I0352626

SAPUI5

SAP PRESS is a joint initiative of SAP and Rheinwerk Publishing. The know-how offered by SAP specialists combined with the expertise of Rheinwerk Publishing offers the reader expert books in the field. SAP PRESS features first-hand information and expert advice, and provides useful skills for professional decision-making.

SAP PRESS offers a variety of books on technical and business-related topics for the SAP user. For further information, please visit our website: *www.sap-press.com*.

Rene Glavanovits, Martin Koch, Daniel Krancz,
Maximilian Olzinger

SAPUI5

The Comprehensive Guide

Editor Meagan White
Acquisitions Editor Hareem Shafi
Copyeditors Julie McNamee
Cover Design Graham Geary
Photo Credit iStockphoto: 103914263/© Philip Meyer
Layout Design Vera Brauner
Production Eric Wyche
Typesetting III-satz, Germany
Printed and bound in the United States of America, on paper from sustainable sources

ISBN 978-1-4932-2660-3
2nd edition 2025

Rheinwerk Publishing, Inc.
2 Heritage Drive, Suite 305
Quincy, MA 02171
USA
info@rheinwerk-publishing.com
+1.781.228.5070

Represented in the E.U. by:
Rheinwerk Verlag GmbH
Rheinwerkallee 4
53227 Bonn
Germany
service@rheinwerk-verlag.de
+49 (0) 228 42150-0

Library of Congress Cataloging-in-Publication Control Number: 2024060910

Contents at a Glance

Contents

Preface

Welcome to *SAPUI5: The Comprehensive Guide*! Developers who want to enter the world of SAPUI5 or deepen their knowledge will find everything they need in this book. SAPUI5 is a powerful framework for developing user-friendly, modern, and scalable web applications in the SAP environment. To provide a comprehensive view of the topic, this book is divided into several parts that accompany the journey from basic understanding to advanced concepts.

Aim and Structure of the Book

The book is divided into five main parts, which build on each other and provide a clear structure. This way, it offers added value to both beginners and experienced developers. The first part covers the basics. Here, you'll learn how SAPUI5 was developed, what the differences and similarities are compared to related frameworks such as OpenUI5 and UI5 Web Components, and how the SAP Fiori design guidelines and the SAP Fiori launchpad are used (**Chapter 1**). Then you'll get to know the infrastructure components necessary for development, such as SAP Business Technology Platform (SAP BTP), the cloud connector, and SAP Business Application Studio or Visual Studio Code (VS Code) development environments (**Chapter 2**). Finally, the importance of TypeScript is emphasized, which makes development more secure and efficient through modern language features (**Chapter 3**).

The second part of the book focuses on practical development with SAPUI5. You'll learn how to create a SAPUI5 application, what its basic structure looks like, and how the model view controller (MVC) approach is implemented (**Chapter 4**). In addition, we'll discuss the use of SAPUI5 modules, including libraries such as `sap.m` or `sap.ui`, as well as various layouts and managed objects (**Chapter 5**). After that, you'll learn about the different types of data binding, such as property or aggregation binding, to dynamically link UI elements with data (**Chapter 6**). Finally, **Chapter 7** shows how to integrate routing and navigation patterns into your application.

The third part is dedicated to data integration. You'll learn how to connect RESTful services using the JSON model or Axios (**Chapter 8**). Next, **Chapter 9** introduces OData service integration, including create, read, update, and delete (CRUD) operations; query options; and advanced features such as function exports. Next, we discuss the differences and specific strengths of the OData V2 and V4 models, such as batch groups or binding contexts. You'll learn how to use them when it comes to CRUD operations and function imports/actions, which need to be executed (**Chapter 10** and **Chapter 11**). In

Chapter 12, you'll learn how to implement advanced concepts such as modularization using fragments and base controllers, including how to integrate functions such as localization, drag and drop, and input validation. In **Chapter 13**, the focus is on developing custom controls. **Chapter 14** covers SAP Fiori elements, in particular, prebuilt floorplans, the flexible programming model, and extension possibilities.

In the fifth and final part of the book, we look at administration and productivity. In **Chapter 15**, you'll learn how to use Git to effectively implement version control. **Chapter 16** is all about deployment. This chapter shows you how to deploy applications both manually and automatically (via continuous integration/continuous delivery [CI/CD]). **Chapter 17** explains the configuration of the SAP Fiori launchpad (on-premises) and the SAP Build Work Zone, standard edition (cloud). **Chapter 18** and **Chapter 19** cover debugging, test-driven development, code quality in general, and security aspects of both the SAPUI5 framework and the corresponding components. Finally, **Chapter 20** briefly introduces how the use of generative AI will optimize development processes in the future.

Note

A PDF of full-size color versions of all figures in this book is available for download at *https://sap-press.com/6012* under the **Product supplements** section.

Why This Book?

SAPUI5 is a comprehensive framework that requires in-depth knowledge and practical experience. This book offers a structured, hands-on approach to learning both basic and advanced topics. Developers are given the tools to create modern, high-performance, and secure SAPUI5 applications.

The aim is to delve into the fascinating world of SAPUI5 and discover the possibilities of this framework!

PART I

Basics

Chapter 1
Introduction to SAPUI5

SAPUI5 is a modern user interface (UI) framework that has become the central technology for the development of user-friendly and scalable business applications thanks to its flexibility, cross-platform use, and seamless integration into SAP systems.

The world of SAP UIs and technologies has evolved considerably over the years. This introduction provides a comprehensive overview of the various aspects and developments that have made *SAPUI5* and *SAP Fiori* what they are today. In the following, we'll discuss the history of SAPUI5, the differences and similarities between SAPUI5, SAP Fiori, *OpenUI5*, and *UI5 Web Components*, as well as the SAP Fiori design guidelines and the SAP Fiori launchpad in detail. We'll also refer to important documentation and additional resources that are crucial for developers. Finally, we'll summarize the most important findings.

1.1 History of SAPUI5

SAPUI5 is a framework from SAP for developing web applications. It offers developers the opportunity to create user-friendly and appealing UIs that work on different devices and platforms. Some important milestones in the history of SAPUI5 are covered in this section.

Legend says that it all began in November 2008, when five or six people from different teams were crammed into a room that was only designed for four people. Their task was to develop a new UI technology. This was to be flexible, expandable, modern, and independent of the backend technology. At the time, no one had any idea where this would ultimately lead. The goal was to develop a framework that would make it possible to create applications that work equally well on different platforms and devices. This was a response to the growing need for flexible and modern UIs that meet the standards and expectations of the 21st century. But first, let's look back at the evolution of SAP's UIs.

The development of UIs for SAP applications has a long and varied history spanning several decades. Over time, SAP has introduced and developed various technologies to meet changing requirements and technological advances.

SAP UIs began with *SAP GUI*, which was one of the first graphical user interfaces (GUIs) for SAP applications. It's based on the Dynpro framework, which is used for the development of dialog-based applications in SAP systems. Dynpro, short for dynamic programming, refers to the dynamic creation of UIs in ABAP. SAP GUI offers a classic, transaction-based UI that is still widely used in many companies. It enables access to SAP ERP systems and other SAP applications via a desktop application. However, SAP GUI wasn't particularly user-friendly compared to modern web-based interfaces and required a local installation of the GUI application.

In the early 2000s, SAP introduced the *Business Server Pages* (BSP) technology, which provided a platform for the development of web-based applications. BSP applications are developed in ABAP and use HTML and JavaScript for display in the web browser. This technology enabled the creation of dynamic web applications running on the SAP web application server and provided a way to develop more user-friendly and accessible applications. However, BSP had limitations in terms of flexibility and performance and was relatively complicated to develop and maintain.

In the mid-2000s, SAP set another milestone with the introduction of *Web Dynpro for Java* and *Web Dynpro for ABAP*. Both technologies are based on the *model view controller* (MVC) design pattern and offer a declarative development environment for the creation of web applications. Web Dynpro for Java, an implementation of Web Dynpro in the Java programming language, was mainly used for the SAP NetWeaver platform. Web Dynpro for ABAP, the implementation in ABAP, is deeply integrated into SAP ERP and other SAP products and enables the development of business applications directly in the ABAP environment. The advantages of these technologies lie in the separation of business logic and UI, support of reusable UI components, and a consistent user experience (UX). However, development and administration proved to be complex, and Web Dynpro for Java lost popularity over time and was increasingly replaced by more modern technologies.

A key turning point in the evolution of SAP UIs was the introduction of SAPUI5 in 2012 and subsequently SAP Fiori in 2013. SAP Fiori is a collection of design guidelines and application development principles aimed at modernizing the UX of SAP applications. SAP Fiori apps are role-based and offer an intuitive, consistent, and responsive UI. The underlying JavaScript framework for the development of these applications is SAPUI5, which uses modern web technologies such as HTML5, cascading style sheets (CSS), and JavaScript. SAPUI5 was developed as a modern and lightweight successor to Web Dynpro for Java and Web Dynpro for ABAP. As smartphones were becoming increasingly popular at the time and business users were moving away from feature phones, another requirement was that applications developed with SAPUI5 should also run on mobile devices. SAP sent out a host of developers and UX designers to create a framework that was on par with Web Dynpro in terms of functionality.

The first official version of SAPUI5 was released in 2012. This was a significant step, as it provided SAP with a powerful tool for developing UIs for its customers and partners.

The release marked the beginning of a new era for SAP applications, which now had a modern and intuitive UI. In contrast to traditional SAP development, which had previously always taken place in SAP GUI, the development of SAPUI5 UIs took place in Eclipse.

SAPUI5 is based on open web standards such as HTML5, CSS, and JavaScript. The framework uses the MVC design pattern to ensure a clear separation between data, logic, and presentation. This makes the development and maintenance of applications much easier. Another important element of SAPUI5 is the integration of libraries such as jQuery, which was the de facto standard at the time. These libraries offer additional functionalities and ensure compatibility with various web browsers on both stationary computers and mobile devices, which simplified cross-device development and significantly improved the UX.

As expected, SAPUI5 is particularly widespread in the SAP ecosystem. It's the preferred UI framework for applications on SAP S/4HANA, on SAP Business Technology Platform (SAP BTP), and in software as a service (SaaS) offerings from SAP. The breakthrough for SAPUI5 applications in the mainstream began in May 2013 with the announcement of SAP Fiori. SAP Fiori began as a collection of 25 new applications that were characterized by their appealing design, user-friendliness, and responsiveness on various devices; the apps were also developed with SAPUI5. This milestone marked the beginning of a revolution in the UX of SAP applications. As mentioned earlier, the design guidelines and application development principles in of SAP Fiori were focused on improving and updating the UX of SAP applications. SAP Fiori is intended to serve as a guide for developers and aims to provide users with a consistent and intuitive UX.

Another defining moment in the history of SAPUI5 was the launch in 2013 of OpenUI5, the open-source version of SAPUI5. OpenUI5 offers the same core functionality as SAPUI5, allowing developers outside the SAP ecosystem to use and contribute to the framework. This significantly boosted the distribution and acceptance of the framework. An active developer community and comprehensive documentation are key factors for the success of SAPUI5. The community provides support, shares knowledge, and contributes to the further development of the framework. SAP supports these efforts with training, certifications, and numerous resources to help developers improve their skills in using SAPUI5.

SAPUI5 has been continuously developed since its introduction. Regular updates bring new functions, new UI components, performance improvements, and support for the latest web technologies. SAPUI5 offers an extensive library of UI components that help developers create responsive and customizable UIs. The framework supports data binding and internationalization, which facilitates the development of applications for different languages and regions.

However, the development of SAP applications has also evolved and changed considerably over the years. While development was traditionally carried out in SAP GUI and later in *Eclipse*, SAP now provides more modern tools to meet the requirements of developers. SAP now offers specialized development environments such as SAP Web IDE and SAP Business Application Studio for development on SAP BTP. In addition, SAP provides plug-ins for Visual Studio Code (VS Code) to provide a flexible and powerful environment for developers who want to work locally on their computers. These tools provide integrated development environments (IDEs) in which developers can efficiently develop, test, and deploy SAPUI5 applications.

SAP Business Application Studio offers a modern and advanced development environment specifically designed to meet the needs of developers in SAP BTP. It supports a variety of programming languages and frameworks, including SAPUI5, SAP Fiori, SAP Cloud Application Programming Model, and more. SAP Business Application Studio enables the integration of *DevOps* tools and processes for an end-to-end development and deployment pipeline and supports the use of Docker containers for isolated and reproducible development environments. The extensible architecture allows developers to develop their own extensions and plugins and then integrate them into the development environment. This modern and flexible development environment offers strong integration with SAP BTP and other SAP services while supporting modern development practices and tools.

For developers who prefer to work locally, SAP offers special plugins for VS Code. These plugins, particularly the *SAP Fiori tools*, support the development of SAP Fiori and SAPUI5 applications with advanced code completion and linting for improved code quality. They also provide integrated debugging tools and testing frameworks such as *QUnit* and *OPA5*, as well as support for connecting to SAP systems and using OData services. Using VS Code with these plugins gives developers the flexibility to develop locally and use preferred tools while remaining seamlessly integrated with existing development workflows and tools. This also facilitates the transition from local development environments to SAP BTP.

1.2 Comparing SAPUI5, SAP Fiori, OpenUI5, and UI5 Web Components

The development of modern web applications is crucial for companies to ensure user-friendly and efficient interaction with their systems. Several technologies are available in the SAP ecosystem for this purpose: SAPUI5, OpenUI5, SAP Fiori, and UI5 Web Components. These technologies offer different approaches and tools to meet the diverse requirements for UIs. Understanding the differences and application areas of these technologies is essential for developers and companies who want to optimize and extend SAP applications.

SAPUI5 and OpenUI5 are two closely related frameworks based on modern web standards such as HTML5, CSS, and JavaScript. SAPUI5 is a proprietary solution from SAP that is deeply integrated into the SAP ecosystem and offers commercial support.

OpenUI5, on the other hand, is the open-source version of SAPUI5, which was published under the Apache 2.0 license and can therefore also be used and modified outside the SAP ecosystem. Both frameworks offer a comprehensive library of UI components and enable the development of complex, responsive web applications.

SAP Fiori differs from SAPUI5 and OpenUI5 in that it's not a technology or framework in the traditional sense, but a collection of design guidelines and predefined applications. SAP Fiori aims to create a consistent and intuitive UX by providing best practices and standardized design principles for SAP applications. Many of the applications developed with SAP Fiori are based on SAPUI5, which underlines the close connection between these technologies.

UI5 Web Components are another important technology in the SAP ecosystem. They are based on web standards and offer lightweight, reusable UI components that can be integrated into various web frameworks such as React, Angular, or Vue. UI5 Web Components are particularly suitable for projects that only require individual UI components without taking on the complexity of a complete framework such as SAPUI5.

1.2.1 SAPUI5

As discussed in the previous section, SAPUI5 is a proprietary JavaScript framework designed specifically for the development of SAP applications. Its core functions are as follows:

- **Extensive UI component library**
  Provides ready-made components for various use cases.
- **Data binding and modeling**
  Supports the synchronization of data between the UI and the backend.
- **Internationalization**
  Facilitates the adaptation of applications to different languages and regions.
- **Theming and customization**
  Enables the appearance to be adapted to the company's corporate identity.

SAPUI5 is supported and documented by SAP. It's mainly used in enterprise applications within the SAP ecosystem, such as SAP S/4HANA, SAP Fiori launchpad, and SAP Business Suite. The flexibility and adaptability of SAPUI5 are high, but mainly within the SAP environment.

1.2.2 OpenUI5

OpenUI5 is the open-source version of SAPUI5 and offers the same core functionality as SAPUI5, including an extensive UI component library, data binding, internationalization, and theming. It's released under the Apache 2.0 license and is supported by the community. OpenUI5 enables the development of web applications both within and outside the SAP ecosystem. Information on OpenUI5 can be found at *https://openui5.org/* (see Figure 1.1).

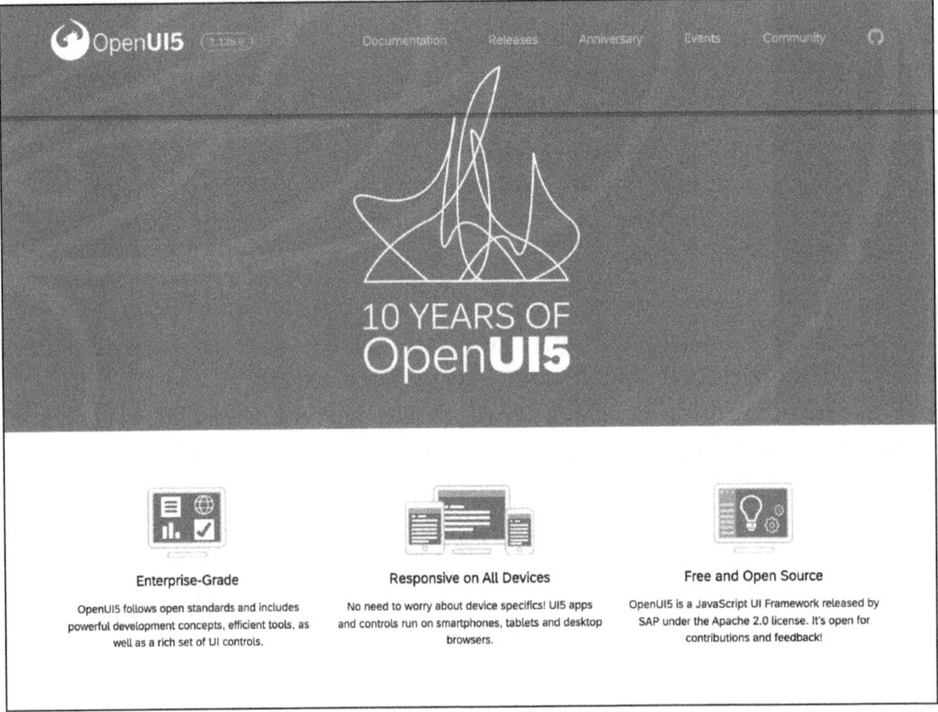

Figure 1.1 OpenUI5 Website

OpenUI5 is released under the Apache License Version 2.0. The Apache 2.0 License is a permissive open-source software license that allows users to freely use, modify, and distribute licensed software. Key features of the license include a requirement for proper attribution to the original authors and a disclaimer of warranties. It allows for both open-source and proprietary usage, meaning the software can be used in closed-source projects. The license also includes patent protections, ensuring that contributors can't sue users over patent infringement claims related to their contributions. This flexibility and legal protection make the Apache 2.0 License a popular choice for software developers and organizations.

The core functions of OpenUI5 are as follows:

- **Same core functions as SAPUI5**
  Offers the same UI component library and data binding functions.
- **Community-driven**
  Open-source community contributes to further development and improvement.
- **Flexibility**
  Suitable for a wide range of web development projects.

OpenUI5 receives community support and no official support contracts. It's suitable for the development of web applications in various contexts, including customizable business apps, and offers a high degree of flexibility that can also be used outside the SAP environment.

1.2.3 SAP Fiori

SAP Fiori has grown from a small initiative into a comprehensive design system for all SAP products, culminating in the new Horizon theme. This latest evolution offers an enhanced visual experience, emphasizing improved accessibility, focus, and branding across all platforms. This development marks a significant milestone in SAP Fiori's journey, reflecting its ongoing commitment to innovation and user-centric design. If you'd like to see a visual representation of the evolution of SAP Fiori, you can check out *https://experience.sap.com/fiori-design-web/sap-fiori/*. Here's a quick recap:

- **SAP Fiori 1.0**
  SAP Fiori, introduced in 2013, redefined business software by adopting a mobile-first approach focused on simplicity and responsiveness. Its design principles—role-based, task-oriented, and device-agnostic—broke down monolithic transactions into streamlined apps. Initially targeting smartphones, SAP Fiori expanded to accommodate more complex functionality on larger devices such as tablets and desktops. SAP Fiori for Web was built on SAPUI5, ensuring that its adaptive design supported both touch and traditional inputs across devices. The framework evolved in SAPUI5 (versions 1.26–1.38) and is available as the open-source OpenUI5.
- **SAP Fiori 2.0**
  Building on the success of SAP Fiori's initial version, SAP Fiori 2.0 expanded into a comprehensive design system capable of handling complex business scenarios. As the interface for SAP S/4HANA, it introduced new floorplans for managing large amounts of information intuitively and added powerful productivity features such as notifications and enhanced navigation in the SAP Fiori launchpad. It also introduced *SAP CoPilot*, a conversational digital assistant, which has since been retired. SAP Fiori 2.0 was integrated into multiple SAP UI technologies and won accolades such as the Red Dot Award, emphasizing its innovative approach.
- **SAP Fiori 3**
  SAP Fiori 3, announced in 2018, marked a significant evolution in SAP's design system aimed at creating a unified UX across all SAP products for the intelligent enterprise. With the introduction of the Quartz theme and new features, it enhanced navigation, homepages, and integrated machine intelligence. The redesigned shell bar supports multi-product navigation, while flexible homepages and spaces offer improved access to various business domains. SAP Fiori 3 also expanded support for modern web technologies, including Angular, React, Vue, and UI5 Web Components, ensuring broader adoption across platforms.

- **SAP Fiori 3 with Horizon**
  The Horizon theme is a modern evolution of the SAP Fiori design system, featuring a vibrant color palette, rounded corners, and softer shapes. It introduces a new icon font, optimized typography, and enhanced accessibility in line with the Web Content Accessibility Guidelines 2.2 (WCAG 2.2). Stronger contrasts, clearer visual hierarchy, and improved spacing make the design intuitive for all users. Horizon offers four themes: Morning Horizon, Evening Horizon, High Contrast White, and High Contrast Black. Horizon is available for web apps built with SAPUI5 (version 1.102) and native mobile apps.

The core functions of SAP Fiori are as follows:

- **Role-based applications**
  Adaptation of the UI to the specific needs of the user roles.
- **Consistent design**
  Uniform appearance and operation across different applications.
- **Responsive design**
  Optimal display and functionality on desktops, tablets, and smartphones.
- **Intuitive UI**
  Reduced complexity and user-friendly interactions.

SAP Fiori is supported and documented by SAP. It's used to optimize the UX in SAP applications such as the SAP Fiori launchpad, SAP S/4HANA, and SAP BTP. Customizations and enhancements are made by following the SAP Fiori design guidelines.

1.2.4 UI5 Web Components

UI5 Web Components are lightweight, reusable UI components based on web standards. They are independent of special frameworks and can be integrated into various web application frameworks such as React, Angular, and Vue. UI5 Web Components offer a flexible and modern way to create reusable UI elements and use them in different projects. Information on UI5 Web Components can be found at *https://sap.github.io/ui5-webcomponents/* (see Figure 1.2).

Core functions of UI5 Web Components are as follows:

- **Reusable UI components**
  Provide prebuilt, lightweight components for different use cases.
- **Integration into various frameworks**
  Support for React, Angular, Vue, and other web frameworks.
- **Lightweight and flexible**
  Minimal dependencies and easy integration into existing projects.

These open-source components are supported by the community. They are suitable for use in various web application frameworks and enable the creation of user-defined dashboards as well as integration into existing web applications. UI5 Web Components are highly flexible and adaptable, making them ideal for a wide range of development requirements.

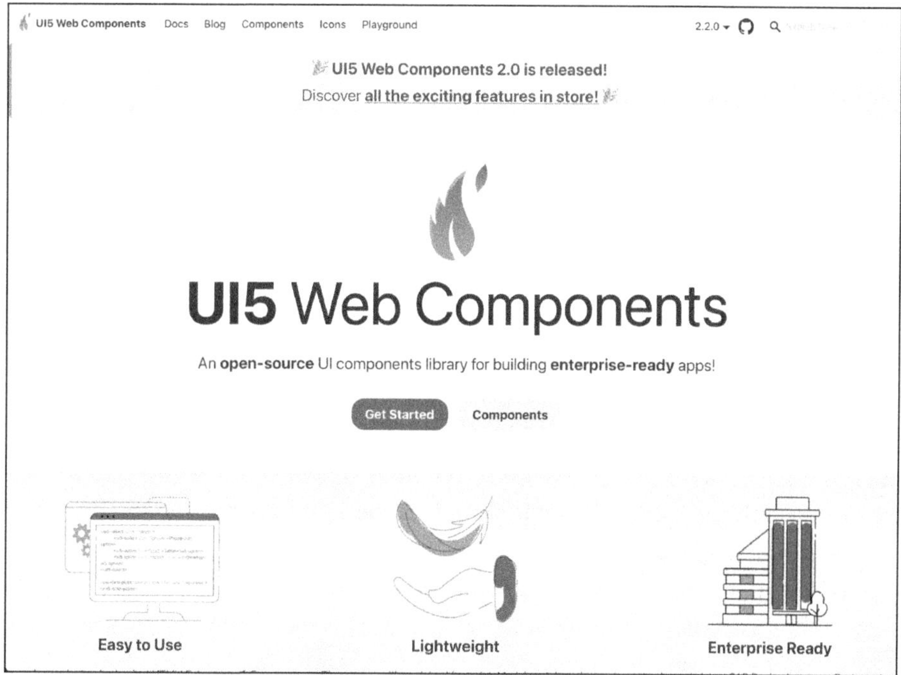

Figure 1.2 UI5 Web Components Website

1.3 SAP Fiori Design Guidelines

The SAP Fiori design guidelines are a comprehensive set of rules that enable the development of user-friendly, consistent, and modern business applications. These guidelines have been developed to provide developers and designers with clear instructions on how to create applications that are both functional and aesthetically pleasing. Consistent application of these guidelines ensures that all SAP Fiori apps provide a consistent and intuitive UX.

A deeper understanding of the structure and the individual components of the SAP Fiori design guidelines is crucial for the successful implementation of projects. In the following section, we'll look at the various building blocks of the guidelines in detail. These building blocks include the design language, SAP Fiori elements, floorplans, interaction patterns, and specific guidelines for particular use cases. Each of these building blocks plays an essential role in creating applications that meet the high standards of SAP Fiori. Before we get to the building blocks, however, we'd first like to introduce you to the SAP Fiori design guidelines at a high level and discuss responsive versus adaptive design.

1.3.1 Core Principles

Following are the main goals and benefits of the SAP Fiori design guidelines:

- **User-centric design**
  The SAP Fiori design guidelines aim to create applications that meet the needs and working methods of users. This is achieved through extensive user research and the consideration of user feedback.
- **Consistency**
  By using standardized design principles and UI components, the guidelines ensure that all SAP Fiori apps offer a consistent UX. This reduces the learning curve and facilitates the use of applications.
- **Role-based applications**
  SAP Fiori apps are specially designed for different user roles. This means that users only see the information and functions that are relevant to their tasks, which increases efficiency.
- **Responsive design**
  The applications automatically adapt to different screen sizes and devices so that users can seamlessly access the applications from desktops, tablets, and smartphones.
- **Simplicity and intuitiveness**
  A central goal of the SAP Fiori design guidelines is to make applications as simple and intuitive as possible. This includes providing a clear and concise UI, using easy-to-understand navigation, and minimizing unnecessary complexity.

The SAP Fiori design guidelines were originally published for web development to ensure a consistent and user-friendly experience on desktop and browser applications. Over time, SAP recognized the need to extend these principles to mobile platforms. The first step was to adapt and extend the guidelines for native iOS apps with SAP BTP SDK for iOS. This enabled the creation of SAP Fiori–based applications that consider the specific design and interaction standards of iOS devices. This was shortly followed by support for Android through SAP BTP SDK for Android, which enabled developers to create SAP Fiori apps for the widely used Android platform. These enhancements ensured that users could enjoy a consistent and optimal UX on all major mobile platforms. The SAP Fiori design guidelines can be found at *https://experience.sap.com/fiori-design/* (see Figure 1.3).

If you're wondering why developers should adhere to the SAP Fiori design guidelines, that's easy to answer. The SAP Fiori design guidelines offer numerous benefits that significantly improve both development and UX. Here are some key reasons to follow these guidelines:

- **Improved user satisfaction**
  Applications developed according to the SAP Fiori design guidelines offer an outstanding UX. User-friendly and intuitively designed applications lead to higher end-user satisfaction and productivity.

- **Reduced training costs**
  Consistent and intuitive UIs mean that users need less time for training and familiarization. This saves companies time and money and enables users to become productive faster.
- **Efficiency and productivity**
  Role-based and easy-to-use applications allow users to complete their tasks faster and more efficiently. This leads to overall higher productivity and better use of business resources.
- **Reusability and maintainability**
  The use of standardized UI components and design principles facilitates code reusability and application maintenance. This leads to lower development costs and better scalability of applications.
- **Future-proofing**
  SAP is continuously developing the SAP Fiori design guidelines to meet the latest technological developments and user requirements. Applications that follow these guidelines therefore remain future-proof and benefit from continuous improvements.
- **Brand image and consistency**
  A consistent look and feel and a uniform UX contribute to a company's positive brand image. This strengthens user confidence in the applications and the company.

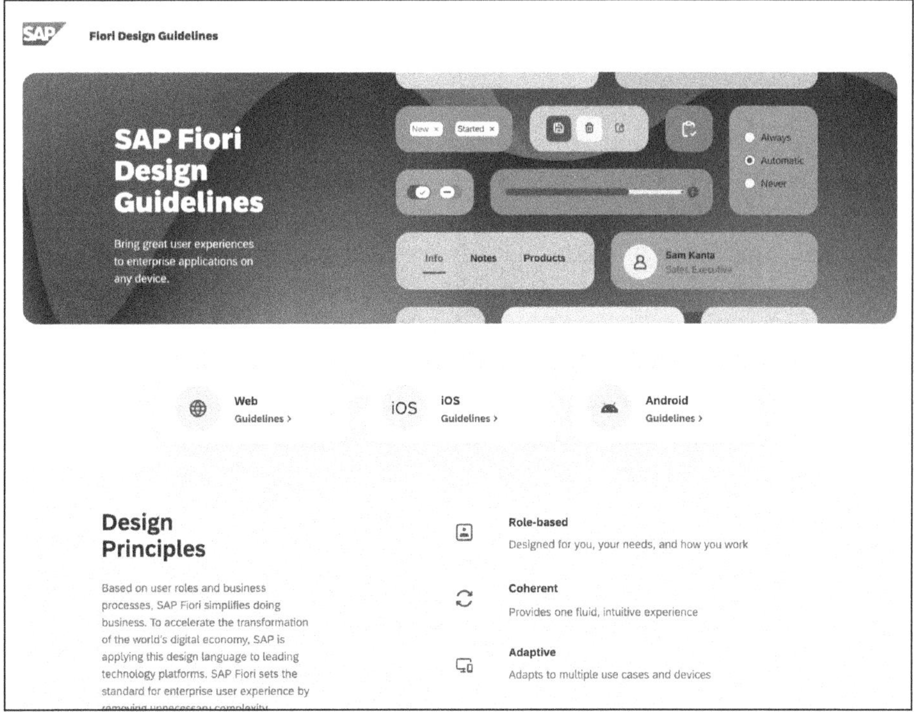

Figure 1.3 SAP Fiori Design Guidelines Overview

The SAP Fiori design guidelines were developed to ensure a user-friendly and consistent experience for SAP applications. These guidelines are based on extensive research and best practices in the field of UX design. They provide developers with clear instructions on how to design applications that are both aesthetically pleasing and functionally efficient. Following are the core principles of the SAP Fiori design guidelines that serve as the foundation for the design and development of these modern UIs:

- **Role-based**
  Applications are developed specifically for the needs of user roles to provide relevant information and functions. This reduces complexity and increases efficiency.
- **Adaptive**
  SAP Fiori apps are designed to work well on different devices and screen sizes. This allows applications to be accessed from both desktops and mobile devices.
- **Simple**
  The UIs are designed to be clear and uncluttered, with a focus on the essentials. Superfluous information and functions are avoided to increase user-friendliness.
- **Coherent**
  Consistent designs and interactions across all applications ensure a familiar and predictable UX. This is achieved using consistent UI components and patterns.
- **Delightful**
  Applications should be not only functional but also aesthetically pleasing to increase user satisfaction and engagement.

1.3.2 Responsive vs. Adaptive Design

SAP supports not only responsive design but also adaptive design. One of the biggest advantages of SAP Fiori is that apps only need to be developed, configured, and maintained once so that all changes take effect on all devices. This gives end users a consistent experience regardless of which device they are using. SAPUI5 facilitates responsive design by providing a variety of UI controls that automatically adapt to different form factors and interaction styles. In addition, SAPUI5 controls can be customized to the type of interaction, be it touch device or keyboard/mouse. The advantage of *responsive design* is that applications can adapt to different screen sizes without the need for additional programming and maintenance. Figure 1.4 shows a simple form displayed on a desktop browser. As you can see in the image, there are three columns. Based on the space available, columns 1 and 2 are rendered side by side in the first row, and column 3 is rendered in the second row.

If you now open the same app on a smartphone, you'll see that the columns are displayed one below the other. It even goes so far as to display the labels in a separate row and the corresponding input fields below them (see Figure 1.5).

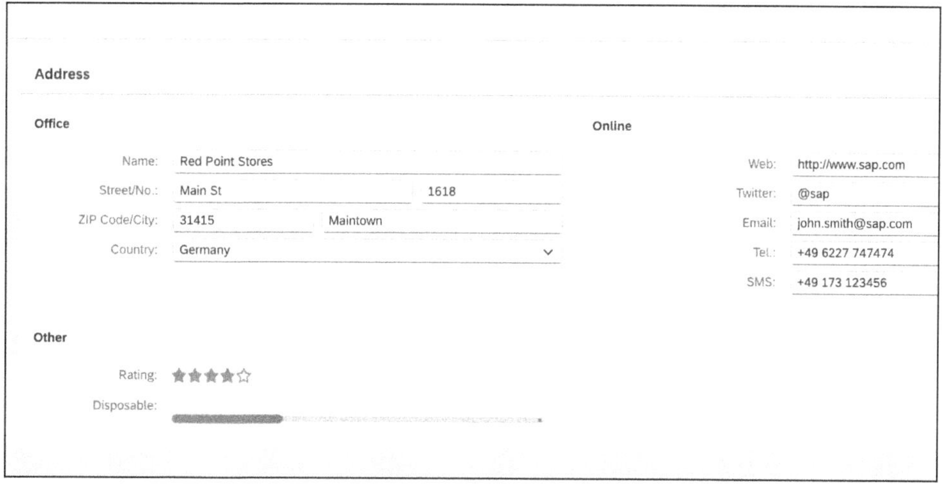

Figure 1.4 Responsive Design: Displayed on a Desktop Browser

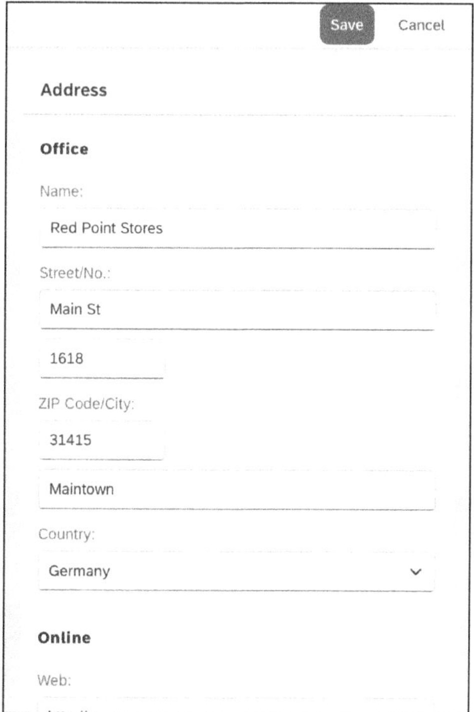

Figure 1.5 Responsive Design: Displayed on a Smartphone

However, there are certain scenarios where a responsive approach isn't optimal. For example, users would probably prefer to enter large amounts of data on their desktop, while on their tablet or smartphone they may only want to view the data or enter a small subset. In such cases, it makes sense to develop for different devices and adapt the complexity of the use case to the respective device. This is known as *adaptive design*. With this approach, app developers must manually define specific designs for the different form factors. Although this requires a little more effort, it also enables more targeted support for device-specific use cases. In Figure 1.6, you can see an example of an adaptive design. As the width of the table isn't sufficient to display all fields, the **Supplier Name** field is rendered in a second row.

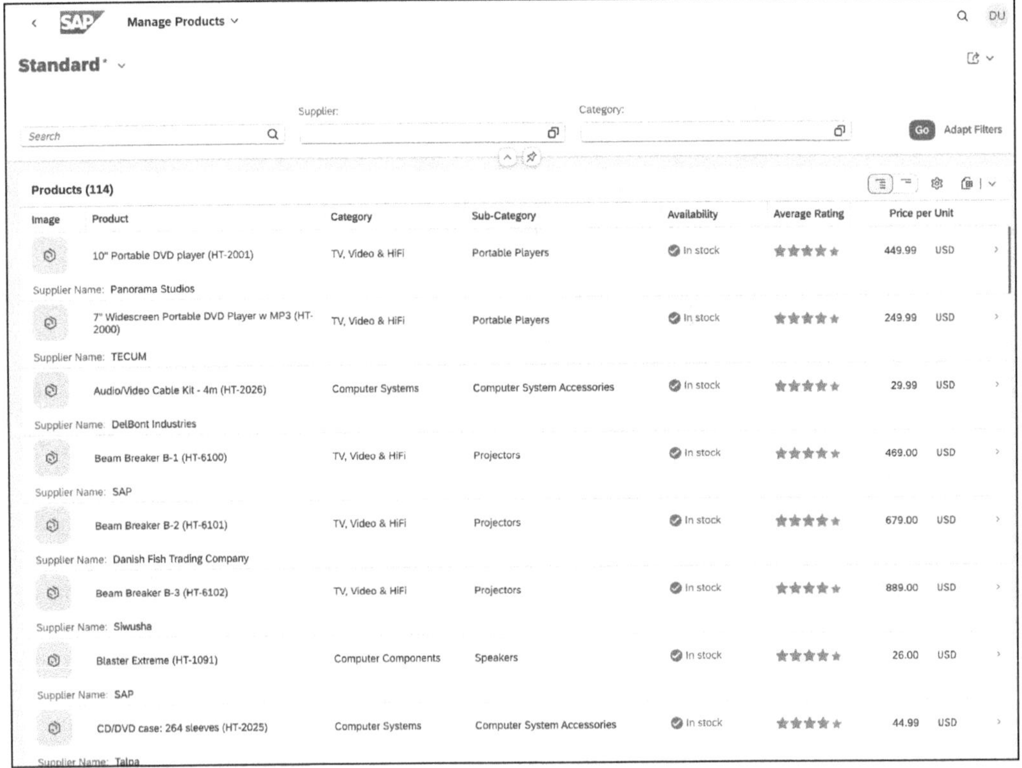

Figure 1.6 Adaptive Design: Displayed on a Desktop Browser

If you now open the same application on a smartphone, you'll see a completely different picture. As you can see in Figure 1.7, all columns are displayed unbroken and row by row. This is a perfect example of adaptive design. The only thing you need to do as a developer is to provide a little information so that the framework knows that it will behave accordingly when displaying the content.

A responsive design adapts dynamically to all screen sizes and uses flexible grids, layouts, and CSS media queries. The advantage is a standardized code base, which makes maintenance easier, but development is more complex. Adaptive design uses fixed layouts for specific screen sizes and uses breakpoints to select the appropriate layout. The advantage is an optimized UX for specific devices, whereas the disadvantage is a higher development effort due to the maintenance of multiple layouts.

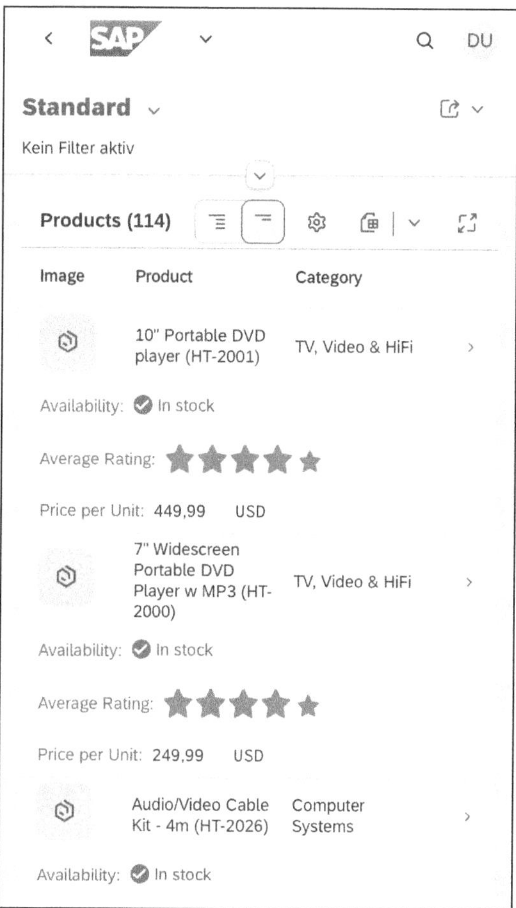

Figure 1.7 Adaptive Design: Displayed on a Smartphone

Content Density

In SAPUI5, content density refers to the density of the display of UI elements that can be adapted to the requirements of different devices and user preferences. This is a property that controls how much space is left between the elements on a UI to achieve either a more compact or a more generous display. The flexible design enables SAPUI5 applications to be optimized for different devices and user contexts.

There are two types of content density:

- **Cozy**
  Suitable for touch devices such as tablets and smartphones by default. Elements have larger gaps to make it easier to use with your fingers.
- **Compact**
  For desktop applications or environments where the mouse and keyboard are used. Elements are packed more densely so that more information can be displayed at once.

The right choice of content density improves user-friendliness and efficiency. Cozy ensures comfortable operation on touch devices. Compact allows more information to be displayed on a limited screen area on desktops.

1.3.3 Design Language

The *design language* of SAP Fiori defines the basic visual and interactive elements that ensure a consistent UX. The most important aspects of the design language include the following:

- **Color scheme**
  Defines the color palette used in all SAP Fiori apps. This includes primary, secondary, and accent colors that provide a consistent visual identity. Color communicates importance and association, as well as provides direction to users. By applying the color palette of the themes, UIs guarantee a clean and lightweight design that is consistent and coherent across all SAP Fiori apps.
- **Typography**
  Defines the fonts and typography styles used in SAP Fiori apps. This includes the selection of fonts, font sizes, and text styles to optimize readability and aesthetics. SAP Fiori uses SAP's proprietary typeface 72, with a fallback to a sans serif system font if 72 can't be loaded.
- **Icons and graphics**
  Determines the use of icons and graphic elements to create a visual hierarchy and guide users through the application. In addition to the general icon semantics and grid system, each article covers the line thickness and corner radius values defined for the respective theme. These styling differences explain why the same icons can look slightly different, depending on the theme you're using.
- **Layouts and spacing**
  Defines the arrangement of UI components and the use of spacing to ensure a clean and uncluttered UI.

1.3.4 SAP Fiori Elements

SAP Fiori elements are prebuilt UI components and templates that provide frequently used patterns and functions. They facilitate the development of applications by providing reusable building blocks that fit seamlessly into the SAP Fiori design guidelines. SAP Fiori elements is a framework that comprises the most-used floorplan templates, is designed to speed up development by reducing the amount of frontend code needed to build SAP Fiori apps, and is used to drive UX consistency and compliance with the latest SAP Fiori design guidelines. The most important SAP Fiori elements floorplans include the following:

- **Lists**
  Components for displaying data records in list form, including filter and sorting functions.
- **Forms**
  Components for entering and editing data, including validation and error-handling mechanisms.
- **Tables**
  Components for the structured presentation of data in tabular form, including functions for paginating and exporting data.
- **Charts**
  Components for the visual representation of data in the form of charts and graphs.

SAP Fiori elements support both OData V2 and OData V4. Each version of SAP Fiori elements is designed to be compatible with a specific version of the OData standard, as reflected in their respective names:

- **SAP Fiori elements for OData Version 2 (V2)**
  This version is tailored for applications using OData V2, offering features and capabilities that align with this protocol's specifications and limitations.
- **SAP Fiori elements for OData Version 4 (V4)**
  This variant is optimized for OData V4, incorporating advancements and enhancements introduced in the later version of the protocol, enabling more modern and efficient data handling.

The choice between these versions depends on the specific requirements and compatibility of the underlying OData services being used in your SAP environment.

1.3.5 Floorplans

Floorplans are predefined layouts for typical business scenarios that provide a structured arrangement of UI components and data. The individual floorplans are briefly explained in the following sections.

Overview Page

The *overview page* is a data-driven SAP Fiori app type and floorplan designed to consolidate all essential information for a user on a single page, tailored specifically to the user's domain or role. This layout enables users to concentrate on the most critical tasks, offering a streamlined experience where they can quickly view, filter, and respond to relevant information. In this floorplan, each task or topic is encapsulated within a card, also known as a content container. The overview page serves as a UI framework that organizes and displays multiple cards in a cohesive, single-page view, making it easy for users to interact with various data points efficiently. The overview page leverages annotated views of application data, allowing for content to be dynamically customized based on the specific domain or role of the user.

A variety of card types are available, each designed to present information in a visually appealing and efficient manner, ensuring that users can quickly grasp and act upon the information most pertinent to their needs (see Figure 1.8).

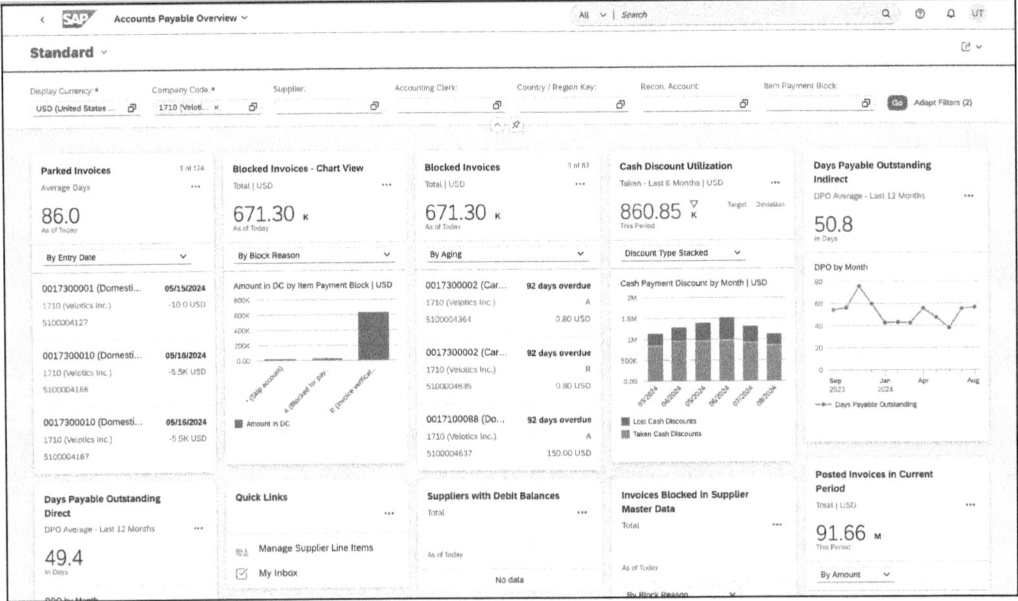

Figure 1.8 Overview Page

In an overview page, the following card types can be used:

- Analytical card
- List card
- Bar chart list card
- Link list card
- Table card
- Stack card
- Quick view card
- Custom card

SAP Fiori Launchpad vs. Overview Page

The SAP Fiori launchpad homepage serves as a central hub that displays all of a user's favorite apps, accessible through tiles. This page covers the full spectrum of roles a user might hold, such as employee, manager, production worker, or quality manager, providing easy access to a wide range of applications relevant to these roles.

In contrast, the overview page is more focused, targeting the key tasks associated with a specific role. It contains only the most frequently used apps relevant to that role, organized through cards instead of tiles. Cards offer more detailed preview information than tiles due to their larger size, enhanced properties, and interactive areas. Some card types even allow users to perform simple actions directly from the card. In essence, cards provide an entry-level view of application content, making it easier for users to engage with critical data and actions at a glance.

List Report

A *list report* is a versatile SAP Fiori floorplan that allows users to efficiently view, filter, and manage large sets of items. Designed with robust features, it enables users to quickly locate and act on the most relevant items within the list (see Figure 1.9). This floorplan is particularly useful as an initial entry point, from which users can navigate to more detailed information about a specific item, typically presented on an object page. In addition to basic viewing and navigation, the list report supports various interactive elements, such as sorting, grouping, and filtering, empowering users to tailor their experience according to their specific needs. This makes it an ideal solution for scenarios where users need to manage complex datasets while maintaining the flexibility to drill down into item details as needed.

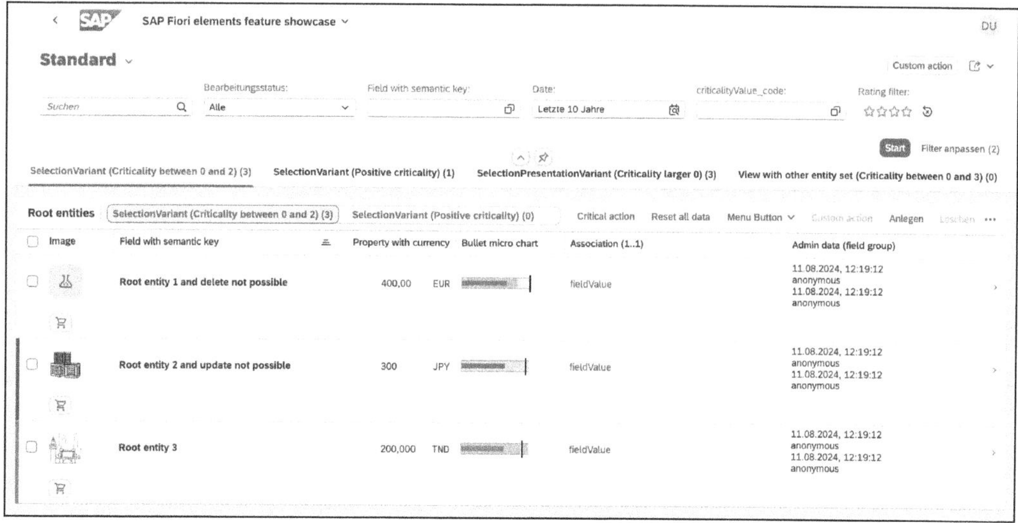

Figure 1.9 List Report

Object Page

The *object page* floorplan is designed to present and organize all pertinent information related to a specific business object. It categorizes content efficiently, allowing users to quickly access different sections via anchor or tab navigation (see Figure 1.10). Users can easily toggle between display mode to view information, edit mode to modify existing content, and create mode to generate new objects. This floorplan features a flexible and responsive layout, making it adaptable to various screen sizes and devices. The dynamic page header is customizable, accommodating both simple and complex business objects, which ensures that the layout can be tailored to a broad range of use cases. The object page is ideal for scenarios where detailed object information needs to be accessed, modified, or created within a structured and user-friendly interface.

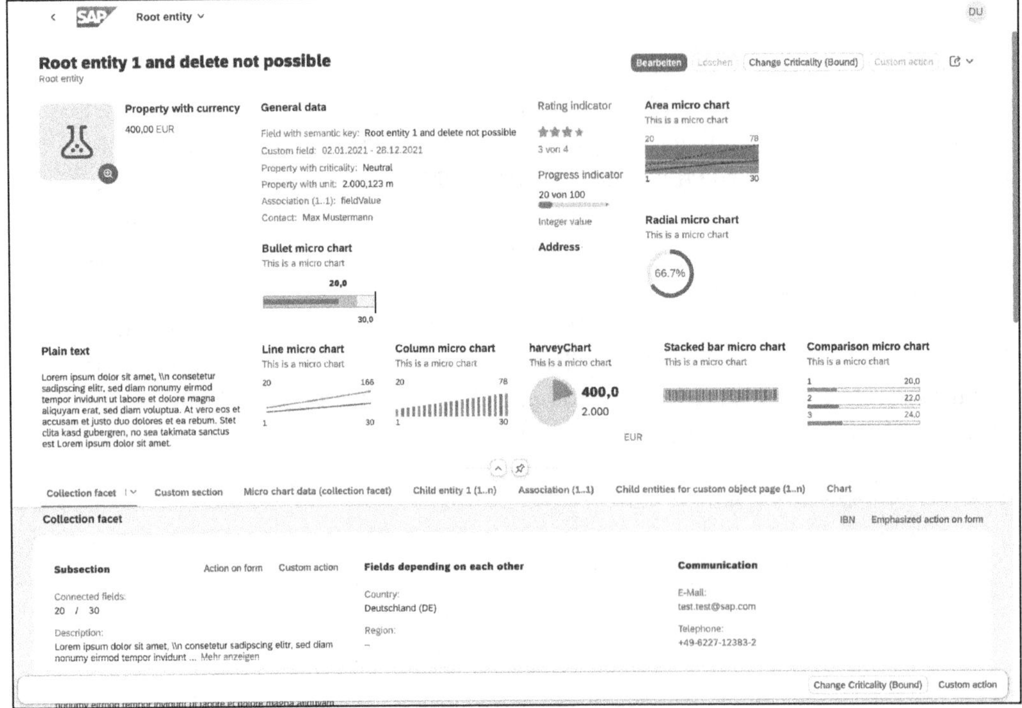

Figure 1.10 Object Page

Worklist

The *worklist* floorplan is designed to present a collection of items that require user attention and action. It serves as a task management tool where users can review item details, take necessary actions, and typically either complete the task or delegate it to others. This versatile floorplan offers three main variants to cater to different user needs and scenarios, as follows:

- **Simple worklist**
  A straightforward page featuring a table that lists the tasks.
- **Worklist with tabs**
  A more organized approach that categorizes tasks into different tabs, making it easier to manage tasks across various categories.

- **Worklist with KPI tags**
  This variant includes one or more key performance indicator (KPI) tags, providing users with immediate insight into critical metrics associated with the tasks.

Each variant is tailored to specific use cases, allowing users to select the format that best meets their workflow requirements. For more detailed information on customization and additional features, refer to the options under **Components**.

Analytical List Page

The *analytical list page* provides a powerful and intuitive platform for users to analyze data progressively from multiple perspectives, enabling root cause investigation through drilldown features and facilitating action on transactional content—all within a single, seamless page (see Figure 1.11). The analytical list page is designed to help users identify key areas of interest within datasets or pinpoint significant individual instances by leveraging data visualization and business intelligence.

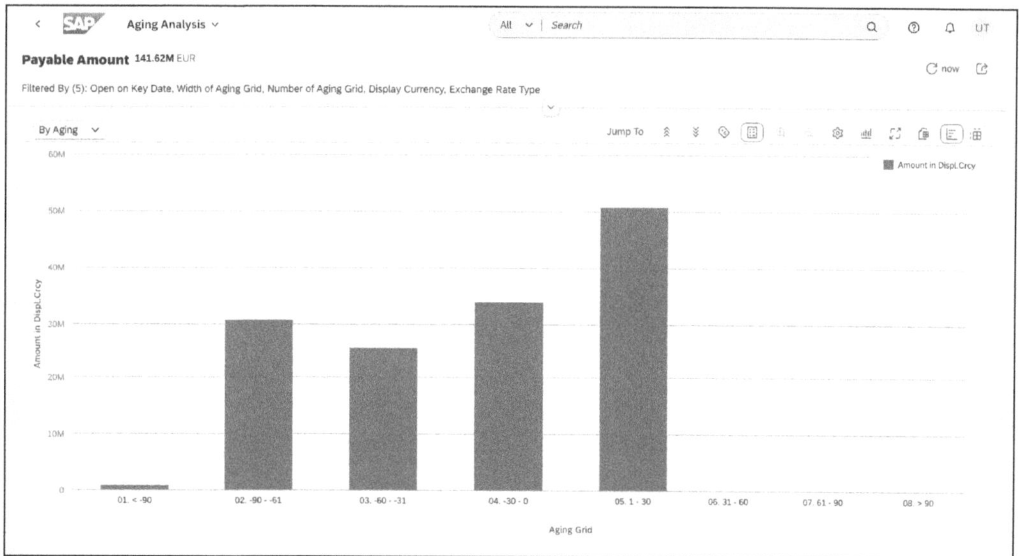

Figure 1.11 Analytical List Page

Data visualizations play a crucial role in this process, allowing users to quickly recognize important facts and situations, thereby minimizing the number of interaction steps required to gain insights or identify critical instances. The use of chart visualizations not only enhances the UX but also enables faster identification of relevant data.

The primary audience for the analytical list page consists of users engaged in transactional content who benefit from transparent access to business object data and direct integration with business actions. The analytical list page offers analytical views and functions without the need to switch between systems, making it a highly efficient tool. Key features include KPIs, a visual filter enriched with measures and visualizations, and a hybrid table/chart view with drill-in capabilities. Users can interact directly with charts to delve deeper into the data, allowing them to quickly spot spikes, deviations, and abnormalities, and take immediate, informed action.

1.3.6 Flexible Programming Model

While SAP Fiori elements is great for most use cases, some applications require greater flexibility that predefined templates can't offer. This is where freestyle development comes into play. Developers can build custom UIs, incorporating unique workflows or business logic that aren't easily accommodated by standard templates. Freestyle development in SAP Fiori follows the same core design principles as template-based development, but gives developers more control over the UI layout, behaviors, and interactions. This approach is ideal for more complex or niche scenarios that require tailored interfaces and features.

The SAP Fiori *flexible programming model* is a dynamic framework that enables developers to create scalable, user-friendly applications in line with the SAP Fiori design principles. This programming model merges the benefits of both SAP Fiori elements and the freestyle development approach, giving developers flexibility while still adhering to SAP's standards for UX. The model is designed to streamline development by offering reusable components, predefined templates, and the ability to introduce custom logic when needed. This section explores the various aspects of the SAP Fiori flexible programming model and its significance in modern application development. The SAP Fiori flexible programming model is currently designed for OData V4. It leverages the benefits of OData V4, such as improved query handling, batch processing, and enhanced metadata annotations. OData V4 is a key aspect of the model as it facilitates the creation of metadata-driven UIs, which are integral to SAP Fiori elements. OData V2 is supported for existing applications, but the flexible programming model itself primarily targets OData V4 to utilize its advanced features and capabilities.

The flexible programming model explorer (*https://sapui5.hana.ondemand.com/test-resources/sap/fe/core/fpmExplorer/index.html*) demonstrates how you can extend SAP Fiori elements apps and custom apps to fit your specific needs. It highlights the full scope of the publicly available flexible programming model, offering live examples in a sandbox environment that allow users to experiment with different extensions and modifications. This hands-on approach provides practical insights into how the model can be adapted for various business use cases, showcasing the versatility and customization options available within the SAP Fiori framework. The flexible programming model explorer can be used by developers to experiment in a sandbox and directly see the impact on code changes (see Figure 1.12).

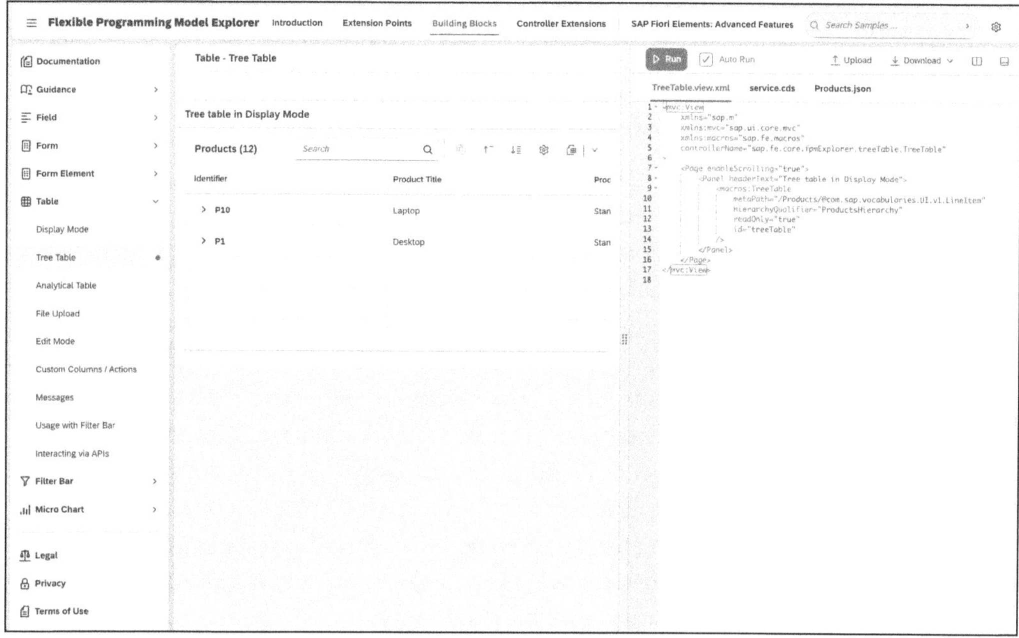

Figure 1.12 Flexible Programming Model Explorer

1.3.7 Draft Handling

Draft handling in SAP Fiori is a powerful feature used in transactional applications to manage uncommitted work. Users can save a draft of their current transaction, allowing them to pause, leave, or come back and continue their task without losing data. Drafts are saved temporarily in the backend and can be resumed later. This feature supports a nonblocking UX through optimistic concurrency control, meaning multiple users can work on the same object, and potential conflicts are resolved only at the final submission phase. Key features of draft handling include the following:

- **Automatic saving**
  Drafts are saved automatically after each user interaction or manually when needed. This minimizes the risk of data loss during lengthy transactions.
- **Optimistic locking**
  Multiple users can edit a record without locking the entire record for others, as updates are merged and conflicts are handled when the draft is submitted.
- **Version control**
  Each draft is treated as a version that can be updated or discarded. Users can choose to resume working on their draft or reset it if changes are no longer needed.
- **Separation of draft and active data**
  Draft data is separated from active records in the system, ensuring that incomplete changes don't affect the operational data until they are finalized.

1.3.8 Interaction Patterns

Interaction patterns describe how users should interact with the applications. These patterns ensure that user interactions are intuitive and consistent. The most important interaction patterns include the following:

- **Navigation**
  Navigation in SAP Fiori mirrors common web application practices, allowing users to navigate through different screens based on their browser settings, with options to open screens in new tabs or windows. SAP Fiori employs a hybrid of the hub-and-spoke and application network models. Users start at the central homepage (SAP Fiori launchpad or spaces) and move through app screens that may form part of larger processes. The back navigation allows easy return to previous screens, and clicking the logo always takes users back to the homepage. Tiles on the SAP Fiori launchpad homepage act as navigation anchors to individual apps. By selecting a tile, users access the corresponding application. These tiles can also integrate legacy UI technologies, allowing for the seamless inclusion of non-SAP Fiori apps. Non-SAP Fiori UI technologies, however, will open in a new tab or window, but this functionality is limited to supported desktop operating systems. This approach ensures compatibility with both modern and older systems while maintaining a consistent UX.
- **Search and filtering**
  The SAP Fiori launchpad provides an enterprise search feature that allows users to search across apps and business objects such as customers, materials, and plans. The search icon, located in the shell bar, is always visible, and users can filter search results by object type before or after searching. Results can be refined through type, app, or term suggestions, and they can be personalized based on search behavior. A filter panel offers advanced filtering options with visualization tools such as bar or pie charts. Personalized search can be turned on or off in user settings, and system configuration may limit some features.

  The filter functionality in SAP Fiori's enterprise search offers advanced options to narrow down search results efficiently. After performing a search, users can open the filter panel to refine results by object type, with filters presented in a hierarchical structure when applicable. Result-specific filters are offered based on context, ensuring only relevant options are displayed (e.g., if all results are from the same country, a country filter won't appear). Filters can be visualized as lists, bar charts, or pie charts, supporting multiple selections. Once applied, filters are visible in the info bar, and users can easily clear them with a single click without reopening the filter panel. This dynamic filtering system adapts to the content being searched, allowing users to efficiently refine and focus on the most relevant results. Additionally, up to five one-click options are provided for each filter for quicker navigation, offering flexibility and clarity in managing search outputs.

- **Data processing**
  One of the key interaction patterns in SAP Fiori design is data processing with validation and error handling. SAP Fiori validates user input in real time, providing instant feedback when data doesn't meet required formats or business rules. Constraints, such as dropdowns and mandatory field indicators, prevent invalid entries. Upon submission, the system checks for errors, displaying detailed messages and offering corrective guidance. Backend validation ensures compliance with organizational standards, and any submission errors are clearly communicated, promoting data integrity and user efficiency.
- **Notifications and messages**
  SAP Fiori provides notifications and messages to inform users of critical events, errors, or updates. Notifications are displayed in real time and can be accessed through the notification center, allowing users to take immediate action. Messages are contextually displayed on relevant screens, offering information or warnings related to the current task. These notifications and messages are designed to be clear, actionable, and nonintrusive, ensuring that users are kept informed without disrupting their workflow.

1.3.9 Guidelines for Specific Use Cases

In addition to general design principles, the SAP Fiori design guidelines offer detailed recommendations for specific scenarios, ensuring a consistent and user-friendly experience across various application types:

- **Dashboard design**
  Dashboards are designed to summarize KPIs, reports, and actionable insights. They provide a real-time overview of important metrics and are optimized for visual clarity, with responsive layouts, color-coded data, and drilldown capabilities. Best practices include focusing on essential information and making it easily digestible at a glance, while enabling users to explore details through interactive features.
- **Analytical applications**
  These guidelines are crafted for data-heavy applications, focusing on tools for visualizing and interacting with large datasets. The design supports interactive charts, graphs, filters, and drilldowns, enabling users to analyze data dynamically. Clear navigation between datasets and a user-friendly interface for sorting and comparing data are emphasized. SAP Fiori also ensures that analytical apps are responsive, perform at a high level across devices, and are adaptable to user roles for personalized data views.
- **Mobile apps**
  Mobile app design guidelines focus on delivering a seamless experience across different screen sizes, with emphasis on touch-friendly interfaces and gesture-based

interactions. The design ensures quick access to key functionalities while maintaining optimal performance. Features such as responsive layouts, minimalistic design, and the use of intuitive icons help enhance usability. Special consideration is given to network connectivity and data synchronization, ensuring that apps work efficiently both online and offline.

1.4 SAP Fiori Launchpad

The SAP Fiori launchpad is the central hub for accessing SAP Fiori apps, providing a role-based, customizable interface that enhances user productivity. It serves as a shell that hosts various SAP Fiori apps and offers services such as navigation, personalization, embedded support, and application configuration. The SAP Fiori launchpad's design focuses on improving the UX by providing a personalized, role-specific workspace that adapts to the needs of each user, whether they are accessing the system via desktop or mobile devices. In this section, we'll first take a look at the key components in Section 1.4.1. Then we'll introduce tiles—a central component of the SAP Fiori launchpad—in Section 1.4.2. In Section 1.4.3, we'll tackle the theming concept and introduce the theme designer. Finally, in Section 1.4.4, you'll learn about UI adaptations.

1.4.1 Key Components

The SAP Fiori launchpad contains the following key components:

- **Homepage**
  As the core of the launchpad, the homepage features tiles that represent business applications (see Figure 1.13). These tiles can display live data, such as open tasks, and are customizable based on the user's role.

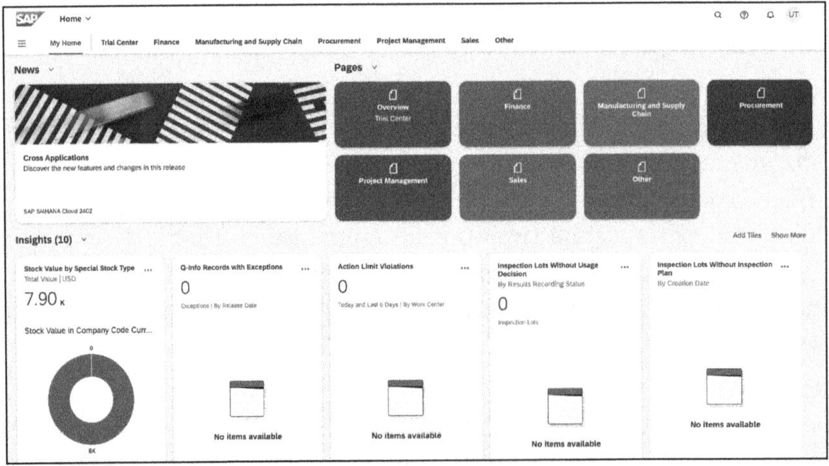

Figure 1.13 SAP Fiori Launchpad Homepage

- **Spaces**
  A new feature introduced with SAPUI5 1.75, spaces allow for the organization of apps based on specific business roles. This structured approach ensures that users only see the most relevant apps and information needed for their daily tasks (see Figure 1.14).

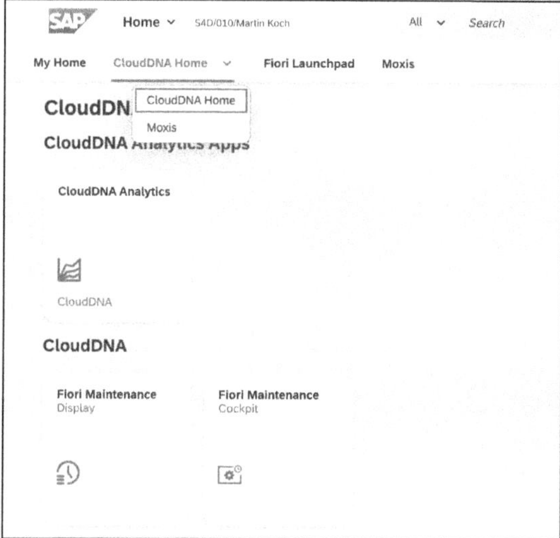

Figure 1.14 SAP Fiori Launchpad Spaces and Pages

- **User actions menu**
  Accessible from the right side of the shell bar, this menu provides a range of user-specific services, including general settings, a catalog of available apps (*App Finder*), recently visited apps, and options for personalization and support (see Figure 1.15).

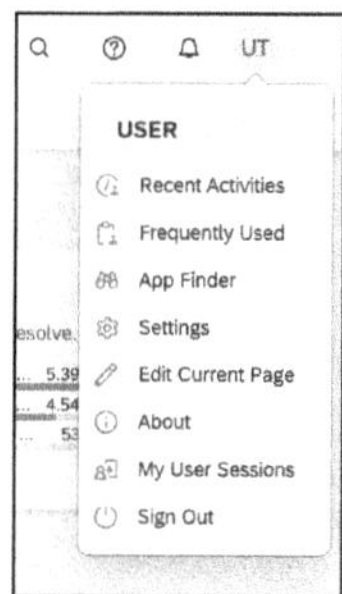

Figure 1.15 SAP Fiori Launchpad User Menu

- **Notifications**
  Users can access a notification list from the shell bar, which aggregates system-generated alerts from various sources, such as workflow inboxes or chat notifications. These notifications can be grouped and prioritized, and they offer direct links to the relevant apps (see Figure 1.16).

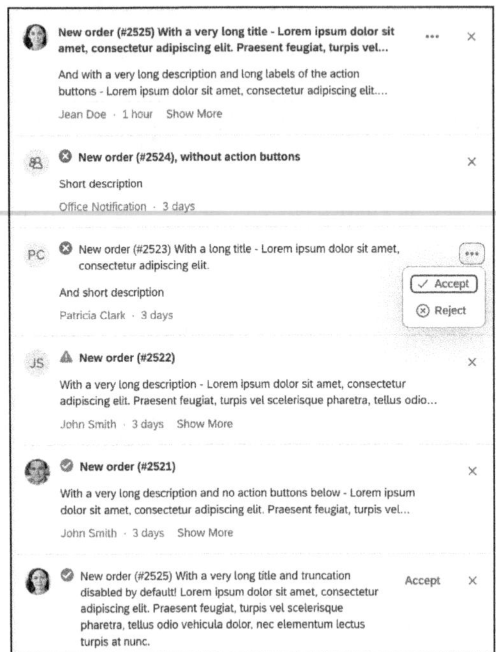

Figure 1.16 SAP Fiori Launchpad Notifications

- **Settings**
  The SAP Fiori launchpad can be customized to the user's needs via a settings dialog. This is where you can, for example, set default settings (see Figure 1.17).

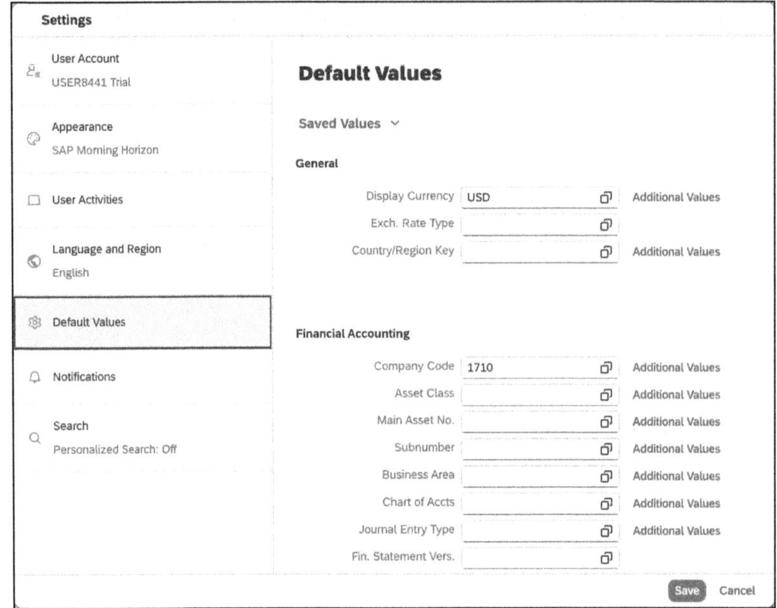

Figure 1.17 Settings: Default Values

1.4.2 Tiles

Tiles are a central element in the SAP Fiori launchpad, as they serve as containers that display applications on the start page. Each tile can display different content provided by the respective application. This includes icons, titles, informative texts, KPIs, counters, and even diagrams. This visual variety allows users to grasp important information briefly and quickly access relevant applications. Figure 1.18 shows the SAP Fiori launchpad for an SAP S/4HANA Cloud system with different types of tiles.

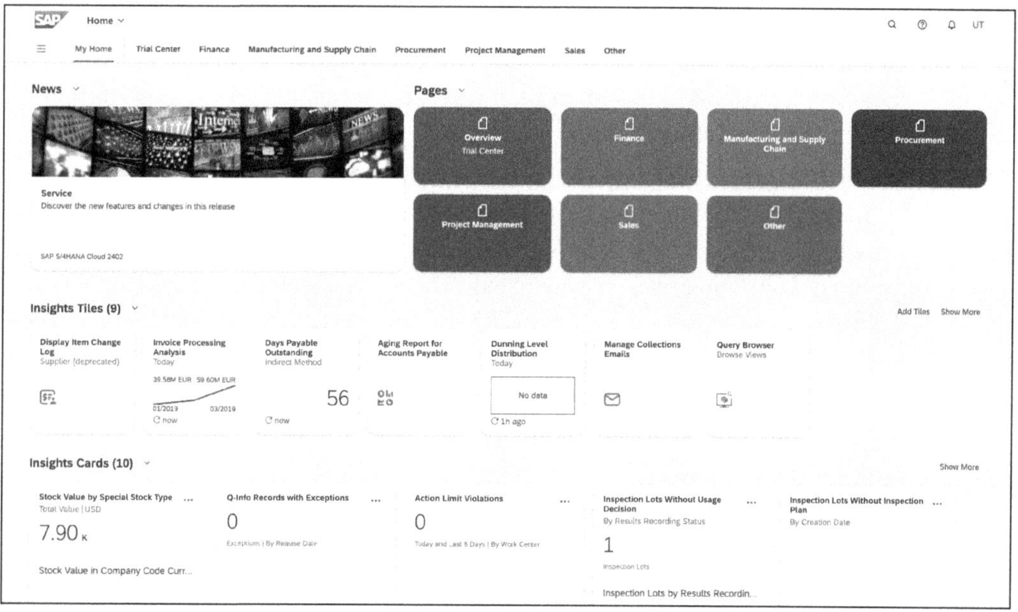

Figure 1.18 SAP Fiori Launchpad Tiles

SAP offers various types of tiles. It's also possible to develop your own tiles if required. The following tile types are provided by SAP:

- **Basic launch tile**
  A basic launch tile typically includes a title, subtitle, and icon. However, it can also be configured as a text-only tile, providing flexibility in how information is presented on the SAP Fiori launchpad.
- **KPI tiles**
  KPIs are crucial for assessing a company's performance at both strategic and operational levels. On SAP Fiori tiles, KPI values are prominently displayed as large, easily recognizable digits. The tile can also include deviation arrows, negative values, and scaling factors, with semantic colors to emphasize key data points. The 4 × 2 tile format allows you to combine content from two tiles, such as pairing a KPI with a comparison chart, while maintaining a single headline, subtitle, clickable area, and target.

- **Comparison chart (micro chart)**
  Comparison charts are ideal for displaying detailed comparisons with semantic coloring, especially in a "Top *N*" list. These charts allow you to highlight key differences using color-coded indicators. You can choose between two different layouts to best suit the data presentation, ensuring clarity and ease of analysis for the user.
- **Bullet chart (micro chart)**
  A bullet chart is a specialized form of a bar chart designed to compare a single primary value against one or more target values. The primary value is displayed within qualitative ranges (thresholds) that often represent levels such as poor, satisfactory, and good. This format helps to provide context by showing how the primary value measures up against predefined performance criteria, making it a useful tool for tracking progress and performance against goals.
- **Trend chart/area chart**
  Trend charts, also known as area charts, are used to display cumulative totals over time, typically based on amounts or percentages. These charts can also show trends for related attributes, with the area below the plotted line filled to indicate volume. A stacked trend chart is one example, where multiple data series are layered to show their collective trends. Trend charts are like plot charts, but with the added visual element of the colored area to represent volume, providing clearer insight into data progression over time.
- **Column chart (micro chart)**
  Column charts, also known as bar charts, are useful for comparing different categories using vertical bars. One axis represents the specific categories, while the other displays a discrete value. Additionally, you can cluster multiple bars into groups within the chart, allowing for comparisons within and across categories.
- **Monitoring tile**
  The monitoring tile is ideal for displaying status updates or object counts. It allows you to apply semantic colors to the status bar, visually indicating whether the status is positive, negative, or critical. This feature helps users quickly assess the situation with a quick visual and small amounts of text.
- **SAP Jam tile**
  If an organization integrates SAP Jam, users can add SAP Jam tiles to the SAP Fiori launchpad. These tiles display new notifications, updating every 10 seconds and cycling through up to 10 notifications. The tile refreshes its content every five minutes, showing the most recent notification if no new ones are available. If the tile lacks an icon, the headline spans the entire width of the 4 × 2 container. This feature helps keep users informed of important updates directly from the launchpad. For further details, see the collaboration documentation at *http://s-prs.co/v601200*.
- **Feed tile**
  The feed tile in SAP Fiori is a larger, specialized tile that displays a rotating news feed, updating every three to five seconds. It includes a headline, background image, news

source, and timestamp. Users can control the feed with start and pause buttons, which are always visible on touch devices and appear on hover on desktops. The number of news items is indicated by dots at the bottom of the tile.

The SAP Fiori launchpad homepage can be customized by the user (see Figure 1.19).

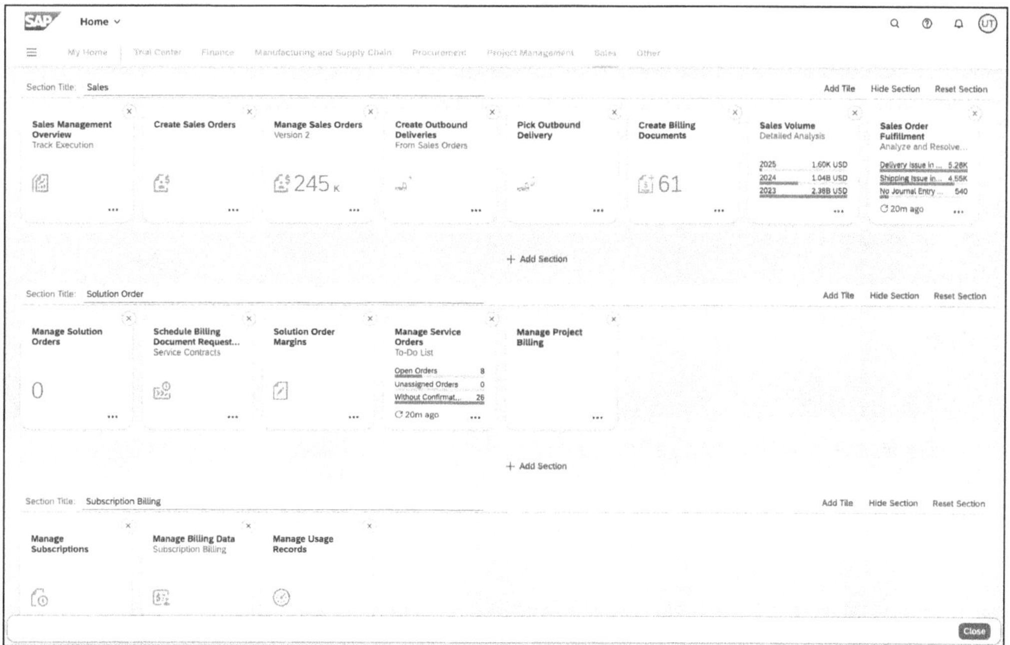

Figure 1.19 SAP Fiori Launchpad Customizing

1.4.3 Theming

You can personalize the SAP Fiori UX by customizing the standard SAP Fiori theme to reflect your organization's brand identity. This can range from simple changes such as replacing the logo or updating the primary UI colors to match their corporate palette, to more advanced customization of UI control colors for greater visual control. Some organizations take this a step further by creating multiple themes to represent different sub-brands or divisions, ensuring consistency across various departments while maintaining distinct visual identities. This flexibility allows businesses to tailor SAP Fiori to their branding needs.

The latest theme for SAP applications is Morning Horizon, which offers a fresh and modern design. SAP also provides alternative themes, including Evening Horizon, a dark theme option, along with Quartz Light and Quartz Dark. For accessibility needs, SAP offers High-Contrast Black (HCB), as shown in Figure 1.20, and High-Contrast White (HCW) themes, designed to improve readability for users with visual impairments. These themes offer flexibility in design to cater to different user preferences and accessibility requirements across various SAP environments.

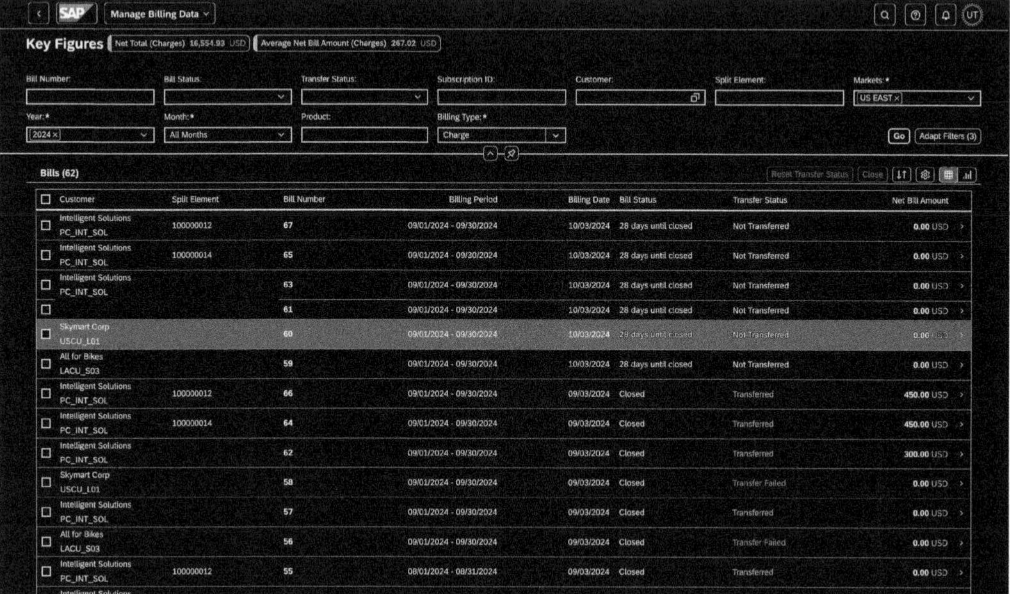

Figure 1.20 High-Contrast Black Theme

Theming is done in a tool called the *UI theme designer* (see Figure 1.21). This is the entry point for creating new themes, importing themes, and editing existing themes.

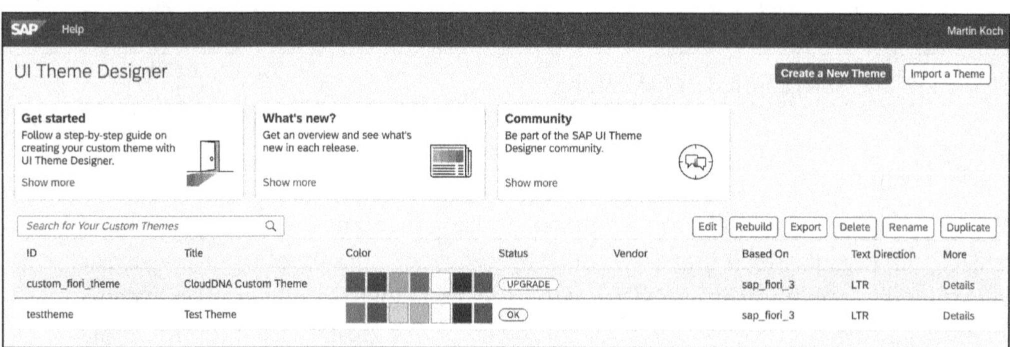

Figure 1.21 UI Theme Designer Overview

Theming in SAP Fiori is based on your organization's requirements and follows a top-down approach, allowing for varying levels of customization:

- **Quick theming**
  Involves simple changes such as replacing the logo and adjusting primary brand colors.
- **Detailed theming**
  Allows for more granular adjustments, such as modifying specific color values for individual UI controls (see Figure 1.22).

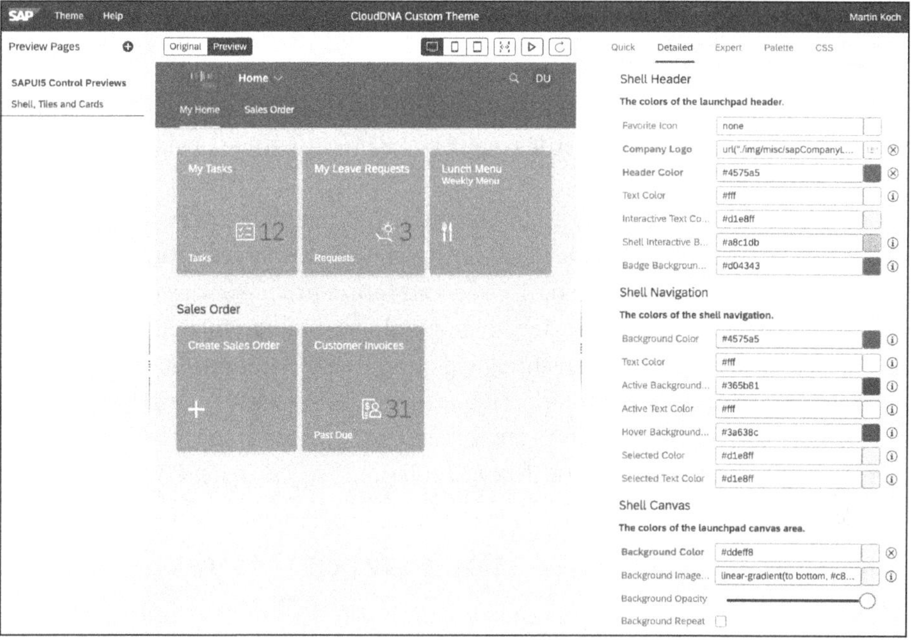

Figure 1.22 UI Theme Designer: Theme Editor

1.4.4 User Interface Adaptations

Both application developers and key users have roles in customizing SAP Fiori apps. Developers can adapt the UI during design time using SAP Fiori tools or manual adjustments for OData V4. For OData V2, they follow specific UI adaptation guides for components such as list reports or object pages. On the other hand, key users can make runtime changes via the **UI Adaption** option to tailor apps for end users (see Figure 1.23). The SAPUI5 flexibility framework supports these customizations, allowing seamless personalization and adaptability across applications. Detailed documentation is available via the SAP Fiori elements feature map at *http://s-prs.co/v601201*.

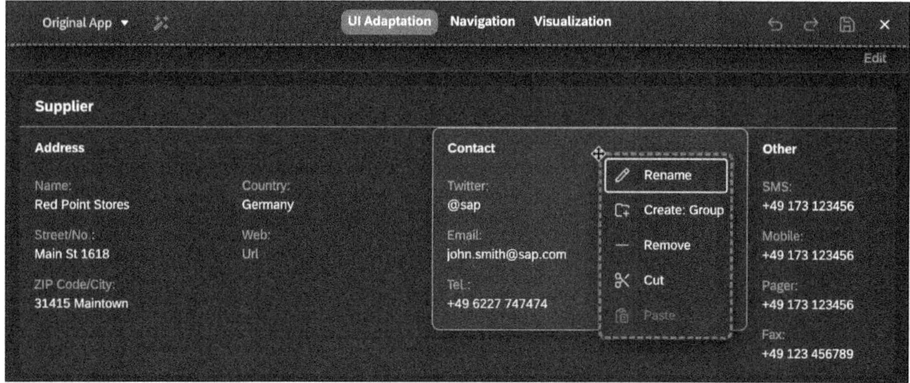

Figure 1.23 Key User Adaptations

1.5 Documentation and Additional Resources

Good documentation serves as a road map for problem-solving, enabling developers to quickly find answers and avoid potential pitfalls. Whether it's accessing SAP's official Software Development Kits (SDKs), leveraging community-driven insights, or finding practical examples in online forums, knowing where to find trustworthy information saves time and effort. Moreover, as SAP technologies rapidly evolve, staying current with the latest updates and enhancements through solid resources ensures that developers can keep their applications optimized and future-proof. For developers, having access to high-quality documentation is essential to building efficient, scalable, and maintainable applications. These technologies are at the forefront of SAP's approach to delivering intuitive, role-based interfaces, but they also come with complexity. Understanding how to navigate the intricate application programming interfaces (APIs), design principles, and best practices is key to ensuring success in development.

1.5.1 General Learning Platforms for SAPUI5, SAP Fiori, and SAP Fiori Elements

This section provides a collection of key learning platforms that offer structured courses and learning journeys for SAPUI5, SAP Fiori, and SAP Fiori elements. These platforms cater to both beginners and experienced developers who want to learn the fundamentals or deepen their expertise in SAP's UI technologies. These resources, like SAP Learning Hub and SAP Fiori design guidelines, help learners follow a guided path from initial concepts to advanced development practices, making it easier to build professional-grade applications. These resources focus on SAP's learning journeys, courses, and developer resources for building apps using SAPUI5, Fiori, and related technologies. They are ideal for beginners or those looking to deepen their knowledge:

- **SAP Learning Hub** (*https://learning.sap.com/*)
  A central platform offering courses and certifications on various SAP technologies.
- **SAP Fiori design guidelines** (*https://experience.sap.com/fiori-design-web/*)
  This site covers best practices for designing SAP Fiori apps, from layout to UX.
- **SAP Learning Journey for developing apps with SAPUI5** (*https://learning.sap.com/learning-journeys/develop-sapui5-applications*)
  An official SAP journey guiding you through SAPUI5 app development.
- **Learning the basics of SAP Fiori** (*https://learning.sap.com/learning-journeys/learn-the-basics-of-sap-fiori*)
  A learning pathway that introduces the fundamental principles and features of SAP Fiori apps.

1.5.2 Best Practices and Documentation

For developers aiming to create high-quality, maintainable applications, knowing best practices and having access to detailed documentation is essential. This section curates official documentation and best practices guides, such as the UI5 best practices and SAPUI5 API documentation, which provide step-by-step guidance on how to structure your projects, use libraries, and implement responsive design. By following these guidelines, developers can ensure that their applications are efficient, scalable, and aligned with SAP standards. These are essential resources for developers seeking best practices and technical documentation to enhance their development process:

- **Flexible programming model explorer** (*https://sapui5.hana.ondemand.com/test-resources/sap/fe/core/fpmExplorer/index.html*)
  A site and virtual playground and sandbox where the flexible programming model is explained in detail.
- **SAPUI5 API documentation** (*https://ui5.sap.com/*)
  Comprehensive API documentation for SAPUI5, with details on various libraries, modules, and UI components.

1.5.3 Tools for SAP Fiori Development

In the world of SAP Fiori, having the right tools can significantly enhance your productivity. This section highlights tools such as SAP Fiori tools documentation and SAP Fiori theme designer, which simplify the development and customization of SAP Fiori apps. These tools allow developers to quickly prototype, extend, and style applications according to specific business requirements, reducing development time, and ensuring a more intuitive UX. This set of tools tailored for developing and customizing SAP Fiori apps include tools for app design, theme development, and UI inspections:

- **SAP Fiori tools documentation** (*https://help.sap.com/docs/SAP_FIORI_tools*)
  Detailed documentation for SAP Fiori tools, used for app development and extension in the SAP ecosystem.
- **UI5 Tooling** (*https://sap.github.io/ui5-tooling/stable/*)
  The UI5 Tooling resources that provide the build and development environment for SAPUI5 and OpenUI5 applications.
- **UI theme designer** (*https://pages.community.sap.com/topics/ui-theme-designer*)
  A tool used to customize and style SAP Fiori apps to meet specific corporate branding needs.

1.5.4 Open-Source and Inspection Tools

Developers working with SAPUI5 and SAP Fiori often need tools to inspect, analyze, and debug their applications effectively. This section focuses on open-source projects such as UI5 Inspector and UI5 Tooling, which provide vital resources for ensuring that applications function correctly. These tools help developers dive deeper into the internal workings of their apps, offering insights into performance optimization and debugging best practices. These resources focus on open-source projects and tools to inspect and analyze SAPUI5 components:

- **UI5 Inspector** (*https://sap.github.io/ui5-inspector/*)
  A Chrome DevTools extension to inspect, analyze, and debug SAPUI5 apps.
- **UI5 Tooling** (*https://sap.github.io/ui5-tooling/v3/*)
  Tools for setting up and managing development environments for SAPUI5 apps.

1.5.5 Git Tutorials

These three YouTube tutorials offer comprehensive *Git* training, providing developers with the knowledge and skills necessary to effectively use Git in their version control and project collaboration workflows. The tutorials cover everything from Git fundamentals to more advanced features, helping viewers understand how to manage code changes, collaborate on projects, and streamline their development processes.

- **Intermediate Git features** (*www.youtube.com/watch?v=tHsZtNqIOAk*)
  This video dives into more intermediate topics such as branching, merging, and resolving conflicts. It helps developers understand how to manage multiple versions of a project simultaneously.
- **Advanced Git techniques** (*www.youtube.com/watch?v=2oH-Me2ERtA*)
  The third tutorial takes things further by introducing advanced features such as rebasing, squashing commits, and setting up Git hooks. It's aimed at those who are already familiar with Git basics and want to take their skills to the next level.

1.5.6 Additional Resources and Community Contributions

These three SAP PRESS books are essential resources for developers and consultants working with SAP Fiori elements and SAP technologies. Each book delves into different aspects of the SAP landscape, providing in-depth knowledge, practical examples, and best practiceYs for building robust applications:

- *SAP Fiori Elements* (*www.sap-press.com/sap-fiori-elements_5641/*)
  This book focuses on SAP Fiori elements, guiding readers through the creation of consistent, user-friendly applications using predefined templates. It covers everything from setup to customization, offering a step-by-step approach to implementing SAP Fiori elements in real-world projects.
- *Full Stack Development with SAP* (*www.sap-press.com/full-stack-development-with-sap_5733/*)
  For developers aiming to master the full spectrum of SAP development, this book is an indispensable guide. It covers frontend, backend, and database integration, providing readers with comprehensive insights into developing full-stack applications within the SAP ecosystem.

- *Git: Project Management for Developers and DevOps Teams* (*www.sap-press.com/git_5555/*)
  This book serves as a complete reference for those looking to enhance their development skills using the powerful Git source code management.

1.5.7 Versioning

The official SAPUI5 documentation includes a section on versioning and backwards compatibility. This documentation describes the guidelines and practices of SAPUI5 regarding introducing new features, maintaining existing APIs, and dealing with deprecated properties, events, and methods:

- **Versioning**
  - Semantic versioning: SAPUI5 follows the semantic versioning scheme, where version numbers are specified in the form MAJOR.MINOR.PATCH.
  - MAJOR: Major versions with changes that are potentially not backward compatible.
  - MINOR: Minor versions with new, backward-compatible features.
  - PATCH: Bug fixes and minor enhancements that are backward compatible.
- **Backward compatibility**
  SAPUI5 places a high value on ensuring that existing applications continue to work even after updates. Any changes that could affect backward compatibility are carefully checked and avoided whenever possible.
- **Deprecated properties, events, and methods**
  If certain API elements are marked as deprecated, they will usually remain available for a transitional period before being removed. Developers are encouraged to switch to the recommended alternatives to avoid future compatibility issues.

For detailed information and specific guidelines, it's recommended that you read the full documentation directly on the SAPUI5 website.

1.6 Summary

In this chapter, we explored the evolution, comparison, design guidelines, and resources relevant to SAPUI5 and SAP Fiori technologies, offering a comprehensive foundation for developers and designers working within the SAP ecosystem.

Section 1.1 traced the origins of SAPUI5. It began as a tool to address SAP's need for a modern, responsive UI toolkit, and, over time, it has grown into a core component of SAP's broader strategy, particularly within the SAP Fiori environment. Moving to Section 1.2, we provided a comparison of the different technologies under the SAP UI umbrella. SAPUI5 is the commercial product used for SAP's business solutions, while

OpenUI5 is its open-source sibling, offering similar capabilities for broader use. SAP Fiori, while using SAPUI5, goes beyond just a framework, offering a comprehensive design system focused on UX. Lastly, UI5 Web Components enable developers to use SAPUI5 elements in non-SAP environments, making them adaptable for integration with other frontend technologies such as React and Angular.

In Section 1.3, we detailed how the guidelines define the UX for SAP Fiori apps. These guidelines emphasize simplicity, consistency, and efficiency, ensuring that SAP Fiori apps offer a seamless and intuitive experience across devices, adhering to a role-based design principle.

Section 1.4 explored the centralized hub that acts as the entry point for users to access their SAP Fiori apps. SAP Fiori launchpad personalizes user access to various applications and services based on roles, creating a streamlined, user-focused experience essential to any enterprise environment using SAP.

Finally, Section 1.5, covered the vital role of accessible and high-quality resources in development. Proper documentation such as the SAPUI5 SDK and SAP Fiori design guidelines, alongside community platforms and official learning tools, were emphasized as key to successful SAP UI development, ensuring that developers can keep their knowledge up-to-date and their projects aligned with best practices.

Chapter 2
Infrastructure

SAP Business Technology Platform (SAP BTP) plays a central role in the development of SAPUI5 applications. In addition to the development environment, SAP BTP also provides you with services for continuous integration/continuous delivery (CI/CD) and the provision of applications.

To successfully develop and deliver SAPUI5 and SAP Fiori applications, it's essential to choose the right development infrastructure. Of course, it's possible to perform simple development tasks in a conventional text editor. However, this approach has significant disadvantages that can severely impact developer efficiency. A text editor offers very limited support, and the loss of convenience and functionality is significant, especially in complex development projects.

A fully *integrated development environment* (*IDE*), on the other hand, offers significant advantages that go far beyond mere code writing. Modern IDEs, such as SAP Business Application Studio or Visual Studio Code (VS Code) with the corresponding SAP plug-ins, are specifically designed to optimize the development process.

A particularly critical point when accessing on-premise systems is security. In most companies, SAP systems aren't allowed to be accessed directly from the internet. This means that a secure connection between the cloud development environment and the on-premise system is required. This is where the *cloud connector* comes into play, a software component that acts as a reverse proxy or virtual private network (VPN). It allows you to securely connect cloud applications to on-premise systems without putting sensitive company data at risk. The cloud connector integrates fully with SAP Business Technology Platform (SAP BTP) and enables developers to securely and efficiently connect their applications to on-premise data sources. At this point, as a developer, administrator, or architect, you're faced with an important decision: Develop in the cloud or locally?

In this chapter, we provide you with a comprehensive overview and an introduction to the infrastructure for developing SAPUI5 applications. We start in Section 2.1 with a general overview of SAP BTP. Here, you'll learn about the role this platform plays in SAP's modern cloud ecosystem and how it supports the development and operation of applications through its diverse components, including data management, integration, extensibility, and artificial intelligence (AI). SAP BTP offers a variety of services and tools that enable developers to create and deliver innovative solutions quickly and efficiently. You'll also learn how SAP BTP acts as a bridge between existing on-premise systems and cloud environments to enable seamless integration and extension. SAP BTP is SAP's strategic platform for application development, integration of SAP and non-SAP applications, and extension of SAP software as a service (SaaS) products such as SAP S/4HANA Cloud or SAP SuccessFactors.

In the rest of the chapter, we'll look at the various development environments available for SAPUI5 development. In Section 2.2, we start with SAP Business Application Studio, the IDE developed specifically for SAP BTP that is ideally suited for cloud-based development. In Section 2.2.1, you'll learn how to use this modern IDE for your development processes, from creating a project to deploying the finished application. You'll see how SAP Business Application Studio helps you manage the entire lifecycle of an application in a unified and user-friendly environment.

For developers who prefer to work locally, VS Code is a strong alternative. In Section 2.2.2, we show you how to use this popular and versatile development environment for SAPUI5 development. You'll learn which plugins and configurations you need to optimize VS Code for your needs and how you can link local development processes with the services of SAP BTP.

By taking a holistic view of the development infrastructure and environments in conjunction with SAP BTP, this chapter not only provides you with the necessary knowledge but also with practical tools to implement your development projects efficiently and successfully. Regardless of whether you're developing in the cloud or on-premise, by the end of this chapter, you'll be able to select and optimally set up the development environment that suits your needs.

2.1 SAP Business Technology Platform

SAP BTP is a comprehensive and integrated platform as a service (iPaaS) offering that includes several key technologies: application development, process automation, integration, data and analytics, enterprise planning, and AI. With SAP BTP, companies can transform data into real business value, design end-to-end business processes, and link entire IT landscapes. In addition, the platform offers the option of customizing, extending, and personalizing existing SAP applications. This significantly reduces the total cost of ownership (TCO) by reducing the maintenance effort for SAP landscapes and third-party software over the entire lifecycle.

In Section 2.1.2, we delve deeper into the architecture of SAP BTP. You'll gain detailed insights into the platform's fundamental building blocks, such as subaccounts, the global account, and providing services through service plans and entitlements. You'll also understand how SAP BTP's multicloud strategy enables you to run your applications on different cloud infrastructures such as Amazon Web Services (AWS), Microsoft Azure, or Google Cloud.

Another key building block for success when developing in SAP BTP is the cloud connector, which we discuss in Section 2.1.3. Here you'll learn how the cloud connector establishes a secure and high-performance connection between your on-premise systems and the cloud. This service enables you to seamlessly integrate cloud applications into existing enterprise landscapes without putting sensitive data or systems at risk. Finally, in Section 2.1.4, we'll cover key aspects of SAP BTP security.

2.1.1 SAP BTP for Intelligent Enterprises

In today's business world, companies need access to real-time data to make informed decisions and integrate advanced technologies into agile, integrated business processes. SAP's strategy is aimed at integrating end-to-end processes, whether the solutions are from SAP, SAP partners, or third-party providers. This integration of processes and systems is of central importance for creating an intelligent company in which business processes run smoothly while drawing on the latest technologies and best practices. An integral part of the SAP strategy is to support end-to-end business scenarios, such as the following:

- Lead to cash (from initial customer contact to invoicing)
- Source to pay (from purchase to payment)
- Design to operate (from design to operation)
- Hire to retire (from the hiring process to the retirement of an employee)

SAP BTP provides integration capabilities that ensure end-to-end business processes are seamlessly connected across both SAP and non-SAP solutions. The following key characteristics, also known as suite qualities, support a consistent and continuous user experience (UX). Important properties of SAP BTP are as follows:

- **Seamless UX**
  With SAP Fiori, SAP offers a unified user interface that can be used across many SAP solutions. This ensures a consistent look and feel, improves the UX, and reduces interruptions in the workflow.
- **Consistent security and identity management**
  With SAP Cloud Identity Services, companies can centrally manage identities and use the single sign-on (SSO) functionality across various end-to-end processes. This ensures consistent and secure management of access rights and user data.
- **Coordinated domain models, APIs, and events**
  SAP Master Data Integration, as a cloud service, provides a unified view of master data in hybrid IT landscapes. APIs can be used to automate integration processes between SAP solutions, SAP, and third-party providers. Similarly, an event-based

integration strategy, based on predefined integration content, enables fast and efficient networking. All relevant APIs are available in SAP Business Accelerator Hub.

- **Embedded analytics across different solutions**
  SAP BTP enables embedded analytics capabilities that are available in various SAP solutions such as SAP S/4HANA or SAP SuccessFactors. These analytics provide in-depth insights into business processes and support informed decision-making based on real-time data. Embedding SAP Analytics Cloud enables the creation of reports and dashboards directly in business processes.
- **Unified workflow inbox**
  The workflow inbox in SAP BTP provides a unified overview of upcoming tasks from various SAP solutions, both on mobile devices and on desktops. This speeds up the completion of tasks because all relevant workflows are available in one place.
- **Coordinated lifecycle management**
  SAP BTP provides harmonized solutions for application deployment, setup, and operation. These include automated deployments and guided integration setups that enable companies to carry out implementation projects more efficiently and reduce manual configurations. This is particularly true for end-to-end scenarios such as lead to cash, source to pay, recruit to retire, and design to operate.
- **End-to-end process blueprints**
  SAP provides process blueprints based on the Industry Reference Architecture standard. These support you both in planning implementation projects and in making architectural decisions. These blueprints make it easier for companies to plan and build their IT landscapes faster, which reduces the planning effort.

SAP BTP not only offers a comprehensive platform for companies to automate, integrate, and analyze business processes, but it also supports developers—especially those who create SAPUI5 applications—in a variety of ways. SAP BTP provides tools and services that optimize the entire development cycle, from planning and programming to application deployment and maintenance. The architecture of SAP BTP is designed to provide organizations with a highly flexible and scalable environment for developing and operating cloud applications. It enables a clear separation and management of projects, resources, and environments through a multilevel structure of global accounts and subaccounts. We'll take a detailed look into the architecture in Section 2.1.2.

SAP BTP supports developers of SAPUI5 applications in various areas. SAP BTP offers SAP Business Application Studio, an IDE, which will be covered in Section 2.2.1. This modern tool enables developers to create, test, and deploy SAPUI5 applications in the cloud. Features such as code completion, syntax highlighting, debugging, and intelligent error detection significantly speed up the development process. For developers who prefer a cloud-based solution, SAP Business Application Studio offers seamless integration with SAP BTP services and simplified access to the development resources they need.

To help you navigate the world of SAP BTP, which currently offers around 90 different services, SAP provides SAP Discovery Center (*https://discovery-center.cloud.sap/*) as a central point of contact. This not only offers a comprehensive overview of the available services but also provides detailed reference architectures that show you how these services can be optimally used in real business scenarios. In addition, SAP Discovery Center offers *missions*—guided learning paths that have been specially developed for self-study. These missions enable developers and companies to get to know the various services step-by-step and to use them in projects in a practical way. Whether you're implementing a new application or enhancing existing processes, SAP Discovery Center offers a wealth of information and practical guidance to help you get started with SAP BTP and use the services effectively. In addition, best practice examples and technical guides are provided to ensure that both beginners and advanced users can use the services successfully.

Figure 2.1 shows the service details of SAP Business Application Studio in SAP Discovery Center. Service descriptions are usually structured identically. You find relevant information under the **Overview**, **Pricing**, **Related Missions**, **Roadmap**, and **Customer Stories** tabs. On the right side, you'll find **Tools**, **Resources**, and **Tutorials & Learning** areas, along with a link to the **Community** homepage.

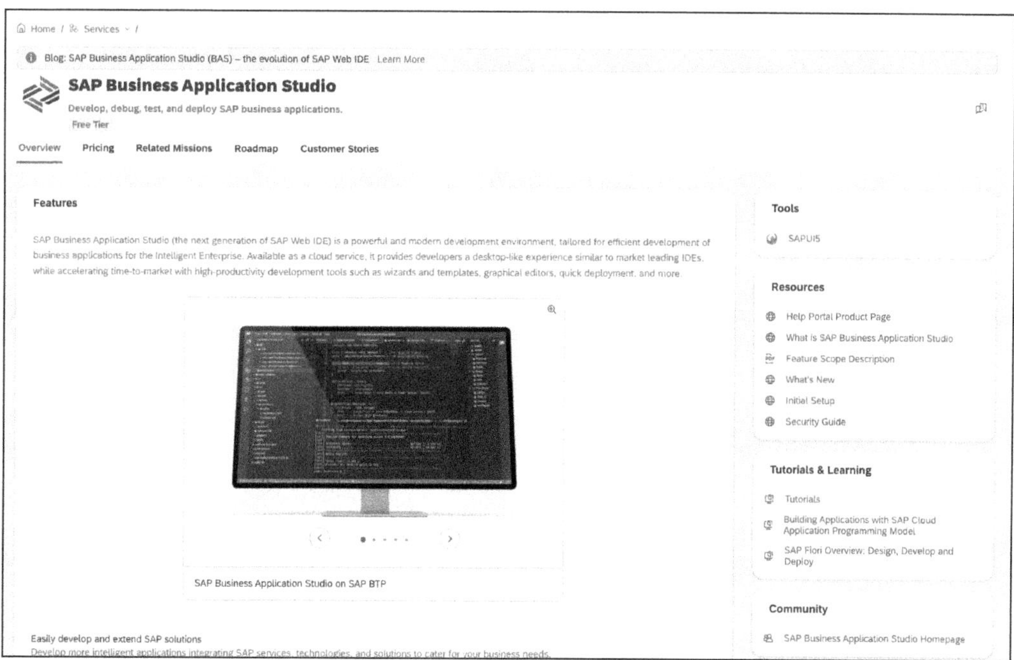

Figure 2.1 SAP Discovery Center

SAP BTP provides comprehensive APIs and OData services that are seamlessly integrated with SAP backend systems. Developers of SAPUI5 applications can use these APIs to retrieve data from SAP S/4HANA, SAP SuccessFactors, and other SAP applications and use that data in their applications. SAP Business Accelerator Hub (*https://api.sap.com/*) provides predefined APIs that enable developers to integrate business processes across cloud and on-premise systems (see Figure 2.2). This eliminates the need to create complex backend integrations yourself.

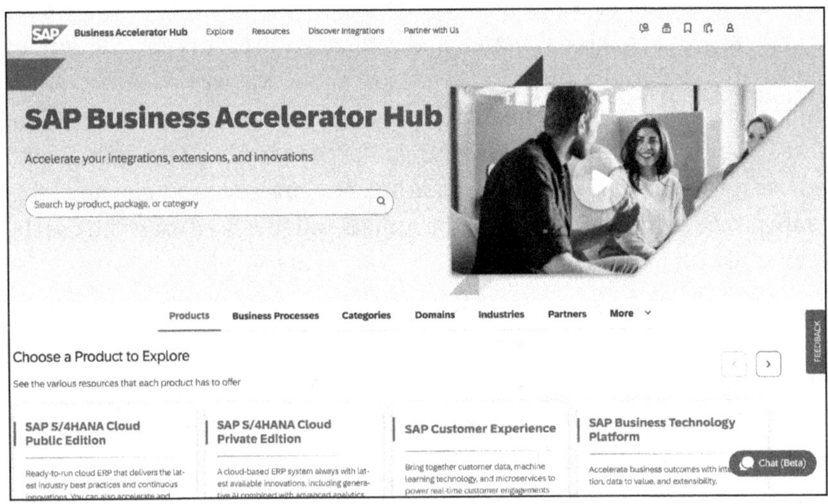

Figure 2.2 SAP Business Accelerator Hub

A key feature of SAP BTP is the cloud connector (covered in Section 2.1.3), which allows developers to access on-premise data sources securely without compromising the security of the company's systems. This is particularly important for SAPUI5 developers who create applications that need to access both cloud and on-premise data sources. The cloud connector acts as a secure bridge between the cloud environment and internal systems, providing a flexible and secure development environment.

A major advantage of SAP BTP is the ability to extend and personalize SAP applications. Developers can seamlessly integrate SAPUI5 applications into existing SAP solutions or create specific extensions for existing business applications to meet the specific needs of their organization. The flexibility of SAP BTP makes it possible to develop customized solutions without affecting the standard of SAP applications.

SAP BTP also offers developers the option of integrating AI and machine learning into SAPUI5 applications. By providing ready-made AI services, developers can easily integrate functions such as text recognition, image recognition, chatbots, or prediction models into their applications. This enables developers to equip SAPUI5 apps with advanced capabilities to make automated decisions or forecasts and improve the UX.

One of the biggest challenges for developers is deploying and scaling applications. SAP BTP provides scalable cloud infrastructures that enable developers to quickly and reliably deploy their SAPUI5 applications worldwide. With automated deployment capabilities and support for CI/CD pipelines, developers can seamlessly move their applications from the development environment to the production environment.

2.1.2 Architecture

The architecture of SAP BTP is specifically designed to provide companies with a powerful, flexible, and scalable cloud environment for application development, integration, and operation. The modular structure of SAP BTP enables companies to manage their IT resources efficiently while developing customized solutions for specific business needs. A key concept here is the organization into global accounts and subaccounts. This structure allows companies to optimally segment their cloud resources and clearly separate specific deployment and development environments, such as for development, testing, and production.

The global account acts as a high-level management and structural object to set central policies and quotas for resource usage. This allows organizations to keep track of their overall SAP BTP usage, and efficiently control how different projects or departments access the platform and which cloud resources they can use. The management of the global account is done in the SAP BTP cockpit (see Figure 2.3).

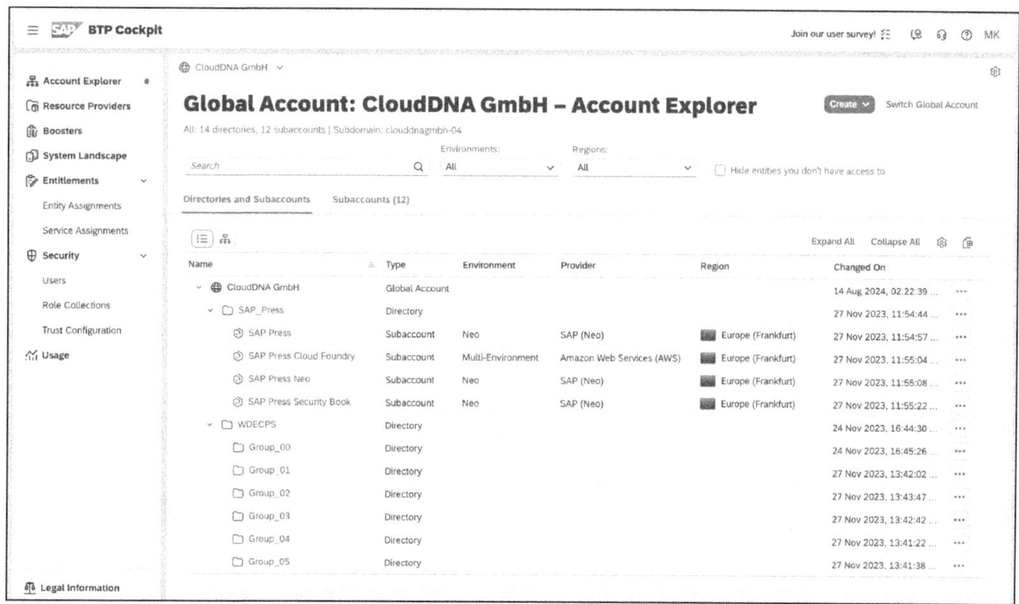

Figure 2.3 SAP BTP Cockpit Global Account Overview

SAP BTP offers two types of global accounts: trial accounts, which are completely free, and enterprise accounts for companies. Within an enterprise account, both free and paid plans can be used, depending on the specific requirements of the company. A global account is the contractual agreement between you or your company and SAP. It serves as a central administration level through which all relevant resources are organized and assigned. In a global account, you manage, among other things, subaccounts, members, entitlements (usage rights), and quotas (resource contingents). You receive a certain amount of entitlements and quotas to use the platform resources. These resources are initially allocated to the global account and then distributed to the various subaccounts as needed to enable actual consumption.

There are two billing models for SAP BTP:

- **Consumption-based model**
  With this model, you pay only for the resources you use. It offers maximum flexibility, as you can scale dynamically and use resources as needed. This is particularly useful for companies that have variable requirements and need to adjust their consumption regularly.
- **Subscription-based model**
  Here, you pay a fixed fee for a specific package of resources. This model is well suited for organizations with stable, predictable requirements, as it offers fixed costs and makes planning easier.

The subaccounts, on the other hand, provide a further level of structuring and flexibility. They make it possible to separate different projects, business areas, or geographical regions so that individual applications or workloads can be assigned to specific teams or environments. Each subaccount can be individually configured and has its own authorizations, services, and data, which offers granular control and flexibility for different business scenarios. This clear separation ensures both better scalability and increased security and compliance, as access to sensitive data and systems can be precisely controlled. Each subaccount must be created within a region and a data center. The *region* reflects the physical location of the data center where applications, data, and services are hosted (see Figure 2.4). The region doesn't have to be the same as the region of your enterprise location.

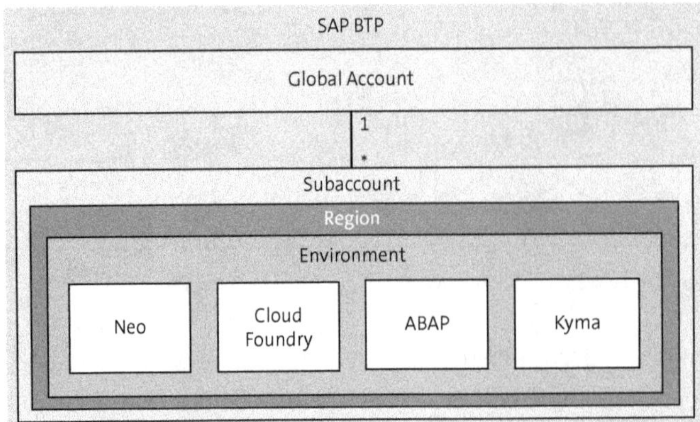

Figure 2.4 Account Structure

The subaccounts structured under a global account are independent of each other. You must take this into account when planning the overall architecture in terms of security, member management, data management, data migration, and integration. The current usage of resources can be monitored on a global account level in the SAP BTP cockpit (see Figure 2.5).

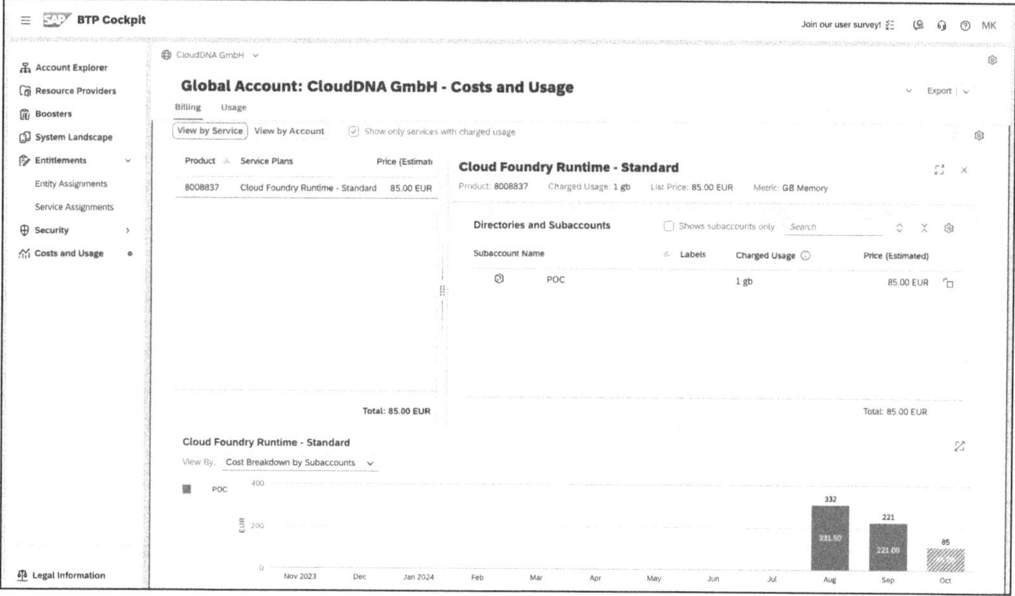

Figure 2.5 Cost and Usage Reporting

In addition, the architecture of SAP BTP supports different runtime environments and multicloud deployments, enabling companies to operate applications in different regions and on multiple cloud infrastructures such as AWS, Microsoft Azure, or Google Cloud. This multicloud capability extends the flexibility of the platform by enabling global scalability and geo-redundant deployment of applications. In the SAP BTP cockpit, all subaccounts assigned to your global account are displayed under the **Subaccounts** menu item (see Figure 2.6).

Choosing a Reasonable Subaccount Structure

At the subaccount level, you can configure the system connections to the on-premise landscape, among other things. Therefore, the subaccounts play an important role in connection with the cloud connector. Separate subaccounts for production systems and development are recommended, if only to separate resources and system connections. Furthermore, you can create additional security by separating them according to the backends used (e.g., SAP ERP Human Capital Management [SAP ERP HCM], SAP SuccessFactors, SAP S/4HANA, etc.) or according to responsible teams.

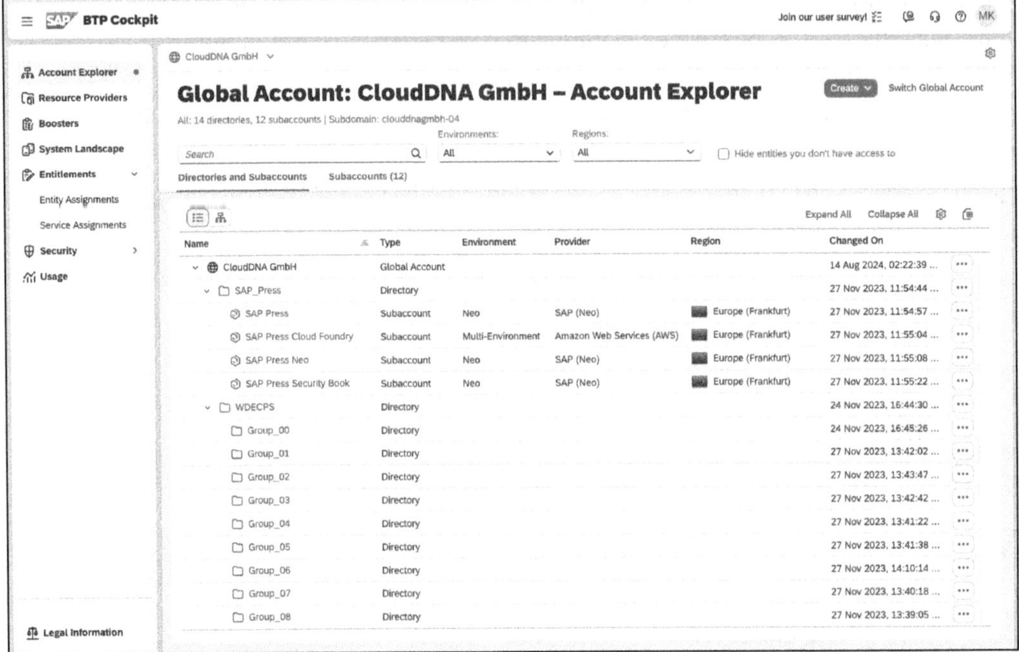

Figure 2.6 Subaccount Overview

Unlike global accounts, you can create subaccounts yourself, by following these steps:

1. Click on the **Create** button in the subaccount overview.
2. In the dialog for creating the subaccount, provide a meaningful **Display Name**, as shown in Figure 2.7.

Create Subaccount
Display Name*
SAPUI5
Description
SAP Press SAPUI5 Book
Region*
Europe (Frankfurt)
Subdomain*
sapui5-66qhdfbc
Parent*
CloudDNA GmbH
Advanced
Used for production
Enable beta features
Labels
Name:
Values:
Enter label name
Create
Cancel

Figure 2.7 New Subaccount Dialog

3. Select a **Region** (i.e., hyperscaler data center) where your subaccount should be hosted. Please don't use the SAP-operated data centers anymore (e.g., SAP BTP, Neo environment, as Neo will be sunset in 2028).
4. In the next step, a **Subdomain** is automatically created. If required, adjust the subdomain. Please be aware that the subdomain must be unique within a datacenter; that is, if the subdomain is already used by another customer, you won't be able to use it.
5. The **Parent** field is set to your global account by default, as SAP allows you to create directories for structuring your subaccounts. You can select any directory as parent.
6. If you select the **Used for Production** checkbox, this won't directly affect your subaccount, but it will help SAP prioritize your support tickets if it's a production subaccount.
7. You can optionally add **Labels** to your subaccount, which make it easier to identify them in the **Subaccount Overview**.
8. Finally, click the **Create** button.

Normally, the next step would be for you to manually assign the relevant entitlements to the subaccount, such as for SAP Business Application Studio. This is necessary if you want to activate and configure the service independently. However, SAP has recognized that this process is often time-consuming and therefore offers boosters for the most frequently used services. Before you consider the manual process, check whether a booster is available for the service you want, such as SAP Business Application Studio. A booster is a guided wizard that takes you step-by-step through the setup and configuration of the respective service. Once all the required information has been provided, the booster takes over the automated assignment of entitlements and the complete configuration of the service. This approach not only saves time but also reduces the potential for configuration errors, as the booster ensures that all the necessary settings are applied correctly. This means you can start actual development faster without having to worry about the detailed configuration.

Open the **Boosters** section in the side menu of the SAP BTP cockpit, and search for the desired service, such as SAP Business Application Studio, to see if a suitable booster is available (see Figure 2.8).

Please be aware that some SAP BTP services, such as SAP Continuous Integration and Delivery, aren't provided in all regions. The regional availability can be checked in the service details in SAP Discovery Center. Figure 2.9 shows the availability of the SAP Continuous Integration and Delivery service.

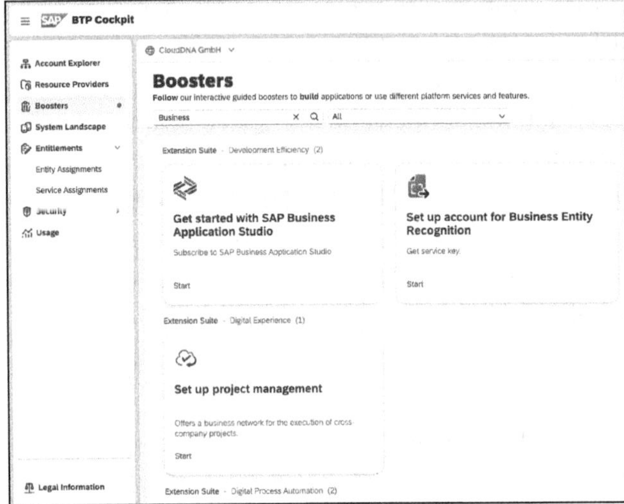

Figure 2.8 Booster Overview

Figure 2.9 SAP Continuous Integration and Delivery Regions

2.1.3 Cloud Connector

The cloud connector is the central component for connecting your on-premise system landscape to the SAP cloud systems. The basic idea of the cloud connector is to be able to administer access centrally and to enable the appropriate monitoring.

In a hybrid system landscape, the complexity of your system architecture inevitably increases. A few years ago, there was already a trend to reintegrate systems, for example, with the methods and tools of system landscape optimization. The integration of the business partner model into the world of SAP ERP was also an example of this trend. Today, cloud solutions add another external layer to these integrated systems. This changes the requirements for system administration. For on-premise systems, issues such as the following play a front-line role:

- Where is the server located?
- How is this server virtualized?
- In which network or zone is this server located?

In the cloud world, this information isn't necessarily transparent.

The cloud connector forms the central interface to the services of SAP BTP within your own system landscape. It acts as a link between cloud and on-premise systems and communicates with SAP BTP via secure connections.

At the architecture level, the goal is for system connections that send data to or receive data from SAP cloud solutions to be handled by the cloud connector. Multiple instances of the cloud connector can also be deployed. Distinct from these integration scenarios are connections from end devices that typically access the cloud applications or services directly.

When a service in SAP BTP, such as SAP Build Apps, requires data from your local SAP systems, the cloud connector comes into play. Because users access SAP BTP services from the internet, you want the accesses into the local system to take place via a central tool rather than directly via the internet. To do this, you use the cloud connector. Figure 2.10 illustrates the architecture of this scenario.

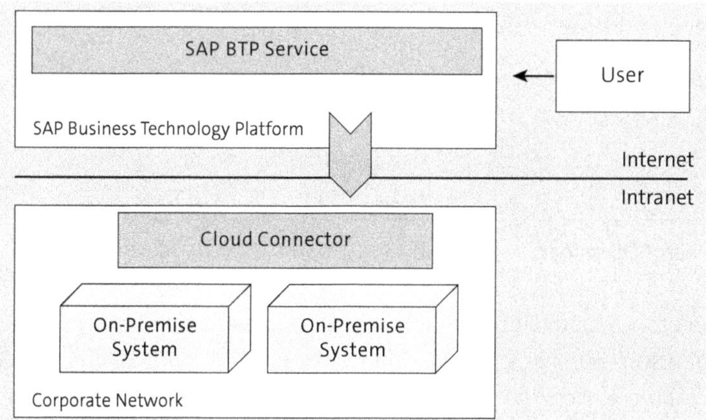

Figure 2.10 Cloud Connector

In this section, we'll show you how to configure the cloud connector. The most common scenario in combination with the SAP Build product portfolio is accessing data from an on-premise ABAP system (e.g., SAP Business Suite or SAP S/4HANA). This configuration is done via a mapping.

The cloud connector is offered for the Linux and Windows operating systems. There is also a version for MacOS, but this isn't intended for productive use.

As illustrated in Figure 2.11, all versions can be downloaded from *https://tools.hana.ondemand.com*. Preconfigured test versions (portable), which are aimed at developers, or versions suitable for production (installer) can be downloaded there. Developer versions enable testing of system connections to the cloud, for example, without having to contact the SAP Basis department. However, it should be noted, for example, that this version can't be run in the background. For productive installations, an installation program is provided via which you execute the installation. To install the cloud connector, a version of the Java Development Kit (JDK) must be installed on the local machine.

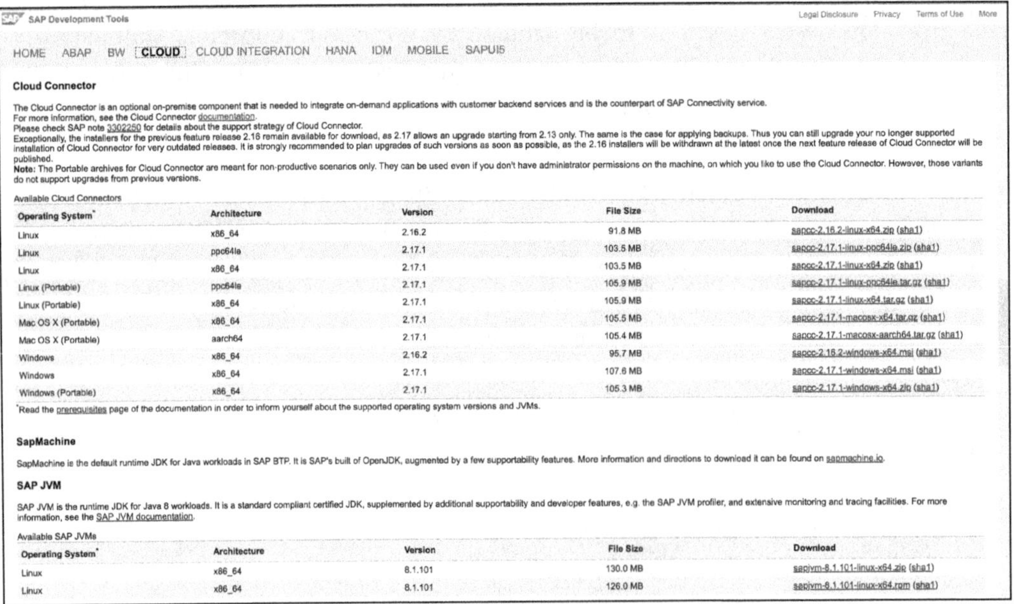

SAP Development Tools

Legal Disclosure Privacy Terms of Use More

HOME ABAP BW CLOUD CLOUD INTEGRATION HANA IDM MOBILE SAPUI5

Cloud Connector

The Cloud Connector is an optional on-premise component that is needed to integrate on-demand applications with customer backend services and is the counterpart of SAP Connectivity service.
For more information, see the Cloud Connector documentation.
Please check SAP note 3302250 for details about the support strategy of Cloud Connector.
Exceptionally, the installers for the previous feature release 2.16 remain available for download, as 2.17 allows an upgrade starting from 2.13 only. The same is the case for applying backups. Thus you can still upgrade your no longer supported installation of Cloud Connector for very outdated releases. It is strongly recommended to plan upgrades of such versions as soon as possible, as the 2.16 installers will be withdrawn at the latest once the next feature release of Cloud Connector will be published.
Note: The Portable archives for Cloud Connector are meant for non-productive scenarios only. They can be used even if you don't have administrator permissions on the machine, on which you like to use the Cloud Connector. However, those variants do not support upgrades from previous versions.

Available Cloud Connectors

Operating System*	Architecture	Version	File Size	Download
Linux	x86_64	2.16.2	91.8 MB	sapcc-2.16.2-linux-x64.zip (sha1)
Linux	ppc64le	2.17.1	103.5 MB	sapcc-2.17.1-linux-ppc64le.zip (sha1)
Linux	x86_64	2.17.1	103.5 MB	sapcc-2.17.1-linux-x64.zip (sha1)
Linux (Portable)	ppc64le	2.17.1	105.9 MB	sapcc-2.17.1-linux-ppc64le.tar.gz (sha1)
Linux (Portable)	x86_64	2.17.1	105.9 MB	sapcc-2.17.1-linux-x64.tar.gz (sha1)
Mac OS X (Portable)	x86_64	2.17.1	105.5 MB	sapcc-2.17.1-macosx-x64.tar.gz (sha1)
Mac OS X (Portable)	aarch64	2.17.1	105.4 MB	sapcc-2.17.1-macosx-aarch64.tar.gz (sha1)
Windows	x86_64	2.16.2	95.7 MB	sapcc-2.16.2-windows-x64.msi (sha1)
Windows	x86_64	2.17.1	107.6 MB	sapcc-2.17.1-windows-x64.msi (sha1)
Windows (Portable)	x86_64	2.17.1	105.3 MB	sapcc-2.17.1-windows-x64.zip (sha1)

*Read the prerequisites page of the documentation in order to inform yourself about the supported operating system versions and JVMs.

SapMachine

SapMachine is the default runtime JDK for Java workloads in SAP BTP. It is SAP's built of OpenJDK, augmented by a few supportability features. More information and directions to download it can be found on sapmachine.io.

SAP JVM

SAP JVM is the runtime JDK for Java 8 workloads. It is a standard compliant certified JDK, supplemented by additional supportability and developer features, e.g. the SAP JVM profiler, and extensive monitoring and tracing facilities. For more information, see the SAP JVM documentation.

Available SAP JVMs

Operating System*	Architecture	Version	File Size	Download
Linux	x86_64	8.1.101	130.0 MB	sapjvm-8.1.101-linux-x64.zip (sha1)
Linux	x86_64	8.1.101	126.0 MB	sapjvm-8.1.101-linux-x64.rpm (sha1)

Figure 2.11 Cloud Connector Download

The actual installation of the cloud connector won't be discussed in detail here. We recommend that you consult the book *Cloud Connector for SAP* (SAP PRESS, 2023, *www.sap-press.com/cloud-connector-for-sap_5683/*) for the installation. After the cloud connector has been installed, it can be accessed via a web browser. For this purpose, a user with the name "administrator" and the password "manage" is created in the standard.

It's mandatory to change the password when logging in for the first time. You must also specify whether it's the **Master** or **Shadow** installation (see Figure 2.12).

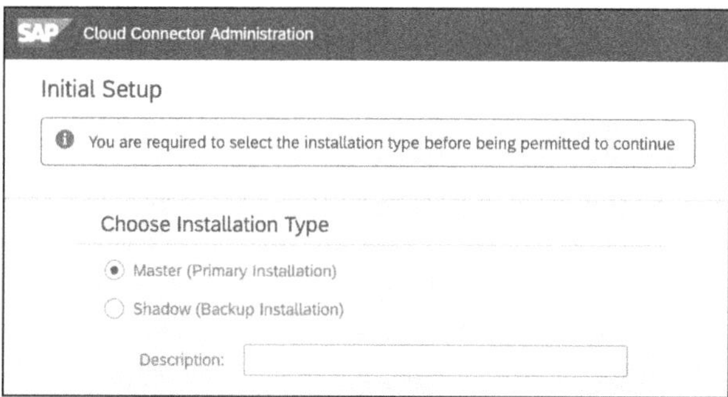

Figure 2.12 Initial Setup

In the next step, you must connect to the desired subaccount in SAP BTP from the cloud connector. Therefore, navigate to the **Define Subaccount** entry in the side menu, and click the **+ Add Subaccount** button (see Figure 2.13).

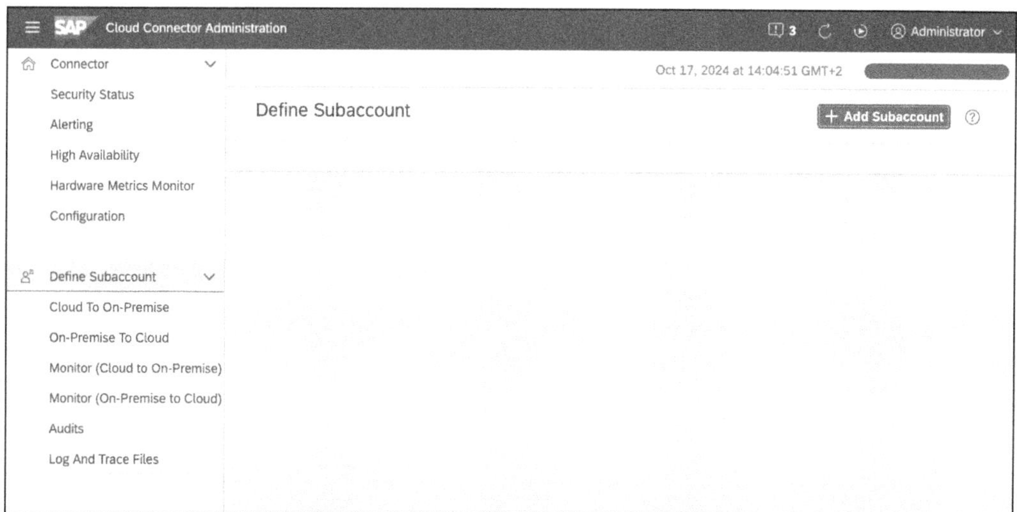

Figure 2.13 Add Subaccount

You can choose between a manual configuration and a configuration based on the authentication data downloaded from SAP BTP. Subsequently, we show the automated configuration, which was introduced in one of the latest versions of the cloud connector. We recommend using this configuration option, as it reduces the number of potential configuration errors. Select the corresponding option, and click the **Next** button (see Figure 2.14).

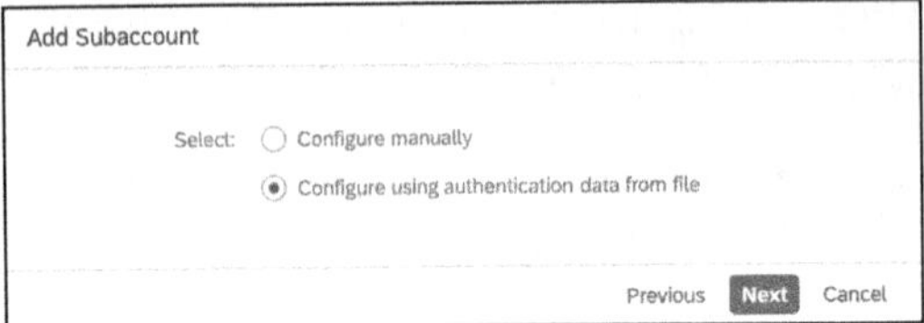

Figure 2.14 Perform Automatic Configuration

You must now download the authentication data from your subaccount. Therefore, open the subaccount in the SAP BTP cockpit, and use the side menu to navigate to the **Connectivity • Cloud Connectors** section, as shown in Figure 2.15. Then, click the **Download Authentication Data** button.

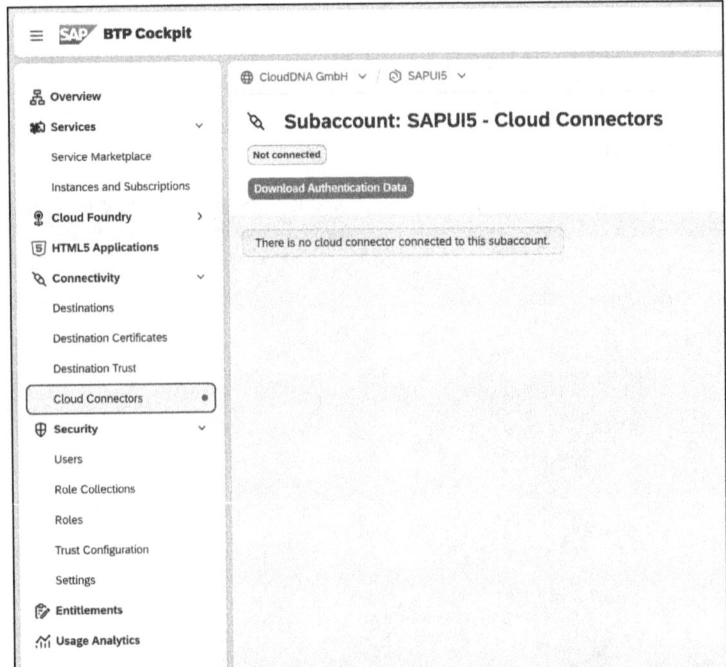

Figure 2.15 Download Authentication Data

Now switch back to the cloud connector setup/configuration, and select the configuration file in the **Add Subaccount** dialog (see Figure 2.16).

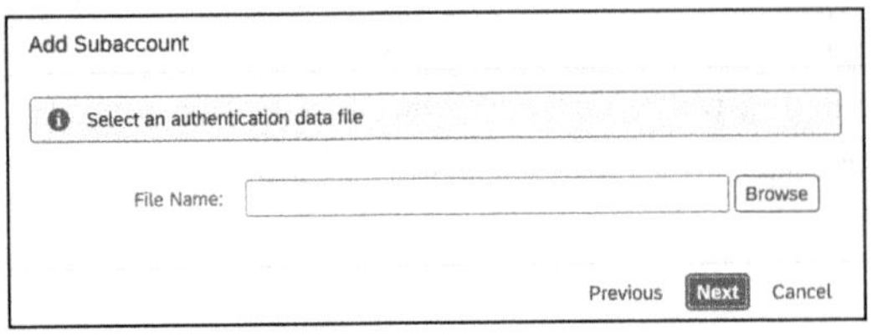

Figure 2.16 Select Authentication Data

You can then optionally add a **Location ID** and **Description** (see Figure 2.17). Finally, click on the **Finish** button.

Add Subaccount

Region: cf-eu10
Subaccount: 86f91ba2-550e-4150-a92d-b707d7b81514
Display Name: sapui5-a29wqi23
Location ID:
Description:

Previous | Finish | Cancel

Figure 2.17 Add Location ID and Description

You'll now find the subaccount in the side menu. The cloud connector should be connected to the desired subaccount (see Figure 2.18).

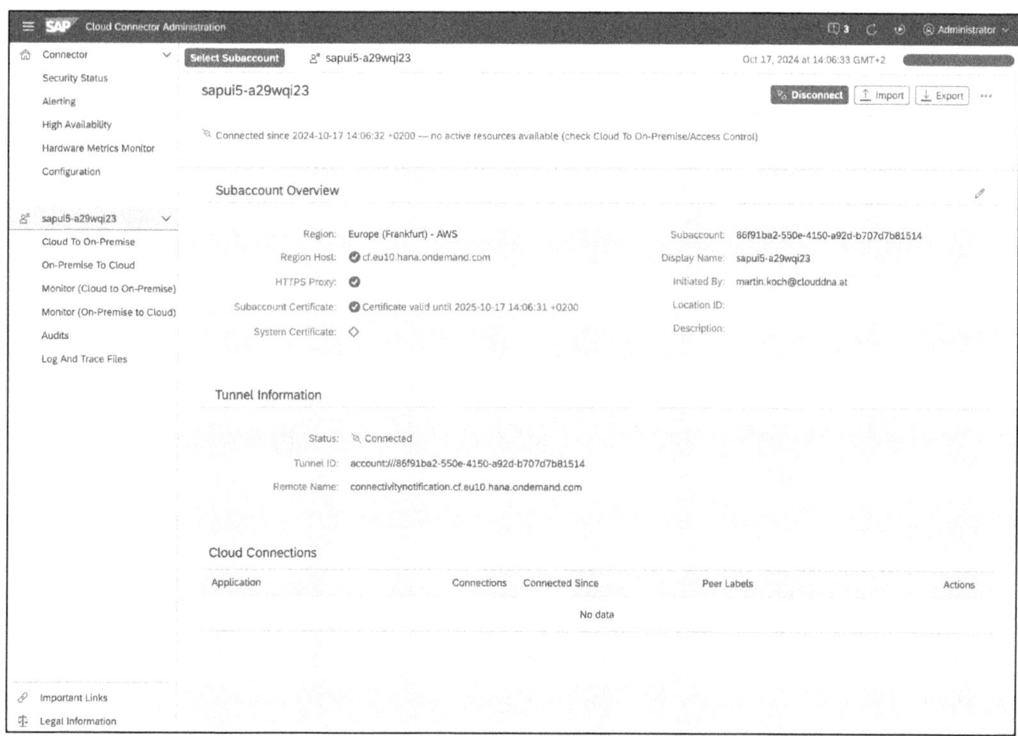

Figure 2.18 Subaccount Connection Status

You should also check the connection status in the subaccount. Therefore, open the subaccount in the SAP BTP cockpit, and navigate to the **Connectivity • Cloud Connector** section in the side menu. You should see that the cloud connector has successfully connected to the subaccount (see Figure 2.19).

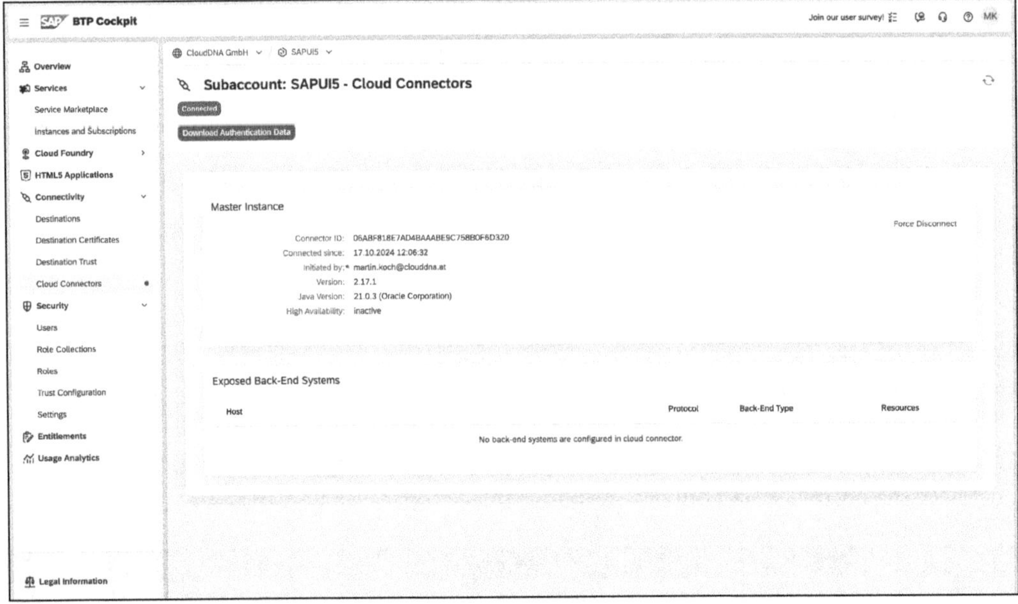

Figure 2.19 Cloud Connector Connection Status

All connections from SAP BTP to your on-premise systems must be explicitly enabled in the cloud connector. In addition to the type of connection (e.g., ABAP system), you specify the connection data of the system and how authentication is to take place. Therefore, you define a mapping. The mapping is used to hide the physical hostname and port in SAP BTP and to use a virtual hostname and port instead. As illustrated in Figure 2.20, the configuration is triggered in the **Cloud To On-Premise** section in the side menu. Click on the **+** button to add a new mapping.

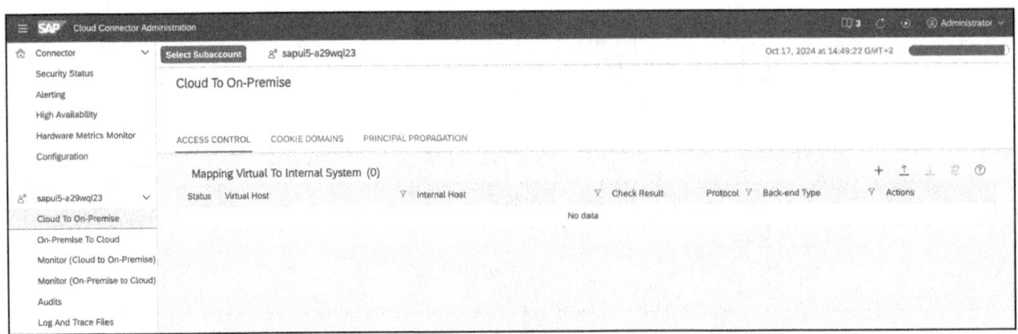

Figure 2.20 Mapping Overview

In the first step, you must select a **Back-end Type**. In the case of an SAP Business Suite or an SAP S/4HANA system, you can select **ABAP System**, as shown in Figure 2.21.

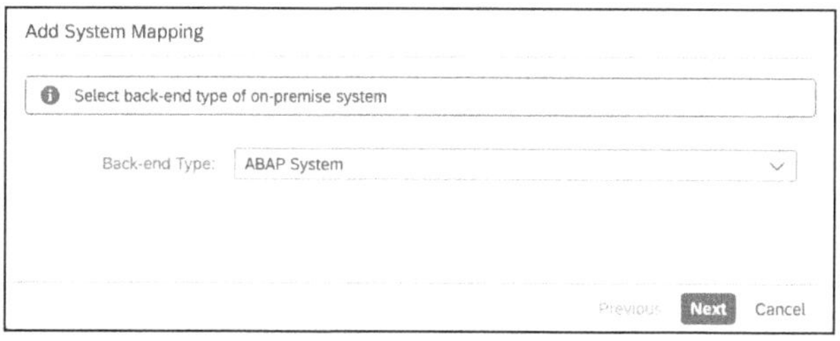

Figure 2.21 Backend Type Selection

In the next step, you need to select the protocol that will be used for communication between the cloud connector and the backend. If OData services or SAP Fiori apps are accessed, this will be **HTTP** or alternatively **HTTPS**, as shown in Figure 2.22.

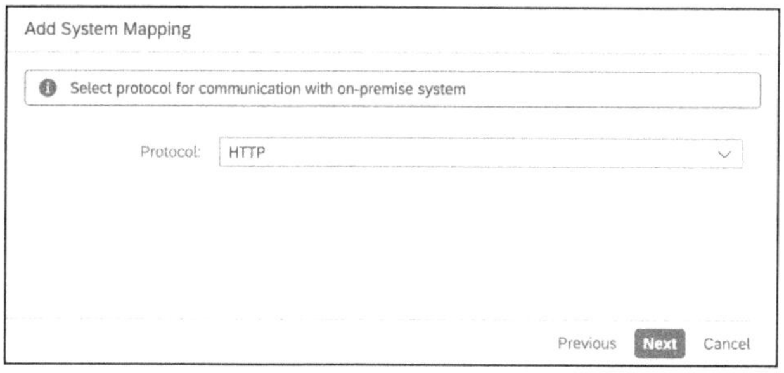

Figure 2.22 Protocol Selection

After that, you must specify the **Internal Host** and the **Internal Port** (see Figure 2.23). This is the hostname and port through which the cloud connector reaches the backend.

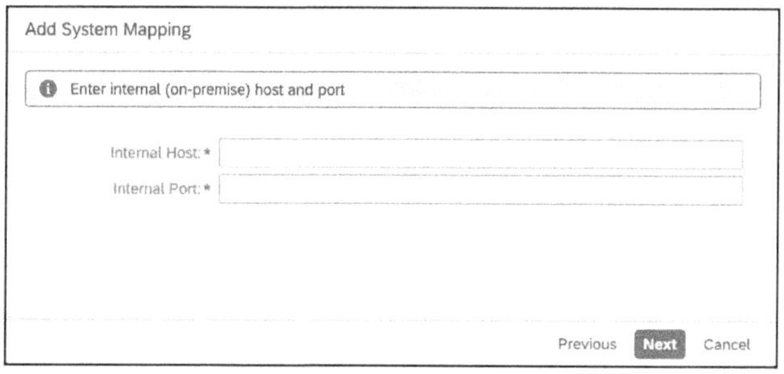

Figure 2.23 Internal Host Configuration

Then, as shown in Figure 2.24, the **Virtual Host** and the **Virtual Port** must be specified. This allows the backend to be addressed in the SAP BTP subaccount. SAP recommends choosing a virtual hostname that is different from the physical hostname. The virtual hostname can't be changed later.

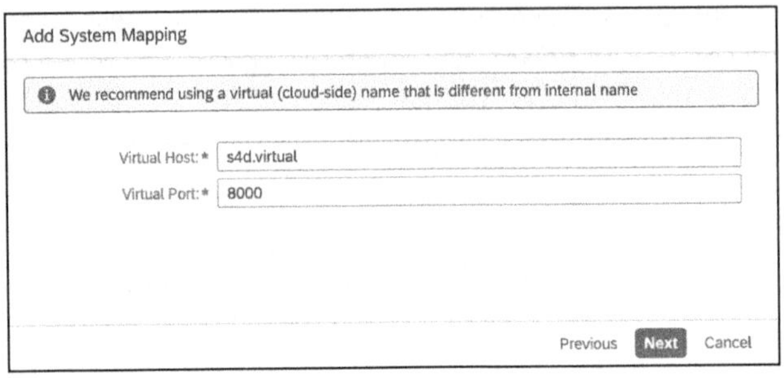

Figure 2.24 Virtual Host Configuration

You can then specify whether principal propagation should be used (see Figure 2.25). Principal propagation allows you to route the identity of the user from SAP BTP through to the backend. This is necessary, for example, when authorization checks take place in the backend.

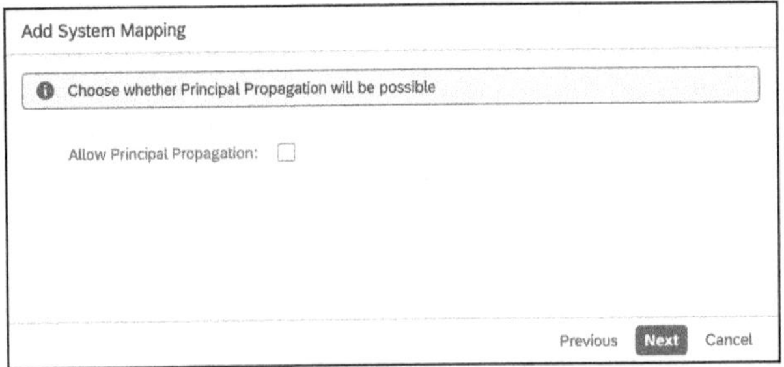

Figure 2.25 System Mapping Principal Propagation

After that, you still need to configure which hostname should be used in the HTTP request header host (see Figure 2.26). In most cases, it's sufficient that the virtual host is used.

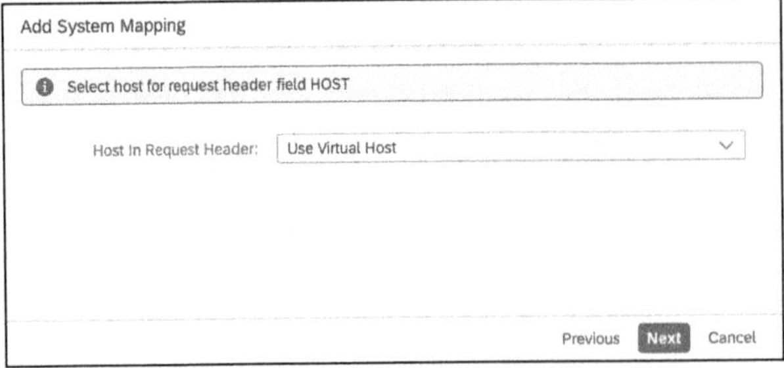

Figure 2.26 System Mapping Request Header Configuration

In the next step, you can optionally assign a **Description** (see Figure 2.27).

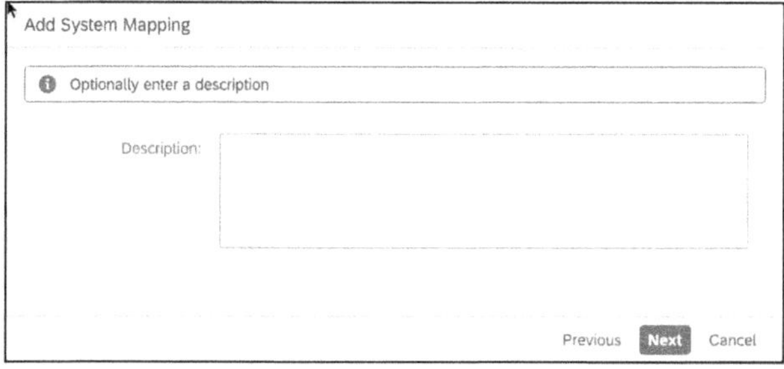

Figure 2.27 System Mapping Description

Finally, as shown in Figure 2.28, you'll again see a summary of the entries made previously. In addition, you can specify whether a check should be made immediately after saving the mapping to determine whether the internal host can be reached from the cloud connector.

Add System Mapping

Summary

Protocol: HTTP (None)

Internal: s4dev.clouddna.at:8001

Virtual: s4d.virtual:8000

Check Internal Host: ✓

Previous Finish Cancel

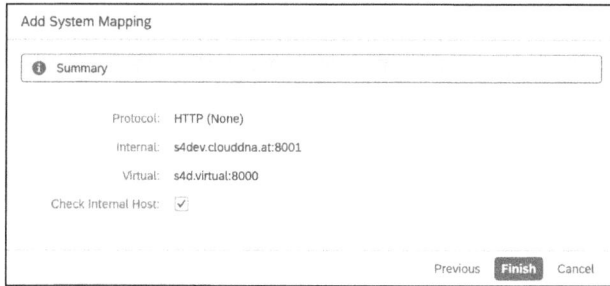

Figure 2.28 System Mapping Overview

If you use HTTPS as the communication protocol, an error will usually occur at this point (see Figure 2.29). The error occurs because the cloud connector checks if the Secure Sockets Layer (SSL) certificate of the backend system is in the allowlist. This requires that either the certificate is imported into the cloud connector or that the check is turned off.

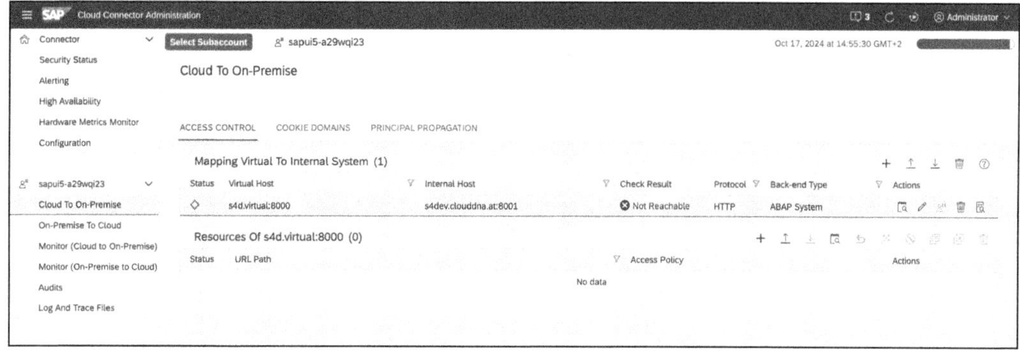

Figure 2.29 Connection Status

In the cloud connector, navigate to the **Connector • Configuration** section in the side menu to turn off the check (see Figure 2.30).

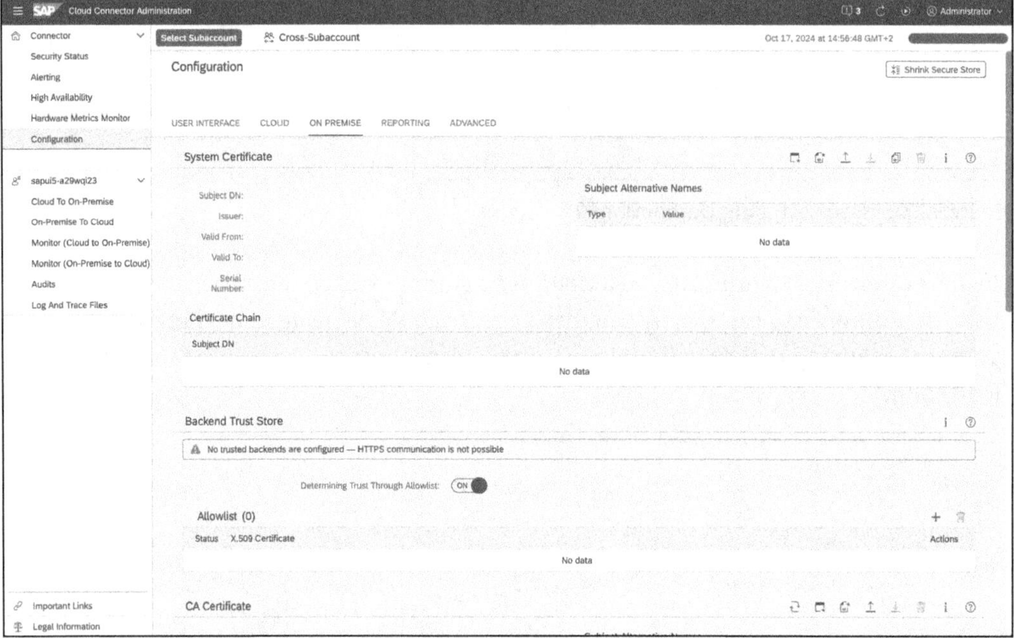

Figure 2.30 Trust Configuration

Disable **Determining Trust Through Allowlist**. Alternatively, you can import the corresponding certificate by clicking the **+** button in the **Allowlist** area.

This means that you've created all the technical prerequisites for accessing the backend. When using OData services and accessing SAP Fiori apps, the Internet Communication Framework (ICF) is used in the backend. Therefore, as shown in Figure 2.31, you still need to configure in the cloud connector which URL paths the cloud connector is allowed to access. To do this, click on the **+** button in the **Resources Of** area.

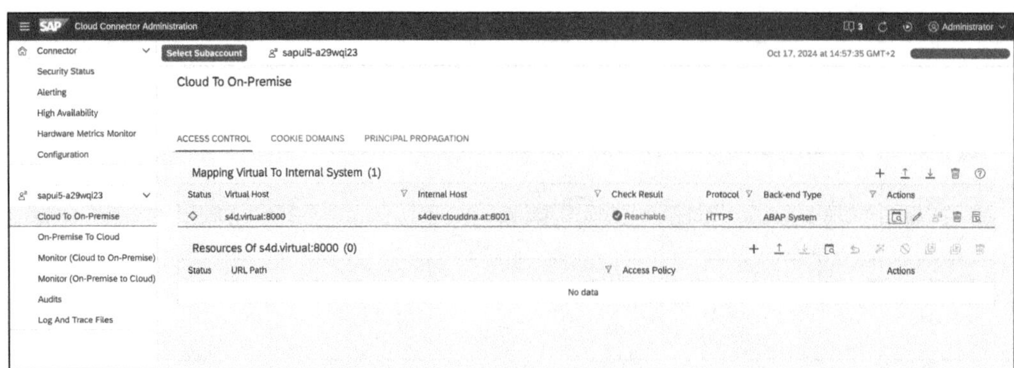

Figure 2.31 Add Resource

You can now specify the desired **URL Path**, for example "/sap". In addition, as shown in Figure 2.32, you can configure whether only this path can be accessed explicitly or also all paths below it.

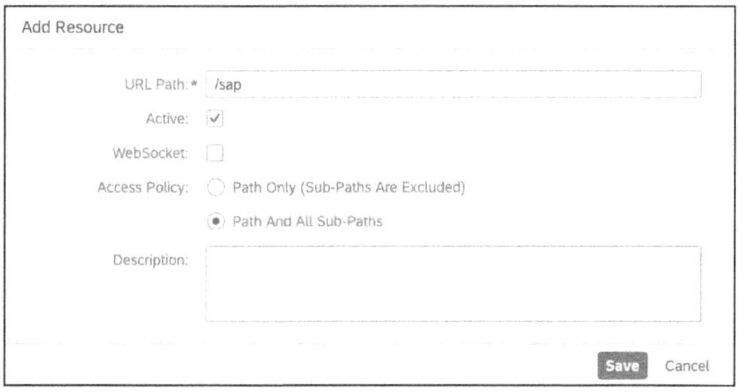

Figure 2.32 Resource Details

2.1.4 Security

In today's digital age, the security of corporate data is at the forefront. Security starts with the selection of secure data transmission technologies and continues with the identification of users through to the assignment of rights. The maintenance and configuration of user management shouldn't be independent of the company's own infrastructure. Users should also be able to log in to the cloud with the usual authorizations, and if a user is no longer valid, the cloud applications and services should also be aware of this. Integration at the technical level is therefore a prerequisite for seamless integration of the applications at the business level. To illustrate the importance of secure communication, we'll use a trivial example that is very commonly used in the literature at this point. Imagine a simple scenario: you want to send a message by mail to a specific recipient. In this scenario, insecure communication is equivalent to sending the message by postcard. A postcard can be read by any person involved in transit, such as the mail carrier. Sending the message to the recipient in an envelope corresponds to secure communication. Only the recipient can read the message, provided the envelope isn't opened in transit.

In this section, we'll take an in-depth look at the security aspects of SAP BTP. Let's start with user authentication and learn how SAP BTP ensures that only authorized users have access to the platform and its services. We'll explore different authentication methods and best practices to verify the identity of users and ensure the integrity of the platform. We'll then go into the authorizations.

User Authentication

Authentication and authorization are the two cornerstones of *access control* and security. In these subsections, you'll learn how these key concepts provide a solid foundation for secure and controlled system access.

Authentication refers to the process of verifying the identity of a user or entity to ensure that they are indeed the person or entity they claim to be. Authentication uses a variety of methods, such as username and password, biometrics, tokens, or two-factor authentication. The goal of authentication is to ensure that the user or entity is legitimate and is allowed to access the desired resources.

Here's an example of authentication in an SAP system: A user wants to log on to an SAP system. He enters his username and password. The system checks this information to determine whether the user has provided the correct credentials. If the authentication is successful, the user is recognized as a legitimate user in the system and can access his personalized data and functions.

There are several ways to authenticate. The most common methods are listed here:

- **Username and password**
  This is the classic method where a user enters a unique username and password to log in. The password is matched against the stored user data to verify authentication.
- **Two-factor authentication (2FA)/multifactor authentication (MFA)**
  This method combines two or more different authentication factors to provide a higher level of security. Typically, a combination of something the user knows (e.g., password) and something the user has (e.g., a one-time password via SMS message or a security token app) is used.
- **Biometric authentication**
  This uses a person's biometric characteristics, such as fingerprint, facial recognition, iris scan, or voice recognition. These characteristics are captured and compared with previously registered biometric data to confirm identity.
- **Certificates**
  This uses a digital certificate issued by a trusted certification authority. The certificate contains information about the user's identity and is used to ensure the authenticity and integrity of data.
- **Social login**
  This method allows users to confirm their identity through their existing social media accounts, such as Facebook, Google, or LinkedIn, instead of creating a separate account and password.

Authorization, on the other hand, refers to the process of setting access rights and permissions for an authenticated user or entity. Once authentication is successfully completed, authorization determines which actions, resources, or functions the user or entity can use. This is done by assigning *roles*, *permissions*, or *access control lists*. Authorization determines what actions a user can perform based on their specific rights and permissions.

Here's an example of authorization in an SAP system: After the user is successfully authenticated, authorization is performed to ensure that the user can only access the resources and functions for which he is authorized. Let's assume that the user has the role of a sales representative. Authorization specifies that the user has access to sales data, customer data, and sales reports. However, he doesn't have authorization to access financial data or personnel files. Authorization thus controls which specific areas of the SAP system the user can use based on his assigned role and defined access rights.

In SAP BTP, a distinction must be made between global account and subaccount regarding authentication. Basically, the *SAP ID service* can be used for authentication at both levels. User identities are used, which you can either create or manage via *SAP for Me* (*https://me.sap.com/*) (see Figure 2.33). However, this approach also has some limitations. One of these drawbacks is that you have limited or no control over certain configuration parameters, such as password policies.

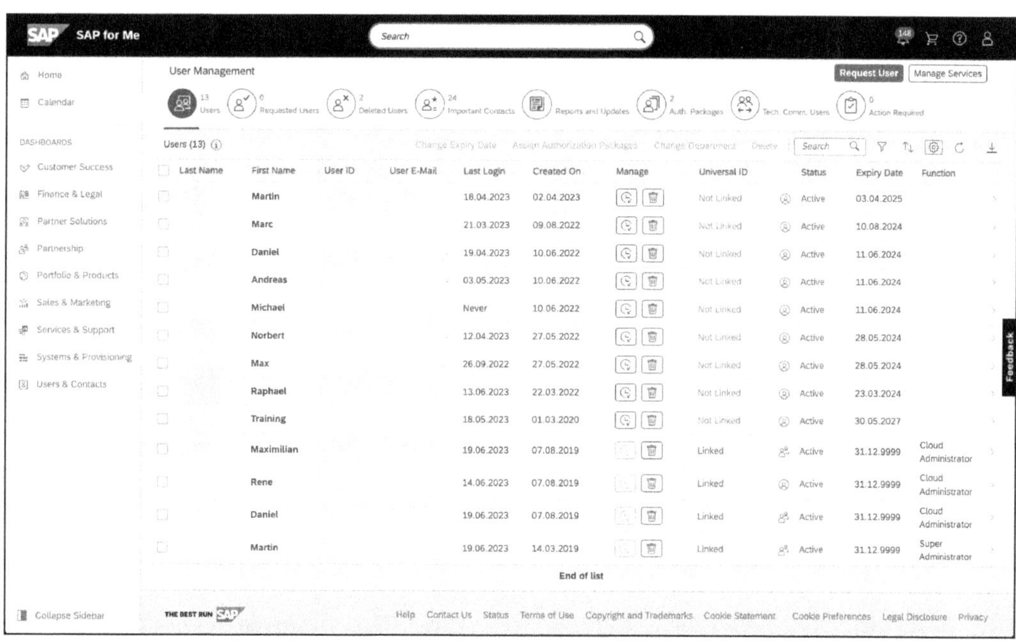

Figure 2.33 SAP for Me User Management

Alternatively, you can use the *Identity Authentication service* instead of the SAP ID service. This gives you control over the administration and maintenance of the users. Identity Authentication is included in the SAP BTP license and can therefore be used without additional licensing costs. Identity Provisioning and Identity Authentication are the most important parts of *SAP Cloud Identity Services* whose goal is to enable seamless SSO across all systems while ensuring secure access to systems and data.

Alternatively, you can use a third-party *SAML identity provider* at the subaccount level to perform user authentication. A common example of this is *Microsoft Azure AD*, which is widely used in practice. By integrating Microsoft Azure AD as a SAML Identity

Provider with your SAP subaccount, users can use their existing Microsoft Azure AD credentials to access resources and services within the SAP system. This seamless integration enables a great UX. You also benefit from the extensive security features and controls that Microsoft Azure AD provides to enhance access security.

User Authorization

As already mentioned, authentication checks the user's credentials. In simplified terms, the system checks whether the user is who he claims to be. Because the user shouldn't be able to access all functions without further checks after successful authentication, appropriate authorization checks are required. Authorizations are managed and assigned to users in SAP BTP based on *role collections*.

A *role* is an instance of a *role template* (see Figure 2.34). You can build a role based on a role template and assign the role to a role collection. The roles are defined in the respective services, such as SAP Integration Suite, and are provided in the subaccount when the services are activated. Roles are assigned to role collections, which are assigned in turn to users or user groups.

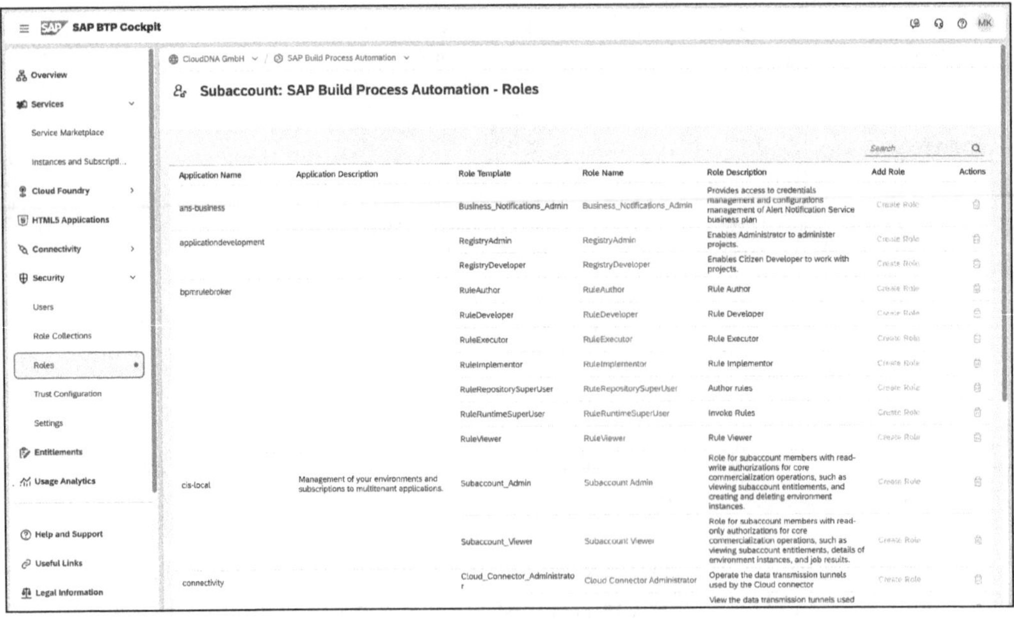

Figure 2.34 Subaccount Roles

By using the SAP BTP cockpit, you can view and access information regarding the role collections that have been established, along with the roles contained within each role collection (see Figure 2.35).

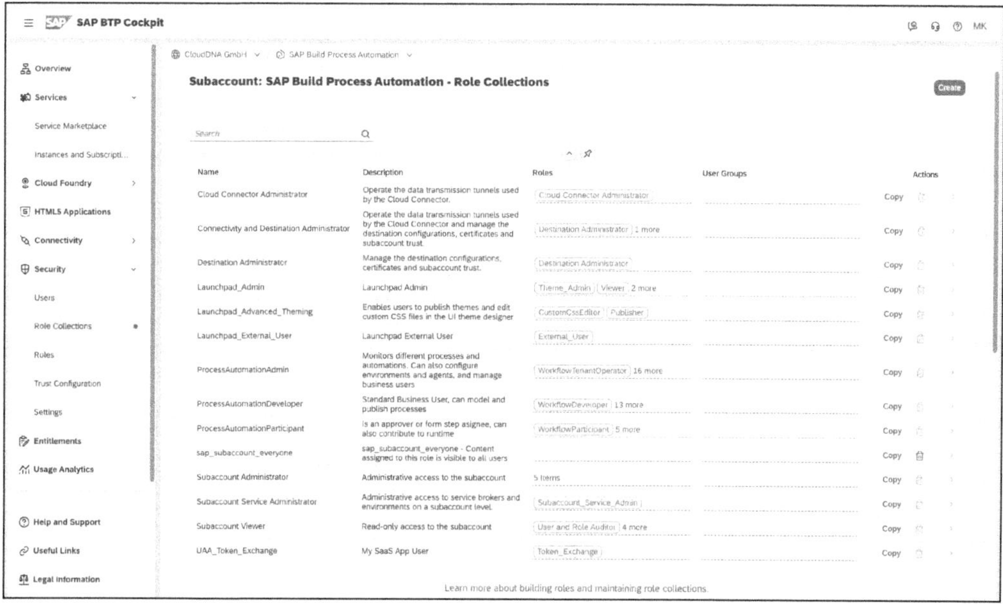

Figure 2.35 Subaccount Role Collections

In the details of the role collection, you can see which roles are assigned to this collection (see Figure 2.36). There, you can also assign users to the role collection or map them to user groups of the underlying identity provider. In addition, you can perform an *attribute mapping*. In doing so, you can derive the assignment to the role collection from an attribute of the identity provider, for example, the department.

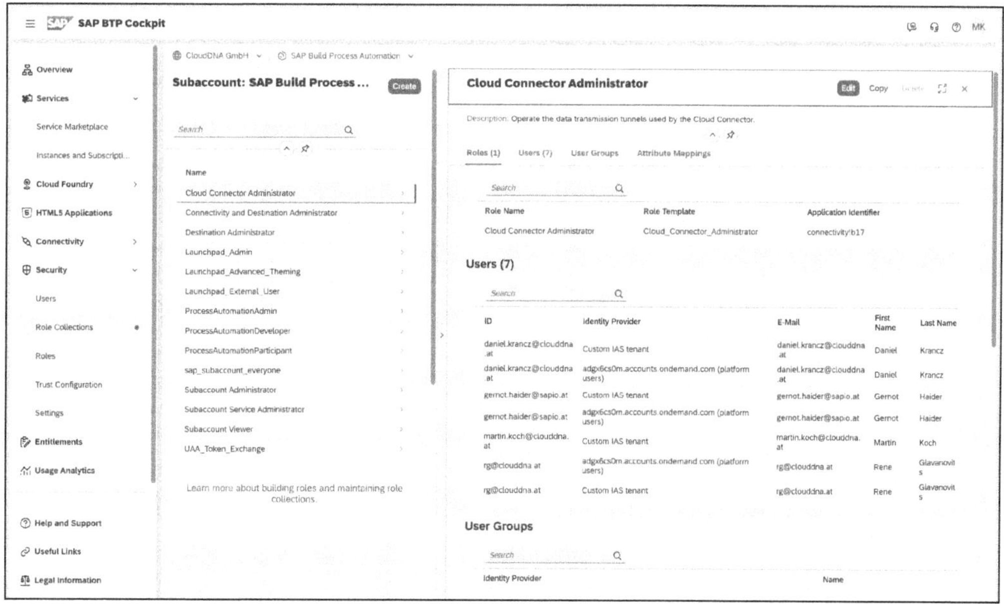

Figure 2.36 User Assignment

A distinction must be made between role collections at the level of global accounts and subaccounts. In addition to the standard role collections delivered by SAP, you can also create your own role collections and tailor them to your requirements on both the global account and subaccount level. The global account comes with two standard role collections named **Global Account Administrator** and **Global Account Viewer** (see Figure 2.37). Developers and end users don't need a user on the global account.

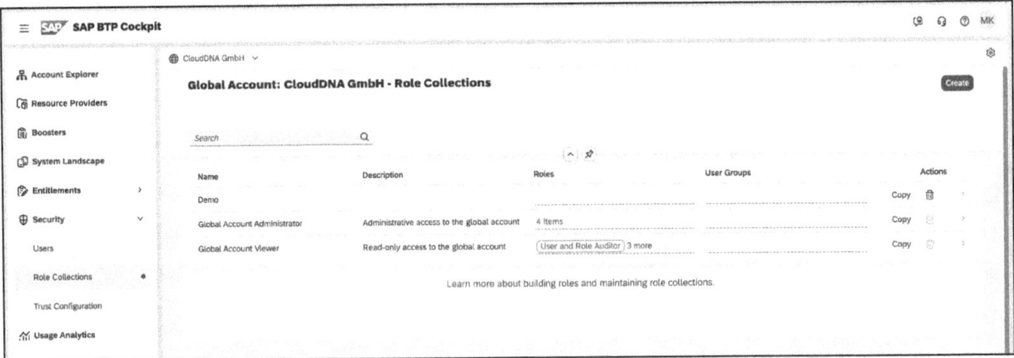

Figure 2.37 Global Account Role Collections

Single Sign-On

Single sign-on (SSO) is an authentication method where a user logs in once and then automatically gains access to multiple connected systems and applications without having to log in again. With SSO, users can access multiple services and resources with an SSO, increasing usability and productivity and reducing the number of passwords.

In the SSO process, logon is usually done via a central *identity provider*, which verifies the user's authentication information and creates an SSO token. This token is then passed to the various connected systems and applications to enable access without the user having to provide additional credentials. Any application that relies on this identity provider is also called a service provider.

SSO enhances the UX by allowing users to log in only once and then seamlessly access multiple applications and systems without having to log in repeatedly. SSO serves to increase efficiency by reducing the number of passwords users need to manage. But it also reduces the time needed to log in and switch between different applications. By centrally managing authentication information and the associated use of modern security standards, SSO can improve security and reduce the risk of password theft and unauthorized access.

In this context, the identity provider and the *service provider* are two important components that play a crucial role in the context of authentication and authorization:

- **Identity provider**
  The identity provider component is responsible for managing and verifying user identities. The identity provider authenticates users and issues identity credentials

in the form of digital tokens. The identity provider assumes responsibility for verifying the user's identity and securely providing authentication information to the service provider. Examples of identity providers include Microsoft Azure AD or Okta.

- **Service provider**
  The service provider is an application, that is, a service, that users want to access. The service provider is responsible for protecting the resources and ensuring that only authenticated and authorized users are granted access. The service provider trusts the identity provider to perform authentication and obtain information about the authenticated user. This allows the service provider to verify access rights and release the appropriate resources. Examples of service providers are web applications, APIs, or cloud services that users want to access.

Various standards and protocols are used for SSO, such as *Security Assertion Markup Language* (SAML) and *OpenID Connect* (OIDC). These established standards ensure seamless collaboration between different systems and facilitate the smooth integration of SSO into existing IT infrastructures.

SAML is an XML-based authentication and authorization protocol commonly used for SSO. It enables the secure transfer of authentication information between an identity provider and a service provider. The identity provider issues a SAML response containing the user's authentication status, while the service provider verifies this response and grants access to the requested resources. SAML is based on the principle of digital signatures and enables secure and trusted communication between the parties involved.

OIDC is an OAuth 2.0–based identity protocol designed specifically for identity management on the web. It allows users to log in to a web application using their existing identity providers (e.g., Google, Facebook, or Microsoft) instead of creating separate accounts and credentials. OpenID Connect uses *JSON Web Tokens* (JWTs) to securely exchange information about the authenticated user between the identity provider and the relying party. It provides a simple and easy-to-use way to implement SSO for web applications and ensure that users are properly authenticated.

Identity Authentication

SAP offers SAP Cloud Identity Services in SAP BTP with different areas of responsibility. With this collection of services, you can map the entire security lifecycle of users or identities. The spectrum ranges from the onboarding of new users and the provision of these users in the various cloud applications to the detection of potential problems in the assignment of critical authorizations. The Identity Authentication service is also part of SAP Cloud Identity Services. This allows you to implement SSO scenarios and delegate authentication to third-party identity providers or identity providers in your on-premise systems.

In connection with SAP Cloud Identity Services, the Identity Directory plays an important role. The *Identity Directory* is the central component for the persistence of users and groups within SAP Cloud Identity Services. It's the central point of contact for users who have access to SAP cloud applications.

Identity Authentication is a SAML 2.0 and OIDC identity provider for authenticating users in SAP BTP applications and in all SAP SaaS products. Optionally, the service can also be used for applications running in your on-premise landscape and for non-SAP applications. The Identity Authentication service enables you to authenticate and SSO users in the cloud. The service is provided for both the SAP BTP, Neo environment and SAP BTP, Cloud Foundry runtime. It's a key element for cloud architecture security. SAP's strategy is to deliver its cloud solutions preconfigured with Identity Authentication.

Identity Authentication provides you with the following core functions (see Figure 2.38):

- **Authentication**
  The Identity Authentication service acts as an identity provider that validates credentials provided by users. Upon successful authentication, it issues a SAML assertion that is used by the target applications.
- **Identity federation and SSO**
  The service can delegate authentication to a third-party identity provider, enabling SSO across on-premise and cloud applications.
- **Risk-based authentication**
  The service enables MFA if stronger means of authentication are required to access specific business applications.
- **User management**
  The service enables management of its users.

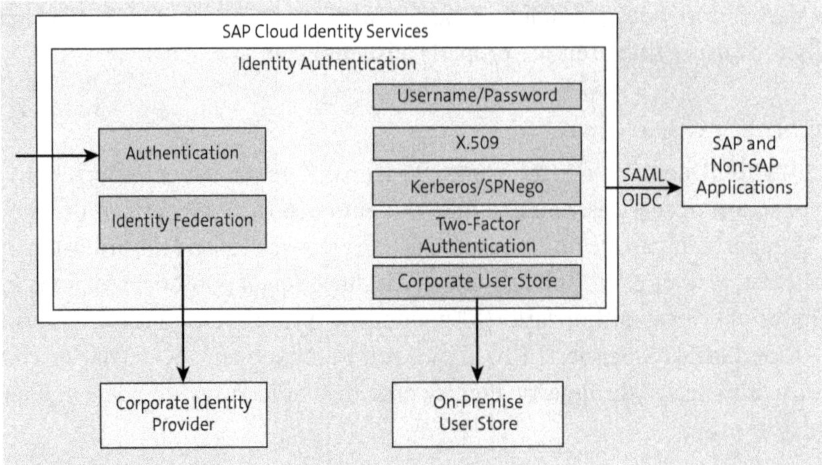

Figure 2.38 SAP Cloud Identity Services

This service plan for the Identity Authentication service is included in the overall SAP BTP contract. You receive two instances of the Identity Authentication service: one for testing purposes and one for production use. These instances are called Identity Authentication tenants. In principle, you have the option to purchase additional tenants, regardless of whether they are used productively or for testing purposes. The license scope of the Identity Authentication service includes all logins from SAP cloud to SAP cloud. If you would also like to connect non-SAP products, then an additional license must be purchased. The metric for this is based on the number of logins.

After you purchase a license for the Identity Authentication service, you receive an email invitation to register the initial administrator. The email contains a link to the initial page of the *Identity Authentication tenant* administration console. There, you can confirm the registration of the initial administration user.

You can log in to the Identity Authentication tenant at *https://<IdentityAuthenticationTenantID>.accounts.ondemand.com*. Replace the placeholder *<IdentityAuthenticationTenantID>* with the ID of your Identity Authentication tenant.

The tenant ID is an ID automatically generated by SAP. The first administration user created for the client receives an activation email with a URL. This URL contains the tenant ID. However, you also have the option to jump directly from the SAP BTP cockpit to the Identity Authentication tenant.

In the Cloud Foundry subaccount, it's a prerequisite that a trust relationship has been established between the subaccount and the Identity Authentication tenant. To do this, navigate to the **Security • Trust Configuration** section, and click the **Establish Trust** button (see Figure 2.39).

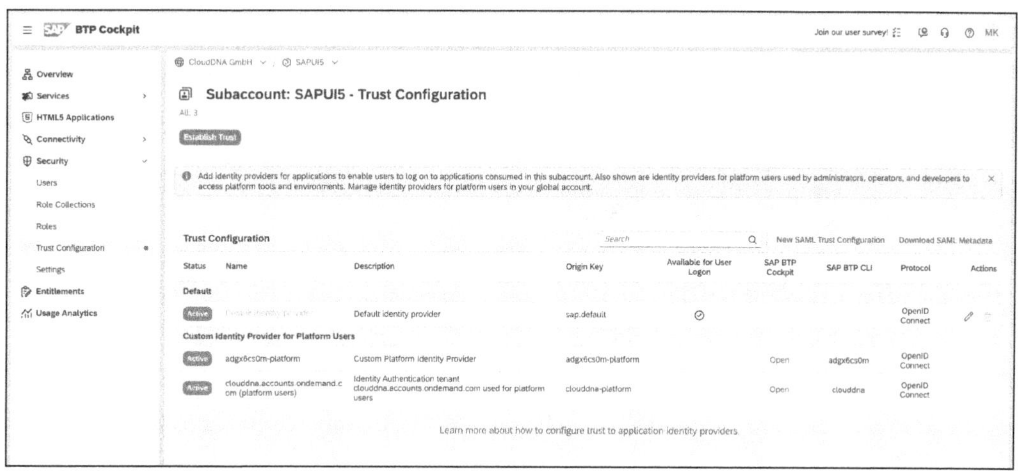

Figure 2.39 Subaccount Establish Trust

You can then select an Identity Authentication tenant assigned to your global account (see Figure 2.40). Additionally, you can perform an advanced configuration.

After the configuration has been performed, you can jump directly to the Identity Authentication tenant by clicking on the link (see Figure 2.41). As you can also see in the figure, both for the connection to the Identity Authentication tenant and for the connection to the default identity provider, the SAP ID service, the previously mentioned OIDC protocol is used, and not SAML 2.0.

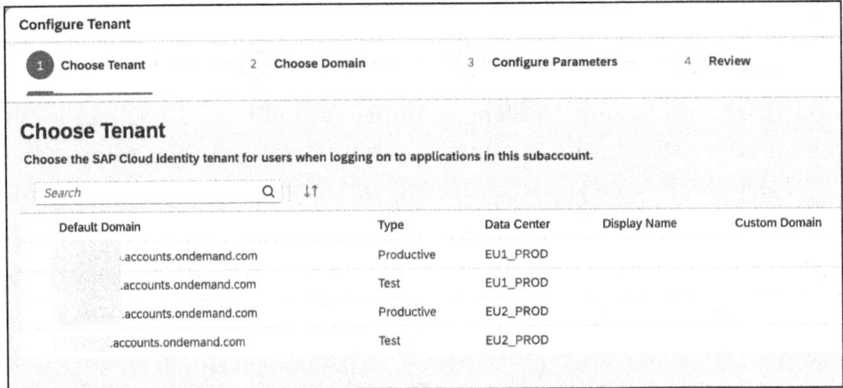

Figure 2.40 SAP Cloud Identity Services Tenant Selection

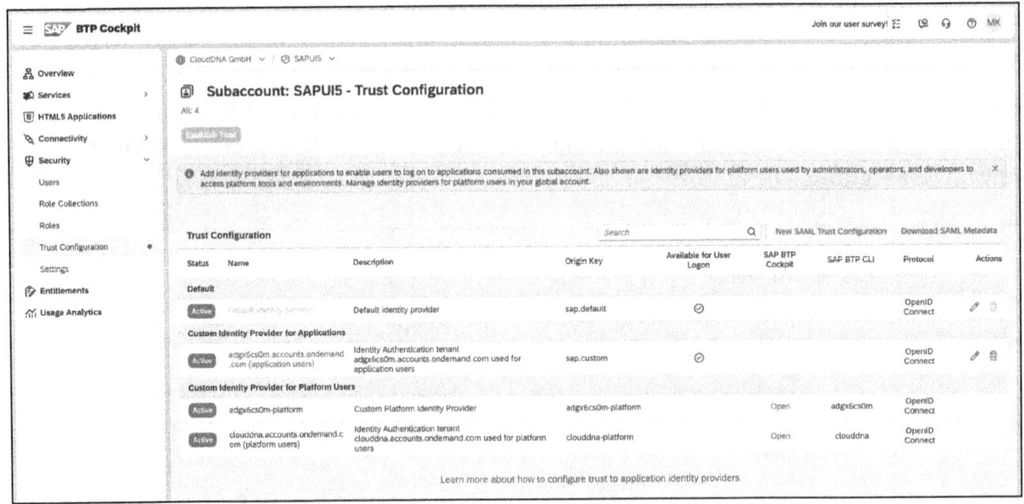

Figure 2.41 Trust Established for Application Authentication

At this point, you can jump to the corresponding Identity Authentication tenant. There, you'll get to the **Home** tab, where all functions can be called directly (see Figure 2.42).

In the Identity Authentication tenant, you can manage users in the **User Management** area. For example, you can create, delete, and delimit users (see Figure 2.43).

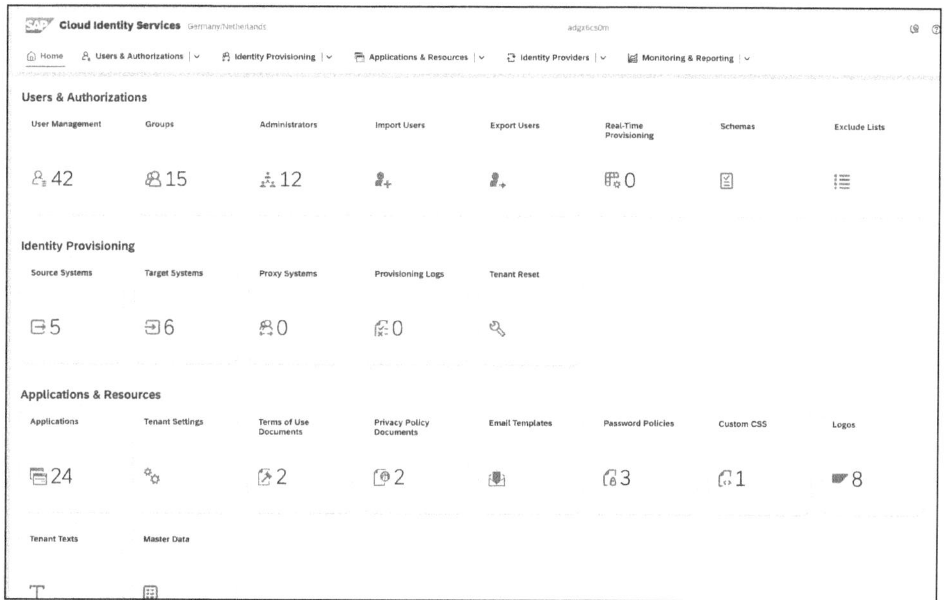

Figure 2.42 SAP Cloud Identity Services Overview

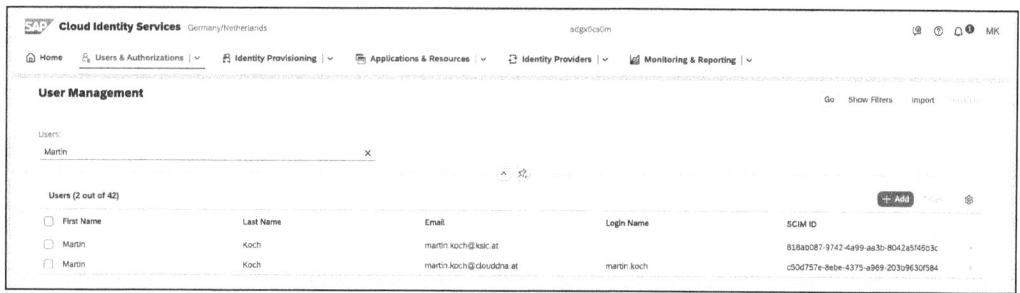

Figure 2.43 SAP Cloud Identity Services User Management

In the Identity Authentication tenant, as shown in Figure 2.44, you can also create groups and assign users to these groups. In SAP BTP, you can map the groups to role collections.

In addition to using the SAP ID service and the Identity Authentication service, it's possible for SAP BTP applications to delegate authentication and identity management to an existing identity provider within your company (a corporate identity provider). This can authenticate your company's employees against a corporate directory service, for example. This allows your employees and, if applicable, customers and partners, to log in to the cloud application using their usual user information. All information about a user required by SAP BTP can be securely passed on with the logon process based on a proven and standardized security protocol.

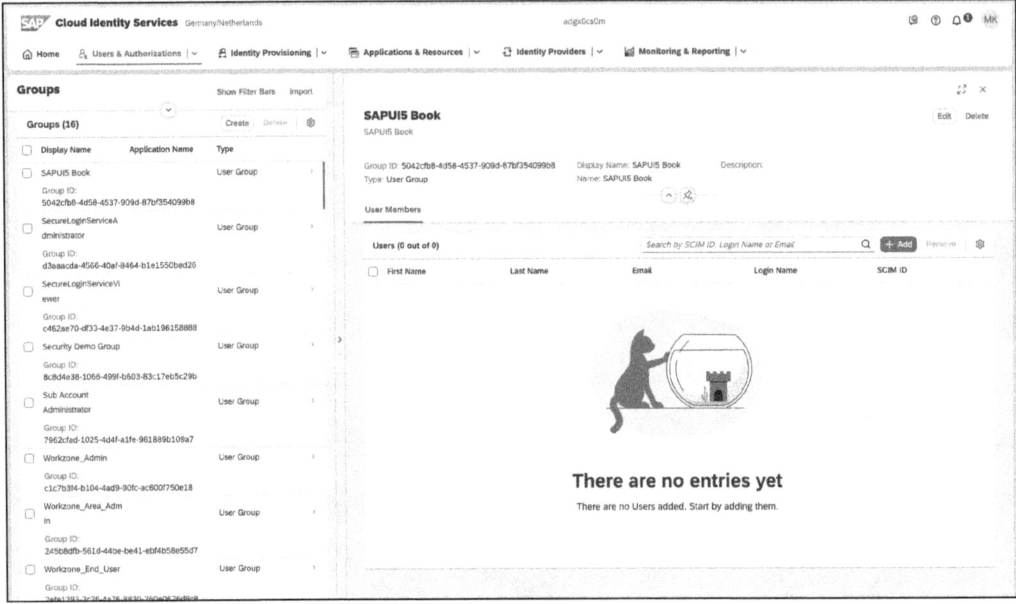

Figure 2.44 SAP Cloud Identity Services Groups

2.2 Development Environments

When developing applications, the goal is to ensure a high level of developer productivity. Of course, you could opt for a classic text editor, but you'll miss some comfort there. A full IDE, on the other hand, offers decisive advantages that go far beyond just writing code. Modern IDEs, such as SAP Business Application Studio or VS Code with the corresponding SAP plugins, are specifically designed for the following:

- **Code completion (auto-completion)**
  This function makes writing code significantly easier by providing suggestions for code snippets in real time based on the syntax and libraries used in your project. This is an indispensable support, especially in extensive frameworks such as SAPUI5, in which numerous modules and components are used.
- **Direct access to documentation**
  In an IDE, you can access official SAP documentation and other relevant sources directly. This not only saves time but also ensures that you always have the latest information available without having to leave your working environment.
- **Local application launch and debugging**
  Full IDEs enable you to launch, test, and debug applications locally before they are transferred to productive systems. This functionality is crucial to identifying errors early on and ensuring the quality of your application.

- **Git integration**
  Direct access to Git repositories is another major advantage of IDEs. Git has established itself as the de facto standard for source code management. Especially in distributed teams, Git offers significant added value with its branching concept. Integration into the IDE not only allows you to manage source code efficiently but also to collaborate directly with other team members via pull requests and branching strategies. This way, multiple developers can work on a project simultaneously without any conflicts arising.
- **Access to different data sources**
  Connecting to backend systems, especially to OData services—whether they are provided in the cloud or by an on-premise system—is a central requirement in SAP development. IDEs offer you the possibility to easily connect and test data sources, which is especially important when developing SAP Fiori applications.

As a developer or as a company you again have an important decision: Development in the cloud or locally?

If you decide to develop in the cloud (SAP BTP), you benefit from the advantages of a fully managed environment. We'll discuss SAP Business Application Studio in more detail in Section 2.2.1. It offers seamless integration into SAP BTP, direct access to SAP BTP services, and rapid application deployment. Furthermore, there is no need to maintain a local infrastructure of any complexity because the entire development and test environment is already provided in the cloud. At this point, the license costs that arise as a result must also be mentioned. This is mainly because SAP has to provide resources for SAP Business Application Studio in SAP BTP. The number of developers is used as the metric for licensing. With SAP Business Application Studio, you can develop not only classic SAPUI5 or SAP Fiori apps but also a variety of other applications and technologies, such as SAP Cloud Application Programming Model or Mobile Development Kit (MDK) applications.

For developers who want to work within the corporate network, VS Code combined with the SAP Fiori plugins offer a flexible and powerful development environment. We'll discuss this in more detail in Section 2.2.2. The advantage of this local development is that you retain full control over your infrastructure and may be able to access internal company resources more quickly. In addition, on-premise development is particularly suitable for companies that don't want to or aren't allowed to use cloud solutions for regulatory or security reasons, as is often the case in a regulated environment. There are no license costs for on-premise development with VS Code. However, you shouldn't forget that this will present you with different challenges. For example, you must ensure that VS Code is provided on the developers' computers. Furthermore, developers must have more extensive permissions at the operating system level, and you must ensure that Node.js or the Node Package Manager (NPM) is available on the computers. To install the plugins, it's also necessary that the developers' computers have an internet connection so that access to the plugin repositories can be established.

Regardless of the development environment you choose, SAP provides you with the tools to develop SAPUI5 and SAP Fiori apps efficiently and securely. The key is to carefully select your development infrastructure to maximize the efficiency of your work and ensure the security of your company data.

2.2.1 SAP Business Application Studio

SAP Business Application Studio is the full IDE provided in SAP BTP in the form of a service. Information about SAP Business Application Studio can be found in SAP Discovery Center (*https://discovery-center.cloud.sap/serviceCatalog/business-application-studio?region=all*). To use SAP Business Application Studio, it must first be provided in a subaccount. This can be done manually by first creating a subaccount, then assigning the corresponding entitlement to it, and finally creating an instance of SAP Business Application Studio in the subaccount. As mentioned earlier, SAP also offers a booster for this (refer to Figure 2.8). You can find the booster at global account level. The booster homepage provides an overview of the architecture and the required services (see Figure 2.45). These may also include optional services, such as SAP Continuous Integration and Delivery. To enable deployment based on the booster, you must click the **Start** button, as shown in Figure 2.45.

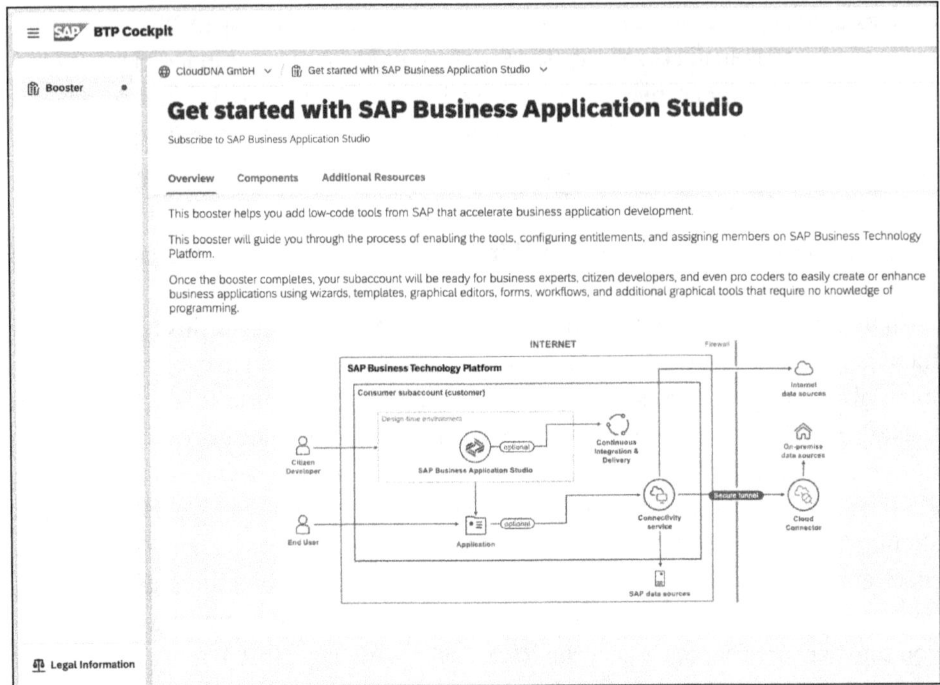

Figure 2.45 Start the Booster

In the first step, the booster checks the prerequisites (see Figure 2.46). This first determines whether you have the necessary permissions to run the booster and create the associated subaccount. At the same time, a check is made to see if the required entitlements (i.e., licenses) are available to use the booster's functions. These checks ensure that all the resources and access rights required for the booster to run successfully are available.

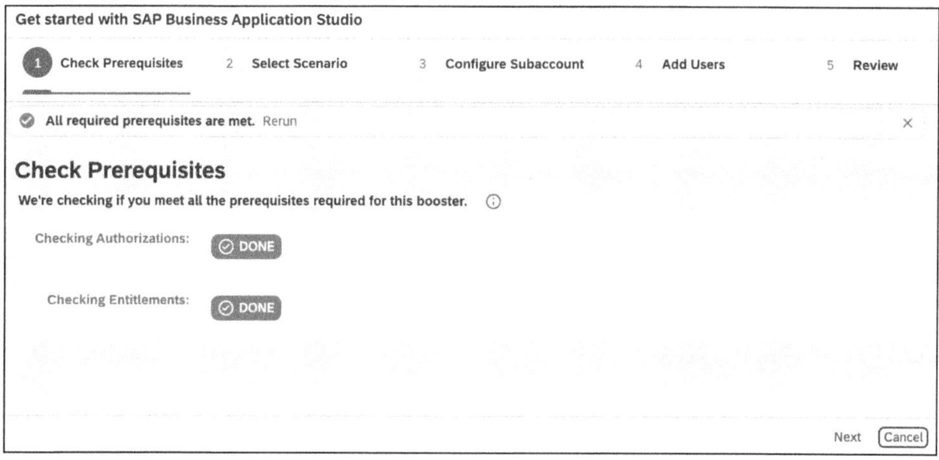

Figure 2.46 Prerequisites Check

In the next step, you need to decide whether you want to use an existing subaccount or set up a new subaccount as part of the installation (see Figure 2.47). This decision is entirely up to you. In principle, we recommend creating a separate subaccount for SAP Business Application Studio. The subaccount itself doesn't incur any additional license costs; only the services used in it, such as SAP Business Application Studio, are licensed and incur corresponding costs. In the example shown, we've opted for an existing subaccount that we previously created in the SAP BTP cockpit at the global account level.

Figure 2.47 Subaccount Selection

In the next step, you have the option to adjust the entitlements. In addition to SAP **Business Application Studio**, the **Continuous Integration & Delivery** service will also be added. However, this is optional and not mandatory for using SAP Business Application Studio. If you don't need this service, you can remove it from the service assignment. If you later decide that you want to add this or other services to your subaccount, you can do so at any time. In addition, the **Subaccount**, the associated Cloud Foundry **Org**, and the underlying **Space** are displayed. These can also be customized here if needed (see Figure 2.48). This flexibility allows you to tailor the environment to your exact needs and to make changes as needed, even after setup.

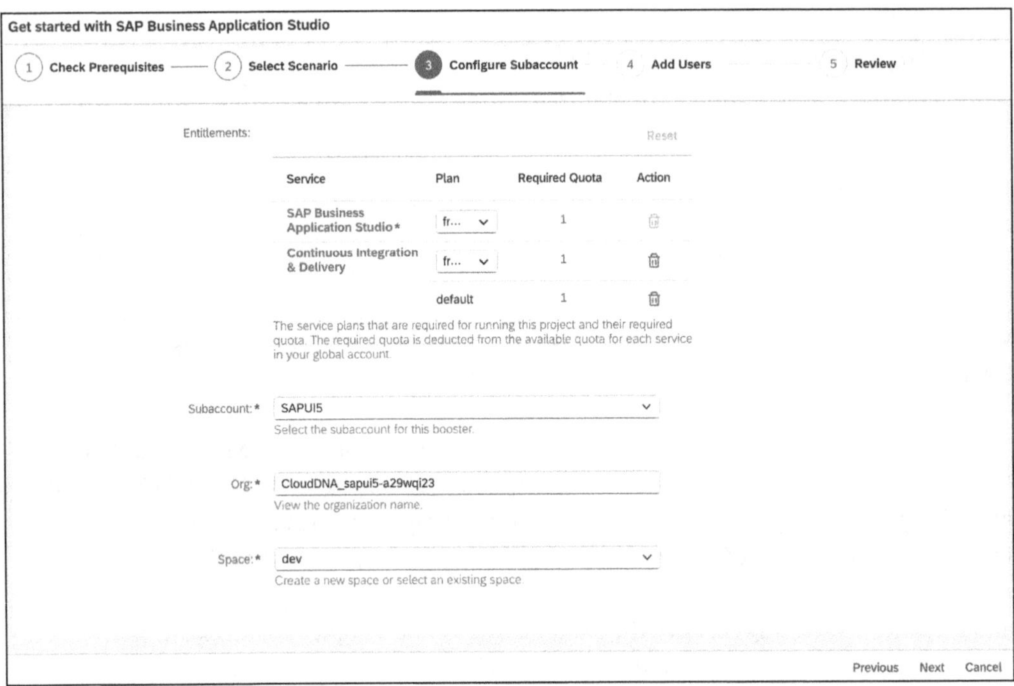

Figure 2.48 Subaccount Configuration

You then have the option to add users (see Figure 2.49). You can define both **Administrators** and **Developers** by entering their email addresses in the corresponding input field. Administrators receive, among other things, the role collections **Subaccount Administrator** and **BAS_Administrator**. These users are authenticated via the **Custom Identity Provider for Platform Users.** Here, you can use either the SAP ID service or your own SAP Cloud Identity Services instance as the identity provider. If you use the SAP ID service, administrators can log in with their S-user ID. If you decide to use SAP Cloud Identity Services, you can manage users directly through SAP Cloud Identity Services or delegate authentication to Microsoft Entra ID.

Developers are assigned the role collections "BAS_Developer" and "Subaccount Viewer". They are authenticated using the **Custom Identity Provider for Applications.**

Here, too, you can choose between the SAP ID service or an SAP Cloud Identity Services instance. We recommend using SAP Cloud Identity Services because it allows you to benefit from advanced features that aren't available in the SAP ID service. These include, for example, defining password policies or the option of designing the logon portal in your own corporate design. This flexibility in the choice of identity provider allows you to optimally adapt user administration and authentication to the requirements of your organization.

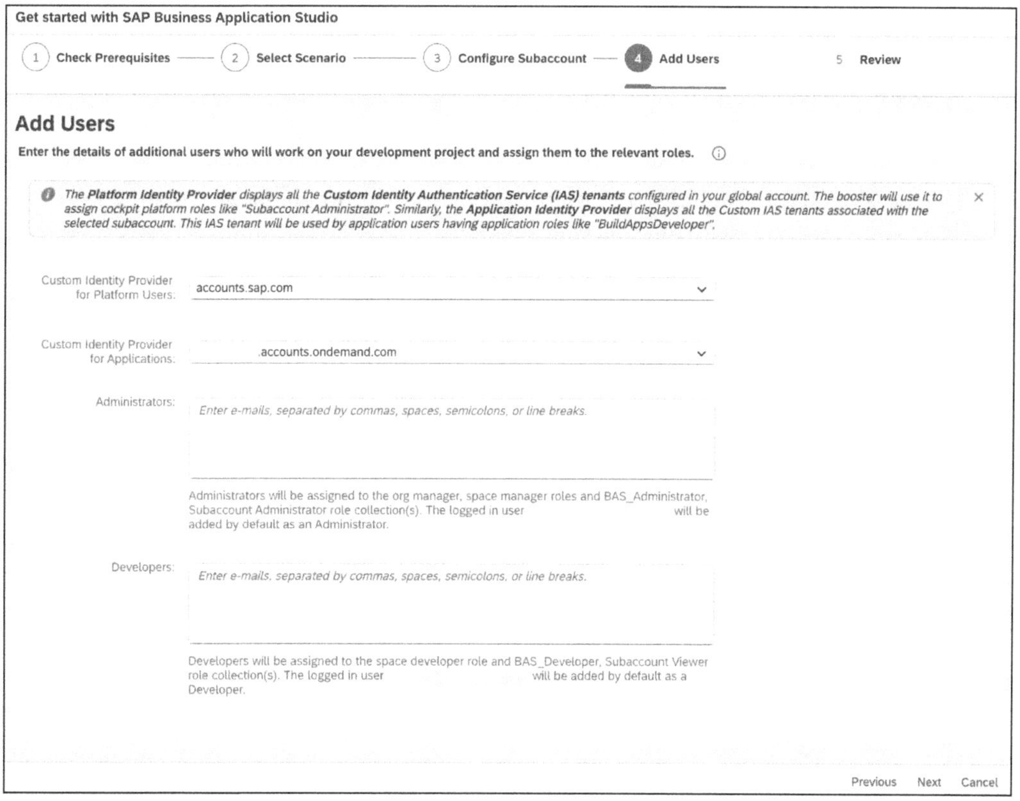

Get started with SAP Business Application Studio

1 Check Prerequisites — 2 Select Scenario — 3 Configure Subaccount — 4 Add Users 5 Review

Add Users

Enter the details of additional users who will work on your development project and assign them to the relevant roles.

The Platform Identity Provider displays all the Custom Identity Authentication Service (IAS) tenants configured in your global account. The booster will use it to assign cockpit platform roles like "Subaccount Administrator". Similarly, the Application Identity Provider displays all the Custom IAS tenants associated with the selected subaccount. This IAS tenant will be used by application users having application roles like "BuildAppsDeveloper".

Custom Identity Provider for Platform Users: accounts.sap.com

Custom Identity Provider for Applications: .accounts.ondemand.com

Administrators: *Enter e-mails, separated by commas, spaces, semicolons, or line breaks.*

Administrators will be assigned to the org manager, space manager roles and BAS_Administrator, Subaccount Administrator role collection(s). The logged in user will be added by default as an Administrator.

Developers: *Enter e-mails, separated by commas, spaces, semicolons, or line breaks.*

Developers will be assigned to the space developer role and BAS_Developer, Subaccount Viewer role collection(s). The logged in user will be added by default as a Developer.

Previous Next Cancel

Figure 2.49 Authentication Configuration

Finally, a summary of the settings you've made so far will be displayed. This overview allows you to review all configurations. Once everything is correct, you can start the installation and configuration process by clicking on **Finish**. This step ensures that all settings are applied and the necessary components are provided so that your environment is ready for immediate use.

You'll then be informed about the progress of the installation and configuration in a popup (see Figure 2.50). If errors occur, you'll normally need to contact SAP Support.

After a successful installation, you'll be informed in a popup. From there, you can jump directly to the subaccount. To do so, click on **Navigate to Subaccount** (see Figure 2.51).

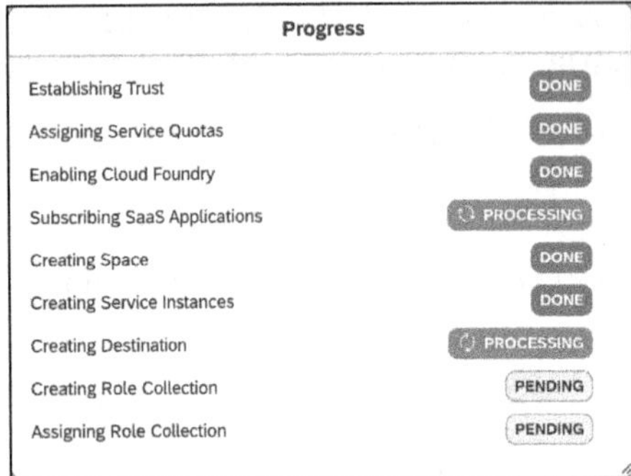

Figure 2.50 Configuration Progress

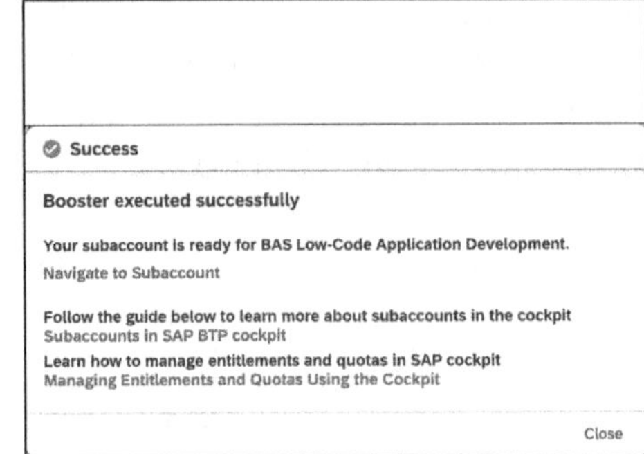

Figure 2.51 Booster Success Message

In the subaccount, several role collections have been created by the booster and the activation of the services. To check this, you can navigate to the **Security • Role Collections** area in the side menu (see Figure 2.52).

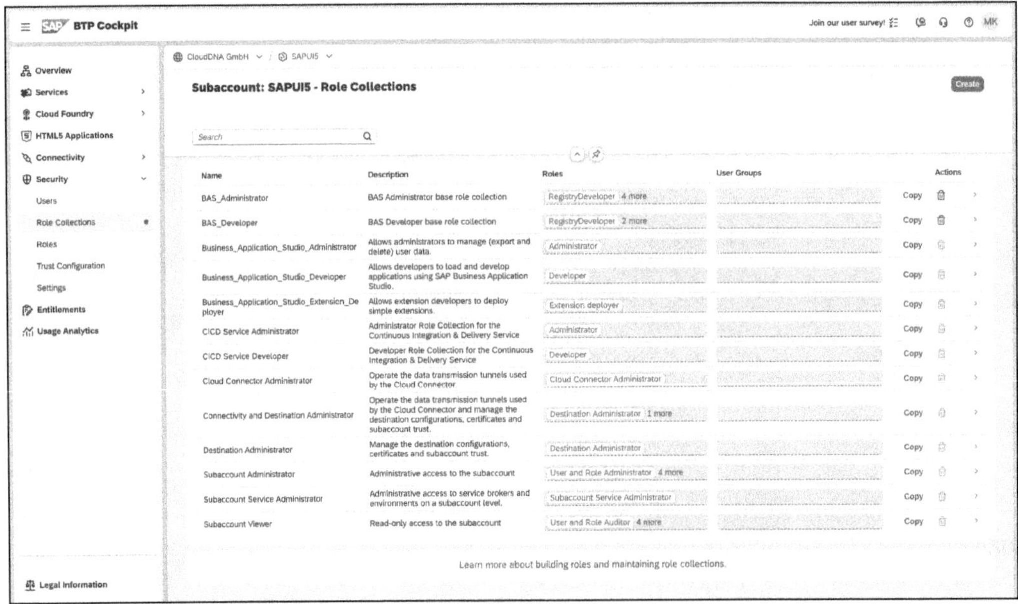

Figure 2.52 Role Collections

Additionally, you can use the page menu **Security • Users** to jump to the user overview and check whether the specified administrator and developer users have been added as desired (see Figure 2.53).

In the example shown here, the SAP Continuous Integration and Delivery service has been installed and configured in addition to SAP Business Application Studio. To get an overview of the application subscriptions and instances, navigate to **Services • Instances and Subscriptions** in the side menu. There, you can view and manage all active services and instances (see Figure 2.54). This view provides a central overview for efficiently managing your installed and used services. In the **Applications** area, click on the **SAP Business Application Studio** link to jump to SAP Business Application Studio.

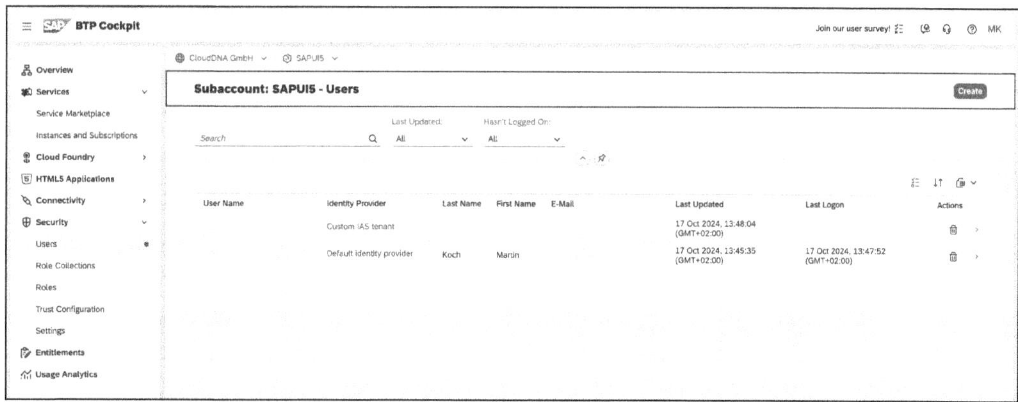

Figure 2.53 Automatically Created Users

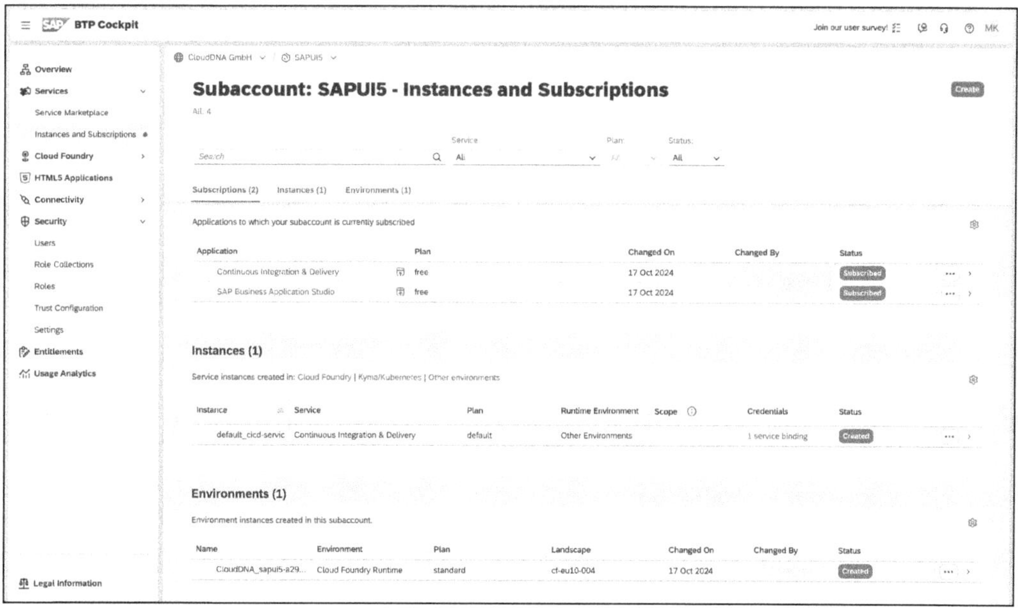

Figure 2.54 Created Instances and Subscriptions

To start SAP Business Application Studio, you have to authenticate yourself accordingly. The available identity providers are displayed, from which you can select the one you want to use for the logon (see Figure 2.55). This is the custom identity provider for applications defined in the booster. If only one identity provider has been configured for the logon, you're automatically redirected to the logon page of this provider. This process ensures that you authenticate yourself using the correct identity provider and that you can use SAP Business Application Studio securely and efficiently.

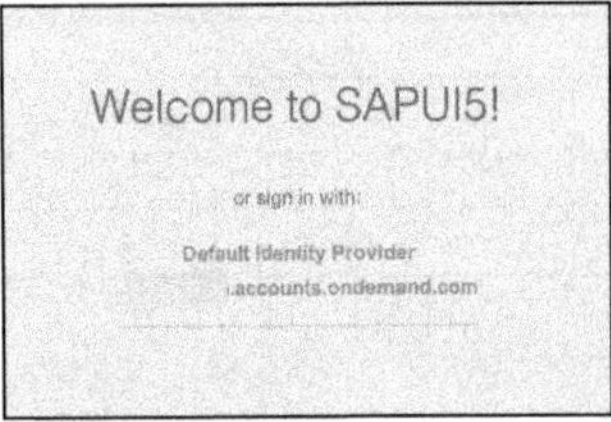

Figure 2.55 SAP Business Application Studio: Authentication Method Selection

When you first start SAP Business Application Studio, you see the screen shown in Figure 2.56. SAP Business Application Studio uses dev spaces, which form the basis for your development environment. A *dev space* defines what kind of developments—such as SAP Fiori, SAPUI5, SAP Cloud Application Programming Model, MDK, and others—can be carried out within this space. Depending on the selected dev space, the corresponding extensions and tools are automatically installed in SAP Business Application Studio to support development. To create a new dev space, click **Create Dev Space**. Here, you can configure the dev space according to your project's requirements, so that all the development tools you need are directly available. This approach provides a flexible and optimized working environment that is specifically tailored to the needs of your SAP development projects.

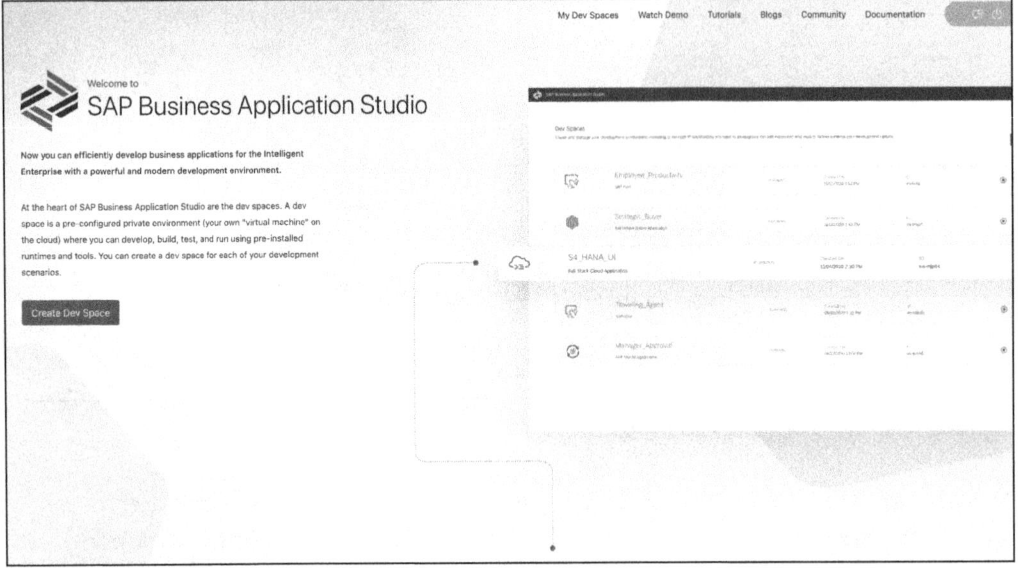

Figure 2.56 SAP Business Application Studio Initial Screen

You must first define a name for the dev space. In addition, you specify what kind of applications you want to develop (see Figure 2.57). We've opted for **SAP Fiori**. Based on your selection, the required extensions are automatically preconfigured so that the development environment is optimally tailored to your requirements. This makes it easier to get started and ensures that all relevant tools and functions for the respective development type are immediately available.

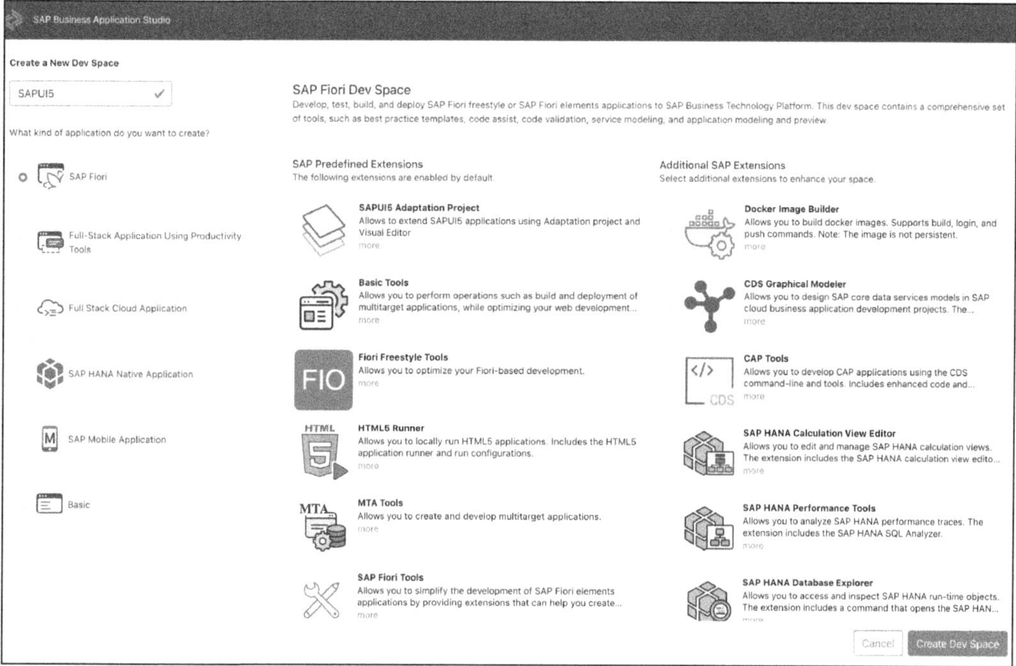

Figure 2.57 Dev Space Creation Process

Once the dev space has been created, it must be started. When it's first created, the dev space is automatically started (see Figure 2.58). You can recognize this from the space displaying the status **RUNNING** in the overview. If the dev space isn't used for a certain period of time, it's stopped to avoid unnecessary use of resources. In such a case, you must restart the dev space manually to continue your development work. This resource-saving feature ensures that only active projects consume computing power. Click the name of the space to open it.

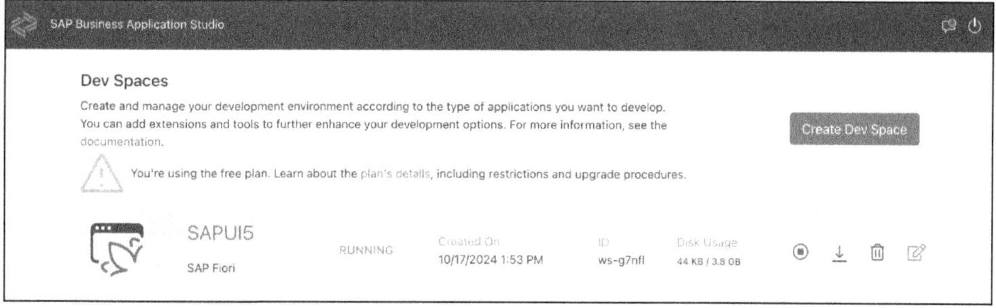

Figure 2.58 Dev Spaces Overview

After the dev space has been opened, the **Get Started** dialog appears in SAP Business Application Studio (see Figure 2.59). This serves as an entry point, and you can now immediately begin with your development. The dialog provides helpful quick links to make your development process efficient. From here, you can either start a new project or import existing projects into the dev space and continue developing them.

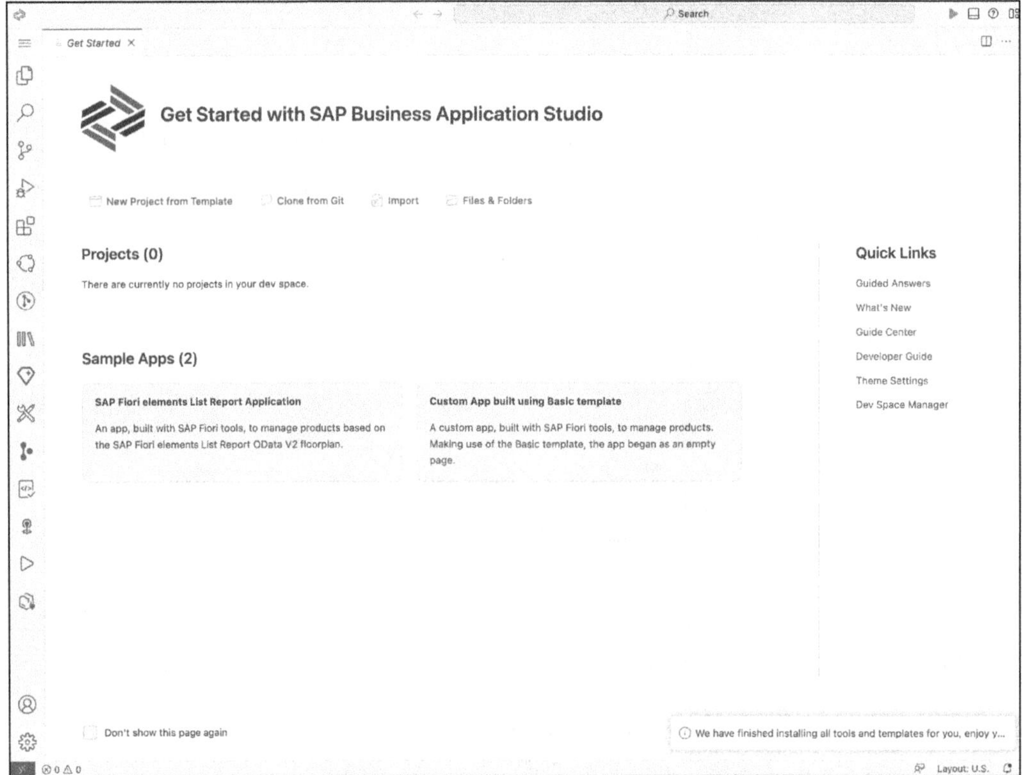

Figure 2.59 SAP Business Application Studio Start Screen

2.2.2 Visual Studio Code

Visual Studio Code (VS Code) is a powerful and popular source code editor developed by Microsoft that is particularly popular in the developer community. It's a lightweight but versatile development environment that supports numerous programming languages and frameworks, including SAPUI5. In recent years, VS Code has become the preferred tool for SAPUI5 development, largely replacing Eclipse, the previously dominant development environment.

One reason for the popularity of VS Code is its flexibility and extensibility. A huge range of extensions is available to customize VS Code to specific requirements. For SAPUI5 developers, they have the advantage, among others, of accessing a wide range of plugins and tools that are specifically optimized for developing SAPUI5-based applications and

SAP Fiori apps. This includes the *SAP Fiori tools* extension, which helps developers quickly and efficiently develop applications in the SAPUI5 context.

Compared to Eclipse, VS Code is characterized by its lean architecture. This leads to a faster way of working. Eclipse has often been criticized for its cumbersome nature, while VS Code impresses with its short start-up time and resource-efficient operation. Additionally, VS Code supports modern web technologies, and it's strongly embedded in the open-source community. This makes it an ideal tool for developing SAPUI5 applications, which are based on HTML5, CSS, and JavaScript.

The first step is to download VS Code for the desired operating system. VS Code is available for all common operating systems, such as Windows, macOS, and Linux, so developers can work independently of their preferred platform. The download is available on the official VS Code website (*https://code.visualstudio.com/Download*), where the appropriate installation file can be found for each operating system (see Figure 2.60).

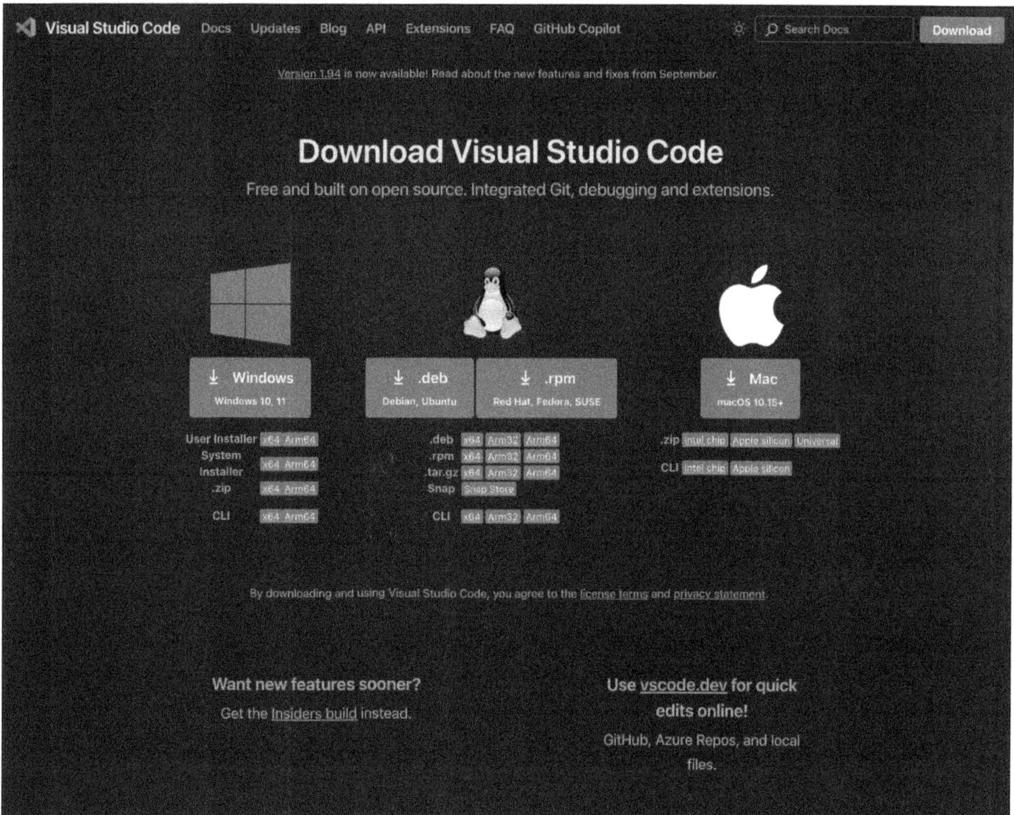

Figure 2.60 Visual Studio Code Download

Before VS Code can be used in the SAPUI5 context, it's necessary to install *Node.js*. Node.js provides a JavaScript runtime environment that is required by many development tools.

As shown in Figure 2.61, Node.js can be downloaded directly from the official Node.js website (*https://nodejs.org/en/download/package-manager*). We recommend choosing an **LTS** (long term support) version, as this version is stable and optimized for long-term use in production environments. After installing Node.js, the package manager *npm* (Node Package Manager) is also available, which is used to install dependencies and tools in SAPUI5 development.

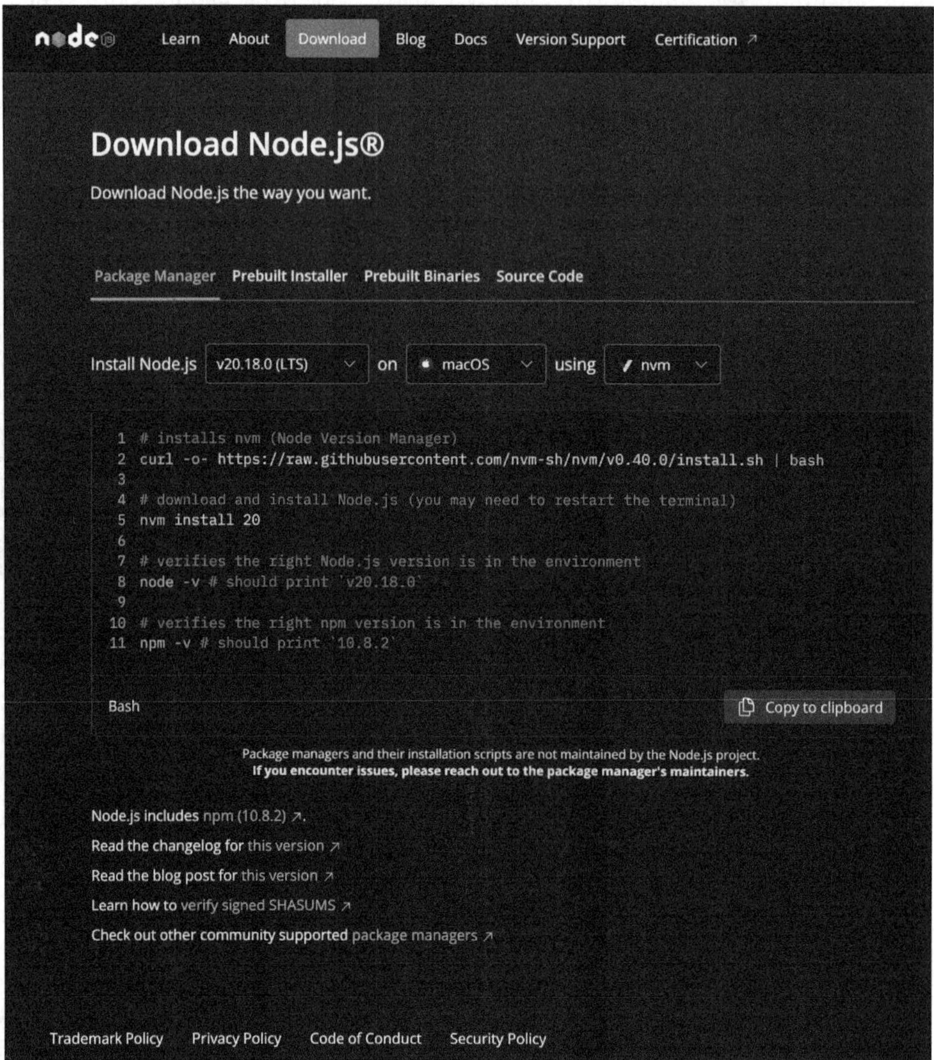

Figure 2.61 Node.js Download

After downloading and installing VS Code and Node.js, VS Code can be used immediately. However, for SAPUI5 development, it's recommended to install some additional extensions that make work much easier. These include, in particular, the SAP Fiori tools extension pack, which offers specific functions for creating and managing SAPUI5 applications. To do this, navigate to the **Extensions** area in the side menu in VS Code and then search for the SAP Fiori tools. Click on **SAP Fiori Tools - Extension Pack** in the list of found extensions to view the details. Then, click on **Install** (see Figure 2.62).

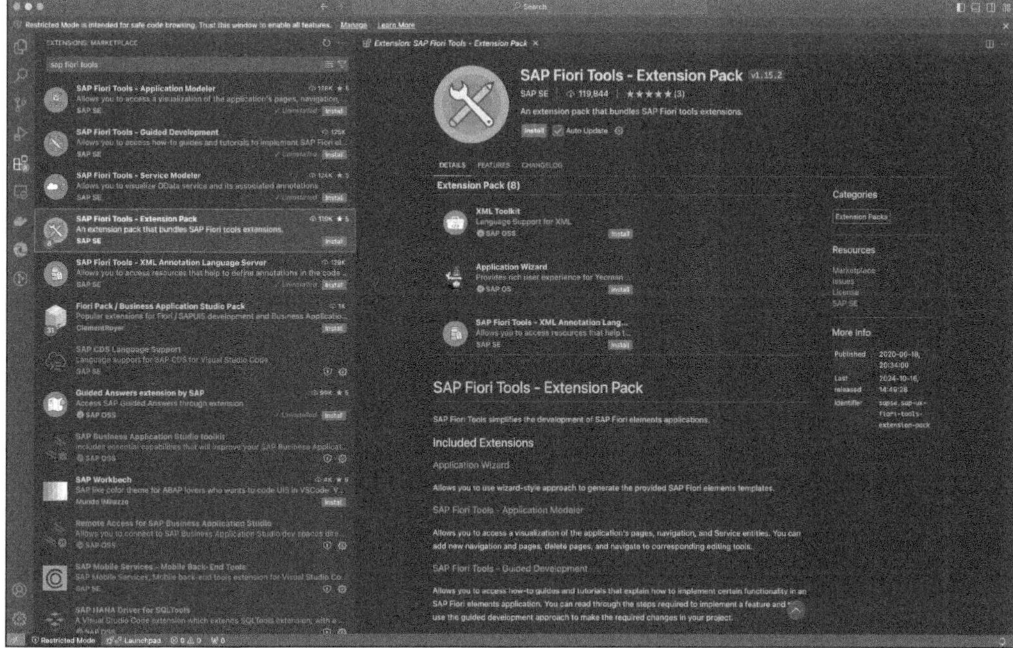

Figure 2.62 SAP Fiori Tools Extension

After the SAP Fiori tools have been installed, the SAP Fiori tools commands are available in the command palette, as shown in Figure 2.63.

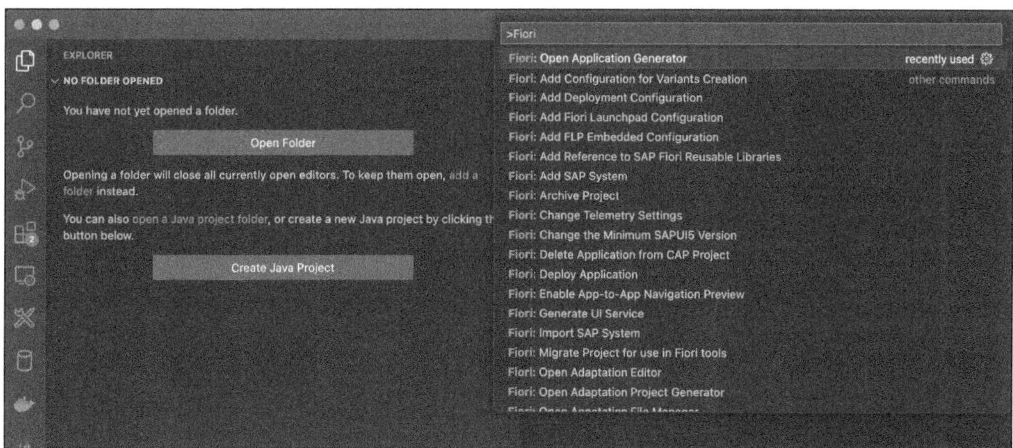

Figure 2.63 SAP Fiori Tools Commands

2.3 Summary

In this chapter, the main infrastructure elements in SAPUI5 development were presented. These include SAP BTP and IDEs. SAP BTP was discussed in Section 2.1. It serves as a central platform for integration between various cloud services as well as between cloud and on-premise solutions. It also plays a key role in the expansion of SAP cloud solutions by offering flexible and powerful expansion options. The architecture of SAP BTP was explained in detail in Section 2.1.2. For the secure connection between SAP BTP and the on-premise landscape, the cloud connector is a key element, which was discussed in Section 2.1.3. The cloud connector not only enables secure integration but also supports principal propagation, that is, the transfer of the user identity to the on-premise system to ensure end-to-end authentication. The security aspects of SAP BTP were described in section Section 2.1.4, including basic concepts for authentication and authorization. In addition, the topic of SSO was briefly addressed, followed by an introduction to SAP Cloud Identity Services, which offer various options for secure identity management and authentication.

Finally, Section 2.2 discussed the available development environments. Developers can choose between SAP Business Application Studio on SAP BTP (Section 2.2.1) and a local development environment such as VS Code (Section 2.2.2), depending on which working environment best suits their needs.

Chapter 3
TypeScript

TypeScript is an extension of JavaScript that enables the development of more robust and maintainable code bases. For this, aspects such as typing, classes, interfaces, generics, and other functions play a major role in simplifying the development process and increasing source code quality. We'll see how TypeScript helps us detect type-related errors early on, reduce code complexity, and increase code readability.

The *TypeScript* scripting language is important for SAPUI5 development because it provides strong typing, better tooling, and enhanced code maintainability, enabling developers to write robust, scalable, and error-free applications while leveraging SAPUI5's type definitions and application programming interfaces (APIs) efficiently. Based on SAP's suggestions, TypeScript is also the way to go when implementing SAPUI5 applications.

TypeScript was developed by Microsoft and released in 2012 based on the ECMAScript 6 (aka ES6) standard and made available under the Apache license. TypeScript is probably most widespread in the field of web development. Due to its numerous advantages over JavaScript, its importance in terms of UI5 development is also increasing enormously. Compared to JavaScript, TypeScript brings many additional features, such as typing.

In this chapter, we'll take a closer look at TypeScript. In Section 3.1 and Section 3.2, we'll look at the basics and the advanced concepts of TypeScript, respectively. Section 3.3 covers classes and interfaces, which are also important in other programming languages such as Java. In Section 3.4, you'll learn about generic, what advantages they offer, and how to use them. Finally, in Section 3.5, we'll discuss developing modules.

3.1 Basics

As already mentioned, TypeScript is based on the ECMA standard, just like JavaScript. The latest version is ECMAScript 2024 (aka ES2024), which was published in July 2024 as the 15th version.

TypeScript is a superset of JavaScript and can therefore use various libraries such as Angular. Initially, TypeScript was only supported by the Visual Studio Code (VS Code) development environment, but this is no longer the case. Almost all common integrated development environments (IDEs) now also support this programming language. Because of the similarities with JavaScript, a combination of both languages is often used. In principle, JavaScript can also be used wherever TypeScript is used; however, the reverse isn't possible. TypeScript code is converted into JavaScript code by a compiler.

ECMA

European Computer Manufacturers Association (ECMA) is a nonprofit organization that develops standards for computer hardware, communications, and programming languages.

Whether TypeScript and JavaScript are two languages in their own right is often debated, as TypeScript contains all of JavaScript. In most cases, however, developers who learn TypeScript are already familiar with JavaScript or learn it at the same time.

So where are the differences between JavaScript and TypeScript? To understand the advantages and disadvantages of the two languages, a direct comparison of features and characteristics is best. In addition, we'll look at possible areas of use and how the languages work. In the following sections, we'll cover the different data types TypeScript provides.

3.1.1 JavaScript

JavaScript is a lightweight scripting language. It can calculate, validate, and change data of various kinds to make web pages interactive. It can also be used to dynamically insert content into HTML and cascading style sheets (CSS) documents. JavaScript is often associated with Java because of its name, but it's important to make a strict distinction here. Apart from the name, the two languages have nothing to do with each other. JavaScript can simply be executed in the browser without any further tools. There's no typing in JavaScript, which is also referred to as *dynamic typing*, meaning that the type of a variable always depends on its content. If you store a string in a variable, it has string as its type. If you now write a date instance into the same variable, the variable no longer has the type string, but date. Although this may be convenient in development because every variable can always be assigned any value or type, it often leads to problems that wouldn't have occurred if languages with strict typing had been used. In addition, the readability of the program code suffers to a certain extent.

3.1.2 TypeScript

In addition to the functions available in JavaScript, TypeScript has all the features and concepts of object-oriented development, so classes, interfaces, inheritance, modules, and so on are available. Unlike JavaScript, TypeScript can't simply be executed in the browser because TypeScript code must first be converted into JavaScript code by a compiler before execution. This process is called *transpiling*. An essential advantage of TypeScript is that it has strict typing. This makes the written code less susceptible to errors because these can often be caught in an earlier phase of development. It also provides better syntax highlighting, which significantly simplifies the development and analysis of source code.

If we want to assign the value Hello World to a variable in JavaScript, it looks like this:

```
let myString = "Hello World";
```

In TypeScript, this looks almost identical. Here, a type is also specified, in our case string:

```
let myString: string = "Hello World";
```

If you were to try to assign a value of the type number to the variable in TypeScript, this would result in a syntax error, whereas you wouldn't receive any feedback in JavaScript. For example, it could look like this:

```
myString = 1;
```

In addition to the already-existing standard types such as string, number, Boolean, and so on, you can also define your own types in TypeScript. We'll see what this looks like in Section 3.3.

However, the question of when to use JavaScript and when to use TypeScript hasn't yet been clarified. In principle, JavaScript is most suitable if you want to write a quick, short script or develop small projects. However, if it's a large project in which several developers are involved, TypeScript is the better decision, especially because source code written in TypeScript is much easier to read. In addition, it's easier to integrate external libraries and various frameworks in TypeScript.

3.1.3 The Purpose of TypeScript

TypeScript allows better quality source code to be written because problems occur much earlier and can therefore also be corrected earlier than is the case in JavaScript. For example, errors caused by incorrect types can already be detected during development through appropriate typing and syntax highlighting. In JavaScript, such errors can usually only be discovered when testing the application. This saves time both in development and in quality assurance.

Not only is the developed code more qualitative, but the development itself is easier. Through appropriate typing, which doesn't exist in JavaScript, convenience functions such as code completion can be used. In this way, available functions or properties are suggested to you, which can be carried out with an object. In this way, you don't have to look up the documentation, but it's integrated into the development environment, so to speak. This aspect of TypeScript is very useful, especially if you use libraries or frameworks.

Often, of course, you want to create program documentation at the end of the development. Here, too, TypeScript provides a remedy through typing. While in JavaScript, documentation has to be added by comments, TypeScript can do this out of the box. Through the types used, it's possible to explain how the source code is to be used. In addition, you can use a generator, such as *TypeDoc*, which is responsible for the output of a documentation.

3.1.4 How to Use TypeScript

While JavaScript can be used without further installation, TypeScript requires installation to use it. To install TypeScript, Node.js and Node Package Manager (NPM) are required.

Node.js and Node Package Manager

Let's look at these tools in slightly more detail:

- **Node.js**
  Node.js is a server-side JavaScript runtime environment that allows JavaScript code to be executed outside of a browser. The asynchronous mode of operation enables high scalability and performance. In addition, developers are also offered numerous modules and packages with which applications can be extended by various functionalities. These packages can be installed with NPM.
- **Node Package Manager**
  NPM is a tool that allows you to install, update, and manage various packages, modules, or libraries for Node.js applications. NPM contains more than a million projects provided by the open-source community. This makes it the largest package repository for JavaScript. The functional scope of these provided packages is very broad. For example, there are complex frameworks that are used for web development as well as packages that only offer simple utility functions.

To integrate such packages into your own applications, the dependencies to these packages must be entered in the *package.json* file, which exists in Node.js projects. NPM ultimately ensures that the defined dependencies are available and installed.

Node.js can be downloaded and installed at *www.nodejs.org/en/download*. By installing Node.js, the Node Package Manager is also installed.

Once this installation is complete, TypeScript can be installed using NPM with the following command via the command line:

```
npm install -g typescript
```

As shown in Figure 3.1, a success message is displayed after successful installation.

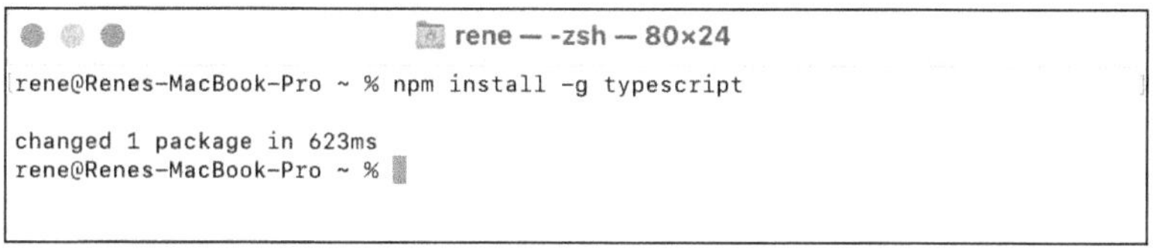

```
rene — -zsh — 80×24
rene@Renes-MacBook-Pro ~ % npm install -g typescript

changed 1 package in 623ms
rene@Renes-MacBook-Pro ~ %
```

Figure 3.1 Installation of TypeScript via NPM

If you now type "tsc" in the console and execute it, you should see a similar output as in Figure 3.2. In this output, you can see which version of TypeScript is installed. Furthermore, you'll also get help on which commands are available.

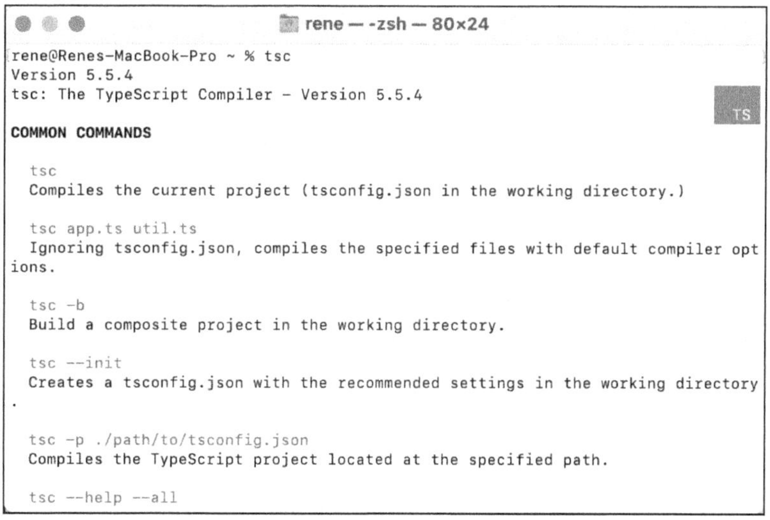

```
rene — -zsh — 80×24
rene@Renes-MacBook-Pro ~ % tsc
Version 5.5.4
tsc: The TypeScript Compiler - Version 5.5.4                        TS

COMMON COMMANDS

  tsc
  Compiles the current project (tsconfig.json in the working directory.)

  tsc app.ts util.ts
  Ignoring tsconfig.json, compiles the specified files with default compiler opt
ions.

  tsc -b
  Build a composite project in the working directory.

  tsc --init
  Creates a tsconfig.json with the recommended settings in the working directory
.

  tsc -p ./path/to/tsconfig.json
  Compiles the TypeScript project located at the specified path.

  tsc --help --all
```

Figure 3.2 Check Installation of TypeScript

To compile TypeScript so that executable JavaScript is created, there's another command called `tsc`, which stands for TypeScript compiler. However, when TypeScript is first set up in a project, a TypeScript configuration file should be created called *tsconfig.json*, which contains, among other things, configurations for the compiler. A default configuration can be created with the following command:

```
tsc -init
```

Initially, this file can look like the one shown in Listing 3.1.

```
{
  "compilerOptions": {
    /* Language and Environment */
    "target": "es2016",
    /* Modules */
    "module": "commonjs",
```

```
    /* Interop Constraints */
    "esModuleInterop": true,
    "forceConsistentCasingInFileNames": true,
    /* Type Checking */
    "strict": true,
    /* Completeness */
    "skipLibCheck": true
  }
}
```

Listing 3.1 tsconfig.json: TypeScript Configuration File

After the configuration file has been created, the code can be compiled by executing command `tsc`. Note that these commands are always executed in the root of the project.

3.1.5 Data Types

Now that typing is supported in TypeScript, let's look at the different data types that are available to us. Of course, the following data types can be supplemented with specially developed ones at any time:

- **Primitive types**
  TypeScript brings with it the following primitive types, which are also the most commonly used:
  - String: The string data type can be used to represent character strings, for example, "Hello World!".
  - Number: The number data type, on the other hand, is used for all numbers. Unlike in programming languages such as Java or C, there's no distinction between integer and float. Numbers such as 38 or 3.9 can be represented with the number type.
  - Boolean: This type is used for Boolean expressions. A variable of the type Boolean can thus assume the values `true` or `false`.
- **Arrays**
  This is a special type for representing lists or arrays. In principle, an array can be created from any type. The syntax is `type[]`, so to define an array of numbers, you would write `number[]`.
- **Any**
  This special type can always be used when you want to do without typing. If a variable with the any type is present, all of its properties can be accessed. Basically, you can do almost anything with it that is syntactically correct. However, this type should only be used if you want to deliberately avoid writing a long type definition just to convince TypeScript of the correctness of the code. The any type is also used if no type is specified.

3.2 Next-Generation TypeScript

Of course, TypeScript also has more advanced concepts and features, which we'll look at in more detail in this section. In the following sections, you'll see how anonymous and arrow functions can be implemented to quickly define and implement certain functions. Furthermore, you'll also learn about function overloading, rest parameters, static members, variable declarations, and intersection and union type application.

3.2.1 Anonymous Functions

Similar to other languages, such as Java, it's also possible to define anonymous functions in TypeScript. Such a function, just like any other function, can have any number of passing parameters and can have any return type or none at all. The syntax for an anonymous function is shown in Listing 3.2.

```
let fnAnonymous = function([args]) {
  // body
}
```

Listing 3.2 Syntax for an Anonymous Function

3.2.2 Arrow Functions

In addition to anonymous functions, arrow functions are also very popular. Basically, these are anonymous functions in a shorter notation. Arrow functions are also known as Lambda functions in other programming languages. The syntax of this function is as follows:

```
(param1, …, paramN) => expression
```

The arrow (=>) used here, from which this type of function also gets its name, replaces the keyword function. In addition, the arrow separates the passing parameters (left) from the function body (right). Just like anonymous or normal functions, this type of function can have any number of transfer parameters and also any return type or none at all.

Passing parameters are passed here in the round brackets (). If you only have one transfer parameter, the brackets can also be omitted. The same applies to the instructions of the function body, which must be placed inside the curly brackets. If the function consists of only one instruction, these brackets can also be omitted. If you now want to define a function that calculates the sum of two passed numbers, for example, this can look like Listing 3.3.

```
let sum = (a: number, b: number): number => {
  return a + b;
```

```
}
sum(3, 5); // returns 8
```

Listing 3.3 Definition of an Arrow Function

Because this function consists of only one instruction, it can be optimized even further, as shown in the following:

```
let sum = (a: number, b: number): number => a + b;
sum(3, 5); // returns 8
```

As you can see, the optimized version omits not only the curly brackets but also the return keyword.

3.2.3 Rest Parameters

Often, it's uncertain or changing how many parameters are actually passed to a function. TypeScript provides a remedy in this case with *rest parameters*. To do this, simply place ... in front of the parameter identifier. If a parameter is defined as a rest parameter, any number of arguments or even no arguments at all can be passed to it. The compiler automatically creates an array that contains the passed parameters. Listing 3.4 provides an example of how rest parameters can be used.

```
function print(…vehicles: string[]) {
  console.log(vehicles);
}
print('Audi', 'BMW', 'VW'); // prints "Audi, BMW, VW"
print('Ford'); // prints "Ford"
print('`'); // prints ""
```

Listing 3.4 Function with Rest Parameters

As shown in the example, any number of strings can be passed to the function, which will ultimately be merged into an array within the function. The only thing to note when using these parameters is that they may only be listed as the last parameter in the function declaration.

3.2.4 Function Overloading

Another helpful concept that TypeScript offers is *function overloading*, in which you can have several functions with the same name. These functions differ only in their parameter types and return types. The number of parameters should be identical. Let's look at this in an example (see Listing 3.5).

```
function add(a: number, b: number): number;
function add(a: string, b: string): string;
```

```
function add(a: any, b: any): any {
   return a + b;
}

add(3, 5); // returns 8
add('Hello ', 'World'); // returns "Hello World"
```

Listing 3.5 Function Overloading

In the example in Listing 3.5, we've defined a function add() with two function declarations and one implementation. The first declaration has two parameters of the type number and a return type number. The second declaration, on the other hand, has two parameters of type string and a string as return type. Furthermore, there's a function implementation that also takes two parameters, but of the type any. The function does nothing other than adding up the two parameters passed. If numbers are passed, only the sum is formed. If the parameters are of the type string, they are concatenated together.

Warning

Function overloading with different numbers of transfer parameters isn't supported.

3.2.5 Static Members

Static members are also included in TypeScript. The special thing about static members is that they exist and can therefore be used without having to create an instance of the class. Static members are accessed via the class name and the name of the member, and they are identified or defined by the keyword static. Auxiliary functions or constants, as well as other variables and functions, are often defined as static members.

Listing 3.6 provides an example of a class with static members and how they can be accessed. Of course, static and nonstatic members can also be mixed.

```
class MathUtil {
   static PI: number = 3.141592;

   static square(a: number): number {
      return a*a;
   }
}
let PI = MathUtil.PI;
let result = MathUtil.square(3);
```

Listing 3.6 Class with Static Members

3.2.6 Intersection and Union Types

In TypeScript, the developer can define intersection or union types. An *intersection type* is created by merging two different types. The new type thus contains all the properties of the other two types. To achieve this, the & operator is used. Here, any number of types can be merged. For the sake of simplicity, we'll create an intersection type with only two, type1 and type2:

```
type type_12 = type1 & type2;
```

In contrast to the intersection type, a *union type* is used to define that different types are allowed. For example, you can define that a variable can be assigned type1 or type2. The | operator is used for this type. In the following example, again we only use two different types, but there's no restriction here:

```
let var1 = type1 | type2;
```

Whether an intersection type or a union type is used, the order in which the individual types are defined is irrelevant.

3.2.7 Variable Declarations

As in JavaScript, TypeScript offers the following possibilities for declaring variables:

- **var**
  var is the longest existing way to declare variables. It's often necessary to declare variables with var, especially if the application will run in old browsers. Variables defined with var can be declared several times, which of course can lead to unintentional errors or misunderstandings. It's important to note that variables defined with var are globally available, so they can be accessed even before they have been declared.
- **let**
  The keyword let is the successor of var. Unlike var variables, variables declared with let are only available in the block in which they were defined. This is also referred to as block scope, meaning if a variable is declared with let within an if statement or a loop, it can no longer be accessed outside the block.
- **const**
  Constants can be created with the keyword const. It's important here that these are assigned a value immediately, as this is no longer possible at a later time. The scope of a const variable is limited to the block in which it's defined.

Some advantages of let over var are as follows:

- Variables with block scope can't be read or modified before they have even been declared (see Listing 3.7).

```
console.log(var1); // Compiler Error
let var1: number = 1;

console.log(var2); // No Error, Ouput: undefined
var var2: number = 1;
```

Listing 3.7 let vs. var

In the case of the declaration with `let`, we would run into an error in this example. However, if the variable is declared with `var`, no error would occur, and `undefined` would be the value in the variable.

- `let` variables can't be declared more than once. While `var` variables can be declared several times with the same name, this isn't possible with `let` variables. With `let` variables, we'd also run into an error here because variables occur several times in one and the same block. Listing 3.8 shows an example of this.

```
var num1: number = 1; // OK
var num1: number = 2; // OK

let num2: number = 1; // OK
let num2: number = 2; // Compiler Error
```

Listing 3.8 Declaring Variables Multiple Times

- `let` variables reduce the susceptibility to errors. The use of `let` not only promotes the maintainability and readability of the code but also ensures that the number of runtime errors is reduced because possible errors are already detected by the compiler.

3.3 Classes and Interfaces

Similar to Java, object-oriented concepts such as classes and interfaces are available to the developer in TypeScript. We'll discuss both of those concepts, in addition to interfaces and abstract classes, in the following sections.

3.3.1 Classes

Basically, classes in TypeScript are similar to other programming languages. Classes offer the possibility to define a template for creating objects. This template ultimately contains various methods and properties. The defined methods don't necessarily have to be related to the defined properties. In addition, classes in TypeScript, just as in Java, can contain properties, methods, constructors, and access modifiers. If an object of a class is created, it's called an instance.

Classes in TypeScript, just as in other object-oriented programming languages, can contain the following:

- Constructors
- Properties
- Methods

Before we discuss these three elements, let's first look at a few examples of a TypeScript class.

Class Examples

A TypeScript class could look like the one shown in Listing 3.9.

```
class Vehicle {
   brand: string;
   color: string;

   constructor(pBrand: string, pColor: string) {
      this.brand = pBrand;
      this.color = pColor;
   }

   getBrand(): string {
      return this.brand;
   }
}
```

Listing 3.9 Example of a Class in TypeScript

However, as you learned in a previous section, TypeScript code is converted to JavaScript code by the compiler. You can see how our example class would ultimately look in JavaScript in Listing 3.10. As we'll see in a moment, the class is much easier to read and understand in TypeScript.

```
var Vehicle = (function() {
   function Vehicle(brand, color) {
      this.brand = brand;
      this.color = color;
   }

   Vehicle.prototype.getBrand = function() {
      return this.brand;
```

```
  };
  return Vehicle;
}());
```

Listing 3.10 TypeScript Class Compiled to JavaScript

Constructor

A constructor is a special type of method of a class. Whenever an instance of a class is created, its constructor is called. In TypeScript, this always has the method name `constructor`. In other programming languages, such as Java, a method with the name of the class represents the constructor.

In our example, the `Vehicle` class contains a constructor with the parameters `pBrand` and `pColor`. In the constructor, the member variables of the class can be accessed directly via the `this` keyword—in this case, `this.brand` and `this.colour`. Nothing happens here other than that the values of the two passed variables are stored in the corresponding member variables. A class doesn't necessarily have to have a constructor. If this is omitted, the default constructor exists, which doesn't accept any parameters.

To now create an instance of a class and thus also call the constructor, the keyword new is used. In our case, we would do this with the following statement:

```
let vehicle = new Vehicle('Audi', 'blue');
```

Visibility of Properties and Methods

Properties and methods of a class can have different visibility levels. The use of such visibilities is due to encapsulation. The following are available:

- `private`
- `protected`
- `public`

We'll look at what the individual visibilities mean in detail in the following.

private

If a property or a method is defined with the visibility `private`, this means that it can only be accessed within the class. It can't be accessed from the outside. As a rule, properties are defined as `private`, but often methods are also defined as private if they are only used within the class. These are mostly auxiliary methods.

As shown in Listing 3.11, one property was defined as `private` and one as public (default value). If an instance is now created outside the class and both properties are accessed, this will sometimes lead to errors. The public property `color` can of course be accessed. However, accessing the private property `brand` will create a compiler error because private properties are only available within the class. In this case, you would either have to change the visibility of this property or, as is usual, define public methods to access this property.

```
class Car {
   private brand: string;
   color: string;
}
let car = new Car();
car.brand = 'Audi';
car.color = 'blue';
```

Listing 3.11 Class with a Private Member

protected

The access modifier `protected` is similar to `private`. However, there's a small difference: members that are defined as `protected` can also be accessed from derived classes. Protected can therefore be seen as a mixture between private and public.

public

The last access modifier, which also specifies the default value, is `public`. No restrictions are placed on the use of public. It's therefore possible to access these members both inside and outside. As a rule, methods are defined as `public`.

In the example from Listing 3.12, we've now defined a class `Car` that contains the properties `brand` and `color`. While `brand` is defined as `public`, no modifier was specified for `color`, which means that `public` is the default value here. Because both properties are public, they can be accessed directly outside the class, as is the case in the example.

```
class Car {
   public brand: string;
   color: string;
}
let car = new Car();
car.brand = 'Audi';
car.color = 'blue';
```

Listing 3.12 Class with Public Members

3.3.2 Inheritance

Just as in other programming languages, *inheritance* plays an important role in TypeScript, especially when it comes to reusing properties and methods from other classes. The class that inherits properties and methods from another class is called a child class. The class that is inherited is called the parent class.

As already mentioned, inheritance achieves a certain reusability and thus also prevents identical code from being defined multiple times. In JavaScript, classical inheritance isn't supported, but prototypical inheritance is. In TypeScript, however, inheritance is supported in the same way as in ES2024.

To define that a child class inherits from a parent class, the `extends` keyword is used. It's best to look at this in a practical example. Suppose we have a `class Vehicle`, as shown in Listing 3.13.

```
class Vehicle {
   private brand: string;
   private color: string;

   constructor(brand: string, color: string) {
      this.brand = brand;
      this.color = color;
   }

   getBrand(): string {
      return this.brand;
   }
}
```

Listing 3.13 Definition of a Parent Class

As you can see here, we have a `class Vehicle` with the attributes `brand` and `color`. A vehicle can basically be many different things—a car, a truck, and so on—but what they all have in common are the attributes `brand` and `color`.

To implement a concrete vehicle, such as a car, we proceed as shown in Listing 3.14.

```
class Car extends Vehicle {
   private registrationDate: date;
   ...

   constructor(brand: string, color: string, registrationDate: date) {
      super(brand, color);
      this.registrationDate = registrationDate;
   }
}
```

Listing 3.14 Definition of a Child Class

As shown in the previous example, the properties `brand` and `color` aren't defined in the `class Car` because they are inherited from the `class Vehicle`. This means that all methods and properties from the `class Vehicle` are taken into the inheriting class.

Nevertheless, the class can be extended by further properties and methods. In our case, the class is extended by another property called `registrationDate`.

Besides the use of the `extends` keyword, there's another conspicuous feature in this class definition. Namely, the function `super` is called in the constructor. This function is the call to the constructor of the parent class. It's important that this call is the first call in the constructor of the child class. Furthermore, the passing parameters must match the parameters as defined in the parent class. In addition to calling the constructor, functions of the parent class can also be called with the `super` keyword. This syntax would look as follows:

```
super.functionName();
```

Because all properties and methods are inherited from the parent class, expressions like the following are allowed and also provide a correct result:

```
let car = new Car('Audi', 'blue', new Date()); car.getBrand(); // returns "Audi"
```

If possible, the parent classes should be designed so that the methods can also be used for all child classes. Often, however, this isn't possible. In this case, the method can simply be redefined in the child class, as shown in Listing 3.15.

```
class Car extends Vehicle{
   getBrand(): string {
      return 'Brand: ' + this.brand;
   }
}
```

Listing 3.15 Override a Method of the Parent Class

If an instance of the `Car` class is now created and the method `getBrand` is called, the implementation from the parent class is no longer used; instead, the overwritten implementation in the concrete `Car` class is used.

3.3.3 Interfaces

While classes represent the structure of an object as well as its functions, an interface is only used to represent the structure of an object. This means that an interface only specifies which properties or methods a class must contain in a later instance. Classes that implement an interface must therefore follow the structure of the interface.

Unlike classes, interfaces aren't converted into JavaScript code by the compiler. Interfaces, on the other hand, are used for *type checking* (aka *duck typing* or *structural typing*). Interfaces are declared with the keyword `interface`. In addition to properties, interfaces can also contain method declarations. In Listing 3.16, we see an example of such an interface.

```
interface IVehicle {
   brand: string;
   color: string;
   getMileage: () => number;
   getBrand(): string;
}
```

Listing 3.16 Interface in TypeScript

The interface in Listing 3.16 contains two properties: `brand` and `color`. In addition, it also has two methods: `getMileage` and `getBrand`. The method `getBrand` is declared as a normal method with a string as return type. The second method `getMileage` is declared as an arrow function with a number as return type. How these methods are declared, however, plays no role here and is a matter of taste for the respective developer. Due to the structure of the interface, a class that implements this interface must define both properties and both methods.

In the following, we'll see how an interface can be implemented. Additionally, we'll also see how interfaces can be used to define certain types such as function types or array types. After that, we'll take a look at how certain interface properties can be defined as optional or read-only.

Implementing an Interface

Similar to other programming languages (e.g., Java), classes can be defined that implement an interface. In Listing 3.17, you can see what a class would look like that implements the interface from Listing 3.16 shown earlier. The keyword used for this is `implements` in the definition of the class.

```
class Vehicle implements IVehicle {
   brand: string;
   color: string;
   getMileage: () {
      return 1000;
   }
   getBrand(): string {
      return this.brand;
   }
}
```

Listing 3.17 Class Implementing an Interface

However, not only can interfaces be defined to be implemented in classes but also for the definition of types. In Listing 3.18, you can see what such an interface might look like.

```
interface MessageStatus{
   status: string;
   message: string;
}
```

Listing 3.18 Interface as Type

As you can see, we've defined an interface with the two properties `status` and `message` of the type string. When creating variables, this interface can now be used in addition to the already-existing types. Because none of the variables is optional, both variables must be specified when creating the object. It's not only important that both variables are supplied, but also that the type matches.

In Listing 3.19, we've now created three variables each with the type `MessageStatus`. We'll now look in detail at which of these are valid and which lead to errors.

```
let msg1: MessageStatus = {status: 'OK', message: 'Ok!'};
let msg2: MessageStatus = {status: 'NOK', msg: 'Not ok!'};
let msg3: MessageStatus = {status: 1, message: 'Not ok!'};
```

Listing 3.19 Create Variables with Interface as Type

The variable `msg1` follows the structure of the interface and is therefore valid. The second variable `msg2` does contain the property `status`, but the property `message` is missing here because the short form `msg` is used instead of `message`. This causes the compiler to run into an error, and the instruction is therefore invalid. If we now look at the third and last example, `msg3`, you might think at first glance that this is a valid instruction. Look more closely, however, and you might notice that although the names of the defined properties are correct, the type of the passed parameter `status` differs from the type in the interface. In the interface, the status is defined with a string, but a parameter of the type number is passed. Thus, this instruction is also invalid, and the compiler will indicate this with an error.

In this example, we've used the interface as a type for a variable. However, it's also possible to use interfaces as types for functions or arrays. The procedure is basically the same.

Interface as Function Type

Listing 3.20 shows how an interface can be used as a function type. Here, a method signature is defined in the interface. Then, a variable can be created with the interface as type. In our case, the definition of the interface ensures that the variable can only contain references to functions that have the same signature as defined in the interface. If one were to try to assign a function with a different signature—a different transfer parameter and a different return type—this would result in an error.

```
interface MessageStatus {
   (status: string, message: string): void;
}
function print(status: string, message: string): void{
   console.log('Status=' + status + ', message=' + message);
}
let processor: MesssageStatus = print;
processor('OK', 'Ok!'); // Output: Status=OK, message=Ok!
```

Listing 3.20 Interface as Function Type

Interface as Array Type

Interfaces can also be used as types for an array. The type of the index can also be defined here. If we want to create an array that uses an index as well as a value of the type string, this is implemented as shown in Listing 3.21.

```
interface IArrayString {
   [index:string]: string;
}
let arrStr: IArrayString = {};
arrStr['OK'] = 'Ok!';
arrStr['NOK'] = 'Not ok!';
```

Listing 3.21 Define Array Type via Interface

In the previous example, the interface `IArrayString` defines an array in which both the index and the value found at an index in the array are of the type string.

Optional Properties

You don't always want to define all properties of an interface as mandatory fields. To avoid having to define separate interfaces for different cases, individual properties can be defined as optional. These are marked with a `?` and therefore don't have to be specified as mandatory when creating the interface.

As shown in Listing 3.22, we've now extended the interface used earlier by another property `errorCode`. Because there can usually only be one if an error occurs, this is marked as optional by the `?`. In this case, both statements—`success` and `error`—are valid, even if the former doesn't define this new property. If the property isn't defined, it can also be accessed, but the value returned will be `undefined`.

```
interface MessageStatus {
   status: string;
   message: string;
   errorCode?: string;
}
```

```
let success: MessageStatus = {status: 'OK', message: 'Ok!'};
let error: MessageStatus = {status: 'NOK', message: 'Not ok!', errorCode:
'500'};
```

Listing 3.22 Interface with Optional Property

Read-Only Properties

If you want to ensure that a property can only be assigned a value once, which can then no longer be changed, TypeScript also provides the `readonly` keyword. An example interface with a `readonly` property is shown in Listing 3.23.

```
interface Person {
   name: string;
   readonly birthday: string;
}
let pers = {name: 'Max', birthday: '2000-01-01'};
pers.name = 'Max II';
pers.birthday = '2000-02-02';
```

Listing 3.23 Interface with readonly Property

In this interface, there are two properties of the type string. The property `name` isn't read-only and can therefore be changed at will, while the second property `birthday` has the keyword `readonly`. As a result, this property can be set initially, but it can't be changed afterward. The earlier instruction to change the value of this property will result in a compiler error.

3.3.4 Abstract Classes

TypeScript also supports the concept of abstract classes. Basically, *abstract classes* are used to define how derived classes should behave. Unlike normal classes, abstract classes can't be instantiated. If a class inherits from an abstract class, it's automatically abstract unless the abstract methods are implemented. The most important keyword in connection with abstract classes is `abstract`.

Typically, one or more abstract methods are defined in an abstract class. However, these don't contain an implementation; they only define the signature. We find the implementation of abstract methods in the concrete class, which inherits the abstract class.

We can see how the definition of an abstract class might look in Listing 3.24.

```
abstract class Vehicle {
   abstract getPrice(): number;
}
```

Listing 3.24 Definition of an Abstract Class

After defining an abstract class with an abstract method in the preceding example, it's now necessary to implement it in another class. In addition to the inheritance of the class, the method implementation must also be defined. In Listing 3.25, we see how this might look in our case.


```
class Car extends Vehicle {
   getPrice(): number {
      return 30000;
   }
}
```

Listing 3.25 Implementing an Abstract Class

3.4 Generics

No matter which area of software development you're in, reusability is a very important factor. This plays a decisive role for UI components, classes, and logic when attempting to write qualitatively good, clean code. In addition, reuse also increases flexibility and avoids having to maintain identical code several times.

Generics exist precisely to achieve this at class level. Generics enable developers to make components reusable in a certain way so that components can handle different data types and aren't limited to a specific one. In the end, generic components can be used with any data types without losing functionality. In TypeScript, generics are similar to those in C#.

In the following sections, we'll first see why generics are important and should be used in development. After that, we'll look at the single methods provided by generics. Last but not least, we'll have a look at the implementation of generics in both classes and interfaces. We'll also take a look at how we can use generic constraints as well.

3.4.1 Generics Example

Let's look at an example of why generics are relevant in development. In Listing 3.26, we have a function `getArray()` that accepts an array of type any. In it, nothing happens except that a new array is created in which the elements from the passed-in array are inserted and the result is finally returned. Because the type any was defined, arrays with any data type can be passed at this point. However, this might not be the desired behavior. In this example, we could insert elements of the type number and string in the two arrays `numberArray` and `stringArray`. As a rule, you wouldn't want this, but an array should only have elements of the same type.

```
function getArray(arr: any[]): any[] {
   return new Array().concat(items);
}
```

```
let numberArray = getArray([10, 20, 30]);
let stringArray = getArray(['Full', 'Stack']);

numberArray.push(40); // OK
stringArray.push('Full-Stack'); // OK
numberArray.push('Test'); // OK
stringArray.push(50); // OK

console.log(numberArray); // [10, 20, 30, 40, 'Test']
console.log(stringArray); // ['Full', 'Stack', 'Full-Stack', 50]
```

Listing 3.26 Why We Need Generics

Generics can be used to solve this problem. Basically, the getArray() method must be changed to a generic type. In our case, this is done with the type variable <T>. This variable is a special case of variables that identifies a type. On the one hand, this type variable must be included in the respective function, which is to be generic, and on the other hand, a type must be passed when it's called. In Listing 3.27, you can see what changes this entails in the implementation.

```
function getArray<T>(arr: T[]): T[] {
  return new Array<T>().concat(items);
}
let numberArray = getArray<number>([10, 20, 30]);
let stringArray = getArray<string>(['Full', 'Stack']);

numberArray.push(40); // OK
stringArray.push('Full-Stack'); // OK
numberArray.push('Test'); // Error
stringArray.push(50); // Error
```

Listing 3.27 Using Generics

Because the getArray() method now accepts a type, this must also be specified when using it. Instead of type any that we used before, the type used is called passed. In addition, the type is used when instantiating the array, which means that it's no longer possible to mix elements of different types in an array. This can also be seen in the example from Listing 3.27.

You can also call generic functions without specifying a type, but this isn't recommended. The compiler determines the type to be used from the type of the parameter passed.

Generics can be applied to the following:

- Parameters of a function
- Return types of a function

- Variables of a class
- Methods of a class

The number of generic data types used isn't limited. Even if we only use one such data type in our example, any number can be assigned. The only important thing is that the names of these are kept unique. Furthermore, generic data types can also be used in conjunction with nongeneric data types.

3.4.2 Methods of Generic Types

If generic type variables are used, they can basically only access methods and properties that are independent of their type. For example, the methods `toUpperCase()` or `toFixed()`, which only exist for strings or numbers, can't be accessed. The following methods are available for every object are, for example:

- `toString`
- `hashCode`
- `equals`

Constraints can be used to bring further flexibility into play here as well.

3.4.3 Generic Constraints

In principle, the generic type allows any type, but often this isn't desired. Therefore, TypeScript provides a remedy here as well, namely by using constraints. Listing 3.28 shows how this could look in practice and how it should be used.

```
class Vehicle {
  brand: string;
  color: string;

  constructor(pBrand: string, pColor: string) {
    this.brand = pBrand;
    this.color = pColor;
  }
}
function print<T extends Vehicle>(v: T): void {
  console.log('Brand=' + v.brand + ' color=' + v.color);
}
let vehicle = new Vehicle('Audi', 'blue');
print(vehicle); // Output: Brand=Audi color=blue
print("Audi"); // Error
```

Listing 3.28 Generic with Constraints

As you can see in Listing 3.28, we have a class Vehicle. In addition, there's a function print(), which is a generic function with a constraint. This constraint is specified in the angle brackets after the generic type. In our case, this is <T extends Vehicle>, which means nothing more than that the generic type must inherit the class Vehicle. In concrete terms, this means that it's no longer possible to pass on type any, but only the class Vehicle or a subclass of it. If an attempt is made to pass types that don't follow this rule, this results in an error.

3.4.4 Generic Interfaces

Generic data types can be applied not only to methods but also to design interfaces generically. Let's take a look at this with a short example. In Listing 3.29, we've defined an interface that contains two properties: a key and a value. The variable key is assigned the type of the type variable K, while the variable value is assigned the type of the type variable V. The designation of these variables can be chosen arbitrarily.

```
interface KeyValue<K, V> {
   key: K;
   value: V;
}
```

Listing 3.29 Define a Generic Interface

Listing 3.30 shows how an object of the type KeyValue is created. When using it, it's again important that the types are specified accordingly and that the specified values correspond to the correct type. In both our cases, it would work every time without any problems. However, if we were to expect a string and pass a number, it would lead to an error.

```
let pair1: KeyValue<string, string> = {key: 'a', value: 'Test I'};
let pair2: KeyValue<string, number> = {key: 'a', value: 1};
```

Listing 3.30 Using a Generic Interface

In this case, we've used the interface as a type definition. As already mentioned, interfaces can also be used in classes, for example. However, nothing else changes here, except that the generic type must of course also be specified there.

A big advantage of this type of implementation is that it offers a lot of flexibility. Basically, all key-value pairs can be mapped with this interface. If you were to do this without generics, you'd have to design separate interfaces for each application.

3.4.5 Generic Classes

In addition to generic interfaces, TypeScript also supports generic classes. With classes, the whole thing looks rather similar to interfaces. Here, too, the generic type is specified in angle brackets in the class definition. Generic classes can have any number of generic types, generic members, and generic methods. If we were to implement the KeyValue example shown earlier as a class, it would look like the one shown in Listing 3.31.


```
class KeyValue<K, V> {
   private key: K;
   private value: V;

   put(key: K, value: V): void {
      this.key = key;
      this.value = value;
   }

   print(): void {
      console.log(this.key + '=' + this.value);
   }
}
let pair1 = new KeyValue<string, string>();
let pair2 = new KeyValue<string, number>();
pair1.put('a', 'Test I');
pair1.print(); // a=Test I

pair2.put('a', 1);
pair2.print(); // a=1
```

Listing 3.31 Implementing a Generic Class

In Listing 3.31, a generic class KeyValue was implemented with the two generic type variables K and V. These two type variables specify the types for the properties key and value. In addition, the types are also used in the function put(), which is passed two parameters and sets the key and value of the object accordingly.

Both classes and generic classes can also implement other generic classes or generic interfaces. The type of these implementing classes or interfaces must also be specified. You can either specify a fixed type or do this via the generic type variables. If we were to use the generic interface from one of the previous examples and implement it in a class, it would look like Listing 3.32.

In Listing 3.32, we see an example where a class implements in a generic interface. However, the types are hardcoded, which diminishes flexibility. Here, you would have to create your own class for the different use cases.

```
class KeyValueStatic implements KeyValue<string, string>{
}
let pair1 = new KeyValueStatic();
pair1.key = 'a';
pair1.value = 'Test I';
```

Listing 3.32 Using a Generic Interface in Class

We can see how to make these facts more dynamic in the example in Listing 3.33.

```
class KeyValueDynamic<K,V> implements KeyValue<K,V> {
}
let pair1 = new KeyValueDynamic<string, string>();
pair1.key = 'a';
pair1.value = 'Test I';
```

Listing 3.33 Using a Generic Interface in a Generic Class

Unlike the previous example, in this one, the type must be specified when instantiating. This is then passed on to the generic interface. Thus, this implementation can be used with any data types without having to change anything.

3.5 Module Development

In JavaScript, various approaches have been tested in the past to modularize code. Over the years, ES modules have finally become established with TypeScript. You often hear about them under the name import-export syntax. In this section, you'll learn what modules are and why to use them. We'll also take a deeper look at the two types of modules: import and export.

3.5.1 What Are Modules?

Any TypeScript file that contains a top-level `import` or `export` statement is a module. Conversely, any file that doesn't contain a top-level `import` or `export` statement is considered a script that is globally available. Files that are globally available are also available for modules.

The big difference with modules is that they are executed in their own scope and not in the global scope. This means that all methods, variables, classes, and so on that are declared within the module aren't visible outside the module. However, there is an exception: they can also be visible outside the module only if they are explicitly provided with an export statement. Conversely, a variable that has been exported from another module must be imported into the desired module with an `import` statement.

3.5.2 Why Use Modules?

Because TypeScript code is by default in the global scope, this means variables, functions, and so on can be accessed from any file. These can be accessed not only by reading but also by writing, meaning that they can be manipulated, which can lead to unexpected behavior. This isn't optimal in terms of encapsulation and error-proneness. The answer to this problem comes via modules.

By using modules, you can achieve a certain local scope within a file. As already mentioned, `import` and `export` are the two most important keywords in connection with modules. We'll look at what this means in detail in the following.

To illustrate this practically, let's imagine we have two files, *file_1.ts* and *file_2.ts*. Let's first look at *file_1.ts*, as follows:

```
let author: string = 'Max Mustermann';
```

Next, let's look at *file_2.ts*, as follows:

```
console.log(author); // Max Mustermann
author = 'Max';
```

The expressions in these two files are valid, allowed, and will also work because, as previously discussed, all variables are in the global scope. Thus, the value of this variable can be overwritten, which can lead to problems in productive code.

3.5.3 Export Modules

To create a module in the first place, the `export` keyword is used. If we now look at *file_1.ts* from the previous example again as a module, it could look like in the following:

```
export let author = 'Max Mustermann';
```

Basically, not much has changed, except that the variable has been prefixed with the keyword `export`. This means that there's now an `export` statement at the top level, which turns this file into a module. Now this variable can be imported from other modules by means of an `import` statement. However, the difference is that you're no longer in the global scope.

3.5.4 Import Modules

The `import` keyword is used to import modules into another module and thus use the functions and variables from it. If we now want to access the property `"author"` as before, it won't work because it wasn't imported as a module. To get the code to work again anyway, we have to change the implementation a bit as follows:

```
import {author} from 'file_1.ts';
console.log(author); // Max Mustermann
```

Often, the identifiers are chosen inappropriately. In this case, an export module can be renamed. To do this, enter as followed by the desired name when importing the module. This feature is very helpful with regard to the readability of the code. In the following, we see how this might look in our case:

```
import {author as authorName} from 'file_1.ts';
console.log(authorName); // Max Mustermann
```

Several variables can be exported in one export module. In this case, each one can be imported separately, or the entire module can be imported into one variable. This variable would then contain all exports. To do this, we extend our *file_1.ts* with another export as follows:

```
export let author = 'Max Mustermann';
export let age = 30;
```

If we were to import this module, we would have to specify what is to be imported. However, we want to import the entire module into a variable. We can see how this looks and how the members can be accessed in Listing 3.34.

```
import * as authorObj from 'file_1.ts';

console.log(authorObj.authorName);
console.log(authorObj.age);
```

Listing 3.34 Import Module into a Variable

Another solution is to import each property separately.

3.6 Summary

In this chapter, we've dealt with the basics of TypeScript as well as more advanced concepts. Especially in view of constantly growing and sometimes changing requirements, it's important to write high-quality code from the outset. This often makes subsequent changes easier to implement. In TypeScript, too, the quality of the code can be significantly increased through the targeted encapsulation of data and separation of logics. Not to be forgotten are concepts such as generics and modules, which help in view of the reusability of the code.

Now that we've dealt with the SAPUI5 framework, the infrastructure and the development environments used, and TypeScript as a programming language in this and the previous chapters, we'll create an initial application with SAPUI5 in the following chapter.

PART II
Development

Chapter 4
First Steps in SAPUI5 Development

SAPUI5 is the framework when it comes to the development of modern web applications in the SAP environment. The development of these apps differs fundamentally from the development of classic applications (ABAP, Web Dynpro, etc.). SAPUI5 apps, as they are called, are supplied with data via OData services.

We've already introduced you to the SAPUI5 framework itself in Chapter 1, but we would like to remind you once again what the architecture of such an application looks like. There are various approaches to architecture to ensure that software is both easy to maintain and easy to extend and reuse. There are *design patterns* that group logically related parts of software and isolate these groups from each other. This idea of a tripartite division also occurs in the *model view controller* (MVC) design pattern, which was used for SAPUI5.

Before we can think about the structure in more detail, we first need to create a project in Section 4.1. We'll show you how to use the state-of-the-art development environments SAP is recommending. In Section 4.2, we'll take a closer look at the application structure and the files, which were generated automatically by using the corresponding template for SAPUI5 basic apps. In Section 4.3, the MVC design pattern will lead us through the development of the UI, the application logic, and the data maintenance. In Section 4.4, we'll go into more detail about how the SAPUI5 application can be opened in a preview.

4.1 Create Your Application

You already know that there are several integrated development environments (IDEs) for this. As SAP Business Application Studio is the recommended IDE from SAP, we'll start with this first. Nevertheless, we'll also show you how to create a project in Visual Studio Code (VS Code).

4.1.1 SAP Business Application Studio

When you start SAP Business Application Studio for the first time, you'll be greeted with a welcome text. A dev space must first be created to be able to advance the developments at all. A *dev space* is an instance for the development environment. In addition to the name, you must select the developments to which the IDE is to be adapted for this instance. We've selected **SAP Fiori** and then clicked on **Create Dev Space** (see Figure 4.1).

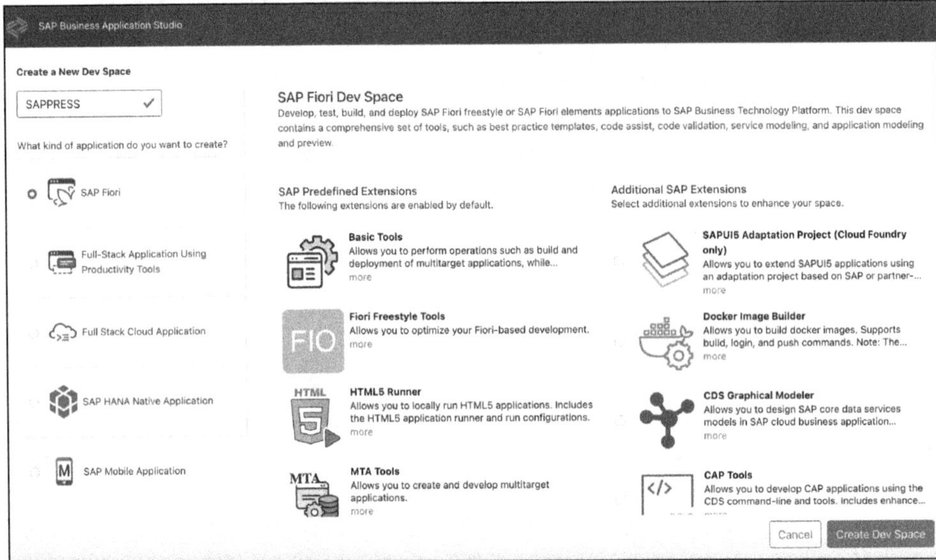

Figure 4.1 Creating a New Dev Space

Creating alone isn't enough. You have to wait until the IDE is up and running. As soon as the name of the dev space appears as a link, you can click on it and navigate to the development environment (see Figure 4.2).

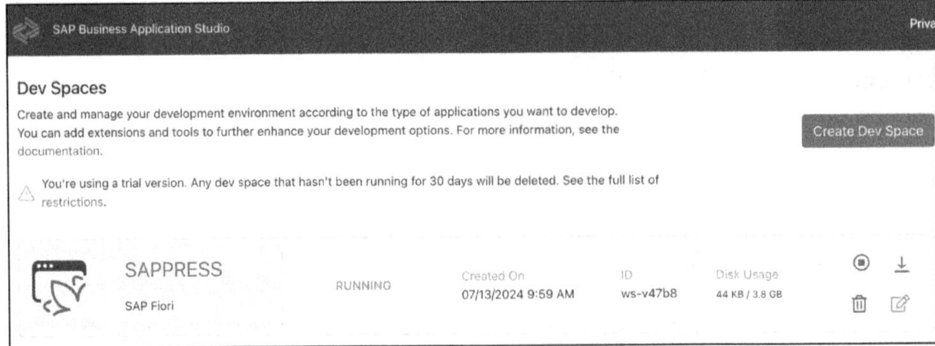

Figure 4.2 Dev Space Started and Ready to Use

After the development environment is started in the browser, you can also save the link to your dev space as a favorite to save you from having to go through the dev space lobby every time. First, we need to create a new project by choosing **File • New Project from Template** from the menu bar.

You can choose from several project types: the **SAP Fiori generator** is of interest for our purposes (see Figure 4.3). **Basic Multitarget Applications** are applications that could contain both frontend and backend components and are usually deployed in SAP Business Technology Platform (SAP BTP). With **SAPUI5 Adaption Projects**, SAP has presented the enhancement concept of SAP Fiori (standard) apps.

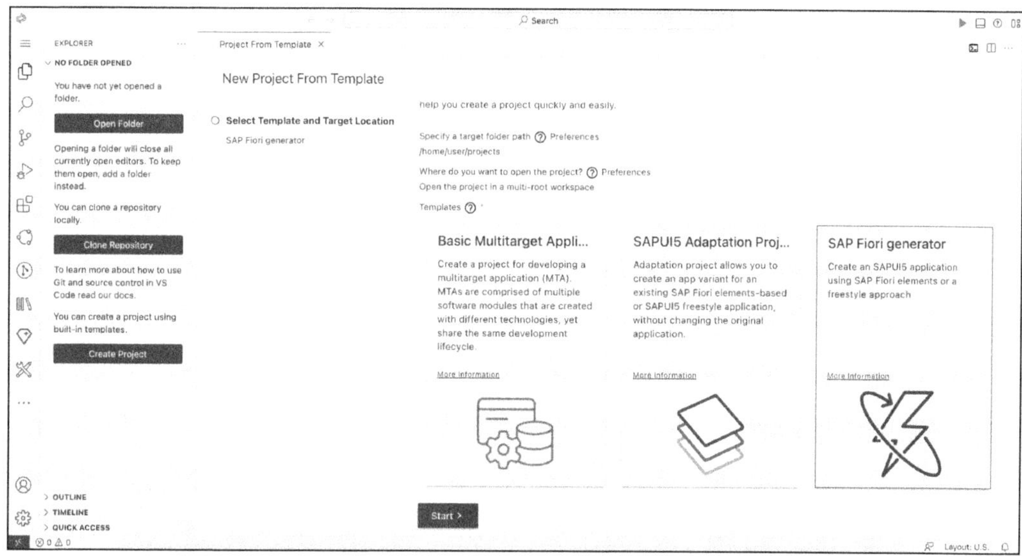

Figure 4.3 Selecting the SAP Fiori Application Template

The SAPUI5 freestyle applications are created using the **Basic** application template (see Figure 4.4). All other templates relate to low-code/no-code developments with the *SAP Fiori elements framework* or the newly introduced *flexible programming model.*

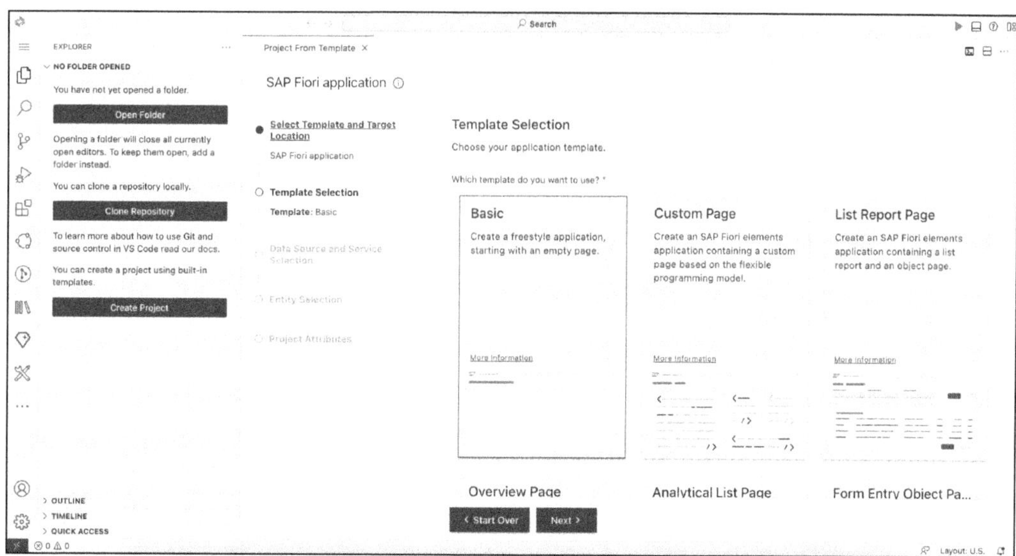

Figure 4.4 Selecting Basic as the Application Type

In the next step, we're asked whether we already want to integrate an OData service. We initially select **None** here, as we'll first get to know the integration of remote data sources (see Figure 4.5). You're not blocking the possibility of integrating data sources into the project at a later point in time.

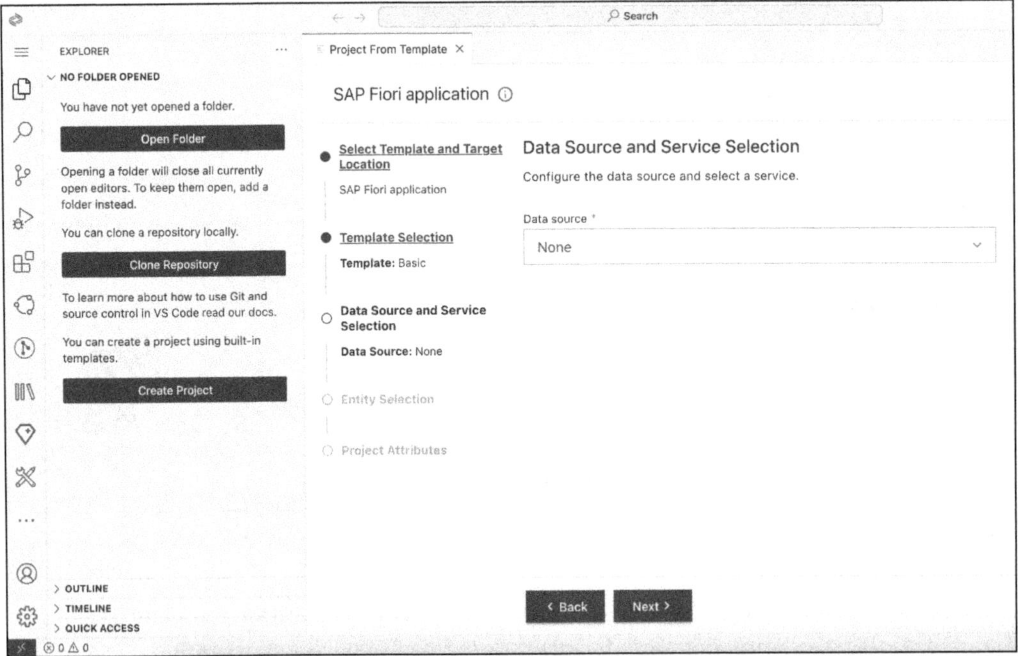

Figure 4.5 Selecting the Data Source and Service to Get the OData Service Connection Created

Because we're creating a SAPUI5 project, the wizard assumes that we'll also need a view for displaying the UI components. For this reason, we can assign a name to this initial view. We've entered "Main" as the **View name**, as shown in Figure 4.6.

In the next step (see Figure 4.7), we need to enter the project-specific details:

- **Module name**
  Name of the project (can only be changed later with considerable effort).
- **Application title**
  Title of the application, which can be easily changed later.
- **Application namespace**
  Namespace of the app, whereby the reverse domain name of the company is often chosen here (can only be changed again later with considerable effort).

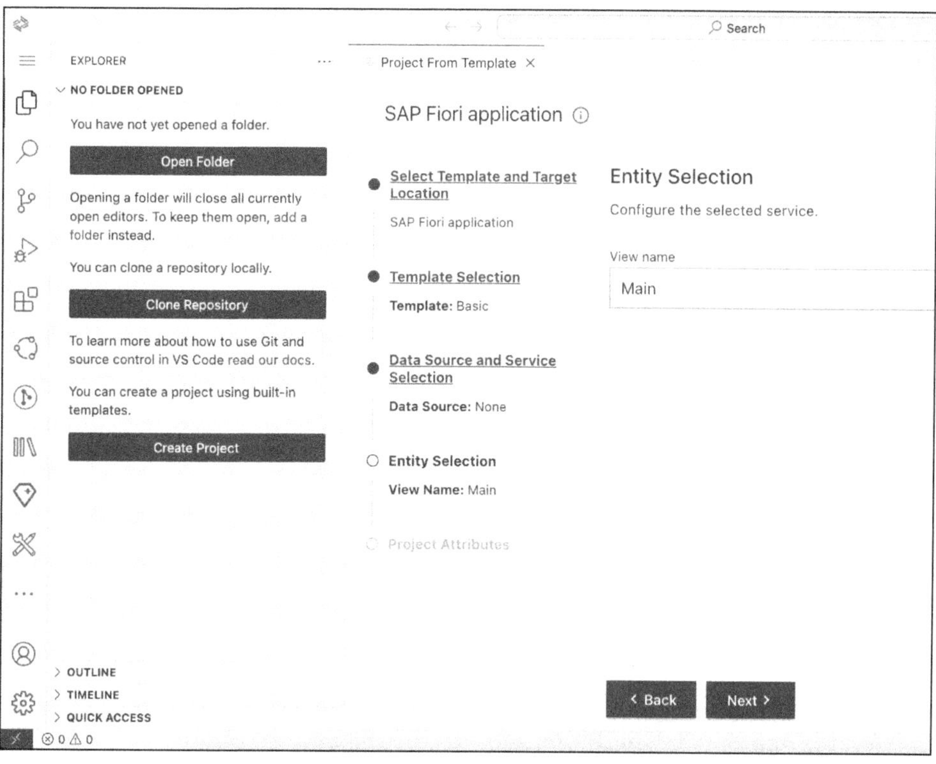

Figure 4.6 Selecting a Name for the Initial View, Which Will Be Created Automatically

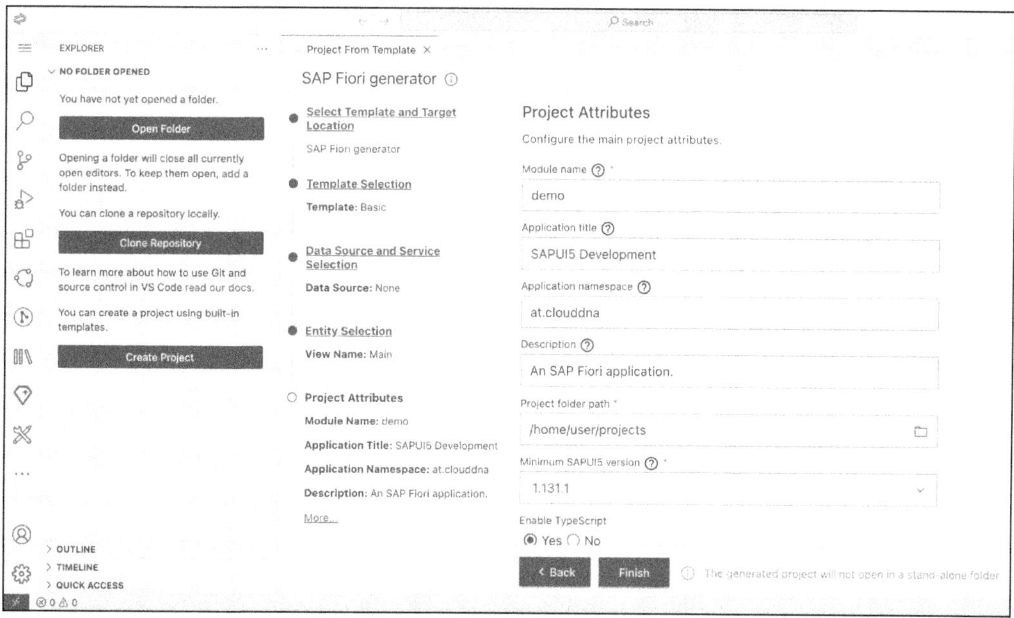

Figure 4.7 Project Attributes for Module Name, SAPUI5 Version, or Project Path

Technical ID of the Application

You need to know that the technical ID of the application is made up of the application namespace and the module name. These are concatenated and separated by a dot. It's important that this ID must later be unique in the SAPUI5 ABAP repository. This uniqueness can't be checked when the app is created, but only when it's first deployed on the SAP system.

- **Description**
  Description of the application, which is only important for the metadata.
- **Project folder path**
  Directory in SAP Business Application Studio in which the project is created. The default projects directory (*/home/user/projects*) is usually used, but you can set up the directory structure as you wish.
- **Minimum SAPUI5 version**
  Version of the SAPUI5 framework used for development. This version won't be used later, as the installed version of the SAPUI5 component of the SAP system plays a role.

Select the Appropriate SAPUI5 Version

The SAPUI5 framework is versioned, which means both that changes may have been made to the components already delivered with higher versions and new components may have been introduced. If you develop with a different SAPUI5 version than will actually be available on the deployed system later, you may only realize at the time of deployment that you're using components that don't yet exist on the SAP system.

In Figure 4.8, you can see that we've activated **Enable TypeScript** via the **Yes** radio button. We're not interested in any of the other options for now. We'll cover the other topics represented by radio buttons, such as deployment (**Add deployment configuration**), SAP Fiori launchpad configuration (**Add FLP configuration**), and in-depth programming concepts (**Configure advanced options**) in Chapter 16, Chapter 17, and Chapter 12, respectively.

The project has been created but not yet opened. In any case, you can open a directory again via **File • Open Folder**. Navigate through the project directory until you can find and select the project, and continue with **OK** or `Enter`, as shown in Figure 4.9.

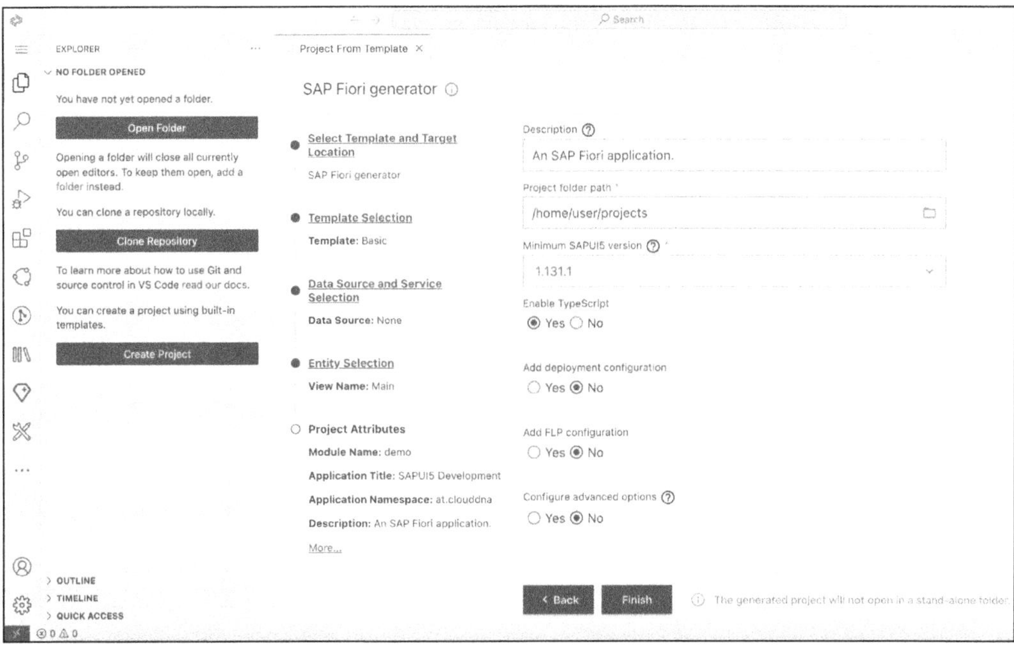

Figure 4.8 Enable TypeScript to Use It Instead of JavaScript

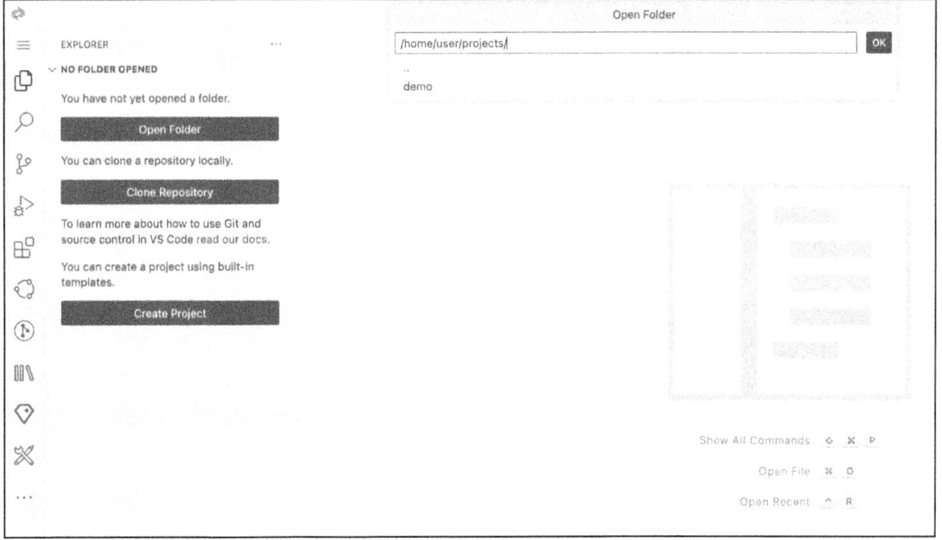

Figure 4.9 Selecting and Opening the Desired Folder

When you open a project, you're greeted with the **Storyboard**. Some project-specific information is displayed here, along with *quick actions*, such as adding an OData service or similar (see Figure 4.10).

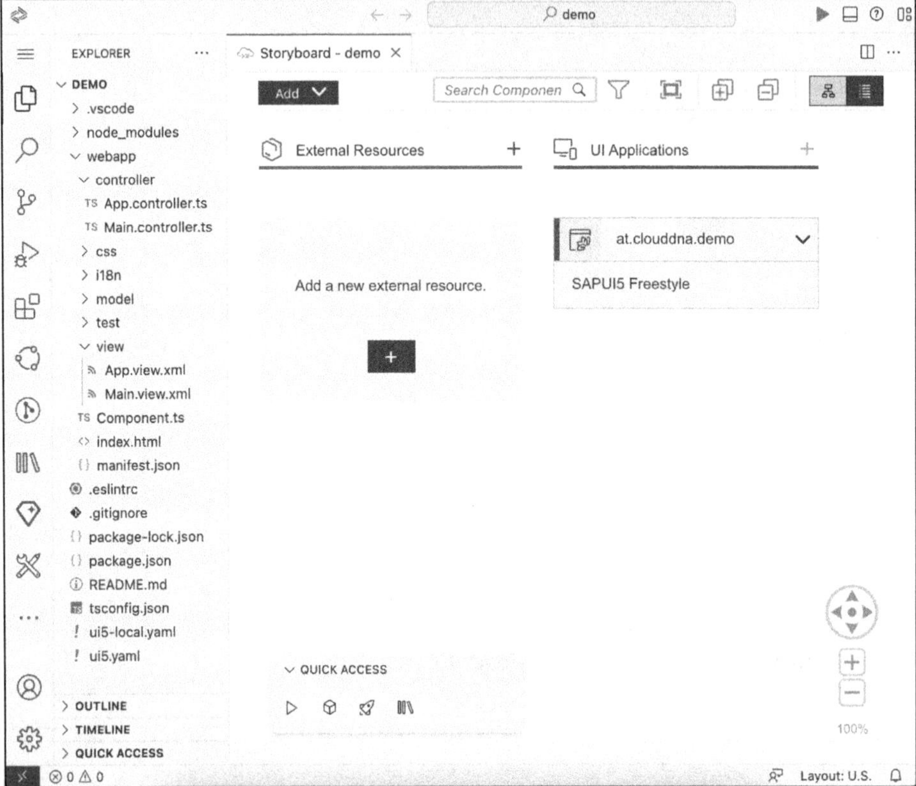

Figure 4.10 Storyboard of an Application Offering Quick Actions

Source Code and Version Management

The SAPUI5 application that has just been created only exists in your development environment. No one else has access to it, nor is the source code distributed, even if it's later deployed to the SAP system. With this new technology, you have to introduce a source code and version management system yourself. The recommendation is to use Git and store the source code in a repository. Although the initial outlay speaks against this new step, it opens up new possibilities in terms of collaboration and development of software and features. In Chapter 15, you'll learn all the necessary details about Git and how this tool can be used.

4.1.2 Visual Studio Code

In this section, we'll show you how to create a SAPUI5 freestyle application in VS Code. The wizard behind the application generator should already be familiar. The design is similar to SAP Business Application Studio, but the headings of the individual areas may have different names. For example, the selection of the template for SAPUI5 freestyle applications is simply called **SAPUI5 Application**. It's also important that you select **SAPUI5 freestyle** as the value for **Application Type** in the select box (see Figure 4.11).

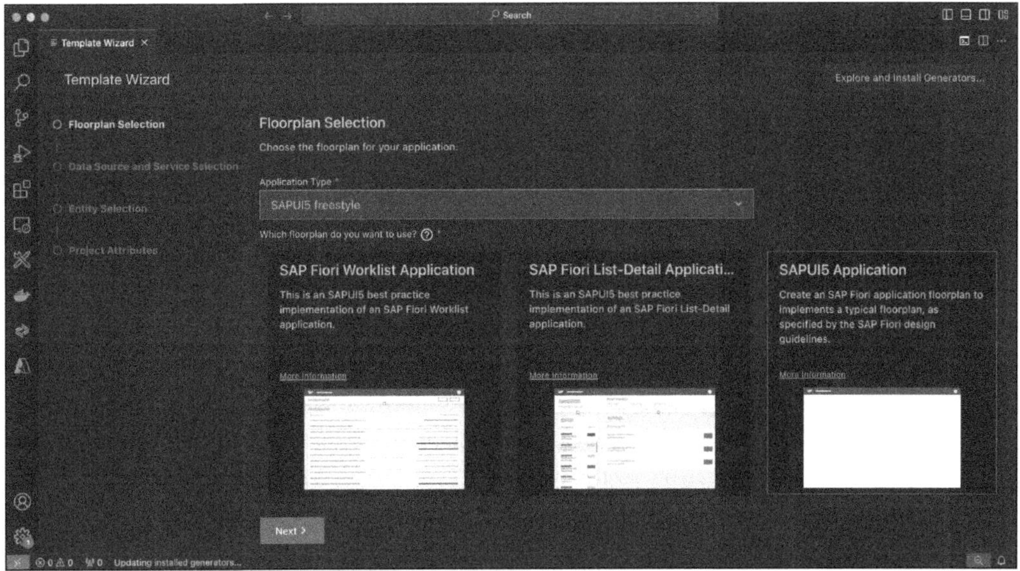

Figure 4.11 Creating SAPUI5 Applications to Achieve the Same Result as in SAP Business Application Studio

All other steps from this step on in the wizard are the same as in SAP Business Application Studio.

4.2 Explore the Application Structure

Let's take a look at the most important files that were created by the wizard and are also essential for operating an SAPUI5 app. We'll distinguish between five important representative files in this section:

- *Component.js*
- *manifest.json*
- *package.json*
- YAML files
- *index.html*

4.2.1 Components and the Component Controller: Component.js

As you can imagine, an SAPUI5 app can be divided into several independently operating parts. Each of these parts can take on a different function in the overall construct and can be separately instantiated and used as appropriate. These parts are called *components*. An SAPUI5 app consists of at least one component but can consist of several independent components. In the standard system, this initial component is located in the *webapp* directory, but the name of this directory has no influence on the component itself. The files in the directory that are mandatory are what make up a component. These include, in particular, the *Component.js* file, which also predominates in the standard system.

A component doesn't always have to have a UI because there are two different types of components. *Faceless components* only contain business logic that doesn't require a UI and can therefore only provide auxiliary classes, convenience functions, formatters, or other things for other components. *UI components*, on the other hand, have a renderer in addition to their own business logic, which ensures that the UI component can be displayed on the UI using HTML in the browser. When you create an SAPUI5 app, a component controller is created, and it's assumed that you want to create a UI component by default. The *Component.js* file is called the component controller, which helps to bring the component to life by the framework as soon as it's called (see Figure 4.12). This `BaseComponent`, which you're extending, defines whether you want a UI component (`sap/ui/core/`**`UIComponent`**) or a faceless component (`sap/ui/core/`**`Component`**).

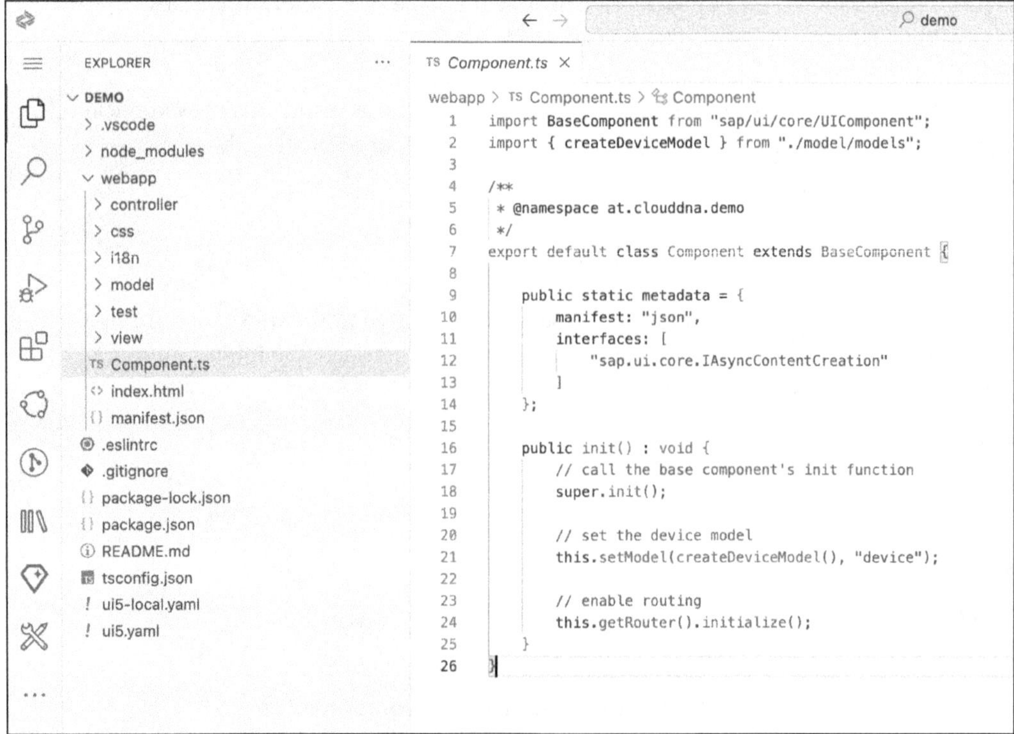

Figure 4.12 Component Controller Created by Default

Although the metadata can be defined in the designated JavaScript object, it's recommended that the metadata be stored in a separate file. This file is called *manifest.json* and is the centerpiece of the application. We'll discuss this file in more detail in a moment, but you can see in the following code snippet that the reference is made when the component is initialized:

```
public static metadata = {
    manifest: "json"
};
```

A specific method can be used to overwrite the constructor inherited from a component. In this init function, you can implement logic that is to be executed once when the respective component is instantiated. In any case, because you're overwriting the standard constructor here, we recommend that you also call the constructor prepared by SAP so that the framework can start up as planned. Otherwise, the functionalities will be affected, and you'll have to manually call the necessary steps when starting the framework. We won't go into the details of navigation until Chapter 7, but you can see in Listing 4.1 that the router is initialized by default. In addition to the router, a model called device is also created, whose logic is stored in another file that we can use to access device-specific information such as operating system, window size, or device features (e.g., touch input).

```
public init() : void {
    // call the base component's init function
    super.init();
    // set the device model
    this.setModel(createDeviceModel(), "device");
    // enable routing
    this.getRouter().initialize();
}
```

Listing 4.1 Logic Called by Default When the Component Is Initialized

4.2.2 Application Descriptor: manifest.json

The centerpiece of the application is the *application descriptor*. The *manifest.json* file stores the most important metadata information at the top level of the component in the project structure (see Figure 4.13). You already know that the reference to this file is made in the component controller.

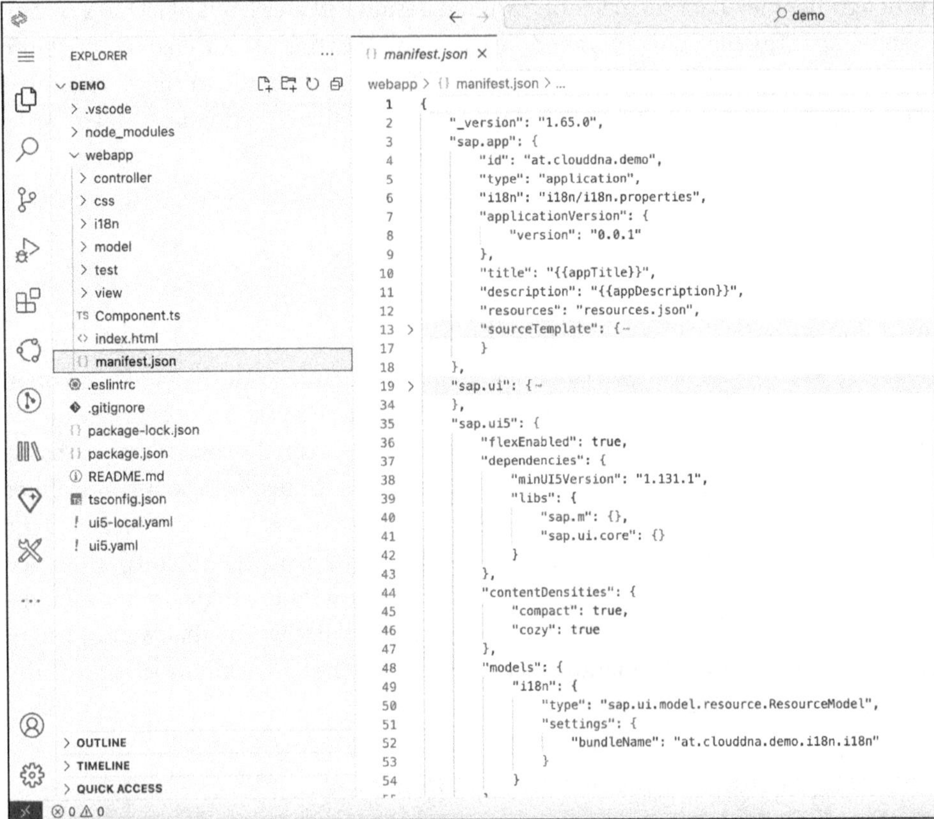

Figure 4.13 The manifest.json File as a Centerpiece for the SAPUI5 Application

There are a number of settings that you can make in this JavaScript Object Notation (JSON) file, but from experience, they are always the same ones that you'll either customize or from which you want to obtain information. The most important properties are discussed in more detail in Table 4.1.

Property in the manifest.json File	Description
`sap.app.id`	The technical ID of the application is stored here, which was assigned when the project was generated. This ID must be unique when it's deployed to the SAP system. This must be used as the namespace throughout the entire application. If you want to change it later, it must be changed not only here, but in every place (*Component.js*, views, controllers, etc.) throughout the entire application.

Table 4.1 Most Important Properties in the Application Descriptor

Property in the manifest.json File	Description
`sap.app.type`	This is where you define whether the component is an `application`, a `library`, or a `card`. By default, `application` will be stored here.
`sap.app.title`	This is the title of the application, which is also displayed at the top of the browser tab. In addition to the title, there are also subtitles and descriptions. Here, you don't enter the text directly, but instead use double curly brackets to refer to a key in i18n, where the application-specific translations are stored as key-value pairs. We'll discuss i18n in more detail in Section 4.3.3 and in Chapter 12.
`sap.app.dataSources`	This is where you maintain the OData and Representational State Transfer (REST) services you want to communicate with. Due to security restrictions in the JavaScript environment, you use relative URLs to the respective services here. We'll explain in Section 4.2.4 how these can still be resolved by the framework.
`sap.ui.fullWidth`	This property can be used if the application is to run later in the SAP Fiori launchpad and you want the application to always take up the full width. Otherwise, the application will have a responsive margin on the left and right, as far as the device width allows.
`sap.ui5.flexEnabled`	With this property, you can use `true` or `false` values to determine whether the app is enabled for adaptation via the SAPUI5 flexibility. This means that other developers will be able to use your project as a basis for *adaptation projects*, and key users will be able to extend your project using *key user adaptations*.
`sap.ui5.dependencies.minUI5Version`	This property indicates the minimum SAPUI5 version of the libraries that must be available on a server, but this version isn't enforced during operation. If this isn't fulfilled, only a warning is issued on the console. It has much more influence on development because if the SAPUI5 libraries are obtained from the content delivery network (CDN), then this SAPUI5 version is used. Make sure that you've entered the same version here as will be used on the server later to avoid developing with a much higher version and thus being tempted to use controls that don't yet exist on the system.

Table 4.1 Most Important Properties in the Application Descriptor (Cont.)

Property in the manifest.json File	Description
`sap.ui5.dependencies.libs`	The SAPUI5 libraries that should definitely be loaded eagerly by the framework when the application is started are specified here. Nonetheless, you can also use other libraries in your application, which are then subjected to lazy loading.
`sap.ui5.models`	The models that you'll get to know in Section 4.3.3 as part of the MVC concept are defined globally here. These can be OData, JSON, XML or resource models, which always have the same function but provide the data in a different format and can communicate with remote data sources depending on their definition.
`sap.ui5.routing`	As promised, we'll deal with routing in Chapter 7. However, know that all the necessary definitions concerning the individual routes within a SAPUI5 app are stored here.
`sap.ui5.rootView`	Every UI component has a root view, which is the first view that opens. Inside, it's either a full-screen application or a container that provides navigation.

Table 4.1 Most Important Properties in the Application Descriptor (Cont.)

4.2.3 package.json

The *package.json* file plays a central role in a JavaScript-based framework in combination with Node.js, such as SAPUI5. This file is used to define the metadata of a project (project name, description, license used, etc.), but more importantly, it defines the dependencies on *node modules* and the scripts offered for starting, building, or previewing an application (see Figure 4.14).

Node modules are an integral part of *Node.js* projects and contain the libraries a project needs to function. These libraries are stored in a public repository and can be downloaded and placed in the application. The dependent modules are versioned in the *package.json* file under `dependencies`. They are stored in the *node_modules* folder, which is generated after dependencies are installed using a package manager such as `npm` or `yarn`. The idea is that these modules aren't distributed among the developers, but that everyone can install the corresponding modules using the distributed definition, for example, with `npm install`. The two properties `dependencies` and `devDependencies` are used to distinguish between modules that are required for operation and those that are only required for development.

In addition, a number of `scripts` are already predefined. These help us to preview the application both with mock data and with a connection to a backend. Others are responsible for executing the tests or performing a build and subsequent deployment.

We'll come back to the most important of these scripts several times during the course of the book.

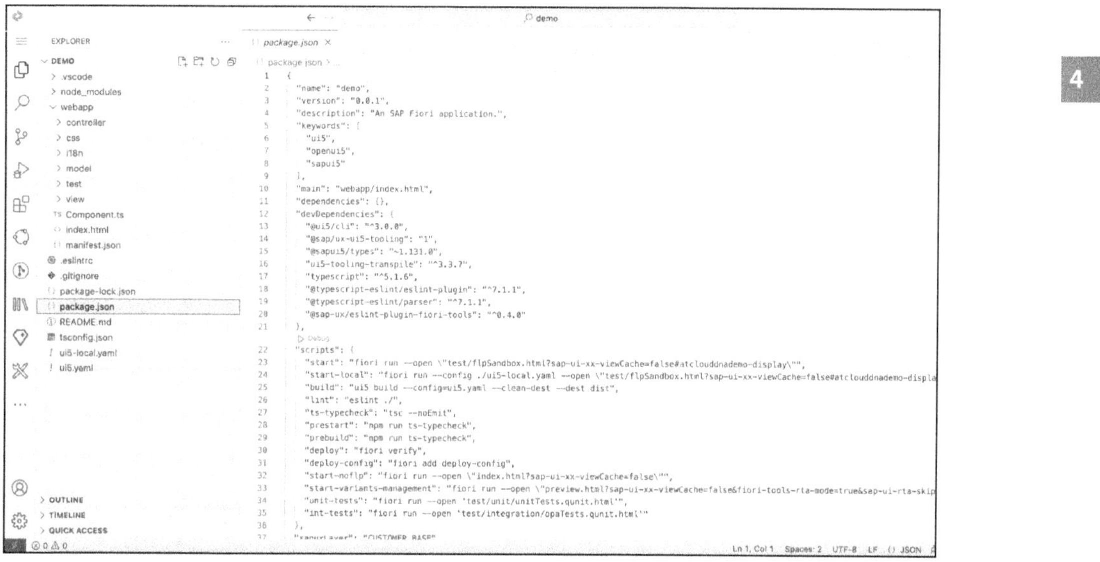

Figure 4.14 The package.json File Stores Dependencies and Scripts

4.2.4 YAML Files

By default, two YAML files exist (see Figure 4.15). The most important one is *ui5.yaml* so that a preview can be started and the application can be previewed locally.

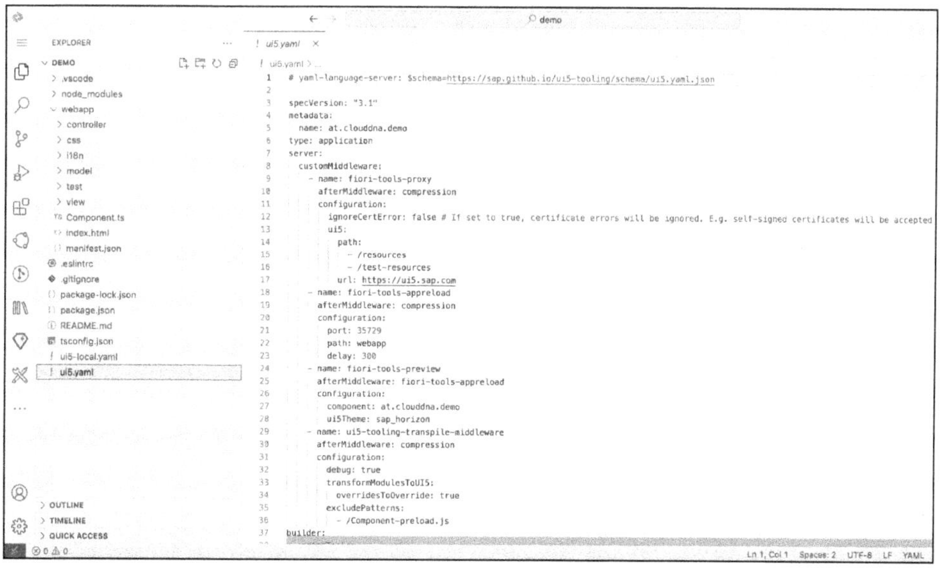

Figure 4.15 YAML Files for Running the Application within a Sandbox, Previewing with Mock Data, and Deploying to a System

Middleware is also defined here, which resolves the relative URLs that are defined in *manifest.json* under `dataSources`. In this way, the middleware routes the requests to the SAPUI5 CDN on the one hand and to the OData service to an SAP S/4HANA system on the other. For security reasons, it's not permitted to send these requests directly to the various hosts, which is why middleware must be used here. You can find more information on the topic of security concepts and requirements in Chapter 19. Further YAML files may (and will) be created during the development process, which we'll discuss in due course.

4.2.5 index.html

The *index.html* file is one of the files that can be used to start the app (see Figure 4.16).

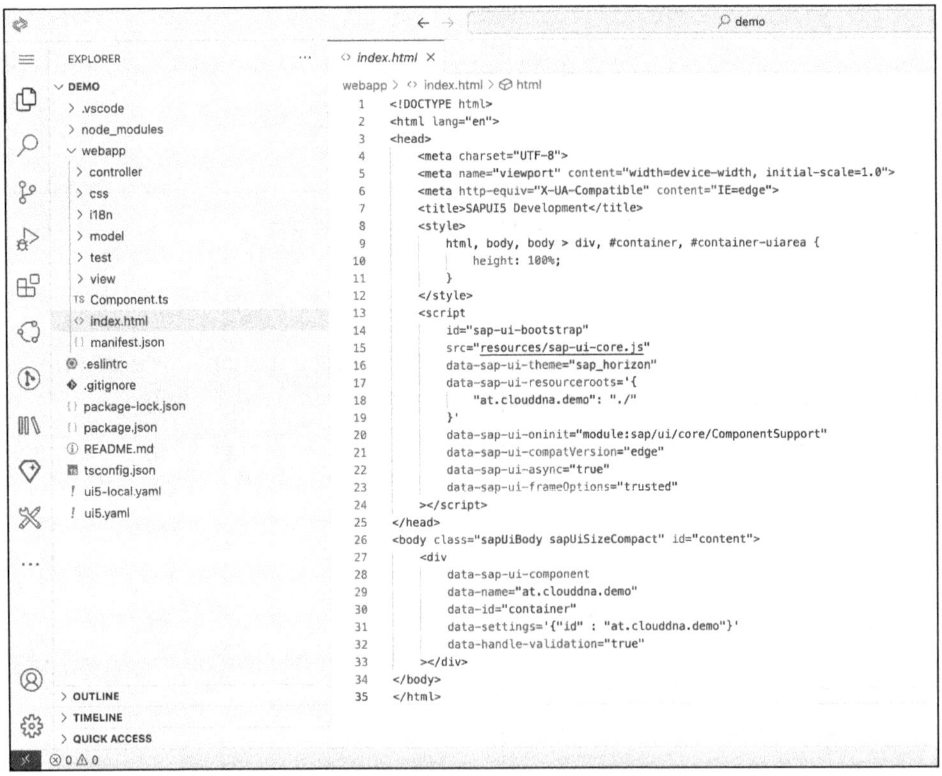

```
<!DOCTYPE html>
<html lang="en">
<head>
    <meta charset="UTF-8">
    <meta name="viewport" content="width=device-width, initial-scale=1.0">
    <meta http-equiv="X-UA-Compatible" content="IE=edge">
    <title>SAPUI5 Development</title>
    <style>
        html, body, body > div, #container, #container-uiarea {
            height: 100%;
        }
    </style>
    <script
        id="sap-ui-bootstrap"
        src="resources/sap-ui-core.js"
        data-sap-ui-theme="sap_horizon"
        data-sap-ui-resourceroots='{
            "at.clouddna.demo": "./"
        }'
        data-sap-ui-oninit="module:sap/ui/core/ComponentSupport"
        data-sap-ui-compatVersion="edge"
        data-sap-ui-async="true"
        data-sap-ui-frameOptions="trusted"
    ></script>
</head>
<body class="sapUiBody sapUiSizeCompact" id="content">
    <div
        data-sap-ui-component
        data-name="at.clouddna.demo"
        data-id="container"
        data-settings='{"id" : "at.clouddna.demo"}'
        data-handle-validation="true"
    ></div>
</body>
</html>
```

Figure 4.16 index.html File Runs the SAPUI5 Application Standalone

This file isn't used if the operation is carried out on an SAP system or, more specifically, is integrated into an SAP Fiori launchpad. *index.html* is only used for standalone operations, for example, if the web application is started on a regular web server. For this reason, we recommend that you don't embed scripts here, as you may be accustomed to doing with other JavaScript-based frameworks.

4.3 Model View Controller Design Pattern

Figure 4.17 shows the architecture of the SAPUI5 framework using the MVC design pattern on the one hand and the specific designs of the individual components on the other. The client-side models are either XML or JSON models. The difference between these models is the format in which the data is stored. The only server-side model delivered by SAP in the SAPUI5 framework is the OData model. You can also use the JSON model for an ordinary REST model, but this is in no way as powerful as the OData model. The OData model expects an OData service for communication on the remote server and takes on the task of communicating with this OData service via HTTP calls.

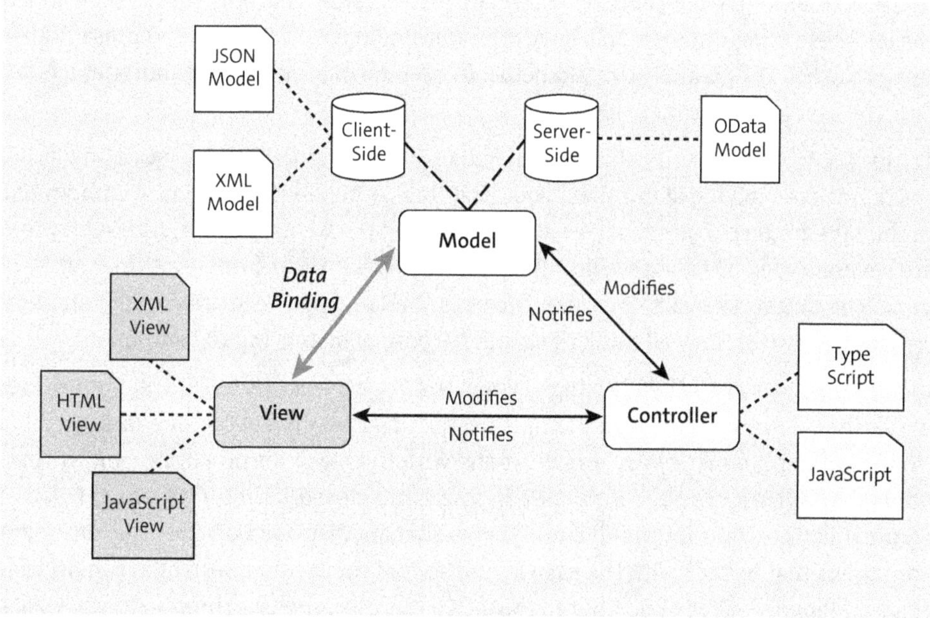

Figure 4.17 Model View Controller Design Pattern Used as the Architecture Pattern in SAPUI5 Apps

The *view* can be structured in three different ways—*JavaScript*, *XML*, and *HTML*—with the most common variant being the use of XML views. Although SAPUI5 is referred to as an HTML5-based framework, it may be unusual to build the UI using XML definitions. This is due to the comparatively short definitions and the better readability compared to HTML. Due to the hierarchical structure of an XML page, it's easy to get an idea of the final layout during development. Furthermore, a page that is structured with XML can be validated more easily by the system.

The entire application logic is in the respective *controllers*, whereby a controller always belongs to a view. A controller is programmed in either JavaScript or TypeScript and supports the view so that the view only must take care of the visualization, but not the business logic, such as the implementation of lifecycle methods, event handlers, or formatters.

These three components work together in a constant exchange of information, as we'll see, to provide a smooth, clearly structured, and modern application for both developers and users.

4.3.1 View

UI components are either delivered by SAP in the SAPUI5 standard or additionally developed by developers. In addition to the components that are to be displayed on the UI and therefore have a renderer, there are also faceless components, as mentioned earlier These components don't have a renderer and aren't intended to present anything to users. They usually contain helper classes, formatters, or other libraries to support the user's own business logic.

To display UI components at all, you need an interface on which they can be placed. We're not even talking about the layout here, but rather about a technical component in the MVC pattern. You may have already guessed it—we need views. Although there are several types of views, we'll limit ourselves to the most important one: an XML view. The difference from the other views is the language with which they are constructed. However, they all fulfill the same purpose of displaying UI components.

In our case, a view was already created when the project was created. If you want to create further views, you can do this simply by creating a new file in the *view* directory. It's important to note that XML views must end with the file extension *.view.xml*. The previous `sap.ui.core.mvc.view` is opened in line 1 and closed in line 7, as you can see in Figure 4.18. This includes the UI components that are displayed on the page. There are containers that help us with the page layout so that the layout doesn't have to be arbitrary and involve a lot of positioning work. In our case, the UI component `sap.m.Page` was already inserted when the project was created (see lines 2–6). This UI component divides the page into the three parts of header, content, and footer and offers us the option of placing any UI components.

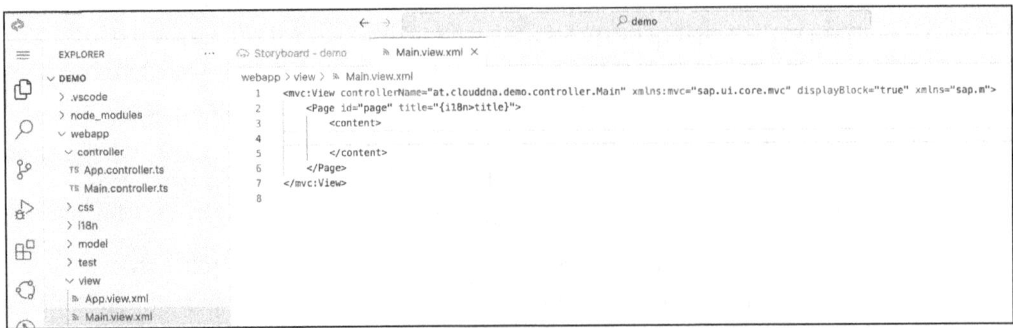

Figure 4.18 A View with One UI Component Inside: An Empty Page

You've already seen from our description that the UI components always originate from a specific namespace/library. These namespaces are integrated with `xmlns` in the view definition. There's always a default import (in our case, `sap.m`), so UI components that originate from this namespace don't have to be explicitly marked in the file. As there can only be one default import, all other namespaces must have an abbreviation set and must always be used in the XML via an abbreviation for the UI component from the namespace (see difference between `mvc:View` and `Page`).

Now let's try to place a first text in the content of the page. In contrast to HTML, we're not allowed to simply write the text, but need a UI component that offers this possibility. In the SAPUI5 documentation (*https://ui5.sap.com/*), we found out through research that there's a UI component called `sap.m.Text`. If we look at the properties of this component, we find that a property called `text` can be set. The description states that this property represents the set text on the UI (see Figure 4.19).

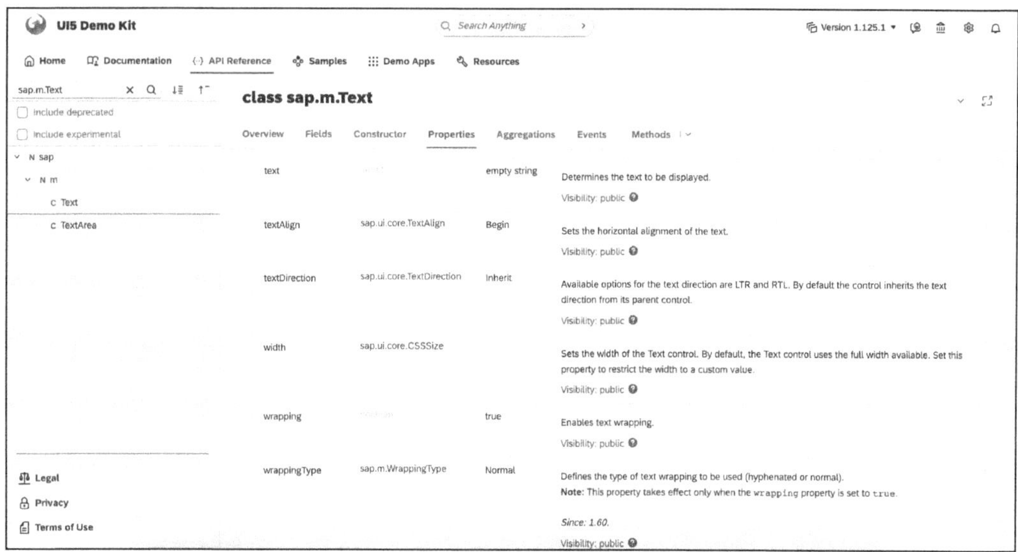

Figure 4.19 UI Component sap.m.Text in the SAPUI5 Documentation

We'll now place this UI component in the content of our page and because `sap.m` is already imported as a namespace by default, we don't need to prepend anything to the `sap.m.Text`. Our text, which we assign as the value of the `text` property, is displayed on the interface. We've inserted the text as shown in Figure 4.20.

Now, let's what the app could look like for the end users. To do this, click on any directory in the project, and select **Preview Application** from the context menu (see Figure 4.21).

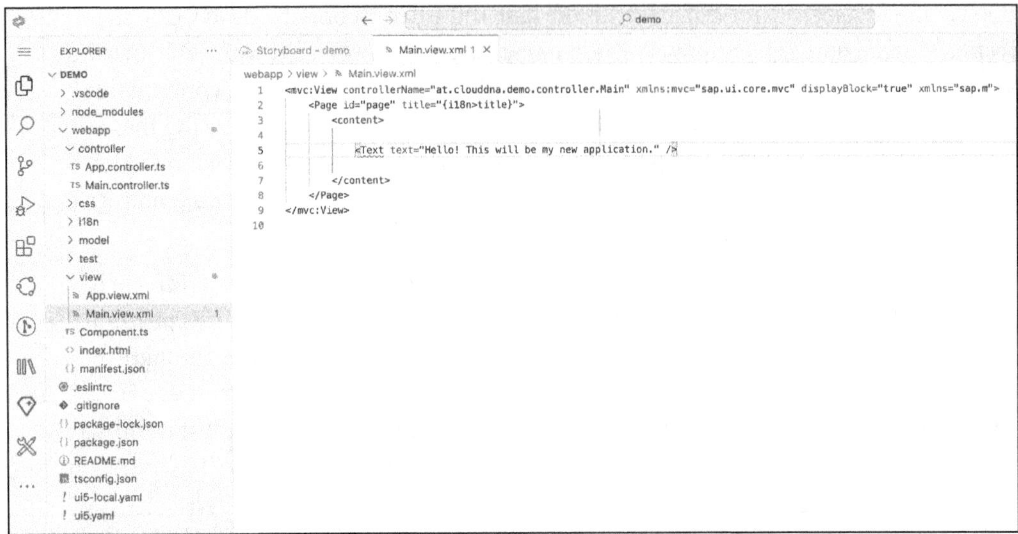

Figure 4.20 UI Component sap.m.Text Inserted into the Content of sap.m.Page

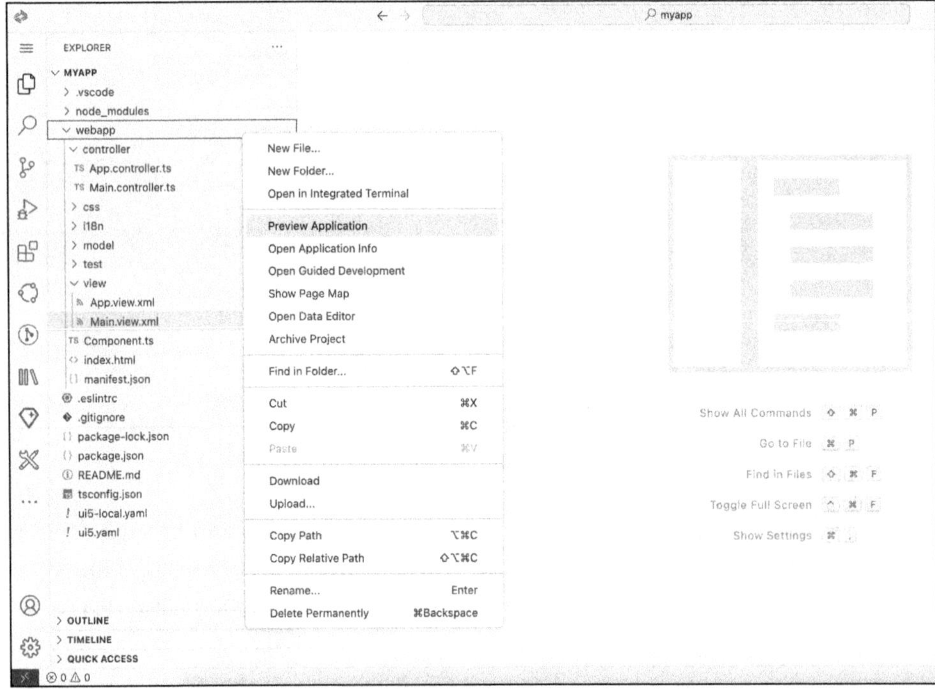

Figure 4.21 Use the Preview to Check How the Application Will Look for End Users

We're offered several scripts from the *package.json* with which we can start a preview (see Figure 4.22). We'll briefly explain the two most important scripts:

- `start`
  Use this script to start an SAP Fiori launchpad sandbox instance in which the app is displayed as if it were running in the launchpad. In this case, the app doesn't start via *index.html* but via the *Component.js* file, which we've already discussed. The configuration for the middleware used by this preview is in the *ui5.yaml* file.
- `start-noflp`
  In this case, the app isn't embedded in an SAP Fiori launchpad sandbox but started directly via *index.html* as if it were being made available on any non-SAP web server. The configuration for the middleware used by this preview is in the *ui5.yaml* file.

We've opted for the variant with the SAP Fiori launchpad sandbox.

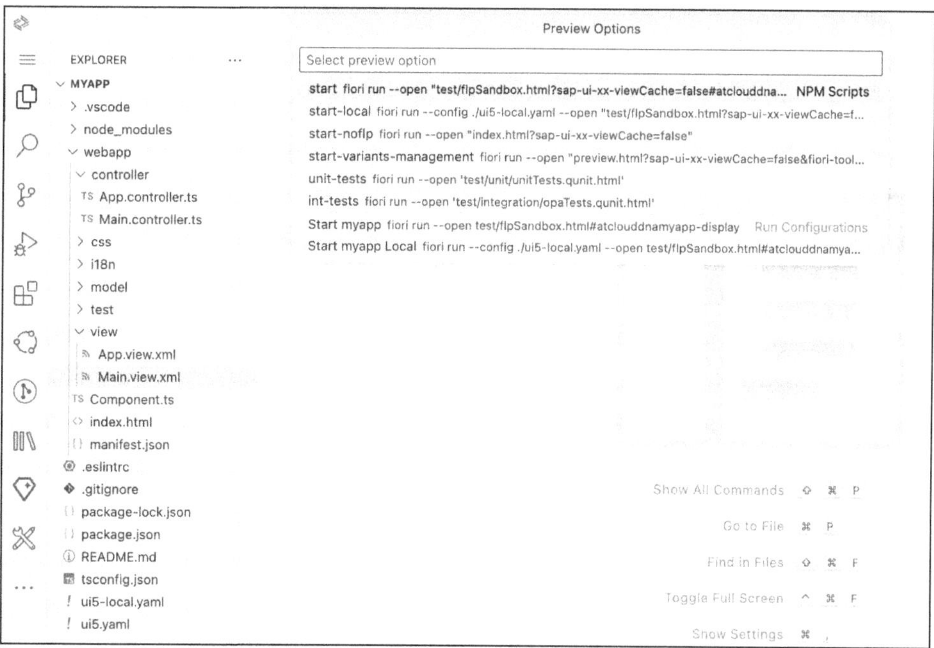

Figure 4.22 Selecting One of the Predefined Scripts to Start the Preview

Popup Blocker May Block the Preview

The first time you start the preview, the popup blocker may prevent it from opening in a new tab. Explicitly allow popups to be opened from this page, and restart the preview.

After all of these steps, you can see in Figure 4.23 that a preview of the application is started directly in the browser. Now, you can follow every change after saving in this additional tab and directly see the effects on the application.

Preview Means Preview

What we mean by this is that this preview doesn't represent the executable app. Although the preview is in a browser, so you could copy and distribute the link, the app will no longer run as soon as you close the terminal where the script was inserted for starting the preview.

As you can see in Figure 4.23, the page is divided into three parts, and we could place our UI components in these three areas (almost) at will. The header ❶ and footer ❷ are limited in height, which is why not all components make sense here. The content area ❸ is the most used area, which can grow and protrude beyond the visible area. In this case, a vertical scroll bar is provided.

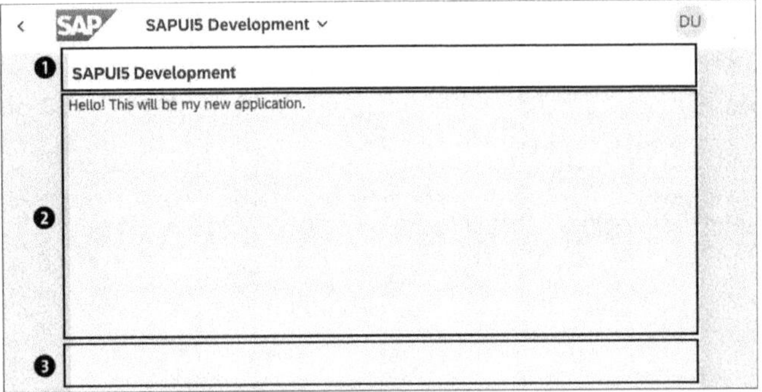

Figure 4.23 Main Parts of the Page Where Content Can Be Placed

If we look in the documentation for `sap.m.Page`, we'll find the corresponding areas `content`, `contentHeader`, or `customHeader`, and `footer` under **Aggregations** (see Figure 4.24).

UI5 Demo Kit | Search Anything | Version 1.125.1

Home | Documentation | API Reference | Samples | Demo Apps | Resources

sap.m.Page

Include deprecated

Include experimental

N sap
N m
C Page
C PageAccessibleLandmarkInfo
E PageBackgroundDesign

Legal

Privacy

Terms of Use

class sap.m.Page

Overview | Constructor | Properties | Aggregations | Events | Methods

Aggregations

Show borrowed aggregations

Name	Cardinality	Type	Alt types	Description
content(default)	0..n	sap.ui.core.Control		The content of this page
customHeader	0..1	sap.m.IBar		The (optional) custom header of this page. Use this aggregation only when a custom header is constructed where the default header consisting of title text + nav button is not sufficient. If this aggregation is set, the simple properties "title", "showNavButton", "NavButtonText" and "icon" are not used.
footer	0..1	sap.m.IBar		The (optional) footer of this page. It is always located at the bottom of the page
headerContent	0..n	sap.ui.core.Control		Controls to be added to the right side of the page header. Usually an application would use Button controls and limit the number to one when the application needs to run on smartphones. There is no automatic overflow handling when the space is insufficient. When a customHeader is used, this aggregation will be ignored.
landmarkInfo	0..1	sap.m.PageAccessibleLandmarkInfo		Accessible landmark settings to be applied on the containers of the

Figure 4.24 Aggregations of the UI Component sap.m.Page in the Documentation

In contrast to properties, aggregations are written between the opening and closing tag of the respective UI component. In Figure 4.25, we've inserted the three areas as an example and marked the places where we could place other UI components with the comments to place them in the header, content, or footer.

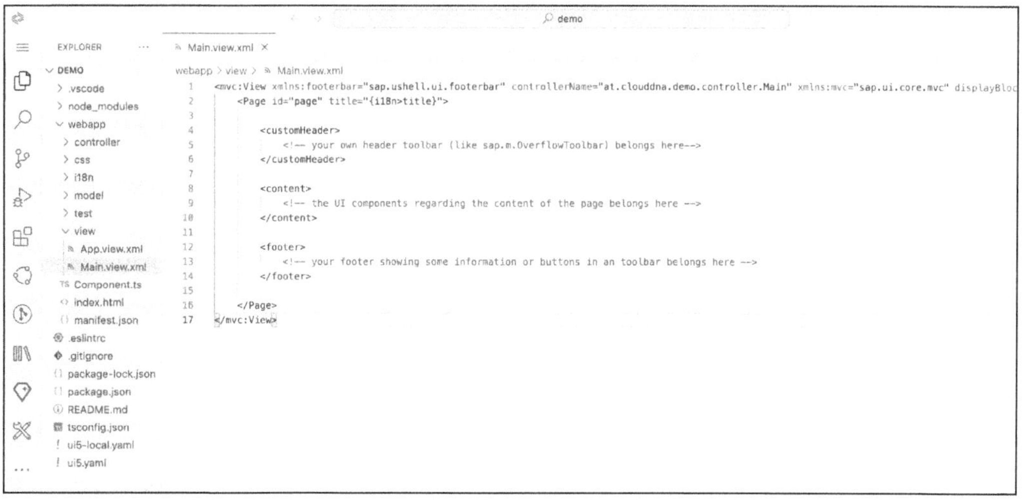

Figure 4.25 Aggregations Written Between the Opening and Closing Tag of a UI Component

But what exactly is an aggregation? *Aggregations* are parent-child relationships between UI components. The aggregation content from `sap.m.Page` means that we can place any UI component that inherits from `sap.ui.core.Control` in the aggregation and display it in the content area of the page. Let's take another example that's possibly more tangible: The UI component `sap.m.List` represents a list and has an aggregation called `items`. List elements that should be displayed in the UI can be placed in this aggregation. Listing 4.2 shows a list with two list items.

```
<List headerText="List Heading">
      <items>
            <StandardListItem title="First Item"/>
            <StandardListItem title="Second Item"/>
      </items>
</List>
```

Listing 4.2 Example: Two-Item List to Show Aggregations

Listing 4.3 shows the syntax for using the information you can extract from the SAPUI5 documentation regarding properties, aggregations, events, and associations of a UI component.

```
<Component property="value" event="listener" association="value" >
      <aggregation>
```

```
            <ChildComponent property="value" />
        </aggregation>
</Component>
```

Listing 4.3 Syntax for Using UI Components with Their Properties, Aggregations, Events, and Associations

IDs Essential for UI Components

The syntax check has always warned us that we should fill the id property with a value for each UI component. If you don't assign IDs, the SAPUI5 framework will do so. This is disadvantageous because you don't have control over the naming and assignment of the IDs yourself and therefore can't access the UI components from other places later.

Another disadvantage is that you won't be able to use the extension concept for this app. Both the *adaptation projects* and the *key user adaptations* live from the fact that the individual UI components have IDs.

Now what if we don't want to display a list but want to offer users a form? As shown in Listing 4.4, we could use a sap.m.VBox, which arranges the UI components in the aggregation items one below the other. For each line we want to display, we could use a sap.m.HBox, causing sap.m.Label and sap.m.Input to be displayed next to each other.

```
<VBox id="vbox1">
    <items>
        <HBox id="hbox1">
            <items>
                <Label id="labelFistname" labelFor="inputFirstname" text="First
name"/>
                <Input id="inputFirstname" />
            </items>
        </HBox>
        <HBox id="hbox2">
            <items>
                <Label id="labelLastname" labelFor="inputLastname" text="Last
name"/>
                <Input id="inputLastname" />
            </items>
        </HBox>
    </items>
</VBox>
```

Listing 4.4 Using Boxes as Containers Impacts the Layout Presented to End Users

But if we look at the result in Figure 4.26, we'll see that the structure isn't very appealing. The labels aren't the same height nor centered like the input fields. In addition, there's no margin to the left- or right-hand side. Of course, we could start to use CSS to achieve our goal, but there are a lot of UI components that can act as a layout container.

Figure 4.26 Opening the Preview to See the Effect of the Boxes

We'll now use a `sap.ui.layout.form.SimpleForm` to provide a simple representation of the data. The `SimpleForm` can be displayed in a user-friendly and responsive way for all device sizes (`S`, `M`, `L`, `XL`, etc.) using several definitions. The `SimpleForm` is normally based on the `ResponsiveGridLayout`. This layout means that the screen is divided into 12 equal columns, regardless of the device and therefore the screen width. These 12 columns can be assigned with the following properties, each of which exists in four versions:

- **`labelSpan`**
  The number of `labelSpan` specifies how many columns a label should occupy in the `SimpleForm`.
- **`emptySpan`**
  The `emptySpan` specifies how many columns in each line are left empty at the end of the line.

Figure 4.27 shows an example of the effects a definition for the device size `L` could have. According to the definition, the labels (in our case, first name and last name) take up the first four columns. We've also defined that three columns should be left blank for each line. If we were to have an input after a label in our XML definition, this input field would be assigned the remaining columns and therefore take up space.

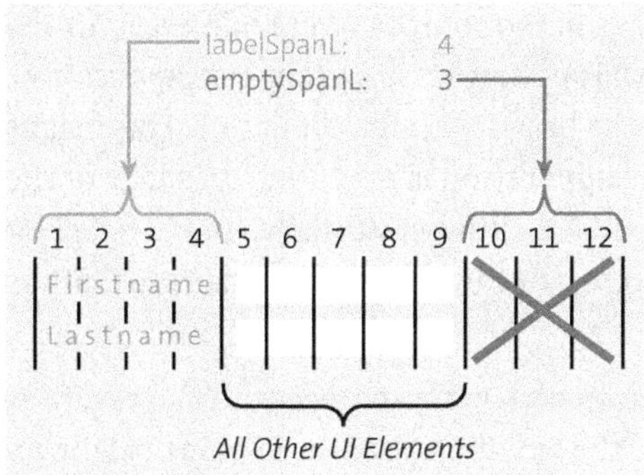

Figure 4.27 Theory of the SimpleForm Regarding the Responsive Grid Layouts

These rules are also very important for the SimpleForm:

- A sap.m.Label always starts a new line.
- A sap.ui.core.Title or sap.m.Toolbar starts a new group.
- All UI elements that come after a sap.m.Label use the available space (12 – labelSpan – emptySpan = Available columns). An attempt is made to distribute this space evenly among all UI elements in the same ratio. If the number is odd, the first UI element is credited with one more column.

We now start by inserting a SimpleForm in the aggregation content in our main view. You can see the XML definition for this in Listing 4.5.

```
<f:SimpleForm id="detail_simpleform" editable="true" layout=
"ResponsiveGridLayout" labelSpanXL="3" labelSpanL="3" labelSpanM="3"
labelSpanS="12" emptySpanXL="4" emptySpanL="4" emptySpanM="4" emptySpanS="0"
columnsXL="1" columnsL="1" columnsM="1">
    <f:content>
        <!-- Content of the SimpleForm belongs here -->
    </f:content>
</f:SimpleForm>
```

Listing 4.5 Add the Definition for the SimpleForm into the Content Aggregation of the Page

If we add an sap.m.Label to the SimpleForm, we start a new line. Let's try this out by displaying a line for the first name and one for the last name (see Listing 4.6).

```
<Label id="labelFistname" labelFor="inputFirstname" text="First name" />
<Input id="inputFirstname" />
<Label id="labelLastname" labelFor="inputLastname" text="Last name" />
<Input id="inputLastname" />
```

Listing 4.6 Add the Content of the SimpleForm to Display the Labels and Input Fields

Interestingly, when we switch back to the preview, we only see a blank page. We see no syntax errors in our IDE and no error messages in the console where our script for the preview was started. This is because the SAPUI5 application is a client-side application. The application is only built and compiled in the browser on the client side. This means that we only see the runtime errors when the application is executed. Let's look at the console of the *developer tools* of the respective browser (usually, press [F12] or [Ctrl]+[Shift]+[I]). Figure 4.28 shows the browser's console where we can see that the SimpleForm couldn't be loaded.

This is because the SimpleForm is loaded from the namespace sap.ui.layout.form, which we haven't imported, or we've prefixed the opening tag of the SimpleForm with an f as an abbreviation, which isn't even defined. Let's adjust the definition of the view as follows to carry out the import:

```
<mvc:View xmlns="sap.m" xmlns:f="sap.ui.layout.form" controllerName=
"at.clouddna.demo.controller.Main" xmlns:mvc="sap.ui.core.mvc" displayBlock=
"true">
```

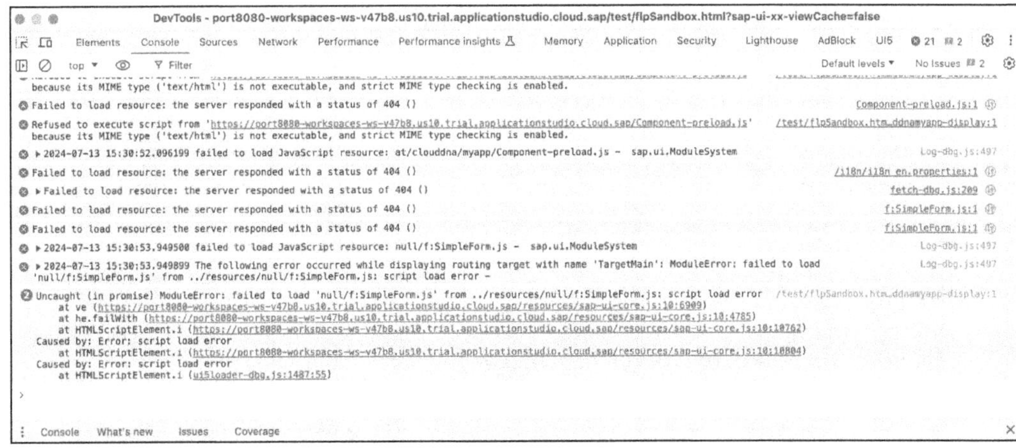

Figure 4.28 SimpleForm Can't Be Loaded and Raises a Runtime Error

Our SimpleForm already appears in its full splendor and displays the respective labels and inputs arranged in a user-friendly way (see Figure 4.29).

Figure 4.29 SimpleForm Displays Two Lines with Input Fields

As a final step, before we move on to the application logic, we add a footer to the page. Here we'll insert two instances of sap.m.Button in a sap.m.OverflowToolbar as a container (see Listing 4.7). Thanks to the sap.m.ToolbarSpacer, the buttons are moved to the far right, as with a tab. These buttons won't be able to do anything in the first step, apart from being visible and clickable. We'll learn how to define the effect of the buttons in the next section when we look at the business logic.

```
<footer>
    <OverflowToolbar id="footerToolbar">
        <ToolbarSpacer id="footerToolbarSpacer"/>
        <Button id="saveButton" text="Save" type="Accept" icon="sap-icon://
save" press="onSavePressed" />
        <Button id="cancelButton" text="Cancel" type="Reject" icon="sap-icon://
cancel" press="onCancelPressed" />
```

```
    </OverflowToolbar>
</footer>
```

Listing 4.7 Adding Buttons in an OverflowToolbar to the Footer Aggregation

Thanks to the properties we've assigned for formatting, our buttons are already displayed in color and with icons on our page (see Figure 4.30).

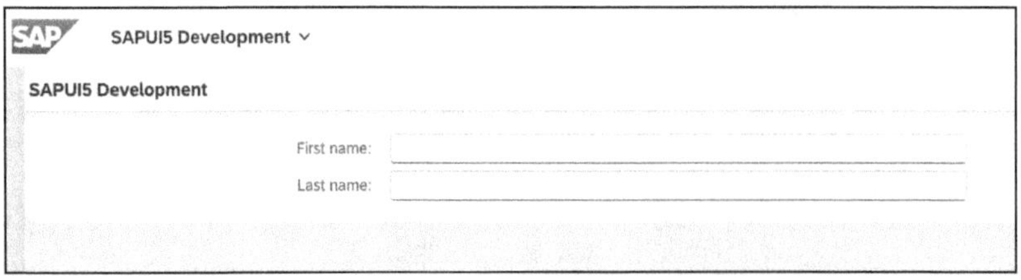

SAP SAPUI5 Development ∨

SAPUI5 Development

First name:

Last name:

Figure 4.30 Buttons Displayed in the Footer Aggregation of the Page

Icons Can Be Used Out of the Box

Wherever you have the option of assigning icons, you can use this list of icons available and supplied as standard: *https://sapui5.hana.ondemand.com/sdk/test-resources/sap/m/demokit/iconExplorer/webapp/index.html#/overview.*

4.3.2 Controller

In this section, we'll deal with the application logic. Before we can start developing with TypeScript, it's important to clarify where the controllers are to be found and what their function is.

The controllers are located in the *controller* directory and, depending on whether you're developing with JavaScript or TypeScript, the files here end with *.controller.js* or *.controller.ts*. To avoid the view having to handle all the logic, the lifecycle methods, formatters, event handlers and all other business logic are outsourced to a controller for each view. Thus, a controller is always assigned to a view (even if there are several technical possibilities). This assignment is initiated by the view because the reference to the controller is stored in the view definition with the `controllerName` property:

```
<mvc:View xmlns="sap.m" xmlns:f="sap.ui.layout.form" controllerName=
"at.clouddna.demo.controller.Main" xmlns:mvc="sap.ui.core.mvc" displayBlock=
"true">
```

Lifecycle Methods

The associated controller, as shown in Figure 4.31, is initially empty and only contains the *lifecycle method* `onInit()`, which is implicitly called due to the name equivalence.

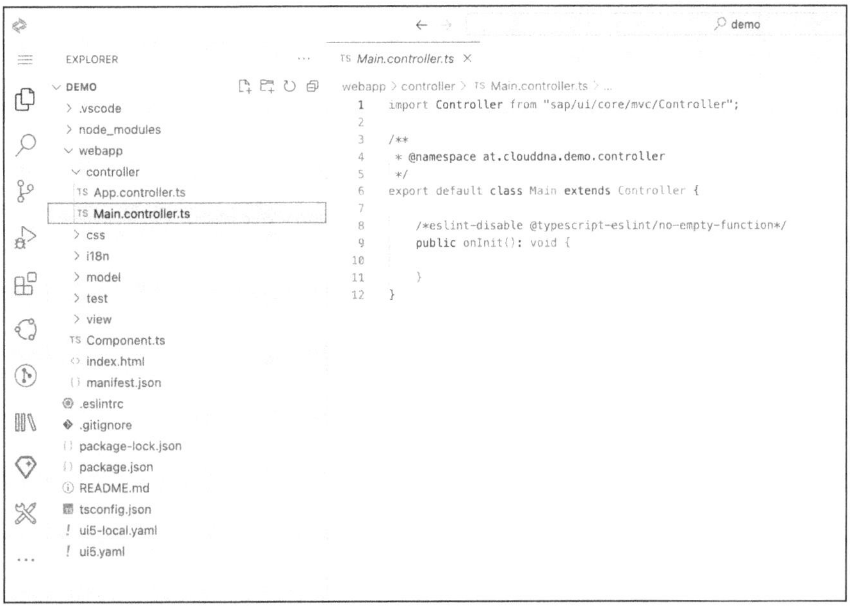

Figure 4.31 An Empty Controller Created by the Wizard

There are four lifecycle methods that are implicitly called by the framework. These methods can be predefined in the template, but they don't have to be. In any case, you can insert or remove these methods as you like, as they will always be found by the framework due to their identical names. The following four methods can be used in any controller:

- `onInit`
  This method is automatically run by the framework when the view is initialized for the first time together with the controller.
- `onBeforeRendering`
  This method is called before the view starts rendering the UI components.
- `onAfterRendering`
  This method is called when the rendering has been completed.
- `onExit`
  This method is automatically run by the framework when the view is destroyed. This is the case when the view or component itself is destroyed during operation, either by triggering the `destroy()` method in the business logic or by the app simply being exited by the user.

These methods are all only called once during the lifecycle of a view. Business logic that is called repeatedly shouldn't be placed in these functions. For this, a suitable listener must be found. In Chapter 7, we'll present some of these listeners that are frequently used in practice.

View Manipulation

The relevance of the IDs becomes clear when we want to access UI components in the view from the controller. First of all, thanks to the 1:1 relationship between controller and view, we can access the view instance in each method with the following method call:

```
this.getView()
```

Because we now receive an instance of sap.ui.core.mvc.View, we can look at the SAPUI5 documentation to see which methods are available to us (see Figure 4.32).

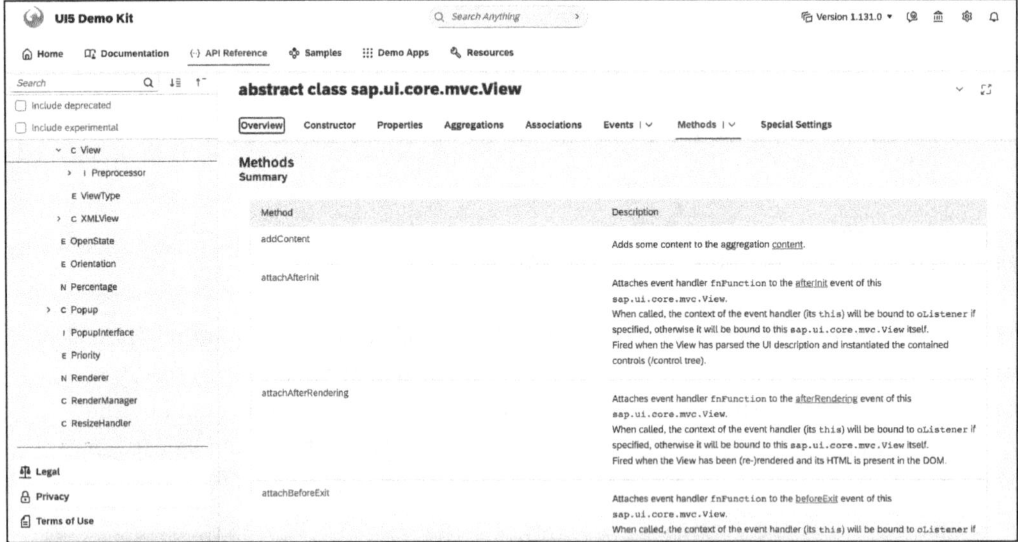

Figure 4.32 Available Methods on an Instance of a sap.ui.core.mvc.View

We find what we're looking for in the documentation. We can use the method byId(sId) to get a UI component with the desired ID, so to access the input for the first name, it would look like this:

```
this.getView().byId("inputFirstname")
```

Now the process starts all over again. We know that we get our sap.m.Input returned here, so we look again in the documentation to see which methods are available for this UI component. The documentation shows we can write a value in the input field with the method setValue(sValue). Putting all these puzzle pieces together, we can access the first name input in the onInit() method and write any value to the input from the controller (see Listing 4.8).

```
public onInit(): void {
  let oView = this.getView(),
    oInputFirstname = oView?.byId("inputFirstname") as Input;
```

```
  oInputFirstname.setValue("Daniel");
}
```

Listing 4.8 Accessing an Input Field to Use a Method of That UI Component

As we're using the UI component `Input` via this alias here at the first time, we need to do an `import` at the top of the controller so our TypeScript controller knows this UI module:

```
import Input from "sap/m/Input";
```

As you can see in Figure 4.33, the input field is now filled with the text we've set.

Figure 4.33 Accessing a UI Component via the ID throughout the Controller

Access via IDs in the Lifecycle Methods

Care should be taken when accessing the UI components by their ID in the lifecycle methods. Depending on the scenario, the view might be built first, and, especially in the `onInit()` method, an `undefined` is returned instead of the instance of a UI component because the UI component is technically not yet available. Any further method chaining at this point will result in a runtime error.

Among the lifecycle methods, `onAfterRendering()` is the most appropriate, but even this isn't always 100% reliable. When we discuss routing in Chapter 7, we'll show you what we consider to be the most appropriate listener.

Event Handler

The controller can be used not only to access the UI components from the lifecycle methods but is also available to the events as an *event handler*. We've already defined two buttons that have an event called `press`. We can examine this event more closely by looking at the definition of `sap.m.Button` (see Figure 4.34).

Thanks to the documentation, we now know that we've not stored anything for our two buttons in the XML view except for the two method names that are called when the `press` event is triggered by the user clicking on the UI. So, we prepare these two methods in our controller and write the method header into the body of the controller:

```
private onSavePressed(){}
private onCancelPressed(){}
```

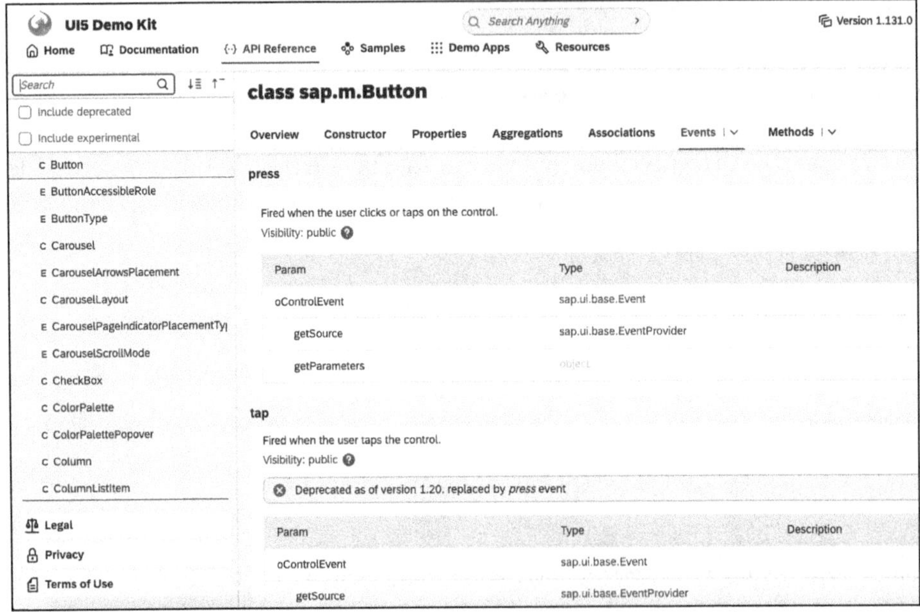

Figure 4.34 Looking Up the Events of a UI Component

In the next step, we add a logic to the event handler. When the user clicks on **Save**, a toast message **Successfully saved** should appear. When the user clicks on **Cancel**, a prompt should appear asking whether the user really wants to cancel along with a message stating that data could be lost. This is easy to do because SAP provides the two standard components, `sap.m.MessageToast` and `sap.m.MessageBox`, which are equipped with static methods and can perform these tasks for us. Let's adjust our methods as shown in Listing 4.9.

```
private onSavePressed(){
  MessageToast.show("Successfully saved");
}
private onCancelPressed(){
  MessageBox.warning("Are you sure you want to cancel?", {
    actions: [MessageBox.Action.YES, MessageBox.Action.NO],
    emphasizedAction: MessageBox.Action.YES,
    onClose: (sSelectedAction: String) => {
      if(MessageBox.Action.YES === sSelectedAction){
        MessageBox.success("Successfully cancelled.");
      }
    }
  });
}
```

Listing 4.9 Two Event Handlers Using the Default MessageToast and MessageBox UI Components

Don't forget to import these two UI components; otherwise, you won't be able to access them in the corresponding controller:

```
import MessageBox from "sap/m/MessageBox";
import MessageToast from "sap/m/MessageToast";
```

If we now click **Cancel**, the `MessageBox` appears with the definition that we stored in the `onCancelPressed` method (see Figure 4.35).

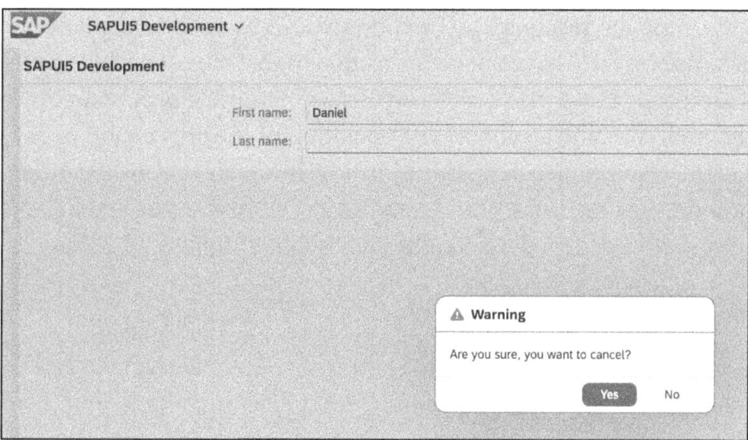

Figure 4.35 MessageBox Used to Show a Dialog Shipped by SAPUI5

4.3.3 Model

The *model* is responsible for holding the data in an SAPUI5 application at runtime and making it available to the view and the controller. If the model is connected to a remote data source (OData service or any REST service), our model also handles the communication with the service. Especially when it comes to remote data sources, we can't emphasize enough that the model acts as a kind of cache for the runtime in this case. Operations such as reading, filtering, sorting, grouping, and many more are executed (in a standard use case) on the server side and returned to the model. Our model stores the data as long as it's not emptied by the business logic or by a refresh of the web application.

Figure 4.36 shows a dream scenario for every SAPUI5 developer because here we have either one view with one controller or several such pairs. These are supplied with data by a single, globally known model. In this case, "global" means that all views and controllers in this (UI) component know the model and can therefore interact with it. This one model is a local model, which doesn't need a remote data source to get the data. For example, the prefilling will happen from a JSON or XML file. This may happen, but it's rather unlikely. More often, you'll have at least one model that communicates either with an OData service (as is common in the SAP environment) or with any REST service (e.g., developed with Java or Python).

In this case, if the view wants to display data or our controller needs data for processing in the business logic, our model triggers the corresponding HTTP request. Our model waits for the HTTP response and, in most cases, provides us with a corresponding listener so that we can also react to it in our business logic. One thing is very important here: the communication is asynchronous. In this request-response cycle, several hundred milliseconds to seconds can pass, depending on the user's hardware specs, the user's internet connection, the firewall and security on the server side, the hardware specs on the server side, and the business logic in the backend that is reading/processing the data and returning the response. We mentioned this as an ideal scenario because an ideal-typical use case will have one or more models that are connected to a remote data source. Alongside these, you'll have smaller, sometimes global, sometimes more specific, local models that will support the operation of the application or support specific parts of the application such as one specific view or one specific UI component of the application. So, as you can see, we need to learn more about the model, the instantiation, and types of such a model, and especially about visibility.

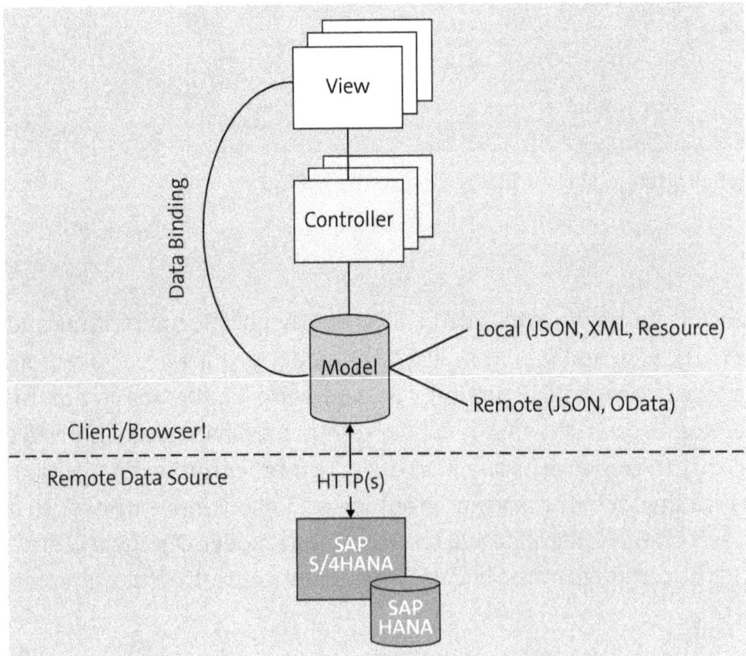

Figure 4.36 Global Model Providing Data to Views and Controllers

As you already learned in the introduction, the MVC concept provides a connection between the model and the view. This connection is called *data binding* and helps us access data directly without having to ask the controller each time. Using data binding, attributes in a view can be bound to the data in the model. There are different types of this connection that describe the data flow. This can either exist from the model to the view or be bidirectional, so that the view is also able to write data back to the model through user input and interaction. When the data in the model changes, the model can notify the view if it's bound to attributes at any point. This notification tells the view that parts of the UI need to be re-rendered. Because data binding is a simple yet powerful concept, we've devoted all of Chapter 6 to the topic.

In the following sections, we'll take a first look at the models. As the topic of models requires comprehensive knowledge, we'll only give you a brief introduction in this chapter and explain the most important details. You can find further information on the JSON model in Chapter 8, on the OData model in Chapter 10 and Chapter 11, and on the resource model and especially localization in Chapter 12.

JavaScript Object Notation Model

A *JSON model* must first be instantiated before it can be accessed. This instantiation can either be done in the application descriptor, specifically in the *manifest.json* file, or anywhere in the application logic, specifically from a controller. If you want to instantiate it from a controller, this is done in JavaScript or TypeScript. According to the official documentation, instantiation in TypeScript looks like this:

```
let oMyJsonModel = new sap.ui.model.json.JSONModel();
```

After instantiating a JSON model, we have to take care of its visibility. It's not enough to have created a JSON model if we don't make it known to any component. The following question can help us decide: "Do I only want to use a JSON model with a specific component (e.g., a view or a simple form), or should it be globally accessible in the application?" If we want to use a JSON model globally in the application, we can define it in *manifest.json* under the property `models` (see Listing 4.10).

```
{ //...
    "models": {
        "myGlobalModel": {
            "type": "sap.ui.model.json.JSONModel"
        }
    }, //...
}
```

Listing 4.10 Defining a Global JSON Model in the Application Descriptor

As you can see in this snippet, a model of the type `sap.ui.model.json.JSONModel` is defined in *manifest.json* with the name `myGlobalModel`. Optionally, you could also define the property `uri` in addition to the `type`, which could refer to a JSON file in our project. If this property is defined, the JSON model is prefilled with the data from the file.

Default and Named Models

As you can see in the last code snippet, you can give such models names. This name then also appears everywhere when gaining access to this model. If you leave the name empty (although this isn't allowed with the classic JSON), this model becomes the default model and is always addressed when no name is specified. Even if a default model is already defined (either in the application descriptor or somewhere in the code), this can be overwritten by using the JavaScript methods shown in Table 4.2.

If you define a global model with the owner component, this is done in the corresponding JavaScript or TypeScript code. This can be done in the *Component.js* or *Component.ts* file (where the owner component is also created and instantiated) or in any controller. The methods for this are almost identical, with the only difference being that when accessing the owner component in a controller, the instance of the owner component must be returned via the method `getOwnerComponent()`. In both cases, however, the method `setModel()` is called via the method chain, which receives the JSON model as a parameter. Optionally, a `name` for the model can be passed as a second parameter. If no name is provided, it becomes the default model. These two options are shown in Table 4.2.

Type of Assignment	Corresponding Method Call
Assignment in *Component.ts*	`this.setModel(oMyJsonModel);`
Assignment from any controller	`this.getOwnerComponent().setModel(oMyJsonModel);`

Table 4.2 Creating a Global JSONModel via JavaScript or TypeScript

Creating a Global Model

The two options shown ensure that the OData model created is globally available in the application. However, we recommend entering the global models in the application descriptor. This has two advantages:

- **Clean code**
  With this, you've defined the models that are to be globally available in a central location and can more easily keep track of them.
- **Instantiation**
  The application descriptor handles not only the global visibility setting but also the instantiation. This means that you don't have to worry about the model being created in JavaScript or TypeScript code.

If a JSON model is only intended to supply a certain component (e.g., a view) with data, then the model is assigned to the desired component in a controller. In the following code snippet, you can see that setting the model to an individual component is similar to setting it to the owner component. To do this, the method `setModel()` is called up, and the instance of the model is passed as a parameter (see Listing 4.11).

```
public onInit(): void {
    let oMyJsonModel = new sap.ui.model.json.JSONModel();
    this.getView().setModel(oMyJsonModel);
}
```

Listing 4.11 Instantiate a JSON Model in the Controller and Assign It to a Specific UI Component to Achieve Limited Visibility

In summary, Figure 4.37 shows that assigning a model to a component affects visibility. If the model is set via the application descriptor or the owner component, it becomes globally visible. Otherwise, it's only known to the specific UI component.

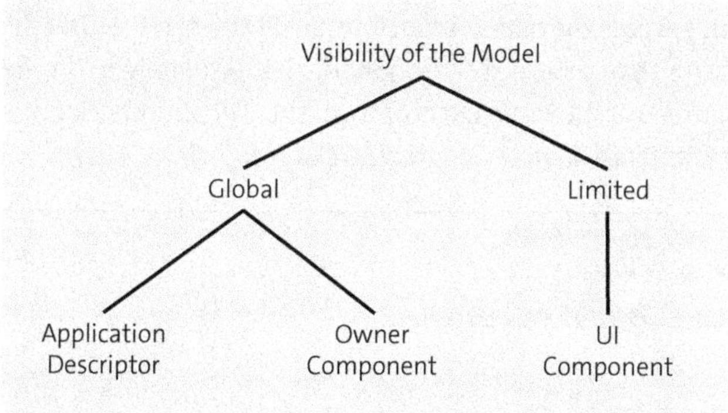

Figure 4.37 Visibility of the Model Depends on the Assignment

In this section, we wanted to present the most important information about the JSON model so that we can gain the necessary basic knowledge about all parts of the MVC design pattern. The possibilities don't stop there, which is why we'll discuss the JSON model in more detail in Chapter 8 and show how you can use it both as a local model and for a remote data source.

OData Model

As you already know, OData services are required to communicate with the SAP system. For a long time, only OData Version 2 (V2) services were available in the SAP environment. At that time, SAPUI5 was also launched, which is why the SAPUI5 framework with the component `sap.ui.model.odata.v2.ODataModel` (also known as *OData V2 model*) supports this protocol perfectly. There's already a successor under OData V4, but the migration of existing OData services isn't possible out of the box, especially if you've developed it with Transaction SEGW projects in the backend. For this reason, it will continue to be necessary for development and, above all, maintenance to know the OData V2 model and to be able to develop applications with it.

The *OData model* handles communication with an OData service for us. Because the OData service can be available in several versions, there are also several OData model components. In the component that SAP delivers as standard, operations such as create, read, update, and delete (CRUD) are encapsulated using methods. In contrast to the JSON model, if the data hasn't yet been loaded, the OData model automatically sends a request to the OData service and requests the data with a `GET` request. The OData model also has a number of convenience functions, such as tracking data changes at runtime, which can be sent to the backend at once. In this section, we'll get to know the basics of an OData model so that we can get to know all parts of the MVC design pattern. Because communicating with an OData model is a specialized topic, we've prepared the information for you in Chapter 10 (OData V2 model for OData V2 services) and Chapter 11 (OData V4 model for OData V4 services).

In the application itself, we don't access the data provided by an OData service directly. An OData model is nothing more than a faceless component that SAP delivers in the SAPUI5 component library. Figure 4.38 shows an excerpt from the API reference of this class (see *https://ui5.sap.com/#/api/sap.ui.model.odata.v2.ODataModel*).

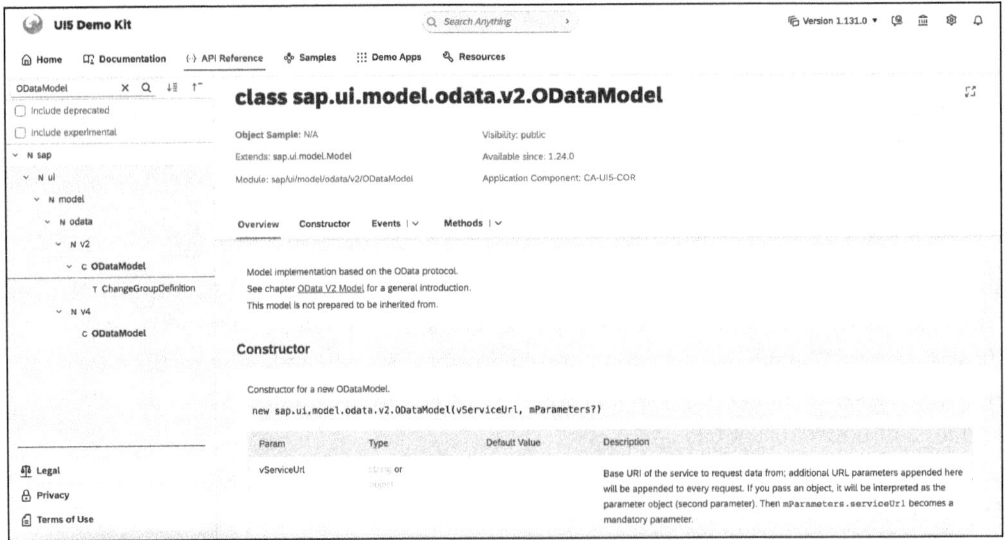

Figure 4.38 OData V2 Model Faceless Component in the SAPUI5 Documentation

An OData model of this kind must first be instantiated before we can access it at all. This instantiation can be carried out either in the application descriptor (i.e., in the *manifest.json* file) or at any point in the application logic (i.e., in a controller). If you want to carry out this instantiation from a controller, it takes place in JavaScript or TypeScript. You can also see what the instantiation looks like in the official documentation:

```
let oMyODataModel = new sap.ui.model.odata.v2.ODataModel(vServiceUrl,
mParameters?)
```

The vServiceUrl parameter must be replaced here by the path pointing to the OData service. In the second optional parameter mParameters, we can define a JavaScript object and assign further settings for this OData V2 model. We'll discuss some of these parameters in the chapters already mentioned. In any case, however, you can refer to the documentation, where all possible parameters are listed with their respective default values, data types, and descriptions.

The OData model is responsible for storing all data accessed during communication with an OData service for the runtime. It is, so to speak, a cache that we can access. The OData model doesn't store this data in any old way, but structures it according to the paths accessed at runtime. Listing 4.12 shows a simple example to illustrate how the data might be stored, even though this isn't how things work behind the scenes.

```
data: [
  "/Customer(1)": {
    Name: "Mary Sample"
  },
  "/Customer(2)": {
    Name: "Max Sample"
  },
  "/Product('ABC123')": {
    Title: "Flatscreen Monitor",
    Price: 399.99
  }
```

Listing 4.12 Example of How an OData Model Caches the Data

When the runtime is over, whether because the application was closed and reopened, the application was refreshed in the browser by the user, the developer as the author of the application logic simply wanted it over, our OData model starts empty again with the next instantiation.

OData Model Is Like a Cache

Before we go into more detail about the theory and functionality of the OData model, we want to make a comparison, which we also like to use in SAPUI5 workshops.

The OData model automatically triggers the read operations when data binding is used. In other words, if we use binding for the first time at runtime and bind to "/Customer(1)", then this data binding sends a GET request to "/Customer(1)", retrieves the entity, and stores it in the cache of the OData model for this specific binding. If we now access the name of this customer with ID 1 a second time in the same runtime, the

OData model recognizes that this data has already been loaded and accesses it directly instead of reading or requesting it again from the OData service.

However, this also means that if the name of the customer with ID 1 has changed in the meantime in the backend, our application won't notice this until the data in the OData model's cache has been updated. You'll learn how to perform this update yourself or when it's done automatically by the model in the OData model chapters already mentioned.

It's still not enough to have created an OData model if we don't make it known to any components or set it. The following question can help you decide: "Do I only want to use an OData model in a specific component (e.g., in a view), or should it be globally accessible in the application?" If you want to use an OData model globally in the application, you can define it in the `models` property in the *manifest.json* file, as you would for a JSON model (see Listing 4.13).

```
{ //...
  "models": {
    "myODataModel": {
      "dataSource": 'mainService',
      "preload": true,
      "settings": { /* ... */ }
    }, //...
  }
}
```

Listing 4.13 Definition of an OData Model in the Application Descriptor

As you can see in this code snippet, an OData model named `myODataModel` is defined in *manifest.json*. If you hadn't assigned a name to this entry, the OData model would be the default model. The same rule applies here regarding default and named models, which we've already presented in the JSON models.

The OData service isn't defined directly under `models`, but specified via a data source. The data source is defined earlier in the file under `dataSources` and could look like Listing 4.14.

```
{ //...
  "dataSources": {
    "mainService": {
      "uri": '/sap/opu/odata/IWFND/RMTSAMPLEFLIGHT',
      "type": "OData",
      "settings": {
        "localUri": "localService/metadata.xml",
        "odataVersion": "2.0"
      }
```

```
      }
    }, //...
}
```

Listing 4.14 Connections Defined in the Application Descriptor

This data source is named `mainService` and stores a relative path in the `uri` property that refers to the OData service. The `type` property defines that it's an OData service. In other settings, that is, in `settings`, there's a reference to a local copy of the metadata file and to the version of the OData service. You already know that this `uri` stores a relative URL. During development, this can be forwarded thanks to the middleware in *ui5.yaml* and resolved locally, as well as locally, on the SAP system anyway—without middleware after deployment.

The good news is that if we've already specified a data source when creating the project, this OData service is entered under `dataSources` and an associated global default OData model is entered under models in *manifest.json*. You can also add an OData service later. There are several ways to do this. Ultimately, each option will result in exactly these entries being added to *manifest.json* under `dataSources` and `models`. This means that you could also maintain or adapt these entries manually. If you don't want to do this, you can access the **Service Manager** via **View • Command Palette • Fiori: Open Service Manager** and then manually add further OData services to the project there (see Figure 4.39).

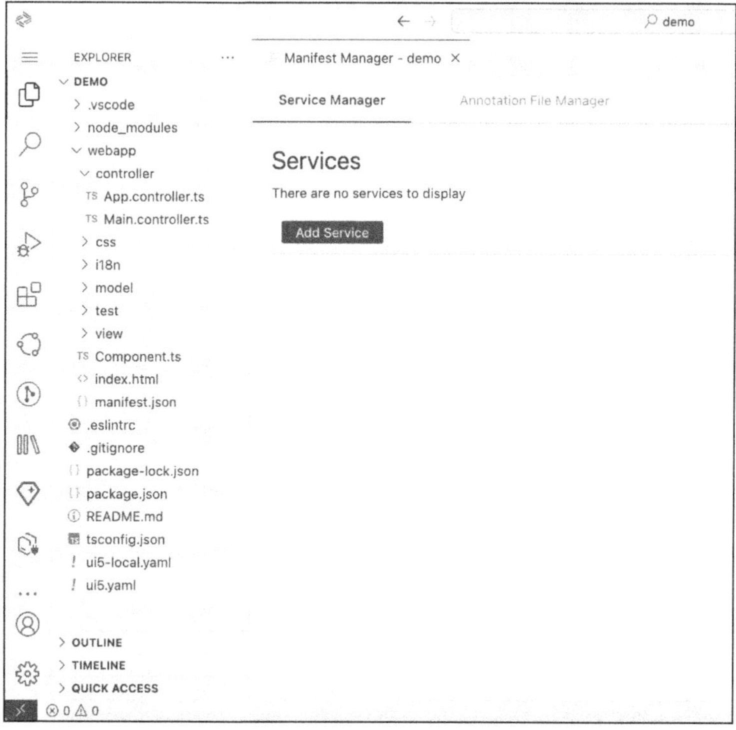

Figure 4.39 Add OData Services Manually via the Service Manager

Resource Model

A *resource model* stores language-dependent information, either as key-value pairs or in XML format, if you want to achieve a more granular and hierarchical storage. The resource model first determines the user's logon language and stores it under `locale` for the runtime. We'll discuss how this works later on in this section. Furthermore, a *resource bundle* is linked to the resource model. This can be either in the form of a file path or a URL. This resource bundle must contain JavaScript properties files with the names defined in the resource model and the ISO639 language codes of the languages that are managed. This is the only way the resource model can later find the corresponding file to get the language-specific texts and provide them to either the view or the controller. This process is shown in Figure 4.40.

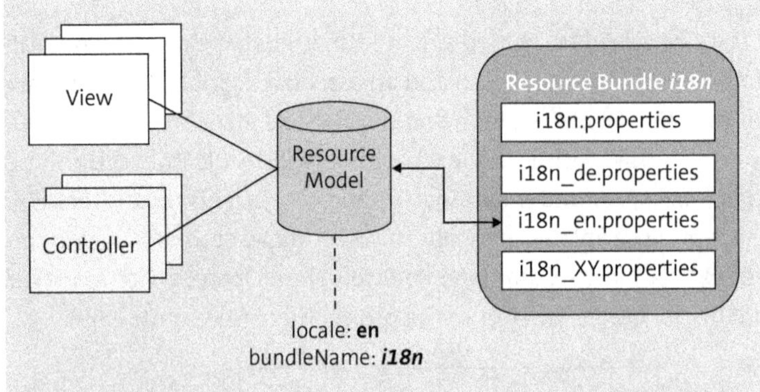

Figure 4.40 Resource Model Determining the User's Locale and Searching for Corresponding Files in the Resource Bundle

In the SAPUI5 framework, a resource model for translations is already included in the standard version. The corresponding entry can be found in `models` in the *manifest.json* file (see Listing 4.15).

```
"models": {
  "i18n": {
    "type": "sap.ui.model.resource.ResourceModel",
      "settings": {
        "bundleName": "at.clouddna.demo.i18n.i18n"
      }
    }
  }
},
```

Listing 4.15 A Resource Model for Translations Is Predefined in Every SAPUI5 Project

This global resource model can be accessed from all views and controllers with the name `i18n`. But what values does this model store? When the project was created, a directory named *i18n* was also created. It contains Java properties files that store texts using *key-value pairs*. The names of the files indicate the language for which they are responsible. Table 4.3 shows a few examples of how the naming is crucial for the target language. There's a fallback if no i18n has been maintained for the user's determined language.

i18n Filename	Purpose
i18n.properties	Default/fallback i18n
i18n_en.properties	English
i18n_de.properties	German
i18n_en_US.properties	American English
i18n_en_GB.properties	British English

Table 4.3 Examples for Different i18n Files to Cover (Specific) Languages

The way in which the user's language is determined is fixed and is processed using a checklist. If no answer is found for a question, the process moves on to the next point until a language is determined:

1. Was the language set programmatically in the application with `sap/base/i18n/Localization.setLanguage`?
2. Was the query parameter `sap-locale` set in the URL?
3. Was the query parameter `sap-language` set in the URL?
4. Was the query parameter `sap-ui-language` set in the URL?
5. What language is set in the browser?

SAPUI5 fallback is English if you couldn't find an answer to all the other questions.

As we've already mentioned, these files store key-value pairs separated by an equal sign. So, with the same key, you define the respective value in all the other i18n files in the different languages. You can see a few simple examples of this in Table 4.4.

Key	Value (en_US)	Value (en_GB)	Value (de)
`appTitle=`	Manage Bookings	Manage Bookings	*Buchungen verwalten*
`saveBtn=`	Save	Save	*Speichern*
`mailNotificationText=`	You have new messages in your mailbox.	You have new messages in your postbox.	*Sie haben neue Nachrichten in Ihrem Postfach.*

Table 4.4 Examples for i18n Entries for Translation Purposes

Accessing the resource model is covered in detail in Chapter 12, but a basic knowledge of data binding from Chapter 6 will be necessary for this.

4.4 Application Launch

You already know that we can preview an application by executing certain scripts. In this section, we focus on the possibilities of such scripts, which are predominant by default. As shown in Figure 4.41, these scripts are defined in the *package.json* file and can either be called directly in the console or selected via the context menu. Both result in the same functionalities.

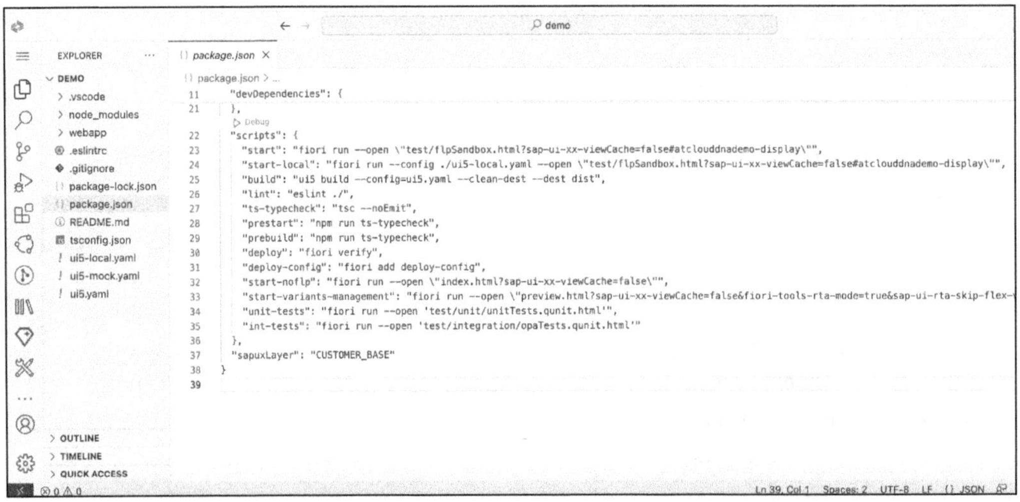

Figure 4.41 Scripts Predefined in the package.json File to Preview, Test, Build, or Deploy an Application

To preview the application with live data/connection in an SAP Fiori launchpad sandbox, simply execute the following command in the terminal by using the shortcut you've set up in your IDE, or go to **Terminal • New Terminal** in the menu, and enter the following command:

```
npm run start
```

You can also access this and all other scripts by right-clicking on the *webapp* folder and selecting the **Preview Application** option in the context menu.

What do live data and live connection mean? To explain this, we've added an OData service to our project via the service manager shown previously. Now we have to look at the *ui5.yaml* file, which is used for configuration when this script is executed. A middleware has been defined there (see Listing 4.16). This middleware specifies which requests should go to a specific local path or should be redirected to somewhere else. There's a `backend` entry which states that all calls attempting to execute on *https://localhost:<port>/sap* should be redirected to another URL. This URL isn't entered directly here, but is linked to a `destination` in SAP BTP, where the URL is stored.

However, we could have used the property url instead of destination and entered the URL of our system there, if accessible. So, we can see here that when we preview the application, the OData service and thus the data are accessed by an underlying SAP system at runtime. This means that we're previewing with a live connection. You can also see that when the framework tries to access /resources or /test-resources, the request is forwarded to *https://ui5.sap.com*. This is necessary because the SAPUI5 libraries aren't stored directly in the application, but have to be fetched from a CDN. If you want to use a packaged, custom CDN version instead of SAP's CDN, you could point to your own CDN here.

```
customMiddleware:
  - name: fiori-tools-proxy
    afterMiddleware: compression
    configuration:
      ignoreCertError: false
      ui5:
        path:
          - /resources
          - /test-resources
        url: https://ui5.sap.com
      backend:
        - path: /sap
          destination: S4D
          authenticationType: BasicAuthentication
```

Listing 4.16 Excerpt from the ui5.yaml File Regarding the Middleware

As you already know, if the application is running in the SAP Fiori launchpad, it's not started via *index.html* but via *Component.js* or *Component.ts*. This is also the case if you use this sandbox, which is designed to allow you to preview the app without a functional SAP Fiori launchpad. However, if you want to preview the app without this sandbox and actually start the application standalone via *index.html*, you can also use this script:

```
npm run start-noflp
```

In this case, too, an attempt is made to establish a live connection based on the definition in *ui5.yaml*. But what if the backend isn't available at the moment, but we still want to preview the application locally? With a live connection, this is only possible to a limited extent because the connection attempts result in dozens of errors each time. So, in this case, we have to preview with *mock data*. To make this possible, we have to configure a *mock server* and create a suitable script for this possibility. This all sounds very complicated, but don't worry, it just takes a few clicks.

You can configure the mock server in a guided way by opening the *application info* file of your project. This option, **Open Application Info**, is also available in the context menu when you right-click on the *webapp* folder level (see Figure 4.42).

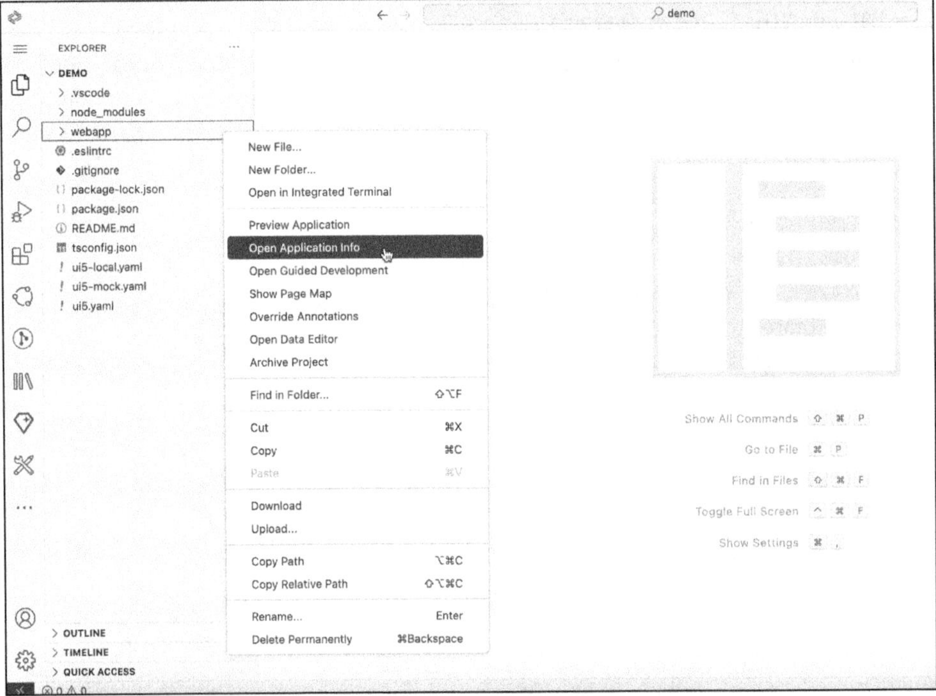

Figure 4.42 Opening the Application Info of a SAPUI5 Application

In addition to project-specific information, you'll also see a few buttons that can be used as quick actions. Along with previewing the app, opening the **Service Manager**, and checking the node modules and reinstalling them, you can also add the mock server with **Add Mockserver Config** (see Figure 4.43).

This option does the same as the following command, which you could also have entered manually:

```
npx --yes @sap-ux/create@latest add mockserver-config
```

After that, you'll see that another script called `start-mock` has been added to the `scripts` in your *package.json*. You can also run this script in the terminal:

```
npm run start-mock
```

The *data editor* is used to ensure that the application doesn't start with any data, but that you can determine the mock data yourself. Here, you can create, edit, and even import and export sample data from your real system to make previewing with mock data even more realistic. You can find this data editor in the context menu when you right-click on the *webapp* directory or in the application info tab with the name **Maintain Mockdata**. In either case, you'll end up in the data editor, as shown in Figure 4.44.

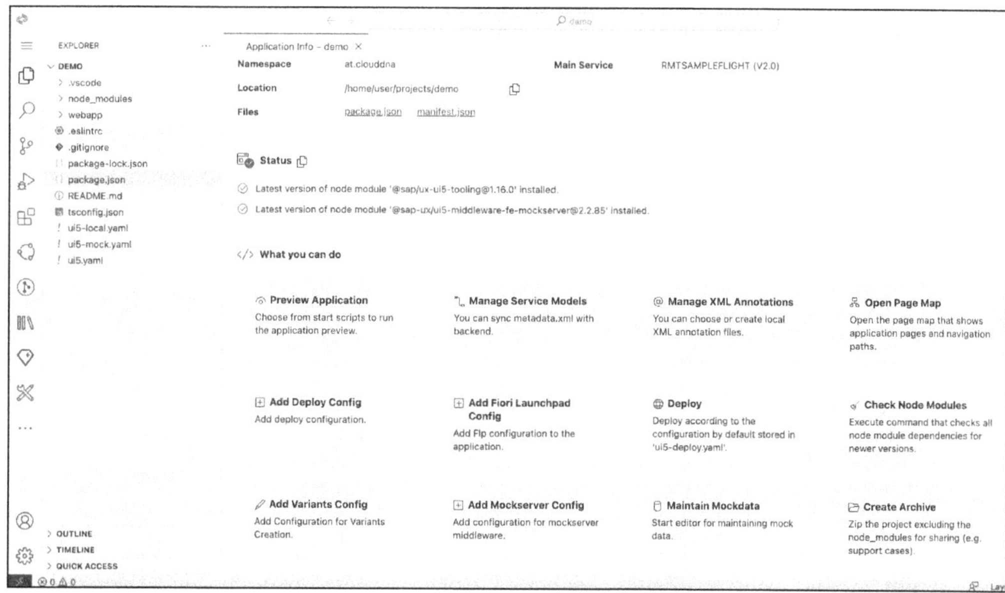

Figure 4.43 Application Info Used to Check Project-Specific Information and to Execute Quick Actions

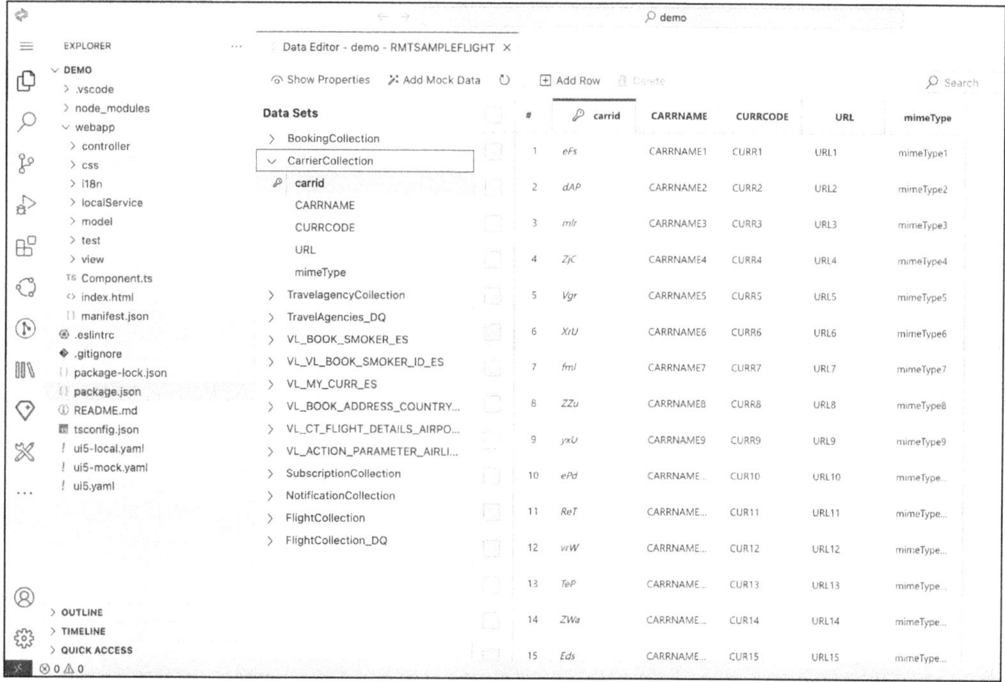

Figure 4.44 Data Editor to Maintain Your Mock Data

4.5 Summary

In this chapter, we took our first steps toward SAPUI5 development. We saw how to create a new application in both the in-house SAP Business Application Studio and VS Code. We clarified that a SAPUI5 app is nothing more than at least one faceless component or a UI component. In addition to the metadata and application logic, the UI component also has a renderer to display the component in the browser. The metadata of each component is stored in the application descriptor (*manifest.json*). The most important and essential component controller (*Component.js/Component.ts*) must not be missing because this is what brings a component to life.

Because we've built a UI component, we got to know the individual parts of an application within the framework of the MVC design pattern. In the view, we displayed the first UI controls, which was more about generic approaches: How do I work with the SAPUI5 documentation? How do I find the individual properties, aggregations, and events of a UI component? After that, we took a first look at the controller and wrote the first event handlers in addition to the lifecycle methods. In the MVC concept, we saw the model as the last piece of the puzzle. This can take several forms and can also communicate with a remote data source. It can be defined in a variety of places, and the subsequent assignment decides whether it will be globally visible or only seen and used by certain components. With this chapter, we've laid the foundation for learning more about the other standard components, data binding, routing, different models and how they work, and more advanced topics in the next chapters of this book.

Chapter 5
Modules

Modules are SAPUI5's way of loading only the things it needs during runtime. They are the foundation of the SAPUI5 library and provide a vast majority of an enterprise's needed functionalities such as UI elements, routing, runtime functionalities, and so on.

5

SAPUI5 modules are fundamental to the SAPUI5 framework, allowing you to break down your application into smaller, more manageable and reusable pieces of code. Modules help you to structure your code logically, making it easier to understand, maintain, and debug. You can create reusable modules that encapsulate specific functionalities, reducing code duplication and promoting consistency. By loading only the necessary modules at runtime, you can improve the initial load time and overall performance of your application.

In this chapter, we'll take a deep dive into the most-used SAPUI5 modules and how they enable you to build enterprise-grade applications. We start with a brief introduction to defining a module in Section 5.1 and then discuss the most common libraries in Section 5.2. Here, we give a first glimpse into how SAPUI5 is structured and what base functionalities you can use.

Then, we transition over to the controls in Section 5.3. This will be your first go-to reference on the most commonly used controls to provide UI functionality. Section 5.4 explains how to generally structure the layout of applications. You'll see which containers you can use to build neat and routable single-page apps. Finally, Section 5.5 provides an overview of how controls are generally built in SAPUI5 and why concepts such as data binding work. These are managed objects.

5.1 Define a Module

To define a module in SAPUI5, you use the `sap.ui.define()` function in JavaScript or the export class statement in TypeScript. The `sap.ui.define` function takes three arguments:

- **Dependencies**
  An array of module IDs that the current module depends on.

- **Factory function**
  A function that defines the module's logic and returns the export value.
- **Export value**
  The value that is exported from the module and made available to other modules.

An example of this in JavaScript is shown in Listing 5.1.

```
sap.ui.define([
    "sap/m/MessageToast"
], function(MessageToast) {
    "use strict";
    return {
        sayHello: function(message) {
            MessageToast.show('Hi dear reader of this book. How ya doin?');
        }
    };
});
```

Listing 5.1 A Module Definition in JavaScript

An example of this in TypeScript looks like Listing 5.2.

```
import MessageToast from "sap/m/MessageToast";

 export default function sayHello(){
    MessageToast.show('Hi dear reader of this book. How ya doin?')
 }
```

Listing 5.2 A Module Definition in TypeScript

To use a module in another part of your application, you can import it in the first argument of the `sap.ui.define` function in JavaScript or use the `import` statement in TypeScript. An example in JavaScript looks like Listing 5.3.

```
sap.ui.define([
    "namespace/sayHello"
], function(sayHello) {
    "use strict";
    sayHello();
});
```

Listing 5.3 Module Usage in JavaScript

An example in TypeScript looks like Listing 5.4.

```
...
import sayHello from "./controls/SayHello";
```

```
...
export default class Main extends Controller {
    public onInit(): void {
        sayHello();
    }
}
```

Listing 5.4 Module Usage in TypeScript

5.2 Libraries

SAPUI5 libraries are collections of reusable UI components that you can use to build your applications. They provide prebuilt controls such as buttons, input fields, tables, charts, and more, saving you time and effort in developing your own custom UI elements. If you want to create your own libraries, see Chapter 13. Here, are the key libraries we'll talk about:

- `sap.base`
  This provides general functionality of the framework such as internationalization, logging, and security.
- `sap.ui`
  This is the foundation library that provides the basic building blocks for all SAPUI5 applications, including models, views, controllers, and routing.
- `sap.m`
  This library is specifically designed for mobile devices and tablets, providing touch-optimized controls and responsive layouts.
- `sap.ushell`
  This library bundles the access to services of the SAP Fiori launchpad.
- `sap.uxap`
  Here, all controls necessary for building SAP Fiori object pages are located.
- `sap.f`
  This library includes special controls for SAPUI5 development that aren't normally found in other frameworks.

5.2.1 sap.base

The `sap.base` library contains, as the name suggests, base functionality that the SAPUI5 framework uses internally. The most-used component of this library is the internationalization concept of SAPUI5. See Chapter 12, Section 12.2, for detailed examples of this internationalization. But other useful implementations are also included in this library. Important components of this library are as follows:

- **sap.base.assert**
  The sap.base.assert method adds a basic assertion mechanism where you can check if a value meets a certain condition. This can also be used for testing purposes.
- **sap.base.i18n.ResourceBundle**
  The sap.base.i18n.ResourceBundle class stores text in separate files based on language. You can use the ResourceBundle module to access these files and display the appropriate text for the user's language. This provides a create method where you provide the base URL of your resource bundle file and optionally the user's preferred language. If the specific language file isn't found, the resource bundle tries alternative versions in a specific order:
  - Locale without region code
  - Fallback language (defaults to English)
  - Base URL file (the original file without any language information)
- **sap.base.Logging**
  The SAPUI5 Logging API with the module sap.base.Logging helps you monitor your application's behavior by creating log entries. These entries contain details such as timestamps, severity levels (debug, info, warning, error, fatal), messages, and component information. You can control which log messages are recorded by setting the minimum level using setLevel. By default, optimized builds only record errors, while debug builds show all messages.
- **sap.base.security: sap.ui.security**
  This offers functions for encoding cascading style sheets (CSS), JavaScript, URLs, URL parameters, and XML, whereas sap.ui.string offers string manipulation functions.
- **sap.base.util**
  The module sap.base.util and its submodules provide functionalities for programmers. This offers methods for working with arrays, cloning data, looping over arrays, generating unique IDs, and more.

5.2.2 sap.ui

The sap.ui namespace is one of the core namespaces relating to everything needed to build web applications. It's structured into sub-namespaces, clustering the different parts required to not only run SAPUI5 applications but also for lifecycle management, core functionalities, and UI elements.

sap.ui.core

The sap.ui.core namespace contains the core functionalities of SAPUI5, as the name suggests. The following components are of most importance:

- **sap.ui.core.Control, sap.ui.core.Element**
  In this namespace, the sap.ui.core.Control and sap.ui.core.Element modules are

placed. They are the main concepts of building UI elements and are also handled in Chapter 13.

- `sap.ui.core.Component`
  The module `sap.ui.core.Component` is the base class in SAPUI5 for defining reusable UI components.
- `sap.ui.core.BusyIndicator`
  The `sap.ui.core.BusyIndicator` can be used to indicate that work is currently done or something is being processed. This blocks the UI part, to disable users' interactions with it.
- `sap.ui.core.dnd`
  This namespace bundles the necessary technology to enable drag and drop in SAPUI5. This is shown in Chapter 12.
- `sap.ui.core.EventBus`
  The `sap.ui.core.EventBus` is an observer pattern that can be used to trigger events and react to these events throughout your application. From experience, this implementation might lead to some problems, so it's advised to create your own event bus.
- `sap.ui.core.Fragment`
  Fragments are a big part of SAPUI5, enabling you to create reusable XML snippets that can be loaded when needed. This is also shown in Chapter 12.
- `sap.ui.core.Icon`
  This control is responsible for displaying icons from the SAP Fiori icon pool.
- `sap.ui.core.Item`
  This is the default implementation of an item that can be used in dropdowns or lists.
- `sap.ui.core.Routing`
  Here, the implementation of the routing concept is defined. An instance of this class is normally automatically created via its definition in *manifest.json* and the instantiate routing method in *Component.js*.
- `sap.ui.core.Title`
  This title can be used to render emphasized text that should serve as a title.

sap.ui.Device

The `sap.ui.Device` API offers insights into the user's browser and device, as well as cross-platform support for events such as media queries, orientation changes, and resizing. It operates independently of the rest of the SAPUI5 framework, enabling its preloading for dynamic SAPUI5 bootstrapping based on device or browser capabilities.

The data of the Device API is normally put into a global named `JSONModel` during the instantiation of *Component.js*.

sap.ui.layout

This namespace bundles the different layout options, SAPUI5 comes shipped with. This includes horizontal and vertical box layouts, different grids, responsive layouts, and so on. We'll talk about this in Section 5.4.1.

sap.ui.model

The `sap.ui.model` namespace brings you the different models that you can use for holding and managing data. This includes the following:

- **`sap.ui.model.json.JSONModel`**
  This client-side model stores data in JSON.
- **`sap.ui.model.odata.v2.ODataModel`**
  This server-side model is for integrating OData Version 2 services.
- **`sap.ui.model.odata.v4.ODataModel`**
  This server-side model is for integrating OData Version 4 services.
- **`sap.ui.model.resource.ResourceModel`**
  This client-side model works with resource data that get bundled like the internationalization files of your application.

In addition to the models, these namespaces also provide classes to work with these models:

- **`sap.ui.model.Filter`**
  This class is responsible to create filters for bindings and will handle their logic and combination.
- **`sap.ui.model.FilterOperator`**
  This enum provides you with valid filter operations.
- **`sap.ui.model.Binding`**
  This is the base class for all binding information.
- **`sap.ui.model.Context`**
  This holds the context of a binding for relative bindings.
- **`sap.ui.model.PropertyBinding`**
  This is the binding between one property of an element and one property of a model.
- **`sap.ui.model.ListBinding`**
  This is the binding between a controls aggregation and a list of model data.
- **`sap.ui.model.TreeBinding`**
  This is like the `ListBinding`, but for hierarchical data.
- **`sap.ui.model.Sorter`**
  This class is responsible for sorting bindings.

5

sap.ui.richtexteditor

The sap.ui.richtexteditor.RichTextEditor control is an advanced input control that allows you to input formatted text. It uses third-party components and might therefore run into certain restrictions when used in specific scenarios. In addition, from experience, this control doesn't handle binding very well, so new values should always be set via its dedicated setter methods and retrieved via its getter methods. In addition, when changing the editable property, it can lead to rendering issues. To solve that, new instances of the control should be created dynamically with the correct editable value provided instead of changing it during runtime via the setter method. Besides that, it's a great control if you want to provide input functionality for things such as notes or comments.

sap.ui.table

The sap.ui.table namespace provides you with the sap.ui.table.Table control, also known as the grid table. It was made to make working with large sets of data possible by offering horizontal and vertical scrolling, fixed row heights, and so on. We'll look into this in detail in Section 5.3.3.

5.2.3 sap.m

The contents of the sap.m namespace were created with the mobile aspect in mind, meaning that every control here supports an adaptive and responsive design. Some of these responsive controls are as follows:

- **sap.m.App**
  This control is the base of a SAPUI5 application and serves as a single page container and a navigation container.
- **sap.m.Avatar**
  The avatar is a round image or text used for displaying users.
- **sap.m.Breadcrumbs**
  The breadcrumbs provide easy navigation functionality by displaying the previous navigation steps as clickable links.
- **sap.m.Button**
  The sap.m.Button is the main control for displaying buttons that users can interact with. It supports different button types for different scenarios.
- **sap.m.Carousel**
  The carousel displays a list of items that can be shuffled from left to right. It can be used to always bring one item into the spotlight.
- **sap.m.CheckBox**
  This displays a checkbox that can be selected or deselected.

- **sap.m.ComboBox**
  The sap.m.ComboBox can be used to display a dropdown that includes between 13 and 200 items. It supports search completion and adding new dropdown items on the fly.
- **sap.m.DatePicker**
  This control offers the input of dates by providing a calendar popover when interacting with it.
- **sap.m.DateTimePicker**
  The DateTimePicker adds time to the DatePicker control.
- **sap.m.Dialog**
  The sap.m.Dialog control provides a modular dialog that can be opened dynamically. This can be used when you want to interrupt the user's current interaction and display data that should be processed now.
- **sap.m.FlexBox**
  This layout control enables you to use the flexible box layout of CSS in SAPUI5.
- **sap.m.HBox**
  The sap.m.HBox layout control allows you to align content horizontally.
- **sap.m.IconTabBar**
  With the IconTabBar, you can cluster content into different tabs that can be selected via a toolbar, displaying icons for each tab. This is great for structuring content and adding visual appeal to them.
- **sap.m.Image**
  This is the general implementation with which you can add images to your apps.
- **sap.m.Input**
  The sap.m.Input control allows you to input data. It can be enhanced with placeholder texts to provide information on what users need to input there.
- **sap.m.Link**
  With the sap.m.Link control, you can add hyperlinks to your apps. This is the pendant to the HTML anchor tag.
- **sap.m.List**
  The list control allows you to display items in a structured list, supporting performance techniques such as lazy loading. You can use the list with different list item types for displaying the list data in different ways such as custom layouts, standard layouts, and so on.
- **sap.m.Menu**
  The sap.m.Menu control makes creating menus possible that can then be structured into submenus. This is for clustering actions into one place.
- **sap.m.MessageBox**
  With the sap.m.MessageBox control, you can display different predefined severities of

messages to users. This will open a dialog that depending on the severity shows a different color and icon scheme.

- **sap.m.MessageStrip**
  The MessageStrip shows messages directly in the UI, also coming with predefined severities.
- **sap.m.MessageToast**
  With the MessageToast control you're able to display messages that pop up and are automatically set invisible after a few seconds. This doesn't interrupt the users interaction flow as hard as a MessageBox, for example.
- **sap.m.OverflowToolbar**
  The OverflowToolbar can be used as a responsive toolbar. When the space becomes smaller, otherwise impacted items in the toolbar will be shifted into a menu in the toolbar. This allows you to tidy up the UI in responsive scenarios.
- **sap.m.Page**
  The sap.m.Page control gives you a general layout consisting of a header, content area, and footer area, and it takes up 100% of the screen.
- **sap.m.PDFViewer**
  With the PDFViewer, you can display files with an application/pdf MIME type in your application.
- **sap.m.PlanningCalendar**
  The PlanningCalendar works as a calendar, showing appointments for different entities in rows. These appointments can be visually enhanced. The calendar itself supports different views and time intervals such as weekly or monthly views.
- **sap.m.Popover**
  The popover can be used to display additional information opened by a button click, for example.
- **sap.m.RadioButton**
  Radio buttons work like checkboxes that are grouped so that only one can be selected.
- **sap.m.RangeSlider**
  A RangeSlider enables users to select a numeric value based on an interval by using a slider.
- **sap.m.SegmentedButton**
  With the SegmentedButton, you can add multiple buttons together to give them the same context. This is often done to display filters for tables such as "own" and "others".
- **sap.m.Select**
  The select works similarly to the ComboBox but should be used for fewer entries.

- `sap.m.StepInput`
  The `StepInput` enables numerical input that can be incremented or decremented by predefined steps.
- `sap.m.Switch`
  With the switch control, you can add a toggle control for setting true/false values.
- `sap.m.Table`
  This table control, also called responsive table, is the main table type that should be used to display tabular data in responsive scenarios. It supports line breaking, device-specific visibility of columns, and grouping table data.
- `sap.m.Text`
  This is the main control for displaying texts in your UI.
- `sap.m.TextArea`
  This works like the input control, but adds multiline support instead of a single line of data.
- `sap.m.VBox`
  The `VBox` control places its content vertically.
- `sap.m.Wizard`
  The wizard is a great control for taking a user step-by-step through a process such as the creation of new data. It should include at least three steps, which can be activated individually.

5.2.4 sap.ushell

The `sap.ushell` namespace holds relevant APIs for working with the SAP Fiori launchpad from SAPUI5 applications. First, the `sap.ushell.Container` API can be used to develop plugins in the SAP Fiori launchpad. A plugin is nothing more than an SAPUI5 component that retrieves the renderer of the Container API during initialization and adds new content to the SAP Fiori launchpad with the renderer's methods.

Then, this namespace holds services provided by the SAP Fiori launchpad. These services can be retrieved via the method `sap.ushell.Container.getServiceAsync("<Service name>")`. This returns a promise with the requested service. The following services might be important for you to know:

- `sap.ushell.services.Configuration`
  This service is relevant when you program custom tiles because it allows you to listen to size changes from the SAP Fiori launchpad.
- `sap.ushell.services.Navigation`
  This service is the main way of implementing app-to-app navigation in the SAP Fiori launchpad. It allows you to navigate from one app to another using intent-based navigation, listen to navigation events, retrieve the intent information of the current application, and much more.

- `sap.ushell.services.UserInfo`
  The user info service retrieves the information of the current SAP Fiori launchpad user. You can fetch its email, name, and ID.
- `sap.ushell.services.BookmarkV2`
  The relatively new bookmark service makes it easy to add shortcuts in the SAP Fiori launchpad.
- `sap.ushell.services.NotificationsV2`
  The notification service is the entry point for retrieving information about the SAP Fiori launchpad notifications. You can't add new notifications, but you can fetch them and execute their actions.

5.2.5 sap.uxap

The `sap.uxap` namespace bundles everything needed to create an SAP Fiori object page layout. We'll discuss this in more detail in Section 5.2.5.

5.2.6 sap.f

The `sap.f` namespace in SAPUI5 houses a collection of UI controls specifically designed for building applications adhering to the SAP Fiori design guidelines. These controls are built to align with the visual style, interaction patterns, and user experience principles of the SAP Fiori design system. They often provide features that improve usability and accessibility for SAP Fiori apps. While adhering to SAP Fiori standards, they offer customization options to tailor the look and feel to specific application requirements. The following controls are important:

- `sap.f.Card`
  A card is a container with predefined header and content, as well as a predefined visual style. It can be used with a list, table, contact information, charts, and so on.
- `sap.f.DynamicPage`
  The dynamic page is the base for the object page. It provides a title, dynamic header that can be expanded or collapsed, and a content area. If you want to have this behavior but don't need all the features from the object page, use the dynamic page. It has significantly fewer dependencies and less rendering time than an object page.
- `sap.f.FlexibleColumnLayout`
  This control functions similarly to `sap.m.SplitContainer`, but with the key distinction of supporting three columns (begin, mid, and end) instead of the two typically found in `sap.m.SplitContainer` (master and detail). The width of each of these three columns can be dynamically adjusted. The control offers a variety of possible layouts, which can be modified programmatically via the control's API or interactively by the user. Draggable column separators enable users to customize column widths

within the current layout. Moreover, dragging a separator past a predefined breakpoint can trigger a transition to a different layout.

- `sap.f.GridContainer`
  The grid container is a layout control that places its content in a grid. This grid is sectioned into rows and columns and can vary in CSS size. Its content then gets the information on how many columns and rows it should occupy.
- `sap.f.GridList`
  The grid list works similarly to the grid container but should be used for displaying its content in the same size. It can also use different grid layouts.
- `sap.f.ProductSwitch`
  The product switch displays a popover that displays a list of items. These items represent different apps, navigation targets, or other items.

5.3 Controls

In this section, we'll now investigate the controls from the namespaces we just read about. This section's target is to give an understanding of how to use the most-used controls in SAPUI5. We'll give you a short introduction to the control, explain how you can define it, and show what it should look like. In addition, we won't really talk about layout controls because this will be the content of the next section. This isn't an exhaustive list of controls because that would require its own book. Some controls will also be used in other chapters when showing things such as routing, validation, and so on. We just want to give you a basic set of controls to work with here so you can start programming. Let's jump right into the basic controls.

5.3.1 Basic

In this section, we'll talk about the basic controls of SAPUI5. This includes basic texts, different input controls, interactive controls such as buttons, and so on.

sap.m.Text

The `sap.m.Text` control simply displays text in your UI. The text itself can be formatted in simple ways, for example, to set its alignment and maximum displayed lines. The following metadata is important:

- `text`
  Sets the displayed text.
- `textAlign`
  Sets the text alignment to left or right.
- `wrapping`
  Enables text wrapping when the space is too short.

- `maxLines`
  Determines the maximum displayed lines.

A sample implementation can look like the following:

```
<Text text="Hello World" maxLines="1" wrapping="true"/>
```

This will produce text that looks like Figure 5.1.

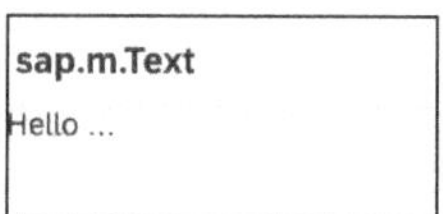

Figure 5.1 The sap.m.Text Control

sap.m.Title

The `sap.m.Title` control displays a text with a higher semantical meaning. It should be used to introduce upcoming content and is divided into the different HTML5 title styles. The following properties are important:

- `text`
  The text of the title.
- `titleStyle`
  The HTML5 title level from H1 to H6. This will affect the styling.
- `level`
  This HTML5 title level is only for the semantic meaning.

A sample implementation can look like this:

```
<Title text="This is a title" titleStyle="H1"/>
```

This will produce the output shown in Figure 5.2.

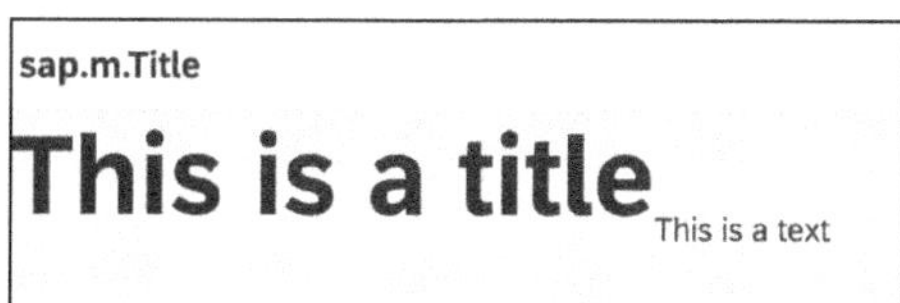

Figure 5.2 The sap.m.Title Control Next to a sap.m.Text Control for Comparison

sap.m.Input

The `sap.m.Input` control is the SAPUI5 equivalent to the HTML5 input tag. It not only allows the input of text but can also provide you with suggestions that you can choose from. Like any interactive control, it can either be enabled or disabled. The following properties are important:

- **value**
  The value of the input is stored here, and will trigger the change event when the context of the input is left or Enter is pressed.
- **description**
  Adds a unit at the end of the control.
- **maxLength**
  Sets the maximum input length.
- **valueLiveUpdate**
  Triggers the liveChange event whenever a new character is put in.

The following events are important:

- **change**
  Triggered by changes of the value property.
- **liveChange**
  Triggered by changes if the valueLiveUpdate is set to true.

sap.m.InputBase

The sap.m.InputBase class is the base class for input controls, including the ones in this chapter. It manages the property value, which holds input field values. The InputBase also provides the change event, which is triggered when the value property is changed. Therefore, all input controls can use these and have a relatively similar implementation.

An example implementation might look like this:

```
<Input placeholder="Provide something" description="Unit"/>
```

This produces the result shown in Figure 5.3.

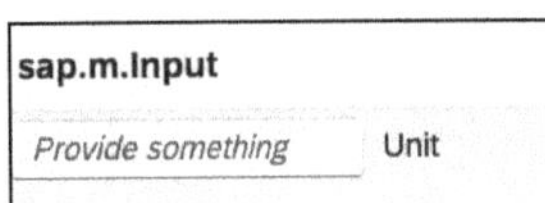

Figure 5.3 The sap.m.Input Control

If you want to add the suggestion feature, you need to set the showSuggestion property and set it to true. An additional icon to clear the input can added with the showClearIcon property. The suggestion aggregation is also suitable for aggregation binding, for example, to fetch the suggestion items from an OData service. An example is shown in Listing 5.5 that uses the sap.ui.core.Item element to add suggestion items.

```
<Input showSuggestion="true" showClearIcon="true">
      <suggestionItems>
```

```
            <core:Item key="1" text="first"/>
            <core:Item key="2" text="second"/>
            <core:Item key="3" text="third"/>
      </suggestionItems>
</Input>
```

Listing 5.5 Add Suggestions to a sap.m.Input

This adds an input with suggestion items that looks like Figure 5.4.

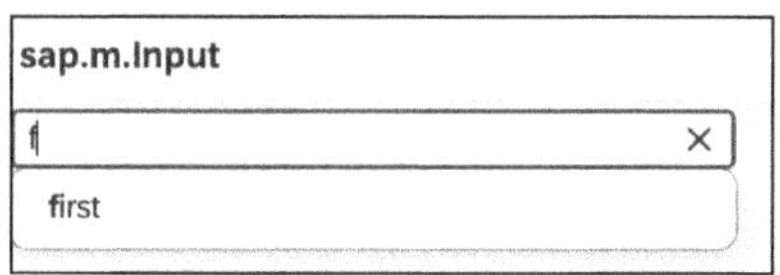

Figure 5.4 The sap.m.Input Control with Suggestion Items

sap.m.StepInput

The `StepInput` works like the input control but only for numerical values that can increment its value in steps. Therefore, you should use the following properties:

- `max`
  The max value.
- `min`
  The min value.
- `step`
  The value that should be incremented or decremented.
- `stepMode`
  Either subtract/add or multiply/divide the value with the step value.

An implementation might look like this:

```
<StepInput value="10" step="10" max="100" min="0" />
```

This produces the result as shown in Figure 5.5.

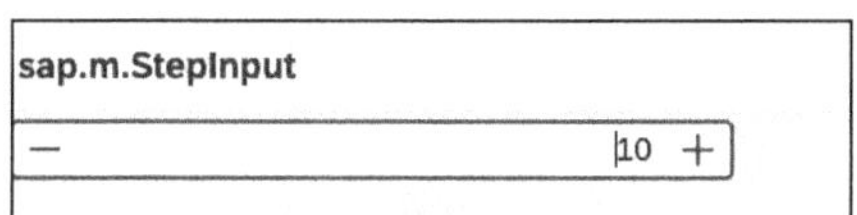

Figure 5.5 The sap.m.StepInput Control

sap.m.MaskedInput

The `MaskedInput` again works as an input control but with a predefined space in which to put data by defining an input mask such as International Bank Account Numbers (IBANs), IDs, phone numbers, or anything that has a predefined format. To declare an input mask, you can use the mask property, which is where you provide an input mask. The character C allows character input, 9 allows number input, and ~ allows all. An implementation might look like this:

```
<MaskInput mask="+43 9999 9999999999"/>
```

This allows for the creation of an input control for Austrian telephone numbers where only numbers can be entered after the +43. This result looks like Figure 5.6.

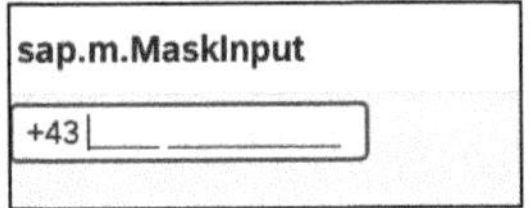

Figure 5.6 The sap.m.MaskInput Control

sap.m.TextArea

The TextArea again works like a normal input control, but it enables multiline text input, which can be controlled via the property rows. An implementation might look like this:

```
<TextArea value="I'm a really long text" rows="6"/>
```

This produces the text area shown in Figure 5.7.

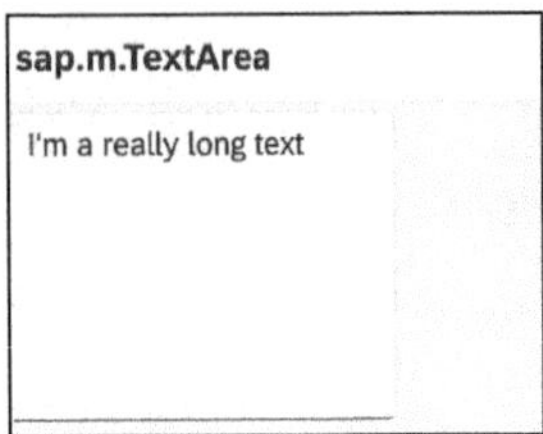

Figure 5.7 The sap.m.TextArea Control

sap.m.DatePicker

The DatePicker is the last input control we're going to look at and the most complex. It allows the input of localized dates that are stored as a string and a JavaScript date object. A popover that displays a calendar for choosing a date is provided out of the box, so users don't need to type in the date manually. By default, it formats the date based on your current locale.

This control works best if a binding is used because a binding to the value property allows you to add a sap.ui.model.type to it which includes formatting and constraint options. The most important properties of the DatePicker are listed here:

- maxDate
  The maximum date that can be chosen.

- `minDate`
  The minimum date that can be chosen.
- `displayFormatType`
  Defines the output format.
- `showCurrentDateButton`
  Displays a button in the calendar popover where you can choose today's date.
- `value`
  Holds the data as string.
- `dateValue`
  Holds the data as a JavaScript date object.

An implementation of this control can look like this:

```
<DatePicker id="datePicker"/>
```

To add some constraints, we can use TypeScript:

```
const datePicker: DatePicker = this.getView()?.byId("datePicker") as
DatePicker;
datePicker.setMinDate(new Date("2024-12-31"));
datePicker.setMaxDate(new Date("2025-01-15"));
```

This produces a date picker with a minimum and maximum date as shown in Figure 5.8.

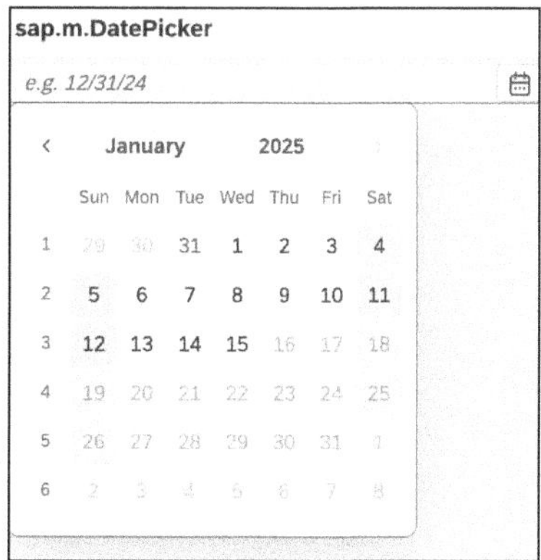

Figure 5.8 The sap.m.DatePicker Control

sap.m.CheckBox

The `CheckBox` allows the input of a binary value. It unfortunately doesn't have a label property so it should always be pared with a labeling control. The most important property here is the `selected` property, which holds the Boolean value of the `CheckBox`. Interacting with this control triggers the `select` event. An implementation can look like the following:

```
<Label text="Do you like this book?"/>
<CheckBox selected="true"/>
```

This produces a checkbox, as shown in Figure 5.9.

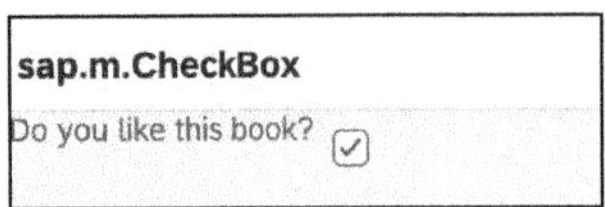

Figure 5.9 The sap.m.CheckBox Control

sap.m.Switch

The switch works like a checkbox that was optimized for mobile devices and touch input. It also adds a label that represents the current state of the `Switch`. This control has a more appealing design in comparison to the checkbox. The most important properties are listed here:

- `customTextOn`
  Displays a custom text if the `Switch` is set to `true`.
- `customTextOff`
  Displays a custom text if the `Switch` is set to `false`.
- `state`
  Defines the switch's Boolean value.

An implementation can look like this:

```
<Label text="Do you like this book?"/>
<Switch state="true" customTextOn="Sure" customTextOff="Meh"/>
```

This produces the switch shown in Figure 5.10.

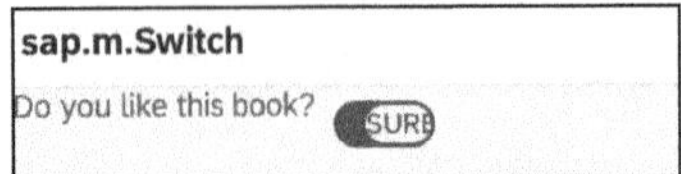

Figure 5.10 The sap.m.Switch Control

sap.m.RadioButton

A radio button again works like a checkbox, meaning it stores a Boolean value. However, it can be grouped into a `sap.m.ButtonGroup`. This makes sure that only one radio button of this group can be selected. Use it if you want to give users multiple options where only one can be chosen. The most important properties are as follows:

- `text`
  Adds a label to the radio button.
- `selected`
  Holds the Boolean value.

- `groupName`
  All radio buttons of the same group name belong to one group if not already in one.

An implementation looks like Listing 5.6.

```
<RadioButtonGroup selectedIndex="1" >
      <RadioButton text="I like it"/>
      <RadioButton text="I like it very much"/>
      <RadioButton text="I like something else"/>
</RadioButtonGroup>
```

Listing 5.6 The sap.m.RadioButton in sap.m.RadioButtonGroup

This adds a new radio button group with three radio buttons, as shown in Figure 5.11.

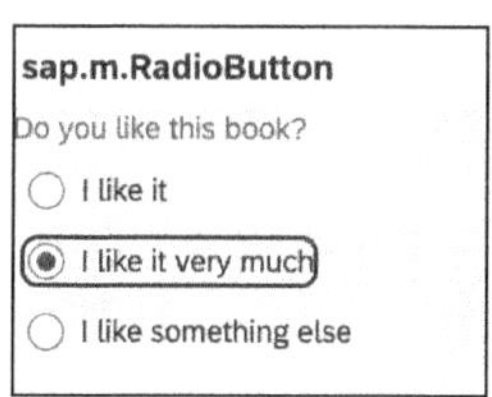

Figure 5.11 The sap.m.RadioButton Control

sap.m.RangeSlider

A range slider lets users choose a ranged numerical value. This is very useful if, for example, you want to add a price filter choice for products. Here, a user can choose in which price range to filter products. The most important properties are as follows:

- `range`
  Holds the range as a float array.
- `value`
  Holds the first value of the selected range.
- `value2`
  Holds the second value of the selected range.
- `min`
  Defines the minimum value.
- `max`
  Defines the maximum value.
- `enableTickmarks`
  Enables the visualization of tick marks.
- `step`
  Defines the number of steps required to handle the range slider.
- `showAdvancedTooltip`
  Shows the selected range value as a popover.

An implementation can look like this:

```
<RangeSlider min="0" max="200" range="0,150" showAdvancedTooltip="true"
enableTickmarks="true"/>
```

This produces a range slider that looks like Figure 5.12.

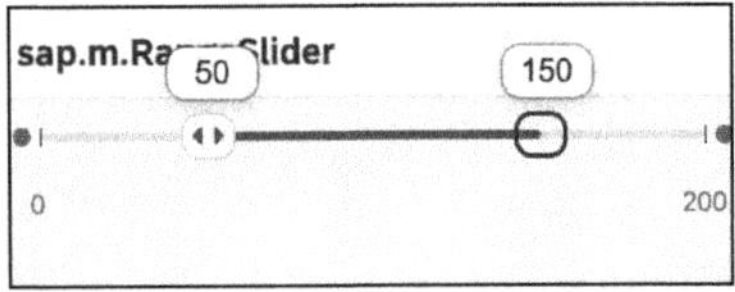

Figure 5.12 The sap.m.RangeSlider Control

sap.m.Select

The select control allows you to choose a predefined value from a list of 2–12 items. It looks like an input control that opens a dropdown when interacted with. It has an `items` aggregation where `sap.ui.core.Items` can be added to define the values of the dropdown. This aggregation is often used with data binding to display values from an OData service. The most important properties are listed here:

- `forceSelection`
  Automatically selects the first entry.
- `selectedKey`
  Holds the key of the selected item.
- `seleectedItemId`
  Holds the ID of the selected item.

An example implementation is shown in Listing 5.7.

```
<Label text="My favorite SAP Technology" />
<Select forceSelection="false" selectedKey="">
      <items>
            <core:Item key="rap" text="ABAP RESTful Application Programming
Model"/>
            <core:Item key="segw" text="SAP Gateway Service"/>
            <core:Item key="cap" text="Cloud Application Programming Model"/>
      </items>
</Select>
```

Listing 5.7 The sap.m.Select Control with Items

This adds a select that looks like Figure 5.13.

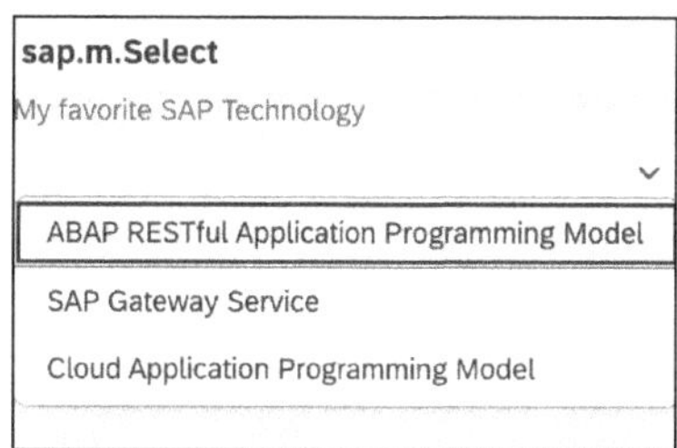

Figure 5.13 The sap.m.Select Control

sap.m.ComboBox

The combo box enhances the basic features from the select control and should be used for 13 to 200 items. It adds the functionality of input search, meaning that you can put in text that is then searched in the dropdown of the combo box. It's also possible to add new entries to the items of the combo box by typing them into the input field. An implementation can look like Listing 5.8.

```
<Label text="My favorite SAP Technology" />
<ComboBox>
       <items>
              <core:Item key="rap" text="ABAP RESTful Application Programming
Model"/>
              <core:Item key="segw" text="SAP Gateway Service"/>
              <core:Item key="cap" text="Cloud Application Programming Model"/>
       </items>
</ComboBox>
```

Listing 5.8 The sap.m.ComboBox Control

This adds a combo box, as shown in Figure 5.14.

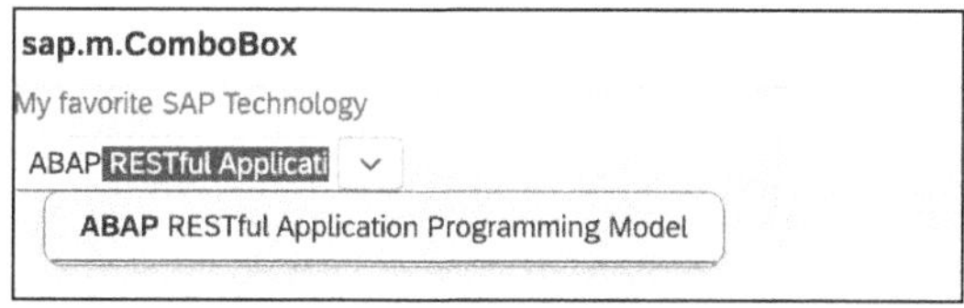

Figure 5.14 The sap.m.ComboBox Control

sap.m.Label

Labels provide a title or text to controls. They should give information about what the labeled control is about. Its usage can be seen in some examples in this chapter, such as in Section 5.3.2.

sap.m.Button

The button control is the pendant to the HTML5 button tag, enhancing it with out-of-the-box button type for different scenarios and semantic meanings. The button can display text, an icon, or both at the same time. When a button is clicked, it triggers its `press` event. Its most important properties are listed here:

- `text`
  Defines the displayed text.
- `icon`
  Defines the displayed icon.
- `buttonType`
  Defines the different button types. Some are as follows:
  - Accept
  - Back
  - Emphasized
  - Reject
  - Transparent

An implementation regarding different button types might look like Listing 5.9.

```
<Button text="I'm a default button"/>
<Button text="Accept" type="Accept"/>
<Button text="Reject" type="Reject"/>
<Button text="I'm transparent" type="Transparent"/>
<Button type="Back"/>
<Button text="I'm important" type="Emphasized"/>
<Button text="I have an icon" icon="sap-icon://education"/>
```

Listing 5.9 Different sap.m.Buttons

This produces different buttons, as shown in Figure 5.15.

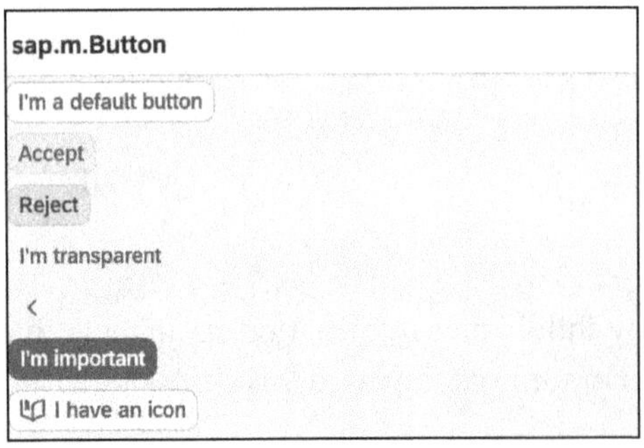

Figure 5.15 The sap.m.Button Control with Different Types

sap.m.MenuButton

With the `MenuButton`, you can create interactive menus. The `MenuButton` renders a button with a dropdown that shows menus (control `sap.m.Menu`) and submenus (control `sap.m.MenuItem`) when clicked. These menu entries can either have their own dedicated `press` event handler methods or keys. When you use keys and press an entry, the `itemSelected` event of the menu is triggered. The most important properties are listed here:

- **`buttonMode`**
  Defines whether you want to have a separator between the button text and the dropdown indicator.
- **`menuPosition`**
  Specifies the placement of the menu.

An implementation might look like Listing 5.10.

```
<MenuButton text="Im a MenuButton" menuPosition="BeginBottom" buttonMode="Split">
        <menu>
                <Menu>
                        <MenuItem text="I'm a MenuItem" icon="sap-icon://edit"/>
                        <MenuItem text="I'm a SubMenu">
                          <items>
                                <MenuItem text="I'm a MenuItem"/>
                                <MenuItem text="I'm a MenuItem too"/>
                          </items>
                        </MenuItem>
                </Menu>
        </menu>
</MenuButton>
```

Listing 5.10 The sap.m.Menu Control

This produces a new menu button that looks like Figure 5.16.

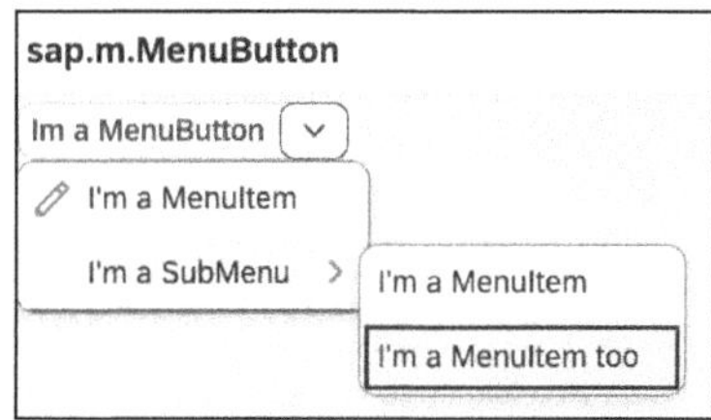

Figure 5.16 The sap.m.MenuButton Control

sap.m.SegmentedButton

The segmented button groups multiple buttons into one group where only one of them can be selected at a time. Each button can have a `key` that is then stored into the selectedKey property of the segmented button. An implementation can look like Listing 5.11.

```
<SegmentedButton selectedKey="second">
      <items>
            <SegmentedButtonItem key="first" text="I'm first"/>
                  <SegmentedButtonItem key="second" text="I'm second"/>
            <SegmentedButtonItem key="third" text="I'm third"/>
      </items>
 </SegmentedButton>
```

Listing 5.11 The sap.m.SegmentedButton Control

This produces a segmented button comprised of three buttons, as shown in Figure 5.17.

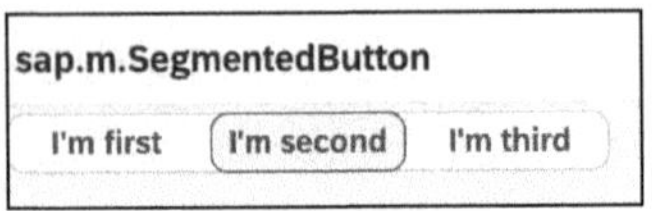

Figure 5.17 The sap.m.SegmentedButton Control

sap.m.Link

The link is the pendant to the HTML5 anchor tag. It's used to navigate in your browser, either between apps or different web pages. The most important properties are as follows:

- **href**
  Defines the target of the navigation.
- **target**
  Defines whether the page should be opened in the same window or in a new.
- **validateUrl**
  Defines whether the value of the href property should be validated.

An implementation can look like this:

```
<Link text="I open SAPUI5" href="https://ui5.sap.com" target="_blank"/>
```

This adds a new link that opens a target in a new window, as shown in Figure 5.18.

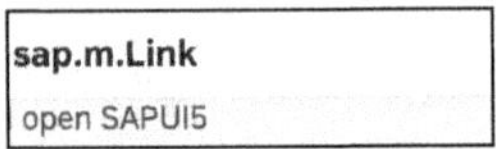

Figure 5.18 The sap.m.Link Control

sap.ui.core.Icon

With this control, you can, as the name states, display icons from the SAPUI5 icon pool (available icons can be found at *https://sapui5.hana.ondemand.com/sdk/test-resources/*

sap/m/demokit/iconExplorer/webapp/index.html). An icon is interactive, meaning it can be clicked and offers a `press` event. The icon control also supports accessibility features for screen readers. Here are some important properties:

- `activeBackgroundColor`
  Defines the background color of the icon.
- `activeColor`
  Defines the color of the icon when it's clicked.
- `alt`
  Defines an alternative text for screen readers.
- `decorative`
  Defines whether the icon should be ignored in screen readers.
- `hoverBackgroundColor`
  Defines the background color of the icon when the mouse hovers over it.
- `hoverColor`
  Defines the color of the icon when the mouse hovers over it.
- `color`
  Defines the main color.
- `src`
  Defines the source of the icon from the icon pool.

An implementation can look like this:

```
<core:Icon src="sap-icon://course-book" color="blue"/>
```

This produces a simple icon, as shown in Figure 5.19.

Figure 5.19 The sap.ui.core.Icon Control

5.3.2 Form-Based

To display data, you can just use a label and a text or input control. Sometimes, this is just enough to provide information or get users to enter something. Most of the time, however, you need to display multiple pieces of information, need to have complex input possibilities, and so on. That's where a form comes into place.

sap.ui.layout.Form

A form groups labels and controls, here called fields, into specific groups. This provides better visibility and a coherency that can't be achieved via simply displaying data. In

SAPUI5, this is realized by the `sap.ui.layout.Form` control. Let's look at an example of how a form works:

```
xmlns:f="sap.ui.layout.form"
<f:Form editable="true">
      <f:title></f:title>
      <f:layout></f:layout>
      <f:formContainers></f:formContainers>
</f:Form>
```

The form control has three main aggregations: a `title` aggregation for defining the main title of the form, a `layout` aggregation for using a certain layout, and a `formContainers` aggregation for the content of the form. In the `title` aggregation, you can add a `sap.ui.core.Title` control like this:

```
<f:title><core:Title text="Book Data"/></f:title>
```

The `formContainers` aggregation then holds your labels and fields as shown in Listing 5.12.

```
<f:formContainers>
       <f:FormContainer>
              <f:formElements>
                     <f:FormElement label="Book name">
                          <f:fields>
                                 <Input placeholder="Book name"/>
                          </f:fields>
                   </f:FormElement>
                   <f:FormElement label="ISBN">
                          <f:fields>
                                 <Input placeholder="ISBN"/>
                          </f:fields>
                   </f:FormElement>
                   <f:FormElement label="Author">
                          <f:fields>
                                 <Input placeholder="First name"/>
                                 <Input placeholder="Last name"/>
                          </f:fields>
                     </f:FormElement>
              </f:formElements>
       </f:FormContainer>
</f:formContainers>
```

Listing 5.12 A Simple Form Container

Here, we first need a `FormContainer` to hold our form data. Each label and field combination is then encapsulated with the `FormElement` control. The `FormElement` holds the label, and the fields go into the `fields` aggregation, where you can, for example, add an `Input` control to create an input form.

In the layout aggregation, you need to add a layout definition. This defines how the labels and fields are placed and adds responive features. There are two main layouts:

- **`sap.ui.layout.form.ColumnLayout`**
  The column layout divides the form into columns. Each cluster of labels and fields is placed into a column. This also adds responive features, as it allows you to define how many columns you want to display on extra large, large, medium, and small displays.
- **`sap.ui.layout.form.ResponsiveGridLayout`**
  The `ResponsiveGridLayout` uses the `sap.ui.layout.Grid` control to add a grid with 12 columns to your form. Each label and fields combination is placed into a row of this grid, where each label has a certain span of columns and the fields have a certain span of columns. You have separate properties to define how many columns a label takes and how many columns the fields take on extra large, large, medium, and small displays. You can also define an empty span, meaning that you leave certain columns empty on different screen sizes for better placement and visibility.

So let's first use the `ColumnLayout`, as follows:

```
<f:layout>
      <f:ColumnLayout columnsM="1" columnsL="2" columnsXL="2"/>
</f:layout>
```

This sets the `ColumnLayout` to be the layout of our form, which displays two columns on large (Figure 5.20) and extra large displays and one column on medium (Figure 5.21) displays.

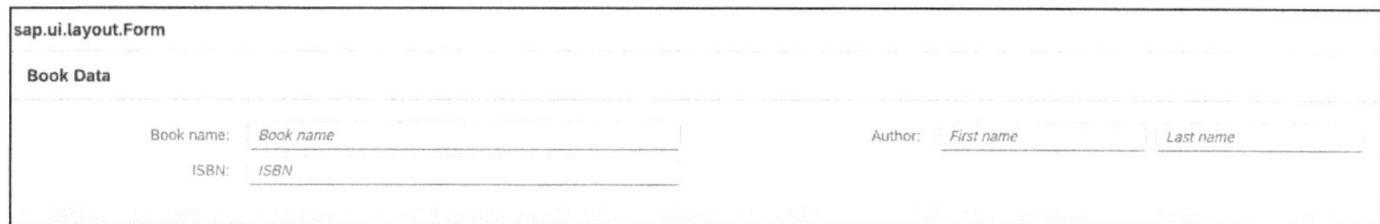

Figure 5.20 The ColumnLayout on Large Displays

Figure 5.21 The ColumnLayout on Medium Displays

Now replace the `ColumnLayout` with a `ResponsiveGridLayout`, as follows:

```
<f:ResponsiveGridLayout labelSpanXL="3" labelSpanL="3" labelSpanM="3"
labelSpanS="12" adjustLabelSpan="false" emptySpanXL="4" emptySpanL="4"
emptySpanM="4" emptySpanS="0" columnsXL="1" columnsL="1" columnsM="1"
singleContainerFullSize="false"/>
```

This defines a grid that offers responsive features with both a three-column-wide label span and a four-column empty span on large (Figure 5.22) and medium displays (Figure 5.23), leaving the remaining five columns for the fields.

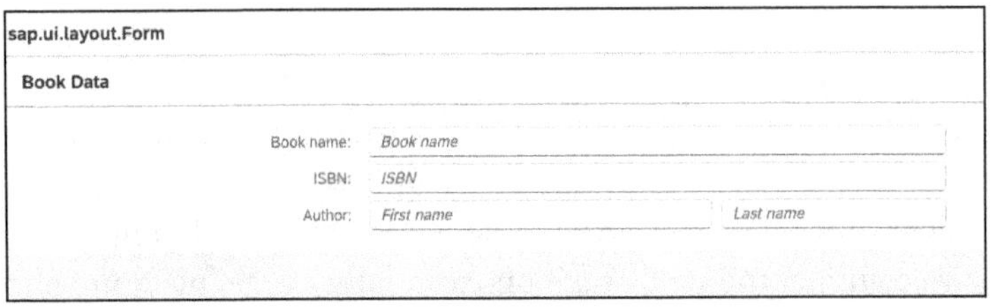

Figure 5.22 The ResponsiveGridLayout on Large Displays

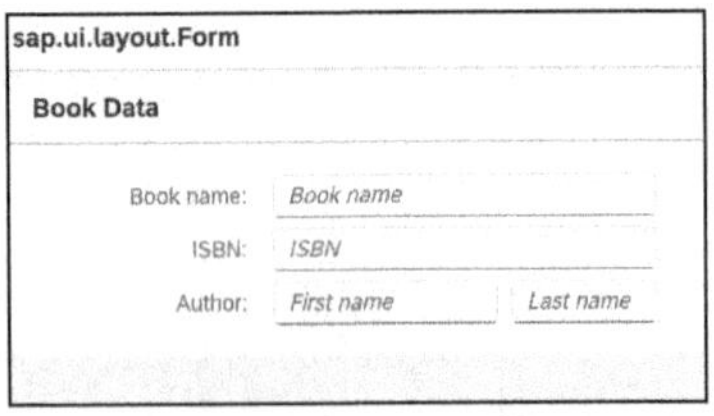

Figure 5.23 The ResponsiveGridLayout on Medium Displays

Maybe you've noticed that the input fields for the **First name** and the **Last name** aren't the same size. If you want to redefine the behavior of how a field should be placed and how many columns it should take, you can use the `sap.ui.layout.GridData` element. With this element, you can specify how a control should behave in a form when using the `ResponsiveGridLayout`. Let's try this, as shown in Listing 5.13.

```
<Input placeholder="First name">
        <layoutData>
                <l:GridData span="XL2 L2 M2 S12" />
        </layoutData>
</Input>
<Input placeholder="Last name">
        <layoutData>
                <l:GridData span="XL2 L2 M2 S12" />
        </layoutData>
</Input>
```

Listing 5.13 Adding the GridData Element to Fields

The GridData element can be added to controls that have the layoutData aggregation. All controls that extend the sap.ui.core.Control class include this aggregation. In this example, we added the GridData element to both input fields. Using the span properties of the GridData element, we defined that the input fields should take two columns on all device sizes exept small displays. As we have two input fields spanning both columns, we take a total of four columns, leaving one empty. This produces the result shown in Figure 5.24.

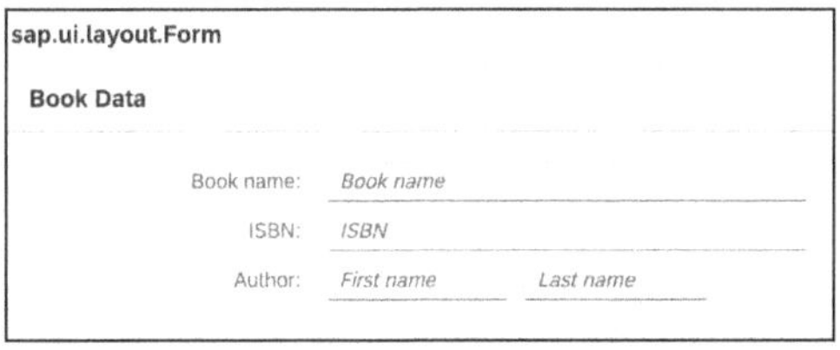

Figure 5.24 The GridData Element in Action

It's also possible to add multiple form containers to a form. This adds new forms to a form with a title for clustering your data, where each form container is basically a new section. Let's try this by adding a second form container to our form's formContainers aggregation, as shown in Listing 5.14.

```
<f:FormContainer title="General Information">
        ...
</f:FormContainer>
<f:FormContainer title="Purchase Information">
        <f:formElements>
                <f:FormElement label="Price">
                     <f:fields>
                             <Input placeholder="Price" type="Number" description=
"€"/>
                     </f:fields>
</f:FormElement>
<f:FormElement label="Purchasing Date">
                     <f:fields>
                             <DatePicker />
                     </f:fields>
                </f:FormElement>
        </f:formElements>
</f:FormContainer>
```

Listing 5.14 Adding a Second FormContainer

We start by adding a title to our first form container. Then, we add a second one, including form elements for purchasing information. This produces two sections that look like Figure 5.25.

Figure 5.25 Two FormContainers in a Form

sap.ui.form.SimpleForm

The simple form simplifies the creation of forms by eliminating the need to declare form containers and form elements. An implementation of our form as a simple form can look like Listing 5.15.

```
<f:SimpleForm editable="true" labelSpanXL="3" labelSpanL="3" labelSpanM="3"
labelSpanS="12" adjustLabelSpan="false" emptySpanXL="4" emptySpanL="4"
emptySpanM="4" emptySpanS="0" columnsXL="1" columnsL="1" columnsM="1"
singleContainerFullSize="false" title="Book Data">
      <core:Title text="General Information"/>
      <Label text="Book name"/>
      <Input placeholder="Book name" />
      <Label text="ISBN"/>
      <Input placeholder="ISBN" />
      <Label text="Author"/>
      <Input placeholder="First name">
          <layoutData>
                <l:GridData span="XL2 L2 M2 S12" />
          </layoutData>
      </Input>
<Input placeholder="Last name">
           <layoutData>
                <l:GridData span="XL2 L2 M2 S12" />
           </layoutData>
      </Input>
      <core:Title text="Purchasing Information"/>
      <Label text="Price"/>
      <Input placeholder="Price" type="Number" description="€"/>
      <Label text="Purchasing Date"/>
      <DatePicker />
</f:SimpleForm>
```

Listing 5.15 The SimpleForm Control

As you can see, there's no need to add form containers and form elements. Each `sap.ui.core.Title` control adds a new form container. Each label definition adds a new form element that includes all the controls until the next label definition appears. In addition, the layout definition is done as properties of the simple form. This produces the result in Figure 5.26, looking the same as in the previous form example but with a simplified development process.

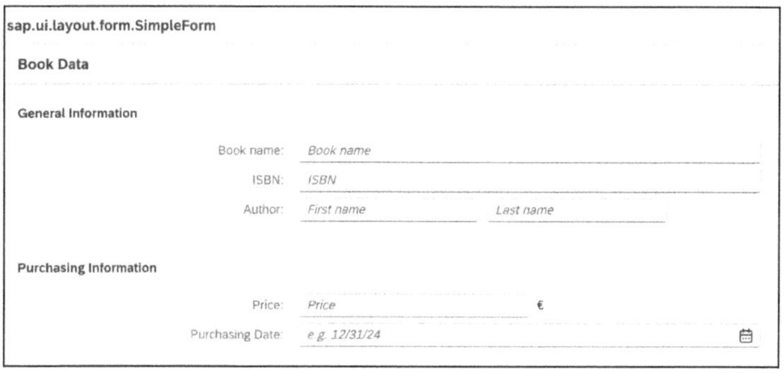

Figure 5.26 The SimpleForm Control: Simplified

5.3.3 List-Based

List-based controls display several items in a normally vertical layout that is scrollable. Often, they are combined with a list or aggregation binding to display data from an OData services entity set. In SAPUI5, you have three basic types of list controls:

- **List**
  A list displays data in a very homogenous way. You don't want to display a lot of information about your data but the most important should be visible. You can either use predefined item templates such as `sap.m.StandardListItem` or `sap.m.DisplayListItem`, or your own layout.
- **Table**
  A table displays data in rows and columns. This should be used when you want to display a lot of structured information. Tables can either be very responsible such as the `sap.m.Table`, provide features to work with a lot of columns such as the `sap.ui.table.Table`, or provide analytical functionalities such as the `sap.ui.table.AnalyticalTable`.
- **Tree**
  A tree displays hierarchical data that is structured into nodes with underlying subnodes. This enables you to drill down into information. Trees can either be displayed as a list or a table.

sap.m.List

The `List` is the first list-based control we're going to investigate. It serves mainly as a container for different list item layouts. As a container, it provides aggregation binding, lazy loading, and header and footer contents. The most important properties are as follows:

- `headerText`
  Defines the title of the list.
- `footerText`
  Defines the label in the footer of the list.
- `mode`
  Defines the mode of the list, meaning its selection mode and how items of this list can be selected.
- `growing`
  Enables lazy loading of items, meaning that only a certain amount of list items is loaded.
- `growingThreshold`
  Defines the certain amount of list items to be loaded and displayed.
- `growingScrollToLoad`
  Automatically loads the next set of items when you scroll to the bottom of the list.

Lazy Loading with the Growing Property

Lists and tables support growing functionalities on their aggregation binding of items, which is also called *pagination*. This was added to improve the performance of applications by reducing the load on services and rendering time.

The client-side JSON model automatically provides this feature. When using server-side OData models though, this behavior needs to be implemented on the server side. The OData protocol provides two pagination request parameters for this:

- `$skip`
  Defines the number of items that should be skipped.
- `$top`
  Defines the number of items that should be fetched after the skipped items.

These parameters are automatically added to OData requests that are sent when using an OData model as the source of an aggregation binding. It's always recommended to use these pagination parameters in combination with sorting parameters and the count parameter to retrieve the total amount of entries that the entity set provides.

Let's implement a list. To use the growing functionality and aggregation binding, we need to define a JSON model somewhere in our app that holds data (see Listing 5.16).

```
const model: JSONModel = new JSONModel({
            books: [{
                name: "Call of Cthulhu",
                author: "H. P. Lovecraft",
                price: 25.00,
                priceUnit: "€",
                releaseYear: 1928,
                inStock: true
            },{
                name: "W40K: Rise of Horus",
                author: "Dan Abnett",
                price: 12.99,
                priceUnit: "€",
                releaseYear: 2006,
                inStock: true
            },{
                name: "Catilinarian orations",
                author: "Marcus Tullius Cicero",
                price: 18.99,
                priceUnit: "€",
                releaseYear: -63,
                inStock: false
            },{
                name: "The Brothers Karamazov",
                author: "Fjodor Dostojewski",
                price: 10.66,
                priceUnit: "€",
                releaseYear: 1880,
                inStock: true
            },]
        });

        this.getView()?.setModel(model);
```

Listing 5.16 Set Book Data as a Model

Here, we set a simple JSON model holding book data to our view. Now, we can add a simple list in a view as in Listing 5.17. First, we start with a `sap.m.StandardListItem`. The `StandardListItem` is the most basic list item, offering a `title`, `description`, `info`, and `infoState` that you can define.

```
<List items="{/books}" headerText="Books">
                <items>
                    <StandardListItem
                        title="{name}"
```

```
                    description="{author}"
                    info="{= ${inStock} ? 'In Stock' : 'Not in Stock'}"
                    infoState="{= ${inStock} ? 'Success' : 'Error'}"
                />
            </items>
        </List>
```

Listing 5.17 A List with a StandardListItem

This produces a simple list, as shown in Figure 5.27.

sap.m.List	
Books	
Call of Cthulhu H. P. Lovecraft	In Stock
W40K: Rise of Horus Dan Abnett	In Stock
Catilinarian orations Marcus Tullius Cicero	Not in Stock
The Brothers Karamazov Fjodor Dostojewski	In Stock

Figure 5.27 A List with a StandardListItem

Next, we look at a list with an ObjectListItem, as shown in Listing 5.18, which offers more ways to display data by adding attributes to the list item. This is really neat when you want to display object data, as the name states.

```
<List items="{/books}" headerText="Books">
                <items>
                    <ObjectListItem
                        title="{name}"
                        number="{price}"
                        numberUnit="{priceUnit}">
                        <firstStatus>
                            <ObjectStatus
                                text="{= ${inStock} ? 'In Stock' : 'Not in
Stock'}"
                                state="{= ${inStock} ? 'Success' : 'Error'}"/>
                        </firstStatus>
                        <ObjectAttribute title="Author" text="{author}" />
                        <ObjectAttribute title="Release Year" text=
"{releaseYear}" />
```

```
                </ObjectListItem>
            </items>
        </List>
```

Listing 5.18 A List with an ObjectListItem

This produces the result shown in Figure 5.28.

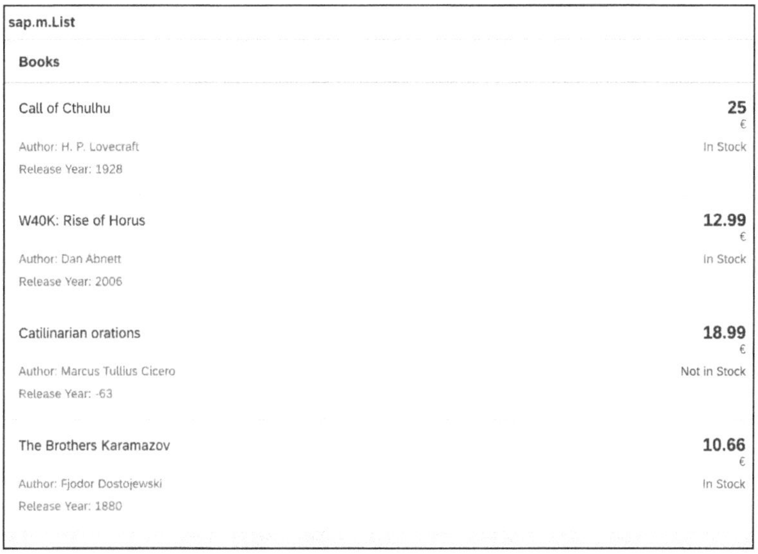

sap.m.List

Books

Call of Cthulhu	**25** €
Author: H. P. Lovecraft	In Stock
Release Year: 1928	
W40K: Rise of Horus	**12.99** €
Author: Dan Abnett	In Stock
Release Year: 2006	
Catilinarian orations	**18.99** €
Author: Marcus Tullius Cicero	Not in Stock
Release Year: -63	
The Brothers Karamazov	**10.66** €
Author: Fjodor Dostojewski	In Stock
Release Year: 1880	

Figure 5.28 A List with an ObjectListItem

There are multiple additional list items that you can choose for your use case, but these two should give you a basic set to play around with.

sap.m.Table

A `sap.m.Table`, also called a *responsive table*, displays complex data structured into columns and rows. It offers responsive features by either setting certain columns invisible on small screens or moving them into a separate row when the horizontal space isn't sufficient to display every column. The most important properties of the table are listed here:

- `headerText`
  Defines the title of the table.
- `mode`
  Defines the selection mode of the table such as single select or multiselect.

We start by creating a basic table definition:

```
<Table headerText="Books" items="{/books}">
<columns></columns>
```

```
<items></items>
</Table>
```

The table mainly has two aggregations: an item aggregation holding all the rows and a column aggregation used to define the dedicated columns. In our case, we want to display each property of our book JSON model as its own column (see Listing 5.19).

```
<columns>
                    <Column>
                        <Text text="Name"/>
                    </Column>
                    <Column>
                        <Text text="Author"/>
                    </Column>
                    <Column>
                        <Text text="Price"/>
                    </Column>
                    <Column>
                        <Text text="Release"/>
                    </Column>
                    <Column>
                        <Text text="Availability"/>
                    </Column>
                </columns>
```

Listing 5.19 Adding Columns to a Table

We add multiple `sap.m.Column` controls to the `column` aggregation of the table with each column containing a `sap.m.Text` control for the column header text. Then, we need to add a template for our `items` aggregation. Typically, you add a `sap.m.ColumnListItem` as a template. The `ColumnListItem` needs one control per column in its `cells` aggregation, as shown in Listing 5.20.

```
<ColumnListItem >
                        <cells>
                            <ObjectIdentifier title="{name}"/>
                            <Text text="{author}"/>
                            <ObjectNumber number="{price}" unit="{priceUnit}"/>
                            <Text text="{releaseYear}"/>
                            <ObjectStatus text="{= ${inStock} ? 'In Stock' :
'Not in Stock'}" state="{= ${inStock} ? 'Success' : 'Error'}" />
                        </cells>
                    </ColumnListItem>
```

Listing 5.20 Adding Items to a Table

Here, we add one control per column. We use the `sap.m.ObjectIdentifier` as our first cell because the first cell should contain identification information such as the name of the book. Then, we add a `sap.m.Text` as our second cell, a `sap.m.ObjectNumber` for the pricing information, a `sap.m.Text` for the release year, and a `sap.m.ObjectStatus` for status information. This produces a table that looks like Figure 5.29.

sap.m.Table

Books

Name	Author	Price	Release	Availability
Call of Cthulhu	H. P. Lovecraft	**25** €	1928	In Stock
W40K: Rise of Horus	Dan Abnett	**12.99** €	2006	In Stock
Catilinarian orations	Marcus Tullius Cicero	**18.99** €	-63	Not in Stock
The Brothers Karamazov	Fjodor Dostojewski	**10.66** €	1880	In Stock

Figure 5.29 The sap.m.Table Control

Let's enhance our table with some responsive features. This can be done at the column control level, where the property `minScreenWith` and `demandPopin` are available. `minScreenWith` declares up to which device size a column should be hidden. If this is combined with `demandPopin`, the column isn't hidden but moved to the next row (see Listing 5.21).

```
<columns>
                <Column>
                    <Text text="Name"/>
                </Column>
                <Column>
                    <Text text="Author"/>
                </Column>
                <Column demandPopin="true" minScreenWidth="Desktop">
                    <Text text="Price"/>
                </Column>
                <Column minScreenWidth="Tablet">
                    <Text text="Release"/>
                </Column>
                <Column>
                    <Text text="Availability"/>
                </Column>
            </columns>
```

Listing 5.21 Adding Responsive Features to the Table

If we make our screen tinier to, say, a phone size, the column **Release** is hidden, and the column **Price** gets moved to a new row, as shown in Figure 5.30.

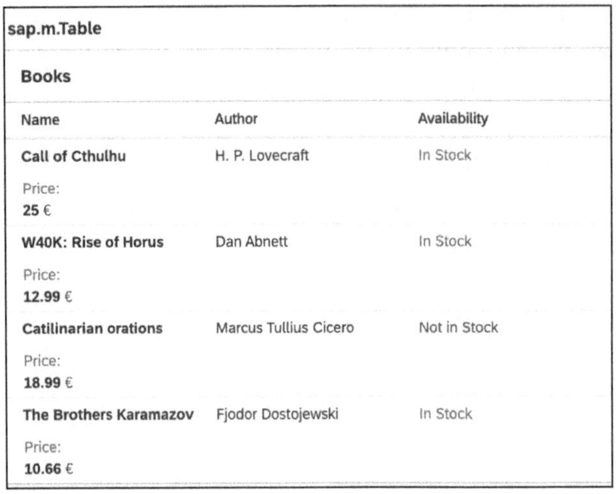

sap.m.Table		
Books		
Name	Author	Availability
Call of Cthulhu Price: **25** €	H. P. Lovecraft	In Stock
W40K: Rise of Horus Price: **12.99** €	Dan Abnett	In Stock
Catilinarian orations Price: **18.99** €	Marcus Tullius Cicero	Not in Stock
The Brothers Karamazov Price: **10.66** €	Fjodor Dostojewski	In Stock

Figure 5.30 The Table on a Small Device

sap.ui.table.Table

The `sap.ui.table.Table`, also called a *grid table*, offers displaying large datasets with better support for desktop and tablet devices. The grid table adds a horizontal scroll functionality to the table, making it possible to display more columns than the screen width can hold. It also works by defining columns that directly include the cell definition. The most important properties are listed here:

- `fixedColumnCount`
  Defines how many columns on the left side should be fixed when scrolling.
- `scrollThreshold/threshold`
  Defines how many rows should be loaded when scrolling.
- `selectionMode`
  Defines the selection mode, such as multiselect.
- `visibleRowCountMode`
  Defines whether the height of the table should be adjusted to fill the screen.
- `visibleRowCound`
  Defines how many rows should be displayed.

We can try this by creating the same table as from the responsive table example but now with a grid table, as shown in Listing 5.22.

```
xmlns:table="sap.ui.table"
<table:Table
                rows="{/books}">
                <table:extension>
                    <OverflowToolbar style="Clear">
                        <Title text="Books"/>
```

```
                </OverflowToolbar>
            </table:extension>
            <table:columns>
                <table:Column width="20rem">
                    <Label text="Name" />
                    <table:template>
                        <ObjectIdentifier title="{name}"/>
                    </table:template>
                </table:Column>
                <table:Column width="20rem">
                    <Label text="Author" />
                    <table:template>
                        <Text text="{author}"/>
                    </table:template>
                </table:Column>
                <table:Column width="10rem">
                    <Label text="Price" />
                    <table:template>
                        <ObjectNumber number="{price}" unit="{priceUnit}"/>
                    </table:template>
                </table:Column>
                <table:Column width="10rem">
                    <Label text="Release" />
                    <table:template>
                        <Text text="{releaseYear}"/>
                    </table:template>
                </table:Column>
                <table:Column width="20rem">
                    <Label text="Availability" />
                    <table:template>
                        <ObjectStatus text="{= ${inStock} ? 'In Stock' :
'Not in Stock'}" state="{= ${inStock} ? 'Success' : 'Error'}" />
                    </table:template>
                </table:Column>
            </table:columns>
        </table:Table>
```

Listing 5.22 Defining a New GridTable

The grid table defines its own columns that can have a fixed width. Each column includes a label that defines its title and a cell template, stating which data you want to display in which form in the cell. This produces the table shown in Figure 5.31.

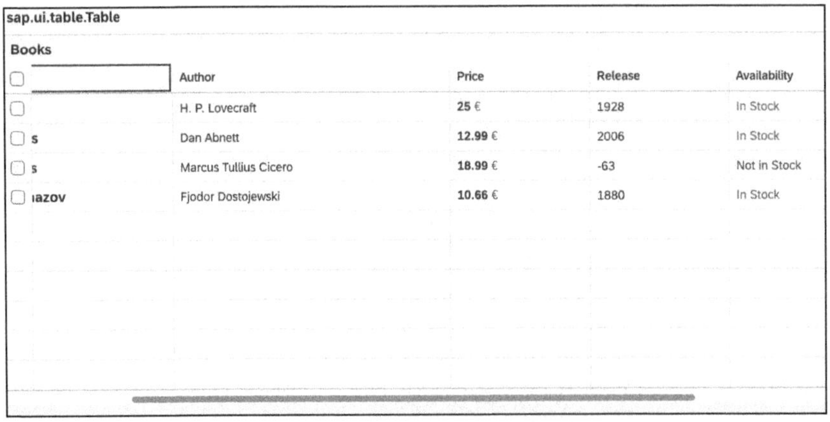

sap.ui.table.Table

Books

	Author	Price	Release	Availability
	H. P. Lovecraft	25 €	1928	In Stock
s	Dan Abnett	12.99 €	2006	In Stock
s	Marcus Tullius Cicero	18.99 €	-63	Not in Stock
ıazov	Fjodor Dostojewski	10.66 €	1880	In Stock

Figure 5.31 The sap.ui.table.Table Control

5.3.4 Miscellaneous

We also want to show you a few more controls that we think are quite handy and useful.

sap.m.OverflowToolbar

The first of these useful controls is the overflow toolbar. It's similar to a normal sap.m.Toolbar, meaning it provides a horizontal layout that typically holds actions. Most SAPUI5 controls offer toolbars out of the box or dedicated aggregations where custom toolbars can be added. The overflow toolbar adds responsive features to a toolbar by moving items to a separate popover menu when the screen space isn't sufficient to display them all. Let's try this with a simple example, as shown in Listing 5.23.

```
<OverflowToolbar width="100%">
                <Title text="I'm a Title"/>
                <ToolbarSpacer />
                <Button text="Press me"/>
                <ToolbarSeparator />
                <Button text="Don't Press me" type="Negative"/>
                <Button icon="sap-icon://settings"/>
            </OverflowToolbar>
```

Listing 5.23 An OverflowToolbar with Items

Here, we add multiple items to an OverflowToolbar, such as a title, a ToolbarSpacer that places everything after it on the right side, buttons, and a ToolbarSeparator that separates the item around it with a small line. This produces a toolbar that looks like Figure 5.32 on large displays.

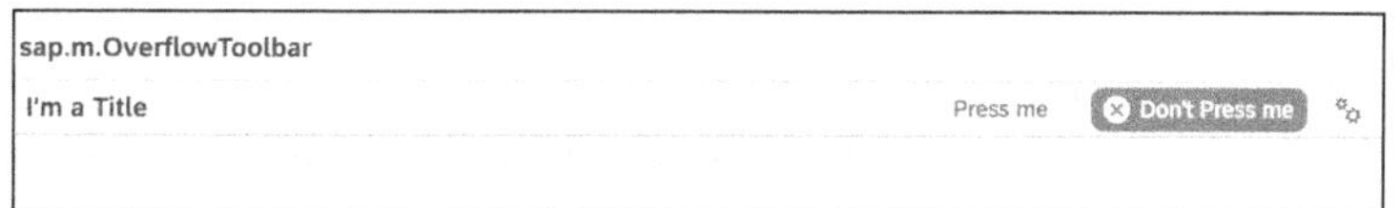

Figure 5.32 An OverflowToolbar on Large Devices

When the screen gets smaller, the rightmost items get moved to a popover, as shown in Figure 5.33.

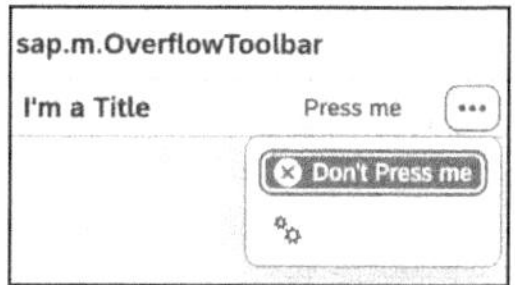

Figure 5.33 An OverflowToolbar on Small Devices

sap.m.PDFViewer

If you want to display PDF documents directly in your application, SAPUI5 offers the PDF viewer. You can either integrate this control in your view or open it in a dialog. The most important properties of this control are listed here:

- `displayType`
  Defines whether the PDF should either be embedded into a view, downloadable via a link, or opened in a new tab. This can also be defined in correspondence to your device type.
- `isTrustedSource`
  Defines whether the source of the PDF can be trusted. This means either directly embedding the PDF or opening it in a new tab outside of the application.
- `source`
  Defines the URI of the PDF.
- `showDownloadButton`
  Defines whether a download button should be shown.
- `title`
  Defines whether the title of the PDF should normally be the title of the PDF document.

Let's try this in a short example:

```
<PDFViewer title="Sample PDF" source="https://www.w3.org/WAI/ER/tests/xhtml/
testfiles/resources/pdf/dummy.pdf"/>
```

This adds an embedded PDF viewer to our application that enables us to directly download the PDF shown in Figure 5.34.

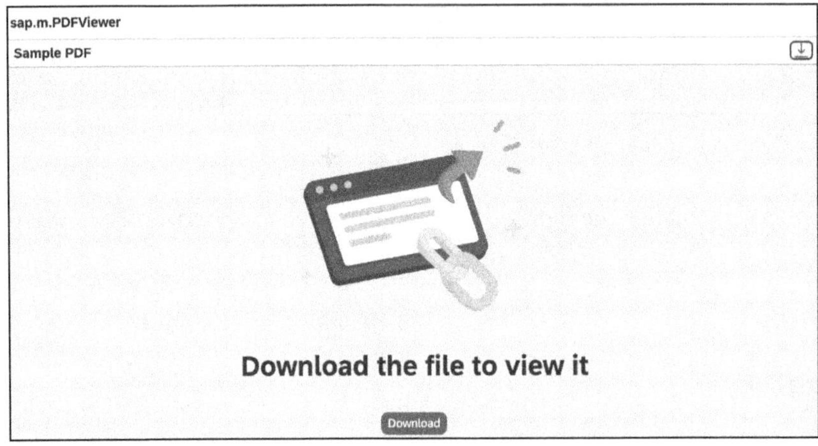

Figure 5.34 The sap.m.PDFViewer Control

sap.m.ProgressIndicator

This control is used to indicate progress by means of displaying a bar that fills up to a percentage value, which can also be animated to fill up smoothly. The most important properties are as follows:

- `displayAnimation`
  Defines whether the progress should be animated.
- `displayValue`
  Defines the displayed text.
- `percentValue`
  Defines the percentage of the fill state.
- `state`
  Defines the semantic color.

Let's try this in a short example:

```
<ProgressIndicator percentValue="45" displayValue="45 Percent" state="Warning"/>
```

This produces a neat way of displaying progress, as shown in Figure 5.35.

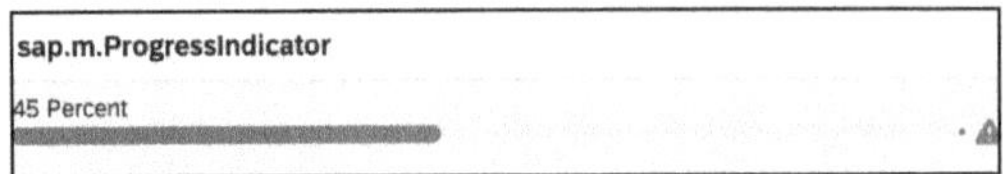

Figure 5.35 The sap.m.ProgressIndicator Control

5.4 Layouts

The next section of this chapter discusses the different options you must have to add a layout to your application. Most SAPUI5 controls offer their own layout to aggregated subcontrols. However, it's often the case that you need to manually place controls into a dedicated layout to structure them not only statically but also for different device sizes.

We first start with some general layouts for structuring your controls. After that, we investigate dedicated single-page containers. These container controls can be used as root controls, meaning that they are the upmost element in your control tree. They enable the display of whole views or multiple views at the same time.

5.4.1 General Layouts

Let's start with some general layouts for providing a layout to structure our controls on a very low level.

sap.m.FlexBox

The flex box is the SAPUI5 implementation of the CSS flexible box layout. A flexible box is a box layout where its children can grow or shrink in size and can be placed freely. The most important properties are as follows:

- **alignContent**
  Defines the controls behavior across the cross-axis.
- **alignItems**
  Defines the items behavior across the cross-axis.
- **direction**
  Defines the direction of the layout, either row or column.
- **fitContainer**
  Defines whether the control should fill its parent's space.
- **justifyContent**
  Defines the control's behavior across the main axis.

If you want to specifically define how an item inside a flex box should behave, you can add a sap.m.FlexItemData element to its layout aggregation if available. You can then add a flex box to try out some different layouts (Listing 5.24).

```
<FlexBox justifyContent="Center">
                <Button text="first"/>
                <Button text="second"/>
                <Button text="third"/>
                <Button text="fourth"/>
            </FlexBox>
            <FlexBox width="200px" direction="Column" alignItems="End">
                <Button text="first"/>
                <Button text="second"/>
```

```
        <Button text="third"/>
        <Button text="fourth"/>
    </FlexBox>
```

Listing 5.24 Different FlexBox Layouts

This adds two flex box layouts that display their items where the items of the first one are centered and the second one's items are displayed in one column that aligns them on the right side with the end of the words lined up, as shown in Figure 5.36.

Figure 5.36 The sap.m.FlexBox Control

sap.m.VBox

The vertical box (VBox) extends the flex box to make it easier to display its items vertically, so it's a vertical flexible box layout. It also shares the same properties with the flex box, meaning you can define alignments and justify its content. Following is an easy example of the VBox:

```
<VBox >
        <Button text="first"/>
        <Button text="second"/>
        <Button text="third"/>
        <Button text="fourth"/>
    </VBox>
```

This adds a simple VBox that displays its content vertically, as shown in Figure 5.37.

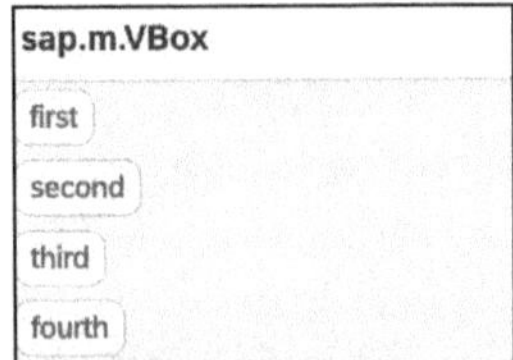

Figure 5.37 The sap.m.VBox Control

sap.m.HBox

The horizontal box (HBox) again works as an extension of the flex box, displaying its items horizontally. An easy example is shown here:

```
<HBox >
                <Button text="first"/>
                <Button text="second"/>
                <Button text="third"/>
                <Button text="fourth"/>
            </HBox>
```


Working with Box Layouts

Most of the time, layout controls are combined with one another. You often see VBoxes that include multiple HBoxes to define some sort of table-like layout, or you see an HBox that includes multiple VBoxes to add a column-based layout.

The default aggregation of these box layouts is the items aggregation that is perfectly capable of having an aggregation binding. This is great for adding dynamic content to a box.

From practical experience, we strongly recommend that you don't use the box layouts to add a form. New SAPUI5 programmers often do this, but there are dedicated controls, such as the simple form, that offer many more capabilities in displaying form data, especially in a responsive scenario.

This adds a simple horizontal layout, as shown in Figure 5.38.

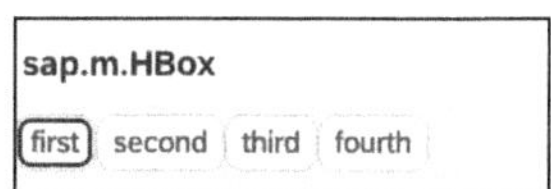

Figure 5.38 The sap.m.HBox Control

sap.m.IconTabBar

The icon tab bar defines a layout that works by displaying a tab bar that shows one tab's content like maybe your browser does. These tabs (sap.m.IconTabFilter) display an icon and a text, and they can be selected to display its content. The icon tab bar can also be used to, for example, filter a table based on a state by adding one tab for each state, such as the SAP Fiori elements worklist floorplan. Let's try this out in Listing 5.25.

```
<IconTabBar >
                <items>
                    <IconTabFilter text="First" icon="sap-icon://bookmark"
iconColor="Neutral">
                        <Text text="I'm in the first Tab"/>
                    </IconTabFilter>
                    <IconTabSeparator />
                    <IconTabFilter text="Second" icon="sap-icon://account"
iconColor="Critical">
```

```
                    <Text text="I'm in the second Tab"/>
                </IconTabFilter>
                <IconTabFilter text="Third" icon="sap-icon://work-history"
iconColor="Positive">
                    <Text text="I'm in the third Tab"/>
                </IconTabFilter>
            </items>
        </IconTabBar>
```

Listing 5.25 Adding an sap.m.IconTabBar

Here, we use an icon tab bar with three icon tab filters, each displaying text, an icon, and a semantic color for the icon. You can also use an icon tab separator to add a line between your tabs. When selecting an icon tab filter, its content is displayed. This produces the result shown in Figure 5.39.

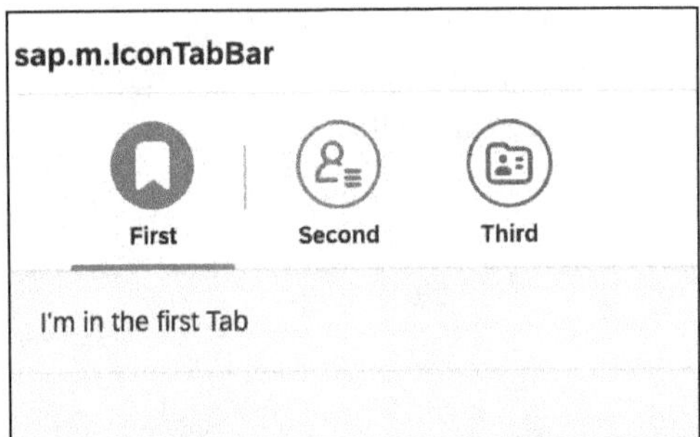

Figure 5.39 The sap.m.IconTabBar Control

sap.m.Panel

A panel is a grouping layout control that has a header and a content area. The content area can either be always visible or expanded and collapsed. The most important properties are listed here:

- **expandable**
  Defines whether the content area can be expanded and collapsed.
- **expanded**
  Defines whether the content is expanded or collapsed.
- **headerText**
  Defines the text in the header area.

An example might look like this:

```
<Panel expandable="true" expanded="true" headerText="I'm a Panel">
                <Text text="I'm the content"/>
            </Panel>
```

This adds a new panel whose content can be expanded or collapsed by clicking the arrow button on the left side of the panel's header, as shown in Figure 5.40.

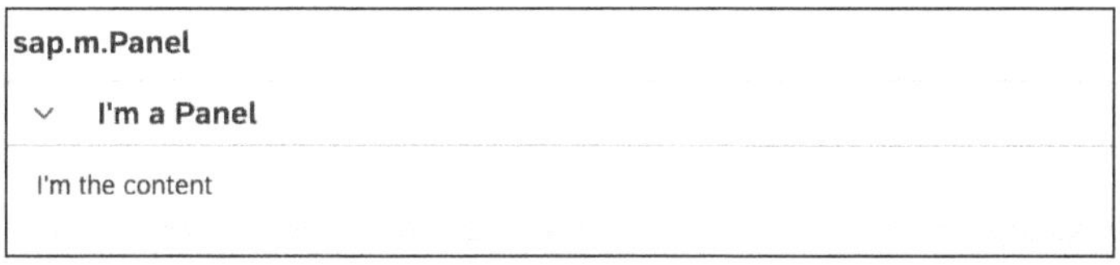

sap.m.Panel

I'm a Panel

I'm the content

Figure 5.40 The sap.m.Panel Control

sap.m.Wizard

A wizard is a more complex layout control that displays three to eight subtasks that depend on one another and should be completed in steps. You might know this from installation wizards when installing new applications on your PC. There might be different steps from choosing storage, accepting the terms and conditions to adding additional software, and so on. In SAPUI5, wizards are often integrated into the creation of new business entities, where the creation might differ from one setting to another and should be done in steps. The most important properties of the wizard are as follows:

- `enableBranching`
  This enables multibranching, where the outcome of one step might lead to different steps afterward.
- `finishButtonText`
  Defines the text of the last button to finish the wizard.
- `showNexButton`
  Defines whether the button to go to the next step should be visible. This can be set invisible when a step isn't completed.

The wizard's steps are implemented with the `sap.m.WizardStep` control. Here, the content of the step is defined, and the logic for when a step is completed, what the next step is, and if a step is valid is contained. The most important properties of a wizard step are as follows:

- `icon`
  Defines the icon of the step.
- `optional`
  Defines whether the step is optional.
- `Validated`
  Defines whether the step is validated and the button to go to the next step should be shown.
- `nextStep`
  Defines the next step (association).

- `subsequentStep`
  Defines the next steps if the `enableBranching` property of the wizard was set (association).

An example of the wizard is shown in Listing 5.26.

```
<Wizard finishButtonText="Complete" >
                <WizardStep title="Buying" icon="sap-icon://currency" id=
"firstStep" nextStep="secondStep">
                    <Text text="Buy our book"/>
                </WizardStep>
                <WizardStep title="Reading" icon="sap-icon://course-book" id=
"secondStep" nextStep="thirdStep">
                    <Text text="Read our book"/>
                </WizardStep>
                <WizardStep title="Rating" icon="sap-icon://favorite" id=
"thirdStep">
                    <VBox >
                        <Text text="Rate our book"/>
                        <RatingIndicator />
                    </VBox>
                </WizardStep>
```

Listing 5.26 A Small Wizard on How to Handle This Book

This wizard includes three steps that need to be completed after one another. Each step has a title, an icon, and an association to the next step. This produces a simple wizard that looks like Figure 5.41.

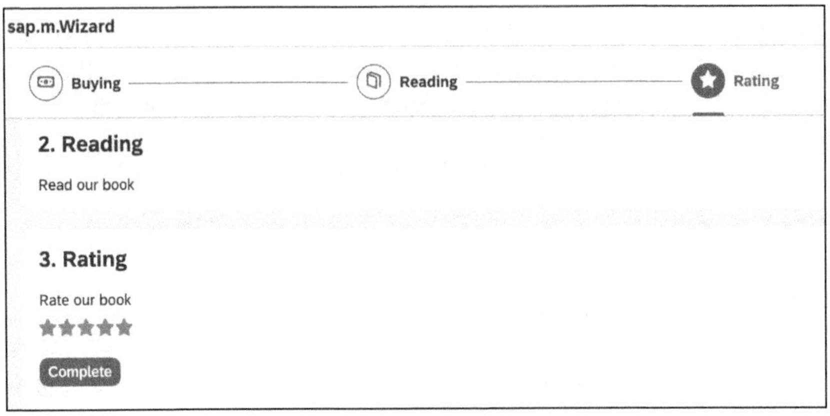

Figure 5.41 The sap.m.Wizard Control

5.4.2 App

The first container control we're going to look into is the `sap.m.App` control. As already mentioned, this is a navigation container, meaning it can be used as the root control for the SAPUI5 routing. This means, that the app control is the root of the SAPUI5 control tree, whereas all other pages or container controls are inserted into its `pages` aggregation through routing. It also automatically adds HTML5 header tags such as for adding a home icon. This makes it useful for mobile apps.

The app control is also automatically added to the root view of your application when you generate a new one and is referenced in the routing configuration of the *manifest.json* file. The most important properties are listed here:

- `autoFocus`
  Automatically focuses the current displayed page.
- `backgroundColor`
  Defines the background color.
- `defaultTransitionName`
  Defines how new pages should be made visible.
- `homeIcon`
  Defines the icon when an app is saved to the home screen of mobile devices.

Although the app control is a control, it doesn't have a visible UI and serves more as a container, so we won't provide an example. As already mentioned, always use app as your upmost control in your control tree, except when using another root container such as a split app.

5.4.3 Page

The next control, the `sap.m.Page`, is also a container control that can be used to display a single page. It can be placed inside, for example, an app control to define a single page with a dedicated purpose. Each page control represents a single page of your application, further including the contents of your page. A page is separated into three areas, a header toolbar, a content area, and a footer toolbar.

The header toolbar should include a title, an optional back-navigation button, and actions that are relevant to the whole page. It's also possible to add a subheader, mostly including things such as breadcrumbs.

The content area is scrollable and holds the content of your page. The footer area is optional and is fixed to the bottom of the content area. It mostly holds finalizing actions when, for example, building a page where you can edit something. The most important properties of the page control are as follows:

- `enableScrolling`
  Enables vertical scrolling.

- `floatingFooter`
  Defines whether the footer should always float on the bottom of the screen.
- `showFooter`
  Defines whether the footer should be shown.
- `showNavButton`
  Defines whether a back-navigation button should be shown in the header.
- `title`
  Defines the title of the page.

The most important aggregations are as follows:

- `customHeader`
  Defines a custom header toolbar.
- `subHeader`
  Defines a custom subheader toolbar.
- `content`
  Defines the content of the page.
- `footer`
  Defines the footer toolbar of the page.

Let's try out all of these aggregations to build a nice page for adding books, as follows:

```
<Page title="Add a new Book" showNavButton="true" >
    <content></content>
</Page>
```

Here, we add a new page with a title and a back-navigation button that looks like Figure 5.42.

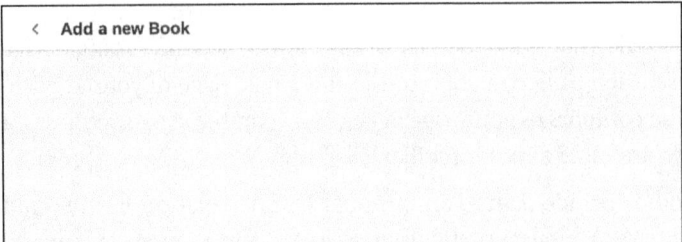

Figure 5.42 Adding a New sap.m.Page Control

Then, we want to add a custom header for a new action in our header area, as shown in Listing 5.27.

```
<Page>
        <customHeader>
            <OverflowToolbar >
```

```
                <Button type="Back"/>
                <Title text="Add a new Book"/>
                <ToolbarSpacer/>
                <Button text="Seach on Rheinwerk"/>
            </OverflowToolbar>
        </customHeader>
        <content>
        </content>
    </Page>
```

Listing 5.27 Adding a Custom Header to a Page

When you add a customHeader aggregation, you redefine the whole header, meaning you need to add your own toolbar with a button for back navigation and a title. We also add a button with text as a sample. This produces a custom toolbar that looks like Figure 5.43.

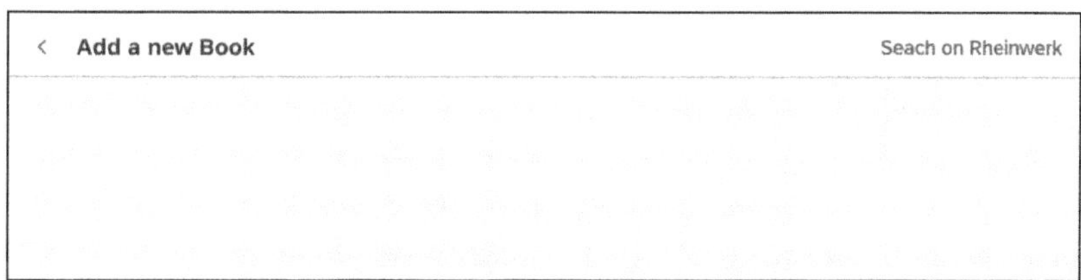

Figure 5.43 Adding a Custom Header

If we also want to create an optional subheader, we need to add it to the subHeader aggregation, as shown in Listing 5.28.

```
<Page showSubHeader="true" >
        <customHeader>
            ...
        </customHeader>
        <subHeader>
            <OverflowToolbar width="100%">
                <Breadcrumbs currentLocationText="New Book">
                <Link text="Books"/>
            </Breadcrumbs>
            </OverflowToolbar>
        </subHeader>
        <content>
        </content>
    </Page>
```

Listing 5.28 Adding a Subheader to a Page

Here, we add a new toolbar to the subheader aggregation with the `sap.m.BreadCrumb` control. This allows us to show a list of previous navigations that offer easy navigation back. Our page now looks like Figure 5.44.

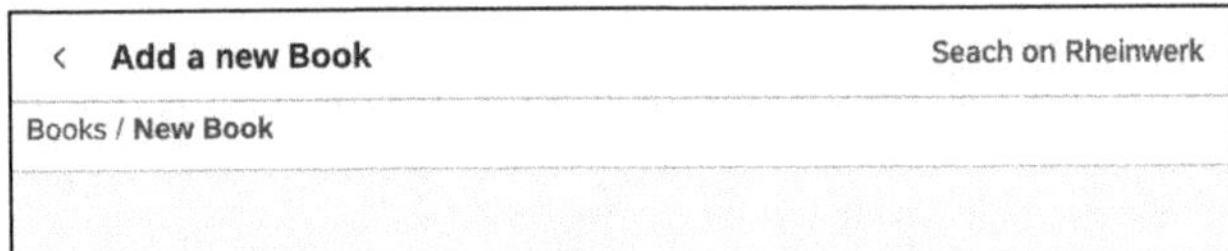

Figure 5.44 Adding a Subheader to a Page

Now we can populate the content area of our page. Here, we add a simple form to provide input for our new book, as shown in Listing 5.29.

```
<content>
            <f:SimpleForm title="Your Book">
                <Label text="Title"/>
                <Input />
                <Label text="Price"/>
                <StepInput description="€"/>
            </f:SimpleForm>
        </content>
```

Listing 5.29 Adding Content to a Page

Now, our page also has content and looks like Figure 5.45.

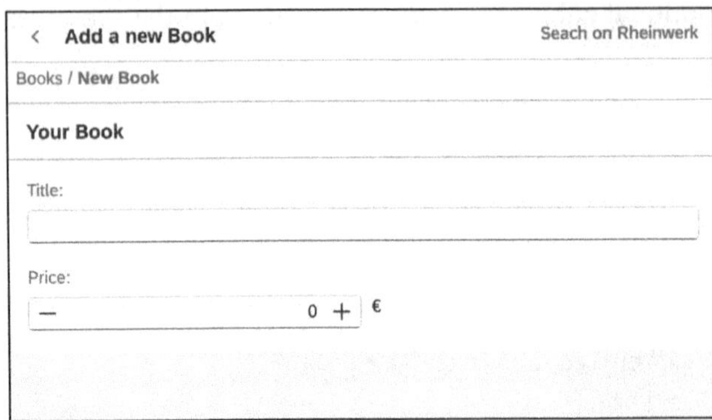

Figure 5.45 Defining the Content of a Page

The last thing to do is add a footer with finalizing actions for our page, as shown in Listing 5.30.

```
<Page showSubHeader="true" showFooter="true">
        <customHeader>
...
        </customHeader>
```

```
        <subHeader>
...
        </subHeader>
        <content>
...
        </content>
        <footer>
            <OverflowToolbar >
                <ToolbarSpacer />
                <Button text="Save" type="Accept"/>
                <Button text="Cancel" type="Reject"/>
            </OverflowToolbar>
        </footer>
    </Page>
```

Listing 5.30 Adding a Footer to a Page

We add a new toolbar with two typed buttons for saving or canceling the addition of a book. Now our page is finished and provides everything the page provides for users. The result should look like Figure 5.46.

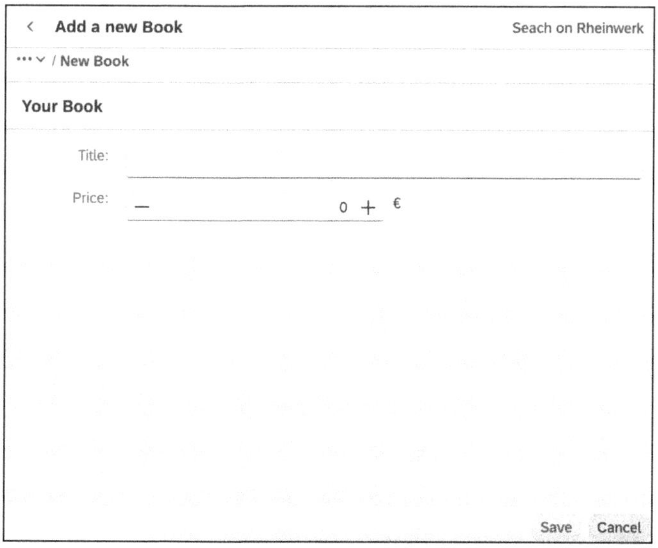

Figure 5.46 The Finished Page

5.4.4 Object Page

The next container control, `sap.uxap.ObjectPageLayout`, is the SAPUI5 control for building SAP Fiori object page layouts, meaning a UI that is tailored for displaying business objects. The object page layout shares some similarities to the page, having a header, content area, and footer area, but it enhances these areas with some features.

The collapsable header area of the object page layout is sectioned in the title and header content area. The title area, built with a sap.uxap.ObjectPageDynamicHeaderTitle, offers a title, breadcrumbs, actions, and different headings dependent on whether the header itself is collapsed or not. Below the title area goes the header content. Facets such as images, icons, graphs, or any other short information items are placed there.

The content area itself is portioned into sections, sap.uxap.ObjectPageSection, and subsections, sap.uxap.ObjectPageSubSection. Each section should contain one group of information that can be further broken down into subsections. To enable easy navigation between these sections, each section title is placed in an anchor bar that includes a dropdown for the subsections. An optional footer area is also available, mostly containing finalizing actions.

Let's try this out by first declaring a new object page layout in a view as a root control, as shown in Listing 5.31.

```
xmlns:uxap="sap.uxap"
    <uxap:ObjectPageLayout
        showTitleInHeaderContent="true"
        showFooter="true"
        upperCaseAnchorBar="false">
        <uxap:headerTitle>
        </uxap:headerTitle>

        <uxap:headerContent>
        </uxap:headerContent>

        <uxap:sections>
        </uxap:sections>

        <uxap:footer>
        </uxap:footer>
    </uxap:ObjectPageLayout>
```

Listing 5.31 Defining a New ObjectPageLayout

Here, we add a new ObjectPageLayout with all the necessary aggregations. We add the headerTitle aggregation, where we later place our ObjectPageDynamicHeaderTitle. Then, we add the headerContent aggregation, where we can place the contents of the header, such as facets. The sections aggregation will later hold our ObjectPageSections and the footer aggregation as an optional footer toolbar.

We can start filling our ObjectPageLayout by adding a new ObjectPageDynamicHeaderTitle control to the headerTitle aggregation, as shown in Listing 5.32.

```
<uxap:headerTitle>
            <uxap:ObjectPageDynamicHeaderTitle>
                <uxap:breadcrumbs>
                    <Breadcrumbs currentLocationText="New Book">
                        <Link text="Books"/>
                    </Breadcrumbs>
                </uxap:breadcrumbs>

                <uxap:expandedHeading>
                        <Title text="Your new Book" wrapping="true"/>
                </uxap:expandedHeading>

                <uxap:snappedHeading>
                        <Title text="Your new Book" wrapping="true"/>
                </uxap:snappedHeading>

                <uxap:expandedContent>
                    <Text text="Add a book"/>
                </uxap:expandedContent>

                <uxap:snappedContent>
                    <Text text="Add a book"/>
                </uxap:snappedContent>

                <uxap:snappedTitleOnMobile>
                    <Title text="Add a book"/>
                </uxap:snappedTitleOnMobile>

                <uxap:actions>
                    <Button text="Seach on Rheinwerk"/>
                </uxap:actions>
            </uxap:ObjectPageDynamicHeaderTitle>
        </uxap:headerTitle>
```

Listing 5.32 Adding an ObjectPageDynamicHeaderTitle

The ObjectPageDynamicHeaderTitle defines the title area of the header of an ObjectPageLayout. Here, you can add breadcrumbs and actions. The title of the ObjectPageLayout itself goes into the aggregations expandedHeading and snappedHeading. With these two aggregations, you can define how the title should look if the header area of the ObjectPageLayout is collapsed or expanded. This is also available for the content of the title with the expandedContent and snappedContent aggregations. Alternatively, a title for mobile devices can be added into the snappedTitleOnMobile aggregation.

We can then add content to the headerContent aggregation, as follows:

```
<uxap:headerContent>
       <FlexBox wrap="Wrap">
            <Avatar class="sapUiSmallMargin" src="sap-icon://add-coursebook"
       displaySize="L" />
       </FlexBox>
</uxap:headerContent>
```

You can put in any content you like, but keep it short, and only include relevant data, such as status information, images, prices, or anything important. In our case, we add a flex box containing a sap.m.Avatar control for displaying user images. Next, we add sections to our object page layout, as shown in Listing 5.33.

```
<uxap:sections>
            <uxap:ObjectPageSection titleUppercase="false" title="I'm a
section">
                <uxap:subSections>
                    <uxap:ObjectPageSubSection title="I'm a subsection">
                        <uxap:blocks>
                            <Text text="I'm a text"/>
                        </uxap:blocks>
                    </uxap:ObjectPageSubSection>
                    <uxap:ObjectPageSubSection title="I'm a subsection too">
                        <uxap:blocks>
                            <Text text="I'm a text too"/>
                        </uxap:blocks>
                    </uxap:ObjectPageSubSection>
                </uxap:subSections>
            </uxap:ObjectPageSection>
            <uxap:ObjectPageSection titleUppercase="false" title="I'm a second
section">
                <uxap:subSections>
                    <uxap:ObjectPageSubSection title="I'm a subsection">
                        <uxap:blocks>
                            <Text text="I'm a text"/>
                        </uxap:blocks>
                    </uxap:ObjectPageSubSection>
                    <uxap:ObjectPageSubSection title="I'm a subsection too">
                        <uxap:blocks>
                            <Text text="I'm a text too"/>
                        </uxap:blocks>
                    </uxap:ObjectPageSubSection>
                </uxap:subSections>
```

```
            </uxap:ObjectPageSection>
        </uxap:sections>
```

Listing 5.33 Add Sections to an ObjectPageLayout

An ObjectPageSection gets a title, which is placed for easy navigation between sections into an anchor bar on top of the content area. The section itself can have multiple ObjectPageSubSections, each with its own title where the content then is placed. But a section must always contain at least one subsection. The ObjectPageSubSection has a blocks aggregation where you add the content. Each block should be one type of control such as a form, a table, a list, or something else. In our case, we just add a text control for demonstration purposes. Then, we can add a footer toolbar into the footer aggregation of our ObjectPageLayout, as follows:

```
<uxap:footer>
            <OverflowToolbar >
                <ToolbarSpacer />
                <Button text="Save" type="Accept"/>
                <Button text="Cancel" type="Reject"/>
            </OverflowToolbar>
        </uxap:footer>
```

This sample here adds an overflow toolbar with two finalizing actions in the form of buttons. Now we can test our object page layout, and we should be presented with a result that looks like Figure 5.47.

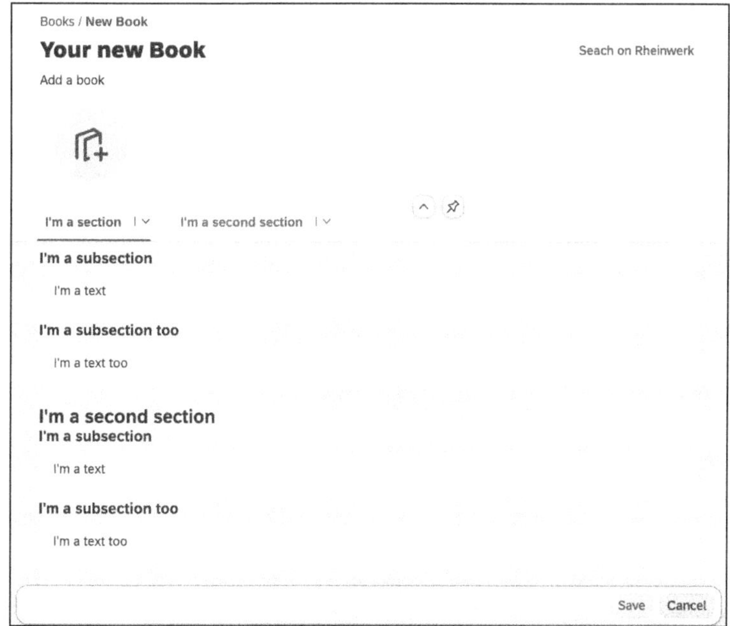

Figure 5.47 A Finished ObjectPageLayout

If you click on the title, scroll down, or click on the little up arrow icon, you can collapse the header. The screen will now look like Figure 5.48.

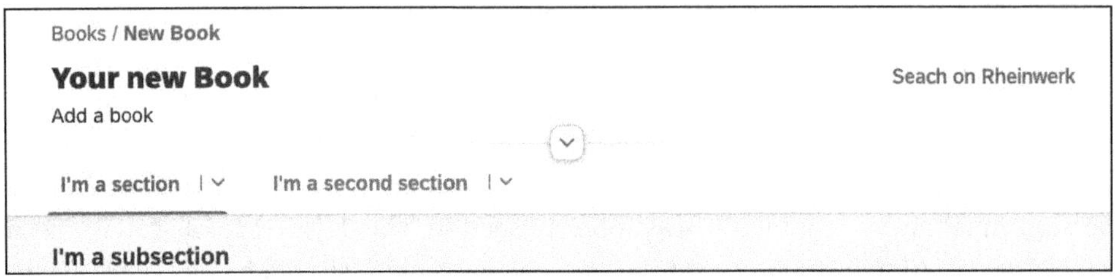

Figure 5.48 A Collapsed ObjectPageLayout Header

5.4.5 Split App

The next container control on our list is the sap.m.SplitApp control. This control extends the sap.m.SplitContainer control and splits the available screen into two fixed areas: master area and details area. The master area mostly contains lists that can be filtered and searched. The details area should then display a selected item of the master area. With this, you can enable master/detail scenarios that always display always the list and the details in the same screen without the need of backwards navigation.

Each area of the split app should include a dedicated page. It's often combined with routing to dynamically add the pages on startup and during navigation. It also supports different modes, meaning you can hide the master area when not needed or on mobile scenarios. We can try this in a short example, as shown in Listing 5.34.

```
<SplitApp mode="ShowHideMode">
        <masterPages>
            <Page title="Master Page">
                <content>
                    <Text text="I'm in the master Page" />
                </content>
            </Page>
        </masterPages>
        <detailPages>
            <Page title="Detail Page">
                <content>
                    <Text text="I'm in the detail Page" />
                </content>
            </Page>
        </detailPages>
    </SplitApp>
```

Listing 5.34 Defining a Simple SplitApp

Here, we add a new split app control and add one page into the `masterPages` and one into the `detailPages` aggregation. When you run this, you'll get the result shown in Figure 5.49.

Master Page
I'm in the master Page
Detail Page
I'm in the detail Page

Figure 5.49 The SplitApp Control

You can see that the screen has been split into two fixed-sized sections. As we've also defined the `mode` property of the split app to be `ShowHideMode`, the master page gets hidden on small screens, as shown in Figure 5.50. In addition, a **Navigation** button is added to the detail page that opens the master page within an overlay. This is only available if you add a `sap.m.Page` as the detail page, however.

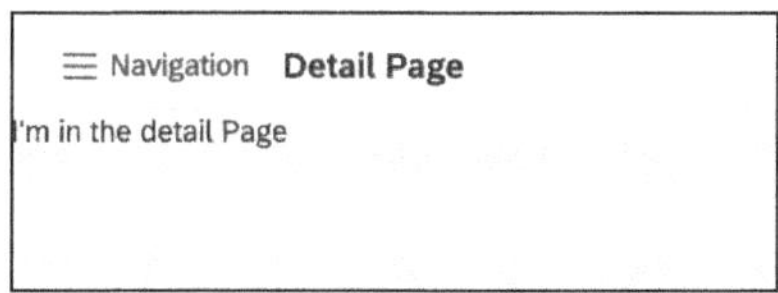

Figure 5.50 The masterPage Hidden in a SplitApp

5.4.6 Flexible Column Layout

The last layout on our list goes one notch up compared to the split app. The `sap.f.FlexibleColumnLayout` layout enhances the split app functionality by making the sections resizable and flexible, as well as by adding a third section.

`FlexibleColumnLayout` is separated into three sections for the pages, each offering a dedicated aggregation. The `beginColumnPages` aggregation holds the left section, the `midColumnPages` aggregation the middle section, and the `endColumnPages` aggregation the right section.

Which section should be expanded and shown is defined by the layout property of the `FlexibleColumnLayout` control. The following properties are available:

- **`OneColumn`**
  Only the left section is shown.
- **`MidColumnFullScreen`**
  The middle section is in full screen mode.
- **`EndColumnFullScreen`**
  The right section is in full screen mode.

- **TwoColumnsBeginExpanded**
  The left section is expanded, and the middle one is small. There's no right section.
- **TwoColumnsMidExpanded**
  The left section is small, and the middle one is expanded. There's no right section.
- **ThreeColumnsBeginExpandedEndHidden**
  The left section is expanded, the middle one is small, and the right one is hidden.
- **ThreeColumnsEndExpanded**
  The right section is expanded, and the other ones are small.
- **ThreeColumnsMidExpanded**
  The middle section is expanded, and the other ones are small.
- **ThreeColumnsMidExpandedEndHidden**
  The middle section is expanded, and the right one is hidden.

`FlexibleColumnLayout` works mostly in combination with routing, where the different views are loaded into their sections and the mode is added as a query parameter to the routing to ensure that the same layout is shown when refreshing the window. We try out a simple example by first declaring the `FlexibleCloumnLayout` as a root in a view:

```
<f:FlexibleColumnLayout>
        <f:beginColumnPages />
        <f:midColumnPages />
        <f:endColumnPages />
</f:FlexibleColumnLayout>
```

Then, we can add pages to our aggregations, as shown in Listing 5.35.

```
<f:FlexibleColumnLayout>
        <f:beginColumnPages>
            <Page title="I'm begin">
                <Text text="I'm a Text"/>
            </Page>
        </f:beginColumnPages>
        <f:midColumnPages>
            <Page title="I'm mid">
                <Text text="I'm a Text"/>
            </Page>
        </f:midColumnPages>
        <f:endColumnPages>
            <Page title="I'm end">
                <Text text="I'm a Text"/>
            </Page>
        </f:endColumnPages>
</f:FlexibleColumnLayout>
```

Listing 5.35 Adding Pages into the Aggregations

Here, we add one page into each section aggregation. The last thing now is to add a layout mode for our `FlexibleColumnLayout`, as follows:

```
<f:FlexibleColumnLayout layout="ThreeColumnsMidExpanded">
    ...
</f:FlexibleColumnLayout>
```

We try out the `ThreeColumnsMidExpanded` layout mode next. This sections our `FlexibleColumnLayout` into three sections where the middle one is expanded, as shown in Figure 5.51.

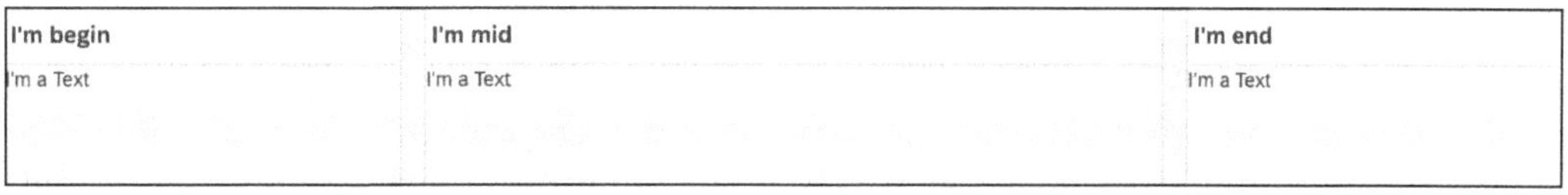

Figure 5.51 FlexibleColumnLayout with Layout Mode ThreeColumnsMidExpanded

Now, you can change the layout mode to make `FlexibleColumnLayout` fit your needs such as using layout mode `TwoColumnsBeginExpanded`, as shown in Figure 5.52.

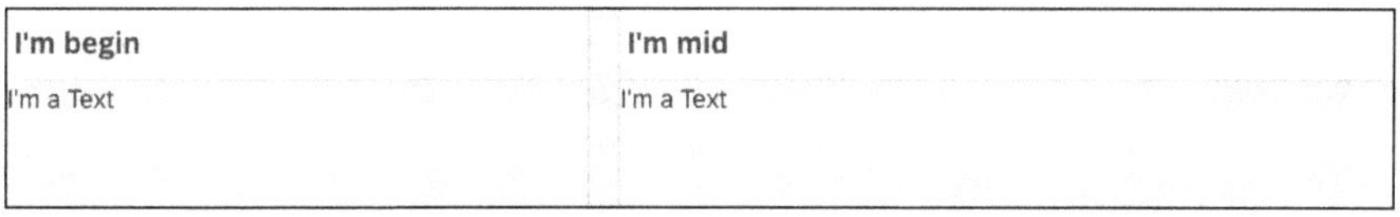

Figure 5.52 FlexibleColumnLayout with Layout Mode TwoColumnsBeginExpanded

5.5 Managed Objects

Managed objects, which we'll also discuss in Chapter 13, are the base class of all SAPUI5 controls and elements. They introduce core concepts such as state management or data binding and provide the following managed features:

- **Properties**
  Properties represent the state of a managed object and store either primitive JavaScript data types or custom SAPUI5 data types. They can be accessed via automatically added setter- and getter-methods. Properties also support data binding out of the box.

  An example of a property is the `text` property of a `sap.m.Button` control, defining the button-displayed text.

- **Aggregations**
  Aggregations store references to other managed objects, either in a one-to-one or one-to-many relationship. The aggregated managed objects are then controlled by the managed object they are placed in. Aggregations, like properties, support data binding for the dynamic managing of the aggregated managed objects.

An example of an aggregation is the `items` aggregation of a `sap.m.List`, defining the displayed list items.

- **Associations**
  Associations define the relationships between managed objects without controlling the associated managed object. Associations are made by storing the ID of the associated managed object.

  An example is the `labelFor` association of the `sap.m.Label` control, defining which control the label is labeling. This is only relevant for accessibility purposes and doesn't affect the labeled control.

- **Events**
  Events enable the communication between controls and JavaScript or TypeScript based on an observer pattern. Each event of a managed object is triggered programmatically at a specific point, which notifies all listeners that are registered to this event. When triggering an event, parameters must be added to it so they can then be consumed by the listeners.

  An example of an event is the `press` event of a `sap.m.Button`. This event is triggered when a user clicks the button, notifying all the listeners that the button has been pressed and something needs to be done.

- **Low-level APIs**
  Managed objects provide methods for accessing properties, manipulating aggregations, changing associations, and using the observer pattern for events.

  An example of this is the method `setText` of the `sap.m.Button` control, which sets a new string value for the displayed button text, re-rendering the button afterward.

If you want to work with managed objects, the good news is you already have without knowing it! As already stated, all SAPUI5 controls are managed objects at the core. Managed objects are the reason we can access the SAPUI5 controls in our controllers and manipulate them, binding OData entity sets to a table or doing something when a button is clicked.

The only time you might need to create your own managed objects is when creating custom controls (see Chapter 13) or when you want to create complex modules that serve a dedicated purpose, such as a managed object that manages the lifecycle of a custom dialog.

5.6 Summary

In this chapter, we looked into the modules of SAPUI5. Modules are reusable, on-demand, loadable JavaScript or TypeScript snippets and are the implementation of the SAPUI5 library.

The different libraries shipped out by SAPUI5 bundle everything included into SAPUI5. They include basic controls, list-based controls, form controls, and a huge variety of different UI and non-UI building blocks and classes you can use to build state-of-the-art enterprise applications.

These controls can then be placed into layout controls for placing them in a fixed or responsive layout. For defining the base layout of your application, you can use container layouts. They declare how your app should look and behave overall—from simple pages to complex flexible column layouts.

We hope to have given you a good overview of the most-used modules that you can use either for your applications or for understanding later chapters.

Chapter 6
Data Binding

Data binding plays an important role when it comes to synchronizing data between the model and the view. In this chapter, we examine the different types of binding detail.

6

Data binding is a central and powerful functionality in SAPUI5 development that makes it possible to flexibly and efficiently connect data sources with user interface (UI) elements. It's a concept that ensures the synchronization of data between the data model and the UI elements so that changes on one side (e.g., in the backend or user input) are automatically visible on the other side. This greatly facilitates the development of modern, dynamic, and responsive applications as it enables a clear separation between data logic and the presentation layer. Data binding allows developers to easily integrate data from different sources (e.g., JavaScript Object Notation [JSON] files, OData services, or XML data) into the UI without having to manually ensure that the display remains up-to-date. At the same time, the concept enables efficient management and reuse of data, as the binding ensures automated updating when the underlying data changes. Data binding in SAPUI5 supports different binding types, models, and modes to cover different application scenarios. For example, developers can connect individual properties of a UI element to a data field, bind entire lists or tables with data collections, or even define complex expressions directly in the UI. Thanks to this flexibility, data binding is one of the most important foundations for creating user-friendly and maintainable applications.

This chapter introduces basic techniques such as property binding, aggregation binding, element binding, and expression binding. First, property binding is explained, which allows UI elements to be dynamically linked to model data. Then, the aggregation binding is examined, which shows how data collections are seamlessly bound to UI controls. This is followed by an explanation of element binding, which focuses on binding individual elements from a data model to a control. Finally, the chapter provides an overview of expression binding, which enables flexible calculation of dynamic values directly in the binding expressions.

6.1 Binding Modes

In addition to the different types of bindings, SAPUI5 also offers different binding modes. These differ in terms of how the data flow between the model and view (UI) takes place. The following modes are supported:

- One-way binding
- Two-way binding
- One-time binding

The binding mode to be used depends on the requirements of the application. Flexible and efficient data management can be implemented by combining the various binding modes with other binding functionalities. The individual binding modes are described in more detail in the following subsections. Furthermore, the default binding mode depends on which type of model is used. For example, the *JSON model* has a different default binding mode than the *OData models*. Nevertheless, the binding mode to be used can be set programmatically.

6.1.1 One-Way Binding

In the *one-way binding* mode, the data flow only takes place in one direction: from the model to the view. If the data in the model changes, this change is synchronized in the view. However, a change to the data in the view has no effect on the data in the model.

One-way binding is particularly suitable when data from a model is only to be displayed without the user being able to edit it. The following scenarios could be examples of this:

- **Display of labels**
  This refers to the display of static or dynamic information, such as labels for defined input fields, which can change.
- **Data display**
  Content such as tables, lists, or diagrams that are used purely for display and can't be changed by the user via the UI, for example, values in a `ComboBox`.
- **Display of status information**
  One-way binding can also be used if error messages, success messages, or status information about an object will be displayed.

Advantages of One-Way Binding

Some advantages of one-way binding are as follows:

- **Simple and efficient**
  The implementation is lean because no changes have to be written from the view back into the model.

- **Clarity about data flow**
  As the data flow only goes in one direction—from the model to the view—it's clear and easy to understand.
- **Performance**
  As no resynchronization of data from the view to the model is required, performance is improved.

Listing 6.1 shows how the binding mode of a specific binding can be set to the one-way type.

```
<Text text="{path: '/bindingPropertyOneWay', mode: 'OneWay'}"/>
```

Listing 6.1 Binding with Binding Mode OneWay

6.1.2 Two-Way Binding

In contrast to one-way binding, the *two-way binding* mode enables a bidirectional data flow. This means that changes are not only synchronized from the model to the view but also in the opposite direction. If the user changes a value on the UI, this change also affects the data in the model.

This type of mode is particularly suitable when the UI needs to allow some form of interaction. Some use case examples include the following:

- **Interfaces with form input**
  If the user makes entries in forms on the UI, these are automatically written to the underlying model and the data is available there.
- **Provision of dynamic UI components**
  Changes to various input fields, checkboxes, dropdown menus, or sliders affect the data in the model.
- **Real-time data acquisition**
  Two-way binding is also helpful when user input needs to be immediately available for validations, calculations, or other logic.

Advantages of Two-Way Binding

Some advantages of two-way binding are as follows:

- **Interactivity and dynamics**
  The data displayed in the view and the data in the model are constantly synchronized. This has a positive effect on the user experience, as there can be no unwanted inconsistencies in the data storage.
- **Efficient development**
  There's no need to implement your own methods or logic to react to user interactions. This makes the application leaner and less complex.

- **Adaptability**
  Changes on both sides—view and model—automatically affect the other side.

Listing 6.2 shows how the binding mode of a binding can be set to the two-way type.

```
<Text text="{path: '/bindingPropertyTwoWay', mode: 'TwoWay'}"/>
```

Listing 6.2 Binding with Binding Mode TwoWay

6.1.3 One-Time Binding

Unlike with one-way or two-way binding, changes aren't synchronized multiple times with *one-time binding*. Instead, the data is only synchronized once from the model to the view—this happens when the model is initially bound. Once the model has been initially bound, no further data transfers are carried out. This type of binding mode is therefore the one with the lowest dynamics. For example, this binding mode is used for the resource model.

One-time binding is particularly suitable for scenarios in which it's ensured that data doesn't need to be synchronized during runtime. Common examples of this include the following:

- **Initial values**
  These values are used for displaying static configuration data or default values that are determined once.
- **Static content/i18n texts**
  If labels, headings, or descriptions will be displayed that can't change under any circumstances, this binding mode is the preferred variant.
- **High-performance applications**
  As there's no need to synchronize data between the view and model, this binding mode is well suited to working with large amounts of static data where synchronization isn't necessary.

Advantages of One-Time-Binding

Some advantages of one-time binding are as follows:

- **High performance**
  One-time binding reduces the overhead of constantly checking for changes in the data, which is common with two-way or one-way binding mechanisms.
- **Easy to implement**
  One-time binding is straightforward to set up. You simply bind the data once, and the value is rendered without worrying about updates.

- **Is well suited for static, nonchanging data**
  It's ideal for data that doesn't change after being initially set, such as static configurations, labels, or information that is fetched and displayed without modification.

Listing 6.3 shows how the binding mode of a binding can be set to the one-time type.

```
<Text text="{path: '/bindingPropertyOneTime', mode: 'OneTime'}"/>
```

Listing 6.3 Binding with Binding Mode OneTime

As already mentioned at the beginning, the default binding mode depends on the type of model used. Table 6.1 lists the individual models with their default mode and the supported mode.

Model	One-Way	Two-Way	One-Time	Default
JSON model	Supported	Supported	Supported	Two-way
OData model (V2 and V4)	Supported	Supported	Supported	One-way
XML model	Supported	Supported	Supported	Two-way
Resource model	Not supported	Not supported	Supported	One-time

Table 6.1 SAPUI5 Binding Modes

6.2 Property Binding

Property binding is one of the central and most frequently used forms of data binding in SAPUI5. It allows the direct linking of properties of UI elements with data from a model. This means that changes in the model are automatically applied to the UI without the need for manual updating. With property binding, numerous properties of UI elements such as texts, numbers, colors, visibility, and status can be dynamically controlled. This makes applications more flexible and data-driven, as the display of the UI is always based on the current data. At the same time, the application remains clear, as data and display are clearly separated from each other. The property binding syntax is simple but flexible. In addition to basic bindings that directly access properties in the data model, it also supports advanced use cases such as the use of formatting functions or complex expressions. Property binding also offers different modes (e.g., one-way or two-way) to adapt the data exchange to the requirements of the application.

There are two ways to apply property binding to a UI control:

- As part of the controls definition in an XML view
- Using JavaScript/TypeScript, either within the settings object in the constructor of a control or in specific scenarios using the control's `bindProperty` method

In the following sections, we'll take a closer look at how to apply property binding, and then we'll see in detail which function the SAPUI5 framework provides to realize property binding.

6.2.1 Property Binding with Named Models

Once the property binding has been defined, the property is automatically updated whenever the value of the property in the bound model changes and vice versa. For our example, let's assume the following data is available in a JSON model (see Listing 6.4) and the `JSONModel` has the name `employeeModel`.

```
let data = {
   "employee": {
      "firstName": "Max",
      "lastName": "Mustermann",
      "age": 40
   }
}
let model = new JSONModel(data);
this.getView().setModel(model, "employeeModel");
```

Listing 6.4 JSON Demo Data

If you now want to bind the data from the preceding listing directly in the view, this looks like Listing 6.5.

```
<mvc:View controllerName="" xmlns="sap.m" xmlns:mvc="sap.ui.core.mvc">
   <Input value="{employeeModel>/employee/firstName}"/>
   <Input value="{employeeModel>/employee/lastName}"/>
   <Input value="{employeeModel>/employee/age}"/>
</mvc:View>
```

Listing 6.5 Property Binding in XML View

The *complex syntax* can also be used in a property binding. This syntax makes it possible to define additional binding information, such as a formatting function. Listing 6.6 shows, for example, how a formatting function can be applied to a binding.

```
<mvc:View controllerName="" xmlns="sap.m" xmlns:mvc="sap.ui.core.mvc">
   <Text text="{parts: [{path: 'employeeModel>/employee/firstName'}, {path:
'employeeModel>/employee/lastName'}], formatter: '.employeeFormatter'}"/>
</mvc:View>
```

Listing 6.6 Apply the Formatter Function to Property Binding

However, if you want to create the UI elements in the controller and set the binding programmatically, this has to be done as shown in Listing 6.7.

```
let oInputFirstName = new Input({value: "{employeeModel>/employee/
firstName}"});
let oInputLastName = new Input({value: "{employeeModel>/employee/lastName}"});
let oInputAge = new Input({value: "{employeeModel>/employee/age}"});
```

Listing 6.7 Property Binding in Controller

6

The bindProperty function offers another option for programmatically setting the binding to a control. This is particularly helpful if bindings are only to be set at runtime at a certain point in time. This method receives the name of the UI control property to which a binding will be set as the first parameter. The second parameter is an object that contains the following information on the property from the model that will be bound:

- path
  Path of the property.
- value
  Defines a static value.
- model
  Name of the model; if not present, the default is considered.
- suspended
  Defines whether the binding is suspended and therefore not executed automatically.
- useRawValues
  Defines whether the value of the binding should be passed as raw values. In this case, the type for the binding parts isn't taken into account.
- useInternalValues
  Defines whether the parameters to the formatter function should be passed as the related JavaScript primitive values.
- type
  Defines the type that should be used for the value of the binding property.
- targetType
  Defines the type that should be used when a formatter function is called.
- formatOptions
  Defines the formatting options, which should be applied to the value. The available formatting options differ depending on the type that is used.
- constraints
  Defines the constraints against which the value should be checked. The available constraints differ depending on the type used.

- `Mode`
  Defines the binding mode that should be used (one-way, two-way, or one-time).
- `parameters`
  Map of additional parameters for this specific binding. The name and value ranges depend on the model implementation used.
- `events`
  Map of event handler functions.
- `parts`
  Array of binding info objects for the parts of a composite binding.

Now, let's assume there's an input like the one shown in Listing 6.8.

```
<mvc:View controllerName="" xmlns="sap.m" xmlns:mvc="sap.ui.core.mvc">
   <Input id="inputFirstName"/>
</mvc:View>
```

Listing 6.8 Input without Binding

To set a binding to this input, you can proceed as shown in Listing 6.9.

```
let oInputFirstName = this.getView().byId("inputFirstName");

oInputFirstName.bindProperty("value", {path: "employeeModel>/employee/
firstName"});
```

Listing 6.9 Set Binding on Input Using bindProperty

A binding can also be removed from a UI element via the `unbindProperty` method. This method receives the name of the UI control property for which the binding will be removed as the first parameter. In addition, a Boolean can be passed as the second parameter, which defines whether the default value should be restored after the binding is removed. If you want to remove the binding from the input used earlier, this can be done as in Listing 6.10.

```
let oInputFirstName = this.getView().byId("inputFirstName");

oInputFirstName.unbindProperty("value");
```

Listing 6.10 Unbind Property from Input

6.2.2 Formatting of Property Values

Values in bindings are usually displayed in an internal format. However, this format often isn't very user-friendly. It's therefore possible to implement formatting functions and apply them to the binding. These functions make it possible to carry out any formatting based on the initial value and ultimately display the formatted value. This is particularly useful when numerical or date values need to be displayed.

SAPUI5 offers two different options for carrying out this formatting:

- Using formatting functions for one-way conversion
- Using data types in two-way bindings

Data types can also be used to validate user input based on defined constraints.

Using Formatting Functions

To use a formatting function, it must be explicitly specified in the property binding in the XML view and implemented in the controller. An example implementation of a formatting function that formats a date value can look like Listing 6.11.

```
sap.ui.define([
   "sap/ui/core/mvc/Controller",
   "sap/ui/model/json/JSONModel"
], function (Controller, JSONModel) {
      "use strict";

      return Controller.extend("at.clouddna.demo.App", {
         ...
         formatDate: function(fValue) {
            if (fValue) {
               return new Date(fValue).toLocaleDateString();
            }
            return "";
         }
      }
   ));
});
```

Listing 6.11 Definition of a formatter Function in Controller

Listing 6.12 shows how the formatting function can be used in the binding in an XML view. To do this, the function name must be specified in the binding in the `formatter` property. The `this` context of a `formatter` function is usually set to the control (or managed object) that has the binding. In XML views, however, the `formatter` function is generally defined in the controller of the view. To refer to this function, the formatting name is preceded by a dot (.) (e.g., `{formatter: '.formatDate'}`). This ensures that this context of the `formatter` is bound to the controller.

```
<mvc:View controllerName="" xmlns="sap.m" xmlns:mvc="sap.ui.core.mvc">
   <Text text="{path: 'model>/someDateField', formatter: '.formatDate'}"/>
</mvc:View>
```

Listing 6.12 Property Binding with the formatter Function

If you're using JavaScript, you can either pass the `formatter` function as the third parameter of the `bindProperty` method or define it in the binding information with the `formatter` key. The `formatter` function receives the value to be formatted as the only parameter and is executed in the context of the control. This gives it access to other properties of the control and to data from the model. Listing 6.13 shows how a `formatter` can be integrated using the `bindProperty` or directly in the constructor.

```
let formatterFunction = function(sValue){…};
oInput.bindProperty("value", "model>/someDateField", formatterFunction);
let oControl = new Input({
   value: {
      path: '/someDateField',
      model: 'model',
      formatter: formatterFunction
   }
});
```

Listing 6.13 Apply formatter via Controller

Because the `formatter` function can contain any code, it can be used for formatting as well as for type conversions. The return type of the function doesn't necessarily have to be identical to the type of the value to be converted. For example, `formatters` can also be used to return a corresponding status or icon based on a Boolean flag.

Refreshing Bindings

The framework only updates a binding if one of the properties contained in the binding definition changes. However, if the `formatter` accesses additional properties that aren't defined in the binding, the framework doesn't recognize this dependency and could therefore overlook necessary updates. To avoid this, a composite binding should be created that explicitly includes all relevant properties, even if they come from different models.

Using Data Types

The use of data types allows data to be formatted, parsed, and validated to check whether the entered data is subject to defined constraints. SAPUI5 already offers a large number of data types as standard. Following are the data types from the `sap.ui.model.type` namespace:

- Boolean
- Currency
- Date
- DateInterval
- DateTime
- DateTimeInterval
- FileSize
- Float
- Integer
- String
- Time
- TimeInterval
- Unit

The following data types from the namespace `sap.ui.model.odata.type` are also available:

- Boolean
- Byte
- Currency
- Date
- DateTime
- DateTimeBase
- DateTimeOffset
- DateTimeWithTimezone
- Decimal
- Double
- Guid
- Int
- Int16
- Int32
- Int64
- ODataType
- Raw
- SByte
- Single
- Stream
- Time
- TimeOfDay
- Unit

The data types from the `sap.ui.model.odata.type` namespace support OData V2 and V4. At the same time, these data types represent the OData EDM (entity data model) data types.

If the data types contained in the standard aren't sufficient, you can also implement your own data types. To do this, a class must be created that inherits from the base type `sap.ui.model.SimpleType` and implements the corresponding methods of `formatValue`, `parseValue`, and `validateValue`.

To define a data type in a binding within an XML view, this must be specified in the corresponding property in the binding (see Listing 6.14).

```
<mvc:View controllerName="" xmlns="sap.m" xmlns:mvc="sap.ui.core.mvc">
   <Text text="{path: 'model>/someFloatField', type:
'sap.ui.model.type.Float'}"/>
</mvc:View>
```

Listing 6.14 Define Type in Binding in XML View

The formatOptions and constraints parameters can be supplied accordingly so that values can be validated or formatted in a binding. The possible parameters that can be defined within the two parameters depend on the types used. Listing 6.15 shows an example in which both formatOptions and constraints are defined for the sap.ui.model.type.Float. In this case, the formatOptions define that the value is always displayed with a minimum of two and a maximum of three decimal places. The constraints define that only values from 5 to 10 are valid.

```
<mvc:View controllerName="" xmlns="sap.m" xmlns:mvc="sap.ui.core.mvc">
   <Text text="{path: 'model>/someFloatField', type: 'sap.ui.model.type.Float',
formatOptions: {minFractionDigits: 2, maxFractioDigits: 3}, constraints:
{minimum: 5, maximum: 10}}"/>
</mvc:View>
```

Listing 6.15 Using formatOptions and Constraints in Binding

The same behavior as in the XML coding can also be achieved by corresponding coding in the controller (see Listing 6.16). The formatOptions must be passed as the first parameter and the constraints as the second parameter in the constructor of the type.

```
let control = new Input({
   value: {
      path: "model>/someFloatField",
      type: new sap.ui.model.type.Float({minFractionDigits: 2,
maxFractioDigits: 3}, { minimum: 5, maximum: 10})
   }
});
```

Listing 6.16 Using formatOptions and Constraints via Controller

To implement a custom type, as already mentioned, a subclass of sap.ui.model.SimpleType must be created, and this custom type can then be used in the same way as the predefined types. A schematic implementation of a custom type can be seen in Listing 6.17.

```
var MyCustomType =
sap.ui.model.type.SimpleType.extend("at.clouddna.MyCustomType", {
   formatValue: function(oValue) {
      return oValue;
   },

   parseValue: function(oValue) {
      return oValue;
   },

   validateValue: function(oValue) {
      if (oValue < 0) {
```

```
            throw new sap.ui.model.ValidateException("Invalid Input!");
        }
    }
});
```

Listing 6.17 Example for Implementation of Custom Type

Resuming Suspended Bindings

To check whether a binding is suspended, you can use the `isSuspended` method of the `sap.ui.model.Binding` class. If the binding is suspended, the `resume` method can be used to trigger the binding. This triggers the corresponding request, causing the binding to no longer be suspended.

6.3 Aggregation Binding

Aggregation binding—or list binding—can be used to automatically create corresponding child elements in the UI based on data contained in a model. This can be done either by defining a template, which is then cloned for each entry, or by using a factory function. To bind aggregations, it's essential that they are bound to lists in the model, for example, collections in an OData model or arrays in a JSON model. Important components for the use of an aggregation binding are as follows:

- **Data model**
  Contains a collection of entries with certain properties.
- **Path**
  Refers to the collection of the model.
- **Template**
  Defines how each element of the collection should look in the UI.
- **Binding**
  Links data of the model with the aggregation of the control.

For our example, let's assume there's a list, as shown in Listing 6.18, in a JSON model.

```
{
   employees: [
      {
         "employeeId": 1,
         "employeeName": "Max Mustermann I"
      },
      {
         "employeeId": 2,
         "employeeName": "Max Mustermann II"
```

```
        }
    ]
}
```

Listing 6.18 JSON Demo Data for Aggregation Binding

To use this list in an aggregation binding, it's possible to do this either directly in the XML view or from a controller. The following sections cover the `AggregationBindingInfo` object, various approaches for implementing list bindings in both XML views and controllers, removing bindings from controls, and the use of factory functions for aggregation binding.

6.3.1 AggregationBindingInfo Object

The object `sap.ui.base.ManagedObject.AggregationBindingInfo` defines which information is required to link an aggregation with data from a model. For example, information on the data source, the template, and other configuration parameters are stored here. An instance of this class must be specified when defining a binding—regardless of whether the binding is defined in the XML view or via the controller.

The class has the following parameters, whereby the `path` parameter and either the `template` or `factory` parameter must be specified:

- `path`
  Defines the path in the model to bind to and can either be a relative or absolute path. In case of a relative path, the binding context of the corresponding model defines the root. If the path contains a >, the part before the > defines the model name and overwrites the model property; the rest of the path is used as the binding path. This is a mandatory parameter of type `string`.
- `model`
  Defines the name of the model, which is used for the binding. If the property isn't present or is undefined, the default model is used. This is an optional parameter of type `string`.
- `template`
  Defines the template, which is cloned for each entry in the model. This is an optional parameter (either `template` or `factory` must be present) of type `ManagedObject`.
- `templateShareable`
  Indicates whether the application manages the lifecycle of the template. If the value `true` is defined, the template can be reused across multiple bindings, and the framework won't clone it when the managed object is cloned. On the other hand, if the value `false` is defined, the template's lifecycle is bound to the lifecycle of the binding. So, if the binding is unbound or the `ManagedObject` is destroyed, the `template` itself will also be destroyed. This is an optional parameter of type `boolean`.

- `factory`
  Defines a factory function that determines how the created objects for the aggregation should look; it can be used as an alternative to a template. The functions can be called with two parameters: an ID for the object and the binding context. The return value of the function must be an object suitable for the bound aggregation. This is an optional parameter (either `template` or `factory` must be present) of type `function`.
- `suspended`
  Defines whether the binding should be suspended and not executed initially. This is an optional parameter of type `boolean`.
- `startIndex`
  Defines the first entry by index of the list, which should be created. This is an optional parameter of type `int`.
- `length`
  Defines the number of entries the aggregation should contain. It can exceed the model's size limit. This is an optional parameter of type `int`.
- `sorter`
  Defines the initial sorters of the aggregation. This is an optional parameter of type `sap.ui.model.Sorter` or an array of `sap.ui.model.Sorter`.
- `filters`
  Defines the initial filters of the aggregation. This is an optional parameter of type `sap.ui.model.Filter` or an array of `sap.ui.model.Filter`.
- `key`
  The name of the key property or a function that takes the context as a parameter to compute a key for entries. This helps optimize update behavior in models without predefined keys. This is an optional parameter of type of `string` or `function`.
- `parameters`
  Defines a map of additional parameters for the binding. The supported parameter names and value ranges depend on the model implementation and should be documented in the `bindList` method of the respective model class or its `sap.ui.model.ListBinding` subclass. This is an optional parameter of type `object`.
- `groupHeaderFactory`
  Defines an optional factory function to create a custom group header visualization. It should return a control, such as a `sap.m.GroupHeaderListItem` for a `sap.m.List`, suitable for displaying the group header. This is an optional parameter of type `function`.
- `events`
  A map of event handler functions, with each function associated with a specific binding event. The events and their possible values depend on the model implementation and should be documented in the corresponding subclass of `sap.ui.model.ListBinding`. This is an optional parameter of type `object`.

Filters Are Linked by AND by default

By default, the filters passed in the filter array are linked with logical AND, regardless of whether you filter on the same properties or not. To also achieve an OR link, a filter instance must be created in which the filter array is passed. In addition, you can specify whether AND or start Index should be linked by setting the and property:

```
let arrayOfFilters = [new sap.ui.model.Filter("Property1", "EQ", "value1"),
new sap.ui.model.Filter("Property1", "EQ", "value2")];
new sap.ui.model.Filter({filters: arrayOfFilters, and: false});
```

6.3.2 List Binding in an XML View

To apply the list binding directly in a view, coding such as that in Listing 6.19 can be implemented. In this case, we assume that the JSON model with the data is the default model of the application, so no model name needs to be specified in the binding. Furthermore, the entries are displayed in the form of a list (sap.m.List). A standard list item is used within the list, which is therefore generated for each list entry. Depending on which parent control is used, the aggregations and child elements that can be used for the display differ.

```
<mvc:View controllerName="" xmlns="sap.m" xmlns:mvc="sap.ui.core.mvc">
   <List id="employeeList" items="{/employees}">
      <items>
         <StandardListItem title="{employeeName}" description="{employeeId}"/>
      </items>
   </List>
</mvc:View>
```

Listing 6.19 Definition of List Binding in an XML View

As can be seen in Listing 6.19, the sap.m.List provides both an items attribute and a nested items element, as follows:

- The items attribute binds the child elements of the JSON model to the list. However, this alone isn't sufficient to display the individual employees in the list. The parent path (/employees) is used to bind all contained list elements. In addition, a nested element must be defined, which represents the template for all list entries.
- In our case, the nested element is a StandardListItem. This forms the template that is used to create the individual list lines.

Relative bindingpath in Aggregation

As shown in the example, the binding paths for the title and description properties of the StandardListItem are relative to employees. This means that the entire binding path (/employees/employeeId or /employees/employeeName) doesn't have to be specified. The fact that the path is relative is indicated by omitting the / at the beginning of the path.

If an sap.m.List is used, the following other row templates can be used in addition to the StandardListItem:

- ActionListItem
- DisplayListItem
- CustomListItem
- ObjectListItem

As already mentioned, the controls that can be used in an aggregation depend on the parent container. A list of which controls are permitted and when is available in the SAPUI5 API Reference (*https://ui5.sap.com/#/api*).

Size Limitation in List Bindings

For performance reasons, list bindings have a size limit, which is 100 entries by default and can be overridden programmatically. The reason for the limit is to prevent too much data from being rendered in the UI, as this would have a negative impact on performance.

The limitation means that only the defined number of elements are displayed in the UI, even if more entries become available in the model. This can be changed by setting the sizeLimit with the setSizeLimit method of the model or by specifying the length properties in the binding.

6.3.3 List Binding via a Controller

In addition to the option of defining the binding directly in the XML view, the aggregation binding can also be set from the controller in a similar way to the property binding. To do this, the binding can either be set directly in the settings object of the constructor or by calling the bindAggregation method. It's also necessary to define a template, which is then used to create the individual entries. Each clone of this template receives the corresponding binding context of the respective list entry, which in turn allows all bindings within this template to be resolved relatively. If the list in the model on which the parent control is based changes, all aggregated elements are destroyed and recreated.

Listing 6.20 shows how a list can be created from the controller using the constructor.

```
let employeeTemplate = new sap.m.StandardListItem({
   title: "{employeeName}",
   description: "{employeeId}"
```

```
});

let employeeList = new sap.m.List({
   items: {
      path: "/employees",
      template: employeeTemplate
   }
});
```

Listing 6.20 Define the List Binding in the Controller Using a Constructor

Another way to apply the aggregation binding is by calling the `bindAggregation` method. The parent control (in our case the `sap.m.List`) must already exist. Listing 6.21 shows how the corresponding method can be called. Note that the `bindAggregation` method contains the name of the aggregation to be bound as the first parameter and the second parameter is the `BindingInfo`.

```
let employeeTemplate = new sap.m.StandardListItem({
   title: "{employeeName}",
   description: "{employeeId}"
});

let employeeList = this.getView().byId("employeeList");
employeeList.bindAggregation("items", {
   path: "/employees",
   template: employeeTemplate
});
```

Listing 6.21 Define the List Binding Using bindAggregation

Furthermore, various controls—such as the `bindAggregation` method—also have methods that enable specific aggregations to be addressed directly. In the case of the `sap.m.List`, the `bindItems` method exists here. This basically does nothing more than the `bindAggregation` method, but the "items" aggregation is addressed directly and doesn't have to or can't be explicitly specified in the method. Listing 6.22 shows how this method will look.

```
let employeeList = this.getView().byId("employeeList");
employeeList.bindList({
   path: "/employees",
   template: employeeTemplate
});
```

Listing 6.22 Define List Binding Using bindItems

6.3.4 Remove Binding from Control

To remove an aggregation binding from a control, the `unbindAggregation` method can be used. This receives the name of the aggregation to be removed as a parameter. In our case, this will look like Listing 6.23.

```
let employeeList = this.getView().byId("employeeList");
employeeList.unbindAggregation("items");
```

Listing 6.23 Unbind Aggregation from sap.m.List Using unbindAggregation

Analogous to the `bindAggregation` method, methods are also offered for unbinding in various controls that address the aggregation directly. In this case, this would be the `unbindItems` method (see Listing 6.24).

```
let employeeList = this.getView().byId("employeeList");
employeeList.unbindItems();
```

Listing 6.24 Unbind Aggregation from sap.m.List Using unbindItems

6.3.5 Factory Function for Aggregation Binding

If a factory function is used instead of a template, the coding must be adapted accordingly. To achieve the same end result as in our earlier examples, the coding should look like Listing 6.25.

```
let employeeList = this.getView().byId("employeeList");

employeeList.bindList({
   path: "/employees",
   factory: (sId, oContext) => {
      return new sap.m.StandardListItem({
         id: sId,
         title: oContext.getProperty("employeeName"),
         description: oContext.getProperty("employeeId")
      });
   }
});
```

Listing 6.25 Define Aggregation Binding Using the Factory Function

6.4 Element Binding

Element binding, also known as context binding, enables elements to be bound to a specific object in the model, creating a binding context. This context allows relative bindings within the control and all its subordinate elements. This is particularly useful in scenarios with lists and detailed views, as it enables a clear structure and simple data binding between the different levels of the UI.

As a result, subordinate UI elements—such as templates within an aggregation—don't have to use absolute paths but can bind to the corresponding properties from the model using relative paths. This significantly improves the complexity and readability of the code. In this case, we again assume there's a model with the data shown in Listing 6.26.

```
{
   "employee": {
      "firstName": "Max",
      "lastName": "Mustermann",
      "age": 40
   }
}
```

Listing 6.26 JSON Demo Data for Element Binding

To apply an element binding to a UI control, the `binding` property must also be specified in the UI control. It's important that the specified path points to the entire object and not to a specific property. There are again two ways to do this:

- Element binding directly in the XML view
- Element binding via the controller

6.4.1 Element Binding Directly in the XML View

To perform an element binding on a control in the XML view, the `binding` parameter must be defined. Based on our demo data, the binding should be `employee`. Listing 6.27 shows how this could look for the properties `firstName`, `lastName`, and `age`.

```
<mvc:View controllerName="" xmlns="sap.m" xmlns:mvc="sap.ui.core.mvc">
   <Input id="firstNameInput" binding="{/employee}" value="{firstName}"/>
   <Input id="lastNameInput" binding="{/employee}" value="{lastName}"/>
   <Input id="ageInput" binding="{/employee}" value="{age}"/>
</mvc:View>
```

Listing 6.27 Apply Element Binding via the XML View

This type of syntax creates a binding context on the `employee` property from the model on the input. The property binding in the `value` attribute can then be used to access the corresponding property relatively. If you were to dispense with the binding element in this case, the path in the value attribute would look like `{/employee/<propertyName>}`—where `<propertyName>` is a placeholder for `firstName`, `lastName`, or `age`.

Element binding is also inherited downward. In other words, if the binding element is applied to the XML view, the binding property can be omitted, and the value of the value property remains identical (see Listing 6.28).

```
<mvc:View controllerName="" xmlns="sap.m" xmlns:mvc="sap.ui.core.mvc" binding=
"{/employee}">
   <Input id="firstNameInput" value="{firstName}"/>
   <Input id="lastNameInput" value="{lastName}"/>
   <Input id="ageInput" value="{age}"/>
</mvc:View>
```

Listing 6.28 Apply Element Binding via the XML View with Inheritance

6.4.2 Element Binding via the Controller

Similar to property and aggregation binding, an element binding can also be created directly from the controller. The bindElement method can be used for this purpose. If the XML view looks like the one in Listing 6.29, you can see how this can be done for an input in Listing 6.30.

```
<mvc:View controllerName="" xmlns="sap.m" xmlns:mvc="sap.ui.core.mvc">
   <Input id="firstNameInput"/>
</mvc:View>
```

Listing 6.29 XMLView with Input without Binding

It's important that a binding context is first created using the bindElement method of the input. Various properties can then be bound using the bindProperty method (see Listing 6.30).

```
let firstNameInput = this.getView().byId("firstNameInput");

firstNameInput.bindElement("/employee");
firstNameInput.bindProperty("value", "firstName");
```

Listing 6.30 Apply Element Binding via Controller

Here, too, it's possible to create the element binding to higher-level elements, for example, to the XML view directly. The individual properties can then be bound relatively to the inputs (see Listing 6.31).

```
let firstNameInput = this.getView().byId("firstNameInput");

this.getView().bindElement("/employee");
firstNameInput.bindProperty("value", "firstName");
```

Listing 6.31 Apply Element Binding via the Controller with Inheritance

In contrast to the previous example, the binding element is applied directly to the XML view here (`this.getView().bindElement(...)`). This type of binding is used particularly often in detail pages. It's sufficient to create the element binding once on the view—or another type of superordinate container—and then define all other property bindings relatively.

Element binding can also be used to bind individual entries from lists. To do this, the index or key of the respective entry must be specified in the binding. The syntax of the `bindElement` method is `bindElement(vPath, mParameters)`, as explained here:

- `vPath`
  Defines the path that will be used for the element binding.
- `mParameters`
  Defines additional parameters to be used for the binding. These should match the properties from class `sap.ui.base.ManagedObject.ObjectBindingInfo`. However, the supported parameters depend on the model used.

Basically, the following parameters are contained in the `sap.ui.base.ManagedObject.ObjectBindingInfo` class:

- `path`
  Defines the path in the model to bind to, which can either be an absolute or relative path. Should the path contain a >, the first parts before the > sign will be used as the model name, so the model property will be overwritten. The part after the > sign will then be used as the binding path. This is a mandatory parameter of type `string`.
- `model`
  Defines the name of the model that should be used for the binding. The default model is used if the value is not present or undefined. This is an optional parameter of type `string`.
- `suspended`
  Defines whether the binding is initially suspended and therefore not executed. This is an optional parameter of type `boolean`.
- `parameters`
  A map of additional parameters for this binding. The names and value ranges of supported parameters depend on the model implementation and should be documented in the `bindContext` method of the corresponding model class or in the model-specific subclass of `sap.ui.model.ContextBinding`. This is an optional parameter of type `object`.
- `events`
  A map of event handler functions, keyed by the names of the binding events they are attached to. The names and value ranges of supported events depend on the model implementation and should be documented in the model-specific subclass of `sap.ui.model.ContextBinding`. This is an optional parameter of type `object`.

Remove Element Binding via Controller

An element binding can also be removed again via the controller. The `sap.ui.core.Element` class provides the `unbindElement` method for this purpose. If the element binding is removed using this method, this means that all bindings are then released relative to the parent context. This method has the name of the model for which an element binding will be removed as the only transfer parameter. A possible call of this method is shown in Listing 6.32.

```
this.getView().byId("firstNameInput").unbindElement();
```

Listing 6.32 Unbind Element Using Method unbindElement

If you want to retrieve the binding context of a single control, the method `getBindingContext` of the class `sap.ui.base.ManagedObject` can be used. This is the parent class of UI elements. Retrieving the binding context of a UI element can be especially useful if, for example, you click on a list item in a table and want to retrieve further information of the selected entry. This method returns an object of type `sap.ui.model.Context`. Via the instance of `sap.ui.model.Context`, either the whole object (`getObject`) or single properties (`getProperty`) can be retrieved.

To retrieve the binding context of a UI element, the method call can look like Listing 6.33.

```
let bindingContext = this.getView().byId("uiElement").getBindingContext();
let oObject = bindingContext.getObject();
let sProperty = bindingContext.getProperty("SomeProperty");
```

Listing 6.33 Retrieve the Binding Context of a UI Element with the Default Model

This coding works only for default models. However, if you want to use it together with named models, the name of the model must be supplied in the method call of `getBindingContext` (see Listing 6.34).

```
let bindingContext =
this.getView().byId("uiElement").getBindingContext("namedModel");
let oObject = bindingContext.getObject();
let sProperty = bindingContext.getProperty("SomeProperty");
```

Listing 6.34 Retrieve the Binding Context of a UI Element with the Named Model

6.5 Expression Binding

Expression binding is an extension of the SAPUI5 binding syntax. This makes it possible to use expressions instead of specially written `formatter` functions. The use of this special syntax eliminates the overhead that would otherwise arise from defining a separate formatter function. The use of this binding is recommended if the formatter function only has very trivial syntax—such as the display of an indicator based on a property from the model, where the comparison condition is very simple. There are two options for using an expression binding:

- {=expression}
  This type uses one-way binding (defined by the = before the expression). This enables automatic recalculation of the value if it changes in the model.
- {:=expression}
  This type uses one-time binding (defined by := before the expression). This means that the value is only calculated once and doesn't react to changes in the model. This type requires fewer resources but is also less dynamic.

The syntax of the expression is similar to the syntax in JavaScript, but has some restrictions with regard to the comparison operations to be used. The supported operations are shown in Table 6.2.

Syntax Element	Symbol
Literal	Number, for example, 42, 6.022e+23, or -273.15 Object, for example, {foo: 'bar'} String, for example, 'foo' null true false
Grouping	(…), for example, 3 * (4 + 10)
Unary operator	! + - Typeof
Multiplicative operator	* / %
Additive operator	+ -
Relational operator	< > <= >=

Table 6.2 Supported Syntax in Expression Bindings

<table>
<tr><th>Syntax Element</th><th>Symbol</th></tr>
<tr><td>Strict equality operator</td><td>===
!==</td></tr>
<tr><td>Binary logical operator</td><td>&&
||</td></tr>
<tr><td>Conditional operator</td><td>?</td></tr>
<tr><td>Member access operator with the . operator</td><td>{/firstName}.indexOf('M')</td></tr>
<tr><td>Function call</td><td><function name>(...)
Example: text="{=Math.max(%{val1}, %{val2})}"</td></tr>
<tr><td>Array literals</td><td>[...]
Example: [2, 3, 5, 7]</td></tr>
<tr><td>Property/array access</td><td>obj[...]
Example: 'foo/bar'.split('/')[1]</td></tr>
<tr><td>In operator</td><td>'PI' in math, or 0 in []</td></tr>
<tr><td>Global symbols</td><td>Array, Boolean, encodeURIComponent, Infinity, isFinite, isNaN, JSON, Math, NaN, Number, Object, odata.collection, odata.compare, odata.fillUriTemplate, odata.uriEncode, parseFloat, parseInt, RegExp, String, undefined</td></tr>
</table>

Table 6.2 Supported Syntax in Expression Bindings (Cont.)

You can also define in the expression how the type of expression will be cast. A distinction is made between the following two types:

- **${binding}**
  When using this syntax, the resulting value is cast to the target type behind the control property (for example, this would be the Boolean type for a visible property).
- **%{binding}**
  When using this syntax, the type any is used as the resulting type. This type of definition is therefore a shortcut for the following coding: "${path: 'binding', targetType: 'any'}". It's recommended to use this type of syntax; otherwise, it can lead to unwanted problems, especially when using OData V4 models.

Suppose you want to display a specific icon based on an attribute age (if age is less than 20, then use icon *sap-icon://employee*; in all other cases, then use the icon *sap-icon://manager*). To achieve this, coding as shown in Listing 6.35 can be used.

```
<mvc:View controllerName="" xmlns="sap.m" xmlns:core="sap.ui.core" xmlns:mvc=
"sap.ui.core.mvc">
   <core:Icon src="{= ${age} &lt; 20 ? 'sap-icon://employee' : 'sap-icon://
manager'}"/>
</mvc:View>
```

Listing 6.35 Using Expression Binding

Escaping of Some Special Characters

Some special characters must be escaped if they are to be used in the expression binding. An example of this is the concatenation operator: `&`. This must be represented by `&`. Similarly, the character < must be escaped with `<`. Although this makes it possible to represent such expressions, it makes the code difficult to read.

The example just shown in Listing 6.35 is a very simple example of an expression binding. However, it's also possible to map far more complex expressions. Nevertheless, expression bindings should be avoided for very complex expressions and a `formatter` function should be implemented instead. For example, several expressions can also be logically linked together. Furthermore, various functions can be called within the expression, such as functions for converting strings, regular expressions, or mathematical functions.

6.6 Summary

Now that we've dealt with the different types of binding—property binding, aggregation binding, element binding, and expression binding—and the individual binding modes of one-way, two-way, and one-time, we'll take a closer look at topics relating to routing and navigation in SAPUI5 applications and the integration of REST services in the following chapters. Furthermore, we'll see how OData services can be integrated and will also look at the different OData model types—OData model V2 and OData model V4.

Chapter 7
Routing and Navigation

In practice, complex business applications are becoming increasingly common. These should be logically subdivided and kept efficient, which argues against a single-page application for large-scale projects. This is where routing and navigation come into play because it should be possible to navigate between the individual views using different layout controls with user-friendly and responsive designs.

7

Navigation will play a major role in this chapter, because you'll rarely get by with a single-page application. The idea is that we can navigate between several views. This navigation can take place with or without parameters, so that data can be passed between views during navigation. When navigating, we get a lot of things already prepared by the SAPUI5 framework, but nevertheless, we have to pay close attention to a few details and not lose sight of the fact that the web application is running in the client's browser, that client-side application generation is taking place, and that both URL changes and active user navigation requests can be called up.

In this chapter, we'll start by introducing you to our example scenario in Section 7.1. From there, we'll move on with the basics in Section 7.2 and clarify the most important terms and components of a router. Furthermore, we'll discuss how the router is defined and at which point it's initialized. In Section 7.3, we'll already start with the patterns. We'll discuss the three well-known ones, namely full-screen navigation (Section 7.3.1), split app (Section 7.3.2), and the famous flexible column layout (Section 7.3.3). Even if you're only interested in a specific pattern, we recommend that you also look at all the other patterns and stick to the order. This is because we'll develop the knowledge and practices around the various navigation concepts one after the other and build on this knowledge pattern by pattern. Section 7.4 will cover the individual routing events where you can register a listener to react to both intentional and unintentional changes. Finally, Section 7.5 will cover navigation from one SAP Fiori app to another SAP Fiori app (or even legacy user interfaces [UIs]), which takes place in the SAP Fiori launchpad.

7.1 Example Scenario

We'll acquire the necessary knowledge step-by-step by drawing on the data model familiar in the SAP environment: the *Flight Reference Scenario* (see Figure 7.1). The advantage is that this data model, including sample data, can be activated on any SAP system. This allows you to follow the same steps on your SAP system and go through the exercises with us step-by-step. You can find out how to activate the model in the official SAP Help Portal.

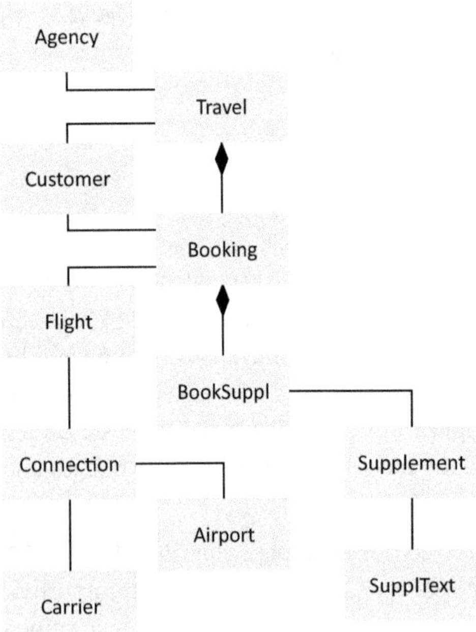

Figure 7.1 Data Model of the Flight Reference Scenario (Source: SAP Help Portal)

If you're not yet familiar with this data, we've shown the most important entities and attributes we'll access in our exercises in Figure 7.2. We'll always start from the `Carrier`. In addition to the `carrid`, which is the primary key, a carrier also has a name (`CARRNAME`), the currency of the airline (`CURRCODE`), and the `URL` to the website. A carrier can have `0` to `*` connection or `Flight`. A flight always belongs to a carrier and is also identified by the connection ID (`connid`) and the flight date (`fldate`). Two counters always tell us the maximum number of seats available (`SEATSMAX`) and how many are already occupied (`SEATSOCC`). These counters are recorded in the backend thanks to the bookings. A flight can have `0` to `*` `Booking`, but a booking always belongs to a flight. The composite primary key contains all the flight information plus a booking ID (`bookid`). In addition to the booking class (`CLASS`), the amount and currency paid in (`FORCURAM` and `FORCURKEY`), the booking date (`ORDER_DATE`), whether the booking has already been canceled (`CANCELLED`), and the passenger's name and date of birth (`PASSNAME` and `PASSBIRTH`) are also stored.

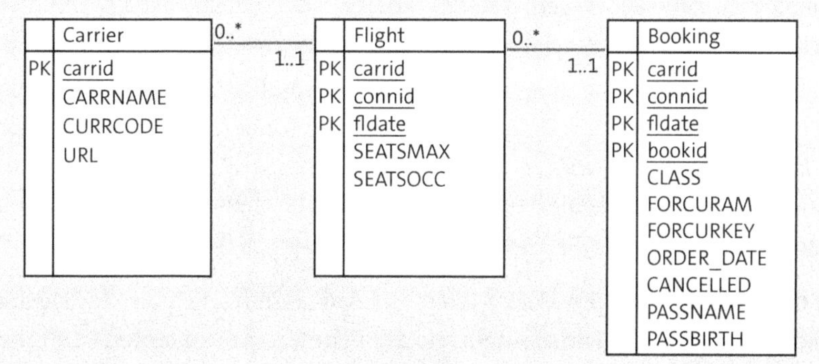

Figure 7.2 Data Model of the Flight Data Example with the Attributes We Need

7.2 General Concepts

Before we implement the navigation, we must look at the theory behind it. We'll guide you step-by-step through the processes shown in Figure 7.3 to give you an understanding of the background of the navigation:

- **View**
  For a view to be navigated to, it must of course exist. We'll also create new views, but there's no automated process for this, so we'll simply create new files and adhere to the naming conventions so that these files are recognized as views.
- **Target**
  A target explicitly refers to a view. However, a view can be used by multiple targets. We need these targets because we won't refer directly to a view during navigation but will need targets as an intermediate step.

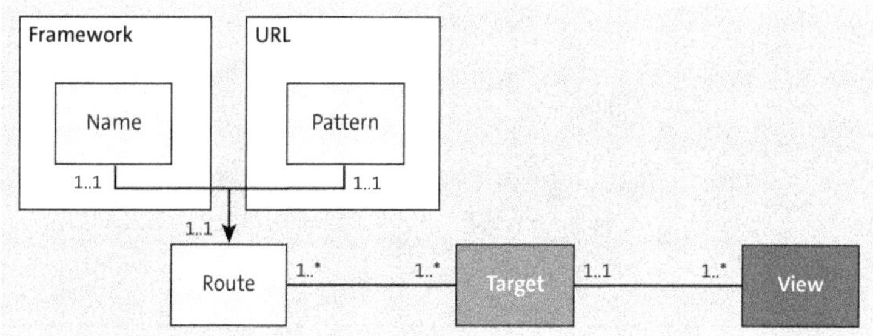

Figure 7.3 Components That Contain a Routing Concept

- **Route**
  A route can address one or more targets. Routes are the objects that are now used for the actual navigation. How these routes can be addressed is explained here:

- Name: For a route to be addressed, it needs a name. This name is used within the framework to load and address the route (e.g., in TypeScript from the controller).
- Pattern: Because we're in the browser and the web applications are opened via URLs, the navigation is also part of this URL. If we use a route for the navigation or the URL changes in general, this can also affect the navigation. If the patterns in the URL are recognized and can be assigned to a route, then the corresponding route is used.

As already mentioned, a router is provided in the standard system and is already stored with some basic configurations. The configuration for the router can be found in the application descriptor (*manifest.json*) under the item `sap.ui5` (see Figure 7.4). This includes the config, which defines which router class should be used, where the views can be found, what the animation should look like when navigating, and which container should be used for the navigation. We'll explain the container and the associated attributes `controlId` and `controlAggregation` in more detail in Section 7.3.1. Another important definition is the `rootView` also under `sap.ui5`. This object defines which view should be opened first, when the application starts up by defining a `viewName`, `type` and assigning an `id` to it. This view is called the *root view*.

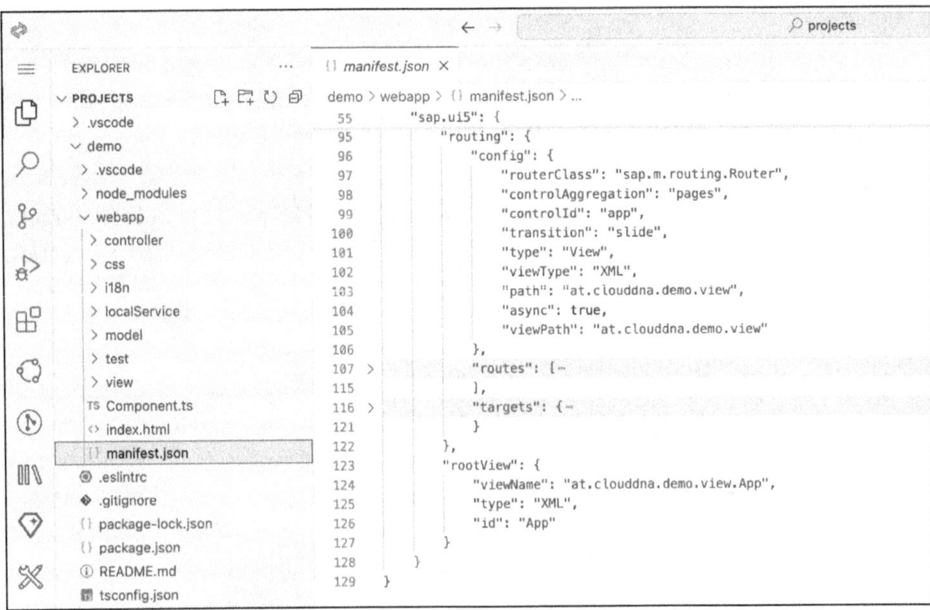

Figure 7.4 Router Configuration in the Application Descriptor

SAP delivers a number of faceless components in the SAPUI5 framework that can be used for navigation. We'll get to know the two important representatives of the routing classes, starting with the `sap.m.routing.Router` (see Figure 7.5). This is provided and used by default. We'll also work on the individual methods and events of this router in the hands-on examples in Section 7.3.

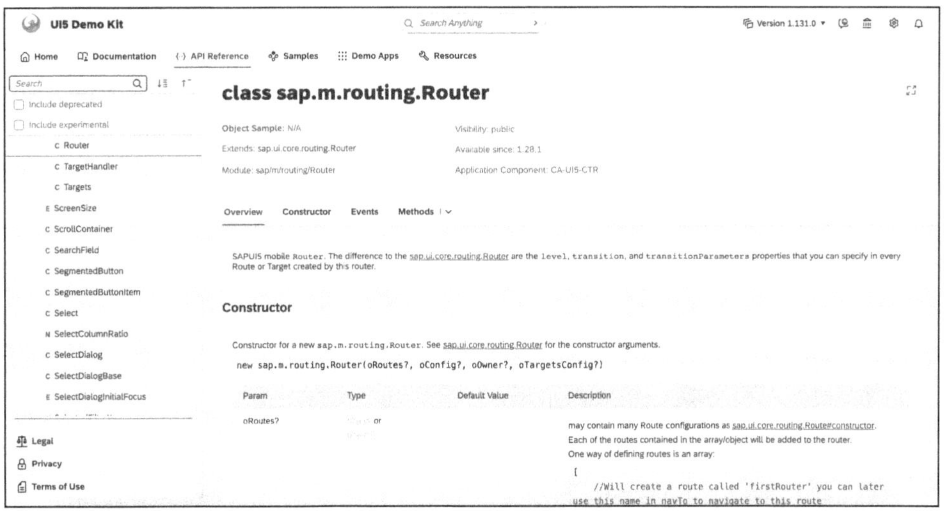

Figure 7.5 Faceless Component sap.m.routing.Router Used by Default

The router must be initialized when the application starts up. If you haven't already noticed, this happens in the component controller (*Component.ts* or *Component.js*) in the `init` method (see Listing 7.1). By default, the call `this.getRouter().initialize()` is stored here, which is used to access the router and prepare it for use.

```
import BaseComponent from "sap/ui/core/UIComponent";
export default class Component extends BaseComponent {
  public static metadata = { manifest: "json" };
  public init() : void {
    // call the base component's init function
    super.init();
    // enable routing
    this.getRouter().initialize();
  }
}
```

Listing 7.1 Router Initialized in the Component Controller's init Method

In simple scenarios, the router only exists once and is available globally in a component, so we'll always be able to access the router via the owner component, as follows:

```
this.getOwnerComponent().getRouter()
```

7.3 Patterns

This brief introduction will be enough to get us started with the first hands-on exercises on patterns. You'll learn step-by-step and develop your knowledge from pattern to pattern, building on it in subsequent exercises. For this reason, we recommend that you start with full-screen navigation, even if you're only interested in the flexible column layout, followed by the split app to refresh your knowledge of the methods and events of routing.

7.3.1 Full-Screen Navigation

Full-screen navigation describes the navigation within a UI component when we want to navigate from view A to view B. Both views always take up the full width and height of the screen. This process is illustrated in Figure 7.6.

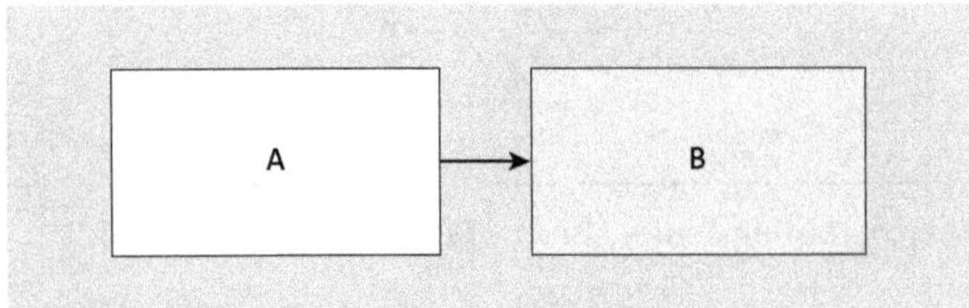

Figure 7.6 Full-Screen Navigation Illustrated

This navigation is designed so that nothing else happens behind the scenes other than a container being filled with new XML elements and then being displayed. Figure 7.7 shows that users always see the *App.view.xml* (because this is the root view defined in the application descriptor), but the content of this view replaces the content, depending on which view is called up by a navigation.

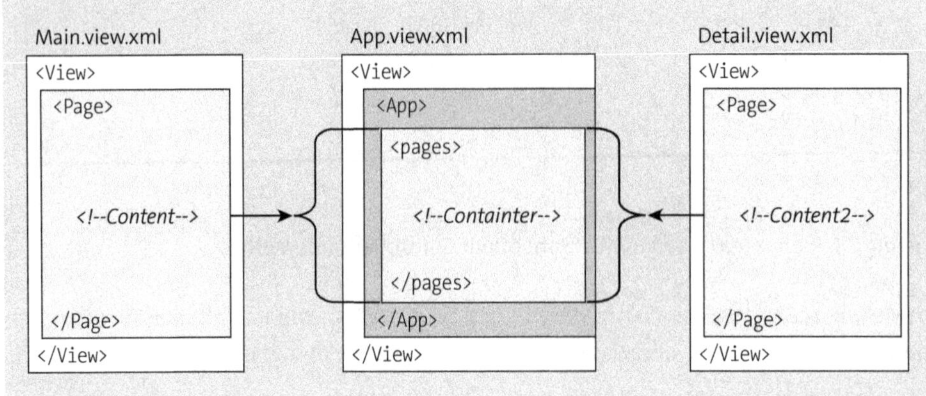

Figure 7.7 Full-Screen Navigation in Theory

Now we know that there's a root view, but how does the router find the UI component `App`, and how does the router know that the contents of the two views `Main` and `Detail` should be inserted exactly into the aggregation `pages`? Our router takes this from the config, which is stored in the application descriptor. If you remember, the `controlId` is stored there. Exactly the same ID is assigned here to the UI component App in the view *App.view.xml*. If we now look at the documentation for sap.m.App and in particular the aggregations, we understand that there's an aggregation here called pages (see Figure 7.8). This aggregation can contain 0 to *n* of any UI components. For this reason, the other views no longer have the same structure with app and pages, but only a Page. The sap.m.Page has the peculiarity that it always takes up 100% of the width and height on a page and can have a maximum of one header, content area, and footer.

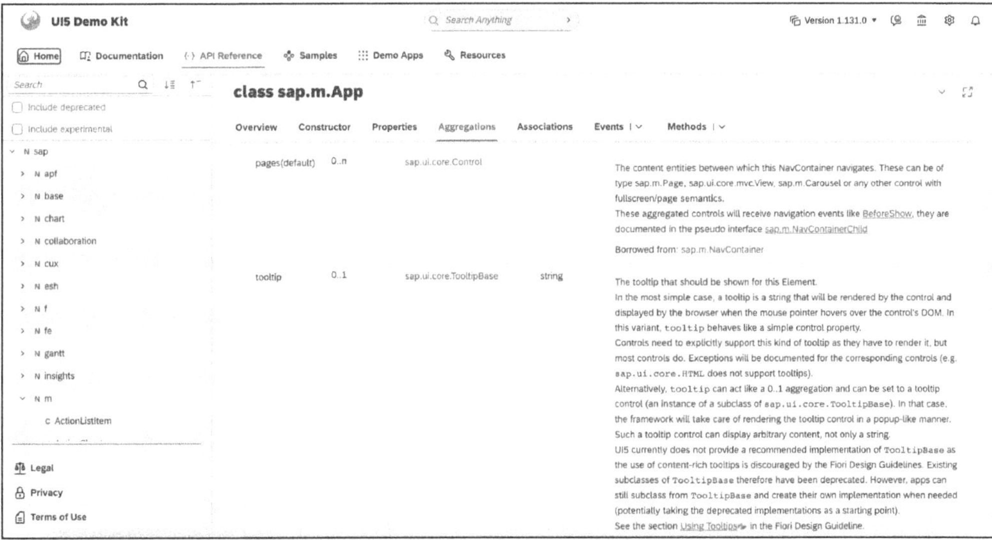

Figure 7.8 Documentation of the sap.m.App Used as a Container in the App View

We've created a new SAPUI5 application at.clouddna.fullscreennav for full-screen navigation, which consumes the OData service RMTSAMPLEFLIGHT. Via this OData service, we can access the aforementioned data model and consume and display carriers, flights, and bookings. We've named the initial view Main and activated TypeScript in the project details.

In the following sections, we'll present the individual options for defining a parameter and using this parameter to transfer data from one view to another during navigation. We'll distinguish between hard-coded routing (= without parameters) and routing with optional parameters and with mandatory parameters. For the first time, we'll show a best practice for routing in combination with element binding.

Hard-Coded Routing

Let's start with *hard-coded routing* (i.e., navigation without parameters) to give you a feel for the steps necessary to implement routing in general. To do this, we need to create another view. Open the context menu at the folder level, and click on **New File**. Name this new file *Detail.view.xml*. This naming convention will identify the file as an XML view. Fill this view with a sample implementation, as shown in Listing 7.2.

```
<mvc:View xmlns="sap.m" xmlns:f="sap.ui.layout.form" controllerName=
"at.clouddna.fullscreennav.controller.Detail" xmlns:mvc="sap.ui.core.mvc"
displayBlock="true">
  <Page id="pageDetail" title="Details">
    <content>
      <Text text="This is your detail page."/>
    </content>
  </Page>
</mvc:View>
```

Listing 7.2 Detail View Showing Just a Text at First

A view always has a controller. For this reason, we're also creating a new file named *Detail.controller.ts* in the *controller* directory. You can fill this file with the code shown in Listing 7.3.

```
import Controller from "sap/ui/core/mvc/Controller";
/**
 * @namespace at.clouddna.fullscreennav.controller
 */
export default class Detail extends Controller {
  /*eslint-disable @typescript-eslint/no-empty-function*/
  public onInit(): void {}
}
```

Listing 7.3 Empty Detail Controller Created

To navigate to this view, we need to make further entries in the application descriptor, that is, in *manifest.json*. If you scroll down to the bottom of this file, you'll find an entry called `targets`. We need to add another entry to this object, in addition to the existing entry for our main view, to create a new target for the detail view. As you can see in Listing 7.4, this new target points to our detail view.

```
"targets": {
  "TargetMain": {
    "viewType": "XML",
        "transition": "slide",
        "clearControlAggregation": false,
        "viewId": "Main",
        "viewName": "Main"
    },
    "Detail": {
        "viewType": "XML",
        "transition": "slide",
        "clearControlAggregation": false,
        "viewId": "Detail",
```

```
        "viewName": "Detail"
    }
}
```

Listing 7.4 New Target Added to the "targets" Object in manifest.json

Furthermore, there's an entry called `routes` just above the targets, which manages the routes of the router. Let's add an entry to this array as well and thus create a route that points to our new target belonging to the detail view. You can see this entry in Listing 7.5.


```
"routes": [
    {
        "name": "RouteMain",
        "pattern": "",
        "target": [
            "TargetMain"
        ]
    },
    {
        "name": "Detail",
        "pattern": "staticDetail",
        "target": [
            "Detail"
        ]
    }
],
```

Listing 7.5 New Route Added to the Array of Routes

This route can now be addressed in two ways: (1) from the framework, for example, if we access the name of the route `Detail` in TypeScript; and (2) from the pattern `/staticDetail` if it's recognized in the URL (the view will also be addressed).

Default Route Called First at Application Startup

If we now compare the two routes between `RouteMain` and `Detail`, we notice that the route from `Main` has no `pattern` stored. This makes this route the *default route*. This route is then called if no other path or parameter is recognized in the URL. This is also the case when the application is started because the start usually takes place without a defined pattern.

Let's try out the navigation by allowing a list element to be clicked in a list. When a list element is clicked, we want to trigger the navigation, that is, react to it with an event handler. In the `content` aggregation, insert the list in Listing 7.6 into your *Main.view.xml*.

```
<List id="carrierList" headerText="Carrier" items="{/CarrierCollection}">
  <StandardListItem id="listItemTemplate" title="{CARRNAME}" description=
"{carrid}"
    icon="/img/{carrid}.png" type="Navigation" press="onNavToDetail"/>
</List>
```

Listing 7.6 List Showing the Carrier Collection with a Standard List Item

The value `Navigation` for the property `type` of a `StandardListItem` also displays a different cursor so that users know that the list element is clickable (see Figure 7.9). If you don't specify either `Navigation` or `Active` here, then `Inactive` remains set by default, and the list element can't be clicked.

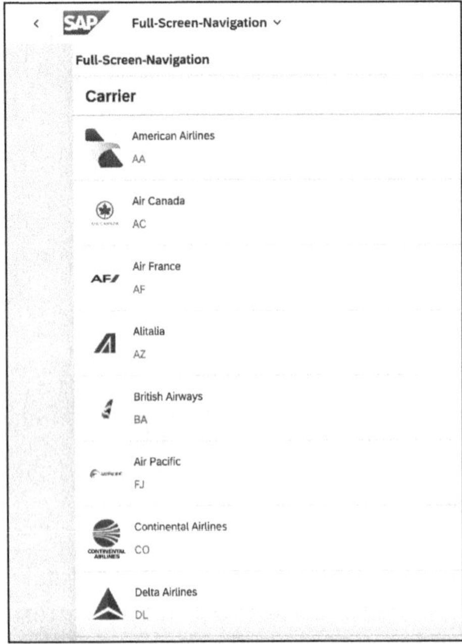

Figure 7.9 A List of the Carrier Collection Shown in the Main View

In a corresponding event handler in the controller, we react to the click and have the default `sap.m.routing.Router` delivered to us. With this router, we have the option of calling up a route using the method `navTo(sRouteName, oParameters?, oComponentInfo?, bSkipHistory?)`. As you can see in Listing 7.7, we use this method to navigate to the route called `Detail`.

```
private onNavToDetail() :  void {
  let oRouter = (this.getOwnerComponent() as UIComponent).getRouter();
  oRouter.navTo("Detail");
}
```

Listing 7.7 Method navTo of a Router Used to Navigate to the Route Called Detail

Now we can click on a list entry in the list and should be redirected to another page (see Figure 7.10).

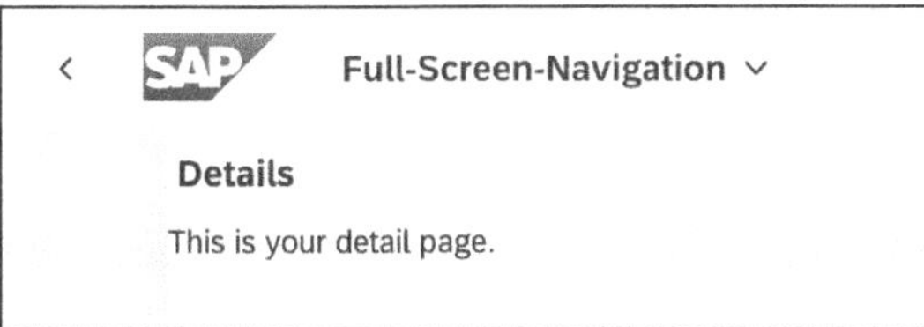

Figure 7.10 Successful Navigation to the Detail View

You can now see that the URL has changed. The pattern that is also linked to the route `Detail` has become part of the URL: *https://port8080-workspaces-... &/staticDetail.*

You can navigate back to the list using the standard browser navigation or the **Back** button of the SAP Fiori launchpad sandbox. This is also an advantage because the navigation in SAPUI5 is part of the browser navigation and thus automatically ends up in the history.

Routing with Mandatory Parameters

The option to navigate to another view is nice, but not very helpful in practice. In practice, you'll often want to display details about a selected object. In this case, you also need to pass the information about which object the details should be displayed for. Of course, you could also use runtime variables for this, but then you would have the problem that the information is lost after a refresh. Let's now see how navigation with a parameter can help here as well.

Let's make an adjustment to the route to the detail in *manifest.json* and specify a parameter between the curly brackets in the definition for the `pattern`. You can see this adjustment in Listing 7.8.

```
{
  "name": "Detail",
  "pattern": "mandatoryParamDetail/{param1}",
  "target": [ "Detail" ]
}
```

Listing 7.8 Curly Brackets Used for Mandatory Parameters in the Application Descriptor

This change will ensure that our previous navigation no longer works. You'll also find a note to this effect in the browser console. Figure 7.11 shows and indicates that we want to call a route, but the mandatory parameter named `param1` isn't filled in this navigation.

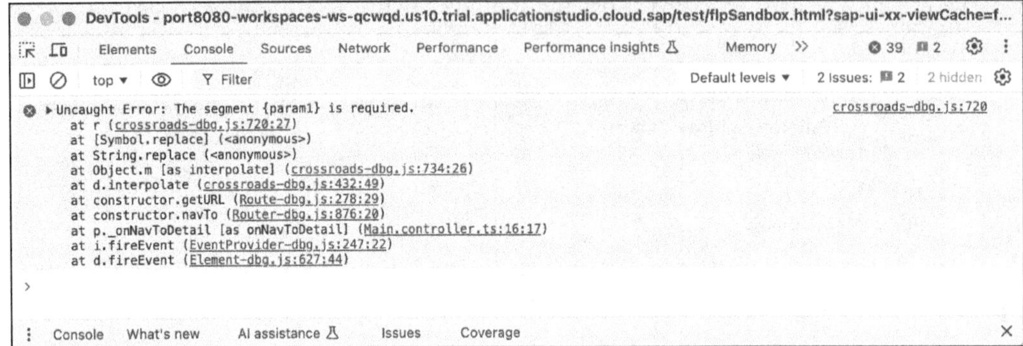

Figure 7.11 Error If We Navigate without Having Filled the Parameter

We can do this manually in the URL, but we won't be spared the task of also adapting the navigation logic in *Main.controller.ts*. You can see how we can fill this parameter in Listing 7.9.

```
private onNavToDetail() :  void {
  let oRouter = (this.getOwnerComponent() as BaseComponent).getRouter();
  oRouter.navTo("Detail", { param1: "Value1234" });
}
```

Listing 7.9 Routing Parameters Given in the navTo Method as the Second Parameter

Now, in the `navTo` method, we provide an object as the second parameter that populates this parameter. The parameter can be received, read, and used in the detail view. Unfortunately, the lifecycle methods aren't suitable for reading the parameters because they are only called once. However, we can register a listener in one of these lifecycle methods that's called every time the detail view is navigated to. This listener for the `patternMatched` event of a route is now called every time we navigate to the **Details** page and the pattern is found (see Listing 7.10).

```
import MessageBox from "sap/m/MessageBox";
// ...
public onInit() : void {
  let oRouter = (this.getOwnerComponent() as UIComponent).getRouter(),
    oRoute = oRouter.getRoute("Detail");
  oRoute?.attachPatternMatched(this.onPatternMatched, this);
}
private onPatternMatched(oEvent: Event) {
  let oArguments = (oEvent as any).getParameters().arguments,
    sParam1 = oArguments.param1;
  MessageBox.show("Param1: " + sParam1, { title: "Routing Parameter" });
}
```

Listing 7.10 Using the patternMatched Event to Extract the Parameter after Navigating

If we now navigate to the **Details** page, a `sap.m.MessageBox` is opened, and the parameter we passed is displayed as a string based on the logic in our `onPatternMatched` method. This is the parameter that was passed during navigation (see Figure 7.12).

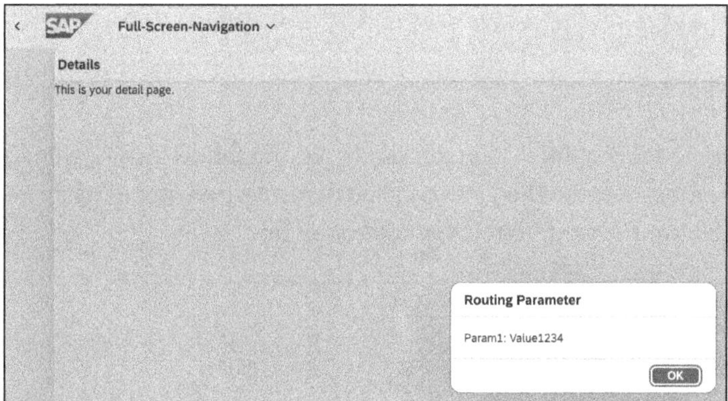

Figure 7.12 Parameter Extracted in the Target Route View and Displayed in a Message Box

Parameter Value Visible in URL

The advantage of the information that was passed to the detail view using parameters in the URL can also be a disadvantage. This parameter is part of the URL, which is why a refresh would already work in our case without us losing the information about which parameter was used to navigate to the detail view. However, don't forget that this information is visible in the URL and can therefore also be seen by the user. In our case, the URL was changed as follows after navigation:

```
https://port8080-workspaces-...&/mandatoryParamDetail/Value1234
```

Of course, you can also define more than just one parameter in the pattern. These can also have a wide variety of separators because all other parts of the pattern are then static and must appear exactly in the URL. Here's an example of how we could define three different mandatory parameters for a route:

```
"pattern": "manyMandatoryParamDetail/{param1}-{param2}/{param3}",
```

Before we use this navigation to display flight data in the detail view in the next section, we want to point out that there are several types of parameters. Not only can you define mandatory parameters but also optional parameters. Furthermore, you can also allow query parameters and read them in the detail view.

Routing with Optional Parameters

To enable optional parameters to be defined, the desired parameter in the pattern must be marked with colons at the beginning and end instead of curly brackets (see Listing 7.11).

```
{
  "name": "Detail",
  "pattern": "optionalParamDetail/:param1:",
  "target": [ "Detail" ]
}
```

Listing 7.11 Optional Parameters Defined in the Pattern

In Figure 7.13, you can see that even if we no longer send a parameter with navTo, this no longer results in an error, but instead the view is called up as desired. In our patternMatched event, we do read out the value, but it contains undefined.

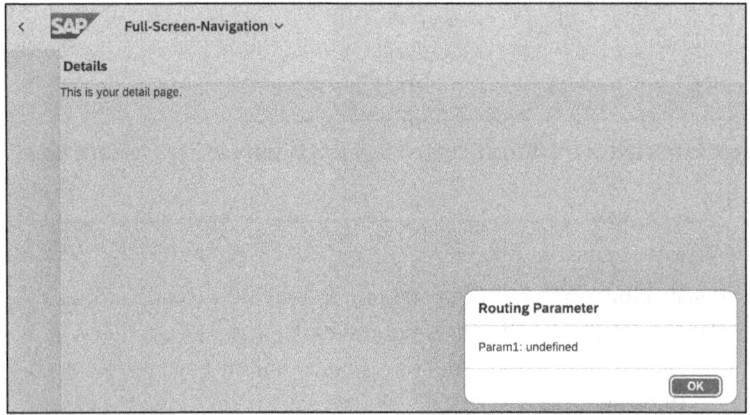

Figure 7.13 Empty Parameter Doesn't No Longer Resulting in an Error

Routing with Query Parameters

If we want to use query parameters, we add a question mark to the curly brackets for mandatory parameters and colons at the beginning and end for optional parameters, as we've already learned. This question mark marks the query parameter. This way, you can define a mandatory query parameter with {?paramName} and an optional query parameter with :?paramName:. We've also defined a mandatory query parameter in our example (see Listing 7.12).

```
{
  "name": "Detail",
  "pattern": "mandatoryParamDetail/{?queryParam}",
  "target": [ "Detail" ]
}
```

Listing 7.12 Mandatory Query Parameter Defined in the Pattern

If we now want to navigate to the detail view, we also have to structure the parameters here differently. We can store any number of properties within the query parameter because now we're not tied to the number of parameters in the pattern (see Listing 7.13).

```
private onNavToDetail() :  void {
  let oRouter = (this.getOwnerComponent() as UIComponent).getRouter();
  oRouter.navTo("Detail", {
    queryParam: {
      value1: "Asdf",
      value2: 1234
    }
  });
}
```

Listing 7.13 Query Parameters Passed as an Object at the navTo Method

In the `patternMatched` event in *Detail.controller.ts*, we can also access these query parameters and display them (see Listing 7.14).

```
private onPatternMatched(oEvent: Event) {
  let oArguments = (oEvent as any).getParameters().arguments,
    oQueryParams = oArguments["?queryParam"];
  MessageBox.show("Value1: " + oQueryParams.value1 + " | Value2: " +
    oQueryParams.value2, { title: "Routing Parameter" });
}
```

Listing 7.14 Query Parameters Extracted in the patternMatched Event Handler

These query parameters appear in the URL as follows:

https://port8080-workspaces-...&/mandatoryParamDetail/?value1=Asdf&value2=1234

The corresponding result is shown in Figure 7.14.

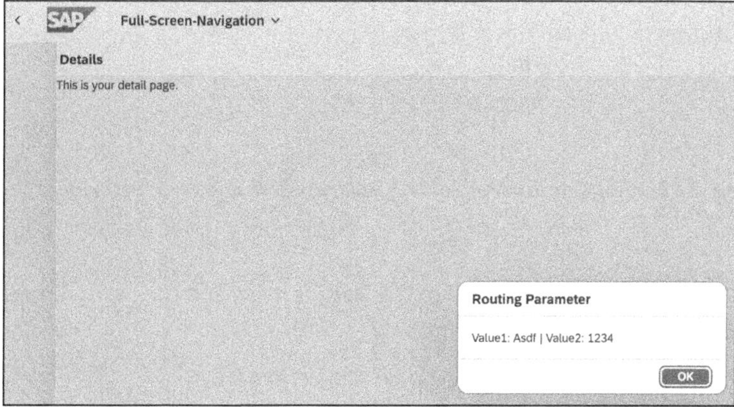

Figure 7.14 Query Parameters Extracted in the Detail View and Shown in a MessageBox

Now you understand why the route `RouteMain`, which is created by default, has a pattern of `:?query:` but can still be the default route. In addition, you now know that optional query parameters can be filled, but don't necessarily have to be.

Navigation Together with Element Binding

Whether full-screen navigation, flexible column layout, or split app, it often happens in practice that when an element is selected in the table/list, its detailed information should be displayed on the next page. For this reason, we want to provide you with a blueprint for this task. You've already learned everything we need to know about this throughout the previous chapters in this book: aggregation binding, event handler, navigation with parameters, and element binding.

Let's switch back to a mandatory parameter, but this time with the name `path` (see Listing 7.15).

```
{
  "name": "Detail",
  "pattern": "detail/{path}",
  "target": [ "Detail" ]
}
```

Listing 7.15 Pattern Switched Back to a Mandatory Parameter Called Path

In the event handler that is triggered when a row in the list is clicked, we adjust the logic so that the path of the selected element is read. We can access the path via the event object, which is written to the `oEvent` parameter. We specify the path when navigating to the detail view. We recommend using an `encodeURIComponent` here so that you don't have to worry about special characters that end up in the URL due to parameter-based navigation and thus falsify it (see Listing 7.16).

```
private onNavToDetail(oEvent: Event) :  void {
  let oRouter = (this.getOwnerComponent() as UIComponent).getRouter(),
    oBindingContext = (oEvent as any).getSource().getBindingContext(),
    oObject = oBindingContext?.getObject(),
    sPath = oBindingContext?.getPath();
  oRouter.navTo("Detail", { path: encodeURIComponent(sPath) });
}
```

Listing 7.16 Extracting the Binding Context Path of the Selected Item and Pass It as a Parameter

If we now click on a list item, we're navigated to the **Details** page, and the path that we have URI-encoded is entered in the URL like this:

https://port8080-workspaces-...&/detail/%252FCarrierCollection('AA')

In the detail view, in the listener of the `patternMatched` event, we'll receive this path, restore it to its original state using `decodeURIComponent`, and then bind the path to the detail view using `bindElement` (see Listing 7.17).

```
private onPatternMatched(oEvent: Event) {
  let oArguments = (oEvent as any).getParameters().arguments,
    sPath = decodeURIComponent(oArguments.path);
  this.getView()?.bindElement(sPath);
}
```

Listing 7.17 The Extracted Path Used for Element Binding after Decoding

Now we use our knowledge of `sap.ui.layout.form.SimpleForm` to incorporate it into the detail view with labels and texts based on our data from a carrier. There we access the individual properties via relative paths (see Listing 7.18). These can be resolved because a specific entity has been bound to the page via the element binding.

```
<f:SimpleForm id="detail_simpleform" editable="true" layout=
"ResponsiveGridLayout" labelSpanXL="3" labelSpanL="3" labelSpanM="3"
labelSpanS="12" emptySpanXL="4" emptySpanL="4" emptySpanM="4" emptySpanS="0"
columnsXL="1" columnsL="1" columnsM="1">
  <f:content>
    <Label id="labelLogo" labelFor="avatarLogo" text="Logo"/>
    <Avatar id="avatarLogo" src="/img/{carrid}.png" displaySize="S"
displayShape="Square" showBorder="true"/>
    <Label id="labelCarrid" labelFor="textCarrid" text="ID"/>
    <Text id="textCarrid" text="{carrid}"/>
    <Label id="labelCarrname" labelFor="textCarrname" text="Name"/>
    <Text id="textCarrname" text="{CARRNAME}"/>
    <Label id="labelWebsite" labelFor="linkWebsite" text="Website"/>
    <Link id="linkWebsite" text="{URL}"/>
    <Label id="labelCurr" labelFor="textCurr" text="Currency"/>
    <Text id="textCurr" text="{CURRCODE}"/>
  </f:content>
</f:SimpleForm>
```

Listing 7.18 SimpleForm Showing the Details of a Carrier

The result after navigation is shown in Figure 7.15.

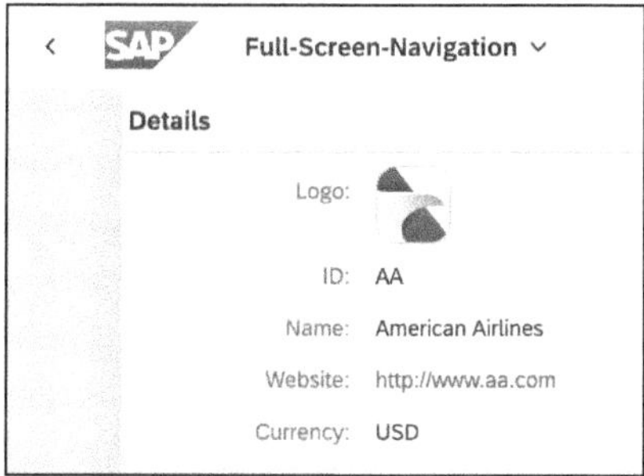

Figure 7.15 SimpleForm Showing the Details of a Carrier

A Proper Routing Configuration Allows Deep Links

The advantage of this approach is that this **Details** page of the application can be saved as a *deep link* via a bookmark or used for sharing. In doing so, the full UI state of the page, which is contained in the URL, should be restored. For example, flexible layouts such as list detail views receive a new URL when another element is selected, which can be used as a deep link.

When a deep link is accessed, the full-page state should be restored as far as technically feasible. If the target object doesn't exist or isn't accessible, a blank page or a bypassed target such as a **Not Found** page should be displayed. For shared links, displayed content may vary depending on user permissions.

But what happens if the URL is changed manually, or the app is customized in such a way that a previously saved URL no longer works? We want to react to this event as well, so we prepare a new view called *NotFound.view.xml*. You can see the content of this view in Listing 7.19.

```
<mvc:View controllerName="at.clouddna.fullscreennav.controller.NotFound"
xmlns:mvc="sap.ui.core.mvc" displayBlock="true" xmlns:f="sap.ui.layout.form"
xmlns="sap.m">
  <IllustratedMessage id="illustratedMessageNotFound" illustrationType=
"sapIllus-PageNotFound">
    <additionalContent>
      <Button text="Take me back to the list" press="onNavToMain"/>
    </additionalContent>
  </IllustratedMessage>
</mvc:View>
```

Listing 7.19 Not Found View Created to Show an Illustrated Message

To go with this, we create a controller `NotFound.controller.ts` that doesn't do much except provide a method for navigating back to the list of carriers. As you can see in Listing 7.20, we've used the fourth parameter in the `navTo` method to determine that this navigation shouldn't be included in the browser history.

```
import Controller from "sap/ui/core/mvc/Controller";
import UIComponent from "sap/ui/core/UIComponent";
/**
 * @namespace at.clouddna.fullscreennav.controller
 */
export default class NotFound extends Controller {
  /*eslint-disable @typescript-eslint/no-empty-function*/
  public onInit(): void {}
  private onNavToMain(){
```

```
    (this.getOwnerComponent() as UIComponent).getRouter().navTo("RouteMain",
undefined, undefined, true);
  }
}
```

Listing 7.20 Navigating Back to the List without Saving This Navigation to the Browser History

The name of the target pointing to this NotFound view is stored as bypassed in the sap.ui5.router.config. This bypassed target is then automatically called by the framework when an invalid pattern is recognized in the URL (see Listing 7.21).


```
"routing": {
  "config": {
    // ...
    "bypassed": { "target": "NotFound" }
  }
},
```

Listing 7.21 Bypassed Target Defined at the Router "config" in the Application Descriptor

Let's assume that the URL used to look like the one following this paragraph, and a user saved this URL as a bookmark in their browser. After a long period of absence, the application was further developed, and the pattern of the URL was changed from detail-PageCarrier/{path} to our current detail/{path}. So, the URL that was saved by the user could no longer be resolved by the framework:

https://port8080-workspaces-...&/detailPageCarrier/%252FCarrierCollection('AA')

To ensure that the application doesn't end up in an error state, but that this unwanted navigation can also be intercepted, we've stored the bypassed target, which now automatically redirects the user to the NotFound view in such a case (see Figure 7.16).

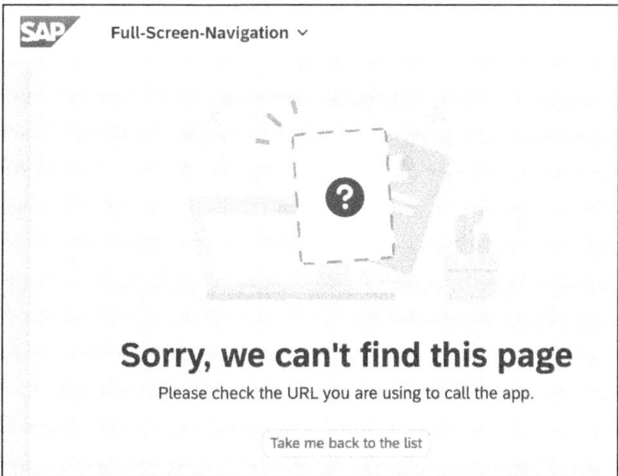

Figure 7.16 Not Found View Shown if the Pattern Doesn't Match Any Route

But what happens if a valid URL is called up, but the parameter contains an invalid path? This can happen, for example, because the carrier that was stored in the bookmark no longer exists or because the user enters a carrier in the URL by hand in the path that doesn't exist in this form:

https://port8080-workspaces-...&/detail/%252FCarrierCollection('ASDF')

We can also react to this, but in this case, a router event won't help us. Instead, we have to register an event handler for the `dataReceived` event in the `bindElement`. To access this event, we have to use a JavaScript object instead of a parameter as a string in the `bindElement` and make the definition a little more complex. This listener checks whether data is coming back, which, in our case, means that the carrier was found and the path was valid. If no data comes back, we navigate to the `NotFound` route and thus to the `NotFound` view. This navigation shouldn't be included in the history either. You can see this adjustment to the bind element in Listing 7.22.

```
private onPatternMatched(oEvent: Event) {
  let oArguments = (oEvent as any).getParameters().arguments,
    sPath = decodeURIComponent(oArguments.path);
  this.getView()?.bindElement({
    path: sPath,
    events: {
      dataReceived: (oResult: any) => {
        if(!oResult.getParameters().data){
          (this.getOwnerComponent() as
UIComponent).getRouter().navTo("NotFound", undefined, undefined, true);
        }
      }
    }
  });
}
```

Listing 7.22 If No Data Received in the Element Binding, Navigate to the Not Found View

But remember our adjustments regarding a bypassed target? For a bypassed target, we didn't need a route, only the target. So, we still need to create a route for our `NotFound` target:

```
{ "name": "NotFound", "pattern": "notFound", "target": [ "NotFound" ] }
```

Now the not found view is displayed here as well if the URL itself is correct, but the path of a carrier is invalid.

7.3.2 Split App

With the split app, we want to display two views at the same time. This UI component was formerly known as the master-detail page, which is reflected in some property, event, and aggregation names to maintain backward compatibility. View A usually has a selection (e.g., a list), with view B, which is wider, displaying the details of this selection (see Figure 7.17). In our example, we'll initially limit ourselves to an XML view that contains both pages A and B.

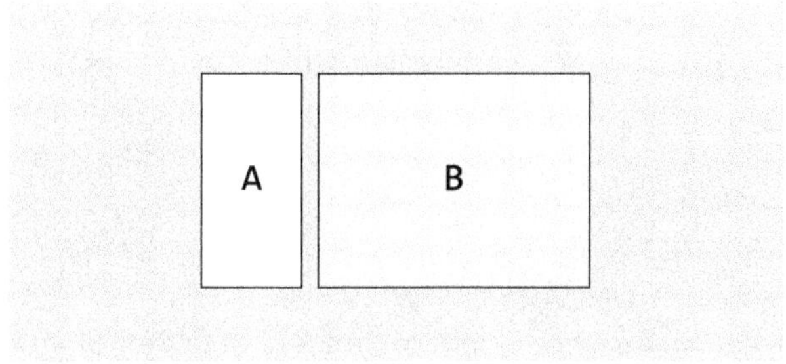

Figure 7.17 Split App Showing Two Pages at the Same Time

We've created a new SAPUI5 application at.clouddna.splitapp. This has again integrated the OData service from RMTSAMPLEFLIGHT and created an initial view called Main. As usual, we've activated TypeScript.

In this created Main view, we now insert the sap.m.SplitApp UI component. This component has two important aggregations: masterPages and detailPages. The masterPages are for the sap.m.Page elements that will be displayed in the left-hand area. In our case, the list that you already built in the full-screen navigation exercise is inserted here. The detailPages aggregation contains the page elements that will display the actual content. Here, we copy the simple form that you've already built for the full-screen navigation in the detail view. We could have done without the main view entirely and done everything in the app view, but we want to keep all options open for the future, which is why we're working with this main view as the default route. You can see the content of the view in Listing 7.23.

```
<SplitApp id="splitAppContainer" initialDetail="detail" initialMaster=
"listPage">
  <masterPages>
    <Page id="listPage" title="Carrier" backgroundDesign= "List">
      <!-- List belongs here -->
    </Page>
  </masterPages>
  <detailPages>
    <Page id="detail" title="Details" backgroundDesign= "Solid">
      <!-- SimpleForm belongs here -->
    </Page>
  </detailPages>
</SplitApp>
```

Listing 7.23 Main View Containing a SplitApp UI Component

We've now completed the first step of setting up the basic layout (see Figure 7.18). You'll see that when you click on a list element, nothing is displayed in the **Details** page. This is understandable because our old logic with the navigation no longer works.

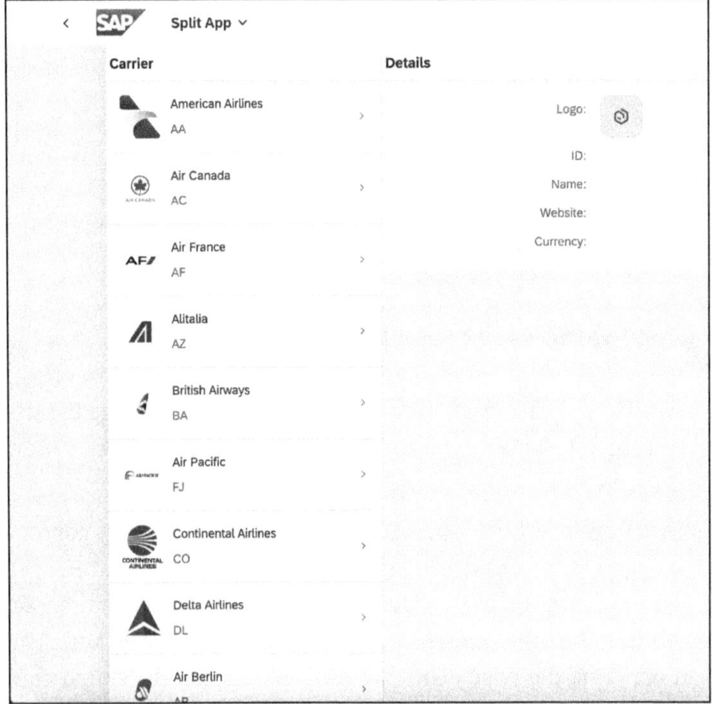

Figure 7.18 Split App Showing a List of Carriers But No Details of a Selected One

To display the selected list items on the left in the content area on the right, we have to adjust our `onNavToDetail` method, but instead of a navigation, simply set the element binding for the **Details** page. Because we're in the same view, we can easily access this page using `byId("detail")`. The entire method is shown in Listing 7.24.

```
private onNavToDetail(oEvent: Event) {
  let oBindingContext = (oEvent as any).getSource().getBindingContext(),
    sPath = oBindingContext?.getPath(),
    oDetailPage = this.getView()?.byId("detail") as Page;
    oDetailPage.bindElement(sPath);
}
```

Listing 7.24 Method onNavToDetail Adjusted So Element Binding Is Set to the Details Page

If you now select an element on the left, the binding is set on the right via element binding. After that, the relative paths can be resolved automatically, and the data of a carrier can be displayed (see Figure 7.19).

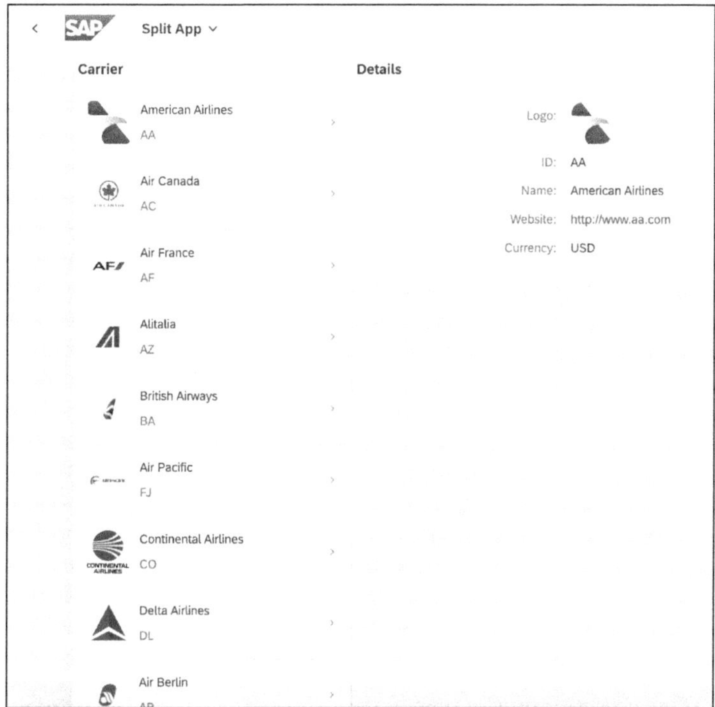

Figure 7.19 Thanks to Element Binding, the Detail Page Already Works

We now want to adjust the app so that the first element in the list is always preselected when the application is initially started. We can achieve this by registering an event handler for the `dataReceived` event when data binding the aggregation `items`:

```
<List id="carrierList" items="{ path: '/CarrierCollection', events: {
dataReceived: '.onDataReceived' } }">
```

The implementation of this method is shown in Listing 7.25. First, we check whether data has been returned at all or whether the list is empty. If there are elements in the list, we take the first element, construct the path, and also create an element binding to the details page.

```
private onDataReceived(oEvent: Event) {
  let oFirstItem = (oEvent as any).getParameters().data.results[0];
  if(oFirstItem){
    let oDetailPage = this.getView()?.byId("detail") as Page,
    oModel = this.getView()?.getModel() as ODataModel,
    sPath = oModel.createKey("/CarrierCollection", { carrid: oFirstItem.carrid
});
```

```
    oDetailPage.bindElement(sPath);
    }
}
```

Listing 7.25 DataReceived Event Used to Preselect the First Item in the List

But this event has a shortcoming. To demonstrate it, we add a refresh button to our list. This button is placed in the aggregation customHeader and replaces the default header. The button is aligned to the right thanks to the ToolbarSpacer. When this button is clicked, the onRefreshList method is called (see Listing 7.26).

```
<Page id="listPage" title="Carrier" backgroundDesign= "List">
  <customHeader>
    <Toolbar id="toolbarMaster">
      <Title id="titleCarrier" text="Carrier"/>
      <ToolbarSpacer id="toolbarSpacerMaster"/>
      <Button id="buttonRefresh" icon="sap-icon://refresh" type="Transparent"
press="onRefreshList"/>
    </Toolbar>
  </customHeader>
  <!-- ... -->
</Page>
```

Listing 7.26 Toolbar Added as a Custom Header to the List Page

This method does nothing more than access the items binding of the list and use the refresh() method to update the list data (see Listing 7.27).

```
private onRefreshList(){
  let oListBinding = this.getView()?.byId("carrierList")?.getBinding("items");
  oListBinding?.refresh();
}
```

Listing 7.27 Refreshing the List Using the Refresh Method of the Aggregation Binding

The problem now is that a refresh reloads the data, and the dataReceived event is triggered again. Our event handler ensures that the first element in the list is always selected with each refresh. In doing so, we don't take into account whether the user may have already made a different selection. To prevent this from happening, we create a member variable in our controller. This variable bInitial is initially set to true. We also request the value in the onDataReceived event handler, as well as the content of the first element. If we've set the element binding initially, we set the member variable to false so that this logic is no longer executed when a refresh occurs (see Listing 7.28).

```
private bInitial = true;
private onDataReceived(oEvent: Event) {
```

```
  let oFirstItem = (oEvent as any).getParameters().data.results[0];
  if(oFirstItem && this.bInitial){
    let oDetailPage = this.getView()?.byId("detail") as Page,
      oModel = this.getView()?.getModel() as ODataModel,
      sPath = oModel.createKey("/CarrierCollection", { carrid:
oFirstItem.carrid });
      //sPath = "/" + oEvent.getSource().aKeys[0];
    oDetailPage.bindElement(sPath);
    this.bInitial = false;
  }
}
```

Listing 7.28 Member Variable Used to Decide Whether the dataReceived Event Is Fired the First Time

A further improvement is to use the `SingleSelectMaster` mode of the list instead of making the list element clickable to load the detail view. This marks the selection visually and retains it until the element is deselected by the user making a new selection. However, we now also have to use a list event, which is the `selectionChange` event:

```
<List id="carrierList" items="{ path: '/CarrierCollection', events: {
dataReceived: '.onDataReceived' } }" mode="SingleSelectMaster" selectionChange=
"onNavToDetail">
```

You now have to remove the `type` property and the `press` event from the list item; otherwise, these two event handlers will interfere with each other.

The only problem we'll have as a result of the change is that we have to access the selected element differently in the event handler. `oEvent.getSource()` no longer contains the selected element, but an instance of the list itself. Of course, this is because the list is the trigger for the event. However, we can use `oEvent.getParameters().listItem` to access the element that is entered as the selected item in the list. We only have to adjust the first line and everything else can remain the same:

```
let oBindingContext = (oEvent as
any).getParameters().listItem.getBindingContext(),
```

As you can see in Figure 7.20, this mode preserves the selection in the list, and, apart from the detail view, the user knows which element is currently selected.

To top it all off and not only set the element binding for the first element at the initial start but also mark it as selected in the list, we adjust our `dataReceived` event handler again. There, we now access the list directly and use `getItems()` to get a list of all elements in the list. This helps us because we now not only need the binding context path of the first element but also an instance of this item.

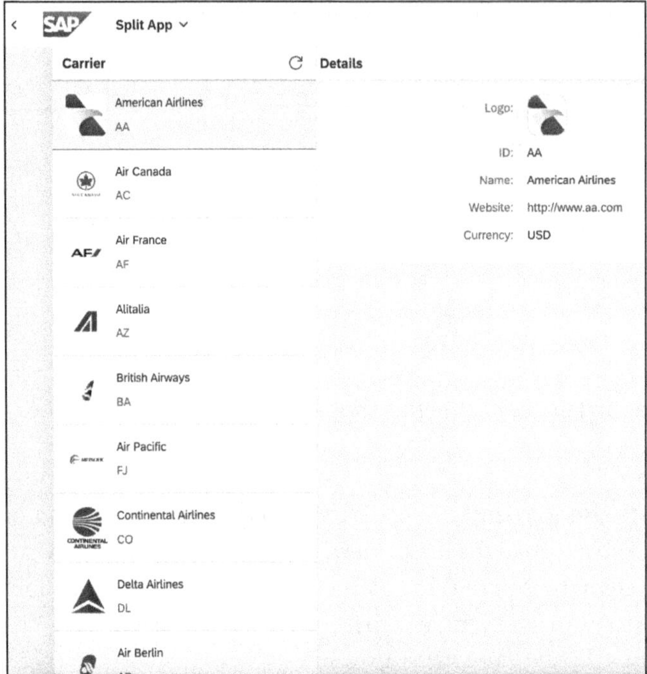

Figure 7.20 Using the SingleSelectMaster Mode of the List

With this instance, we can also call the `setSelectedItem` method in addition to the previous logic and pass the instance of this item there (see Listing 7.29).

```
private onDataReceived(oEvent: Event) {
  let oList = this.getView()?.byId("carrierList") as List,
    aItems = oList.getItems();
    if (aItems.length > 0 && this.bInitial) {
      let oDetailPage = this.getView()?.byId("detail") as Page,
        oModel = this.getView()?.getModel() as ODataModel,
        oFirstItem = aItems[0],
        sPath = oFirstItem.getBindingContext()?.getPath() as string;
      oDetailPage.bindElement(sPath);
      oList.setSelectedItem(oFirstItem);
    this.bInitial = false;
  }
}
```

Listing 7.29 Marking the Item as Selected in the dataReceived Event

We have one more improvement to offer. If we've selected an element in the list (not the first one) and refresh the page, this selection is lost. We could also not take advantage of deep links. We'll now use the query parameter already defined by default. You can see a section of `RouteMain` in Listing 7.30.

```
"routes": [
  {
    "name": "RouteMain",
    "pattern": ":?query:",
    "target": [ "TargetMain" ]
  }
],
```

Listing 7.30 Default Route Called RouteMain with an Already-Defined Optional Query Parameter

First, we adjust the event handler for the `selectionChange` event. In addition to the previous logic, in which we read the path of the selected element and set the element binding, we'll insert a navigation. In this navigation, we always navigate to the `RouteMain` in the `navTo`, but we set the selected path as a query parameter (see Listing 7.31).

```
private onNavToDetail(oEvent: Event) {
  let oBindingContext = (oEvent as
any).getParameters().listItem.getBindingContext(),
    oObject = oBindingContext?.getObject(),
    sPath = oBindingContext?.getPath(),
    oDetailPage = this.getView()?.byId("detail") as Page,
    oRouter = (this.getOwnerComponent() as UIComponent).getRouter();
  oDetailPage.bindElement(sPath);
  oRouter.navTo("RouteMain", {
    query: { path: encodeURIComponent(sPath) }
  });
}
```

Listing 7.31 Setting the Binding Context Path as a Query Parameter if a List Item Selected

Next, we register an event handler for the `patternMatched` event in the `onInit` lifecycle method (see Listing 7.32).

```
public onInit(): void {
  let oRouter = (this.getOwnerComponent() as UIComponent).getRouter(),
    oRoute = oRouter.getRoute("RouteMain");
  oRoute?.attachPatternMatched(this.onPatternMatched, this);
}
```

Listing 7.32 Registering a Listener to the patternMatched Event

In this `patternMatched` event handler, we read the query parameters. If a parameter named `path` is set in the `query` parameter, we store this path in a member variable named `sInitialPath` so that we can access and apply it later (see Listing 7.33).

```
private sInitialPath = "";
private onPatternMatched(oEvent: Event) {
  let oArguments = (oEvent as any).getParameters().arguments;
  if (oArguments["?query"] && oArguments["?query"].path) {
    this.sInitialPath = decodeURIComponent(oArguments["?query"].path);
  }
}
```

Listing 7.33 Query Parameter Extracted in the patternMatched Event

We then adjust the event handler for the `dataReceived` event again. Now we take into account the member variable `sInitialPath` in our `IF` queries. If this is set, it takes precedence after a refresh and is used for the initial setting of a selected element. If it's not set, the first element from the list is initially taken (see Listing 7.34).

```
private onDataReceived(oEvent: Event) {
  let oList = this.getView()?.byId("carrierList") as List,
    aItems = oList.getItems();
  if (aItems.length > 0 && this.bInitial) {
    let oDetailPage = this.getView()?.byId("detail") as Page,
      oModel = this.getView()?.getModel() as ODataModel,
      oFirstItem: ListItemBase,
      sPath = "";
    if (this.sInitialPath) {
      let aFilteredItems = aItems.filter((oItem) => oItem.getBindingContext()?
.getPath() === this.sInitialPath);
    if (aFilteredItems.length > 0) {
      oFirstItem = aFilteredItems[0];
      sPath = this.sInitialPath;
    } else {
      oFirstItem = aItems[0];
      sPath = oFirstItem.getBindingContext()?.getPath() as string;
    }
  } else {
    oFirstItem = aItems[0];
    sPath = oFirstItem.getBindingContext()?.getPath() as string;
  }
  oDetailPage.bindElement(sPath);
  oList.setSelectedItem(oFirstItem);
  this.bInitial = false;
  }
}
```

Listing 7.34 Adjusting the dataReceived Event Handler to Take into Account the Member Variable

If you now select an entry in the list, you won't notice anything of the navigation because we navigate from the same view to the same view, but you'll see a change in the URL. You'll see that the query parameter is set:

https://port8080-workspaces-...&/?path=%25252FCarrierCollection('DL')

What would a detailed view of a carrier be if we didn't show the associated flights? For this reason, we insert a sap.m.Table below the SimpleForm that loads the flights of a carrier in descending order by flight date using the navigation property carrierFlights (see Listing 7.35).


```
<Table id="tableFlights" headerText="Flights" items="{ path: 'carrierFlights',
sorter: { path: 'fldate', descending: true } }" alternateRowColors="true"
growing="true" growingThreshold="200">
  <columns>
    <Column id="columnConn">
      <Label id="columnLabelConn" text="Connection"/>
    </Column>
    <Column id="columnFldate">
      <Label id="columnLabelFldate" text="Flight date"/>
    </Column>
    <Column id="columnSeatsMax">
      <Label id="columnLabelSeatsMax" text="Max. capacity"/>
    </Column>
    <Column id="columnSeatsOcc">
      <Label id="columnLabelSeatsOcc" text="# of occ. seats"/>
    </Column>
  </columns>
  <items>
    <ColumnListItem id="columnListItemTemplate">
      <cells>
       <ObjectIdentifier title="{connid}" />
       <ObjectIdentifier title="{ path: 'fldate', type:
'sap.ui.model.odata.type.Date'}" />
       <Text text="{SEATSMAX}"/>
       <Text text="{SEATSOCC}"/>
      </cells>
    </ColumnListItem>
  </items>
</Table>
```

Listing 7.35 A Table Showing All the Flights of a Selected Carrier

But, unfortunately, as also shown in Figure 7.21, we don't yet see any flights by the individual carriers. What could be the reason for this?

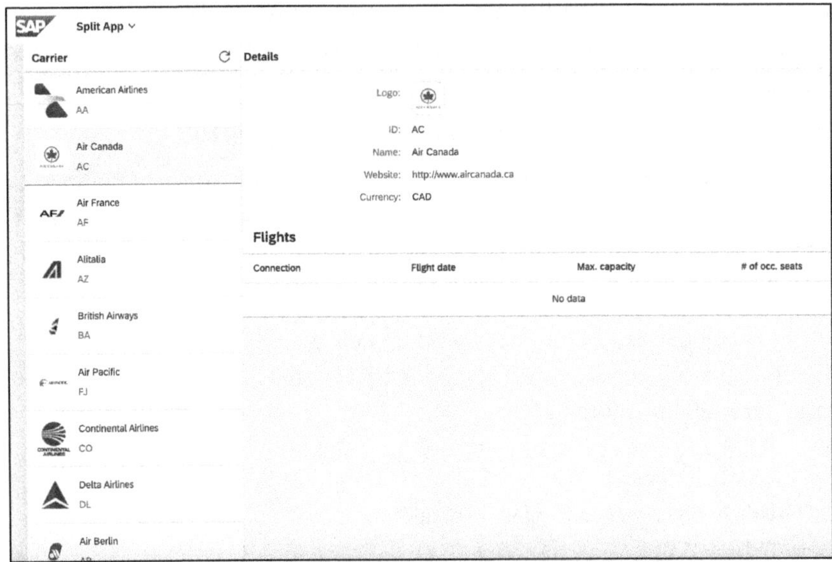

Figure 7.21 Table Showing No Data

This is because we haven't specified the navigation property that we're using in the element binding. These aren't called up implicitly, but must be defined explicitly. Now insert the following wherever you use the `bindElement` method:

```
oDetailPage.bindElement({ path: sPath, parameters: { "$expand":
'carrierFlights' } });
```

Now we can also see the flights of a selected carrier, as shown in Figure 7.22.

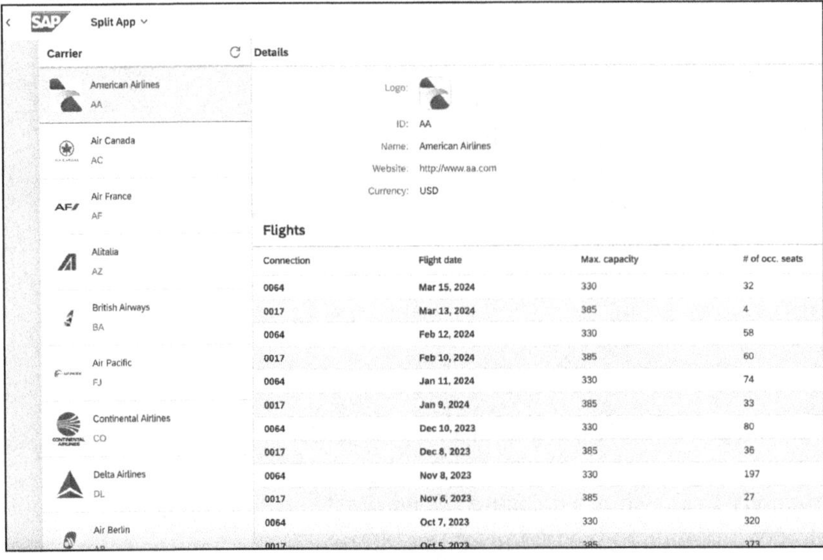

Figure 7.22 Table Showing All the Flights of a Specific Carrier

We make use of expression binding and formatters to provide a user-friendly and visually appealing way to show the remaining seats available. To do this, we replace the two existing columns for the seats with a single column:

```
<Column id="columnSeatsAv">
  <Label id="columnLabelSeatsAv" text="Seats available"/>
</Column>
```

The cell associated with this new column calculates the number of seats still available and uses two formatters that we'll implement in a moment:

7

```
<ObjectStatus id="main_objectstatus_seatsavtemplate text="{=${SEATSMAX}-
${SEATSOCC}}"
  state="{parts: ['SEATSMAX','SEATSOCC'], formatter: '.stateAvSeats'}"
  icon="{parts: ['SEATSMAX','SEATSOCC'], formatter: '.iconAvSeats'}"/>
```

The formatters go into the controller and return the corresponding colors and icons to the object status (see Listing 7.36).

```
private stateAvSeats(seats: int, seatsocc: int): string {
  let iDiff = seats - seatsocc;
  if (iDiff <= 15) { return "Error"; }
  else if (iDiff <= 25) { return "Warning"; }
  else { return "Success"; }
}
private iconAvSeats(seats: int, seatsocc: int): string {
  let iDiff = seats - seatsocc;
  if (iDiff <= 15) { return "sap-icon://message-error"; }
  else if (iDiff <= 25) { return "sap-icon://message-warning"; }
  else { return "sap-icon://message-success"; }
}
```

Listing 7.36 Formatters for Calculating the Icons and Colors Based on the Available Number of Seats

Now we've built the ultimate split app, which both displays a list of all carriers and also visualizes them in the detailed view with the individual flights and their available seats (see Figure 7.23). Even after refreshing or bookmarking the page, the selection isn't lost, which is what we used the routing concepts and query parameters for.

Now we have a list of flights per carrier, but how are the number of seats determined? Which passengers are on a flight? And from which destination to which are they scheduled? To find an answer to all these questions, we'll get to know the flexible column layout and learn how to display not only carriers and carrier details but also flight details in a user-friendly and responsive design.

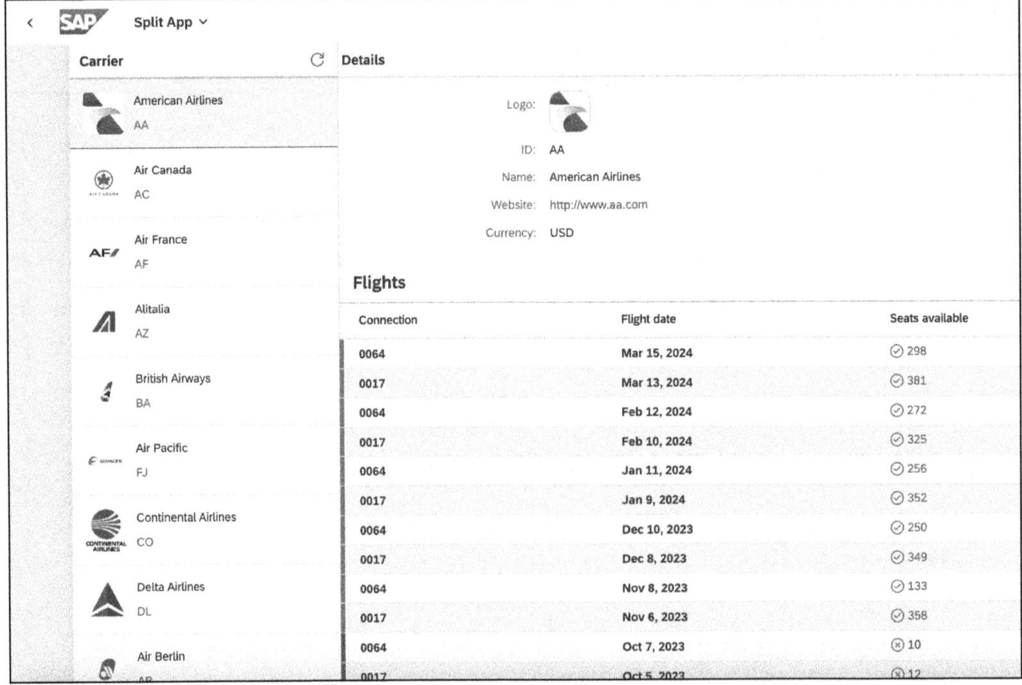

Figure 7.23 Split App Showing a List of Carriers and a Selected Carrier's Details Page with All the Flights

7.3.3 Flexible Column Layout

With the *flexible column layout*, we can display up to three views or columns simultaneously. Each of these columns can be filled with any control in the view, but the classic options are `sap.f.DynamicPage` or `sap.uxap.ObjectPageLayout`. Figure 7.24 shows how these columns are divided by default. If you add the second column, you can split the view between A and B at a ratio of 33:66. If you go down a level and also want to display the third column, the result is a 25:50:25 split. The user can change this split at runtime because there's a slider between each view that can be used to change the size of a column.

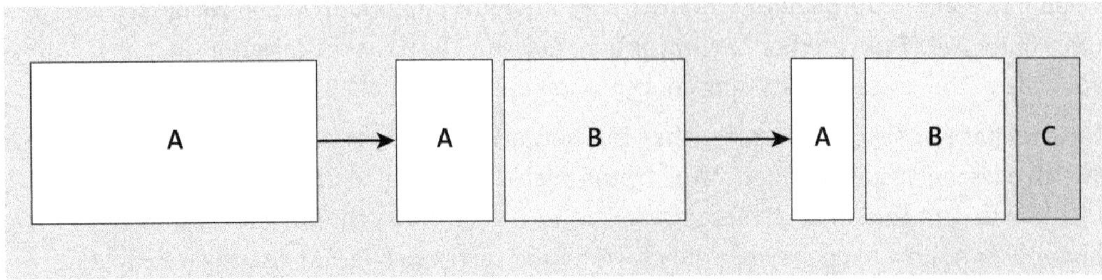

Figure 7.24 Default Ratio between the Columns within the Flexible Column Layout

The change for the user only remains for the runtime. However, the developer can provide other divisions in the standard code. These divisions are shown in Figure 7.25. Finally, as a developer, you offer corresponding buttons so that a single column can be opened in full screen or closed again.

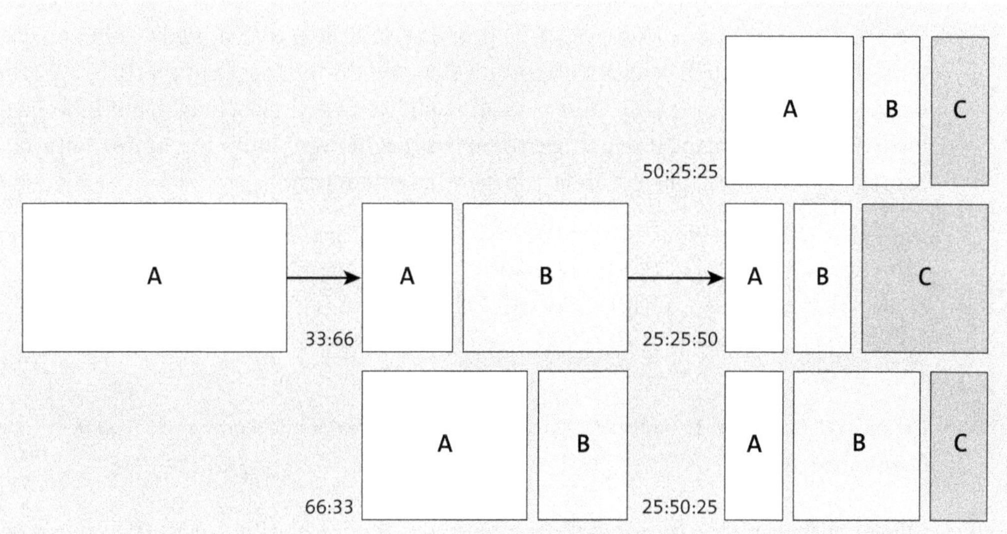

Figure 7.25 Different Possible Ratios for the Columns

The flexible column layout is shipped within the namespace `sap.f` and is described in detail in the official documentation, which also provides important additional information to be aware of during implementation (see Figure 7.26).

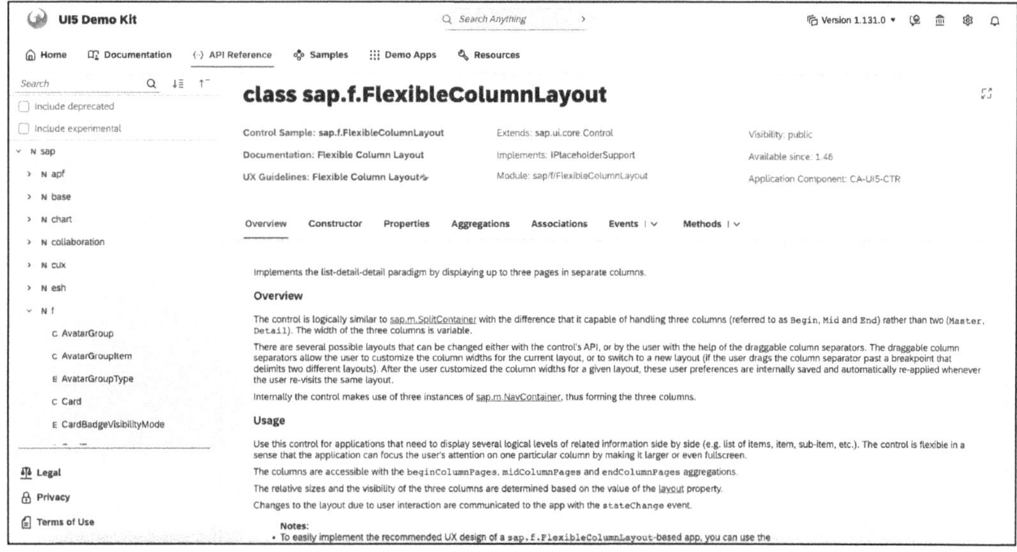

Figure 7.26 The Flexible Column Layout Shipped in the sap.f Namespace

We've created a new SAPUI5 application called `at.clouddna.fcl`. This app in turn consumes the `RMTSAMPLEFLIGHT` and requires TypeScript for the application logic. This time, we have let `Carrier` name the initial view.

Let's start in the root view (*App.view.xml*) by replacing the container from `sap.m.App` to `sap.f.FlexibleColumnLayout` (see Listing 7.37). This `FlexibleColumnLayout` initially gets the property `layout` `OneColumn` set. This property indicates that initially only one column should be opened. It's very important that you set the `height` property to `100%` for the view because the `FlexibleColumnLayout` itself hasn't defined a height, but inherits it from the parent control. If you forget to set the height here, your `FlexibleColumnLayout` container won't be displayed, and the page will remain white.

```
<mvc:View xmlns:f="sap.f" controllerName="at.clouddna.fcl.controller.App"
xmlns:html="http://www.w3.org/1999/xhtml" height="100%" xmlns:mvc=
"sap.ui.core.mvc" displayBlock="true" xmlns="sap.m">
  <f:FlexibleColumnLayout id="fcl" layout="OneColumn" />
</mvc:View>
```

Listing 7.37 Container Used for Navigation in the Root View Replaced by sap.f.FlexibleColumnLayout

When configuring the router in the *manifest.json* file, we have to use a different router class the first time. The router `sap.f.routing.Router` was created specifically for the flexible column layout. This router supports this three-column layout and also offers additional functionalities. Furthermore, we adjust the `controlId` because here we store the ID that we also assigned to the `FlexibleColumnLayout` in our root view. It's also important that the `controlAggregation` is removed here (see Listing 7.38).

```
"routing": {
  "config": {
    "routerClass": "sap.f.routing.Router",
    "viewType": "XML",
    "async": true,
    "viewPath": "at.clouddna.fcl.view",
    "controlId": "fcl",
    "transition": "slide"
  }
},
```

Listing 7.38 Router Configuration Changed to Support the Flexible Column Layout

From now on, we no longer determine globally which view ends up in which aggregation, but rather separately for each target. For the targets, under `controlAggregation`, we insert either `beginColumnPages` (first column), `midColumnPages` (second column), or `endColumnPages` (third column). This way, the router knows in which aggregation the view that is opened via the routing must be inserted. For our initial view, which was created by the wizard, we've specified that it should be embedded in the first column (see Listing 7.39).

```
"targets": {
  "TargetCarrier": {
    "viewType": "XML",
    "viewId": "Carrier",
    "viewName": "Carrier",
    "controlAggregation": "beginColumnPages",
    "clearControlAggregation": true
  }
},
```

Listing 7.39 Property controlAggregation in Each Target Defining in Which Column It Should Be Embedded

Now let's copy the carrier list that we already used in the split app and insert it into our *Carrier.view.xml*. Instead of a `sap.m.P`, we now use the `sap.f.DynamicPage` (see Listing 7.40).

```
<mvc:View xmlns:f="sap.f" controllerName="at.clouddna.fcl.controller.Carrier"
xmlns:mvc="sap.ui.core.mvc" displayBlock="true" xmlns:uxap="sap.uxap" xmlns=
"sap.m">
  <f:DynamicPage id="dynamicPageId" >
    <f:title>
      <f:DynamicPageTitle>
        <f:heading>
          <Title text="Carrier"/>
        </f:heading>
      </f:DynamicPageTitle>
    </f:title>
    <f:content>
      <List id="carrierList" items="{/CarrierCollection}" mode=
"SingleSelectMaster" selectionChange="onNavToDetail">
        <StandardListItem id="listItemTemplate" title="{CARRNAME}" description=
"{carrid}" icon="/img/{carrid}.png" />
      </List>
    </f:content>
  </f:DynamicPage>
</mvc:View>
```

Listing 7.40 Using a List to Display the Carrier Collection in the Carrier View

When we now start the app, we see that the full screen list opens again (see Figure 7.27). The flexible column layout doesn't start with three columns, but only with one column that takes up the entire width, due to our property `layout`.

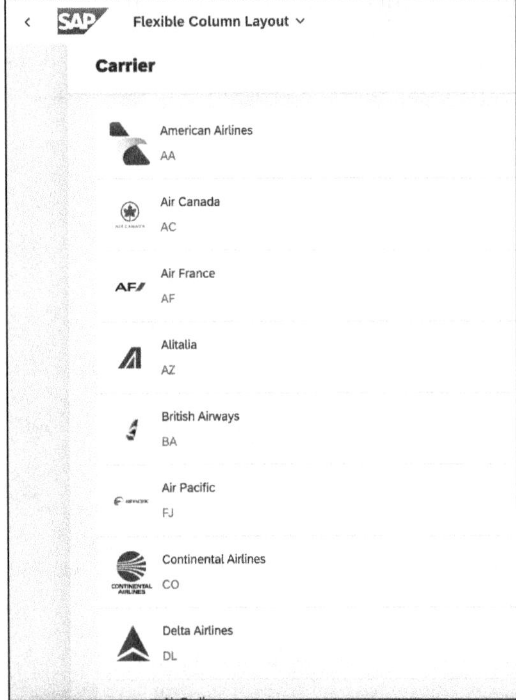

Figure 7.27 Flexible Column Layout Starting with One Column Showing the List of Carriers

Now let's create a second view called *CarrierDetails.view.xml*. In this view, we use a `sap.uxap.ObjectPageLayout` so that we have more options for structuring the data in the header and in the content area. We insert our familiar simple form for a carrier into the first `ObjectPageSection` and its `ObjectPageSubSection`, with the title `General Information`. The second `ObjectPageSection` and its `ObjectPageSubSection` contains the table of flights that we've also already built into the split app (see Listing 7.41).

```
<mvc:View xmlns:f="sap.ui.layout.form" controllerName=
"at.clouddna.fcl.controller.CarrierDetails" xmlns:mvc="sap.ui.core.mvc"
displayBlock="true" xmlns:uxap="sap.uxap" xmlns="sap.m">
  <uxap:ObjectPageLayout>
    <uxap:headerTitle>
      <uxap:ObjectPageDynamicHeaderTitle>
        <uxap:expandedHeading>
          <Title text="{CARRNAME}"/>
        </uxap:expandedHeading>
      </uxap:ObjectPageDynamicHeaderTitle>
    </uxap:headerTitle>
    <uxap:sections>
      <uxap:ObjectPageSection title="General Information">
        <uxap:subSections>
```

```
          <uxap:ObjectPageSubSection>
            <uxap:blocks>
              <!-- SimpleForm belongs in here -->
            </uxap:blocks>
          </uxap:ObjectPageSubSection>
        </uxap:subSections>
      </uxap:ObjectPageSection>
      <uxap:ObjectPageSection title="Flights">
        <uxap:subSections>
          <uxap:ObjectPageSubSection>
            <uxap:blocks>
              <!-- Table belongs in here -->
            </uxap:blocks>
          </uxap:ObjectPageSubSection>
        </uxap:subSections>
      </uxap:ObjectPageSection>
    </uxap:sections>
  </uxap:ObjectPageLayout>
</mvc:View>
```

Listing 7.41 ObjectPageLayout Showing the Carrier Details and Its Assigned Flights

To be able to call this newly created view, we first have to create a target (see Listing 7.42). For the target named `TargetCarrierDetails`, we must make sure that we assign the value `midColumnPages` to the `controlAggregation`. This is how the view is entered in the second column behind this target.

```
"TargetCarrierDetails": {
  "viewType": "XML",
  "viewId": "CarrierDetails",
  "viewName": "CarrierDetails",
  "controlAggregation": "midColumnPages",
  "clearControlAggregation": true
},
```

Listing 7.42 Creating a Target for the Second View for Insertion in the Second Column

Using a new route, namely `RouteCarrierDetails`, we want to address not just one but two targets simultaneously for the first time (see Listing 7.43). Of course, when the URL `carrier/{path}` is opened, both the first and second columns should be displayed at the same time. The mandatory parameter called `path` will contain the path to the carrier to be used in the second column for the element binding.

```
{
  "name": "RouteCarrierDetails",
  "pattern": "carrier/{path}/:?query:",
```

```
  "target": [ "TargetCarrier", "TargetCarrierDetails" ]
},
```

Listing 7.43 Route Contains Two Targets Simultaneously

To enable navigation to this view, we implement the corresponding event handler of the `selectionChange` event of the list. For this navigation, we provide the `layout` as a `query` parameter. With this layout, we specify that two columns should now be displayed, with the middle one taking up more space (`TwoColumnsMidExpanded`) (see Listing 7.44).

```
private onNavToDetail(oEvent: Event) {
  let oBindingContext = (oEvent as
any).getParameters().listItem.getBindingContext(),
    sPath = oBindingContext?.getPath(),
    oRouter = (this.getOwnerComponent() as UIComponent).getRouter();
  oRouter.navTo("RouteCarrierDetails", {
    path: encodeURIComponent(sPath),
    query: { layout: "TwoColumnsMidExpanded" }
  });
}
```

Listing 7.44 The Navigation Itself as Usual

But how is this `TwoColumnMidExpanded` read? To do this, we use an event handler that is called every time any route is matched. We implement this global event handler in the corresponding controller of the root view, that is, *App.controller.ts*. There, in the `onInit` method, we register the event handler for the `routeMatched` event. Furthermore, we prepare a `JSONModel` in which the current `layout` is stored. Last but not least, we store the instance of the `FlexibleColumnLayout` in a member variable so that we can access it more easily in the other methods (see Listing 7.45).

```
private oFCL: FlexibleColumnLayout;
public onInit(): void {
  let oOwnerComponent = (this.getOwnerComponent() as UIComponent);
  oOwnerComponent.getRouter().attachRouteMatched(this.onRouteMatched, this);
  oOwnerComponent.setModel(new JSONModel({ layout: "OneColumn" }), "ui");
  this.oFCL = this.getView()?.byId("fcl") as FlexibleColumnLayout;
}
```

Listing 7.45 Registering a Listener to the Global routeMatched Event

In the corresponding event handler of the `routeMatched` event, we check whether this `query` parameter is set to `layout`. If so, it's stored in the `JSONModel`; otherwise, we assume that only the first column is needed, and we hard-code `OneColumn` (see Listing 7.46).

```
private onRouteMatched(oEvent: Event) {
  let sLayout = (oEvent as any).getParameters().arguments["?query"]?.layout;
  if (!sLayout) { sLayout = "OneColumn"; }
  if (sLayout) { ((this.getOwnerComponent() as UIComponent).getModel("ui") as
JSONModel).setProperty("/layout", sLayout); }
}
```

Listing 7.46 routeMatched Event Handler Extracting the Layout of the Query Parameter If Set

For this value to be applied in the JSONModel of the FlexibleColumnLayout, we have to bind the layout property in the *App.view.xml* to it:

```
<f:FlexibleColumnLayout id="fcl" layout="{ui>/layout}" />
```

However, we don't want to take care of the corresponding strings that contain the ColumnLayout in the future. For this, SAP has created a helper that not only returns the next possible layout (according to the screen width) but will also provide us with further auxiliary functions later. For this reason, we'll provide a method called getHelper in the component controller (*Component.ts* or *Component.js*) that returns this helper sap.f.FlexibleColumnLayoutSemanticHelper and that we can access from anywhere (see Listing 7.47).

```
public getHelper() {
  let oFCL = (this.getRootControl() as any).byId("fcl") as
FlexibleColumnLayout,
    oParams = new URLSearchParams(window.location.search),
    oSettings = {
      defaultTwoColumnLayoutType: LayoutType.TwoColumnsMidExpanded,
      defaultThreeColumnLayoutType: LayoutType.ThreeColumnsMidExpanded,
      maxColumnsCount: oParams.get("max")
    };
  return FlexibleColumnLayoutSemanticHelper.getInstanceFor(oFCL, oSettings);
}
```

Listing 7.47 Method That Returns the FlexibleColumnLayoutSemanticHelper

For example, in onRouteMatched, we use the helper to get back the information for the first column using getNextUIState(0). This returns a JavaScript object that also contains the layout suggested by the helper if no layout has been specified by the query parameter (see Listing 7.48).

```
private onRouteMatched(oEvent: Event) {
  let sLayout = (oEvent as any).getParameters().arguments["?query"]?.layout;
  if (!sLayout) {
   let oNextUIState = (this.getOwnerComponent() as
any).getHelper().getNextUIState(0);
```

```
    sLayout = oNextUIState.layout;
  }
  if (sLayout) {
    this.getOwnerComponent().getModel("ui").setProperty("/layout", sLayout);
  }
}
```

Listing 7.48 The Helper Used to Suggest a Layout Appropriate to the User Screen Width

Using the layout helper makes much more sense from the second and third column onward. This is because, instead of writing our own method to determine a proper ratio based on the user's current screen width and zoom level, the helper we're using does it all for us. In the onNavToDetail event handler, we can use getHelper().getNextUIState(1) to get suggestions for the information that would make sense for the second column according to the helper. This way, we no longer statically assign TwoColumnsMidExpand, but rather the one that best suits the current screen width (see Listing 7.49).

```
private onNavToDetail(oEvent: Event) {
  let oBindingContext = (oEvent as
any).getParameters().listItem.getBindingContext(),
    sPath = oBindingContext?.getPath(),
    oNextUIState = (this.getOwnerComponent() as
any).getHelper().getNextUIState(1),
    oRouter = (this.getOwnerComponent() as UIComponent).getRouter();
  oRouter.navTo("RouteCarrierDetails", {
    path: encodeURIComponent(sPath),
    query: { layout: oNextUIState.layout }
  });
}
```

Listing 7.49 Helper Used When We Need to Get the Current or Next Possible Layout

In CarrierDetails.controller.ts, we'll register a corresponding listener for the patternMatched event. In this listener, we read the path parameter and use it for the element binding, including the $expand parameter for the flights (see Listing 7.50). Don't forget the formatters that we already used in the split app.

```
public onInit(): void {
  let oRouter = (this.getOwnerComponent() as UIComponent).getRouter(),
    oRoute = oRouter.getRoute("RouteCarrierDetails");
  oRoute?.attachPatternMatched(this.onPatternMatched, this);
}
private onPatternMatched(oEvent: Event) {
  let oArguments = (oEvent as any).getParameters().arguments,
    sPath = decodeURIComponent(oArguments.path);
```

```
    this.getView()?.bindElement({
      path: sPath,
      parameters: { "$expand": "carrierFlights" }
    });
}
```

Listing 7.50 PatternMatched Event Handler in the Carrier Detail Controller

Figure 7.28 shows that we can open the second column by clicking on a list element. The `patternMatched` method is used to bind the specified `path` from the carrier there, and this also allows the relative paths in the known `SimpleForm` and in the `Table` to be resolved. Thanks to the global `routeMatched` event, we don't have to read out the `layout` separately in each controller, but this is recorded and set for us. The big difference between this and the split app is that we've separated the pages into two separate views with their own controllers for the application logic and that, later on, the user can flexibly decide whether they want to stick with this ratio or change it.

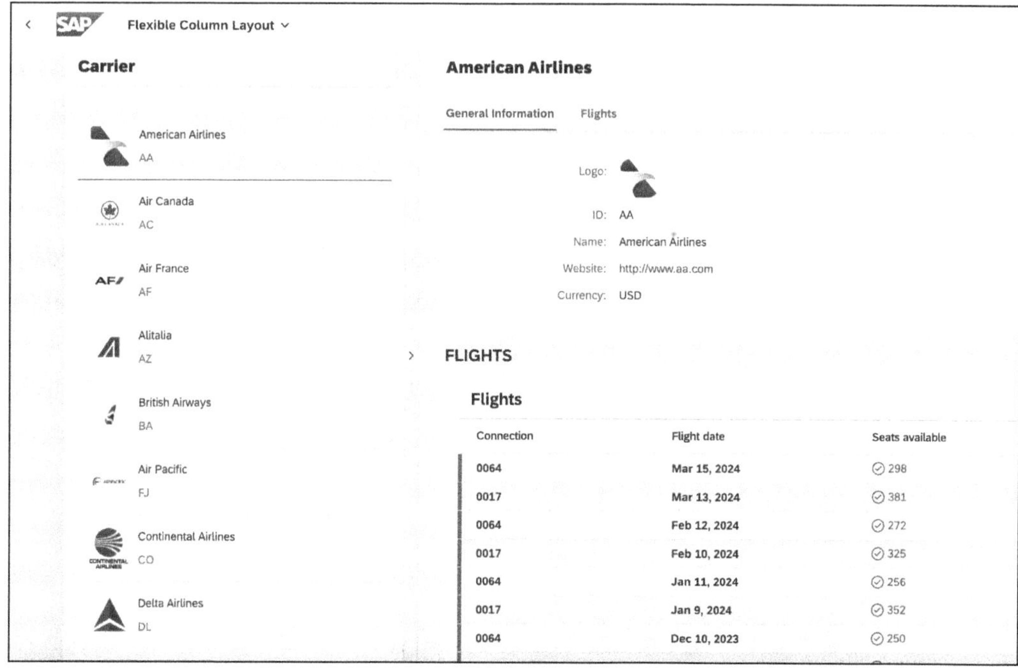

Figure 7.28 Flexible Column Layout Showing Two Columns with the Second Column Expanded

We now come to the last new view—a detail view called *FlightDetails.view.xml*—in which we'll display the flight details and the individual bookings. Let's now create a target for this new view, where `controlAggregation` is used to determine that this view should be inserted into the third column (=`endColumnPages`) (see Listing 7.51).

```
"TargetFlightDetails": {
  "viewType": "XML",
  "viewId": "FlightDetails",
  "viewName": "FlightDetails",
  "controlAggregation": "endColumnPages",
  "clearControlAggregation": true
}
```

Listing 7.51 Target for the Newly Created View Which Should Be Displayed in the Third Column

We also create a route for the flight details. Here, we have to construct our pattern cleverly. It's best to use the same pattern as for the carrier details and expand it to `/flight/{path2}`. It's also important to note that this route will now address three targets for the first time: the two targets for the first and second column, as well as the new one for the flight details in the third column (see Listing 7.52).

```
{
  "name": "RouteFlightDetails",
  "pattern": "carrier/{path}/flight/{path2}:?query:",
  "target": [ "TargetCarrier", "TargetCarrierDetails", "TargetFlightDetails" ]
}
```

Listing 7.52 The Newly Created Route Addressing Three Targets at the Same Time

In this view, we'll again use a `sap.uxap.ObjectPageLayout`. In the first `ObjectPageSection`, we've prepared a `SimpleForm` that shows the details of a connection (connection ID, flight date, departure, and destination). After that, in the second `ObjectPageSection`, we have a `Table` that displays the individual bookings. For each booking, the booking number, class, passenger, and amount are displayed (see Listing 7.53).

```
<mvc:View xmlns:f="sap.ui.layout.form" controllerName=
"at.clouddna.fcl.controller.FlightDetails" xmlns:mvc="sap.ui.core.mvc"
displayBlock="true" xmlns:uxap="sap.uxap" xmlns="sap.m">
  <uxap:ObjectPageLayout>
    <uxap:headerTitle>
      <uxap:ObjectPageDynamicHeaderTitle>
        <uxap:expandedHeading>
          <Title text="{connid}"/>
        </uxap:expandedHeading>
      </uxap:ObjectPageDynamicHeaderTitle>
    </uxap:headerTitle>
    <uxap:sections>
      <uxap:ObjectPageSection title="General Information">
        <uxap:subSections>
```

```
        <uxap:ObjectPageSubSection>
          <uxap:blocks>
            <f:SimpleForm id="flight_simpleform" editable="true" layout=
"ResponsiveGridLayout" labelSpanXL="3" labelSpanL="3" labelSpanM="3" labelSpanS=
"12" emptySpanXL="4" emptySpanL="4" emptySpanM="4" emptySpanS="0" columnsXL="1"
columnsL="1" columnsM="1">
              <f:content>
                <Label id="labelConnid" labelFor="textCarrid" text="Connection"/>
                <Text id="textConnid" text="{connid}"/>
                <Label id="labelFldate" labelFor="textFldate" text="Flight date"/
>
                <Text id="textFldate" text="{ path: 'fldate', type:
'sap.ui.model.odata.type.Date'}"/>
                <Label id="labelCityFrom" labelFor="objectIdentifierCityFrom"
text="Departure"/>
                <ObjectIdentifier id="objectIdentifierCityFrom" title=
"{flightdetails/airportFrom}" text="{ path: 'flightDetails/cityFrom', formatter:
'.toUpper'}"/>
                <Label id="labelCityTo" labelFor="objectIdentifierCityTo" text=
"Destination"/>
                <ObjectIdentifier id="objectIdentifierCityTo" title=
"{flightdetails/airportTo}" text="{ path: 'flightDetails/cityTo', formatter:
'.toUpper'}"/>
              </f:content>
            </f:SimpleForm>
          </uxap:blocks>
        </uxap:ObjectPageSubSection>
      </uxap:subSections>
    </uxap:ObjectPageSection>
  <uxap:ObjectPageSection title="Bookings">
  <uxap:subSections>
    <uxap:ObjectPageSubSection>
      <uxap:blocks>
        <Table id="tableBookings" headerText="Bookings" items="{ path:
'flightBookings', sorter: { path: 'ORDER_DATE', descending: true } }"
alternateRowColors="true" growing="true" growingThreshold="200">
          <columns>
            <Column id="columnBookid">
              <Label id="columnLabelBookid" text="Booking ID"/>
            </Column>
            <Column id="columnClass">
              <Label id="columnLabelClass" text="Class"/>
            </Column>
            <Column id="columnPass" demandPopin="true" minScreenWidth="Desktop">
```

```
            <Label id="columnLabelPass" text="Passenger"/>
          </Column>
          <Column id="columnAmount" demandPopin="true" minScreenWidth=
"Desktop">
            <Label id="columnLabelAmount" text="Amount"/>
          </Column>
        </columns>
        <items>
          <ColumnListItem id="columnListItemBookingTemplate" highlight="{=
${CANCELLED} ? 'Error' : 'Success'}">
            <cells>
              <ObjectIdentifier id="objectIdentifierBookid" title="{bookid}"/>
              <Text id="textClass" text="{CLASS}"/>
              <ObjectIdentifier id="objectIdentifierPass" title="{PASSNAME}"
text="{ path: 'PASSBIRTH', type: 'sap.ui.model.odata.type.Date'}"/>
              <ObjectNumber id="objectNumberAmount" number="{FORCURAM}" unit=
"{FORCURKEY}" />
            </cells>
          </ColumnListItem>
        </items>
      </Table>
    </uxap:blocks>
   </uxap:ObjectPageSubSection>
  </uxap:subSections>
 </uxap:ObjectPageSection>
</uxap:sections>
</uxap:ObjectPageLayout>
</mvc:View>
```

Listing 7.53 ObjectPageLayout Showing the Flight Details and Connections

In the *FlightDetails.controller.ts* controller, we also react here in the listener of the `patternMatched` event and extract the parameter `path2`, which contains the path of the flight (see Listing 7.54).

```
public onInit(): void {
  let oRouter = (this.getOwnerComponent() as UIComponent).getRouter(),
    oRoute = oRouter.getRoute("RouteFlightDetails");
  oRoute?.attachPatternMatched(this.onPatternMatched, this);
}
private onPatternMatched(oEvent: Event) {
  let oArguments = (oEvent as any).getParameters().arguments,
    sPath = decodeURIComponent(oArguments.path2);
  this.getView()?.bindElement({
    path: sPath,
```

```
    parameters: { "$expand": "flightBookings" }
  });
}
private toUpper(sValue: string){ return sValue?.toUpperCase(); }
```

Listing 7.54 PatternMatched Event for the Flight to Extract the Path and Use Element Binding

Now we adjust the table in the second column so that a table entry can be clicked. In the selectionChange event, we call the onNavToFlightDetail method, which we'll develop in a moment:


```
<Table id="tableFlights" headerText="Flights" items="{ path: 'carrierFlights',
sorter: { path: 'fldate', descending: true } }" alternateRowColors="true"
growing="true" growingThreshold="200" mode="SingleSelectMaster"
selectionChange="onNavToFlightDetail">
```

In the event handler just mentioned, we extract from the event parameters which table entry was selected and the binding context path. For the navTo method, we now have to fill two mandatory parameters: the path for the carrier and the path for the flight. We let the helper suggest which layout makes sense in a three-column screen. You can see all of this in Listing 7.55.

```
private onNavToFlightDetail(oEvent: Event) {
  let oBindingContext = (oEvent as
any).getParameters().listItem.getBindingContext(),
    sPath = oBindingContext?.getPath(),
    oNextUIState = (this.getOwnerComponent() as
any).getHelper().getNextUIState(2),
    oRouter = (this.getOwnerComponent() as UIComponent).getRouter();
  oRouter.navTo("RouteFlightDetails", {
    path: encodeURIComponent(this.getView()?.getElementBinding()?.getPath()),
    path2: encodeURIComponent(sPath),
      query: { layout: oNextUIState.layout }
  });
}
```

Listing 7.55 Event Handler for the selectionChange Event in the Flights Table in the Second Column

We must not forget that when the route RouteFlightDetails is called, we must also react to it in the second column, that is, in the *CarrierDetails.controller.ts* (see Listing 7.56). We must also read the parameter path in the corresponding event handler and set the element binding, which is why we register the same listener here. Because we've named the parameters in the patterns of RouteCarrierDetails and RouteFlightDetails the same, we can reuse the method. If we don't create this listener, then in the case of a refresh, the first and third columns can be built correctly, but the second column would lose the binding or the binding wouldn't be determined from the parameter and set.

```
public onInit(): void {
  let oRouter = (this.getOwnerComponent() as UIComponent).getRouter(),
    oRoute = oRouter.getRoute("RouteCarrierDetails"),
  oFlightRoute = oRouter.getRoute("RouteFlightDetails");
  oRoute?.attachPatternMatched(this.onPatternMatched, this);
  oFlightRoute?.attachPatternMatched(this.onPatternMatched, this);
}
```

Listing 7.56 The Second Column Needs to Listen to the patternMatched Event of the Third Column

The result is shown in Figure 7.29. When clicking on a flight, we have a three-column layout where we can simultaneously see the information from a carrier, its flights, and the flight details with the individual bookings. The default ratio is 25:50:25, but the user can adjust this at runtime.

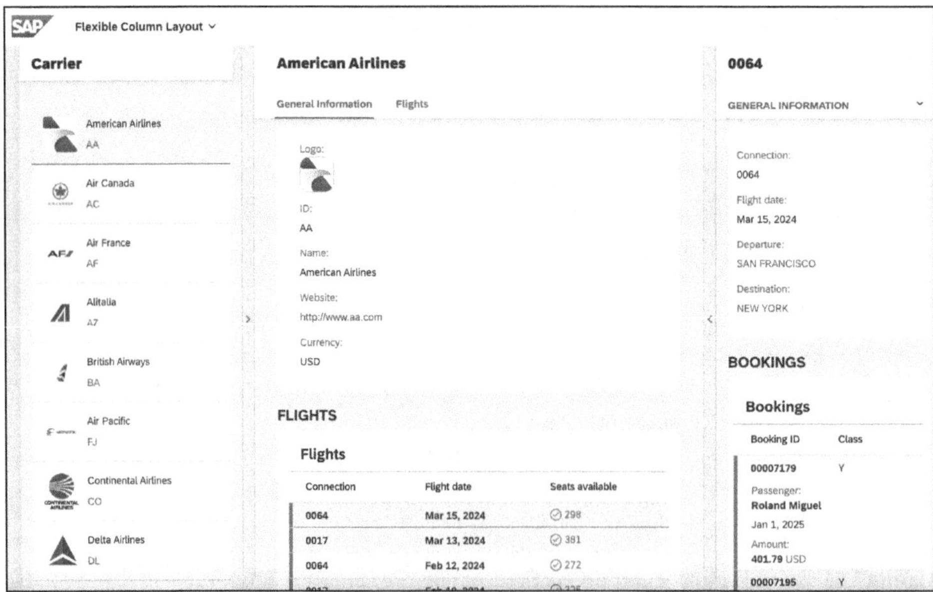

Figure 7.29 Three Columns Showing a Carrier List, the Selected Carrier, and Flight at the Same Time

What the user can't yet do is independently open the second and third columns in full screen or even close them. We therefore provide buttons in `ObjectDynamicHeaderTitle` for this. Whether or not these buttons should be displayed will be saved in the same global `JSONModel` that also saves the current `layout`. See the adjustment for *CarrierDetails.view.xml* in Listing 7.57.

```
<uxap:ObjectPageDynamicHeaderTitle>
  <uxap:expandedHeading>
    <Title text="{CARRNAME}" wrapping="true" class="sapUiSmallMarginEnd"/>
  </uxap:expandedHeading>
  <uxap:navigationActions>
    <OverflowToolbarButton type="Transparent" icon="sap-icon://full-screen"
press="handleFullScreen" id="enterFullScreenBtn" tooltip="Enter Full Screen
Mode" visible="{= ${ui>/actionButtonsInfo/midColumn/fullScreen} !== null }"/>
    <OverflowToolbarButton type="Transparent" icon="sap-icon://exit-full-
screen" press="handleExitFullScreen" id="exitFullScreenBtn" tooltip="Exit Full
Screen Mode" visible="{= ${ui>/actionButtonsInfo/midColumn/exitFullScreen} !==
null }"/>
    <OverflowToolbarButton type="Transparent" icon="sap-icon://decline" press=
"handleClose"  tooltip="Close middle column" visible="{= ${ui>/
actionButtonsInfo/midColumn/closeColumn} !== null }"/>
  </uxap:navigationActions>
</uxap:ObjectPageDynamicHeaderTitle>
```

Listing 7.57 Buttons in the Object Page Header to Allow Opening and Closing in Full Screen

But where is this UI information for controlling the buttons filled? Our `routeMatched` event and the corresponding event handler in *App.controller.ts* are suitable for this because here we can still provide the suggestions for the display control of the buttons in the current `layout` at the end of the helper method. We store this information in the aforementioned `JSONModel` called `ui` (see Listing 7.58).

```
private onRouteMatched(oEvent: Event) {
  // ...
  let oActionButtonsInfo =
this.getOwnerComponent().getHelper().getCurrentUIState().actionButtonsInfo;
  this.getOwnerComponent().getModel("ui").setProperty("/actionButtonsInfo",
oActionButtonsInfo);
}
```

Listing 7.58 Visibility Information for the Buttons Coming from the Helper and Saved into the Global JSONModel Called ui

To ensure that the buttons in *CarrierDetails.view.xml* have the appropriate event handlers, we implement the methods for opening and closing the full screen one after the other. This is simply navigating the same route, only with the `layouts` stored in the buttons. As explained earlier, these layouts were assigned by the helper and stored in the `JSONModel` (see Listing 7.59).

```
private handleFullScreen () {
  let sNextLayout = (this.getOwnerComponent()?.getModel("ui") as
```

```
JSONModel).getProperty("/actionButtonsInfo/midColumn/fullScreen");
   this.getOwnerComponent().getRouter().navTo("RouteCarrierDetails", {
    path: encodeURIComponent(this.getView()?.getElementBinding()?.getPath()),
           query: { layout: sNextLayout }
  });
}
private handleExitFullScreen () {
  let sNextLayout = (this.getOwnerComponent()?.getModel("ui") as
JSONModel).getProperty("/actionButtonsInfo/midColumn/exitFullScreen");
  this.getOwnerComponent().getRouter().navTo("RouteCarrierDetails", {
    path: encodeURIComponent(this.getView()?.getElementBinding()?.getPath()),
    query: { layout: sNextLayout }
  });
}
private handleClose () {
  let sNextLayout = (this.getOwnerComponent()?.getModel("ui") as
JSONModel).getProperty("/actionButtonsInfo/midColumn/closeColumn");
  (this.getOwnerComponent() as UIComponent).getRouter().navTo("RouteCarrier", {
    query: { layout: sNextLayout }
  });
}
```

Listing 7.59 Handling Full Screen, Exiting Full Screen, or Closing the Whole Column in Carrier Details View

In Figure 7.30, you can see that in the second column, as long as no third column is open, the buttons are displayed in the upper-right corner.

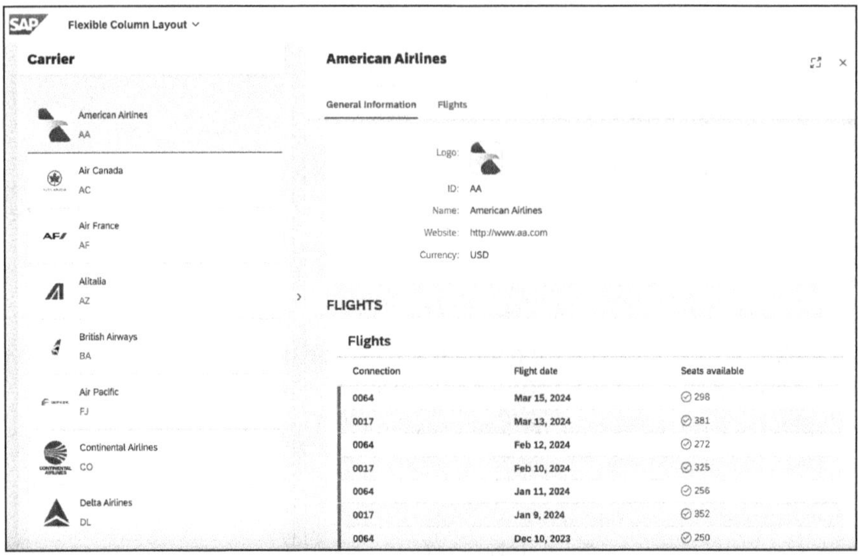

Figure 7.30 Buttons in the Second Column Allow Opening Full Screen or Closing the Column

We want to do the same for the third column, so we insert buttons in the header in *FlightDetails.view.xml* again. However, for the visible property, we use the data that will later be stored for the endColumn (see Listing 7.60).

```
<uxap:ObjectPageDynamicHeaderTitle>
  <uxap:expandedHeading>
    <Title text="{connid}" wrapping="true" class="sapUiSmallMarginEnd"/>
  </uxap:expandedHeading>
  <uxap:navigationActions>
    <OverflowToolbarButton type="Transparent" icon="sap-icon://full-screen"
press="handleFullScreen" id="enterFullScreenBtn" tooltip="Enter Full Screen
Mode" visible="{= ${ui>/actionButtonsInfo/endColumn/fullScreen} !== null }"/>
    <OverflowToolbarButton type="Transparent" icon="sap-icon://exit-full-
screen" press="handleExitFullScreen" id="exitFullScreenBtn" tooltip="Exit Full
Screen Mode" visible="{= ${ui>/actionButtonsInfo/endColumn/exitFullScreen} !==
null }"/>
    <OverflowToolbarButton type="Transparent" icon="sap-icon://decline" press=
"handleClose"  tooltip="Close middle column" visible="{= ${ui>/
actionButtonsInfo/endColumn/closeColumn} !== null }"/>
  </uxap:navigationActions>
</uxap:ObjectPageDynamicHeaderTitle>
```

Listing 7.60 Buttons Placed in the Third Columns Header

Again, we implement the individual event handlers in *FlightDetails.controller.ts*. However, before we can do that, we must store the carrier's path in the patternMatched event handler because we'll need it for both the full screen and the close. For this reason, we've created a member variable called sParentPath and filled it in the onPatternMatched method (see Listing 7.61).

```
private sParentPath : string;
private onPatternMatched(oEvent: Event) {
  let oArguments = (oEvent as any).getParameters().arguments,
    sPath = decodeURIComponent(oArguments.path2);
  this.getView()?.bindElement({
    path: sPath,
    parameters: { "$expand": "flightBookings" }
  });
  this.sParentPath = oArguments.path;
}
```

Listing 7.61 Extracting the Path Parameter in the Third Column for the Carrier

The corresponding event handlers for full-screen mode and closing the column are apparently similar to the methods for the second column, but they are slightly different. It's important that we not only fill path2 with the binding context path of the flight but also fill path with the previously stored sParentPath (see Listing 7.62).

```
private handleFullScreen () {
  let sNextLayout = (this.getOwnerComponent()?.getModel("ui") as
JSONModel).getProperty("/actionButtonsInfo/endColumn/fullScreen");
      this.getOwnerComponent().getRouter().navTo("RouteCarrierDetails", {
    path: encodeURIComponent(this.getView()?.getElementBinding()?.getPath()),
    path2: this.sParentPath,
    query: { layout: sNextLayout }
  });
}
private handleExitFullScreen () {
  let sNextLayout = (this.getOwnerComponent()?.getModel("ui") as
JSONModel).getProperty("/actionButtonsInfo/endColumn/exitFullScreen");
    this.getOwnerComponent().getRouter().navTo("RouteCarrierDetails", {
    path: encodeURIComponent(this.getView()?.getElementBinding()?.getPath()),
    path2: this.sParentPath,
    "?query": { layout: sNextLayout }
  });
}
private handleClose () {
  let sNextLayout = (this.getOwnerComponent()?.getModel("ui") as
JSONModel).getProperty("/actionButtonsInfo/endColumn/closeColumn");
  this.getOwnerComponent().getRouter().navTo("RouteCarrierDetails", {
    path: this.sParentPath,
    query: { layout: sNextLayout }
  });
}
```

Listing 7.62 Event Handlers in the FlightDetails Controller to Open and Close Full Screen

Now, as shown in Figure 7.31, we also have buttons in the third column so that it can be opened or closed in full screen.

We still have a problem that we didn't take into account in this setting. When the third column is closed, the selection remains in the **Flights** table, so the same element can be selected as long as no other entry is selected. This state is shown in Figure 7.32. We also have the same problem regarding the second column and the first column.

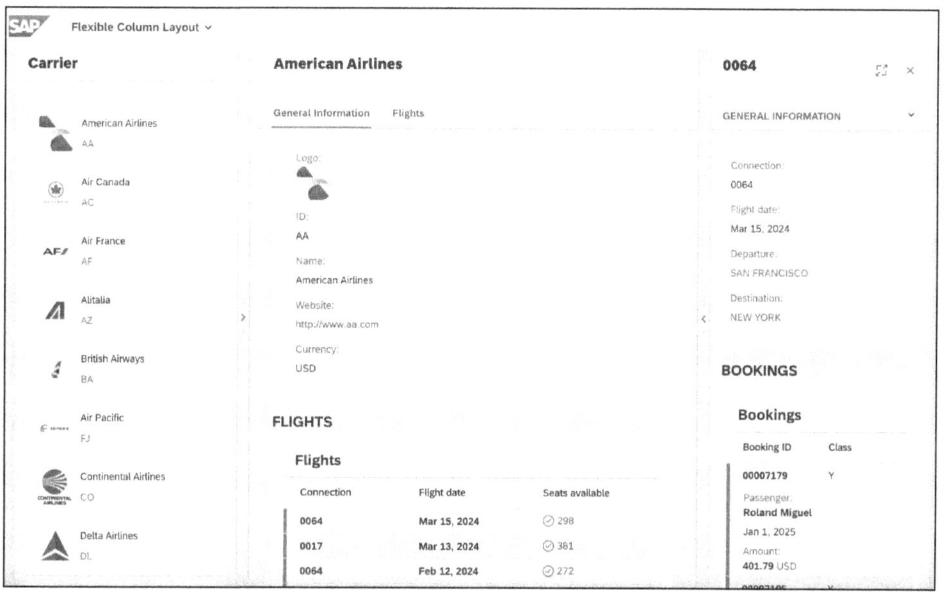

Figure 7.31 Buttons Visible in the Third Column's Header

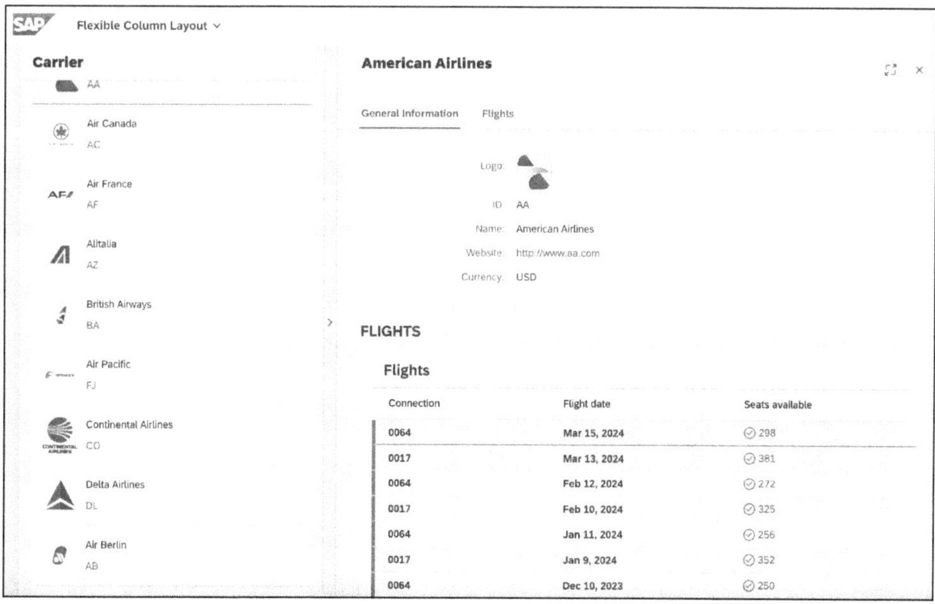

Figure 7.32 After Closing the Third Column, the Selection in the Table Remains

In the *Carrier.controller.ts*, we can provide a method in the lifecycle method `onInit` that can be triggered from any other controller using the combination `"Carrier"` and `"resetSelection"` via the *event bus* delivered by default. The underlying method does nothing other than delete the selection in the list (see Listing 7.63).

```
public onInit(): void {
 this.getOwnerComponent().getEventBus().subscribe("Carrier", "resetSelection",
() => {
    (this.getView()?.byId("carrierList") as List).removeSelections(true);
  });
}
```

Listing 7.63 Using the Event Bus to Offer a Method That Can Be Called from Everywhere

We can trigger this method when the second column is closed in the `handleClose` method (see Listing 7.64).

```
private handleClose () {
  this.getOwnerComponent().getEventBus().publish("Carrier", "resetSelection");
  // ...
}
```

Listing 7.64 Using the Event Bus to Trigger the resetSelection in the Carrier Controller

We prepare a similar method under `"CarrierDetails"` and `"resetSelection"` in *Carrier-Details.controller.ts*. This method removes the selection in the flight table (see Listing 7.65).

```
public onInit(): void {
  //...
  this.getOwnerComponent().getEventBus().subscribe("CarrierDetails",
"resetSelection", () => {
     (this.getView()?.byId("tableFlights") as Table).removeSelections(true);
  });
}
```

Listing 7.65 Event Bus Offering a Method in the Flight Details Controller That Resets the Selection

We also trigger this when the third column is closed (see Listing 7.66).

```
private handleClose () {
  // ...
  this.getOwnerComponent().getEventBus().publish("CarrierDetails",
"resetSelection");
}
```

Listing 7.66 Using the Event Bus to Trigger the resetSelection in the Carrier Details Controller

Now we have a fully functional flexible column layout that can display up to three columns at the same time. This section has also clarified why a route can have more than one target. It's important not to forget the `patternMatched` events in the first and second columns in such a scenario, when a column further to the right is opened, and our logic should also be executed for these routes, for example, to make use of the parameters for data binding.

7.4 Routing Events

You've already become familiar with some routing events in this chapter and have seen them in use in the exercises. We want to provide a brief summary of the events that are used most frequently with the `sap.m.routing.Router` and explain them briefly in Table 7.1.

Event	Description
`beforeRouteMatched`	This event is fired before the corresponding target is loaded and placed, when the current URL hash matches a pattern of a route, subroute, or nested route within the router.
`bypassed`	This event is fired when no route matches the changed URL hash.
`routeMatched`	This event is fired when the current URL hash matches the pattern of a route, subroute, or nested route within the router.
`routePatternMatched`	This event is fired only when a route's own pattern is matched with the URL hash, not its subroutes.

Table 7.1 Common Events of a Router Used to Register Listeners

There are not only events that affect the entire router but also some that go to a specific route. You can see the most important events listed in Table 7.2.

Event	Description
`beforeMatched`	This event is fired before the corresponding target is loaded and placed, and when the current URL hash matches the pattern of the route, its subroute, or its nested route.
`matched`	This event is fired when the current URL hash matches the pattern of the route, its subroute, or its nested route.
`patternMatched`	This event is fired only when the current URL hash matches the pattern of the route.

Table 7.2 Common Events of a Route Used to Register Listener

Additionally, there's an event called `display` regarding a target's events. This event is fired when a target is displayed. It can be triggered by calling the display function or by the router when a target is referenced in a matching route.

7.5 App-to-App Navigation

Navigation in SAP Fiori is based on familiar web application paradigms and is done via a central start page, the *SAP Fiori launchpad*. Users start their journey on the launchpad start page (e.g., **My Home** or *spaces*) and navigate from there to various apps that can be connected to each other across multiple screens (see Figure 7.33). Back navigation is always possible, either via the browser back button, the back arrow icon in the shell bar, or the launchpad logo, which returns you to the homepage.

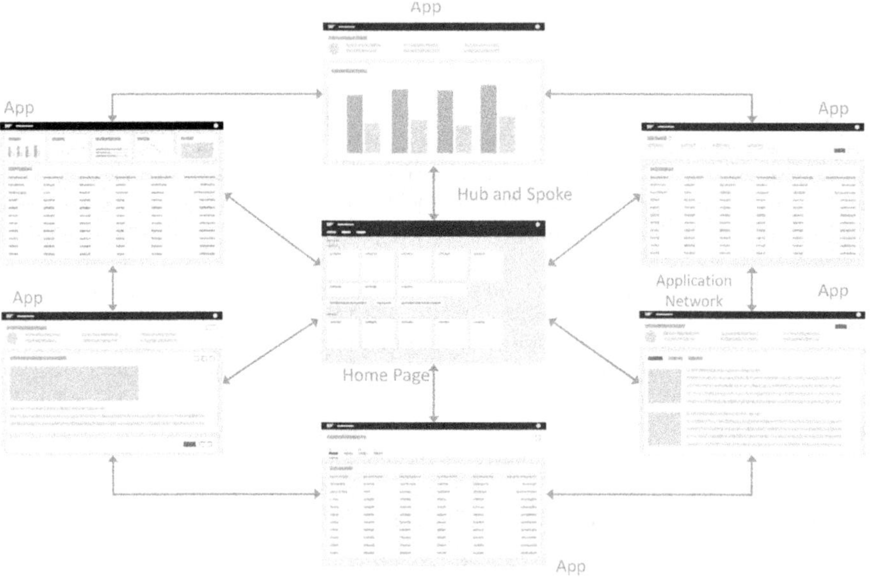

Figure 7.33 Navigation within the SAP Fiori Launchpad

The tiles on the homepage act as anchor points to individual apps. They can be used to access both modern SAP Fiori apps and older UIs (legacy). The latter usually open in new tabs or windows. Within the apps, navigation is done via various elements such as links, table rows, buttons, or other UI elements. Context is passed between apps, enabling flexible reusability and the combination of modular apps into more complex processes.

Links can either lead directly to new pages or open popovers such as quick views or smart links that provide additional navigation targets. This behavior should be consistent within an app, for example, by handling links in tables in a uniform manner. If navigation within the current tab isn't possible, for example, to avoid data loss, apps (especially non–SAP Fiori apps) can be opened in new windows or tabs.

Legacy applications and other non–SAP Fiori apps always open in a new tab or window when launched from the SAP Fiori launchpad. URL-based navigation, which affects non–SAP Fiori apps, also follows this pattern. This ensures a smooth transition between modern and legacy technology.

Let's now assume the following example: we've developed an application called Carrier Flights app that is responsible for displaying the flights of a single carrier. This application is shown in Figure 7.34 and has already been deployed and is running in the SAP Fiori launchpad. You can find the source code of this application in the following Git repository: *https://github.com/clouddnagmbh/sappress-ui5-012025-carrierflights*.

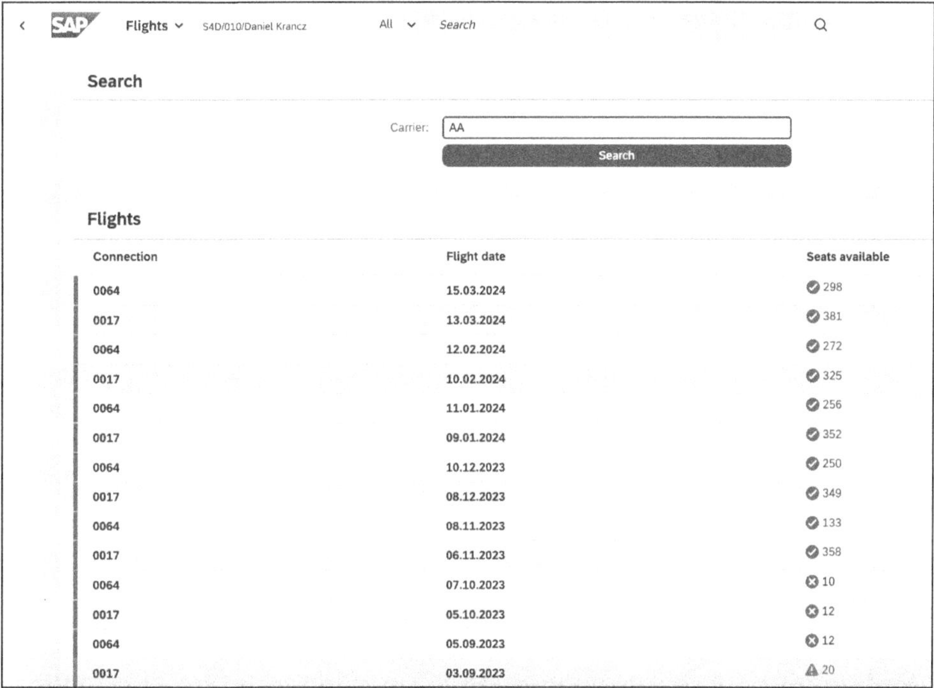

Figure 7.34 Carrier Flights App Developed to Display the Flights of One Carrier

To enable this application to be called up, a **Semantic Object** `CarrierFlight` and **Action** `display` have been defined in the *SAP Fiori launchpad app manager* (see Figure 7.35). The combination of these two strings makes it possible to call up the underlying target application, namely the corresponding Carrier Flights SAPUI5 app.

We now want to implement an app-to-app navigation from our full-screen navigation example to this Carrier Flights app. An app-to-app navigation with the SAP Fiori launchpad configuration is only possible if we deploy the app. However, to enable us to test it without having to deploy the app, we activate the *app-to-app navigation preview* in SAP Business Application Studio.

Before we do this, we'll store the same `inbounds` rule in the *manifest.json* under `sap.app.crossNavigation` in the Carrier Flights app, which displays the flights of a carrier that are also defined in the SAP system in the SAP Fiori launchpad configuration (see Figure 7.36).

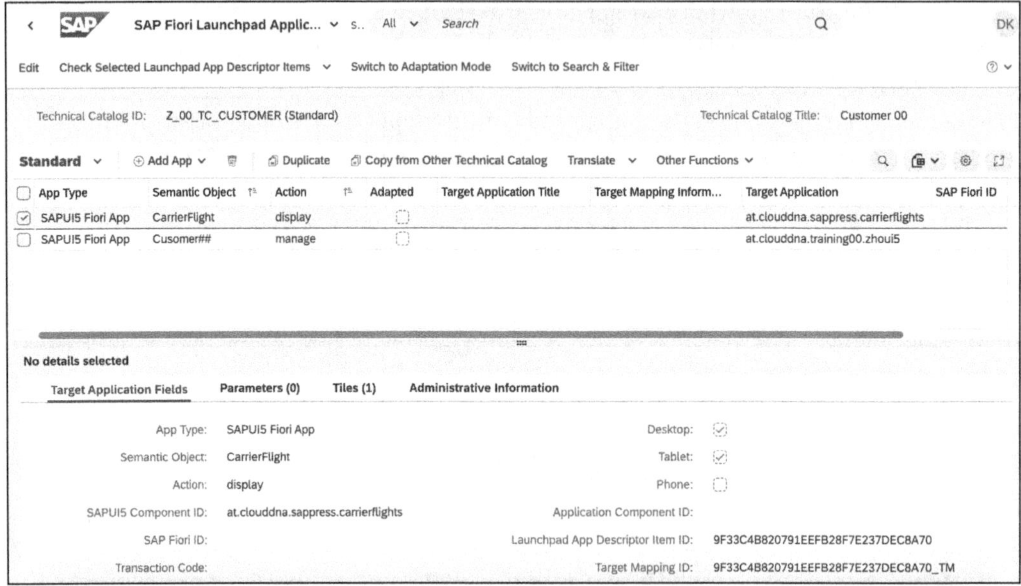

Figure 7.35 SAP Fiori Launchpad App Manager Configuration for Semantic Object and Action

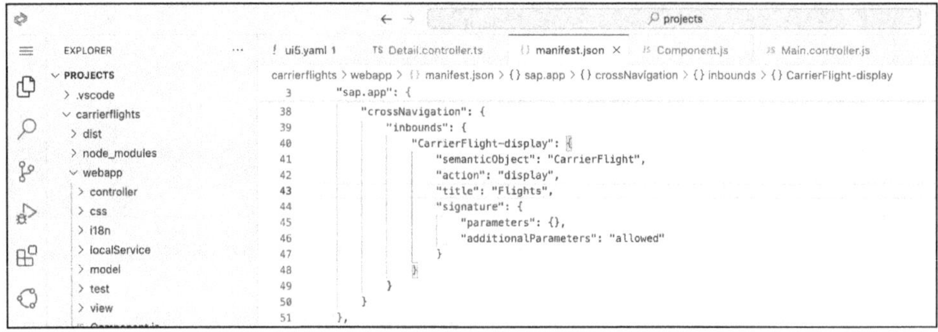

Figure 7.36 Enter the Same Cross-Navigation Inbound Rule as Defined in the SAP Fiori Launchpad Configuration

Now we'll use the command palette by choosing **View • Command Palette** to search for a command called **Fiori: Enable App-to-App Navigation Preview** (see Figure 7.37). When you select this, you must first select the project from which the navigation is to take place (**Source Project**) and then the project to which you want to navigate (**Target Project**).

This command creates a directory *appconfig* with a new file in it called *fioriSandbox-Config.json* within the source project (see Figure 7.38). This definition allows not only the Full-Screen Navigation app but also the Carrier Flights app to be opened in the SAP Fiori launchpad sandbox.

Furthermore, a configuration is stored in *ui5.yaml* so that the source project can access the resources of the target project (see Figure 7.39).

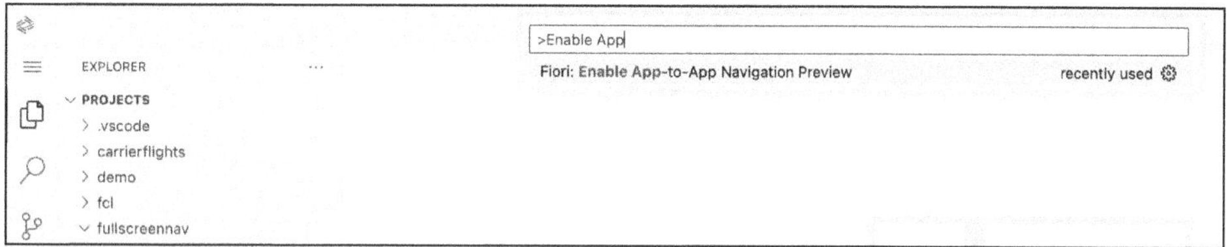

Figure 7.37 Enable the App-to-App Navigation Preview in SAP Business Application Studio

```
{
  "applications": {
    "CarrierFlights-display": {
      "additionalInformation": "SAPUI5.Component=at.clouddna.sappress.carrierflights",
      "applicationType": "URL",
      "url": "../resources/at.clouddna.sappress.carrierflights"
    }
  }
}
```

Figure 7.38 SAP Fiori Sandbox Config Added to the Source Project

```
server:
  customMiddleware:
    - name: ui5-tooling-transpile-middleware
      configuration:
        excludePatterns:
    - name: fiori-tools-servestatic
      afterMiddleware: compression
      configuration:
        paths:
          - path: /resources/at.clouddna.sappress.carrierflights
            src: ../carrierflights/webapp
          - path: /appconfig
            src: appconfig
```

Figure 7.39 Configuration Enabling the Source Project to Access the Resources of the Target Project

Now start the preview of the full-screen navigation example, and click on the **SAP** logo in the top-left corner. You'll then be taken to the homepage of the sandbox. You'll see that not only the Full-Screen Navigation app but also the Carrier Flights app can be opened via a tile separately (see Figure 7.40).

In the Full-Screen Navigation app, we incorporate an `OverflowToolbar` in the detail view in the `headerContent` aggregation, which will ultimately contain a `Button`. This button will later allow the user to navigate from this app to the other (see Listing 7.67).

```
<headerContent>
  <OverflowToolbar>
    <ToolbarSpacer/>
    <Button type="Transparent" text="Carrier Flights" press=
```

```
"onNavToCarrierFlights" icon="sap-icon://forward"/>
  </OverflowToolbar>
</headerContent>
```

Listing 7.67 Button in the Detail View's Header to Execute the App-to-App Navigation

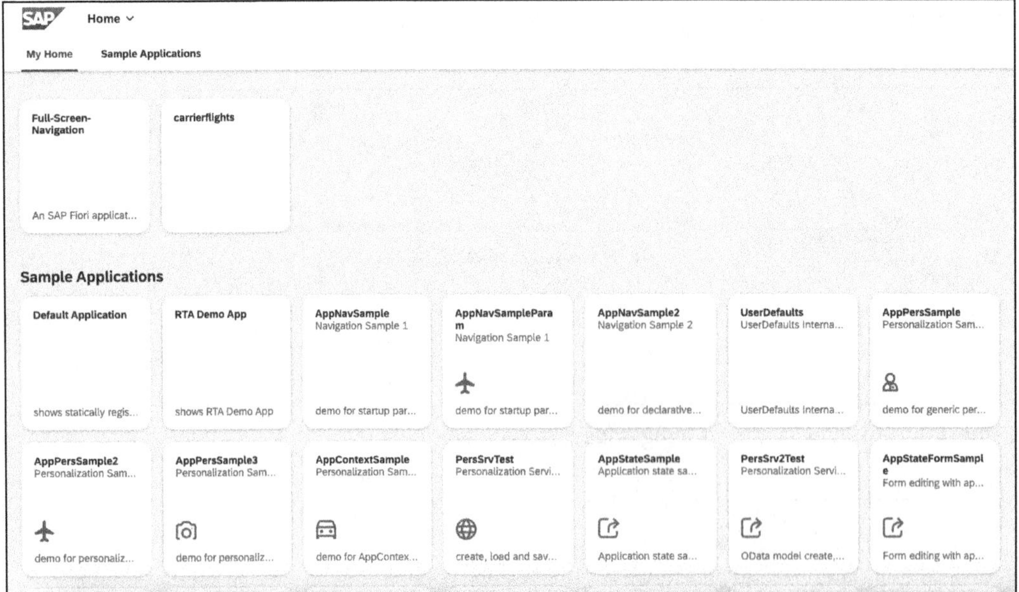

Figure 7.40 Both Apps Able to Be Opened within the Same SAP Fiori Launchpad Sandbox Runtime

This button is shown in Figure 7.41.

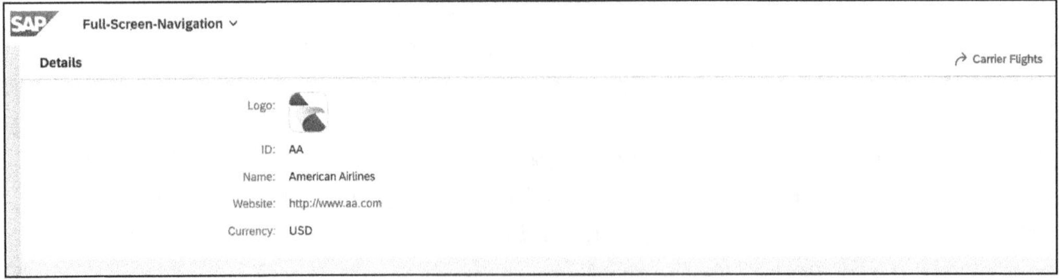

Figure 7.41 Button to Execute the App-to-App Navigation

In *Detail.controller.ts*, we now insert two methods that use app-to-app navigation. This is provided by the `CrossApplicationNavigation` and can be loaded from the `sap.ushell.Container` by default. Ultimately, there's a method called `toExternal`, where you specify the combination of `SemanticObject` and `Action` in the `target` and the parameters to be sent in `params` (see Listing 7.68).

```
protected getCrossApplicationNavigationService():
Promise<CrossApplicationNavigation> {
  return ((sap.ushell as any).Container as
any).getServiceAsync("CrossApplicationNavigation");
}
private crossAppNavigation(oTarget: any, oParams?: any): void {
  this.getCrossApplicationNavigationService().then((service:
CrossApplicationNavigation) => {
    service.toExternal({
      target: oTarget,
      params: oParams
    });
  });
}
```

Listing 7.68 Navigation to an External App Using CrossApplicationNavigationService

After that, we implement the event handler of the button. This event handler calls one of the previously inserted methods and passes as the `target` exactly the combination of the `SemanticObject` and `Action` that also exists in the `crossNavigation` in the application descriptor and later in the SAP Fiori launchpad (see Listing 7.69).

```
private onNavToCarrierFlights(){
  let oTarget = {semanticObject: "CarrierFlight", action: "display"};
  this.crossAppNavigation(oTarget);
}
```

Listing 7.69 Event Handler of the Button Triggering the Cross-Application Navigation

You can now test the app and click on the button. You should then be navigated from the Full-Screen Navigation app to the Carrier Flights app (see Figure 7.42).

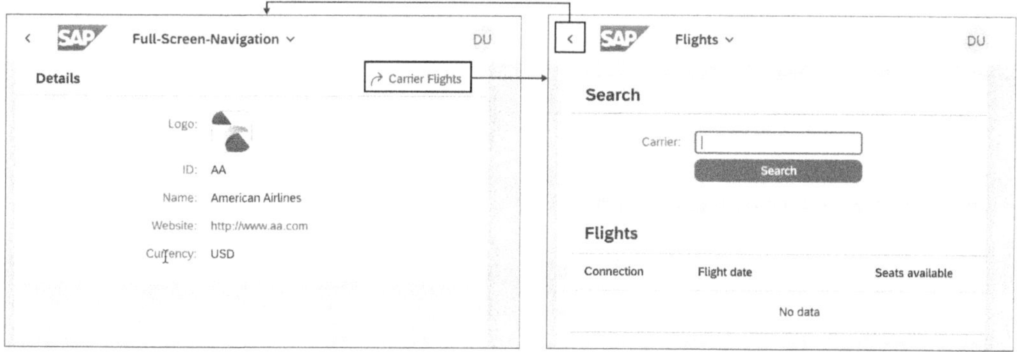

Figure 7.42 App-to-App Navigation

Now we add that in addition to the target consisting of semantic object and action; the carrier ID is also given as a parameter during navigation (see Listing 7.70). In this case, we again have to look in the SAP Fiori launchpad configuration to see which name was set for this *target mapping* as a parameter. There, we find the information that the parameter can be passed to the other app with the name `Carrier`.

```
private onNavToCarrierFlights(){
  let oTarget = {semanticObject: "CarrierFlight", action: "display"},
    oObject = this.getView()?.getElementBinding()?.getBoundContext()?
.getObject(),
    oParams = {Carrier: oObject.carrid};
  this.crossAppNavigation(oTarget, oParams);
}
```

Listing 7.70 Passing Additional Parameters to the Other Application in Cross-Application Navigation

Let's look at the whole thing from the perspective of the other application, that is, from the Carrier Flight app. We could then read the parameters in the `init` method in the *Component.ts* or *Component.js* using the `getComponentData()` method with the `startupParameters` property (see Listing 7.71).

```
let oStartupParameters = this.getComponentData().startupParameters;
if(oStartupParameters && oStartupParameters.Carrier){
  this.Carrier = oStartupParameters.Carrier[0];
}
```

Listing 7.71 Extracting the Startup Parameters Passed in Cross-Application Navigation

We can access the selected parameter via the owner component to apply this parameter in the main controller, for example, insert it into the input field and trigger the search (see Listing 7.72).

```
onInit: function () {
  let oRouter = this.getOwnerComponent().getRouter();
  oRouter.getRoute("RouteMain").attachPatternMatched(this.onPatternMatched,
this);
},
onPatternMatched: function () {
  let sCarrierParam = this.getOwnerComponent().Carrier;
  this.getView().setModel(new sap.ui.model.json.JSONModel({
    carrier: sCarrierParam ? sCarrierParam : null
  }), "input");
  if (sCarrierParam) {
```

```
    this.onSearchCarrierFlights();
  }
},
```

Listing 7.72 Use the Extracted Startup Parameter as a Filter

As you can see in Figure 7.43, on the one hand, the app-to-app navigation takes place, and on the other hand, the parameter `Carrier` is provided as a query or startup parameter when navigating. As already mentioned, the parameter must be read; otherwise, sending the parameter is of no use.

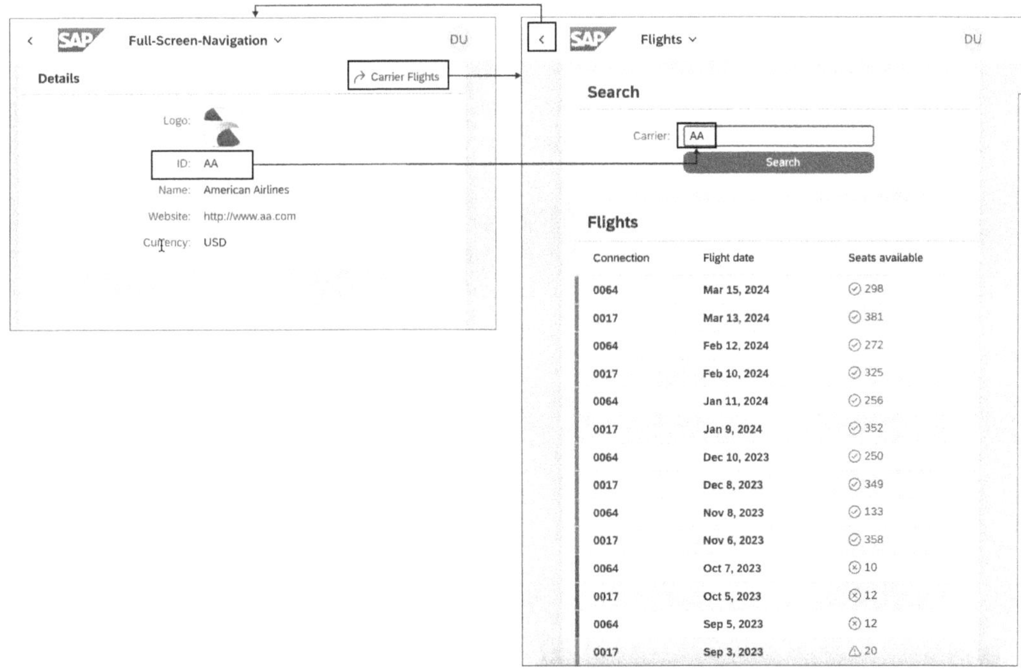

Figure 7.43 App-to-App Navigation with Startup Parameters

This way, we were able to test the app-to-app navigation not only in the SAP Fiori launchpad but also in SAP Business Application Studio. This saves us from having to deploy the app every time a change is made and from having to have the entire configuration set up before we're even sure of the matter. Nevertheless, it's now possible to navigate directly from one application to the other, passing information to the target app that is used to trigger a search, for example.

7.6 Summary

In this chapter, we've learned a great deal about routing and navigation in the SAPUI5 framework. On the one hand, we've seen that the router is already prevalent and is also initialized out-of-the-box. This can be used to navigate from one view to the next. For this to happen, we've learned that there are at least three parts needed: route, pattern, and target. However, the navigation itself is only apparent because the views are inserted into a container designed for this purpose. This container is also referenced in the application descriptor, just like the router information and configurations.

We've looked at the three most important types of such containers in Section 7.2. The full-screen navigation not only showed the navigation from one view to the next but also explained the possible parameters (mandatory and optional parameters or query parameters). In the split app, we displayed two pages simultaneously in a split container and thought about deep links. Finally, we tackled the supreme discipline, the flexible column layout. We took the Carrier Flights Bookings app to the extreme and displayed it in a three-part screen. We provisionally summarized the router, route, and target events, even though we already made use of them in the patterns. Last but not least, we talked about app-to-app navigation, which will later take place in SAP Fiori launchpad. The basis for this is the SAP Fiori launchpad configuration, in which the individual apps consist of target mappings (consisting of semantic object and action, and possibly parameters). As developers, we can also access these target mappings in our applications and enable navigation to our own apps or to the SAP standard apps.

PART III

Data Integration

Chapter 8
REST Integration

JavaScript-based web applications use REST clients to efficiently handle HTTP communication with backend services, abstracting complex tasks such as sending requests, managing data serialization, and handling errors. While they simplify API integration and improve error handling, they also pose challenges such as performance issues, security concerns, and dependency risks.

JavaScript-based web applications require Representational State Transfer (REST) clients to communicate efficiently with backend services. REST is an architectural approach that uses HTTP protocol mechanisms such as `GET`, `POST`, `PUT`, and `DELETE` to exchange data between client and server. REST clients enable frontend applications to abstract this communication and integrate it seamlessly into the application. A REST client serves as a bridge between the web application and the RESTful APIs provided by backends. It helps to send HTTP requests, serialize and deserialize data, and handle errors during communication. Without REST clients, each HTTP interaction must be implemented manually, increasing complexity and the likelihood of errors.

REST clients offer numerous advantages that make them indispensable for modern JavaScript-based web applications. They abstract complex HTTP interactions and provide a standardized, reusable interface that makes it easier for developers to integrate APIs and relieves them of the technical details of data transfer. They also improve error handling, provide centralized logging mechanisms, and make it easier for teams to work together using standardized methods. At the same time, however, they also present challenges. Performance issues due to inefficient requests, security issues such as cross-origin resource sharing (CORS) restrictions, and difficulties with API versioning can lead to errors or increased maintenance. Dependence on external libraries such as Axios also carries risks, as changes in these tools can affect the entire application.

In this chapter, we'll get to know two models. We'll start in Section 8.1 with the JSON model. This is used very often in SAPUI5 applications, but rarely to consume REST services. Due to its client-side, generic mechanisms for sorting and filtering and due to the freedom of structuring the data, it's a perfect addition to server-side models. The JSON model wasn't designed for communicating with a web service, and the framework reflects this in its functionality. In Section 8.2, we'll therefore introduce you to a well-known and important representative from the world of HTTP-based REST clients. We're talking about Axios, which is available under a simple permissive license from MIT. This was forked and further developed as part of a project called *RestModel for SAPUI5*.

8.1 JSON Model

The *JSON model* is used to bind UI components to JavaScript object data, which is usually serialized in JSON format. It's a client-side model intended for small datasets that are fully available on the client. It doesn't support server-side pagination or other loading mechanisms, but it supports all the binding modes, especially two-way binding. There's no built-in function for sending data back to the server. With the JSON model, you can use the `loadData()` method to load data from an external resource. The same can be achieved using Ajax in combination with jQuery. However, these solutions aren't always the best, especially because the JSON model is more tailored to store and process local data. We'll present a suitable alternative to you later in this chapter with Axios.

In the following sections, we'll go into the details of the JSON model. In Section 8.1.1, we'll first show you how to initialize and access it. It will be important here to understand how a JSON model differs from any server-side model, how a JSON model stores the data, and how you can access that data using data binding or predefined methods. This client-side model has many generic methods that we can use, even if the connected data source doesn't provide the functionality. Specifically, we're talking about the options for filtering (Section 8.1.2) and sorting data (Section 8.1.3).

8.1.1 Creating and Using the Model

As you can also see in Figure 8.1, the faceless component `JSONModel` is located in the namespace `sap.ui.model.json`. A JSON model can be created using the following constructor:

```
new sap.ui.model.json.JSONModel()
```

In the constructor, you can already pass data as the first parameter as a JavaScript object:

```
let oMyJsonModel = new sap.ui.model.json.JSONModel({ firstname: "John",
lastname: "Doe" });
```

However, you can also specify a path to a file either in the project structure or on a remote server in the constructor. Assuming that there's a directory named *data* in the *webapp* directory and a file named *testdata.json* in that directory, you can use this file for prepopulating as follows:

```
let oMyJsonModel = new sap.ui.model.json.JSONModel("/data/testdata.json");
```

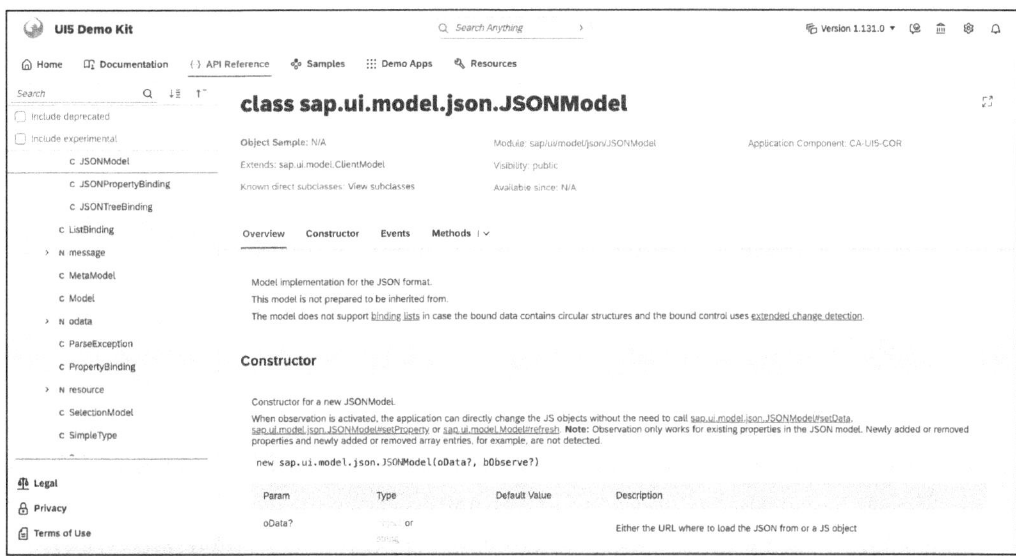

Figure 8.1 Official Documentation of the Faceless Component sap.ui.model.json.JSONModel

Even if you haven't defined anything in the constructor, you can always use the `setData` method to set/change the data retrospectively:

```
oMyJsonModel.setData({ firstname: "Jane", lastname: "Doe" });
```

Finally, you can also load the data retrospectively using a path:

```
oMyJsonModel.loadData("/data/testdata.json");
```

You can also define a JSON model that will be globally visible in the application descriptor under `models`, which saves you the instantiation (see Listing 8.1).

```
"models": {
  "": {
    "type": "sap.ui.model.json.JSONModel",
    "uri": "data/testdata.json"
  },
  "namedJsonModel": {
    "type": "sap.ui.model.json.JSONModel"
  }
},
```

Listing 8.1 Defining Globally Visible JSON Models in the Application Descriptor

A JSON model is mostly used to support the other server-side models, such as an OData model. In the JSON model, you can store data that controls the visibility of buttons or entire toolbars, depending on whether you're in edit mode or not. You can cache input from a simple form before you manipulate this data and finally send it to the server using an OData model. Or you get hierarchical data via the OData model only returned in a flat hierarchy or as a list, so you use a JSON model to transfer the data there and at the same time bring in the hierarchy. Those are just a few examples.

One striking difference to server-side models, however, is where certain operations are processed. We've chosen a very provocative example to illustrate the differences between an OData model and the JSON model. Don't get us wrong—we don't want to put OData services above other REST services because a well-implemented REST service (e.g., in Java, Python, or any other programming language) can provide APIs that are at least as good. In our example, however, we imagine that a REST service uses the /Carriers API to provide a list of carriers that are stored in any database. With our SAP S/4HANA system, the same is possible via an OData service with a /CarrierCollection API. Now our models, which are each linked to a view in different apps, have been given the task of displaying a list of carriers. However, not all of a carrier's data is needed because only the name is needed. Furthermore, the list should be sorted in descending order by the carrier's ID (`carrid`), and you only want to display the first two entries from this result list.

Figure 8.2 shows that the OData model would transform this request into a suitable HTTP `GET` request. The /CarrierCollection API is clear:

- `$top=2` is used to say that only the first two entries are needed.
- The `$orderby=carrid desc` parameter means that the list should be sorted by ID in descending order.
- The URL parameter `$select=name` indicates that only the name should be read, and that no other data such as the airline's currency, website URL, or other information should be included in the response.

The OData service receives this request, extracts the parameters, and returns only the two names of carriers Z and Y in the response.

With the JSON model, the whole thing looks a bit different. As mentioned, we assume that the REST service can't accept such parameters, so we send an HTTP `GET` request to the /Carriers API on the server. This reads all carriers from the table and returns them. Thus, the JSON model gets an entire list of all carriers, although we only need the last two. We also get not only the name, but the entire dataset, provided that we don't encounter any other technical limitations. The JSON model stores all of this data, and then we have the option of sorting and filtering this data locally at the client because the JSON model has generic methods available in the standard that can process the data. We just need to add a small piece of JavaScript or TypeScript code to get only the first two entries.

Now we can see the difference between a client-side and a server-side model more clearly. A JSON model can sort or filter this data itself, but we still need to add the code in JavaScript or TypeScript to get only the first two from the result list. In contrast, the OData model has left these operations to the server regardless of whether the server can resolve and actually apply them. The OData model isn't interested in sorting, filtering, pagination, or other nonfunctional aspects.

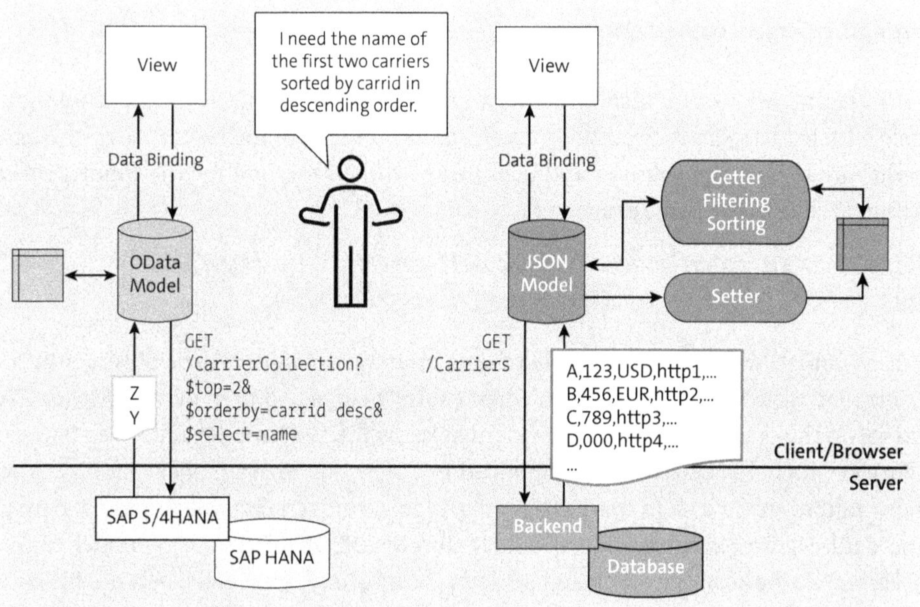

Figure 8.2 Very Simplified Representation of the Differences between Server-Side and Client-Side Models

Before we get too lost in the functionalities that any REST service could provide, let's get back to our JSON model and just concentrate on that. Let's assume that the aforementioned *testdata.json* file contains the data shown in Listing 8.2. This file will continue to be used to fill the default JSON model with data.

```
{
  "firstname": "John",
  "lastname": "Doe",
  "hobbies": [
    {
      "skillLevel": 1,
      "name": "Football"
    },
    {
      "skillLevel": 2,
      "name": "Tennis"
    },
    {
      "skillLevel": 2,
      "name": "Swimming"
```

```
    }
  ]
}
```

Listing 8.2 Content of the JSON Model

Let's assume we want to display the first name stored in the JSON model in an input field with the ID `myInputField`. We could read the contents of the JSON model with `getData()` in a JavaScript method and use the `setValue()` method for the input field to assign a value to the `value` property:

```
let sFirstName = this.getView().getModel().getData().firstName;
this.getView().byId("myInputField").setValue(sFirstName);
```

But, of course, we want to make use of data binding and not play the observer in the controller, which has to notify both when values change and then exchange them. To do so, on the one hand, our input field must know the JSON model. It does so because our JSON model is created in the application descriptor with a global visibility. If we now want to access data in this JSON model, then we must next understand the path to the data. Figure 8.3 shows that the data stored as JSON in the JSON model can be accessed via the absolute paths, and it shows the return type expected behind this path.

Data in the JSON Model	Path to Access the Data	Return Type
`{`	/	Object
`"firstname": "John",`	/firstname	String
`"lastname": "Doe",`	/lastname	String
`"hobbies": [`	/hobbies	Array
`{`	/hobbies/0	Object
`"skillLevel": 1,`	/hobbies/0/skillLevel	Integer
`"name": "Football"`	/hobbies/0/name	String
`},`		
`{`	/hobbies/1	Object
`"skillLevel": 2,`	/hobbies/1/skillLevel	Integer
`"name": "Tennis"`	/hobbies/1/name	String
`},`		
`{`	/hobbies/2	Object
`"skillLevel": 2,`	/hobbies/2/skillLevel	Integer
`"name": "Swimming"`	/hobbies/2/name	String
`}`		
`]`		
`}`		

Figure 8.3 Rules about Accessing the Data via Paths within the JSON Model

So, now we know that if we want to bind the first name to the `value` property in an input field, we can access the data via this path:

```
<Input value="{/firstname}"/>
```

Let's apply our knowledge of SimpleForm and build an appealing form around the input field, where we display the first name and last name (see Listing 8.3).

```
<f:SimpleForm id="detail_simpleform" editable="true" layout=
"ResponsiveGridLayout" labelSpanXL="3" labelSpanL="3" labelSpanM="3"
labelSpanS="12" emptySpanXL="4" emptySpanL="4" emptySpanM="4" emptySpanS="0"
columnsXL="1" columnsL="1" columnsM="1">
  <f:content>
    <Label id="labelFistname" labelFor="inputFirstname" text="First name" />
    <Input id="inputFirstname" value="{/firstname}" />
    <Label id="labelLastname" labelFor="inputLastname" text="Last name" />
    <Input id="inputLastname" value="{/lastname}" />
  </f:content>
</f:SimpleForm>
```

Listing 8.3 Simple Form Used to Display Two Input Fields

The result of our SimpleForm is shown in Figure 8.4.

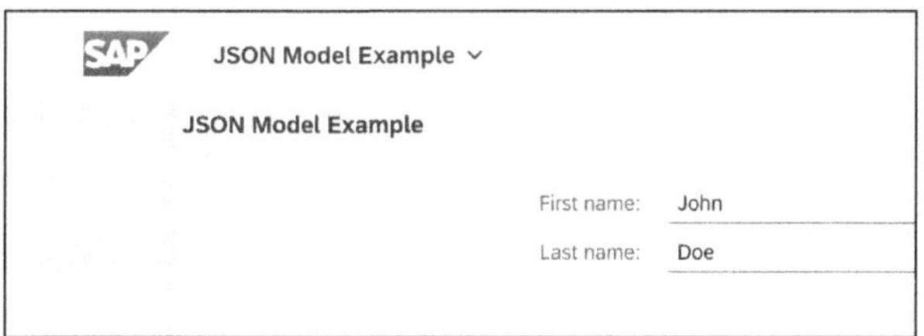

Figure 8.4 Simple Form Visible in the Preview

What happens when we want to access the hobbies? These are stored in an array, which is why we also get an array returned with the /hobbies path. However, you can see from Figure 8.3 that if we want to access the first hobby, we can append the index to the path. So, to display the name of the first hobby in a text control, it looks like this:

```
<Text text="{/hobbies/0/name}"/>
```

However, an array is more suited to a list than to a single text field, so we use a sap.m.List to display several instances of sap.m.StandardListItem. We have three hobbies, so in our list, we also display the StandardListItem, which reads the corresponding names and skill levels of the hobbies using the indices 0, 1, and 2 (see Listing 8.4).

```
<List id="detail_list" headerText="Hobbies">
  <items>
    <StandardListItem title="{/hobbies/0/name}"
      info="Skill level: {/hobbies/0/skillLevel}" />
    <StandardListItem title="{/hobbies/1/name}"
      info="Skill level: {/hobbies/1/skillLevel}" />
```

```
    <StandardListItem title="{/hobbies/2/name}"
      info="Skill level: {/hobbies/2/skillLevel}" />
  </items>
</List>
```

Listing 8.4 List Used to Display the Hobbies with Hard-Coded List Items

Of course, the way we did it in Listing 8.4 is a no-go for possible dynamic data. If we assume that this list can change and that 0 to *n* data could be dynamically loaded into it, then we should resort to another means. Fortunately, we already know about aggregation binding. So, we bind the aggregation `items` of the list to `/hobbies`. The `items` aggregation expects an array, so that basically fits. We now define a single `StandardListItem`, which will act as a template for all entries in the array. This template will be inserted as many times as there's data in `/hobbies`, just like in a `foreach` loop. Because the aggregation binding inserts the elements one after the other, as in a `foreach` loop, and the binding context is set for each element, we can work with relative paths here. For this reason, we don't write `/name` but only `name`. This is because a relative path can be resolved thanks to the binding context set for each element as if we were in the `foreach` loop (see Listing 8.5).

```
<List id="detail_list" headerText="Hobbies" items="{/hobbies}">
  <items>
    <StandardListItem id="detail_listitem" title="{name}"
      info="Skill level: {skillLevel}" />
  </items>
</List>
```

Listing 8.5 Displaying the Hobbies Using Aggregation Binding

Now we have a `List` below our `SimpleForm`, which displays the hobbies from the JSON model (see Figure 8.5).

Figure 8.5 List Uses Aggregation Binding to Show the Hobbies Stored in the JSON Model

8.1.2 Filtering

Filtering is one of the operations that the JSON model performs locally, that is, at the client. All the data required for this is already in the JSON model. We can simply define the filter statically in our XML view if we use complex binding instead of simple binding for the `items` aggregation and specify the desired filter there with `filters`. As you can see from this example, we always filter the list by `skillLevel` equals (`EQ`) to `2`:

```
items="{path: '/hobbies', filters:[{path: 'skillLevel', operator: 'EQ', value1:
2}] }"
```

In Figure 8.6, you can also see that this filtering was successfully performed by the JSON model.

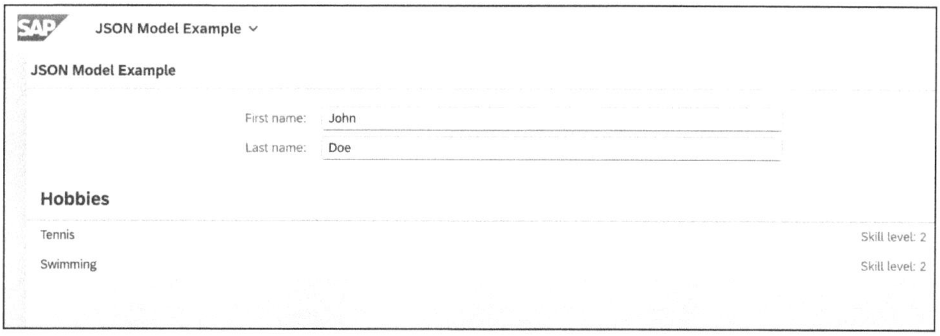

Figure 8.6 Hard-Coded Filter Works

To make the presentation of the filters a bit more exciting, we'll use the UI control `FilterBar` from the namespace `sap.ui.comp.filter`. This not only displays filters in a user-friendly way but also supports variant management and advanced filter configuration. We won't go that far, but in Listing 8.6, you can see that we've stored a `FilterGroupItem` that contains a `Select`. In this select, we've created three items for the three skill levels that we might have. The filter bar has two events, `search` and `clear`, for which we've registered two event handlers and will develop them in a moment.

```
<fb:FilterBar id="filterbar" search=".onSearch" clear=".onClear" showClearOnFB=
"true" showFilterConfiguration="false" >
  <fb:filterGroupItems>
    <fb:FilterGroupItem id="fbItemSkillLevel" label="Skill level" name="Skill
level"
      groupName="Group1" visibleInFilterBar="true">
      <fb:control>
        <Select id="selectSkillLevel" forceSelection="false" resetOnMissingKey=
"true">
          <core:Item id="selectOne" key="1" text="1"/>
          <core:Item id="selectTwo" key="2" text="2"/>
```

```
            <core:Item id="selectThree" key="3" text="3"/>
          </Select>
        </fb:control>
      </fb:FilterGroupItem>
    </fb:filterGroupItems>
</fb:FilterBar>
```

Listing 8.6 Filter Bar Used to Display Filter Items

Let's take a look at the filter bar in action. In Figure 8.7, you can see that the filter bar can be expanded or collapsed by default. The filters are listed one after the other. The search can be triggered using the **Go** button, for which we still need to develop the event handler. The same applies to the **Clear** button.

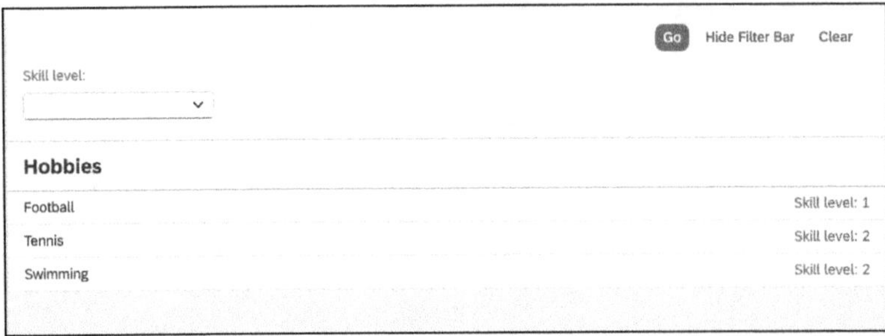

Figure 8.7 Filter Bar Shown in the Preview

Let's start with the onSearch method (see Listing 8.7). First, we get a reference to the aggregation binding from the list with getBinding("items") and store it in oListBinding. We also need the selectedKey from the Select, which can be read with the method getSelectedKey(). Initially, we access the select with an ID. In an IF, we check whether our sSelectedKey is filled. If it is, we add a new sap.ui.model.Filter to our array in which we want to collect all filters. This filter is assigned three parameters: the path of the property to which the filter should be applied, an operator from the FilterOperator enumeration, and the value which should be used for comparison when filtering. In our case, we want to check whether the skillLevel is equal to FilterOperator.EQ, the value stored in the variable sSelectedKey. As a final step, we apply the filter() method to the oListBinding and pass our filter array as the first parameter. The filtering doesn't happen globally in the JSON model, but via the corresponding list binding to which the data is bound.

```
private onSearch() {
  let oList = this.getView()?.byId("detail_list") as List,
    oListBinding = oList?.getBinding("items") as ListBinding,
    oSelectSkillLevel = this.getView()?.byId("selectSkillLevel") as Select,
    sSelectedKey = oSelectSkillLevel.getSelectedKey(),
    aFilter : Array<Filter> = [];
```

```
  if(sSelectedKey){
    aFilter.push(new Filter("skillLevel", FilterOperator.EQ, sSelectedKey));
  }
  oListBinding.filter(aFilter);
}
```

Listing 8.7 Event Handler for Filtering

Filters Objects in the Array Are Linked with AND by Default, Except for Filter Arrays

The individual filter objects in the array are linked with a logical AND. The JSON model handles filter arrays in which the same property is filtered differently from an OData model. Here, the filters that go to the same property are automatically linked with a logical OR.

We don't have to program as much for the onClear method, as shown in Listing 8.8. If we pass an empty array to the filter() method in the oListBinding, the filters are reset. Furthermore, we set the selected key to an empty string with setSelectedKey in the select. Because of the resetOnMissingKey property, the select resets the selection automatically to an empty one. This means that nothing should be selected in the select.

```
private onClear() {
  let oList = this.getView()?.byId("detail_list") as List,
    oListBinding = oList?.getBinding("items") as ListBinding,
    oSelectSkillLevel = this.getView()?.byId("selectSkillLevel") as Select;
    oSelectSkillLevel.setSelectedKey("");
  oListBinding.filter([]);
}
```

Listing 8.8 Clear Method to Reset the Filters

To manage the saving of the selected key and the select items even better, however, we can use a separate JSON model. To do this, we create a JSON model called oSkillSelectModel as a member variable. We initialize this in the onInit lifecycle method and immediately fill it with data. Inside, we'll store the item that was selected in the select behind selectedSkillLevel, and we'll store an array behind skillLevels that will offer the options for the select. We still have to make this JSON model known to the view, but we can't set it as the default model (without a name) because we'll overwrite our JSON model with the personal data and hobbies. For this reason, we assign the name filter as the second parameter in setModel (see Listing 8.9).

```
private oSkillSelectModel : JSONModel;
public onInit(): void {
  this.oSkillSelectModel = new JSONModel({
```

```
    selectedSkillLevel: "",
    skillLevels: [{"key": 1},{"key": 2},{"key": 3}]
  });
  this.getView()?.setModel(this.oSkillSelectModel, "filter");
}
```

Listing 8.9 Using a Separate JSON Model to Store the Filter Data

One Default Model per UI Component, but Several Default Models in a View?

For the more advanced data binding masters among you, here's a little hint: we could also set the setModel only to the FilterBar and then as the default model. We overwrite the default model for the filter bar with the person data and hobbies, but the filter bar doesn't need to know this data at all. So, we separate ourselves from the named model and don't need to use the names in the binding or when accessing the JSON model at all.

In the view, we now adjust our select because the data now comes from the JSON model named filter. For this reason, we bind filter>/skillLevels to the aggregation items of the select. We'll create a property binding for the selectedKey property and bind it to filter>/seletedSkillLevel. The items in the select are now fed from the array behind /selectedSkillLevel, so we only need to store a single item as a template, and this is then inserted based on the array. We can then access the key in the items using a relative path, but we must not forget the named model here and also specify filter> (see Listing 8.10).

```
<Select id="selectSkillLevel" items="{filter>/skillLevels}"
 selectedKey="{filter>/selectedSkillLevel}" forceSelection="false"
 resetOnMissingKey="true">
  <items>
    <core:Item id="selectItemTemplate" key="{filter>key}" text="{filter>key}"/>
  </items>
</Select>
```

Listing 8.10 Select Accessing and Storing the Filter Data via a Named JSON Model

Now we can also adjust our onSearch method. We no longer need to access the select because the select writes the selected key to the JSON model we prepared. So, we only need access to the JSON model to read this value with getProperty("/selectedSkill-Level") (see Listing 8.11).

```
private onSearch() {
  let oList = this.getView()?.byId("detail_list") as List,
    oListBinding = oList?.getBinding("items") as ListBinding,
    aFilter : Array<Filter> = [],
```

```
    sSelectedKey = this.oSkillSelectModel.getProperty("/selectedSkillLevel");
  if(sSelectedKey){
    aFilter.push(new Filter("skillLevel", FilterOperator.EQ, sSelectedKey));
  }
  oListBinding.filter(aFilter);
}
```

Listing 8.11 Method for Filtering: Accessing the Data of Selected Filters via a Separate JSON Model

In our `onClear` method, we're even better positioned. With `setProperty("/selectedSkillLevel", "")`, we reset the path in the JSON model, which stores the `selectedKey` (see Listing 8.12). Thanks to two-way binding, the view is notified of this value change and thus also resets the selection in the `select`.

```
private onClear() {
  let oList = this.getView()?.byId("detail_list") as List,
     oListBinding = oList?.getBinding("items") as ListBinding;
  this.oSkillSelectModel.setProperty("/selectedSkillLevel", "");
  oListBinding.filter([]);
}
```

Listing 8.12 Resetting the Filter Made Even Simpler with a Separate JSON Model

We don't want to withhold from you that the method which takes care of filtering can be overwritten in the `sap.ui.model.Filter`, where you could then store your very own JavaScript/TypeScript logic for filtering:

```
new sap.ui.model.Filter("skillLevel", (iValue: int) => {
    return (iValue > 1);
});
```

8.1.3 Sorting

When sorting, we can, as with filtering, store this hard-coded in an aggregation binding:

```
items="{ path: '/hobbies', sorter: [{path:'name',descending: true}] }"
```

However, we'll choose a more practical example again. So, we start by defining our own `OverflowToolbar` for the list, where we place a `Button` that will later open a dialog for the sort options (see Listing 8.13).

```
<headerToolbar>
  <OverflowToolbar id="hobbiesToolbar" style="Clear">
    <Title id="hobbiesTitle" text="Hobbies"/>
    <ToolbarSpacer id="toolbarSpacer"/>
    <Button id="sortButton" tooltip="Sort" icon="sap-icon://sort"
```

```
      press="openViewSettingsDialog"/>
  </OverflowToolbar>
</headerToolbar>
```

Listing 8.13 Button in a Toolbar That Opens a Dialog for Sorting

When you click on this button, we load a `Fragment` that we'll create later. This fragment will contain a `sap.m.ViewSettingsDialog`. This dialog can be used in the standard system to offer sorting criteria, filter options, or grouping options in a separate dialog. We save this dialog in a member variable so that it doesn't have to be initialized the second time it's opened, but can simply be opened with the `open()` method (see Listing 8.14).

```
private openViewSettingsDialog() {
  if(!this.viewSettingsDialog){
    Fragment.load({
      id: this.getView()?.getId(),
      name: "at.clouddna.jsonmodelexample.view.fragment.ViewSettingsDialog",
      controller: this
    }).then((oDialog: any) => {
      this.getView()?.addDependent(oDialog);
      this.viewSettingsDialog = oDialog;
      this.viewSettingsDialog.open();
    });
  }else{
    this.viewSettingsDialog.open();
  }
}
```

Listing 8.14 Method for Loading the Fragment That Contains the View Settings Dialog

We've stored this *ViewSettingsDialog.fragment.xml* in the */view/fragment* directory, and it's loaded using the method shown in the previous listing. Inside this file, we manage the sorting options that are offered to the user when they click on the button (see Listing 8.15).

```
<core:FragmentDefinition xmlns="sap.m" xmlns:core="sap.ui.core">
  <ViewSettingsDialog id="viewSettingsDialog" confirm=".onConfirm">
    <sortItems>
      <ViewSettingsItem id="sortItemName" text="Name" key="name" selected=
"true" />
      <ViewSettingsItem id="sortItemSkillLevel" text="Skill level" key=
"skillLevel" />
    </sortItems>
```

```
  </ViewSettingsDialog>
</core:FragmentDefinition>
```

Listing 8.15 Content of the Fragment Holds a View Settings Dialog

The result, after clicking on the button in the toolbar, is shown in Figure 8.8. A `ViewSettingsDialog` opens and offers to sort these two sorting criteria in ascending or descending order.

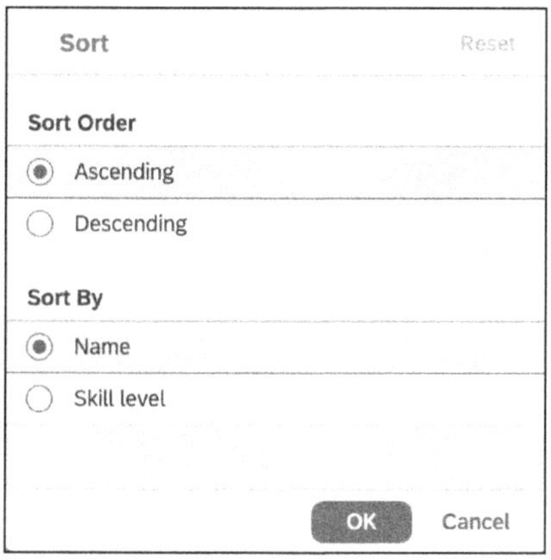

Figure 8.8 View Settings Dialog Offering Several Sort Criteria

This dialog has an event called `confirm`. We've stored our method `onConfirm`, which you can see in Listing 8.16, as the event handler. This method takes the key from the selected `sortItem`. We also use `sortDescending` to determine whether the user wants the sorting to be descending. If not, it can be concluded that it should be ascending. The sorting itself works on the `oListBinding` of the list and is triggered by the `sort()` method. Here, we provide a `sap.ui.model.Sorter` object that receives two parameters: the name of the property to be sorted by and whether this sorting should be descending. We could have also used an array of sorters here, but this time, we passed the object directly.

```
private onConfirm(oEvent: Event) {
  let oList = this.getView()?.byId("detail_list") as List,
    oListBinding = oList?.getBinding("items") as ListBinding,
    bSortDescending = oEvent.getParameters().sortDescending,
    sSortKey = oEvent.getParameters().sortItem.getKey(),
    oSorter = new Sorter(sSortKey, bSortDescending);
  oListBinding.sort(oSorter);
}
```

Listing 8.16 Method to Apply the Sorting Criteria to the List Binding

Order of Sorters in the Array

The order of the sorter objects in the array also determines the order of the sorting operations. So, we could first sort by skill level in descending order and then by the names of the hobbies in ascending order. For this, we need to add two objects to the array.

With `sap.ui.model.Sorter`, you also have the option to overwrite the `fnComparator` (see Listing 8.17). You can imagine that this runs like a bubble sort and always passes two values to the comparator. These two values must be compared with each other in JavaScript, and you can decide how the order should be by returning `-1` (before), `0` (equal), and `1` (after).

```
new sap.ui.model.Sorter("property", (value1, value2) => {
    if (value1 < value2) return -1;
    if (value1 == value2) return 0;
    if (value1 > value2) return 1;
});
```

Listing 8.17 Redefinition of the Default Comparator Method Used for Determining the Sort Order

8.2 Axios

Axios is an HTTP client for web applications and Node.js developed by Matt Zabriskie as an open-source project under the MIT license. As it's also presented in the official documentation (see Figure 8.9), Axios provides promises for asynchronous HTTP calls to a RESTful web service.

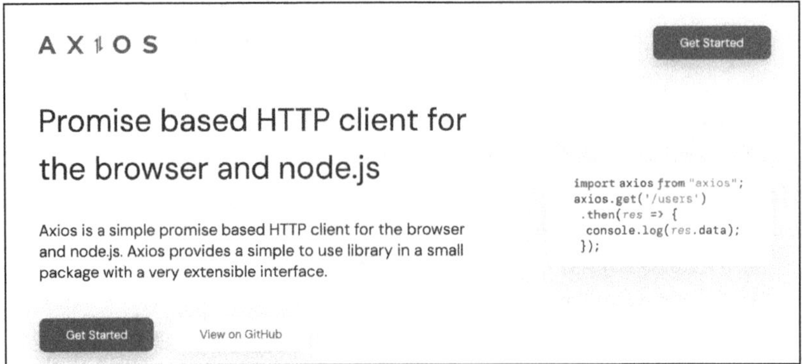

Figure 8.9 Official Website and Documentation of Axios (Source: https://axios-http.com/)

If you're interested in using Axios, you can either fork the GitHub repository and use it as it's provided (see Figure 8.10), or you can make use of the Massachusetts Institute of

Technology (MIT) license and make further developments to the framework. The MIT license is among the best-known simple and permissive licenses, which allows any changes to the source code. The derived work doesn't have to be published under a Free and Open-Source Software (FOSS) license, but copyright notices and liability disclaimers must be included.

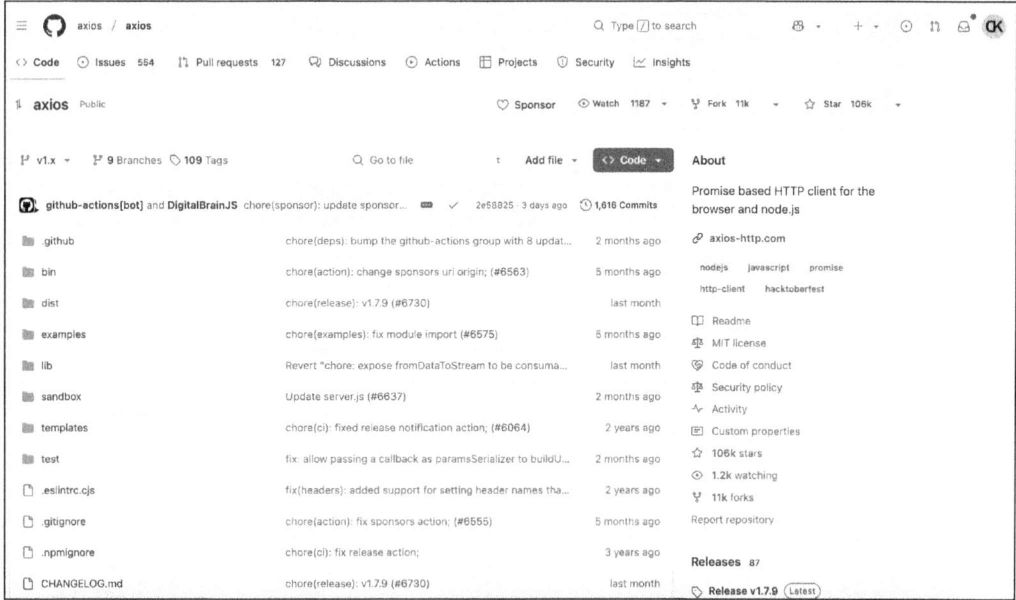

Figure 8.10 Official Git Repository of the Axios Client

We've also added some further developments to the Axios project and published these under the name *CloudDNA RestModel* (see Figure 8.11). The RestModel uses the Axios client to be able to send HTTP calls to a RESTful web service. The RestModel extends the already powerful Axios client and tailors it to an SAPUI5 environment, for example, for the mapping of destinations or self-defined access points in the application.

The RestModel is provided as an open-source project and can be forked at GitHub, where it's updated and provided with new features. Planned features include the storage of requested data in a model similar to the OData model to provide similar features, such as binding and view-triggered requests. The authors would be glad to have motivated contributors to this GitHub project so that together we can make the best out of the RestModel. The GitHub project is located at *https://github.com/clouddnagmbh/RestModel*.

You have two options for installation. We'll show you both, although the first is simpler, and only the second is state-of-the-art.

https://github.com/clouddnagmbh/RestModel/blob/master/webapp/libs/axios.js

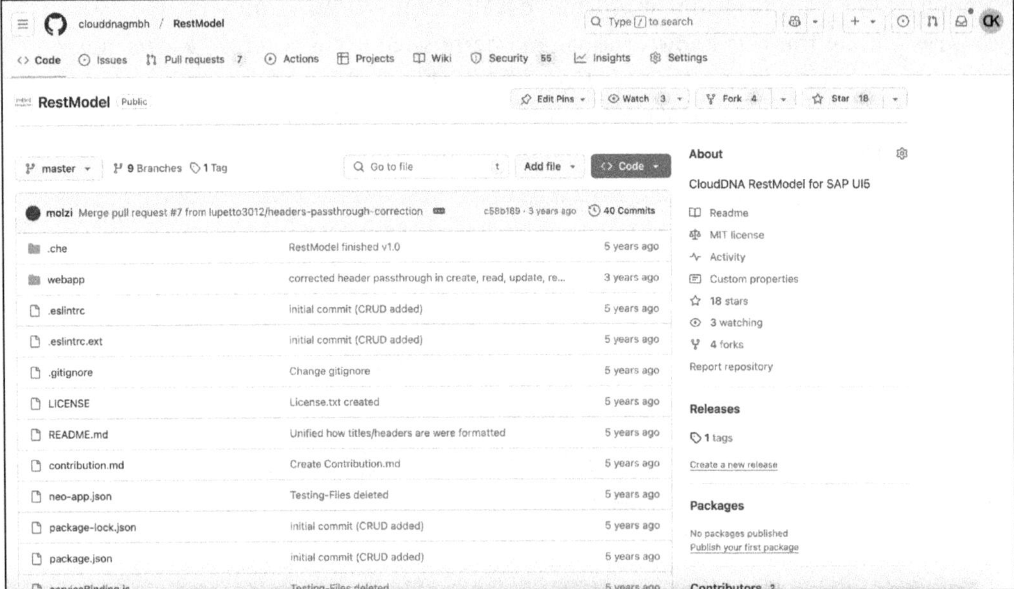

Figure 8.11 CloudDNA RestModel Uses Axios to Provide a Promise-Based HTTP Client in SAPUI5

8.2.1 Installation via Files or the Axios Node Module

In the past, when you were still using SAP Web IDE to develop SAPUI5 applications, it wasn't so easy to install additional node modules. For this reason, the installation was file-based, which means, you had to download the sources and include them in your project structure. These files need to be placed in the *webapp* folder (or in a subordinate folder). The first file that needs to be downloaded is called *axios.js* (see *https://github.com/clouddnagmbh/RestModel/blob/master/webapp/libs/axios.js*), which also holds the core functionalities of the Axios client. After that, the file *RestModel.js* needs to be downloaded (see *https://github.com/clouddnagmbh/RestModel/blob/master/webapp/libs/RestModel.js*) and placed in the same directory where the file *axios.js* was stored.

Now, in *RestModel.js*, you just need to adjust the second parameter in the array from `sap.ui.define` so that the Axios client can be found in your project directory:

```
sap.ui.define([ "sap/ui/base/Object", "myfolderpath/axios", "sap/base/Log"
], function (Object, axios, Log) { //...
```

The disadvantage of this procedure was that *axios.js* had to be updated with every further development, change, and bugfix in the actual Axios client. A lot has changed since then, and with SAP Business Application Studio and the possibility of using Visual Studio Code (VS Code) with the SAP Fiori tools, many new opportunities have arisen.

So, the state-of-the-art option is to either load the source code from the Axios client from a content delivery network (CDN) or integrate it as a node module. We've also listed the associated command for installing the node modules:

```
npm install axios
```

This means that you no longer need the *axios.js* file because the source code of Axios is in the node modules. You can now also use the name `axios` for the import and no longer a file path:

```
sap.ui.define([ "sap/ui/base/Object", "axios", "sap/base/Log"
], function (Object, axios, Log) { //...
```


To use the RestModel, a new instance needs to be created that consumes starting parameters (see Listing 8.18).

```
new RestModel({
  {string} url - URL of REST service.
  {number} [timeout=5000] - Reqest timeout.
  {object} [headers={}] - Request headers.
  {boolean} [xcsrfTokenHandling=false] - Activation of X-CSRF Token Handling.
});
```

Listing 8.18 Constructor of the RestModel That Needs at Least the URL of the REST Service

Here's an example:

```
this.oModel = new RestModel({ url: "https://mysubdomain.domain.com/api" });
```

To make sure that the RestModel is globally visible, we recommend initializing it in the `init` method in the component controller (see Listing 8.19).

```
sap.ui.define([
    "sap/ui/core/UIComponent",
    "at/clouddna/restmodel/model/models",
    "at/clouddna/restmodel/RestModel"
], (UIComponent, models, RestModel) => {
    "use strict";
    return UIComponent.extend("at.clouddna.restmodel.Component", {
        metadata: { … },
        init() {
            …
            // initialize the RestModel
            this.oModel = new RestModel({ url: "https://mysubdomain.domain.com/
api"});
        }
```

```
    });
});
```

Listing 8.19 Initialize the RestModel in the Component Controller to Get Global Visibility

If we start the preview, the RestModel will be initialized thanks to the component controller and available globally with `this.getOwnerComponent().oModel`. However, we'll have the problem that the first time the RestModel is used, this `axios` would be tried to be resolved on the path */resources* at localhost and thus directly in the SAPUI5 framework. There, of course, the request will find nothing, fail, and return an error (see Figure 8.12).

Figure 8.12 Node Module Axios Couldn't Be Loaded by the SAPUI5 Framework

To be able to access node modules directly in SAPUI5, no `require` can be used, but another node module called `ui5-tooling-modules` can, which allows the framework to access the required node modules first. Beforehand, the SAP Fiori tools proxy middleware is activated, and all the requests regarding module loading will be sent to the SAPUI5 framework CDN. This module can be installed with the following command:

```
npm install ui5-tooling-modules --save-dev
```

Further adjustments need to be made, in order so the corresponding `ui5-tooling-modules-middleware` is launched and used before the `fiori-tools-proxy` overtake. You can see these adjustments, which needs to be made in *ui5.yaml*, in Listing 8.20.

```
specVersion: "3.1"
metadata:
  name: test
type: application
builder:
```

```
  customTasks:
    - name: ui5-tooling-modules-task
      afterTask: replaceVersion
server:
  customMiddleware:
    - name: ui5-tooling-modules-middleware
      afterMiddleware: compression
    - name: fiori-tools-proxy
      afterMiddleware: ui5-tooling-modules-middleware
      configuration:
...
```

Listing 8.20 Adjustments to the ui5.yaml in Order So the Node Modules Can Be Loaded

To use an SAP Business Technology Platform (SAP BTP) destination, an entry referring to this `destination` needs to be created in the *ui5.yaml*. This entry serves as a mapping between the destination and an entry path, which is used as a base-path URL in the RestModel (see Listing 8.21).

```
- name: fiori-tools-proxy
      afterMiddleware: ui5-tooling-modules-middleware
      configuration:
        ignoreCertError: false
        ui5:
          path:
            - /resources
            - /test-resources
          url: https://ui5.sap.com
        backend:
          - path: /api
            destination: S4D
            url: https://mysubdomain.domain.com
```

Listing 8.21 Making Use of the Built-In Middleware to Avoid CORS

If you have errors regarding CORS, using a middleware like the one shown in Listing 8.21 is a solution. If you host the app on any web server, then the `destination` entry won't help you. As you can see, we've already added an example of the `url` entry for this reason. If you delete the destination property, the redirection to this URL will take place.

In this case, the built-in middleware takes care of the redirection and all requests going to the relative path */api* are forwarded to the actual host and port. Therefore, when defining the base-path URL in the constructor, we're using a relative path:

```
this.oModel = new RestModel({ url: "/api" });
```

In this case, you can see in Figure 8.13 that the request to get to the Axios is sent to the */resources* path, but this can be intercepted by the presented UI5 Tooling modules middleware and sent to the local copy of the node modules.

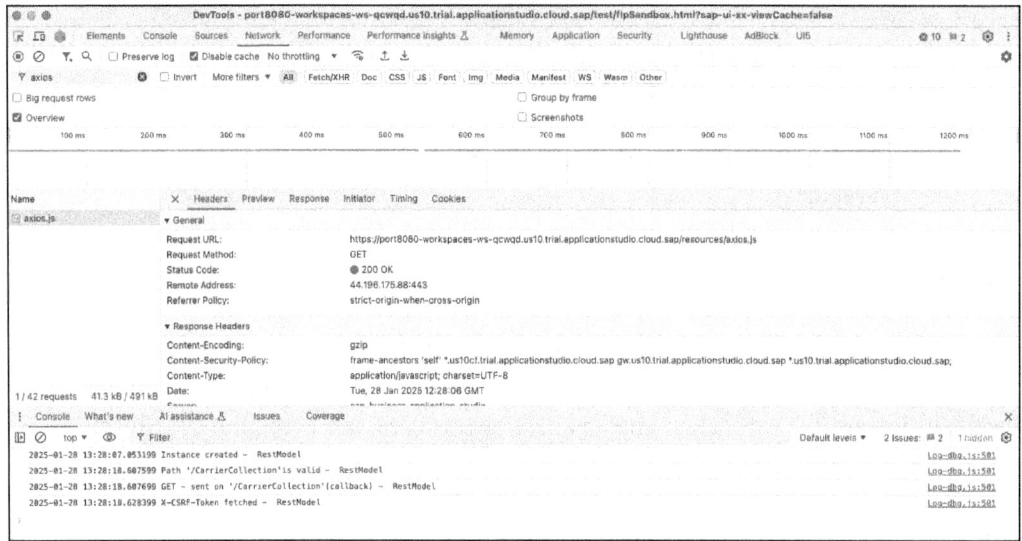

Figure 8.13 Request to /resources/axios Intercepted by the UI5 Tooling Modules Middleware

Nevertheless, in this case, too, the result after a build is that a copy of the Axios source code is stored in the project. You can find these files in the *dist/thirdparty* folder (see Figure 8.14), which is the minified version of the application and will be deployed to the server.

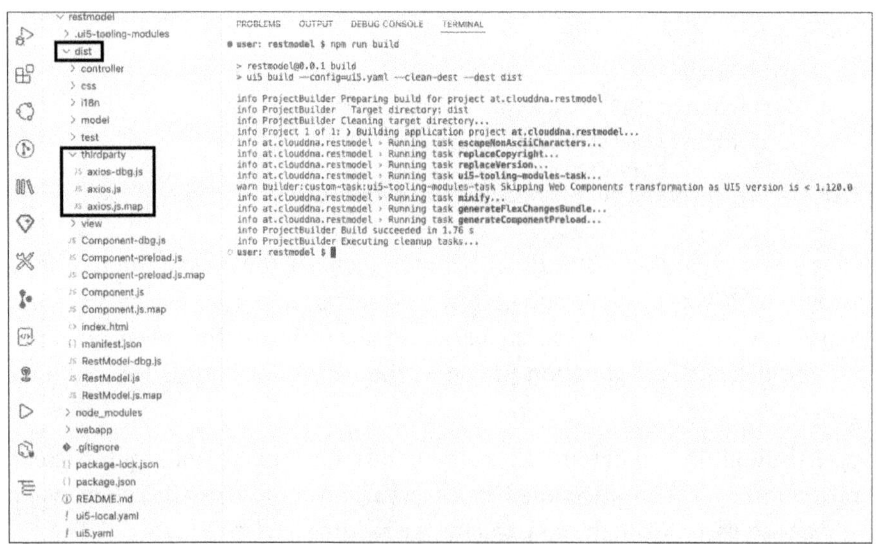

Figure 8.14 The Result after a Build Process: A Copy of the Source Code Stored in the Node Modules

Although Axios is now in a directory called *thirdparty*, you don't need to change your source code. If we look at the built and minified version of the *RestModel* file in the *dist* folder, we can see that the path has also been adjusted by the builder task (Figure 8.15).

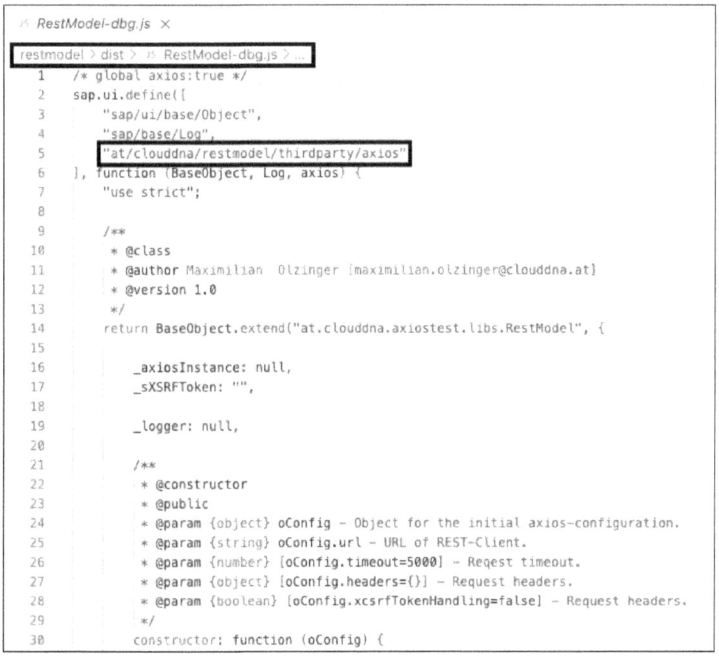

```
RestModel-dbg.js ×
restmodel > dist > RestModel-dbg.js > ...
/* global axios:true */
sap.ui.define([
    "sap/ui/base/Object",
    "sap/base/Log",
    "at/clouddna/restmodel/thirdparty/axios"
], function (BaseObject, Log, axios) {
    "use strict";

    /**
     * @class
     * @author Maximilian  Olzinger [maximilian.olzinger@clouddna.at]
     * @version 1.0
     */
    return BaseObject.extend("at.clouddna.axiostest.libs.RestModel", {

        _axiosInstance: null,
        _sXSRFToken: "",

        _logger: null,

        /**
         * @constructor
         * @public
         * @param {object} oConfig - Object for the initial axios-configuration.
         * @param {string} oConfig.url - URL of REST-Client.
         * @param {number} [oConfig.timeout=5000] - Reqest timeout.
         * @param {object} [oConfig.headers={}] - Request headers.
         * @param {boolean} [oConfig.xcsrfTokenHandling=false] - Request headers.
         */
        constructor: function (oConfig) {
```

Figure 8.15 Builder Task Modified the Paths for the Files to Be Loaded after Deployment

8.2.2 Using CRUD Functionalities

All create, read, updated, and delete (CRUD) functionalities of the RestModel return a promise containing the result or can be called by providing a success callback or an error callback in the parameters of the request. When no callbacks are defined in the parameters, a promise is returned, which can be processed by using `oPomise.then(fnSuccess, fnError)` or `oPromise.then(fnSuccess).catch(fnError)`.

The method `create` of the RestModel uses HTTP `POST` to create new entries. The first parameter is the relative path that will be added to the base-path URL defined in the constructor of the RestModel. The second parameter is a JavaScript object containing the data that should be created. In the third parameter, again a JavaScript object can be passed containing a success function, an error function, and an object containing additional headers. You can examine the syntax in Listing 8.22. If the success and error callbacks aren't used, the method will return a promise.

```
oModel.create(sPath, oObject, oParameters)
//parameters
sPath =  {string} sPath - Path where a new ressource should be added.
oObject = {object} oObject - Data which should be posted.
```

```
oParameters = {
    {function} [success=function(){}] - Success-callback function.
       {function} [error=function(){}] - Error-callback function.
       {object} [headers] - Send additional axios-header-parameters.
}
//returns
returns {promise} [oPromise] - returns promise if no success- or error-callback
was specified.
```

Listing 8.22 Syntax of a "create" with All the Possible Parameters and Its Return Value

In Listing 8.23 you can see an example for a create:

```
this.getOwnerComponent().oModel.create("/Customer",
   { Firstname: "John", Lastname: "Doe" }, {
      success: (oData) => { ... },
      error: (oError) => { ... }
   }
);
```

Listing 8.23 Sample "create" (POST) Request

The method read of the RestModel uses HTTP GET to read external data. The first parameter is the relative path as string, which will be added to the base path of the RestModel. In the second parameter, again a JavaScript object can be passed, which contains the callback methods, additional headers, or additional URL parameters. Again, we've included the syntax of this method in Listing 8.24. If the success and error callbacks aren't used, the method will return a promise.

```
oModel.read(sPath, oParameters)
//parameters
sPath = {string} sPath - Path to a ressource.
oParameters = {
        {function} [success=function(){}] - Success-callback function.
        {function} [error=function(){}] - Error-callback function.
        {object} [headers] - Send additional axios-header-parameters.
        {object} [restUrlParameters] - Additional URL-parameters for REST-
calls.
        {Array} [select] - String-array for $select-parameter.
        {string} [filter] - String for $filter-parameter.
        {number} [skip=0] - Integer for $skip-parameter. Default is 0.
        {number} [top=100] - Integer for $top-parameter. Default is 100.
        {Array} [orderyb] - String-array for $orderby-parameter.
        {boolean} [sendSkipTop=false] - Send default $top=100 and $skip=0.
}
```

```
//returns
returns {promise} [oPromise] - returns promise if no success- or error-callback
was specified.
```

Listing 8.24 Syntax of a "read" with All the Possible Parameters and Its Return Value

An example of a read request is shown in Listing 8.25.

```
this.getOwnerComponent().oModel.read("/Customer", {
  success: (oData) => { ... },
  error: (oError) => { ... }
});
```

Listing 8.25 Sample "read" Request Using the RestModel

The method update of the RestModel uses HTTP PUT to alter external data. The first parameter is the path to the resource, which should be updated. The second parameter is a JavaScript object containing all the data, which should be updated. The third parameter again gives the possibility to send other header parameters or to register a success or error handler within a JavaScript object. If the success and error callbacks aren't used, the method will return a promise. Listing 8.26 shows the syntax of this method.

```
oModel.update(sPath, oObject, oParameters)
//parameters
sPath = {string} sPath - Path where the ressource should be updated.
oObject = {object} sObject - Data which should be posted.
oParameters = {
        {function} [success=function(){}] - Success-callback function.
        {function} [error=function(){}] - Error-callback function.
        {object} [headers] - Send additional axios-header-parameters.
}
//returns
returns {promise} [oPromise] - returns promise if no success- or error-callback
was specified.
```

Listing 8.26 Syntax of an Update with All the Possible Parameters and Its Return Value

In Listing 8.27, you can see an example of a sample update request, where the Firstname is getting updated from John to Jane.

```
this.getOwnerComponent().oModel.update("/Customer/7",
  { Firstname: "Jane", Lastname: "Doe" }, {
    success: function (oData) { … },
    error: function (oError) { … }
});
```

Listing 8.27 Sample Request to Send an Update via the RestModel

The method remove of the RestModel uses HTTP DELETE method. In this case, as with read and update, the path of the resource to be deleted should be entered as the first path. The second parameter again offers the possibility via success and error callbacks to wait for the response. The syntax is again shown in Listing 8.28.

```
oModel.remove(sPath, oParameters)
//parameters
sPath = {string} sPath - Path of the resource to be deleted.
oParameters = {object} {
        {function} [success=function(){}] - Success-callback function.
        {function} [error=function(){}] - Error-callback function.
        {object} [headers] - Send additional axios-header-parameters.
}
//returns
returns {promise} [oPromise] - returns promise if no success- or error-callback
was specified.
```

Listing 8.28 Syntax of an "update" with All the Possible Parameters and Its Return Value

In Listing 8.29, you can see a delete request to delete the customer with the ID 1.

```
this.getOwnerComponent().oModel.remove("/Customer/1", {
    success: function (oData) { … },
    error: function (oError) { … }
});
```

Listing 8.29 Sample Request to Send a Delete Request via the RestModel

By using the method bearerTokenLogin, the user can be authenticated against the provided service. After the authorization is successful, the bearer token is fetched and set for further requests automatically as a header parameter. Listing 8.30 shows the syntax that can be used for this.

```
oModel.bearerTokenLogin: function (sUrl, sUsername, sPassword, oParameters)
//parameters
sUrl = {string} - Path of the auth API.
sUsername = {string} - Username.
sPassword = {string} - Password.
oParameters = {
    success = {function}  - Success-callback function.
    error = {function}  - Error-callback function.
}
```

Listing 8.30 Syntax of the bearerTokenLogin with All Possible Parameters

In addition, the RestModel supports X-CSRF token handling. An X-CSRF token is a security feature designed to protect web applications from *cross-site request forgery* (CSRF) attacks. These attacks occur when a malicious actor tricks a user into performing unintended actions in a web application where the user is authenticated, such as submitting a form or triggering an API request without their consent. The X-CSRF token mitigates this risk by ensuring that each request sent to the server is from an authorized and trusted source. This mechanism is commonly used in web applications and secure systems such as OData services, where a GET request might retrieve an X-CSRF token that is then included in POST, PUT, or DELETE requests to ensure their authenticity. By checking the token, the server can confirm that the request is legitimate and not the result of a CSRF attack. To increase security, the token should be random and unique, transmitted securely over HTTPS, and expire at the end of the session or after a predefined time.

This can be enabled by simply providing the parameter xcsrfTokenHandling when instantiating a new RestModel or by calling the method setXCSRFTokenHandling. Listing 8.31 shows how to change this property of the RestModel at runtime.

```
oModel.setXCSRFTokenHandling(bSetSecurityHeader)
//parameters
bSetSecurityHeader = {boolean} bSetSecurityHeader - Enable X-CSRF-
Tokenhandling
```

Listing 8.31 Method to Change the X-CSRF Token Handling at Runtime

If you want to add request headers that should be applied in general, this can be done via the addHeader method (see Listing 8.32).

```
oModel.addHeader(oHeader)
//parameters
oHeader = {
        {string} name - Header-Name.
        {value} value - Header-Value.
}
```

Listing 8.32 Adding General HTTP Headers to Every Request Sent by the RestModel

To remove a header, just use the removeHeader method (see Listing 8.33).

```
oModel.removeHeader(sHeaderName)
//parameters
sHeaderName = {string} sHeaderName - Header-Name.
//returns
returns {boolean} - Entry deleted or not.
```

Listing 8.33 Removing HTTP Headers That Won't Be Sent by the RestModel Anymore

There's also an additional convenience function that checks if the given path is syntactically correct (see Listing 8.34).

```
oModel.checkPath(sPath)
//parameters
sPath = {string} sPath - Path to a resource.
//returns
returns {string} sPath - Returns the path if no error was thrown.
```

Listing 8.34 Checking the Path of a Resource via the RestModel

8.3 Summary

In this chapter, we got to know two models. The JSON model is delivered and maintained by SAP within the SAPUI5 framework. It has built-in functions to filter or sort the data in the model. Furthermore, it's a great addition to server-side models, such as the OData model, because we can determine the structure of the data ourselves in the JSON model. Using getters and setters, we have access to the data and can make changes as we see fit. Due to its very small feature set, the JSON model isn't suitable for communicating with a web service. It simply lacks the necessary functionalities, such as encapsulated CRUD calls over HTTP, token handling, and CORS support. For this reason, we've brought you one of the many famous representatives from the world of free REST clients. This already has the aforementioned functionalities by default, and a further advantage is that, due to the MIT license, the entire source code of this client can be viewed.

The RestModel for SAPUI5 project forked the Axios project and developed it further, or rather built a wrapper around the Axios client. You've seen how you can integrate Axios into an SAPUI5 project, how you can use the middleware to communicate with a web service, and how the individual calls such as `Create`, `Read`, `Update`, and `Delete` can work using the RestModel.

Chapter 9
OData Service Integration

OData as a data transfer protocol is the connection between our SAPUI5 applications and our SAP system. It allows us to communicate with semantically enriched business data to consume them in our apps and to work with model-based frameworks such as SAP Fiori elements.

In this chapter, we'll look at the OData protocol, which is an ISO/IEC-approved Organization for the Advancement of Structured Information Standards (OASIS) standard that defines the communication and design of Representational State Transfer (*REST*)–based services.

In Section 9.1, we'll get an overview of how the OData protocol is integrated into SAP S/4HANA. Once we understand this, we investigate two technologies mainly used for developing services that run on the OData protocol. Now we can go to the metadata document in Section 9.2, where we see what the technical description of an OData service looks like.

Section 9.3 gives us tools for testing an OData service, so we can look into how one works for the next chapter. In Section 9.4, we'll take our first practical steps by testing an OData service to see how its entities can be addressed to read and manipulate them. Section 9.5 further enhances our capabilities to read entities based on filters, sorters, and so on.

Integrating such an OData service will be the content of Section 9.6. There, we see the approach of adding an OData service into an SAPUI5 application, either during creation or on preexisting applications.

9.1 The OData Protocol in SAP

There are multiple ways of integrating such an OData protocol for a service in SAP. These differentiate foremost between an on-premise SAP S/4HANA system and the cloud environment with SAP Business Technology Platform (SAP BTP). As the cloud is ever-changing, we'll take a look in this chapter at the on-premise side of things. Before we jump into the system architecture, the ABAP RESTful application programming model, and ABAP-based OData services, let's first start with some basics.

9.1.1 Basics

Like REST, OData is based on six architectural layers:

- **Client/server architecture**
  Each layer within an OData architecture must implement a separate client/server model.
- **Caching**
  Each response from a server must indicate whether it can be cached.
- **Statelessness**
  Each request must contain everything the server needs to process it.
- **Multilayered system**
  There must be a clear separation of layers that obscures the lower layers.
- **Uniform interfaces**
  The server must provide the client with information on how to communicate with it and how its communication model is structured.
- **Code on demand**
  On request, the server can provide the client with executable code.

A service that implements the OData protocol specifies a clear communication structure. It's based on an *entity-relationship* (*ER*) model, as shown in Figure 9.1.

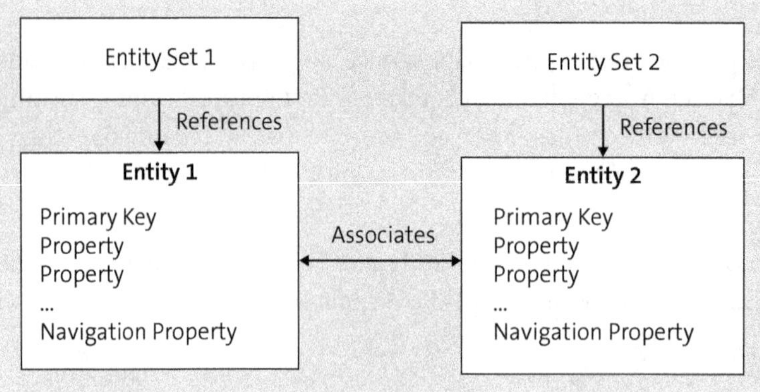

Figure 9.1 The OData ER Model

The ER model is mapped in OData in a metadata document. This metadata document represents entities and their relationship to each other and forms the technical description of the OData service.

Entities consist of properties and relationships to other entities, called associations or navigation properties. Entities are summarized as entity sets.

OData supports two data exchange formats for HTTP communication: the first is JavaScript Object Notation (JSON), and the second is the Extensible Markup Language (XML)–based format, a hierarchical structure language. OData is used in the SAP context where relational business data must be made available for applications or external systems.

9.1.2 System Architecture

In the system architecture for SAP Fiori in an on-premise landscape, a distinction is made between a frontend server and a backend server. This system architecture is shown in Figure 9.2.

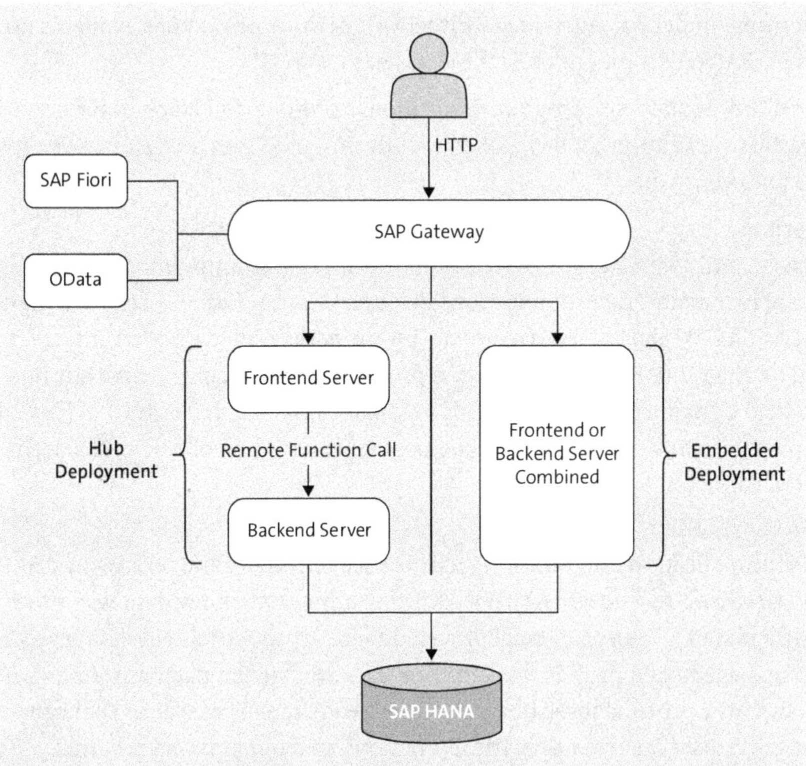

Figure 9.2 The SAP System Architecture for OData

The frontend server doesn't contain any business logic of its own but serves to provide an entry point for users called SAP Fiori launchpad. In addition, the frontend server is used to connect to any systems such as a side-by-side cloud, to provide applications from the SAPUI5 repository or transactions and to serve as a registration point for SAP Gateway.

SAP Gateway is a central interface based on the REST and OData standards. It enables centralized access to an SAP system. Business logic in an OData scenario is consumed via SAP Gateway. SAP applications in different programming languages can be connected to SAP Gateway. The prerequisite is that they can communicate with REST/OData. The backend server, on the other hand, contains the actual business logic and its

ABAP implementation. The backend server is connected to a frontend server via a trusted *Remote Function Call* (RFC), and both systems can be installed on a server as an embedded deployment or be deployed separately on dedicated servers as a hub deployment.

9.1.3 ABAP RESTful Application Programming Model

The ABAP RESTful application programming model is used for ABAP and SAP S/4HANA that makes it possible to build applications in the style of REST. ABAP RESTful application programming model was introduced with SAP S/4HANA 1909 and is available on-premise in the ABAP environment or in the cloud with SAP BTP.

The ABAP RESTful application programming model offers a full-stack approach to developing ABAP applications, which follows uniform guidelines and standards. It's based on the following aspects:

- **User experience**
  ABAP projects built with ABAP RESTful application programming model offer the possibility of metadata-driven applications through the annotations of the core data service (CDS). SAPUI5 and SAP Fiori elements interpret the metadata of an OData service based on the ABAP RESTful application programming model and are thus built and controlled dynamically. This makes it possible to build a uniform UI across all business processes in which the entire business logic and control of the UI is transferred to the backend.
- **SAP HANA capabilities**
  ABAP RESTful application programming model uses CDS views and optionally ABAP Managed Database Procedures (AMDP). AMDP is a framework for database procedures. AMDP and CDS views can be deployed directly to an SAP HANA database via the ABAP managed code pushdown and thus executed at the database level. SAP HANA is not only a pure database but also offers a wide range of services that can be used to map database analyses, analytical functions, and integration scenarios.
- **Efficient development**
  ABAP RESTful application programming model uses Eclipse as the development environment. SAP provides a plugin for the direct connection to the ABAP repository of an SAP system called the ABAP Development Tools (ADT) for Eclipse, which is a prerequisite for the development of certain ABAP development objects, namely CDS. SAP doesn't provide native editors in SAP GUI. Eclipse offers a version management option, documentation aids, and Eclipse-native editors for certain development objects.
- **Cloud**
  The ABAP RESTful application programming model is available on-premise and for the ABAP environment of SAP BTP. SAP BTP is the proprietary SAP cloud environment

and has its own fee-based ABAP environment with an SAP HANA database. ABAP cloud applications are developed using Eclipse.

The ABAP RESTful application programming model is architecturally structured in a layer model, as shown in Figure 9.3. These layers contain different technologies and development objects and thus ensure a separation of tasks in accordance with a layer separation. Each layer has its own task area and development objects.

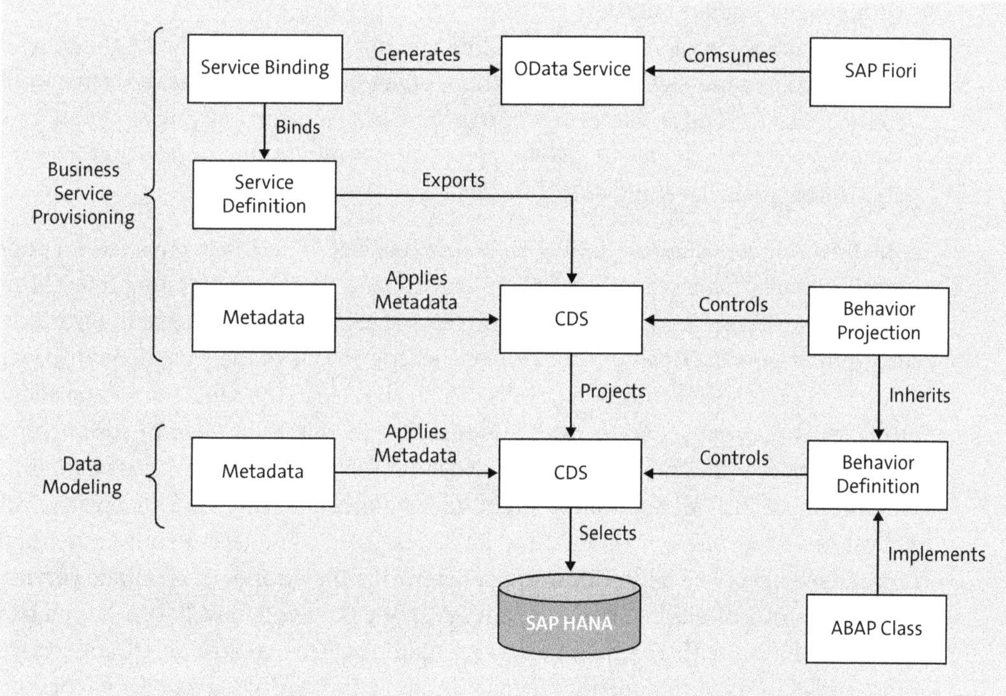

Figure 9.3 The ABAP RESTful Application Programming Model Architectural Layers

The data modeling layer is responsible for the data modeling of business objects, that is, objects of the real world. Here, CDS-based entities are created that read data from database tables and are provided as semantic entities via the Data Definition Language (DDL) of CDS. These CDS entities can be enriched with metadata via annotations. Annotations can influence the runtime behavior or have a purely semantic context.

Behavior definitions are responsible for the exact behavior of an entity. These definitions control write logics, dynamic control of entity fields, and general concepts such as design handling. In addition, an entity can be validated via validators, determine values, and provide actions that can't be mapped via normal write operations. All of this behavior is implemented in ABAP or provided via generic ABAP classes.

The ABAP RESTful application programming model offers two behavioral implementation options:

- **Managed implementation**
  A managed implementation is based on generic ABAP implementation for write logic, locking mechanisms, key assignment, concurrency, and persistence operations, but this requires a certain structure of the CDS entities and their relationship to each other. Managed implementations are suitable for greenfield development, as a CDS entity structure is built from scratch and can be adapted to the assumed structures.
- **Unmanaged implementation**
  The unmanaged implementation requires a manual implementation for the generic operations in a managed implementation. Developments that must be transferred to the ABAP RESTful application programming model on an existing logic can thus be integrated into the ABAP RESTful application programming model architecture style (brownfield development).

In the business service provisioning layer, the business objects are prepared for consumption in a specific scenario. The CDS entities and their behavior are projected in this layer and, if necessary, changed to correspond to the respective consumption scenario. A multiple relationship can arise between the objects of the data modeling layer and several associated projection objects, which all refer to the same object and adapt for different scenarios. CDS entities are enriched in this layer with UI-specific and consumption-specific annotations, provided it's a UI scenario. This makes it possible to consume ABAP RESTful application programming model entities in a wide variety of applications with different UI and functional designs. The projected entities are then summarized in service definitions, which determine the entities of an OData service. The OData protocol, either OData Version 2 (V2) or OData Version 4 (V4), is bound to a service definition with the help of service bindings. This creates an OData service whose CDS-based entities and their behavior-based annotations are available for consumption in an application.

These OData services can be integrated into SAP Fiori apps, provided that the service binding is a UI service. In the case of a web API service, this is consumed on an undefined client, and UI-relevant annotations are excluded.

9.1.4 ABAP-Based OData Services

ABAP-based OData services refers to OData services that are created using SAP Gateway Service Builder. This tool offers options for creating and managing OData services that are provided in the SAP ABAP stack.

SAP Gateway Service Builder (Transaction SEGW) supports the developer when creating a new project. New OData services can be created in versions 2 or 4, but version 3 isn't supported. It's also possible to redefine an existing OData service for extension purposes.

An SAP Gateway Service Builder project serves as a shell/frame for the data model definition, the service implementation, and the service administration of the created OData service.

The data modeling of an OData service in Transaction SEGW enables you to define entity data, entity sets, complex types, and function imports. Entity data can be completely freely modeled or imported based on ABAP structures or RFC modules, and it can be customized. In addition, an associated entity set can be created. After assigning one or more properties of the entity as key attributes, the properties get metadata assigned such as deletability, applicability, and sortability.

In addition, the relationships between the entities with navigation properties can be mapped as an association. This allows a foreign key relationship that can be mapped in an arbitrary cardinality.

In the service implementation, the data procurement of business data and the data model of the OData service are defined as runtime objects. Because this is an ABAP-based service, two ABAP classes are used:

- **Data provider classes (DPC)**
  These classes are responsible for data retrieval and the runtime logic of the entity sets and function imports. Each entity set defines an ABAP method for creating, reading, updating, and deleting (CRUD). These methods must be redefined in the extension class of the DPC and their logic implemented in the extension class called `DPC EXT`. Each change to the data model generates the methods of the DPC and their logic anew, but not the `DPC EXT`.
- **Model provider classes (MPC)**
  These classes provide the definition of the data model of the OData service. They can be extended via an extension class called `MPC EXT`. The data model of the OData service can be adapted in the `MPC EXT` and enriched with metadata or annotations. In this way, annotations can be added that aren't offered by SAP or have to be added dynamically.

A Transaction SEGW–based OData service offers the option of assigning data sources other than the logic of the DPC for entities. A CDS view can be assigned as a source for read operations via the Service Adaptation Definition Language (SADL). For read and write operations, an RFC module can be assigned to an entity set, which can also be located on an external system. After the implementation and definition of a Transaction SEGW–based OData service, it's hosted on a backend server according to the SAP system architecture. This is then registered on a frontend server in SAP Gateway and is available via a node of the Internet Communication Framework (ICF) and can be consumed by a client. The ICF is the programming interface between ABAP and the internet.

The process of performing a query via HTTP is as follows: An HTTP request is sent against SAP Gateway of the frontend server in a hub deployment to the 10ICF node of the OData service. The ICF checks the authentication of the sender and performs basic checks. For example, in the validation of the X-CSRF token, the request is forwarded to the basic software components of SAP Gateway and mapped to an SAP-internal representation. If a hub deployment exists, the OData request is forwarded to the backend server via a trusted RFC. The system alias of the registered OData service is important here, as this points to the backend server, where its logic has been implemented, and the corresponding method of the DPC of the OData service is called, if implemented. The properties of the request are passed to this method. The DPC method calculates the result and sends it back as an HTTP response via the same route.

9.2 Metadata Document

As already mentioned, the metadata document of an OData service is its most important document because it holds the technical information on how this service is composed. In the following sections, we'll first discuss the structure of the metadata document, before discussing some specifics related to OData V2 and OData V4.

9.2.1 Structure

A metadata document of an OData service consists of the following information:

- **Entities**
  Entities are the objects of an OData Service, that is, objects of the real world. They consist of properties that represent an attribute of an entity and have a value, a datatype, and attributes that provide information about this property. One or multiple properties must be assigned as key properties to an entity. Associations between entities are described via navigation properties that are added as a navigational path to the entity.
- **Entity sets**
  Entity sets are collections of entities. They offer access to a list of entities that supports query options. You always address an entity set when working with OData, never the entity directly.
- **Associations**
  Associations are the description of relationships between entities, having a cardinality and a reference between two entities.
- **Complex types**
  With complex types, you can structure and group related properties. They work similar to entities but have no key and are always contained within an entity.
- **Function imports**
  Function imports are nonstandard modifying operations for an entity set. They allow the execution of remote business logic that can't be achieved with the normal CRUD operations and are either bound to a specific entities instance or not.

To view the metadata document of an OData service, you need to understand how the URI of an OData service is structured, as follows: *Scheme://Host:Port/Service Root/ResourcePath?Query Options*.

The URI starts with a scheme such as HTTPS or HTTP. Then, you have your hostname, typically your SAP system where your OData service runs in combination with the port that the scheme runs on. After that, the service root is appended—in an SAP system, this is the ICF node of the OData service in V2 or the service group in V4. Then, you can access the resources of your service, meaning the entity sets. This can be further enhanced with query options.

The metadata document can be accessed via the resource path `$metadata`, which retrieves it as an XML document.

9

9.2.2 OData V2

In OData V2, the metadata document is structured with the root `EDMX` tag. `EDMX` is used for entity data modeling, meaning modeling object relational data. It also includes references to OData and SAP proprietary vocabularies that are used in tags and attributes.

The metadata document holds a `DataServices` tag that gives information about the OData version, which should be set to 2.0 for OData V2 services. The `DataServices` tag holds a schema tag that includes all the OData relevant tags such as `EntityType`, `EntityContainer`, and so on. A sample of an OData service that manages customer data as shown in Figure 9.4.


```
Uri="https://                    /sap/opu/odata/iwfnd/catalogservice;v=2/Vocabularies(TechnicalName='%2FIWBEP%2FVOC_PERSONALDATA',Version='0001',S/
  <edmx:Include Namespace="com.sap.vocabularies.PersonalData.v1" Alias="PersonalData"/>
</edmx:Reference>
▼<edmx:Reference xmlns:edmx="http://docs.oasis-open.org/odata/ns/edmx"
Uri="https://                    /sap/opu/odata/iwfnd/catalogservice;v=2/Vocabularies(TechnicalName='%2FIWBEP%2FVOC_UI',Version='0001',SAP__Origin=
  <edmx:Include Namespace="com.sap.vocabularies.UI.v1" Alias="UI"/>
</edmx:Reference>
▼<edmx:Reference xmlns:edmx="http://docs.oasis-open.org/odata/ns/edmx"
Uri="https://                    /sap/opu/odata/iwfnd/catalogservice;v=2/Vocabularies(TechnicalName='%2FIWBEP%2FVOC_VALIDATION',Version='0001',SAP_
  <edmx:Include Namespace="Org.OData.Validation.V1" Alias="Validation"/>
</edmx:Reference>
▼<edmx:Reference xmlns:edmx="http://docs.oasis-open.org/odata/ns/edmx"
Uri="https://                    /sap/opu/odata/iwfnd/catalogservice;v=2/Vocabularies(TechnicalName='SAP__SELF',Version='0001',SAP__Origin='')/$va
  <edmx:Include Namespace="com.sap.vocabularies.PDF.v1" Alias="SAP__PDF"/>
</edmx:Reference>
▼<edmx:DataServices m:DataServiceVersion="2.0">
 ▼<Schema xmlns="http://schemas.microsoft.com/ado/2008/09/edm" Namespace="cds_z_customer_ui" xml:lang="de" sap:schema-version="1">
   <Annotation xmlns="http://docs.oasis-open.org/odata/ns/edm" Term="Core.SchemaVersion" String="1.0.0"/>
  ▼<EntityType Name="Z_C_CUSTOMERDOCUMENTType" m:HasStream="true" sap:label="CDS Projection - CustomerDocument1" sap:content-version="1">
   ▼<Key>
      <PropertyRef Name="Documentid"/>
    </Key>
    <Property Name="Delete_mc" Type="Edm.Boolean" sap:label="Dyn. Methodensteuerg" sap:creatable="false" sap:updatable="false" sap:sortable="false"
    sap:filterable="false"/>
    <Property Name="Update_mc" Type="Edm.Boolean" sap:label="Dyn. Methodensteuerg" sap:creatable="false" sap:updatable="false" sap:sortable="false"
    sap:filterable="false"/>
    <Property Name="Documentid" Type="Edm.Guid" Nullable="false" sap:label="DocumentID" sap:quickinfo="16 byte UUID in 16 bytes (raw format)"
    sap:creatable="false" sap:updatable="false"/>
    <Property Name="Customerid" Type="Edm.Guid" sap:label="UUID" sap:quickinfo="16 byte UUID in 16 bytes (raw format)" sap:creatable="false"
    sap:updatable="false"/>
    <Property Name="Documentname" Type="Edm.String" MaxLength="257" sap:label="Filename"/>
    <Property Name="Documenttype" Type="Edm.String" MaxLength="128" sap:label="Mimetype" sap:quickinfo="Mimetyp"/>
    <Property Name="Createdby" Type="Edm.String" MaxLength="12" sap:display-format="UpperCase" sap:label="Benutzername" sap:creatable="false"
    sap:updatable="false"/>
    <Property Name="Createdat" Type="Edm.DateTimeOffset" Precision="0" sap:label="Zeitstempel" sap:quickinfo="UTC-Zeitstempel in Kurzform
    (JJJJMMTThhmmss)" sap:creatable="false" sap:updatable="false"/>
    <Property Name="Changedby" Type="Edm.String" MaxLength="12" sap:display-format="UpperCase" sap:label="Benutzername" sap:creatable="false"
    sap:updatable="false"/>
    <Property Name="Changedat" Type="Edm.DateTimeOffset" Precision="0" sap:label="Zeitstempel" sap:quickinfo="UTC-Zeitstempel in Kurzform
    (JJJJMMTThhmmss)" sap:creatable="false" sap:updatable="false"/>
```


Figure 9.4 Metadata Document in OData V2

If you go to the `EDMX->DataService->Schema` tag, you'll first notice your OData services entities. They are described with the `EntityType` tag, as shown in Figure 9.5.

```
▾<EntityType Name="Z_C_CUSTOMERDOCUMENTType" m:HasStream="true" sap:label="CDS Projection - CustomerDocument1" sap:content-version="1">
  ▾<Key>
     <PropertyRef Name="Documentid"/>
   </Key>
   <Property Name="Delete_mc" Type="Edm.Boolean" sap:label="Dyn. Methodensteuerg" sap:creatable="false" sap:updatable="false" sap:sortable="false"
   sap:filterable="false"/>
   <Property Name="Update_mc" Type="Edm.Boolean" sap:label="Dyn. Methodensteuerg" sap:creatable="false" sap:updatable="false" sap:sortable="false"
   sap:filterable="false"/>
   <Property Name="Documentid" Type="Edm.Guid" Nullable="false" sap:label="DocumentID" sap:quickinfo="16 byte UUID in 16 bytes (raw format)"
   sap:creatable="false" sap:updatable="false"/>
   <Property Name="Customerid" Type="Edm.Guid" sap:label="UUID" sap:quickinfo="16 byte UUID in 16 bytes (raw format)" sap:creatable="false"
   sap:updatable="false"/>
   <Property Name="Documentname" Type="Edm.String" MaxLength="257" sap:label="Filename"/>
   <Property Name="Documenttype" Type="Edm.String" MaxLength="128" sap:label="Mimetype" sap:quickinfo="Mimetyp"/>
   <Property Name="Createdby" Type="Edm.String" MaxLength="12" sap:display-format="UpperCase" sap:label="Benutzername" sap:creatable="false"
   sap:updatable="false"/>
   <Property Name="Createdat" Type="Edm.DateTimeOffset" Precision="0" sap:label="Zeitstempel" sap:quickinfo="UTC-Zeitstempel in Kurzform
   (JJJJMMTThhmmss)" sap:creatable="false" sap:updatable="false"/>
   <Property Name="Changedby" Type="Edm.String" MaxLength="12" sap:display-format="UpperCase" sap:label="Benutzername" sap:creatable="false"
   sap:updatable="false"/>
   <Property Name="Changedat" Type="Edm.DateTimeOffset" Precision="0" sap:label="Zeitstempel" sap:quickinfo="UTC-Zeitstempel in Kurzform
   (JJJJMMTThhmmss)" sap:creatable="false" sap:updatable="false"/>
   <NavigationProperty Name="to_Customer" Relationship="cds_z_customer_ui.assoc_9FE20F64318DBBDC74B6B643440E2F64"
   FromRole="ToRole_assoc_9FE20F64318DBBDC74B6B643440E2F64" ToRole="FromRole_assoc_9FE20F64318DBBDC74B6B643440E2F64"/>
 </EntityType>
```

Figure 9.5 Entity in OData V2

Here, you see an entity called `Z_C_CUSTOMERDOCUMENTType`. This entity is labeled via an SAP proprietary attribute called `label`. Each `EntityType` has a `Key` tag, which holds a property reference to the properties that are used as key properties. Then, you have a list of `Properties`, each consisting of a `Name`, a `Type`, and proprietary attributes that further describe the property. Use these attributes to get more information on what you can do with this property. For example, the property `Document` isn't nullable (meaning it can't be empty), can be filtered and sorted because these attributes are missing (only false attributes can be seen here), can't be created, and can't be updated. Be careful about the validity of these attributes because being set to a certain value doesn't mean that the behavior for the attributes was developed in the service. This is often the case for Transaction SEGW OData Services. You can also view the `NavigationProperties` here, which allow you to navigate to get the associated entities.

If you scroll further down, you'll notice the `EntityContainer` tag, as shown in Figure 9.6.

```
▾<EntityContainer Name="cds_z_customer_ui_Entities" m:IsDefaultEntityContainer="true" sap:message-scope-supported="true" sap:supported-
 formats="atom json xlsx pdf">
   <EntitySet Name="SAP__Currencies" EntityType="cds_z_customer_ui.SAP__Currency" sap:content-version="1"/>
   <EntitySet Name="SAP__UnitsOfMeasure" EntityType="cds_z_customer_ui.SAP__UnitOfMeasure" sap:content-version="1"/>
   <EntitySet Name="SAP__MyDocumentDescriptions" EntityType="cds_z_customer_ui.SAP__DocumentDescription" sap:content-version="1"/>
   <EntitySet Name="SAP__FormatSet" EntityType="cds_z_customer_ui.SAP__Format" sap:creatable="false" sap:updatable="false" sap:deletable="false"
   sap:pageable="false" sap:addressable="false" sap:content-version="1"/>
   <EntitySet Name="SAP__PDFStandardSet" EntityType="cds_z_customer_ui.SAP__PDFStandard" sap:creatable="false" sap:updatable="false"
   sap:deletable="false" sap:pageable="false" sap:addressable="false" sap:content-version="1"/>
   <EntitySet Name="SAP__TableColumnsSet" EntityType="cds_z_customer_ui.SAP__TableColumns" sap:creatable="false" sap:updatable="false"
   sap:deletable="false" sap:pageable="false" sap:addressable="false" sap:content-version="1"/>
   <EntitySet Name="SAP__CoverPageSet" EntityType="cds_z_customer_ui.SAP__CoverPage" sap:creatable="false" sap:updatable="false"
   sap:deletable="false" sap:pageable="false" sap:addressable="false" sap:content-version="1"/>
   <EntitySet Name="SAP__SignatureSet" EntityType="cds_z_customer_ui.SAP__Signature" sap:creatable="false" sap:updatable="false"
   sap:deletable="false" sap:pageable="false" sap:addressable="false" sap:content-version="1"/>
   <EntitySet Name="SAP__ValueHelpSet" EntityType="cds_z_customer_ui.SAP__ValueHelp" sap:content-version="1"/>
   <EntitySet Name="Z_C_CUSTOMERDOCUMENT" EntityType="cds_z_customer_ui.Z_C_CUSTOMERDOCUMENTType" sap:content-version="1" sap:deletable-
   path="Delete_mc" sap:updatable-path="Update_mc"/>
   <EntitySet Name="Z_P_CUSTOMER" EntityType="cds_z_customer_ui.Z_P_CUSTOMERType" sap:searchable="true" sap:content-version="1" sap:deletable-
   path="Delete_mc" sap:updatable-path="Update_mc"/>
  ▾<AssociationSet Name="to_formatSet" Association="cds_z_customer_ui.to_format" sap:creatable="false" sap:updatable="false" sap:deletable="false"
   sap:content-version="1">
```

Figure 9.6 Entity Container in OData V2

The `EntityContainer` displays the collections or `EntitySets` for your entities. Always address these when working with an OData service.

9.2.3 OData V4

If you open the same service with an OData V4 protocol, you won't notice many differences at first. The metadata document again is structured with the root `EDMX` tag, under which you have additional references to different vocabularies. You then have your `DataService` tag along with a schema for your ER data like an entity, as shown in Figure 9.7.

```
<EntityType Name="Z_P_CUSTOMERType">
▾<Key>
   <PropertyRef Name="Customerid"/>
  </Key>
  <Property Name="Customerid" Type="Edm.Guid" Nullable="false"/>
  <Property Name="Firstname" Type="Edm.String" Nullable="false" MaxLength="40"/>
  <Property Name="Lastname" Type="Edm.String" Nullable="false" MaxLength="40"/>
  <Property Name="Title" Type="Edm.String" Nullable="false" MaxLength="20"/>
  <Property Name="Phone" Type="Edm.String" Nullable="false" MaxLength="30"/>
  <Property Name="Email" Type="Edm.String" Nullable="false" MaxLength="132"/>
  <Property Name="Gender" Type="Edm.String" Nullable="false" MaxLength="1"/>
  <Property Name="Website" Type="Edm.String" Nullable="false" MaxLength="64"/>
  <Property Name="Createdby" Type="Edm.String" Nullable="false" MaxLength="12"/>
  <Property Name="Createdat" Type="Edm.DateTimeOffset"/>
  <Property Name="Changedby" Type="Edm.String" Nullable="false" MaxLength="12"/>
  <Property Name="Changedat" Type="Edm.DateTimeOffset"/>
  <Property Name="Birthdate" Type="Edm.Date"/>
  <Property Name="__CreateByAssociationControl" Type="com.sap.gateway.srvd.z_customer_ui.v0001.Z_P_CUSTOMERCbAControl"/>
  <Property Name="__EntityControl" Type="com.sap.gateway.srvd.z_customer_ui.v0001.EntityControl"/>
  <Property Name="SAP__Messages" Type="Collection(com.sap.gateway.srvd.z_customer_ui.v0001.SAP__Message)" Nullable="false"/>
▾<NavigationProperty Name="_CustomerDocument" Type="Collection(com.sap.gateway.srvd.z_customer_ui.v0001.Z_C_CUSTOMERDOCUMENTType)"
  Partner="_Customer">
   <OnDelete Action="Cascade"/>
  </NavigationProperty>
</EntityType>
```

Figure 9.7 An Entity in OData V4

In OData V4, entities are more simplified in construction than compared to OData V2. Properties only have a `Name`, a `Type`, and some basic constraints such as `Nullable` or `MaxLength`. SAP proprietary attributes such as labels aren't directly part of the property definition.

> **Annotations in SAP OData Services**
>
> Annotations enhance the semantic meaning of an OData service. They can be either done remotely, meaning they are developed in the system the OData service originates in, or they can be done locally, meaning in the application that the service is consumed in.
>
> When using remote annotations, OData V2 services don't contain them all in the metadata document. Only property constraints, currency information, and labels are directly a part of the document. To retrieve all UI-relevant annotations, SAP provides a dedicated service to retrieve an annotation document as XML, the `CatalogService`, which can be accessed via the following URL:
>
> */sap/opu/odata/IWFND/CATALOGSERVICE;v=2/Annotations(TechnicalName='<OData Service Annotation Name>', Version='<Annotation Version ID>')/$value/*
>
> In OData V4, annotations are directly a part of the metadata document and the `DataService` tag, found with the tag `Annotations`.

Keep this in mind when developing SAP Fiori elements apps. If you consume an OData V2 service, you also need to load the annotations separately. This is normally automatically done during creation of a new SAP Fiori elements app. If you consume an OData V4 service, you don't need to separately load the annotations as they are a part of the metadata document.

9.3 Testing OData Services

If you want to test an OData service, you can either use your web browser with some extension for REST service testing, a REST client such as Postman, or SAP Gateway Client.

SAP Gateway Client, accessible via Transaction /IWFND/GW_CLIENT as shown in Figure 9.8, is a tool that allows you to interact with OData services that are registered on you SAP system.

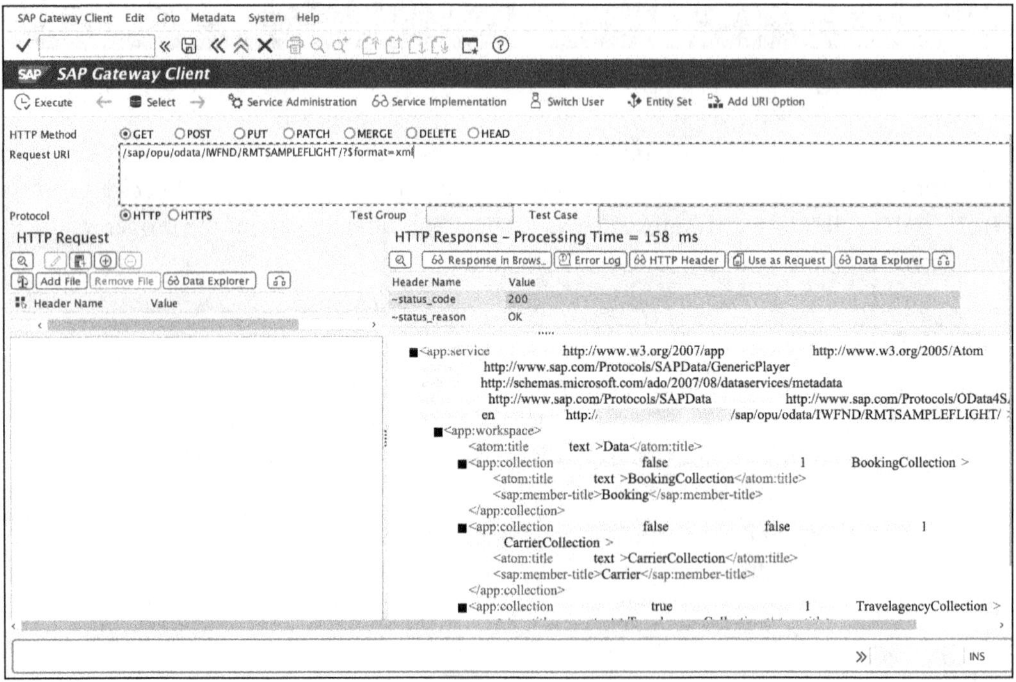

Figure 9.8 SAP Gateway Client

With SAP Gateway Client, you can execute HTTP requests to a certain endpoint. You can choose an **HTTP Method**, enter a **Request URI**, and execute a request. You have a dedicated **HTTP Request** area where you can add request headers and a request body on the left side of the screen. On the right side, you have the response to your request in the **HTTP Response** area, which includes the response headers, HTTP status code, and response content.

From SAP Gateway Client, you can also execute actions such as clearing the metadata cache by going to the **Service Administration** and **Service Implementation** tabs of the registered services.

> **Note**
> Alternatively, you can also use a browser extension such as Yet Another REST Client (YARC), available for Google Chrome.

For testing purposes, you can use the Northwind OData service, available at *https://services.odata.org/*. The OData organization provides simple services for testing purposes, either read-only or with full write access. For the rest of our examples, we're using the provided OData V2 and OData V4 full read-write services because you can easily try them out without needing a certain OData service on your SAP system.

So, go to the Northwind URL, select the **Browse the Full Access (Read-Write) Service** link. You can then manipulate the first part of the resource path in the URL to either use V2 or V4; by default, this is V3, but OData V3 isn't supported at SAP S/4HANA.

9.4 CRUD Operations

As you now understand how an OData service is described through the metadata document and how we're able to test one, we need to take a closer look at how we can communicate with one. As with any REST-compliant protocol, we use CRUD methods to access the resources of such a service. To identify the resources our service provides, we need to look at either the metadata document or the service document. Using the Northwind service, we get the service documents for OData V2 (*https://services.odata.org/V2/OData/OData.svc/*) and OData V4 (*https://services.odata.org/V4/OData/OData.svc/*).

They offer a products-entity set, which we're going to use now in our examples, that has the metadata shown in Figure 9.9 for its product entity.

```
<EntityType Name="Product">
▾<Key>
   <PropertyRef Name="ID"/>
 </Key>
 <Property Name="ID" Type="Edm.Int32" Nullable="false"/>
 <Property Name="Name" Type="Edm.String"/>
 <Property Name="Description" Type="Edm.String"/>
 <Property Name="ReleaseDate" Type="Edm.DateTimeOffset" Nullable="false"/>
 <Property Name="DiscontinuedDate" Type="Edm.DateTimeOffset"/>
 <Property Name="Rating" Type="Edm.Int16" Nullable="false"/>
 <Property Name="Price" Type="Edm.Double" Nullable="false"/>
 <NavigationProperty Name="Categories" Type="Collection(ODataDemo.Category)" Partner="Products"/>
 <NavigationProperty Name="Supplier" Type="ODataDemo.Supplier" Partner="Products"/>
 <NavigationProperty Name="ProductDetail" Type="ODataDemo.ProductDetail" Partner="Product"/>
</EntityType>
```

Figure 9.9 The Product Entity

9.4.1 Read

First, we need to read data by accessing the products-entity set. If we want to get the collection of all products, we address only the entity set with an HTTP GET request:

https://services.odata.org/V4/OData/OData.svc/Products

This retrieves all available products that the service provides with the HTTP status code **200**. When no entries are found, it returns an empty list, but the request will finish with an HTTP **200** too. Normally, accessing the entity set without size limitations or pagination shouldn't be done for performance reasons. We address this in Section 9.5 by using query options.

If you want to address one product, you need to add the key properties to the request's resource path. For the product, the key property is the ID property. This must be added to brackets that are appended to the resource path:

https://services.odata.org/V4/OData/OData.svc/Products(0)

With this addressing of a single resource, you'll get the requested entity based on the provided keys if it exists with the HTTP status code **200**. If it doesn't exist, the service will return an HTTP 404 status code and an error.

To address a single property of your entity, you address the entity and add a slash following the properties name:

https://services.odata.org/V4/OData/OData.svc/Products(0)/Name

This retrieves just the property Name with its corresponding value.

If you want to use a navigation property to get an associated entity, you use the same syntax as when retrieving a single property:

https://services.odata.org/V2/OData/OData.svc/Products(0)/Category

In this case, we want to get the category entity, which has been assigned to the product. Therefore, we address the navigation property Category, as this is, per definition from the metadata document, the linkage to the category entity.

9.4.2 Create

If you want to create new entries of an entity set, we send an HTTP POST request addressing the entity set with the to-be-created data as a JSON object. Properties that aren't creatable, for example, Universally Unique Identifiers (UUIDs) that are generated in the backend, are left out. (See the metadata document for reference.)

So, to create a new product, send a POST request to the product's entity set (make sure you use the Northwind read-write service) with the request body as JSON from Listing 9.1:

https://services.odata.org/V2/<dynamic id>/OData/OData.svc/Products

```
{
"ID": 500,
"Name": "SAPUI5",
"Description": "The Comprehensive Guid",
"ReleaseDate": "2025-04-30",
"DiscontinuedDate": null,
"Rating": 5,
"Price": "85.90"
}
```

Listing 9.1 The POST Request Body

You should get an HTTP **201** status code, indicating that the new product was created, and you should get the newly created product as the response body. For this to work, make sure you provide all necessary properties where their values match the datatypes from the metadata document. If you've made something wrong, the OData service should send a response with the HTTP status code **400** indicating what went wrong while processing this request.

9.4.3 Update

When updating existing entities from an entity set, you can send one of three HTTP methods: PUT, PATCH or MERGE. PUT needs the whole entity with all its properties as a request body, even though only certain properties are changed. It simply puts the request body as the new value. PUT and MERGE work by only sending the properties in the request body that should get a new value.

Let us change the price of our newly created product by addressing the entity set with our ID and then sending a PUT request with the request body as JSON from Listing 9.2:

https://services.odata.org/V2/<dynamic ID>/OData/OData.svc/Products(500)

```
{
"Name": "SAPUI5",
"Description": "The Comprehensive Guid",
"ReleaseDate": "2025-04-30",
"DiscontinuedDate": null,
"Rating": 5,
"Price": "79.90"
}
```

Listing 9.2 The PUT Request Body

Notice that we leave the ID property out, as this is the primary key and can't be changed. We simply take all other properties from our POST request from before and change the Price properties value. When sending this PUT request, we should get an

HTTP status code **204** in return, indicating that the update was successful. We don't get the updated product as a response body, as we already have the updated data.

9.4.4 Delete

To remove a certain entity from an entity set, we send an HTTP `DELETE` request to it. This is quite simple, as this sample shows:

https://services.odata.org/V2/<dynamic ID>/OData/OData.svc/Products(500)

You don't need a request body; you simply send a `DELETE` request to the product. You should get an HTTP status code **204** in return without a response body, as the resource has been deleted.

9.5 Query Options

As previously mentioned, reading an entity set supports a lot of options for defining how the returned result should be structured. You can define filters and sorters, add pagination parameters, use navigation, and so on. The syntax for adding query options is as follows:

https://services.odata.org/V2/OData/OData.svc/Products?<query options>

After your entity set, you add a question mark and then add your query parameters, which are separated by a logical `and` sign. The following is possible via query parameters:

- **Sorting**
  - Parameter name: `$orderby`
  - Syntax: *<service>/<Entity Set>?$orderby=<Property> <asc/desc>*

  The sorting parameter allows you to sort an entity set based on its properties, either ascending or descending. If you want to, for example, sort the products in ascending order based on price, you'll write the code provided at *https://services.odata.org/V2/OData/OData.svc/Products?$orderby=Price asc*.

  You can also make multiple sorters work by writing the code provided at:

  https://services.odata.org/V2/OData/OData.svc/Products?$orderby=Price asc, Name
- **Filtering**
  - Parameter name: `$filter`
  - Syntax: *<service>/<Entity Set>?$filter=<Property> <operation> <value>*

  The filter parameter allows filtering the entity set based on properties and a simple expression syntax. Multiple filters can be combined with either `and` or `or`.

If you want to, for example, filter all products with a price of less than 10, you write the code provided at:

https://services.odata.org/V2/OData/OData.svc/Products?$filter=Price le 10

This can be combined with another filter. For example, to filter products with a rating of 5, you write the code provided at:

https://services.odata.org/V2/OData/OData.svc/Products?$filter=Price le 10 and Rating eq 5

A detailed list of available operations, operands, and string functions for filtering can be found at *www.odata.org/documentation/odata-version-2-0/uri-conventions/* and *https://docs.oasis-open.org/odata/odata/v4.01/odata-v4.01-part2-url-conventions.html.*

- **Pagination**
  - Parameter name: `$skip $top`
  - Syntax: *<service>/<Entity Set>?$<skip/top>*

  The pagination parameters `skip` and `top` allow you to define how many entries you want to receive and how many from the beginning you want to skip. This works best with sorting parameters.

  If you want to, for example, retrieve the first two products, you write the code provided at:

  https://services.odata.org/V2/OData/OData.svc/Products?$skip=0&$top=2

  To get the next two, you skip the first two and get the next top two, you write the code provided at:

  https://services.odata.org/V2/OData/OData.svc/Products?$skip=2&$top=2

  This is great for SAPUI5 tables and lists, where you don't want to load all items, just the first 10 or so. After scrolling down, you load the next 10, and so on.

- **Expanding**
  - Parameter name: `$expand`
  - Syntax: *<service>/<Entity Set>(<keys>)?$expand=<Navigation Property>*

  If you want to retrieve an entity and its associated entities in the same response, you can use the `expand` parameter. This appends the result of a navigation property to your entity.

  If you want to, for example, get a certain product and its category, you write the code provided at:

  https://services.odata.org/V2/OData/OData.svc/Products(0)?$expand=Category

- **Selecting**
  - Parameter name: `$select`
  - Syntax: *<service>/<Entity Set>(<keys>)?$select=<Property>*

If you want to select certain properties of your entity to only retrieve them, you can use the `select` parameter. If you just want to get the ID and the name of your product, for example, you write the code provided at:

https://services.odata.org/V2/OData/OData.svc/Products(0)?$select=ID,Name

This helps in performance-related scenarios, as properties that might be calculated in the backend or that aren't needed can be left out from your request.

9.6 Adding OData Services to SAPUI5 Projects

Adding an OData service to SAPUI5 applications can either be done during the creation of an SAPUI5 app or afterwards (see Chapter 10 and Chapter 11 for more information on this process).

If you create an application with the SAP Fiori generator, one of the first steps is to add a **Data source** to you new app, as shown in Figure 9.10.

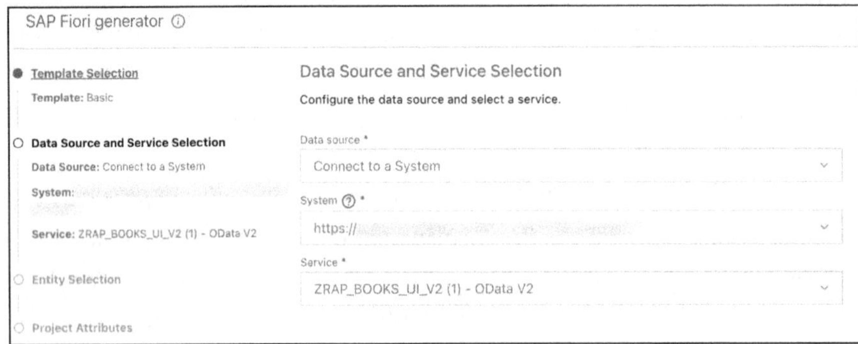

Figure 9.10 Selecting an OData Service during Application Generation

Here, you can either choose **Connect to a System** or **Connect to an OData Service**. The first option allows you to connect to an SAP system based on the SAP Gateway URL and a user. Then, you can choose either an OData V2 or V4 service that is registered on your SAP Gateway. The second option allows you to directly insert a URL of an OData service. Both options add the selected OData service as a new data source and the main model in the application's *manifest.json* file. More information on that can be found in Chapter 10 and Chapter 11.

In addition, a backend configuration is added to the application's *ui5.yaml* file. This file also serves as the request routing. Here, a new node is inserted into `server->customMiddleware->configuration->backend` that might look like the following:

```
backend:
- path: /sap
        url: <system URL>
        client: <system Client>
```

Every request that your app sends to */sap* is redirected to this system. This includes all requests such as OData requests, as an OData services URI starts with the */sap* resource after the system's hostname.

For existing applications, SAP Fiori offers you a few tools for managing your application's OData services that can be either accessed from the **Application Info** screen or from the command palette. The first tool is the **Service Manager**, as shown in Figure 9.11.

Service Manager
Annotation File Manager

Services

Name
URI
mainService
/sap/opu/odata/

Refresh local metadata of service mainService

SAP System
Hostname
Service URI
/sap/opu/odata/
SAP System
https:/
A destination change will only take effect on start of preview
Refresh & Save
Refresh
Cancel

Figure 9.11 SAP Fiori Tools: Service Manager

The **Service Manager** tab allows you to add and configure your OData services. You can also refresh your local copy of the metadata document here, as well as maintain the local annotations for your OData services here.

The next tool is the **Service Modeler**, shown in Figure 9.12. On the **Service Modeler** tab, you can view your entities in a simpler, more intuitive way then looking directly at the metadata document. It also allows you to view the annotations of your OData service and model them to create local annotations.

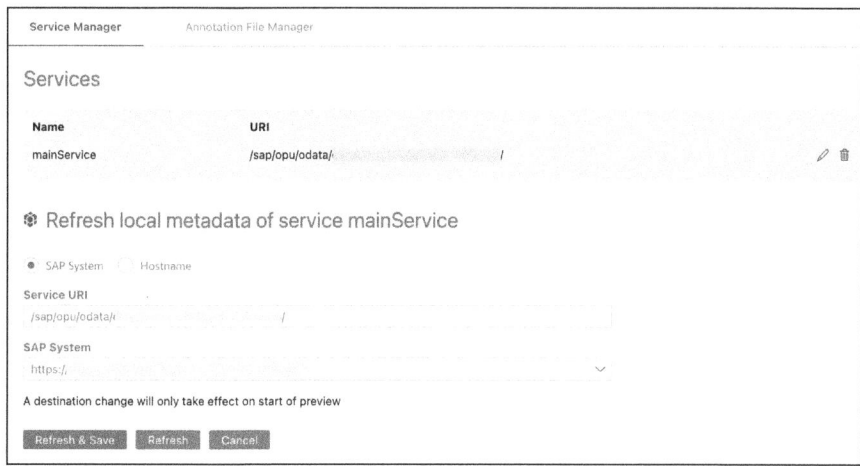

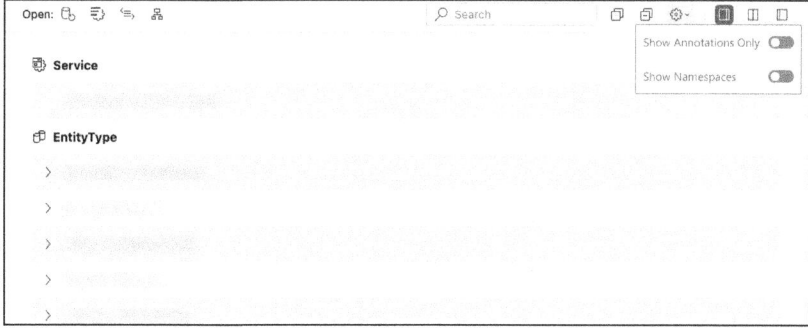

Figure 9.12 SAP Fiori Tools: Service Modeler

9.7 Summary

In this chapter, we looked at the basics of the OData protocol. This REST-based protocol allows the consumption of semantically rich business data that is based on entities. For creating OData services in an SAP S/4HANA system, you can either use the ABAP RESTful application programming model or ABAP-based OData services.

An OData service is described by the metadata document. This document gives a technical description of your entities, their properties and associations to other entities, collection of entities, complex types, and function imports.

If you want to test an OData service, you can use any REST client or the SAP Gateway Client. The OData organization also provides the Northwind service that you can use for testing purposes.

Integrating an OData service can either be done during creation of an SAPUI5 application or afterwards. The SAP Fiori tools can be used for managing OData services that have been added to your applications.

Chapter 10
OData Version 2 Model

Immerse yourself in the world of the OData V2 model and discover how SAPUI5 applications can communicate efficiently with backend systems. This chapter introduces you to the intricacies of the OData model, which is the central interface for working with OData V2 services. You'll learn about the main functions and options for data manipulation and synchronization, as well as how CRUD operations can be used to effectively manage data entities. Finally, the importance of function imports is highlighted, enabling you to seamlessly integrate server-side functions and actions into your applications. This chapter is an essential guide for developers who want to use the full power of the OData V2 model.

The *OData Version 2 (V2) model* provides a powerful way to connect UI components directly to data from OData services. This binding allows SAPUI5 controls, such as tables, lists, input fields, or other UI components, to dynamically access and display data from an OData service. The binding usually takes place via declarative or programmatic data bindings, with the OData V2 model acting as an intermediary between the UI and the backend system. It takes on the task of retrieving the requested data from the server, preparing it, and passing it on to the corresponding UI components. In addition, the model also supports the automatic updating of the UI as soon as the underlying data changes, as well as the synchronization of user interactions back to the backend. In this way, developers can't only implement read data operations but also seamlessly integrate write operations such as updating or creating new datasets into their applications. With support for various binding types, such as list and context bindings, the OData V2 model provides a flexible and scalable foundation for working with complex data structures and enables developers to design data-driven applications in an efficient and user-friendly way.

Because the OData model is indispensable in the development of SAPUI5 interfaces, we'll take a general look at the OData V2 model in Section 10.1. In Section 10.2, we'll look at how create, read, update, and delete (CRUD) operations can be carried out using the OData V2 model. These operations are required for read and write access to data in the backend. We'll then see in Section 10.3 what options the OData V2 model offers, for example, to call up function imports of a service.

10.1 OData Model

The OData model is a *server-side model*, which means that the complete dataset is located exclusively on the server, while the client only knows the currently visible or requested data. All operations, such as sorting and filtering, are carried out directly on the server. The client takes on the role of sending requests to the server to retrieve the required data. As soon as the server processes and returns the requested data, it's displayed to the user on the client side. This keeps data processing lean and efficient, as extensive operations are centralized on the server rather than on the client. This model ensures better performance, especially with large amounts of data, and ensures that the displayed data is always up-to-date and consistent with the server-side information. Requests to the backend can be triggered in various ways, first through list bindings (`ODataListBinding`) or context bindings (`ODataContextBinding`), and second through *create, read, update, and delete (CRUD)* functions provided by the OData model. No backend requests are triggered by property bindings (`ODataPropertyBinding`).

There are two different OData model implementations for OData Services V2: `sap.ui.model.odata.v2.ODataModel` and `sap.ui.model.odata.ODataModel`. However, `sap.ui.model.odata.ODataModel` is deprecated and should no longer be used. New features are only implemented in the other model, and the other model also offers more functionalities. Table 10.1 shows the differences between the two model implementations.

Feature	sap.ui.model.odata.v2.ODataModel	sap.ui.model.odata.ODataModel
OData version support	2.0	2.0
JSON format	Yes (default)	Yes
XML format	Yes	Yes (default)
Support of two-way binding mode	Yes, for property changes only, not implemented for aggregations	Experimental, only properties of one entity can be changed at the same time
Default binding mode	One-way binding	One-way binding
Client-side sorting and filtering	Yes	No
`$batch`	All requests can be batched.	Only manual batch requests are possible.
Data cache in model	All data is cached in the model.	Manually requested data isn't cached.
Automatic refresh	Yes (default)	Yes
Message handling	Yes	No

Table 10.1 Comparison of OData V2 Model Implementations

In the following sections, we'll take a look at how to create an OData V2 model instance, see how additional URL parameters and HTTP headers can be added to requests, and take a look at the creation of entities using the OData V2 model.

10.1.1 Creating a Model Instance

A single instance of an OData model only allows access to a single OData service. If you want to connect several OData services in an application, several instances of an OData model must be created. When creating the model instance, the URL of the OData service must be specified as the only mandatory parameter. The URL can be specified in two different ways within the constructor:

- Passed as a string
- Transfer within the `mParameters` map of the constructor

Both options are shown in Listing 10.1.

```
let oModel_1 = new sap.ui.model.odata.v2.ODataModel("http://services.odata.org/
Northwind/Northwind.svc/");
let oModel_2 = new sap.ui.model.odata.v2.ODataModel({serviceUrl: http://
services.odata.org/Northwind/Northwind.svc/});
```

Listing 10.1 Creation of an OData Model Instance

When an instance of an OData model is created, a request is automatically sent that queries the metadata of the service. For this query, a `$metadata` is appended to the service URL. In our example, the request URL would look like this: *http://services.odata.org/Northwind/Northwind.svc/$metadata*. The metadata of a service is cached via the service URL, and if several model instances are created with the same service URL, the metadata is only loaded from the server once and not for each instance. The `ODataModel` class provides a `getServiceMetadata()` method with which the metadata can be accessed.

Loading of Metadata

In `v2.ODataModel`, the metadata is loaded asynchronously. This behavior can't be changed; that is, the metadata can't be loaded synchronously. However, the OData model offers an event called `metadataLoaded` that is triggered as soon as the metadata has been loaded. This event makes it possible to react to the loading and execute further logic.

Default Binding Mode Depending on the Instantiation

Depending on whether the OData model is instantiated in JavaScript, in TypeScript, or defined in the application descriptor, different binding modes are assigned by default.

If the OData model is instantiated in a controller, a one-way binding is set by default. However, if the OData model has been defined in the application descriptor, it receives a two-way binding by default. Unfortunately, we have to live with this inconsistency. This applies to the OData V4 model as well.

10.1.2 Additional URL Parameters

URL parameters are used for configuration in OData services. Most of the relevant parameters are automatically provided by SAPUI5, depending on the binding. Nevertheless, you can also define and provide your own parameters for general configurations or for authentication purposes, for example. It's often the case that certain parameters, such as `$expand` or `$select`, shouldn't be included in every request, but only in a specific list or context bindings. In this case, the binding methods offer an option to pass these parameters.

There are various options for adding other parameters to the request URL:

- **Appending the parameters to the service URL**
  In this case, the parameters are appended for each request that is sent via the model. The instantiation of the model looks like Listing 10.2.

  ```
  let oModel = new sap.ui.model.odata.v2.ODataModel("http://services.odata.org/
  Northwind/Northwind.svc/?param1=value1&param2=value2");
  ```

  Listing 10.2 Pass Parameters via Service URL

- **Transferring parameters via map**
  If the `mParameters` map is used for parameter transfer instead, parameters can be defined that are only transferred for metadata requests and for any other service requests. The two properties here are `serviceUrlParams` and `metadataUrlParams`. Listing 10.3 shows what these look like.

  ```
  let oModel = new sap.ui.model.odata.v2.ODataModel({
     serviceUrl: "http://services.odata.org/Northwind/Northwind.svc/",
     serviceUrlParams: {
        param1: "value1",
        param2: "value2"
     },
     metadataUrlParams: {
        param1: "value1",
        param2: "value2"
     }
  });
  ```

  Listing 10.3 Pass Parameters via mParameters Map

10.1.3 Custom HTTP Headers

Similar to URL parameters, you can also add your own HTTP headers to the model, which are then sent with every request. For this purpose, the headers can be defined either via the constructor or via a separate method. Listing 10.4 shows how the HTTP headers are passed in the constructor.

```
let oModel = new sap.u.model.odata.v2.ODataModel({
   headers: {
      "myHeader1": "value1",
      "myHeader2": "value2"
   }
});
```

Listing 10.4 Setting Custom HTTP Headers within the Constructor

The setting of the headers using the `setHeaders` method is shown in Listing 10.5.

```
oModel.setHeaders({
   "myHeader1": "value1"
   "myHeader2": "value2"
});
```

Listing 10.5 Setting Custom HTTP Headers Using the setHeaders Method

Custom Headers and Private Headers

If custom headers are added to the OData model, all previously added custom headers will be removed, for example, only the custom headers that are specified in the headers map are applied to the model. However, some headers are private, which means they are set by the model internally and can't be set manually, for example:

- `accept`
- `accept-language`
- `maxdataserviceversion`
- `dataserviceversion`
- `x-csrf-token`

10.1.4 Creating Entities

In addition to the option of creating entities using the `create` method of the OData model (discussed in the following section), entities can also be created in other ways:

- `ODataModel#createKey`
- `ODataModel#createEntry`
- `ODataListBinding#create`
- Deep create

Depending on the use case, it's necessary to consider which method is best suited for the creation of new entities.

ODataModel#createKey

In SAPUI5, the `createKey` method is used with the OData V2 model to generate a URI key segment for a specific entity. This method simplifies accessing or modifying individual entities by dynamically creating the required URL for entity-specific operations such as reading, updating, or deleting.

This method has the following purposes:

- **Generate entity keys**
  Creates a key segment of the URL for a specific OData entity.
- **Handle composite keys**
  Constructs URIs for entities with either single or multiple key properties.
- **Facilitate navigation**
  Provides a standardized way to dynamically build URLs for specific entities.

The syntax of the method looks like `createKey(sEntitySet, oKeyProperties)`. The method returns a string with the key segment of the entities URL. The parameters must be provided as follows:

- `sEntitySet`
  The name of the entity set (e.g., `"Document"`). This is the collection name as defined in the OData service metadata.
- `oKeyProperties`
  An object where each property represents a key field, with the key as the field name and the value as the key's value. If you want to create a key consisting of multiple properties, this object has to contain multiple key-value pairs.

  An example usage looks like the following:

  ```
  let sKey = oModel.createKey("Document", {DocumentId: "4711"});
  ```

  This call of the method returns a string like this: `Document(DocumentId='4711')`.

ODataModel#createEntry

By calling the `ODataModel#createEntry` method, an entry is created, and the corresponding context is returned. It's recommended to use this approach in the following cases:

- You have a form or popup that allows the end user to view and edit the data of a new entry, but there's no table or list control to display the entry.
- You want to create an entry without showing it on the UI.

This method accepts the path of the entity set. In addition, initial properties can optionally be specified that will be used to create the entry. Both the path and the property identifiers must exist in the metadata of the service. Care should also be taken if an object from a context with the getObject method is used for the initial properties. In this case, the _metadata object must be removed from the data, as it must not be part of the payload.

The context returned by this method is transient. This means that the entity initially only exists locally. The client-side creation is only triggered when the submitChanges method is called. In addition, the createEntry method returns a context that is notified as soon as the entity is persisted or reset. The isTransient API can also be used to check whether a context is a transient or persisted context.

The transient context is typically used to bind locally created entities to a form or dialog for further input. This gives the user the opportunity to make changes to the entity before the actual create request is sent. Note also that entries created with createEntry initially receive a temporary ID locally, which is invalid once the entity has been persisted.

Listing 10.6 shows example coding in which a local entry is first created, bound to a view, and then persisted on the server side using the submitChanges method.

```
let oContext = oModel.createEntry("/Document", {
   properties: {
      DocumentName: "Document X"
   }
});

this.getView().bindElement(oContext);

oModel.submitChanges({success: successHandler, error: errorHandler});

oContext.created().then(
   function(){}, // successful creation
   function(){} // creation not successful
);
```

Listing 10.6 Creation of Entity Using ODataModel#createEntry

The createEntry method has an optional refreshAfterChange parameter that controls whether all related bindings are refreshed after the entry is successfully created in the backend. This helps update list bindings to display the newly created entity in table controls. If such an update is needed, it's better to use the ODataListBinding#create API rather than ODataModel#createEntry.

If you want to include navigation properties when the entry is saved, use the optional expand parameter to efficiently request them within the same batch as the POST request for entity creation.

The inactive parameter determines whether an inactive transient context is created, which only becomes active when a property is updated. Until then, it's not considered a pending change, meaning it won't be tracked by `hasPendingChanges` and can't be deleted with `resetChanges`. In addition, `submitChanges` won't trigger a creation request for inactive contexts.

`ODataModel#createEntry` also supports *deep create* for navigation properties with many cardinality.

ODataListBinding#create

The `ODataListBinding#create` method creates a new entry and adds it either at the beginning or end of a list. This entry is directly visible at the corresponding position of the bound control without first having to be saved in the backend and without the binding needing to be updated. This is an advantage over the `ODataModel#createEntry` method. This method is particularly useful if you're using a list or table that displays a collection of entries and one of the following conditions applies:

- New entries should already be visible in the table before they are saved in the backend so that the user can view and edit the data directly.
- The position of the newly created entries in the table should be retained even after saving in the backend, while only the data is updated based on the response to the `POST` request.
- You want to provide inline entry rows that allow new entries to be created quickly.

The `ODataListBinding#create` method internally uses `ODataModel#createEntry` to create a new entry. It supports the same parameters as `ODataModel#createEntry`, but with the following exceptions:

- **`sPath`**
  The path to the entity set is automatically set to the path of the list binding.
- **`mParameters.context`**
  The context for resolving the path is defined by the context of the list binding.
- **`mParameters.created`**
  As `ODataListBinding#create` assumes that the service metadata is already loaded, the method returns the created context synchronously. A callback isn't required.
- **`mParameters.headers`**
  This isn't supported by `ODataListBinding#create`.
- **`mParameters.properties`**
  The initial data for the new entry is provided via the separate parameter `oInitialData`.
- **`mParameters.refreshAfterChange`**
  This isn't supported by `ODataListBinding#create`. The method sets this parameter to `false` by default, as the new entry is already visible in the list before it's saved.

- `mParameters.urlParameters`
  This isn't supported by `ODataListBinding#create`.

New entries are inserted according to the `bAtEnd` parameter. After saving in the backend, they retain their position in the list unless an action occurs that is typically triggered by user interactions, such as `ODataListBinding#filter`, `ODataListBinding#sort`, `ODataListBinding#refresh`, or a rebinding of the list or table control. In these cases, the saved entries are displayed at the position specified by the backend.

Inactive entries can be used to create inline entry rows within a table, allowing new entries to be created quickly without separate forms or popups. Once the table data is loaded, one or more inactive entries can be added. Use `ODataListBinding#isFirstCreateAtEnd` to check if such entries have already been created. When an entry is activated, the list binding triggers the `createActivate` event. You can use this event to add another inactive entry, for example. If the event is canceled with `preventDefault`, the context remains inactive. In this way, you can ensure that all the required properties for creating the entity are set.

10

How a `create` can be called via the `ODataListBinding` is shown in the following code snippets (Listing 10.7 shows the view implementation, and Listing 10.8 the corresponding controller implementation).

```
<Table id="myTable" items="{path: 'ToDocuments', events: {createActivate:
'.onCreateActivateDocument'}}">
   ...
</Table>
```

Listing 10.7 XMLView Implementation for "create" Using ODataListBinding

```
onInit: function(){
   let oItemsBinding = this.getView().byId("myTable").getBinding("items");

   oItemsBinding.attachEvent("dataReceived", function() {
      if(oItemsBinding.isLengthFinal() && oItemsBinding.isFirstCreateAtEnd() ==
= undefined){
         this.onCreateActivateDocument();
      }
   });
},

onCreateActivateDocument: function(oEvent){
   if(!oEvent.getParameter("context").getProperty("DocumentId")){
      oEvent.preventDefault();
      return;
   }
```

```
    let oItemsBinding = this.getView().byId("myTable").getBinding("items");
    oItemsBinding.create({}, true, {inactive: true});
}
```

Listing 10.8 Controller Implementation for "create" Using ODataListBinding

This method also supports the deep create scenario, which is explained in the following section.

Deep Create

One or more subordinate entities can be created for a navigation property of a newly created parent entity in a single OData request. This process is called a *deep create*. The OData V2 model supports this deep create scenario for navigation properties with the cardinality *many*. It's also possible to nest deep create by creating entities for the navigation properties of an already subordinate entity. An entity within a deep create that only contains subentities and doesn't have a temporary parent entity is called a root entity.

To create a subentity, you can use the `ODataModel#createEntry` method. In doing so, you specify a context parameter that represents a transient context and refers to the parent entity, as well as a path parameter that defines the navigation property for the entity type of the context. Alternatively, you can use `ODataListBinding#create`. This uses a list binding that contains a transient binding context and the navigation property as a path. Note that most of the parameters of these APIs aren't applicable when creating a subentity, as they relate to the creation of standalone OData requests. Instead, the subentity data is added as part of the OData payload of the parent root entity, and the request-related settings of the subentities are automatically derived from the API call that creates the root entity.

After a successful deep create request in the OData model, only the transient context of the main entity is updated and can continue to be used. The transient contexts of the subordinate entities, on the other hand, aren't updated and then lose their validity. It's therefore not permitted to save such context references in the application code or to use them after creation as a binding context for a control element, for example.

With a deep create request, the OData service can provide a *deep response*. This means that the response contains not only the data of the root entity (as prescribed by the OData protocol) but also the data of the created subentities. When a deep response is provided, the OData model replaces the transient contexts of the direct subentities of the root entity with new contexts that are created based on the response. This automatically updates controls that are bound to the corresponding navigation property. However, if your service doesn't support deep response, you must manually update the list binding of the control after a successful deep create and retrieve the changed subentities from the backend via a separate `GET` request.

If you use the sap.ui.model.odata.v2.Context#delete method to remove the transient entity to which the context in question points, all associated subentities are also deleted.

In Listing 10.9, we see a controller in which a deep create is shown schematically. In this example, DocumentOwner is the root entity, and Document is the subentity. The table with the ID documentsTable is bound to the navigation property toDocuments. toDocument is a *to many* navigation property in the DocumentOwner entity, which maps the respective documents of the owner.

```
let oDocumentContext,
   oDocumentTable = this.getView().byId("documentsTable"),
   oDocumentItemsBinding = oDocumentTable.getBinding("items"),
   oModel = this.getView().getModel(),
   oDocumentOwnerContext = oModel.createEntry("/DocumentOwnerSet", {properties:
{firstName: "Max", lastName: "Mustermann"}});

oDocumentTable.setBindingContext(oDocumentOwnerContext);
oDocumentContext = oDocumentItemsBinding.create({Name: "Demo Document"});

oDocumentOwnerContext.created().then(function(){
   sap.m.MessageToast.show("Document owner with document created");
});

oModel.submitChanges();
```

Listing 10.9 Sample for Deep Create Using Entities DocumentOwner and "Document"

10.2 CRUD Operations

The OData V2 model allows manual CRUD requests to be sent to the service. A separate method is provided in the OData model class for this purpose. If a manually submitted operation returns data, this is imported into the local model cache and is available there. All operations require at least one path. Optionally, additional parameters can also be specified. In addition, a data object must also be specified for the create and update operation, which is used for processing the manipulating operation.

Each operation returns an object that has an abort function. The request can be aborted by calling this function. If a request is aborted, the corresponding error handler is called. This ensures that for each request—depending on whether it was successful or unsuccessful—either the success handler or error handler is called. It's also possible to include additional URL parameters, header data, or entity tags (ETags).

10.2.1 Create

The OData model provides a `create` method that can be used to create entities. Calling this method results in a `POST` request to the service. This method doesn't return a binding context that can be used to bind the data. In addition, the created entity isn't stored in the model's cache. This approach should only be used if a `create` operation is to be sent to the service and the created entity won't be bound in the UI. In addition, this type of call doesn't support deep creates; that is, only flat entities without child entities can be created.

Listing 10.10 shows what a simple call to the `create` method looks like.

```
let oData = {
   DocumentName: "Demo Document"
};

oModel.create("/Document", oData, {success: successHandler, error:
errorHandler});
```

Listing 10.10 Calling Create Operation of ODataModel V2

In addition to the path, an object with data must also be passed to the `create` method. A map with additional parameters can also be passed here. These are explained in more detail in the following:

- `sPath`
  Defines an absolute path or a relative path to the context given in `mParameters.context`. Should the path contain a query string, the query string will be ignored. For query strings, the parameter `mParameters.urlParameters` must be used.
- `oData`
  Data of the entity that should be created.
- `mParameters`
  Optional parameter map containing the following properties:
  - `context`
    If specified, the path provided in `sPath` must be relative to the path of the given context.
  - `success`
    A callback function that is invoked when the data is successfully retrieved.
  - `error`
    A callback function that is triggered when the request fails. The handler can include the parameter `oError`, which provides additional error details. If the `POST` request was aborted, the error will have the aborted flag set to `true`.
  - `urlParameters`
    Defines a map containing key-value pairs that will be added to the request URL as query parameters.

- **headers**
  A map of headers for this request.
- **groupId**
  Defines the ID of a request group. Requests of the same group will be bundled in one batch request.
- **changeSetId**
  Defines the ID of the ChangeSet that this request should belong to.
- **refreshAfterChange**
  Determines whether all bindings are updated after this change operation is submitted. Refer to #setRefreshAfterChange for details. If set, it temporarily overrides the global refreshAfterChange flag, applying only to this specific operation.

10.2.2 Read

The OData model provides the read method for executing read operations. A call to this method results in a GET request, which is sent to the service. The path is concatenated together depending on the path that was selected when the model was created and the path that is given to the read method. The queried data can be processed in the success handler. A simple call of this method to a defined path "/Document(4711)" could look as follows:

```
oModel.read("/Document(4711)", {success: successHandler, error:
errorHandler});
```

The read method only has two transfer parameters. These are explained in more detail in the following:

- **sPath**
  Defines an absolute path or a relative path to the context given in mParameters.context. Should the path contain a query string, the query string will be ignored. For query strings, the parameter mParameters.urlParameters must be used.
- **mParameters**
  Optional parameter map containing the following properties:
  - **context**
    If specified, the path provided in sPath must be relative to the path of the given context.
  - **urlParameters**
    Defines a map containing key-value pairs that will be added to the request URL as query parameters.
  - **filters**
    Array of filters that will be included in the request URL.

- **sorters**
  Array of sorters that will be included in the request URL.
- **success**
  A callback function that is invoked when the data is successfully retrieved. The handler can have the following parameters: oData and response. The oData parameter contains the retrieved data, while the response parameter provides additional details about the request's response.
- **error**
  A callback function that is triggered when the request fails. The handler can include the parameter oError, which provides additional error details. If the GET request was aborted, the error will have the aborted flag set to true.
- **groupId**
  Defines the ID of a request group. Requests of the same group will be bundled in one batch request.
- **updateAggregatedMessages**
  Specifies whether messages for subordinate entities belonging to the same business object as the requested or modified resources should be updated. This is only considered if sap.ui.model.odata.MessageScope.BusinessObject is set using #setMessageScope and if the OData service supports message scoping.

10.2.3 Update

The OData model provides the update method for executing update operations. A call to this method results in a PUT or MERGE request, which is sent to the service. Whether a PUT or MERGE request is triggered depends on which of the two methods was specified when the model was created. A simple update operation looks like Listing 10.11.

```
let oData = {
   DocumentId: 4711,
   DocumentName: "Demo Document"
};
oModel.update("/Document(4711)", oData, {success: successHandler, error:
errorHandler});
```

Listing 10.11 Calling the Update Operation of ODataModel V2

In addition to the path, an object with data must also be passed to the update method. A map with additional parameters can also be passed here. These are explained in more detail in the following:

- **sPath**
  Defines an absolute path or a relative path to the context given in mParameters.context. Should the path contain a query string, the query string will be ignored. For query strings the parameter mParameters.urlParameters must be used.

- `oData`
  Data of the entity that should be updated. In `PUT` requests, the whole object always must be supplied; in `MERGE` requests, only the properties that changed are required.
- `mParameters`
  Optional parameter map containing the following properties:
  - `context`
    If specified, the path provided in `sPath` must be relative to the path of the given context.
  - `Success`
    A callback function that is invoked when the data is successfully retrieved.
  - `error`
    A callback function that is triggered when the request fails. The handler can include the parameter `oError`, which provides additional error details. If the `PUT`/`MERGE` request was aborted, the error will have the aborted flag set to `true`.
  - `eTag`
    If provided, the `If-Match` header will be set to the specified ETag. Keep in mind that this functionality isn't officially supported, as asynchronous requests can cause data inconsistencies. Should you decide to use it, you must guarantee that the request is fully completed before proceeding with any further data processing.
  - `urlParameters`
    Defines map containing key-value pairs that will be added to the request URL as query parameters.
  - `headers`
    A map of headers for this request.
  - `groupId`
    Defines the ID of a request group. Requests of the same group will be bundled in one batch request.
  - `changeSetId`
    Defines the ID of the `ChangeSet` that this request should belong to.
  - `refreshAfterChange`
    Determines whether all bindings are updated after this change operation is submitted. Refer to `#setRefreshAfterChange` for details. If set, it temporarily overrides the global `refreshAfterChange` flag, applying only to this specific operation.

10.2.4 Delete

The `remove` method initiates a `DELETE` request to the OData service configured when the OData model was created. The application needs to specify the path of the entry to be deleted. A simple delete operation looks like this:

```
oModel.remove("/Document(4711)", {success: successHandler, error:
errorHandler});
```

The `remove` method only has two transfer parameters. These are explained in more detail in the following:

- **`sPath`**
  Defines an absolute path or a relative path to the context given in `mParameters.context`. Should the path contain a query string, the query string will be ignored. For query strings, the parameter `mParameters.urlParameters` must be used.
- **`mParameters`**
  Optional parameter map containing the following properties:
  - **`context`**
    If specified, the path provided in `sPath` must be relative to the path of the given context.
  - **`success`**
    A callback function that is invoked when the data is successfully retrieved. The handler can have the following parameters: `oData` and `response`. The `oData` parameter contains the retrieved data, while the `response` parameter provides additional details about the request's response.
  - **`error`**
    A callback function that is triggered when the request fails. The handler can include the parameter `oError`, which provides additional error details. If the `DELETE` request was aborted, the error will have the aborted flag set to `true`.
  - **`eTag`**
    If provided, the `If-Match` header will be set to the specified ETag. Keep in mind that this functionality isn't officially supported, as asynchronous requests can cause data inconsistencies. Should you decide to use it, you must guarantee that the request is fully completed before proceeding with any further data processing.
  - **`urlParameters`**
    Defines the map containing key-value pairs that will be added to the request URL as query parameters.
  - **`headers`**
    A map of headers for this request.
  - **`groupId`**
    Defines the ID of a request group. Requests of the same group will be bundled in one batch request.
  - **`changeSetId`**
    Defines the ID of the `ChangeSet` that this request should belong to.

- `refreshAfterChange`
  Determines whether all bindings are updated after this change operation is submitted. Refer to `#setRefreshAfterChange` for details. If set, it temporarily overrides the global `refreshAfterChange` flag, applying only to this specific operation.

Furthermore, the `sap.ui.model.odata.v2.Context` class offers a `delete` method. This uses the `remove` method of the OData model internally. In addition, transient and inactive entities are also deleted when this method is called. This method works for all statuses (inactive, transient, and persisted) for a context. The `remove` method of the OData model, on the other hand, only deletes persisted entities.

10.3 Function Import

The OData V2 model provides a `callFunction` function that can be used to call function imports or actions. In addition, the request can also be triggered via the `submitChanges` method if the `callFunction` request is deferred. In the following sections, we'll take a closer look at the method definition of the `callFunction` method. We'll also see how function import parameters can be used in bindings.

> **Note**
> Currently, only `IN` parameters of function imports are supported.

10.3.1 callFunction Method of the OData Model

As already mentioned, the OData model provides a method with which function imports and actions can be called. The syntax of this method is as follows:

```
callFunction(sFunctionName, mParameters?) : object
```

The two parameters are listed here:

- `sFunctionName`
  Defines the name of the function import that should be called. The name must start with a /, for example, `/setStatus`.
- `mParameters`
  Defines a parameter map containing the following properties:
  - `adjustDeepPath`
    Defines a callback to adjust the deep path of an entity returned by a function import call. The deep path represents the entity's resolved path relative to its parent context in the UI hierarchy. For example, for a `ToOwner` context bound to `/Document('47')`, the deep path becomes `/Document('47')/ToOwner`, ensuring messages are correctly assigned and displayed in the UI. This is of type `function`.

- `changeSetId`
  Defines the ID of the changeset that this request belongs to. This is of type `string`.
- `error`
  A callback function triggered in case the request fails. It can include a parameter `oError`, which provides additional details about the error. If the request is aborted, the `oError` will have its `aborted` flag set to `true`. This is of type `function`.
- `eTag`
  If the function import changes an entity, the ETag for this entity can be passed via this parameter. This is of type `string`.
- `expand`
  Defines a comma-separated list of navigation properties that are used to expand the returned entity once the function import is called. The navigation properties are then fetched with a `GET` request within the same `$batch` as the `POST`. For this to work, some prerequisites must be fulfilled, such as batch mode must be enabled, HTTP `POST` must be used as the function import's `requests` method, the function import must return a single entity, and the backend service must support the Content ID header. This is of type `string`.
- `groupId`
  Defines the ID of a request group. Requests belonging to the same group are bundled in one batch request. This is of type `string`.
- `headers`
  Defines a map of headers that are added to this request. This is of type `object`.
- `method`
  Defines the HTTP method that should be used for the function import. `GET` and `POST` are supported. Additionally, the HTTP method of the function import can be found in the metadata of the service. This is of type `string`.
- `refreshAfterChange`
  Specifies whether to refresh all bindings after submitting this change operation. This setting overrides the model-wide `refreshAfterChange` flag but applies only to this specific operation. This is of type `Boolean`.
- `success`
  Defines the callback function that should be called after the function import was executed and data has been retrieved successfully. The handler function can have the following two properties: `oData` and `response`. This is of type `function`.
- `urlParameters`
  Associates the function import parameter name, as defined in its metadata, with its value. The value is formatted according to the parameter's type specified in the metadata. This is of type `object`.

Furthermore, the `callFunction` method returns an object that contains a `contextCreated` function that returns a promise. This promise is resolved with the created context.

An abort function is also available, which can be used to abort the current request. The returned promise of the contextCreated function is rejected if the name of the function import can't be found in the metadata or if expand parameters are used, but the function import doesn't return a single entity.

10.3.2 Binding of Function Import Parameters

Data bindings can be used to change the values of function import parameters. The input parameters of the function import can be changed via the context that is resolved from the contextCreated function. Furthermore, the result of the function call can also be bound via the context. Listing 10.12 shows how this can be implemented.


```
onStartDocumentRetrieval: function(){
   let oView = this.getView();
   let oModel = oView.getModel();
   let oDocumentForm = oView.byId("documentForm");

   let oHandle = oModel.callFunction("/GetDocumentsByOwner", {
      error: function() {
         oModel.resetChanges([oDocumentForm.getBindingContext().getPath()]);
      },
      groupId: "changes",
      urlParameters: {owner: 47}
   });

   oHandle.contextCreated().then(function(oContext){
      oDocumentForm.setBindingContext(oContext);
   });
},

onSubmitDocumentForm: function() {
   this.getView().getModel().submitChanges({groupId: "changes"});
}
```

Listing 10.12 JavaScript Logic for Function Import Binding

The function import returns a single entity or a list of entities; the result data can be accessed via the $result property. The corresponding view implementation is shown in Listing 10.13.

```
<form:SimpleForm id="documentForm">
   <Button text="Submit" press=".onSubmitDocumentForm"/>
   <core:Title text="Results"/>
   <List items="{$result}">
      <StandardListItem title="{Name}"/>
```

```
    </List>
</form:SimpleForm>
```

Listing 10.13 XML View Implementation for Binding the Results of the Function Import

A function import is automatically executed again in the following cases, without a further call to the `callFunction` method:

- An input parameter value is changed via the context, the previous request failed, and `submitChanges` is called.
- The `GroupId` is a nondeferred `GroupId`, and at least one input parameter has been changed via the context.
- The `GroupId` is a deferred `GroupId`, at least one input parameter value has been changed via the context, and `submitChanges` is called for this deferred group.

In addition, failed calls of function imports where input parameter values have changed are automatically repeated. If you want to prevent a function import from being automatically called again, the changes must be reset by calling the `resetChanges` method of the OData model, for example, in the error handler of the function call.

10.4 Summary

After looking at the functionalities of the OData V2 model in this chapter, we'll take a closer look at the OData model V4 in the following chapter. This chapter will therefore also look at the differences between the two models. We'll then look at advanced development techniques, including topics such as custom control development, source control with Git, and the deployment process for SAPUI5 applications.

Chapter 11
OData Version 4 Model

The chapter provides a comprehensive introduction to the use and possibilities of the OData V4 model in SAPUI5 applications. It highlights the core components of this powerful framework and shows how seamless integration and synchronization with backend systems can be achieved. From the basic use of the OData model as an interface to efficient query bundling with batch groups and dynamic data binding using binding contexts, this chapter covers all the essential aspects. You'll also learn how CRUD operations are effectively implemented, how side effects ensure consistency, and how server-side logic is triggered by actions.

The *OData Version 4 (V4) model* is a central component of modern web applications and serves as a powerful interface for data exchange between the frontend and backend. It enables the efficient management of data, supports dynamic data bindings, and promotes consistency through functions such as batch requests and side effects. With its support for create, read, update, and delete (CRUD) operations and server-side actions, the OData V4 model provides a flexible basis for implementing complex business logic and user interactions in SAPUI5 applications.

Therefore, we'll focus on the OData V4 model in this chapter. In Section 11.1, we take a look at the model in general, including how to create an instance of the model and what configuration options are available here. In Section 11.2, we take a closer look at the batch groups, which offer the possibility of handling certain requests within a `$batch` request. This can reduce the number of requests that are sent to the backend and therefore contributes to better performance. In Section 11.3, we look at the different types of binding contexts that the OData V4 model brings with it, and in Section 11.4, we see how CRUD operations can be executed using the OData V4 model. To react to changes on related properties—for example, if another property has to be updated when a property is changed—OData V4 offers side effects, which we'll discuss in more detail in Section 11.5. Finally, in Section 11.6, we'll see how actions or function imports can be used with the OData V4 model. These are primarily used to execute server-side logic and should therefore be considered in addition to the CRUD operations.

11.1 OData Model

The class sap.ui.model.odata.v4.ODataModel provides the implementation of the model, which is intended for the consumption of OData V4 services. The OData V4 model supports the following functionalities:

- Read access
- Updating properties of OData entities (in entity sets and contained entities) via two-way binding
- Deletion of entities
- Operation invocation (functions and actions)
- Grouping of multiple data requests in one single batch request
- Server-side filtering and sorting

Smart Controls vs. Building Blocks

Instead of *smart controls* (sap.ui.comp library) or sap.ui.table.AnalyticalTable (for analytical table scenarios), the *SAP Fiori elements building blocks* (sap.fe.macros) must be used with the OData V4 model in SAPUI5. The tree table isn't supported in combination with the OData V4 model in SAPUI5.

The OData V4 model is primarily designed to work with OData V4 services. Nevertheless, OData V2 services can also be consumed with this model, but this requires an adapter, which is supplied by SAP.

In the following sections, we'll see how an instance of the OData V4 model can be created and also how we can consume OData V2 services using the OData V4 model.

11.1.1 Creating a Model Instance

To instantiate an OData V4 model, only one map must be passed as a mandatory parameter. This map must contain at least the serviceUrl of the service. All other parameters are optional. Listing 11.1 shows a code example in which an OData V4 model is instantiated.

```
let oModel = new sap.ui.model.odata.v4.ODataModel({
   serviceUrl: "/sap/opu/odata4/IWBEP/V4_SAMPLE/default/IWBEP/V4_GW_SAMPLE_
BASIC/0001/"
});
```

Listing 11.1 Instantiation of OData V4 Model

If the OData service and the OData model were integrated via the application descriptor (*manifest.json*), which is also the more common way, this would look like Listing 11.2 for this example.

```
{
   "sap.app": {
      "dataSources": {
         "default": {
            "uri": "/sap/opu/odata4/IWBEP/V4_SAMPLE/default/IWBEP/V4_GW_SAMPLE_
BASIC/0001/",
            "type": "OData",
            "settings": {
               "odataVersion": "4.0"
            }
         }
      }
   },
   "sap.ui5": {
      "models": {
         "": {
            "dataSource": "default",
            "settings": {}
         }
      }
   }
}
```

Listing 11.2 OData V4 Model Creation via manifest.json

The syntax of the OData V4 model's constructor is as follows: `new sap.ui.model.odata.v4.ODataModel(mParameters)`. Here, `mParameters` represents a map that can contain the following properties:

- **`annotationURI`**
  The URL (or an array of URLs) from which the annotation metadata is to be loaded. The specified annotation files are integrated into the service metadata in the order in which they are listed, with the last file loaded taking precedence. If annotations overlap, they are overwritten. However, if an annotation file contains other elements such as type definitions that have already been integrated, this leads to an error.
  - Type: `string` or `string[]`
  - Optional parameter
- **`autoExpandSelect`**
  Determines whether the OData model automatically generates the system query options `$select` and `$expand` based on the binding hierarchy. Dynamic adjustments to the binding hierarchy aren't taken into account. This parameter, which has been available since version 1.47.0 and since version 1.75.0, also allows the inclusion of navigation properties in `$select` via property paths.

- Type: `boolean`
- Optional parameter

- `earlyRequests`
  Specifies whether the following resources are requested as early as possible:
  - The root `$metadata` document and the associated annotation files
  - The security token

  Although the `$metadata` root document and the annotation files are retrieved, they are only converted from XML to JavaScript Object Notation (JSON) if they are actually required. This function has been supported since version 1.53.0. Note that the default value can be changed to `true` in future versions. To additionally speed up the start of an SAPUI5 component, you can also use the *manifest model preload* setting.
  - Type: `boolean`
  - Optional parameter

- `groupId`
  Determines how the model processes batch requests: With `$auto`, the model's requests are combined into a single batch request and sent automatically before rendering. With `$direct`, the requests are sent immediately and without batching. All other values result in an error.
  - Type: `string`
  - Optional parameter

- `groupProperties`
  Defines how batch requests for application groups are handled. This is an assignment of application group IDs to an object that has exactly one property. Permissible values for this property are `API`, `Auto`, and `Direct`.
  - Type: `object`
  - Optional parameter

- `httpHeaders`
  Defines a map of HTTP headers, which should be added to requests.
  - Type: `object`
  - Optional parameter

- `ignoreAnnotationsFromMetadata`
  Specifies whether all annotations from the service metadata and *cross-service references* should be ignored. Only the value `true` is permitted. In this case, only the annotations from the annotation files specified in `annotationURI` are taken into account. This parameter isn't transferred to value list models.
  - Type: `boolean`
  - Optional parameter

- `metadataUrlParams`
  Additional assignment of URL parameters that are used exclusively for `$metadata` requests. Note that the parameter `sap-context-token` is only relevant for the master `$metadata` of the service, but not for cross-service references.
  - Type: `object`
  - Optional parameter
- `odataVersion`
  Defines the version of the OData service. Supported values are `2.0` and `4.0`.
  - Type: `string`
  - Optional parameter
- `operationMode`
  The operation mode for filtering and sorting. As of version 1.39.0, the operation mode `sap.ui.model.odata.OperationMode.Server` is supported. All other operation modes, including undefined, lead to an error if `vFilters` or `vSorters` are specified or if `sap.ui.model.odata.v4.ODataListBinding#filter` or `sap.ui.model.odata.v4.ODataListBinding#sort` is called.
  - Type: `sap.ui.model.odata.OperationMode` (contained values: `Client`, `Default`, `Server`)
  - Optional parameter
- `serviceUrl`
  The root URL of the service from which the data is to be retrieved. The path part of the URL must end with a slash in accordance with the OData V4 specification. It's possible to append custom OData query options to the root URL of the service, separated by a `?`, such as `/MyService/?custom=foo`. However, OData system query options and parameter aliases lead to an error.
  - Type: `string`
  - Mandatory parameter
- `sharedRequests`
  Specifies whether all list bindings for the same resource path share their data so that it's only retrieved once; only the value `true` is permitted. In addition, `sap.ui.model.BindingMode.OneWay` is set as the default binding mode, while `sap.ui.model.BindingMode.TwoWay` isn't allowed. Note that this makes all bindings read-only, which can be particularly advantageous for value list models.
  - Type: `boolean`
  - Optional parameter
- `supportReferences`
  Specifies whether the `<edmx:Reference>` and `<edmx:Include>` directives are supported to load schemas from other `$metadata` documents and integrate them into the current service (cross-service references).

  - Type: boolean
  - Optional parameter
- **updateGroupId**
  The group ID used for update requests. If no specific group ID is specified, the ID from mParameters.groupId is used. Permissible values for update group IDs are undefined, $auto, $auto.*, $direct, or an application group ID.
  - Type: string
  - Optional parameter
- **withCredentials**
  Specifies whether the XMLHttpRequest is called with the withCredentials option so that the user's credentials are included in cross-origin requests from the browser.
  - Type: boolean
  - Optional parameter

11.1.2 Consuming OData V2 Services Using the OData V4 Model

The SAPUI5 framework can consume OData V2 services via an OData V4 model. The framework takes on the task of converting metadata and the data of the V2 service so that it can be used with the V4 model. For the developer, it basically makes no difference whether a V2 or V4 service is behind the V4 model. However, as OData V4 supports some functionalities that aren't supported by OData V2, certain functions of the OData V4 model can't be used. These restrictions are as follows:

- In OData V4 models, OData V2 can only be used for read scenarios.
- The OData V2 services must provide inline type metadata in the responses, that is, the __metadata.__type property. This information is required to convert the data between the OData V2 and OData V4 types.
- System query options such as $orderby, $filter, and $count at the top level, as well as $expand and $select, are supported. All other system query options will result in an exception.
- Not all OData V2 annotations are supported yet.

To use an OData V2 service in an OData V4 model, the service and the model must be integrated via the app descriptor as follows in Listing 11.3. It's important that the odataVersion of the data source is set to 2.0 and that sap.ui.model.odata.v4.ODataModel is selected as the type for the model. In addition, at least SAPUI5 version 1.49 is required.

```
{
   "_version": "1.1.0",
   "sap.app": {
      "dataSources": {
         "default": {
```

```
            "uri": "<ODataV2 Service URL>",
            "type": "OData",
            "settings": {
               "odataVersion": "2.0"
            }
         }
      },
   },
   "sap.ui5": {
      "dependencies": {
         "minUI5Version": "1.49"
      },
      "models": {
         "": {
            "dataSource": "default",
            "settings": {
               "autoExpandSelect": false,
               "operationMode": "Server",
               "synchronizationMode": "None"
            },
            "type": "sap.ui.model.odata.v4.ODataModel"
         }
      }
   }
}
```

Listing 11.3 Service and Model Definition in manifest.json for an OData V2 Service Using the OData V4 Model

As the available types in OData V4 and OData V2 differ, it's necessary to map between the datatypes. The following OData V2 types are supported and mapped:

- `Edm.Binary`, `Edm.Boolean`, `Edm.Byte`, `Edm.Decimal`, `Edm.Double`, `Edm.Guid`, `Edm.Int16`, `Edm.Int32`, `Edm.Int64`, `Edm.SByte`, and `Edm.String` types exist in both OData versions and therefore no type mapping is required.
- `Edm.DatetTime` is mapped to OData V4 type `Edm.Date` if the property has the OData V2 annotation `sap:display-format="date"`. Otherwise it's mapped to `Edm.DateTimeOffset` with the UTC time zone.
- `Edm.Time` is mapped to OData V4 type `Edm.TimeOfDay`.

Some datatypes are represented differently in OData V2 and OData V4. The application developer should use only the OData V4 values. The framework ensures that these values are converted properly before sending the request to the backend (e.g., as a value of a key property) and after receiving the response from the backend.

11.2 Batch Groups

Batch groups are intended to bundle several HTTP requests into one `$batch` request. This means that instead of several individual requests, a collective request is sent to the backend, which has a positive effect on performance. The OData 4 model sends requests to the backend in the following cases:

- **Implicit read requests to retrieve data for a binding**
  For example, if an entity is bound to a list, a `GET` request is triggered to read data.
- **Implicit update requests via two-way binding**
  For example, if the user changes a property via the UI, a `PATCH` request is sent automatically to update the changed property.
- **Explicit requests**
  Those can be initiated via the API, for example, `ODataListBinding.refresh` or `ODataContextBinding.invoke`.

It's possible to define group IDs for each of these cases. Each of these group IDs has a submit mode that affects the batch request. The following submit modes are possible:

- **`sap.ui.model.odata.v4.SubmitMode.API`**
  Requests that are provided with the group ID are sent bundled in a batch request.
- **`sap.ui.model.odata.v4.SubmitMode.Auto`**
  Requests that are assigned to a specific group ID are summarized in a batch request and sent automatically before rendering.
- **`sap.ui.model.odata.v4.SubmitMode.Direct`**
  Requests that are assigned to a group ID are sent immediately and without the use of batch. However, note that certain functions of the OData V4 model are dependent on the backend processing the requests in the correct order. This sequence is only guaranteed for requests within batch requests.

However, the recommendation is to only use the API and auto-submit modes if possible. With regards to group IDs, the following group IDs are possible:

- **`$auto` and `$auto.*`**
  The predefined batch group ID serves as the default if no group ID is specified. It's possible to use different `$auto.*` group IDs to separate different batch requests. The suffix can be any non-empty string consisting of alphanumeric characters of the Latin alphabet and underscores. These groups use the submit mode `sap.ui.model.odata.v4.SubmitMode.Auto`.
- **`$direct`**
  This is a predefined batch group ID that uses the *direct submit mode*.
- **Application group ID**
  An application group ID is any non-empty character string consisting of letters and numbers from the Latin alphabet and underscores. By default, the submit mode of

an application group is set to `sap.ui.model.odata.v4.SubmitMode.API`. However, a different submit mode can also be specified.

- **`$single`**
  With `sap.ui.model.odata.v4.Context#delete`, `sap.ui.model.odata.v4.ODataModel#delete`, and `sap.ui.model.odata.v4.ODataContextBinding#invoke`, you can use the group ID `$single`. This group ID causes a single batch request to be sent directly, similar to groups with the submit mode `sap.ui.model.odata.v4.SubmitMode.Direct`.

To define the group ID for individual requests, the parameter `$$groupId` (for read requests) and the parameter `$$updateGroupId` (for update requests) can be used.

Batch requests for update groups with a submit mode that isn't `$direct` are collected by group ID in a queue. A new batch request with changes is only sent after the previous batch request for the same group ID has been completed and processed. All transmitted changes for this group ID are bundled in a single batch request. Changes submitted by different calls of `ODataModel.submitBatch` are split into separate changesets within the same batch request.

11

Changesets and Order of Requests inside a Batch Request

The OData V4 model automatically organizes all non-`GET` requests into a single change record at the beginning of a batch request. All `GET` requests follow after this. If there's only a single request in the changeset, the batch request is replaced by this single request to minimize communication overhead. `PATCH` requests that are aimed at the same entity are combined into a single request.

If you set the `updateGroupId` to a binding, all changes are collected in a batch queue. These changes are sent to the backend all at once when the `ODataModel.submitBatch` method is called. The OData model also offers the `resetChanges` method, which can be used to reset local changes. These methods can be used, for example, to implement the logic for the `"Save"` and `"Cancel"` actions of a form. The **Save** button calls `submitBatch`, while the **Cancel** button calls `resetChanges`.

Repeating Property Changes

The OData V4 model automatically repeats failed property changes if `API` is defined as the `SubmitMode` and the property change fails on the server side. By calling the `submitChanges` method again, the change would be sent to the backend again.

It's possible to define a submit mode for an application group ID at the time the model is created. To do this, the `groupProperties` must be defined accordingly (see Listing 11.4).

```
"models": {
  "": {
    "dataSource": "default",
```

```
        "settings": {
          "operationMode": "Server",
          "synchronizationMode": "None",
          "groupProperties": {
            "fastGroup": {"submit": "Auto"},
            "slowGroup": {"submit": "Auto"}
          }
        }
      }
}
```

Listing 11.4 Define Submit Mode of Application Group ID on Model Creation

11.3 Binding Contexts

The OData V4 model offers three methods for creating a binding context. Use one of these methods depending on the type of binding you want to create:

- `bindContext`
- `bindList`
- `bindProperty`

11.3.1 bindContext Method

The `bindContext` method can be used to create a context binding (`sap.ui.model.odata.v4.ODataContextBinding`). This can be used to bind to a higher-level container, such as a view or `VBox`, to then be able to bind relatively in nested controls. A context binding represents a single entity (e.g., `"/SalesOrder('4711')"`) or a structural property with a complex type. The syntax of the method is as follows: `bindContext(sPath, oContext, mParameters)`.

`sPath` is the only mandatory parameter and represents the path of an entity. `oContext` is optional and can be used as a parent context if the path is relative. `mParameters` is also optional and is a map that allows you to specify additional configuration parameters. The parameters of `mParameters` are as follows:

- **$expand**
  This is of type `string` or `object`. The object represents an assignment between `expand` paths and their `expand` options. These options are in turn mappings of system query options, which are usually specified as strings. Alternatively, `$count` can be defined as a Boolean value, `$expand` recursively as a map, `$levels` as a numerical value, and `$select` as an array.

- **$select**
  This is of type string or string[] and defines a comma-separated list or an array of items. Property names that are included will be read and be part of the response.
- **$$canonicalPath**
  This is of type boolean and defines if a binding relative to a context uses the canonical path computed from its context's path for data service requests. Only the value true is allowed.
- **$$groupId**
  This is of type string and defines the group ID that will be used for read requests.
- **$$inheritExpandSelect**
  This is of type boolean and defines if $expand and $select are inherited from its parent binding.
- **$$ownRequest**
  This is of type boolean and defines if the binding always uses its own request to retrieve data. Only the value true is allowed.
- **$$patchWithoutSideEffects**
  This is of type boolean and defines if the implicit loading of side effects via PATCH requests is switched off. Only the value true is allowed.
- **$$updateGroupId**
  This is of type string and defines the group ID that will be used for update requests.

11.3.2 bindList Method

The bindList method can be used to create a list binding (sap.ui.model.odata.v4.ODataListBinding). This represents a collection of OData entities—complex or primitive types—for example, "/SalesOrder". This type of binding is used when you want to bind lists or tables, for example. The syntax of the method is as follows: bindList(sPath, oContext, vSorters, vFilters, mParameters).

Here, too, sPath is the only mandatory parameter. This is the path for which the list binding is to be created. If it's a relative path—for example, a navigation property—the parent context can be provided in oContext. vSorters is either an instance or an array of sap.ui.model.Sorter. This can be used to define the sort order. The sorters are provided in the $orderby parameter. vFilters is either an instance or an array of sap.ui.model.Filter, which, like sorters, are provided in the $filter parameter and enable filtering. The mParameters parameter is again a map in which further configurations are possible. The possible values for mParameters are as follows:

- **$apply**
  This is of type string and defines the value of the $apply system query as an alternative to $$aggregation.

- **$count**
  This is of type string or boolean and defines the value for the $count system query.
- **$expand**
  This is of type string or object. The object represents an assignment between expand paths and their expand options. These options are in turn mappings of system query options, which are usually specified as strings. Alternatively, $count can be defined as a Boolean value, $expand recursively as a map, $levels as a numerical value, and $select as an array.
- **$filter**
  This is of type string and defines the value for the $filter system query. This is used for filtering.
- **$orderby**
  This is of type string and defines the value for the $orderby system query. This is used for sorting.
- **$search**
  This is of type string and defines the value for the $search system query. This is used for searching.
- **$select**
  This is of type string or string[] and defines a comma-separated list or an array of items. Property names that are included will be read and be part of the response.
- **$$aggregation**
  This is of type object and defines an object holding information for data aggregation.
- **$$canonicalPath**
  This is of type boolean and defines if a binding relative to a context uses the canonical path computed from its context's path for data service requests. Only the value true is allowed.
- **$$clearSelectionOnFilter**
  This is of type boolean and defines whether the selection state of the list binding should be reset if applied filters are changed.
- **$$getKeepAliveContext**
  This is of type boolean. Specifies whether this binding is taken into account when #getKeepAliveContext is called; only the value true is permitted. A combination with $apply, $$aggregation, $$canonicalPath, or $$sharedRequest isn't permitted. For relative bindings, $$ownRequest must also be activated.
- **$$groupId**
  This is of type string and defines the group ID that will be used for read requests.
- **$$operationMode**
  This is of type sap.ui.model.odata.OperationMode and defines the operation mode used for sorting and filtering. Only the operation mode Server is supported.

- **`$$patchWithoutSideEffects`**
  This is of type `boolean` and defines if the implicit loading of side effects via `PATCH` requests is switched off. Only the value `true` is allowed.
- **`$$ownRequest`**
  This is of type `boolean` and defines if the binding always uses its own request to retrieve data. Only the value `true` is allowed.
- **`$$separate`**
  This is of type `string[]` and defines an array of property names, which will be omitted from the main list request and loaded in separate requests instead.
- **`$$sharedRequest`**
  This is of type `boolean` and defines whether multiple bindings with the same resource path share the data so that this data is loaded only once.
- **`$$updateGroupId`**
  This is of type `string` and defines the group ID, which will be used for update requests.

11.3.3 bindProperty Method

The `bindProperty` method makes it possible to create a property binding (`sap.ui.model.odata.v4.ODataPropertyBinding`). A property binding represents a single, primitive property of an OData entity or a complex type—for example `/SalesOrder('4711')/Name`. The signature of the `bindProperty` method is as follows: `bindProperty(sPath, oContext, mParameters)`.

The `sPath` attribute again defines the path of the property. If the path is a relative one, the parent context can be specified in `oContext`. Here, `mParameters` defines a map of parameters, which are as follows:

- **`$apply`**
  This is of type `string` and defines the value for the `$apply` system query if the path ends with `$count`.
- **`$filter`**
  This is of type `string` and defines the value for the `$filter` system query if the path ends with `$count`.
- **`$search`**
  This is of type `string` and defines the value for the `$search` system query if the path ends with `$count`.
- **`$$groupId`**
  This is of type `string` and defines the group ID that will be used for read requests.
- **`$$ignoreMessages`**
  This is of type `boolean` and defines if the binding should propagate model messages to the control.

- **$$noPatch**
  This is of type boolean and defines if the change of the property triggers a PATCH request. Only the value true is allowed.
- **scope**
  This is of type any and defines the scope for the method requestObject if it's an object; otherwise, it's a custom query option.

11.4 CRUD Operations

The OData V4 model doesn't provide its own methods for executing CRUD operations, as is the case with the OData V2 model. Instead, you have to work with different contexts, which in turn enables you to execute CRUD operations. The exception here is the DELETE operation.

11.4.1 Create

To execute a create operation for an entity—that is, a POST request—a list binding must first be created for this entity. A new instance of the entity can then be created using the create method. Listing 11.5 shows how this can be done.

```
let oModel = this.getView().getModel();
let oListBinding = oModel.bindList("/SalesOrder");

oListBinding.create().created().then();
```

Listing 11.5 Triggering Create Operation Using the OData V4 Model

The create method has the following signature: create(oInitialData, bSkipRefresh, bAtEnd, bInactive). The parameters are as follows:

- **oInitialData**
  This is of type object and contains the initial data used for the creation of entity.
- **bSkipRefresh**
  This is of type boolean and defines if an automatic refresh after the creation is suppressed.
- **bAtEnd**
  This is of type boolean and defines if the new entity should be created at the end of a list or table.
- **bInactive**
  This is of type boolean and defines if an inactive context should be created. Inactive contexts will only be sent to the backend after the first property change.

11.4.2 Read

To perform a read operation—that is, a GET request—three different methods can be called, depending on whether a list of entities, a single entity, or a single property of an entity is to be read, as follows:

- bindContext can be used to read individual entities.
- bindList can be used to read lists of entities.
- bindProperty can be used to read individual properties of a single entity.

Read with bindContext

To read individual entities, the bindContext method of the OData V4 model can be used. To do this, a context must first be created with bindContext. The method requestObject can then be called from this context, which triggers a read request for the corresponding entity (see Listing 11.6).


```
let oModel = this.getView().getModel();
let oContextBinding = oModel.bindContext("/SalesOrder('4711')");

oContextBinding.requestObject().then();
```

Listing 11.6 Trigger a Read Operation via bindContext

The requestObject method returns a promise. When this is resolved, the object read is returned.

Read with bindList

If you want to read a list of entities, you can use the bindList method. This returns a list binding, which can be used with the requestContexts method to determine the contexts for the entities read. These contexts can then be used to query either the entire entity or individual properties of that entity using the requestProperty or requestObject methods. An example of this can be seen in Listing 11.7.

```
let oModel = this.getView().getModel();
let oListBinding = oModel.bindList("/SalesOrder");

oListBinding.requestContexts().then(contexts => {
   contexts.forEach(context => {
      context.requestObject();
      context.requestProperty("PROPERTYNAME");
   });
});
```

Listing 11.7 Triggering a Read Operation via bindList

Read with bindProperty

If you only want to read individual properties of an entity, you can do so using the `bindProperty` method. By calling this method, you get a property binding from which you can read the value of the property using the `requestValue` method. Listing 11.8 shows an example in which the `Name` property of an `SalesOrder` entity is read.

```
let oModel = this.getView().getModel();
let oPropertyBinding = oModel.bindList("/SalesOrder('4711')/Name");

oPropertyBinding.requestValue().then(value => {});
```

Listing 11.8 Triggering a Read Operation via bindProperty

11.4.3 Update

To trigger an update operation, you need an `sap.ui.model.odata.v4.Context`. You get this by calling the `getBoundContext` method of an `ODataContextBinding`. Once you have this context, you can use the `setProperty` method to change the values of properties. Listing 11.9 shows an example of how this can be done.

```
let oModel = this.getView().getModel();
let oContextBinding = oModel.bindContext("/SalesOrder('4711')");
let oContext = oContextBinding.getBoundContext();

oContext.setProperty("Name", "New Name");
```

Listing 11.9 Triggering an Update Operation Using OData V4 Model

The `setProperty` method has the following signature: `setProperty(sPath, vValue, sGroupId, bRetry)`. The individual parameters are explained in more detail in the following:

- **`sPath`**
  This is of type `string` and defines the path of the property that will be changed.
- **`vValue`**
  This is of type `any` and defines the value with which the given property should be updated. This must be primitive.
- **`sGroupId`**
  This is of type `string` and defines the group ID, which will be used for the `PATCH` request.
- **`bRetry`**
  This is of type `boolean`. If this flag is set to `true`, the value isn't reset if the `PATCH` request fails.

11.4.4 Delete

To perform a delete operation for an entity—that is, a `DELETE` request—the `delete` method of the OData V4 model can be used. This method must be passed the path to the entity entry. Listing 11.10 shows how this can be done.

```
let oModel = this.getView().getModel();
oModel.delete("/SalesOrder('4711')").then();
```

Listing 11.10 Triggering the Delete Operation Using OData V4 Model

The signature of the `delete` method of the OData V4 model is `delete(vCanonicalPath, sGroupId, bRejectIfNotFound)`. The individual parameters are explained in more detail in the following:

- **`vCanonicalPath`**
  This is of type string or `sap.ui.model.odata.v4.Context`, defines the path to an entity that will be deleted, and must start with a /. Alternatively, a context instance can also be given.
- **`sGroupId`**
  This is of type `string` and defines the group ID that will be used for the `DELETE` request.
- **`bRejectIfNotFound`**
  This is of type `boolean`. If the flag is set to `true` and the entity isn't found, the deletion fails with an error.

11

11.5 Side Effects

If an end user changes a field by means of an action or entry, this action or these changes could also affect other fields displayed on the UI. These are known as *side effects*. Side effects are basically executed on the backend, but it's still necessary to implement them in such a way that the frontend is also informed of the change. This allows dependent fields to be reloaded and displayed to the user. If this isn't done, the end user will see old data.

Default Side Effects

Default side effects are applicable to most applications. There's no need to annotate these side effects, as they are enabled by default and can't be disabled.

The side effects listed in Table 11.1 are available in SAP Fiori elements out of the box.

User Action	Side Effect
Creating a new entity or draft version, either in the list report or on the object page	The list binding of the parent page is refreshed to show the newly created entity.
Deleting an entity, either in the list report or in the object page	The list binding of the parent page is refreshed to remove the deleted entity.
Creating a draft for an active object	The list binding of the list report page is refreshed to show the new draft.
Discarding a draft version	The list binding of the list report page is refreshed to remove the draft and show the active version.
Activating a draft version	The list binding of the list report page is refreshed to remove the draft and show the active version.
Triggering an action	After a successful execution of an action, the collection for which the action was called is refreshed automatically if the following conditions apply: ■ The action is a bound action. ■ The returned instance doesn't correspond to the bound instance. This applies, for example, to copy actions.

Table 11.1 Default Side Effects

In the following sections, we'll take a closer look at the format of side effect annotations. You'll also see how side effects can be added to SAPUI5 applications.

11.5.1 Side Effect Annotation Format

A side effect annotation consists of the following elements:

- **Side effect source**
  Specifies a modification to a property, the execution of an action, or a structural adjustment (e.g., adding or removing a subitem). As a source, the following annotations can be applied:
  - **Source properties**
    If you use a value help, a combo box, a checkbox, a date picker or a date/time picker, the page effect is triggered directly as soon as the value is set. However, if you enter the value manually, the page effect is only activated when the focus leaves the input field. If the controls are used in conjunction with other input fields as a source for a side effect, this only occurs when the focus leaves the entire group of source fields. Navigation properties can't be used as source properties.

- **Source entities**
  Because these are source entities, navigation properties also can be used. The side effect occurs when structural changes such as the addition or deletion of an element are made. However, changes to a property of any entity don't trigger a side effect. In such cases, the side effect must be defined in the entity type of the corresponding entity.

- **Side effect target**
  Defines a property or structural information that needs to be refreshed. The side effect target consists of the following two elements:
  - **Target properties**
    It's possible to specify a single property or a list of properties to be updated. In addition, 1:1 navigation properties can also be specified. In this case, a request is executed with an $expand on the corresponding navigation property.
  - **Target entities**
    It's possible to specify both 1:1 and 1:n navigation properties. If an empty target is specified, the entire entity is updated. If a trigger action is defined, but neither `TargetProperties` nor `TargetEntities` are specified, only the trigger action is executed.

- **`TriggerAction` (optional)**
  The value of this property defines the path to a function import that should be invoked when the side effect occurs.

11

Side Effects Are Only Triggered If Data Is Sent to the Backend

Normally, no data is sent to the backend as long as UI validation errors exist. This means that a page effect isn't triggered if there are validation errors for the source property or a group of properties.

For example, no page effect is triggered if a value is entered in a data field that refers to the `ProductCategory` or `MainProductCategory` property that exceeds the maximum length of 40 characters (`MaxLength="40"`). For the page effect to be triggered, both properties must be successfully validated.

Listing 11.11 shows how a side effect could be defined. This example is based on the assumption that an entity has the properties `Amount`, `Price`, and `TotalPrice`. When the `Amount` or `Price` fields are changed, the backend recalculates the `TotalPrice` field, which is then updated by the side effect.

```
<Annotations Target="NAMESPACE.ENTITYTYPE">
   <Annotation Term="com.sap.vocabularies.Common.v1.SideEffects" qualifier=
"AmountOrPriceChanged">
      <Record>
         <PropertyValue Property="SourceProperties">
```

```
            <Collection>
               <PropertyPath>Amount</PropertyPath>
               <PropertyPath>Price</PropertyPath>
            </Collection>
         </PropertyValue>
         <PropertyValue Property="TargetProperties">
            <Collection>
               <PropertyPath>TotalPrice</PropertyPath>
            </Collection>
         </PropertyValue>
      </Record>
   </Annotation>
</Annotations>
```

Listing 11.11 Example of Side Effect Annotation

11.5.2 Adding Side Effects to SAPUI5 Applications

There are two ways to add side effects to an SAPUI5 application:

- Manually adding side effects in local annotation file
- Guided adding of side effects via guided development

To add side effects manually, you only need to enter the corresponding annotations in the local annotation file. However, this can be laborious and cumbersome. It's therefore better to add side effects using guided development. To do this, the following steps are required:

1. Open guided development by pressing **View • Command Palette** and searching for "Guided Development".
2. Search for "Side Effect", and select the corresponding entry **Configure side effects** (see Figure 11.1).

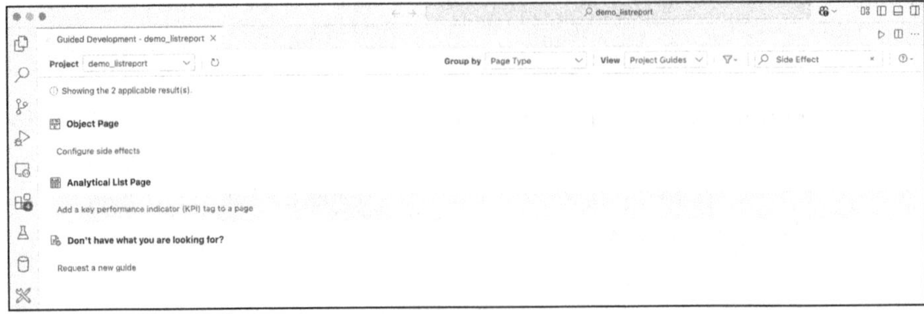

Figure 11.1 Opening the Wizard for Side Effect Configuration

3. A view will then open on the right-hand side in which side effects can be defined. The creation can be started by clicking **Start Guide** (see Figure 11.2).

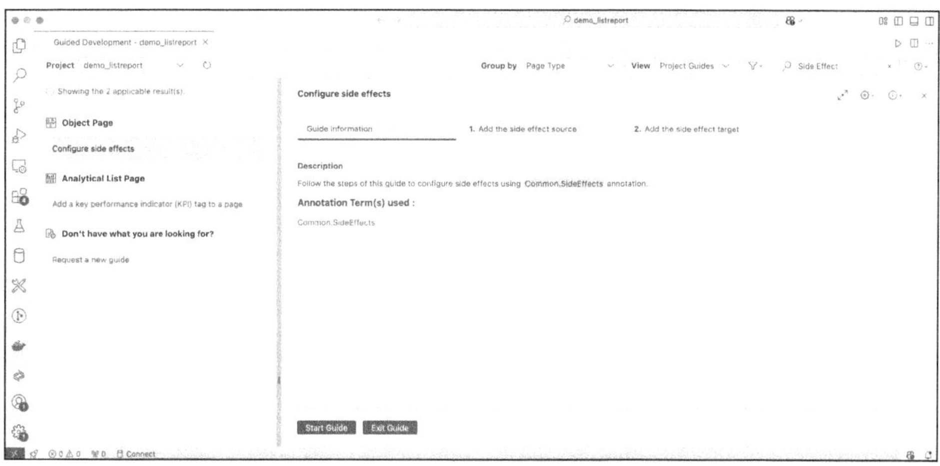

Figure 11.2 Start Guide Button for Side Effect Creation

4. The side effect source must be configured. This involves defining which entity type and which properties or actions should represent a source for the side effect. In this case, select **CustomerType** as the **Entity Type** and **Lastname** as the **Source Property**. Assign the identifier "lastnameChanged" as the qualifier under **Side Effects Qualifier**. These steps are shown in Figure 11.3.

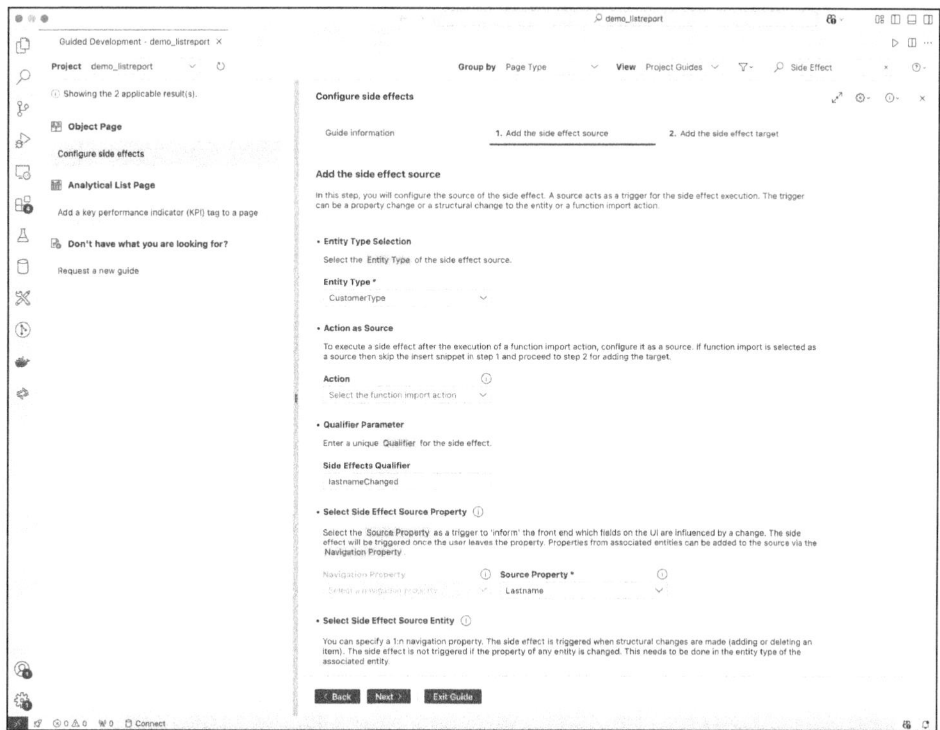

Figure 11.3 Definition of Side Effect Source (1 of 2)

5. A code snippet is generated in the lower area based on the entries made, which can be inserted into the annotation file by clicking on **Insert Snippet** (see Figure 11.4). Then, click on **Next**.

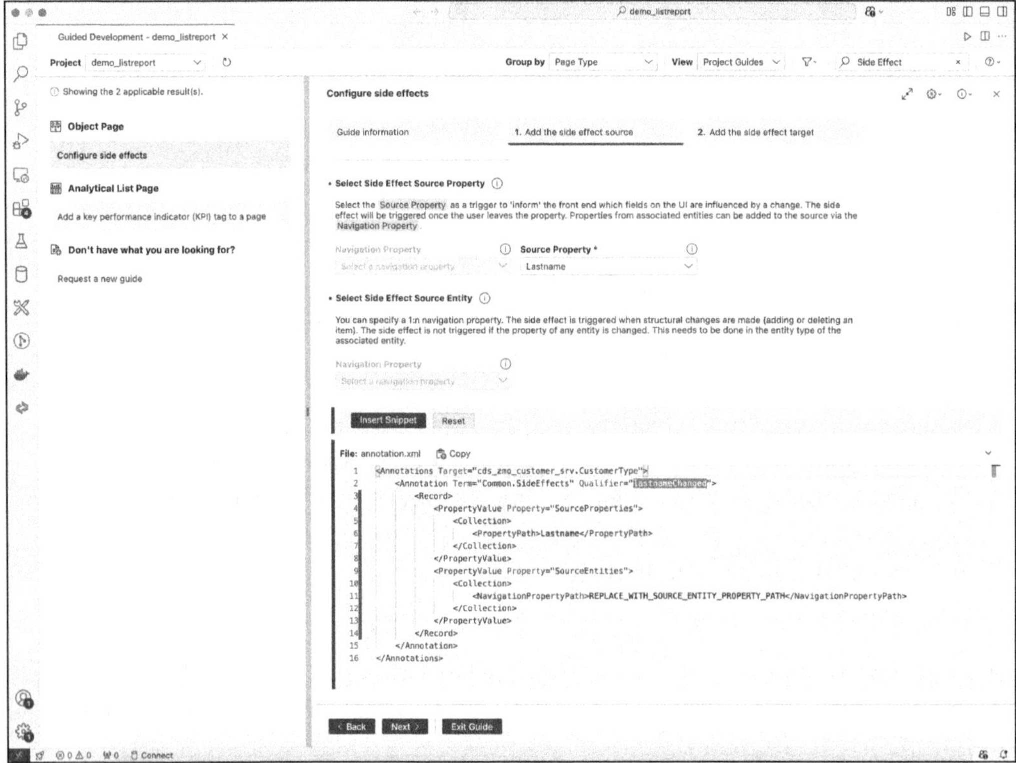

Figure 11.4 Definition of Side Effect Source (2 of 2)

6. Now you're in the mask where the side effect target has to be configured. The target property name must now be specified here, that is, the property that is to be updated by the side effect. Enter "Email" as the **Target Property** (see Figure 11.5).

7. Again, insert the code snippet into the annotation file by clicking on **Insert Snippet**. Then, exit the wizard by clicking **Exit Guide**. In Figure 11.6, we now see the annotation file with the previously configured side effect.

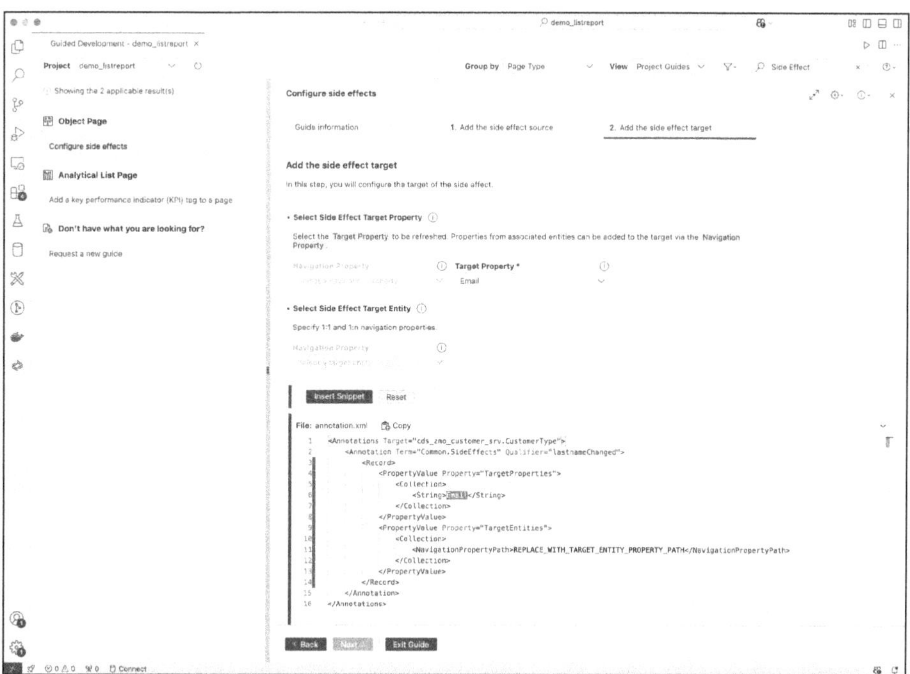

Figure 11.5 Definition of the Side Effect Target

annotation.xml 2 ×

webapp > annotations > annotation.xml

```xml
<edmx:Edmx xmlns:edmx="http://docs.oasis-open.org/odata/ns/edmx" Version="4.0">
    <edmx:Reference Uri="https://sap.github.io/odata-vocabularies/vocabularies/Common.xml">
        <edmx:Include Namespace="com.sap.vocabularies.Common.v1" Alias="Common"/>
    </edmx:Reference>
    <edmx:Reference Uri="https://sap.github.io/odata-vocabularies/vocabularies/UI.xml">
        <edmx:Include Namespace="com.sap.vocabularies.UI.v1" Alias="UI"/>
    </edmx:Reference>
    <edmx:Reference Uri="/sap/opu/odata/sap/ZMO_CUSTOMER_UI02/$metadata">
        <edmx:Include Namespace="cds_zmo_customer_srv"/>
    </edmx:Reference>
    <edmx:DataServices>
        <Schema xmlns="http://docs.oasis-open.org/odata/ns/edm" Namespace="local">
            <Annotations Target="cds_zmo_customer_srv.CustomerType">
                <Annotation Term="UI.LineItem">
                    <Collection>
                        <Record Type="UI.DataField">
                            <PropertyValue Property="Value" Path="Customerid"/>
                            <PropertyValue Property="Label" String="Demo Column"/>
                            <Annotation Term="UI.Importance" EnumMember="UI.ImportanceType/High"/>
                        </Record>
                    </Collection>
                </Annotation>
                <Annotation Term="Common.SideEffects" Qualifier="lastnameChanged">
                    <Record Type="Common.SideEffectsType">
                        <PropertyValue Property="SourceProperties">
                            <Collection>
                                <PropertyPath>Lastname</PropertyPath>
                            </Collection>
                        </PropertyValue>
                        <PropertyValue Property="TargetProperties">
                            <Collection>
                                <String>Email</String>
                            </Collection>
                        </PropertyValue>
                    </Record>
                </Annotation>
            </Annotations>
        </Schema>
    </edmx:DataServices>
</edmx:Edmx>
```

Code snippet has been successfully applied

Figure 11.6 Annotation File with Created Side Effect

11.6 Actions

The OData V4 model supports the following OData operations:

- `ActionImport`
- `FunctionImport`
- Bound actions
- Bound functions

Note that unbound parameters are limited to primitive datatypes. Functions and actions are designed to execute backend logic. The main difference between a function and an action is that a function usually changes data in a backend entity, while an action is only a calculation and has no changing character.

In the following sections, we'll discover topics such as simple function bindings, deferred operations, action bindings, and applying operation parameters. Finally, we'll dive into bound actions and functions.

11.6.1 Simple Function Bindings

A `FunctionImport` can be used in an XML view by binding it to a UI element. If the `FunctionImport` has no parameters and there's no need to control the time of the call, the `FunctionImport` can be bound, as shown in Listing 11.12.

```
<Text text="{path: 'GetNumberOfAvailableItems()', type:
'sap.ui.model.odata.type.Int16'}"/>
```

Listing 11.12 Simple Function Binding

The binding path represents the return value of the `FunctionImport`. The `FunctionImport` is called immediately as soon as the UI element requests the value. In addition, a type must be specified if the return value has a primitive type.

11.6.2 Deferred Operation Bindings

It's often not possible to execute an operation immediately, for example, if the user must first enter parameters. In such cases, an `ODataContext` binding should be used as an element binding on a layout element in the view, such as a `<Form>` or a `<VBox>`. Indicate the operation as deferred by placing an ellipsis in parentheses, for example, `GetNextAvailableItem(...)`. The return value can be read via relative bindings of child elements. In this context, the context binding is referred to as an operation binding, or more precisely as a function or action binding, depending on the type of OData operation used.

If the operation binding defers the execution of the operation, the invoke method must be called to actually execute the operation. In Listing 11.13, we see the view implementation, and Listing 11.14 shows the corresponding controller implementation.

```
<Form id="getNextAvailableItem" binding="{/GetNextAvailableItem(...)}">
   <Label text="Description"/>
   <Text text="{Description}"/>
   <Button text="Call the function" press="onGetNextAvailableItem"/>
</Form>
```

Listing 11.13 Deferred Operation Binding: View Implementation

```
onGetNextAvailableItem: function(oEvent){
   this.getView().byId("getNextAvailableItem").getObjectBinding().invoke();
}
```

Listing 11.14 Deferred Operation Binding: Controller Implementation

The method invoke returns a promise that is resolved if the operation was successful. If the operation fails, the promise is rejected with an error. Note that the result of the function isn't included in the promise.

11.6.3 Action Bindings

Action bindings must be deferred; otherwise, the application has no control over when the action is executed. A deferred action binding is defined in the same way as a deferred function binding (see Listing 11.15).

```
<Form id="Submit" binding="{/Submit(...)}">
   <Button text="Submit the action" press="onSubmit"/>
</Form>
```

Listing 11.15 Action Binding

For actions, a (...) must be appended, even if the resource URL of the action doesn't contain it. This is necessary to mark the binding as deferred. When called via invoke, the binding uses the metadata to distinguish between action and function and to generate the correct resource path for the operation.

11.6.4 Operation Parameters

Within an XML view, parameters of a deferred operation binding can be used. These parameters are identified by the prefix $Parameter. In addition, either each control property can be bound using the prefix $Parameter (see Listing 11.16), or a higher-level binding can be used that binds to the $Parameter path (see Listing 11.17).

```
<Dialog binding="{/ChangeTeamBudgetByID(...)}" id="operation1" title="Change
Team Budget">
   <form:SimpleForm>
      <Label text="TeamID"/>
      <Input value="{$Parameter/TeamID}"/>
      <Label text="Budget"/>
      <Input value="{$Parameter/Budget}"/>
   </form:SimpleForm>
</Dialog>
```

Listing 11.16 Binding Operation Parameters: Option 1

```
<Dialog binding="{/ChangeTeamBudgetByID(...)}" id="operation2" title="Change
Team Budget">
   <form:SimpleForm binding="{$Parameter}">
      <Label text="TeamID"/>
      <Input value="{TeamID}"/>
      <Label text="Budget"/>
      <Input value="{Budget}"/>
   </form:SimpleForm>
</Dialog>
```

Listing 11.17 Binding Operation Parameters: Option 2

In both cases, the parameters are set via the context binding of the control without the need for a controller implementation. Alternatively, the parameters can also be set via a corresponding controller implementation. To do this, the setParameter method of the operation binding must be used (see Listing 11.18).

```
onSubmit: function(oEvent) {
   this.getView().byId("Submit").getObjectBinding().setParameter("Comment",
"Any Comment").invoke();
}
```

Listing 11.18 Setting Parameters Using setParameter of Operation Binding

In addition, the getParameterContext method can be used to access and manipulate the parameters of an operation in the controller. Listing 11.19 shows how this might look.

```
adaptBudgetToTeam: function() {
   let oDialog = this.getView().byId("operation2");
   let oParameterContext = oDialog.getObjectBinding().getParameterContext();

   oParameterContext.setProperty("Budget", 666);
}
```

Listing 11.19 Manipulate Operation Parameters via the Controller Using Parameter Context

11.6.5 Bound Actions and Functions

The previous examples used root-level operations that are invoked via action or function imports. However, it's also possible to bind an action or function to another resource of the service, such as an entity or collection of entities. Bound actions or functions are handled in the same way as unbound operations: add `(...)` to the binding path of the corresponding control property. To call actions or functions bound to a specific entity or navigation property, use a relative binding. In the following example, the `InvoiceCreated` action executes on a selected sales order (see Listing 11.20).

```
let oModel = this.getView().getModel(),
   oTable = this.getView().byId("SalesOrders"),
   oSalesOrderContext = oTable.getSelectedItem().getBindingContext(),
   oActionContext = oModel.bindContext("namespace.InvoiceCreated(...)",
oSalesOrderContext);

oActionContext.invoke().then(
   function() {
      sap.m.MessageToast.show("Invoice created for selected Sales Order");
   },
   function(oError) {
      sap.m.MessageToast.show(oError.message);
   }
);
```

Listing 11.20 Calling a Bound Action from the Controller

Likewise, actions or functions can be called that are bound to a collection of an OData entity set. To do this, a binding context with an absolute path must be created, or a relative path for the operation (e.g., `namespace.DestroyOutdated(...)`) and a list binding context must be provided as the header context. Listing 11.21 shows how an operation can be performed on an entity set.

```
let oModel = this.getView().getModel();
oModel.bindContext("/LeaveRequests/namespace.DestroyOutdated(...)").invoke();
```

Listing 11.21 Call Operation on an Entity Set with an Absolute Path

Listing 11.22 shows how the same operation can be called with a relative path and the additional specification of the list binding context.

```
let oModel = this.getView().getModel(),
   oListBinding = this.getView().byId("leaveRequests").getBinding("items"),
   oHeaderContext = oListBinding.getHeaderContext();

oModel.bindContext("namespace.DestroyOutdated(...)", oHeaderContext).invoke();
```

Listing 11.22 Call Operation on an Entity Set with a Relative Path

11.7 Summary

In this chapter, we provided an overview of the OData V4 model. You've also seen how CRUD operations can be carried out with this model and what role batch groups play in connection with the OData V4 model. Furthermore, we examined in detail the various binding contexts that are relevant for the use of this model, along with actions and side effects.

The following chapters are about advanced development techniques. Among other things, we'll take a closer look at advanced SAPUI5—that is, the modularization of components, localization, the implementation of a file upload, drag and drop, and form validation. We'll also gain insights into the development of custom controls and take a detailed look at SAP Fiori elements.

PART IV

Advanced Development Techniques

Chapter 12
Advanced SAPUI5

In addition to controls for building UIs, routing, and other basic features, SAPUI5 offers more in-depth programming concepts to build enterprise-grade UIs. From modularization and localization to file uploading and validating user input, SAPUI5 provides a set of tools we can use to make our applications more user friendly and better programmed.

In this chapter, we take a deeper look into the capabilities of the SAPUI5 library by focusing more on the advanced features. We start by looking more into modularization in Section 12.1. We talked about basic modularization and what a SAPUI5 module is already in Chapter 5, but we can use them more to reuse certain parts of our XML and controllers.

We then make our application multilanguage ready by implementing the localization and internationalization concepts in Section 12.2. This enables us to make texts that are easily translatable to support multiple languages that are detected automatically. In Section 12.3, we implement the possibility to upload files to an OData service. This goes hand in hand with an OData service that supports media handling, so we'll also investigate the backend side of this.

Drag and drop, as covered in Section 12.4, plays a not-so-small role in modern, mobile-friendly apps. Especially on small displays, dragging and dropping content makes reordering and reorganizing easy. When working with forms, validating the input data can be quite a challenge. So, in Section 12.5, we take a dive into how we can realize this and what possibilities SAPUI5 offers to display hints to help users provide the right data.

This also plays a part in error handling in Section 12.6. When something goes wrong despite validating, we need a way to catch and fix errors, followed by presenting users with what went wrong. In the last Section 12.7, we look at smart controls. These controls work with OData Version 2 (V2) services and are metadata driven, meaning you can build UIs fast and SAP Fiori elements as well.

12.1 Modularization

We start our advanced topics with modularization. As already explained, modules divide our code into smaller, reusable segments. They have a special functionality and are loaded on demand when we need them. This is how the SAPUI5 library was built.

Until now, we've only used JavaScript or TypeScript modules when working with control. But SAPUI5 also allows us to modularize our XML view content, as shown in Section 12.1.1, by using fragments.

One core programming concept is to not only modularize our views but also our controllers. We often call the same methods or coding sequences over and over and over, which goes against the don't repeat yourself (DRY) coding concept. We can handle this with base controllers, as shown in Section 12.1.2.

12.1.1 Fragments and Dialogs

Fragments are lightweight, self-contained UI control trees defined outside of the main view structure. They act as factories, producing the defined UI control tree upon request. When used within declarative views, fragment content is seamlessly integrated.

This works by declaring XML files with the naming convention: *<Fragment Name>.fragment.xml*. Into this file goes a `sap.ui.core.FragmentDefinition`, which holds the UI elements that are a part of this reusable fragment:

```
<core:FragmentDefinition xmlns="sap.m" xmlns:core="sap.ui.core">
    <!-- Content -->
</core:FragmentDefinition>
```

This content can't be loaded dynamically nor placed directly into a view.

Fragments themselves have no controller, as they are placed inside a view that serves as their controller. During loading, however, a dedicated controller can be referenced, which is normally the controller of the view you're adding the fragment into. They also don't automatically inherit the ID of the view, so no unique IDs are guaranteed for the fragment's content.

Fragments are often used when you have content in your application that you're either unsure whether it's seen every time, isn't visible during startup of your view, or should be reused thorough views to get a coherent user experience.

Let's try out a simple example where we can see all the different facets a fragment brings. We'll walk through defining a fragment, loading fragments both statically and dynamically, using dialogs, and using message boxes.

Defining a Fragment

In this example, we want to create a simple form that is either editable or not, meaning we display either text controls or input controls for our fields. We can realize this by creating two fragments, each holding one simple form with either input or text controls. When a user selects the edit mode, we load the fragment with the simple form with input controls; otherwise, we load the simple form with text controls.

Let's start by declaring a small JSONModel that holds the data for our simple forms in a controller that holds book data, as shown in Listing 12.1.

```
const model = new JSONModel({
            title: "SAPUI5",
            description: "A Comprehensive Guide",
            releaseYear: "2025",
            price: 89.99
        });
        this.getView()?.setModel(model);
```

Listing 12.1 Declaring a Simple JSONModel

This JSONModel holds a title, a description, a releaseYear and a price. We then declare our first fragment that displays this data in a simple form.

So, start by adding a new file called *DisplayBook.fragment.xml* to your applications *view* folder. Here, you add a FragmentDefinition and a SimpleForm, displaying your JavaScript Object Notation (JSON) model's data, as shown in Listing 12.2.


```
<core:FragmentDefinition xmlns:form="sap.ui.layout.form"
    xmlns="sap.m"
    xmlns:core="sap.ui.core">
    <form:SimpleForm id="displayForm">
        <Label text="Title"/>
        <Text text="{/title}"/>
        <Label text="Description"/>
        <Text text="{/description}"/>
        <Label text="Release Year"/>
        <Text text="{/releaseYear}"/>
        <Label text="Price"/>
        <ObjectNumber number="{/price}" unit="€" />
    </form:SimpleForm>
</core:FragmentDefinition>
```

Listing 12.2 Defining Our First Fragment

We can now add a second fragment called *EditBook.fragment.xml* to our *view* folder, which contains again a SimpleForm, but now with input controls as shown in Listing 12.3.

```
<core:FragmentDefinition xmlns:form="sap.ui.layout.form"
    xmlns="sap.m"
    xmlns:core="sap.ui.core">
    <form:SimpleForm id="editForm">
        <Label text="Title"/>
        <Input value="{/title}"/>
        <Label text="Description"/>
```

```
        <Input value="{/description}"/>
        <Label text="Release Year"/>
        <Input value="{/releaseYear}"/>
        <Label text="Price"/>
        <Input value="{/price}" description="€"/>
    </form:SimpleForm>
</core:FragmentDefinition>
```

Listing 12.3 Defining Our Second Fragment

Loading Fragments Statically

If we want now to display one of our fragments in our view, we can use the `sap.ui.core.Fragment` class:

```
<core:Fragment fragmentName="<you app id>.view.DisplayBook"/>
```

This instantiates your fragment directly in your view by loading it and inserting the contents of your fragment, meaning your simple form, into your fragment, producing the result shown in Figure 12.1.

Title: SAPUI5
Description: A Comprehensive Guide
Release Year: 2025
Price: 89.99 €

Figure 12.1 The Directly Inserted Fragment

As the fragment is now a part of your view, it's dependent on the view, meaning it also has access to the view's bindings. Therefore, the access to the JSON model in our fragment works. It also inherits the ID of our view, meaning to access the simple form from your controller, you're able to write the following:

```
this.getView().byId("displayForm")
```

It's also possible to add a dedicated ID to the fragment, as follows:

```
<core:Fragment id="displayFragment" fragmentName="<your app
id>.view.DisplayBook"/>
```

Accessing the content like before is no longer possible, as the content now gets the IDs from the fragment. If you want to access contents from your controller, however, import `sap.ui.coreFragment`, and write the following:

```
this.getView().byId(Fragment.createId("displayFragment", "displayForm"))
```

Here, you first create an ID with `Fragment.createId` that combines the ID of the fragment with the ID of the control, which you then pass to your view's `byId` method.

Loading Fragments Dynamically

Let's now load our fragments dynamically, depending on whether we want to display the simple form with text or input controls. We first start by adding a button into a toolbar in our display fragment. With this button, we can switch to the edit fragment as shown in Listing 12.4.

```
<form:SimpleForm id="displayForm">
        <form:toolbar>
            <Toolbar >
                <Title text="Display"/>
                <ToolbarSpacer />
                <Button icon="sap-icon://edit" press="onEditPress"/>
            </Toolbar>
        </form:toolbar>
        ...
    </form:SimpleForm>
```


Listing 12.4 Adding a Button to the Display SimpleForm

Then, we add a button to the edit simple form so we can switch back to the display fragment, as shown in Listing 12.5.

```
<form:SimpleForm id="editForm" >
        <form:toolbar>
            <Toolbar >
                <Title text="Edit"/>
                <ToolbarSpacer />
                <Button icon="sap-icon://save" press="onSavePress"/>
            </Toolbar>
        </form:toolbar>
        ...
    </form:SimpleForm>
```

Listing 12.5 Adding a Button to the Edit SimpleForm

Both of these buttons need an event handler in our controller, where we load the other fragment and insert it into our view instead of the current one displayed. We start with the onEditPress event handler by adding it into our view's controller, as shown in Listing 12.6.

```
private editFragment: Promise<any>;
public onEditPress(){
        if(!this.editFragment){
            this.editFragment = this.loadFragment({
                name: "<your app id>.view.EditBook"
            });
```

```
        }

        this.editFragment.then((fragmentContent: SimpleForm)=>{
            const page = (this.getView()?.byId("myPage") as Page);
            page.removeAllContent();
            page.addContent(fragmentContent);

        })
```

Listing 12.6 Adding the onEditPress Event Handler

Here, we first define a class property called `editFragment`, where we can store a reference to the `EditBook` fragment, as we only need to load a fragment once. In the `onEditPress` method, we check if this property hasn't been initialized yet. If so, we use the `loadFragment` method, provided by the `sap.ui.core.mvc.Controller` instance, which is the base instance of our controller. This method expects the name of the fragment and returns a promise, which is fulfilled when the fragment's content was loaded. The ID of the view is also passed down to the fragment's content when using this method. As this is only available since SAPUI5 1.93, the old and still supported way is by using the `sap.ui.core.Fragment.load` method. It works similar by expecting a `name` and needing a `controller` reference and an optional `ID`.

The same logic using `Fragment.load` is as follows:

```
this.editFragment = Fragment.load({
                name: "<your app id>.view.EditBook",
                controller: this,
                id: this.getView()?.getId()
            })
```

When the promise for loading the fragment is fulfilled, we get the fragment's content in the `then` method of the `editFragment` promise. In this case, we're going to add the content to a `sap.m.Page` that is the root control of the view.

Now if you click on the button when displaying the display fragment, you can switch to the edit fragment, as shown in Figure 12.2.

Now we only need to add the logic for the button in the edit fragment to our controller, as shown in Listing 12.7.

```
private displayFragment: Promise<any>;
public onSavePress(){
        if(!this.displayFragment){
            this.displayFragment = this.loadFragment({
                name: "at.clouddna.customcontrol.view.DisplayBook"
            });
        }
```

```
        this.displayFragment.then((fragmentContent: SimpleForm)=>{
            const page = (this.getView()?.byId("myPage") as Page);
            page.removeAllContent();
            page.addContent(fragmentContent);
        })
    }
```

Listing 12.7 Adding the onSavePress Event Handler

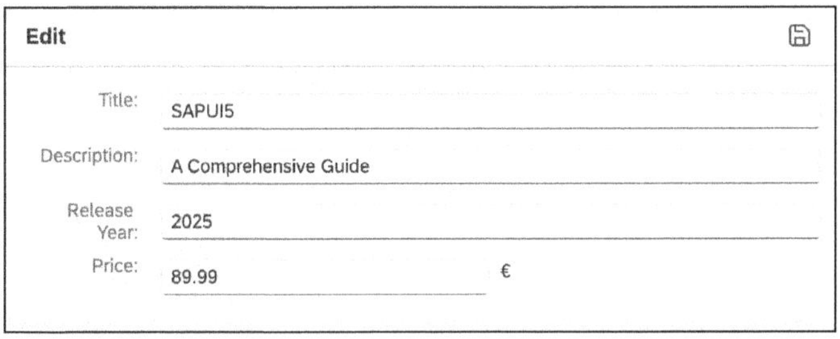

Figure 12.2 The New Fragment Inserted into the View Dynamically

Now we can freely switch between our fragments and only need to load them once when they are needed.

This was just a quick and easy example. We advise using fragments whenever you need to display the same content more than once and especially when you need to display parts of your UI only dynamically. We'll cover this in Section 12.2.

Using Dialogs

In SAPUI5, a *dialog* is a modal window that overlays the main application screen, interrupting the user's current flow to present important information or request an action. Dialogs are modal, meaning they block interaction with the underlying application until the user closes them. This ensures user focus on the dialog's content. You use them when you want to display important information, including messages or warnings that prompt the user to take an action such as confirming, displaying errors, or providing a centralized user input methodology. Dialogs consist of a title area, a content area, and a footer button area. These areas are implemented with the `sap.m.Dialog` control, which has the following important properties:

- **`draggable`**
  Defines if the dialog can be dragged across the screen.
- **`icon`**
  Defines the icon in the header of the dialog.
- **`resizeable`**
  Defines if the dialog can be resized by the user.

- **state**
  Defines the state of the dialog, that is, error, warning, and so on.
- **strech**
  Defines if the dialog should be sized to fill the available scree.
- **title**
  Defines the title of the dialog.

The dialog also provides a customHeader aggregation, in which you can add a toolbar to customize its header area. The content area of a dialog is set with the content aggregation, and buttons that should be displayed in the footer can be added with the buttons aggregation.

Dialogs can either be created in your controller or defined in fragments. When the fragment is loaded that contains your dialog definition, the dialog methods open and close are available to show and hide it, respectively.

We enhance our fragment sample from earlier to show a button that opens a dialog which allows us to buy a book. So, we first need a new fragment called *BuyBookDialog.fragment.xml* in our *view* folder. Here, we declare a fragment that includes a dialog as content, as shown in Listing 12.8.

```
<core:FragmentDefinition xmlns:form="sap.ui.layout.form"
    xmlns="sap.m"
    xmlns:core="sap.ui.core">
    <Dialog title="Buy Book" resizable="true" draggable="true" icon="sap-icon:/
/credit-card">
        <content>
            <form:SimpleForm>
            <Label text="Amount"/>
            <StepInput value="1"/>
        </form:SimpleForm>
        </content>
        <buttons>
            <Button text="Buy" type="Accept" press="onCloseDialogPress"/>
            <Button text="Cancel" type="Reject" press="onCloseDialogPress"/>
        </buttons>
    </Dialog>
</core:FragmentDefinition>
```

Listing 12.8 Defining a New Dialog

This dialog has a title, is resizeable and draggable, and has an icon on top. The content of the dialog is a SimpleForm that holds a StepInput to define how many books you want to purchase. As finalizing buttons, we declare one to buy and one to cancel.

To load the dialog, we need a button to do so. For this, we enhance our *DisplayBook.fragment.xml*, as shown in Listing 12.9.

```
<form:SimpleForm id="displayForm">
        <form:toolbar>
            <Toolbar >
                ...
                <Button icon="sap-icon://credit-card" press=
"onOpenDialogPress"/>
            </Toolbar>
        </form:toolbar>
...
    </form:SimpleForm>
```

Listing 12.9 Adding a Button for Opening a Dialog

12

Now we add the event handler for the `press` event of the button to our controller, as shown in Listing 12.10.

```
private dialogFragment: Promise<any>;

public onOpenDialogPress(){
        if(!this.dialogFragment){
            this.dialogFragment = this.loadFragment({
                name: "<your app id>.view.BuyBookDialog"
            });
        }

        this.dialogFragment.then((dialog: Dialog)=>{
            this.getView()?.addDependent(dialog);
            dialog.open();
        });
    }
```

Listing 12.10 Loading the Dialog

In this handler, we check if we already have loaded the dialog. If not, we use the `loadFragment` method of our controller. When it's loaded, we need to make it dependent on the view. By doing this, the dialog gets the model references of the view and is automatically closed when the view is removed by routing, for example. Always do this, because if you don't, you can't access the models except if you add them manually to your dialog with the `setModel` method.

Then, we call the `open` method of our dialog's instance. Now if you click on the button in the display fragment, a dialog pops up and looks like Figure 12.3.

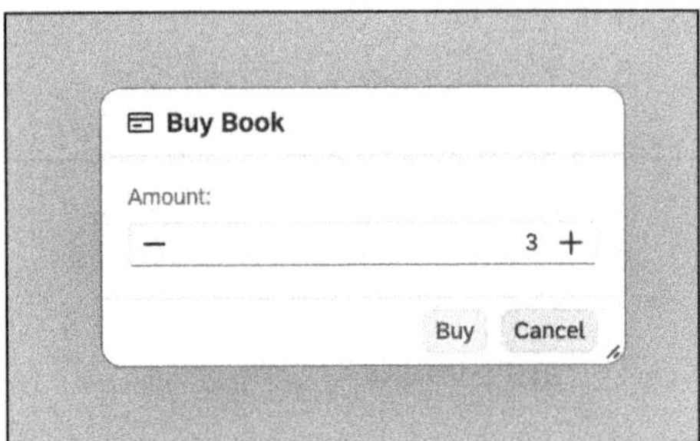

Figure 12.3 The Dialog Now Visible

The last thing now is to be able to close the dialog. We therefore add the event handlers for the two buttons in our dialog to our controller, as follows:

```
public onCloseDialogPress(){
        this.dialogFragment.then((dialog: Dialog)=>{
            dialog.close();
        });
    }
```

Here, we access our `dialog` and call the `close` method, which makes our dialog invisible and removes it. Try it out by clicking on the **Buy** button.

Using Message Boxes

If you want to use predefined dialogs for displaying contents for semantic meaning, such as a success notice, an error message, or a simple confirmation, you don't need to declare a dialog as you can use the `sap.m.MessageBox` control. It provides a standardized, SAP Fiori design guidelines–conforming way of displaying various types of messages. They can be informational messages (type `Information`), success messages (type `Success`), warning messages (type `Warning`), error messages (type `Error`), or a confirmation prompt (type `Confirmation`).

Depending on what type of message box you want to show, you get one dedicated method provided by the `MessageBox` class that gets the to-be-displayed message and optional parameters. For designing your own message box, you can use the `show` method.

If, for example, we want to further enhance our previous sample, we can display a message box when the **Buy** button in our dialog is clicked to notify the user that a new order has been placed. To do so, we add parameters to our press event handlers in our *BuyBookDialog.fragment.xml*, as shown in Listing 12.11.

```
<core:FragmentDefinition xmlns:form="sap.ui.layout.form"
    xmlns="sap.m"
    xmlns:core="sap.ui.core">
```

```
    <Dialog title="Buy Book" resizable="true" draggable="true" icon="sap-icon:/
/credit-card">
        <content>
...
        </content>
        <buttons>
            <Button text="Buy" type="Accept" press="onCloseDialogPress($event,
'buy')"/>
            <Button text="Cancel" type="Reject" press=
"onCloseDialogPress($event, 'cancel')"/>
        </buttons>
    </Dialog>
</core:FragmentDefinition>
```

Listing 12.11 Adding Parameters to the Button Events

Event Handler in XML

In SAPUI5, you can pass optional parameters to event handlers when declaring them in your XML. The syntax looks as follows:

`event="eventhandler(<parameter 1>, ..., <parameter n>)"`

The following parameters can be added when addressing event handler methods:

- **`$event`**
  Provides the default event object of the control event.
- **`${<binding path>}`**
  Provides the value of a data binding.
- **`${$parameters>/<parameter name}`**
  Provides direct access to event parameters.
- **`${$source>/<property name>}`**
  Provides direct access to the control properties.
- **`$controller`**
  Provides the controller reference.
- **JavaScript datatypes**
  Use JavaScript values to provide static values

In this case, we add the default event parameter in the first place, and then the string `'buy'` for the first and `'cancel'` for the second button. This allows us to have one event handler that behaves differently based on the parameters we're giving. So, we enhance our event handler in our controller to display a message box whenever we click on **Buy**, as shown in Listing 12.12.

```
public onCloseDialogPress(event: Event, action: string){
        this.dialogFragment.then((dialog: Dialog)=>{
            dialog.close();

            if(action == 'buy'){
                MessageBox.success("Your Order was placed");
            }
        });
    }
```

Listing 12.12 Opening a MessageBox When Buying a Book

Here, we check if the second parameter has the value `'buy'` after we close our dialog. If so, we use the `MessageBox.success` method (don't forget to import `sap.m.MessageBox`) to display a success message that now looks like Figure 12.4 if you click on the **Buy** button.

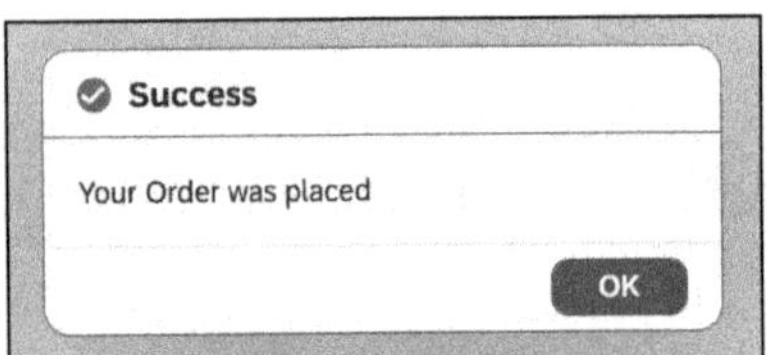

Figure 12.4 A New Message Box Appears

12.1.2 Base Controller

A *base controller* is a controller that your application's controllers can extend to provide more functionality. Methods or properties that are regularly used in your controllers can be added to be defined at a single place, making your application correspond to the DRY concept. You don't want to repeat yourself, writing the same logic repeatedly. This makes it easier to maintain your source code and keeps your coding clean. From practical experience, you write your base controller once and copy it into your new apps, further enhancing it. Often, it's also made generally available at a Git repository to keep track of enhancements and to be able to share your base controller with colleagues.

To create a base controller, you need to add a new TypeScript file to your app. So, let's add a file called *BaseController.ts* to one of our app's *controller* folders, as shown in Listing 12.13.

```
import Controller from "sap/ui/core/mvc/Controller";

/**
 * @namespace <your app id>.controller
 */
```

```
export default class BaseController extends Controller {

}
```

Listing 12.13 Defining a New Base Controller File

We add the module definition of a new controller instance here. To make use of our base controller, we need to extend our existing controllers, as shown in Listing 12.14.

```
import BaseController from "./BaseController";

/**
 * @namespace <your app id>.controller
 */
export default class App extends BaseController {

    public onInit(): void {
      //calling methods from Base Controller instead of redefining the
     //same methods over and over again
    }
}
```

Listing 12.14 Extending from the Base Controller

Then, we can add methods that we regularly use in our applications. We've provided a few examples here.

Let's start with model functions. If you want to access a model in your controller, you always write `this.getView().getModel()`. To make it simpler to access a model, we can add two methods to our base controller, as shown in Listing 12.15.

```
    public getModel(modelName?: string){
        return this.getView()?.getModel(modelName);
    }

    public setModel(model: Model, modelName?: string){
        return this.getView()?.setModel(model, modelName);
    }
```

Listing 12.15 Adding Model Access Methods

These methods allow us to access models in controllers that extend our base controller with just `this.getModel()`.

Another concept that can be moved into the base controller is the setting of the content density. The *content density* is a concept that renders SAPUI5 controls with larger margins and paddings to be optimized for touch devices. We can add two methods to our base controller that allow us to set the content density of our views in our controllers with just one method, as shown in Listing 12.16.

```
private contentDensityClass: string;

    public setContentDensity() {
        this.getView()?.addStyleClass(this.getContentDensityClass());
    }

    private getContentDensityClass(): string {
        if (!this.contentDensityClass) {
            if (!sap.ui.Device.support.touch) {
                this.contentDensityClass = "sapUiSizeCompact";
            } else {
                this.contentDensityClass = "sapUiSizeCozy";
            }
        }
        return this.contentDensityClass;
    }
```

Listing 12.16 Adding Content Density Methods to the Base Controller

We add two methods. The first one, `getContentDensityClass` checks if our device supports touch using the SAPUI5 Device API. If so, it stores the cascading style sheets (CSS) class `sapUiSizeCozy` in a class variable called `contentDensityClass`; otherwise, it stores the CSS class `sapUiSizeCompact`. It then returns the stored CSS class. In the second method, `setContentDensity`, we retrieve this value and set it as a new style class to our view. We can call this method in our controller's `onInit` or `onBeforeRendering` methods to optimize our UI for touch input.

Another handy function is to automatically set the content density whenever you call the router of your application. As getting the router instance is almost done every time in the `onInit` method of your controllers to attach yourself to route pattern matched events, it makes sense to combine these two methods. So, we add a little function to do so in our base controller, as follows:

```
public getRouter () {
        this.setContentDensity();
        return (this.getOwnerComponent() as UIComponent).getRouter();
    }
```

This method first sets the correct content density and then returns our router instance.

Speaking of routing, a nice-to-have method that is especially useful for detail views is to be able to easily route back to the last view. Let's add a method that does this for us, as shown in Listing 12.17.

```
public onNavBack() {
        const previousHash = History.getInstance().getPreviousHash();

        if (previousHash !== undefined) {
            window.history.go(-1);
        } else {
            this.getRouter().navTo("<Default Routing Target>", {});
        }
    }
```

Listing 12.17 Adding Routing Back Functionalities to the Base Controller

With this method, we first check our browser tab's history by using the `sap.ui.core.routing.History` class. If we have a previous routing hash, we simply go one step back in the history by using the JavaScript browser object model. If we haven't navigated yet, for example, we opened a detail view from a browser bookmark, we navigate to our app's default routing target. This method can directly be referenced in a `sap.m.Page` `navButtonPress` event.

12

The last convenience thing we're going to add to our base controller is a centralized logic to handle logging by using the SAPUI5 Logging API, as shown in Listing 12.18.

```
   public logDebug(message: string) {
        const logger = (Log.getLogger(this.getView()?.getControllerName() ||
"your app id") as Log);
        logger.debug(message);
    }

    public logError(message: string) {
        const logger = (Log.getLogger(this.getView()?.getControllerName() ||
"your app id") as Log);
        logger.error(message);
    }

    public logFatal(message: string) {
        const logger = (Log.getLogger(this.getView()?.getControllerName() ||
"your app id") as Log);
        logger.fatal(message);
    }

    public logInfo(message: string) {
        const logger = (Log.getLogger(this.getView()?.getControllerName() ||
"your app id") as Log);
        logger.info(message);
    }
```

```
    public logTrace(message: string) {
        const logger = (Log.getLogger(this.getView()?.getControllerName() ||
"your app id") as Log);
        logger.trace(message);
    }

    public logWarning(message: string) {
        const logger = (Log.getLogger(this.getView()?.getControllerName() ||
"your app id") as Log);
        logger.warning(message);
    }
```

Listing 12.18 Adding Logging Methods to the Base Controller

With these methods, we fully use all logging levels of the `sap.base.Log` class to produce the different kinds of console log messages. We also append our controller's name to the log message to make it easier to identify where this log message was made. If you call `this.logError("I'm an error")`, for example, in one of your extending controllers, you get a nice message in your browsers console, as shown in Figure 12.5.

```
▸ 2025-01-03 16:17:08.245099 I'm an error - at.clouddna.customcontrol.controller.App    Log-dbg.js:497
```

Figure 12.5 Your Own Log Message

Now we have a few nice-to-have methods in our base controller that might make your life programming SAPUI5 a little bit less depraved—trust us, we speak from experience. Feel free to make your base controller truly yours, as it might be the most capable tool you get. A well-curated base controller is your bread and butter. Here are some more ideas on what you could add to it:

- **Message handling**
  You can add methods to work with the SAPUI5 message model here. Add methods for adding new messages, removing duplicates, registering message throwing objects, and so on.
- **OData methods**
  If you always want the same flow when working with OData create, read, update, and delete (CRUD) methods such as logging, caching, and general error handling, you can do so by building your own OData model access API. You can also add a facade to transform the callbacks of the OData V2 model to promises.
- **i18n texts**
  You can add methods to retrieve your localized texts so you don't always need to get the i18n resource model beforehand. We'll discuss this more next.

12.2 Localization

Localization or internationalization (i18n) is the SAPUI5 concept of providing translatable texts for your UI. It uses the `sap.ui.model.resource.ResourceModel` special model to manage your texts and display them using data binding.

The translatable texts are stored in property files in the *i18n* folder of your application. Each file must end with the *.property* filetype and contains key-value pairs. The key is the identifier, and the value is the text in a specific language. In addition to this, each file must contain a locale suffix such as *_de* for German, *_en* for English, and so on. One of these language-dependent property files must contain no suffix. This will be used as the default language file of your application. Therefore, the following naming convention must be used: *i18n_<language key>.property*.

Language Suffixes

The language suffixes that can be used in SAPUI5 are the ISO 639-1 language codes. They offer a two-letter abbreviation for each language, as follows:

- *en*: English
- *de*: German
- *nl*: Dutch
- *hi*: Hindi
- *fr*: French

For SAPUI5 to determine your language, the property files must be named according to this suffix. If you want to add country-specific text, you can add the ISO country code to the suffix such as the following:

- *en_US*: American English
- *en_GB*: British English
- *de_DE*: German
- *de_AT*: Austro-German

So, to add localization support to an application, we need to do a few steps to fully use this technique in SAPUI5:

1. **Setup**
   Add translatable texts, and add a resource model.
2. **Usage in XML**
   Use these texts in our XML views.
3. **Usage in controller**
   Access the texts in JavaScript or TypeScript.
4. **Placeholders**
   Add dynamic texts.
5. **Property metadata binding**
   Access texts from OData metadata.

12.2.1 Setup

We can demonstrate the localization integration by adding two files to the i18n folder into one of our applications. The first one will serve as our main language (English), and the second one should hold the German translation. First add a new file called *i18n.properties* to the *i18n* folder. Here, we add one key-value pair of a text that we want to translate:

```
sample.text=I'm a translateable text
```

Then, we add a second file called *i18n_de.properties* to the *i18n* folder, where we translate this text to German:

```
sample.text=Ich bin ein übersetzter Text
```

As you can see, the key stays the same, only the text changes.

Now you need the resource model to bundle these files into resource bundles. They are the collection of resource files, specifically our property files from the *i18n* folder. Resource models can either be created in the *manifest.json* file or programmatically in your controllers. By default, one resource model is always automatically added to the *manifest.json* that is defined, as shown in Listing 12.19.

```
"models": {
      "i18n": {
        "type": "sap.ui.model.resource.ResourceModel",
        "settings": {
          "bundleName": "<your app id>.i18n.i18n"
        }
      },
      ...
    },
```

Listing 12.19 The Default ResourceModel

This adds a new named global model called `i18n` to your application, which includes a reference to your i18n property files.

For SAPUI5 to determine which language should now be applied and which property file it needs to use, the following logic for determination is included:

1. Hard-coded SAPUI5 default language English
2. Browser-derived language
3. URL parameter `sap-ui-language`
4. URL parameter `sap-language`
5. URL parameter `sap-locale`
6. Locale configured by coding

The last available of these steps is then used to declare the application's language. To add configuration on which languages your application supports and which language should be used as a fallback, two configurations can be added to the settings property of the i18n model definition in *manifest.json*:

- supportedLocales: ['<locale>',…]
- fallbackLocale: '<locale>'

If you've added i18n property files and a model that references these files, the application now on startup determines the language and tries to load the i18n file with the locale and country as a suffix. If it isn't found, the application searches just for the locale suffix; if this isn't found, the application uses English or the default i18n file as language, as follows:

```
de_AT -> de -> en -> ''
```


12.2.2 Usage in XML

Now you can use the previously declared text and bind its key to a control in your view as follows:

```
<Text text="{i18n>sample.text}"/>
```

By using property binding, you can access the keys of the currently used property file of the i18n resource model and access its value. Now your application displays the translated text dependent on your locale.

12.2.3 Usage in a Controller

To load a language-dependent text in your controller, for example, to display a translated error message, you need to access your resource model and load the text, as follows:

```
const resourceModel = (this.getView()?.getModel("i18n") as ResourceModel),
            resourceBundle = resourceModel.getResourceBundle() as
ResourceBundle;
let text = resourceBundle.getText("sample.text");
```

12.2.4 Placeholders

If you want to make your texts somewhat dynamic, you can add placeholders in your property files that can be filled dynamically added. They are added to the text as {index}. So, if you have, for example, a message text that you want to display that includes a message number, you could add an i18n property such as the following:

```
sample.message=Hello {0}, how are you doing?
```

Here, a placeholder with the index 0 was added. This placeholder can now be filled to dynamically add a certain value instead.

To add a value in XML, you can use the sap.base.strings.formatMessage module to format texts and to insert values into the placeholder, as shown in Listing 12.20.

```
<Text core:require="{formatMessage: 'sap/base/strings/formatMessage'}"
            text="{
              parts:[{path: 'i18n>sample.message'}, {path: '<path to the
value to be used instead of the placeholder>'}],
              formatter: 'formatMessage'
           }"/>
```

Listing 12.20 Formatting i18n Texts in XML

Loading Additional Resources

With the core:require attribute, you can load additional modules when needed in your control's definition in XML. This is often done to reference formatters that are outsourced into dedicated modules.

Additionally, with the dependents aggregation of the sap.ui.core.Control, you can add additional controls to your control. These dependent controls aren't rendered but are loaded and receive the binding context of their parent. This can be used to load fragments such as dialogs or popovers. By giving them an ID, you can then access them in your controller and don't need to load them beforehand.

If you want to do so in the controller, you use the getText method of the resourceBundle, as follows:

```
const resourceModel = (this.getView()?.getModel("i18n") as ResourceModel),
            resourceBundle = resourceModel.getResourceBundle() as
ResourceBundle;
resourceBundle.getText("sample.message", [<your values>]);
```

12.2.5 Property Metadata Binding

If you use an OData service, you can reuse the labels that are added to the entity's properties. This is useful if you build your UI and want to be coherent through multiple applications that use the same service or reuse the data dictionary labels. To do so, you bind the label attribute of an OData entity property in your XML, as follows:

```
<Text text="{/#EntityTypeName/PropertyName/@sap:label}"/>
```

Finally, we have some tips for you:

- Always do internationalization in your default language from the start. This makes it easier later for identifying what you still need to translate.
- Always use internationalization. Don't write clear text that isn't translated.
- Do the other languages at the end of the development lifecycle. During development, texts often change. So don't translate your default language in other languages if the application isn't finished. This makes it hard to keep track between your property files.
- Try to use metadata binding for labels that should be abbreviated from an OData service. If the service changes, you don't need to do this in your i18n files.

12.3 File Upload

In SAPUI5, you can upload *media entities* such as documents, pictures, multipart form data items, or any type of file with the help of OData media entities and dedicated SAPUI5 classes and controls. To understand this, we first need to talk about how media entities in OData work. Then, we take a look at how we can implement this in the backend (yes, we need to take a trip onto backend development, but you can skip this if you're already familiar with OData media entities). Then, we can finally implement the necessary functions in SAPUI5.

12.3.1 Media Entities

An OData media entity is a special entity in the data model of an OData service that offers stream methods that must either be implemented manually (in case of Transaction SEGW ABAP-based OData services) or generically provided by a framework (in case of ABAP RESTful application programming model–based services). Such an entity is flagged in the metadata document of an OData service as streamable (attribute `hasStream` is set) and must offer at least the following properties (the naming isn't important):

- `Content`
  The content of the file gets sent, often stored as a raw string on the database level.
- `MimeType`
  Using the MIME type, the content type of the file will be persisted.
- `FileName`
  The file name also must be persisted.

The media content of such media entities can be accessed in OData via the property `$value`, which is automatically added to your entity as in `/MyEntity(keys)/$value`. This property also allows uploading files to this entity.

MIME Types

The internet media type, that is, MIME type, identifies the content type of data on the internet. When using media entities, this information is sent with all requests and must be persistent on a database level. A MIME type consists of a type and a subtype. A type can be an application (for binary data), text (for texts), or image (for pictures). Each type adds subtypes to provide more information, such as `application/pdf` or `text/xml`.

12.3.2 Media Handling in OData V2

In this section, we'll look at how a file upload with media entities can be implemented in OData V2. Here, we create an OData service of the version in an SAP Gateway Service Builder project via the following four steps:

1. Perform the necessary on-premise groundwork.
2. Create an SAP Gateway Service Builder project, and implement the necessary functions in the OData service.
3. Upload file(s).
4. Download file(s).

On-Premise Groundwork

In this step, we'll look at which prerequisites are necessary for the later implementation of the OData service. First, the corresponding structures and database tables must be created, if they don't already exist.

We'll create a database table for this with two important fields: `content` and `mime_type`. The MIME type of the uploaded file as well as its content are stored in these fields. In addition to these fields, the table can also contain other fields.

SAP Data Archiving

Data that takes up a lot of space, such as files or images, shouldn't be stored directly in database tables. They should be moved to *SAP Data Archiving*. This enables storing of data in separate, external locations that are more fitted for storing files. By using this, you improve performance and reduce storage costs.

In a media entity scenario, you can store the information of the file in your database table and the contents via SAP Data Archiving.

To create the table, follow these steps:

1. Open a package in Transaction SE80.

2. To select a package (in our case, **$TMP**), right-click and choose **Create • Dictionary Object • Database Table**.
3. Enter the name of the database table in the dialog that appears, and confirm the input (see Figure 12.6 and Figure 12.7).

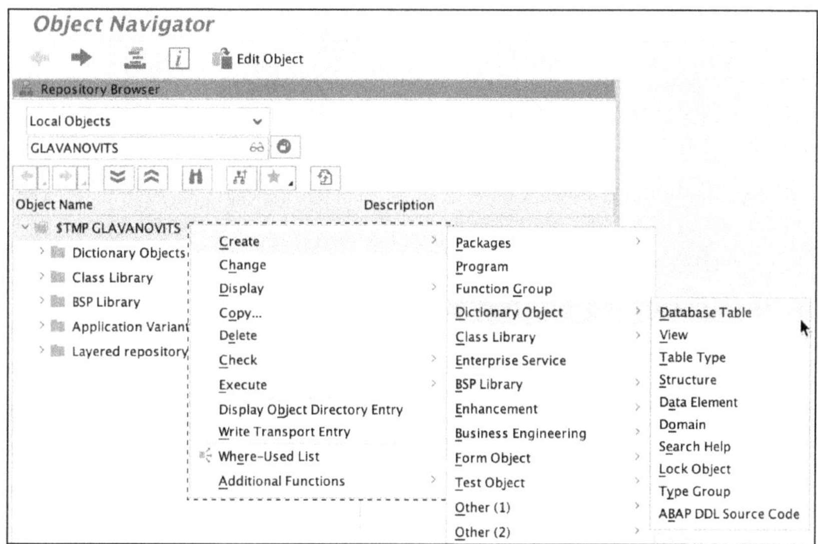

Figure 12.6 Open the Database Table Creation Wizard

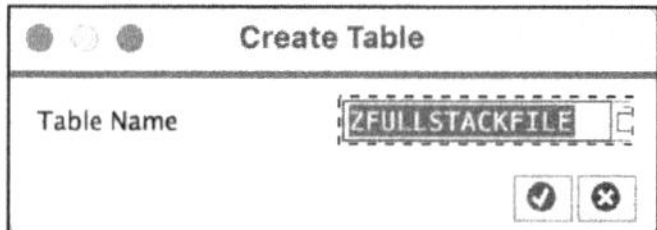

Figure 12.7 Enter the Name of the Database Table

4. Enter a **Short Description**, select "A" as the **Delivery Class**, and set the edit mode via the **Data Browser/Table View Editing** dropdown to **Display/Maintenance Allowed** (see Figure 12.8). Save the database table.

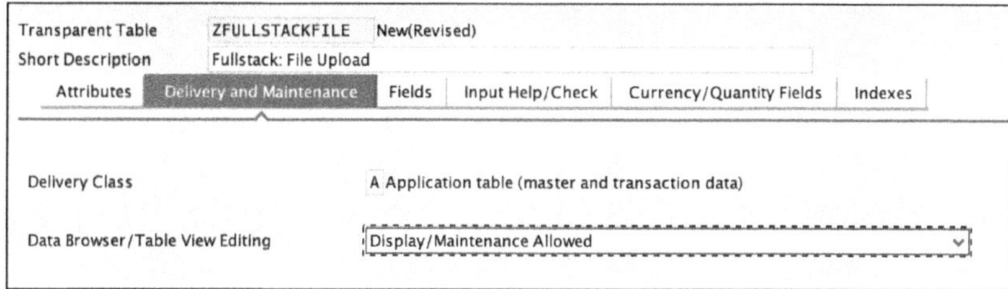

Figure 12.8 Specify the Delivery Class and the Table View Editing Mode

5. Now, the fields can be added to the database table, as shown in Figure 12.9.

Transparent Table ZFULLSTACK_FILE Active
Short Description Fullstack: File Upload
Attributes | Delivery and Maintenance | Fields | Input Help/Check | Currency/Quantity Fields | Indexes

Srch Help | Built-In Type | 1 / 5

Field	Key	Initi...	Data element	Data Type	Length	Decim...	Coordinate	Short Description
MANDT	✓	✓	MANDT	CLNT	3	0	0	Client
ID	✓	✓	SYSUUID_C	CHAR	32	0	0	UUID in character form
FILE_NAME	☐	☐	CHAR20	CHAR	20	0	0	Char 20
MIME_TYPE	☐	☐	SAEMIME	CHAR	128	0	0	MIME Type
CONTENT	☐	☐	XSTRINGVAL	RAWSTRING	0	0	0	XString

Figure 12.9 Create the Database Table

Once this is done, a structure must be created. This is basically a one-to-one copy of the database table. To create a structure, the procedure is similar to the creation of a database table. Follow these steps:

1. Open Transaction SE80 again, and navigate to the package where the structure should be created.
2. Right-click on the package, and select **Create • Dictionary Object • Structure**.
3. When the **Create Structure** dialog opens, enter the name of the structure (see Figure 12.10).

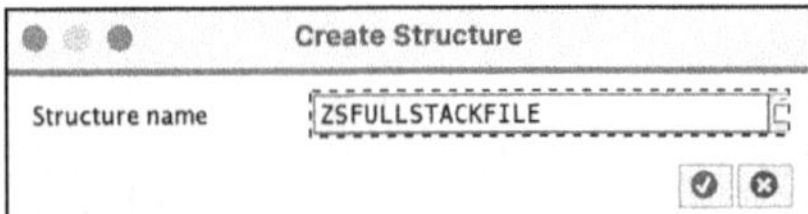

Figure 12.10 Define the Name of Your Structure

4. Enter a **Short Description**, and define the fields of this structure, as shown in Figure 12.11.

Structure ZSFULLSTACK_FILE Active(Revised)
Short Description Structure for Fileupload
Attributes | Components | Input Help/Check | Currency/quantity fields

Built-In Type | 1 / 5

Component	Typing Method	Component Type	Data Type	Length	Decim...	Coordinate	Short Description
MANDT	Types	MANDT	CLNT	3	0	0	Client
ID	Types	SYSUUID_C	CHAR	32	0	0	UUID in character form
FILE_NAME	Types	CHAR20	CHAR	20	0	0	Char 20
MIME_TYPE	Types	SAEMIME	CHAR	128	0	0	MIME Type
CONTENT	Types	XSTRINGVAL	RAWSTRING	0	0	0	XString

Figure 12.11 Create the Structure

Defining the OData Service

After the necessary structure and database table have been created in the previous steps, it's now a matter of creating an SAP Gateway Service Builder project and implementing the necessary functions in the OData service. To do this, follow these steps:

1. Go to Transaction SEGW, and select **Create Project** to create a new project.
2. Enter the following information about the project in the dialog (see Figure 12.12):
   - **Project**: "Z_FULLSTACK_UPLOAD"
   - **Description**: "Fullstack: File Upload"
   - **Project Type: Service with SAP Annotations**
   - **Generation Strategy: Standard**
   - **Package**: "$TMP"
   - When finished, click **Continue**.

12

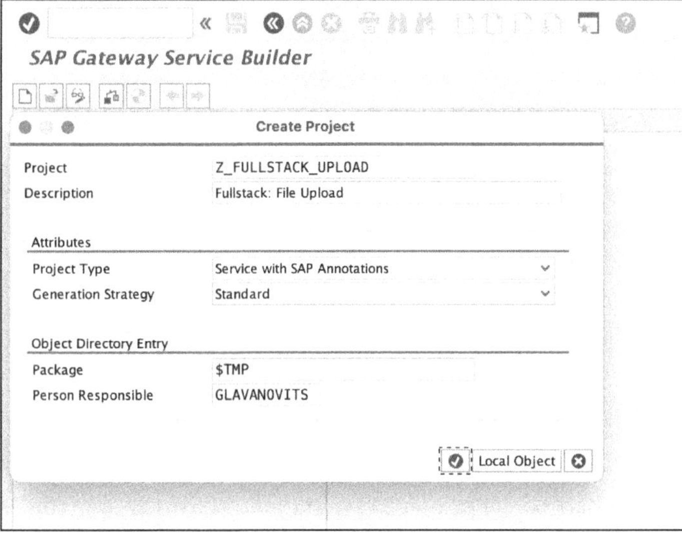

Figure 12.12 Create the SAP Gateway Service Builder Project

3. Now that the SAP Gateway Service Builder project has been created, an entity set must be created. To do this, right-click on **Data Model**, and select **Import • DDIC Structure** in the context menu.
4. In the following dialog, enter a **Name** for the entity type ("File"), and choose a structure in the **ABAP Structure** field (in this case, "ZSFULLSTACK_FILE"), as shown in Figure 12.13. Continue with **Next**.
5. Specify which properties of the structure are to be included in the entity types. In this case, select all available properties, as shown in Figure 12.14, and click **Next**.

Wizard Step 1 of 3: Import from DDIC Structure

Create an Entity Type or Complex Type

Name: File

(•) Entity Type ( ) Complex Type

Import from ABAP Structure

ABAP Structure: ZSFULLSTACK_FILE

[x] Create Default Entity Set

Back | Next | Cancel

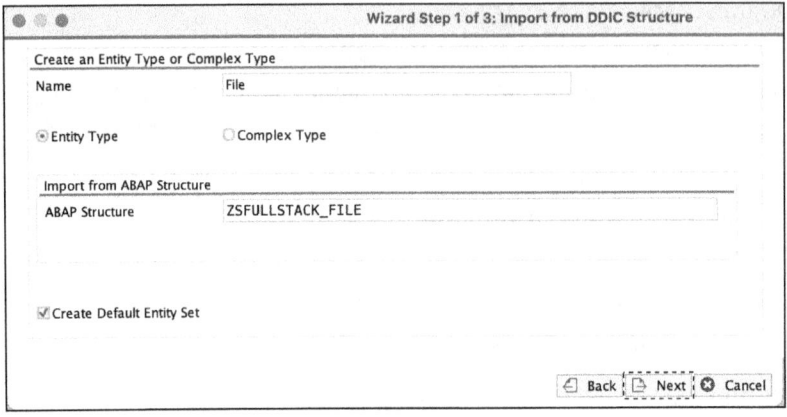

Figure 12.13 Import the Structure as an Entity Type

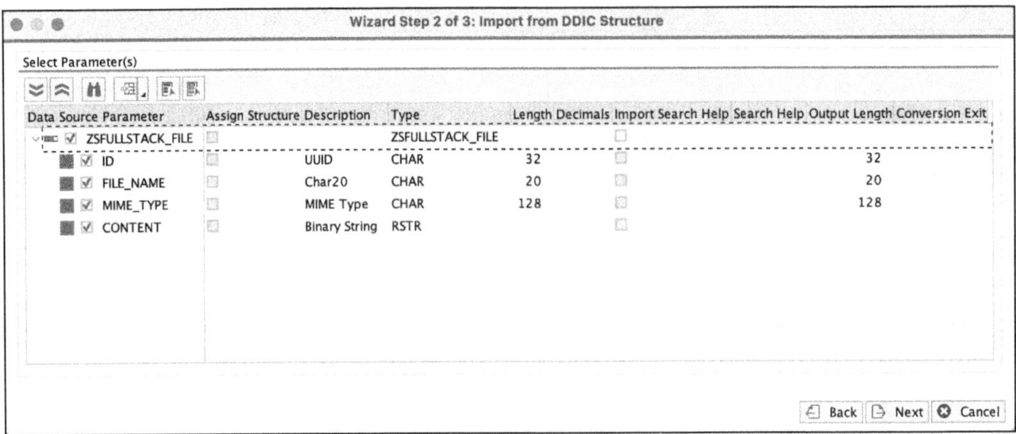

Figure 12.14 Add the Structures Fields to the Entity Type

6. Determine which fields are part of the primary key. In our case, this is only the **ID** field (see Figure 12.15). Click **Finish** to complete the import.

Wizard Step 3 of 3: Import from DDIC Structure

Modify Entity Type

IsEnt.	Complex/Entity Type Name	ABAP Name	Is K..	Type	Name	Label
[x]	File	ID	[x]	CHAR	Id	UUID
[x]	File	FILE_NAME	[ ]	CHAR	FileName	Char20
[x]	File	MIME_TYPE	[ ]	CHAR	MimeType	MIME Type
[x]	File	CONTENT	[ ]	RSTR	Content	Binary String

Back | Finish | Cancel

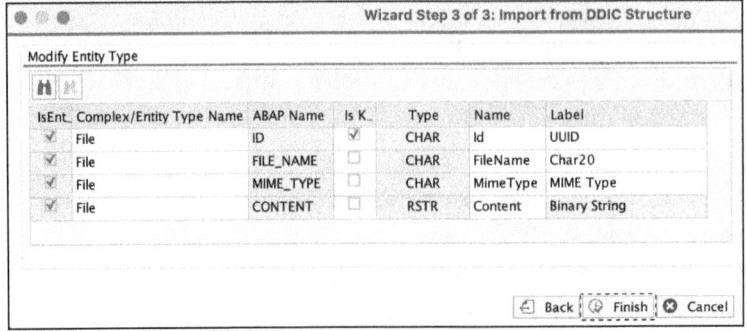

Figure 12.15 Set the Primary Key for Your Entity

Now, both the entity type and the entity set are created. By default, entity types aren't created as media entities, but because this is necessary for our example, this must be set subsequently. To do this, click the entity type, and select the **Media** checkbox (see Figure 12.16). Then, click **Save** and **Generate**.

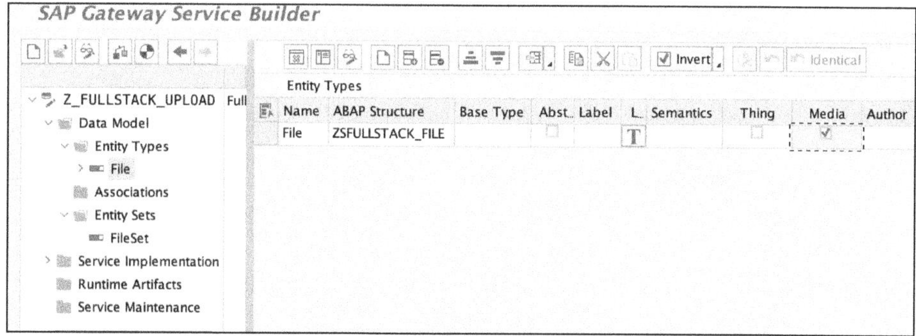

Figure 12.16 Set Your Entity as a Media Entity

After clicking on **Generate**, a dialog like the one shown in Figure 12.17 appears. Here, leave the prefilled values as they are, and click **Continue**. Next, make a package assignment using **$TMP** again. Now, we switch back to Transaction SE80 to implement the generated class accordingly.

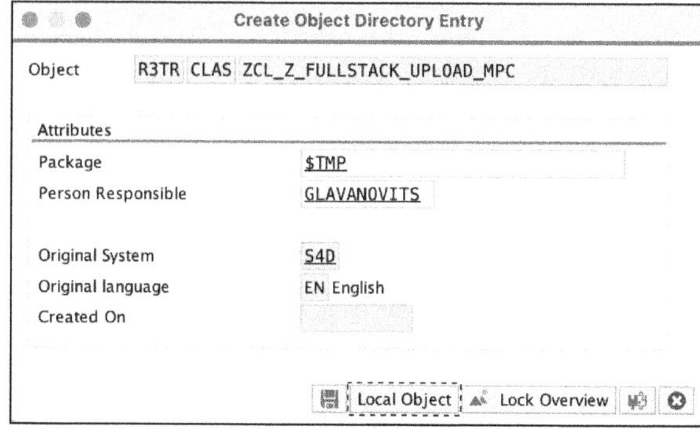

Figure 12.17 Assign the Class to Your Package

First, you must redefine the `DEFINE` method in the model provider class (MPC), which is `MPC_EXT`, for media entities. For more information on this, refer to Chapter 9. This is where you define which entity is a media entity. There are two required methods for this:

- `set_as_content_type`
  Defines the field that holds the MIME type.
- `set_as_content_source`
  Defines the field that contains the content of the file.

In our case, this implementation looks like Listing 12.21.

```
METHOD define.
  DATA: lo_entity   TYPE REF TO /iwbep/if_mgw_odata_entity_typ.
  DATA: lo_property TYPE REF TO /iwbep/if_mgw_odata_property.

  super->define( ).

  lo_entity = model->get_entity_type( iv_entity_name = 'File' ).

  IF lo_entity IS BOUND.
    lo_property = lo_entity->get_property( iv_property_name = 'MimeType' ).
    lo_property->set_as_content_type( ).
    lo_property = lo_entity->get_property( iv_property_name = 'Content' ).
    lo_property->set_as_content_source( ).
  ENDIF.
ENDMETHOD.
```

Listing 12.21 Defining the DEFINE Method

Once this is done, the actual implementation of the upload and download functionality can be started. Again, two methods are needed for this, as follows:

- **GET_STREAM**
  This is used for downloading files.
- **CREATE_STREAM**
  This is used for uploading files.

For the implementation of the GET_STREAM method, a key is passed via which the corresponding data record is read from the table. After the record has been read from the table, the file content is returned in a stream. In addition, a file name can also be returned via the header. Listing 12.22 shows how this looks in our case.

```
METHOD /iwbep/if_mgw_appl_srv_runtime~get_stream.
  DATA: ls_key       TYPE /iwbep/s_mgw_name_value_pair,
  lv_id        TYPE sysuuid_c,
  ls_stream    TYPE ty_s_media_resource,
  ls_header    TYPE ihttpnvp,
  lv_filename TYPE string,
  ls_file      TYPE zsfullstack_file.
*   Read Key
  READ TABLE it_key_tab WITH KEY name = 'Id' INTO ls_key.
  lv_id = ls_key-value.
*   Read file from database
  SELECT SINGLE * FROM zfullstack_file INTO @ls_file WHERE id = @lv_id.
*   Assign values
```

```
    ls_stream-value = ls_file-content.
    ls_stream-mime_type = ls_file-mime_type.
*   Add Headers
    "ls_header-name = |content-disposition|.
    "ls_header-value = |inline; filename={ ls_file-file_name }|.
    "set_header( is_header = ls_header ).
    copy_data_to_ref( EXPORTING is_data = ls_stream CHANGING cr_data = er_stream
).
ENDMETHOD.
```

Listing 12.22 Implementing the GET_STREAM Method

Now that we've implemented the method for the download, the CREATE_STREAM method for the upload is still missing. This implementation looks like this: In addition to the actual file, we also receive the MIME type and the file name. The file name is sent in the slug header. If necessary, this could also be used to send additional information.

In our case, we create a new entry in our database table using the transferred data. It's kept very minimalistic and can still be extended by validations or error handling. Listing 12.23 shows what this implementation looks like.

```
METHOD /iwbep/if_mgw_appl_srv_runtime~create_stream.
  DATA: ls_file TYPE zsfullstack_file.

  ls_file-id = cl_system_uuid=>create_uuid_c32_static( ).
  ls_file-file_name = iv_slug.
  ls_file-content = is_media_resource-value.
  ls_file-mime_type = is_media_resource-mime_type.

  INSERT INTO zfullstack_file VALUES ls_file.

  copy_data_to_ref(
    EXPORTING
      is_data = ls_file
    CHANGING
      cr_data = er_entity ).
ENDMETHOD.
```

Listing 12.23 Implementing the CREATE_STREAM Method

Now that these steps have been completed, the frontend implementation can be started. Don't forget to add your OData service in Transaction /IWFND/MAINT_SERVICE to SAP Gateway.

12

Implementing the SAPUI5 Logic

Now that all the backend arrangements have been made, in this section, we'll look at implementing the file upload in a simple SAPUI5 application called `sap.m.UploadSet`. To do this, an SAPUI5 application must first be created. This can be done as desired in different integrated development environments (IDEs). In our case, we use SAP Business Application Studio. When creating the application, you only need to go through the wizard, and no special settings are required here.

If not specified directly when creating the project, the OData service must be integrated into the application. To do this, a data source and a model (see Listing 12.24 and Listing 12.25) must be defined in your application's *manifest.json* file.

```
"dataSources": {
   "mainService": {
      "uri": "/sap/opu/odata/sap/Z_FULLSTACK_UPLOAD_SRV/",
      "type": "OData",
      "settings": {
         "annotations": [],
         "localUri": "localService/metadata.xml",
         "odataVersion": "2.0"
      }
   }
}
```

Listing 12.24 manifest.json Data Source Definition

```
"models": {
   "": {
      "dataSource": "mainService",
      "preload": true,
      "settings": {}
   }
}
```

Listing 12.25 manifest.json Model Definition

We don't need much in our view; it's sufficient to just include an `UploadSet`. There are a few attributes that need to be set, as follows:

- **`items`**
  This is the aggregation that we must bind the `FileSet` to so that the files are later also visible in the `UploadSet`.
- **`uploadUrl`**
  Here, we must specify which entity set will be called up during the upload.

- `beforeUploadStarts`
  For this event, we have to store a function that sets the slug header in the request and sets the cross-site request forgery (CSRF) token in the request on the other. The latter is a security feature.

In addition, a template must be specified in the aggregation for the items. For now, we only set the file name and the MIME type here. Along with the properties just mentioned, other properties can of course also be specified. For our case, however, those mentioned are sufficient. Figure 12.18 shows what the implementation of the view looks like.

Upload Set with Table Plugin

Since SAPUI5 1.124, you can use the `sap.m.plugin.UploadSetwithTable` plugin for file uploading in combination with tables. When adding this plugin to the dependent aggregation of a table, you get a button to upload files similar to the `UploadSet` control.

On the view side, the implementation is complete. Now, we need to implement the corresponding logic in the controller via a function stored in the `beforeUploadStarts` event.

App.view.xml ×

webapp > view > App.view.xml

```
<mvc:View controllerName="at.clouddna.zfullstackfile.controller.App"
    xmlns:mvc="sap.ui.core.mvc" displayBlock="true"
    xmlns="sap.m" xmlns:upload="sap.m.upload">
    <App id="App">
        <Page id="page" title="{i18n>title}">
            <content>
                <upload:UploadSet
                    id="UploadSet" instantUpload="true" uploadEnabled="true"
                    items="{/FileSet}" beforeUploadStarts="onBeforeUploadStarts"
                    uploadUrl="/sap/opu/odata/sap/Z_FULLSTACK_UPLOAD_SRV/FileSet">
                    <upload:items>
                        <upload:UploadSetItem
                                fileName="{FileName}"
                                mediaType="{MimeType}">
                        </upload:UploadSetItem>
                    </upload:items>
                </upload:UploadSet>
            </content>
        </Page>
    </App>
</mvc:View>
```

Figure 12.18 Implementing the UploadSet in your View

As shown in Listing 12.26, nothing happens in this other than the file name gets passed in the slug header and a CSRF token gets passed in the header to the OData service. This token is necessary to execute the action successfully.

```
onBeforeUploadStarts: function (oEvent) {
  this.oFileUploader = this.getView().byId("UploadSet");
```

```
    let oHeaderSlug = new Item({
        key: "slug",
        text: oEvent.getParameter("item").getFileName()
    });
    let oCSRFHeader = new Item({
        key: "x-csrf-token",
        text: this.getView().getModel().getSecurityToken()
    });

    this.oFileUploader.removeAllHeaderFields();
    this.oFileUploader.addHeaderField(oHeaderSlug);
    this.oFileUploader.addHeaderField(oCSRFHeader);
}
```

Listing 12.26 Implementing the Event Handler for the beforeUploadStarts Event

Initially, the `UploadSet` looks like the one shown in Figure 12.19. To test the upload, click on the **Upload** button, select a file from the file explorer, and confirm the selection. The upload is then initiated (see Figure 12.20).

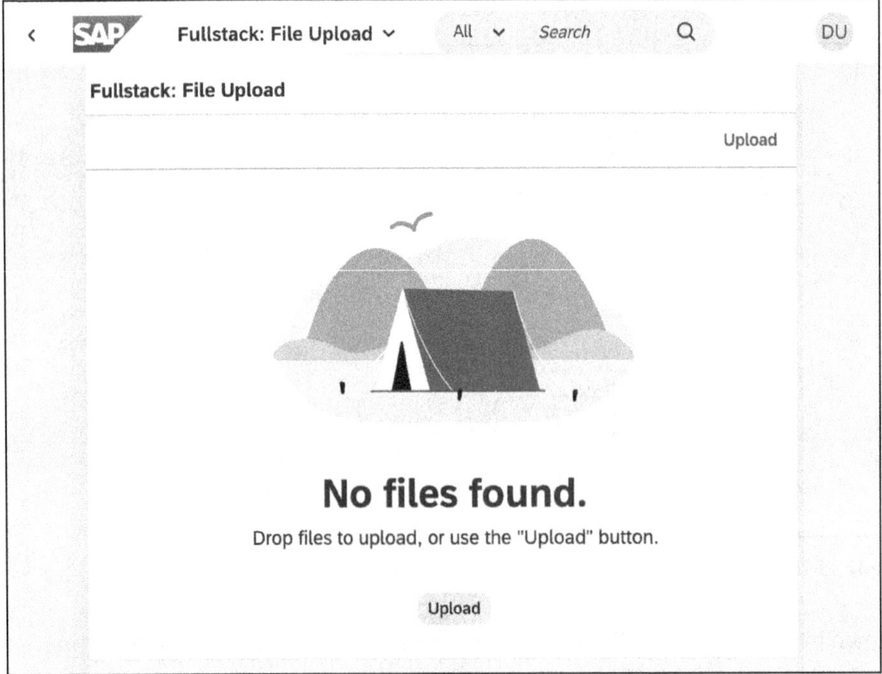

Figure 12.19 The Empty View of an UploadSet

After the upload has gone through successfully, we should see the recently uploaded file, as shown in Figure 12.21. As you can see, the file name has also been taken over correctly.

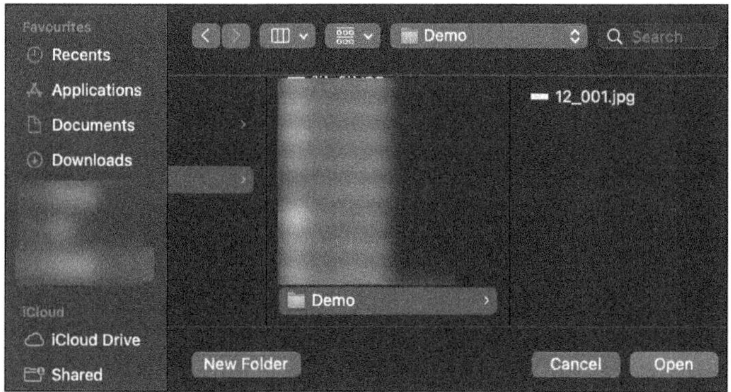

Figure 12.20 Selecting a File from the File Explorer

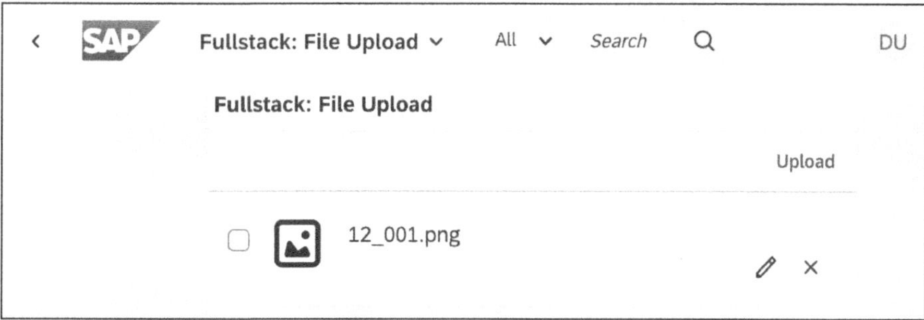

Figure 12.21 The Uploaded File in the UploadSet

Now that we've implemented the file upload, we still need the ability to download the uploaded files again. However, this is relatively easy to implement. We first define a method in our controller that receives the primary key, that is, the ID, of a document. Based on this, we create the corresponding URL and return it. This is needed because in the `UploadSet` control, there's an option for specifying a URL, which is then displayed as a link:

```
getUrl: function (id) {
  return `${this.getView().getModel().sServiceUrl}/FileSet(guid'${id}')/
$value`;
}
```

This completes the controller adjustments. Now this function only has to be called in the view. We enter this directly in the binding to the corresponding property as a formatter function (see Figure 12.22).

If we now switch back to the running application, we can initiate the download by clicking on the link (see Figure 12.23).

```
<upload:UploadSet
    id="UploadSet" instantUpload="true" uploadEnabled="true"
    items="{/FileSet}" beforeUploadStarts="onBeforeUploadStarts"
    uploadUrl="/sap/opu/odata/sap/Z_FULLSTACK_UPLOAD_SRV/FileSet">
    <upload:items>
        <upload:UploadSetItem
                fileName="{FileName}"
                mediaType="{MimeType}"
                url="{path: 'Id', formatter: '.getUrl'}">
        </upload:UploadSetItem>
    </upload:items>
</upload:UploadSet>
```

Figure 12.22 Adding the Formatter Function

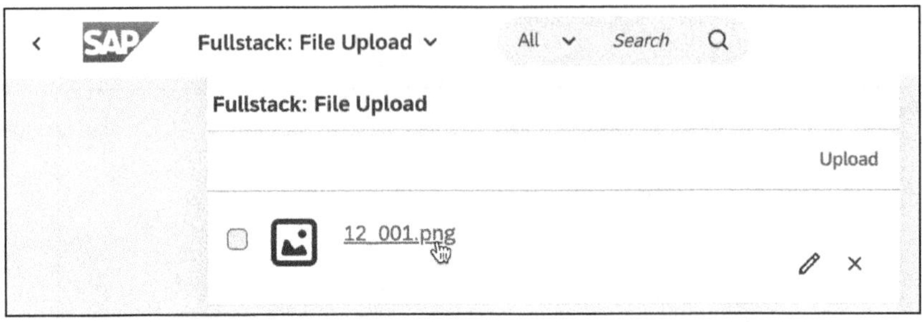

Figure 12.23 Link to the Downloadable File

12.3.3 Media Handling in OData V4

The media handling in OData V4 might differ in the backend because of the different technologies used in development. This is mainly valid for developing ABAP-based services. For developing with the ABAP RESTful application programming model, there's no difference in both protocols.

Let's take a look at a theoretical example to see how media handling can be achieved via the ABAP RESTful application programming model in a managed implementation scenario. First, you need three columns in your ABAP RESTful application programming model business object table, and it doesn't matter if this is your root entity or not:

- `attachment : abap.rawstring(0);`
- `mimetype: mime_typ;`
- `filename: file01;`

You can use the provided datatypes or create your own. These fields now need to be added to your entity that consumes your table with specific annotations, as shown in Listing 12.27.

```
@Semantics.largeObject: {
        mimeType: 'MimeType',
        fileName: 'FileName',
```

```
        acceptableMimeTypes: ['image/png', 'image/png', 'application/pdf'],
        contentDispositionPreference: #ATTACHMENT
    }
    attachment as Attachment,
    @Semantics.mimeType: true
    mimetype as MimeType,
```

Listing 12.27 Adding Large Objects to a CDS Entity

Here, we make the fields available and add the annotations for the ABAP RESTful application programming model to know that a data stream needs to be stored here. Therefore, we add the annotation `Semantics.largeObject` to the `Attachment` field. This annotation makes the `Attachment` field the storage of the file itself. Additionally, we add properties `mimeType` and `fileName`, referencing the fields `MimeType` and `FileName`. This is necessary for the ABAP RESTful application programming model to know where to store this additional information. In addition, we can control which file types can be uploaded by using property `acceptableMimeTypes`. These fields now need to be added to the projection of your entity.

12

Then, you can use the same SAPUI5 logic as previously to implement your `UploadSet`. The only important change is that you can't directly upload in combination with creating new OData entries in OData V4 or OData V2. It's not possible to create new entries directly in combination with a `POST` request. The entity first needs to be created, and then the upload can be done.

To enable this, the `UploadSet` must set the property `instantUpload="false"`. This prohibits the direct upload via a `POST` request. You then need to add the event handler `afterItemAdded="onAfterItemAdded"` to the `UploadSet` and also remove the `onBeforeUploadStarts` handler. In this event handler, you first create your entry and then upload, as shown in Listing 12.28.

```
public onAfterItemAdded(event: Event){
        const uploadSet = (this.getView()?.byId("<id of your upload set>") as
UploadSet);
        const uploadSetItem = event.getParameter("item");

        uploadSet.removeAllHeaderFields();

        this.getView()?.setBusy(true);

        this.getView().getModel().create("<path to your Entity Set>", {}, {
            success: (data, response)=>{
                this.getView().setBusy(false);

                uploadSet.addHeaderField(new Item({
```

```
                key: "X-CSRF-Token",
                text: this.getView().getModel().getSecurityToken()
            }));

            uploadSet.addHeaderField(new Item({
                key: "Content-Disposition",
                text: `filename=${oUploadSetItem.getFileName()}`
            }));

            uploadSet.setUploadUrl(`${this.getModel().sServiceUrl}/<path to
your newly created Entity>/$value`);

            uploadSet.setHttpRequestMethod("PUT");

            uploadSet.uploadItem(oUploadSetItem);
        },
        error: (error)=>{
        }
    });
}
```

Listing 12.28 Uploading a File with ABAP RESTful Application Programming Model

This method first creates a new entry in your OData model. When the creation is successful, it adds the necessary header fields to the `UploadSet` and sets the upload URL to your newly created entity. Afterward, the request method gets changed to a `PUT` request, and the `UploadSet` item that holds your file gets uploaded.

12.4 Drag and Drop

Drag and drop is a native HTML5 feature that allows you to drag content and drop it somewhere. In SAPUI5, this was enhanced to contain more information on what has been dragged, where it has been dropped, where it's allowed to be dropped, and how the drag-and-drop controls are rendered. The namespace `sap.ui.core.dnd` contains all relevant SAPUI5 classes for this handling, as follows:

- **`sap.ui.core.dnd.DragInfo`**
  Class for defining the information, what control was dragged, and how it was dragged.
- **`sap.ui.core.dnd.DropInfo`**
  Class for defining the information, what control was dropped, and how it was dropped.

- `sap.ui.core.dnd.DropEffect`
  Enum for defining the visual drop effect during drag and drop.
- `sap.ui.core.dnd.DropPosition`
  Enum for configuring the drop position.
- `sap.ui.core.dnd.DropLayout`
  Enum for configuring how dropped controls are arranged.

Drag and drop can be used for various reasons. Mostly it's done to provide better usability on mobile devices, as drag and drop works best with touch-enabled devices. It's used to rearrange lists, change items in a control such as a calendar, move items from one place to another, or import and export files such as dropping a document into an `UploadSet`.

Each control in SAPUI5 by default supports drag and drop because of the `dragDropConfig` aggregation of the `sap.ui.core.Element` class. This aggregation needs to contain either a `DragInfo` or a `DropInfo` instance.

12

When creating custom controls or extended controls, you can use the `dnd` aggregation to define if your own control is draggable or droppable by using the metadata definition, as shown in Listing 12.29 (see Chapter 13 for custom controls).

```
metadata : {
          properties : {
               ...
          },
          dnd : { draggable: true, droppable: true },
          aggregations : {
             content: { …, dnd: {draggable: true, droppable: true}}
          }
     }
```

Listing 12.29 The Drag-and-Drop Configuration of a Control

This enables the control to be dragged to a drop target, and makes the `content` aggregation also support drag and drop.

Let's look at a simple example to understand how we can use drag and drop and configure it in our applications. In this sample, we're going to create two lists with items we can move between to rearrange the items of the lists. We start by adding a JSON model in our controller that holds the data for our two lists. The first list contains products, and the second one serves as our shopping cart. See Listing 12.30 for the implementation of the model.

```
const model = new JSONModel({
            products: [{
                key: 1,
```

```
                name: "Apple",
                price: 0.99
            }, {
                key: 2,
                name: "Bread",
                price: 2.99
            }, {
                key: 3,
                name: "Yogurt",
                price: 0.59
            }, {
                key: 4,
                name: "Butter",
                price: 2.99
            }, {
                key: 5,
                name: "Eggs",
                price: 3.49
            }, {
                key: 6,
                name: "Bacon",
                price: 4.59
            }],
            shoppingCart: []
        });
        this.getView()?.setModel(model);
```

Listing 12.30 Adding a Simple Shopping Model

We can then add two lists to an HBox in our view containing two lists that display the model data, as shown in Listing 12.31.

```
        <HBox width="100%" justifyContent="SpaceBetween">
            <List id="productList" items="{/products}" headerText="Products" 
width="250px">
                <items>
                    <ObjectListItem title="{name}" number="{price}"/>
                </items>
            </List>
             <List id="shoppingCartList" items="{/shoppingCart}" headerText=
"Shopping Cart" width="250px">
                <items>
                    <ObjectListItem title="{name}" number="{price}" numberUnit=
"€" />
```

```
            </items>
        </List>
    </HBox>
```

Listing 12.31 Adding Two Lists

This produces the result shown in Figure 12.24.

Products		Shopping Cart
Apple	0.99	No data
Bread	2.99	
Yogurt	0.59	
Butter	2.99	
Eggs	3.49	
Bacon	4.59	

Figure 12.24 Our Two Lists

Now we can add the drag-and-drop configuration to our first list to support dragging an item, as shown in Listing 12.32.

```
<List id="productList" items="{/products}" headerText="Products" width="250px">
            <items>
                <ObjectListItem title="{name}" number="{price}"/>
            </items>
            <dragDropConfig>
                <dnd:DragInfo
                    groupName="productsToShoppingCart"
                    sourceAggregation="items"
                    />
            </dragDropConfig>
        </List>
```

Listing 12.32 Adding the DragInfo to the First List

We add the `DragInfo` class to the list's `dragDropConfig` aggregation, defining a `groupName` for our drag event and the `sourceAggregation`, where our draggable items lay. Now we can drag our items, as shown in Figure 12.25.

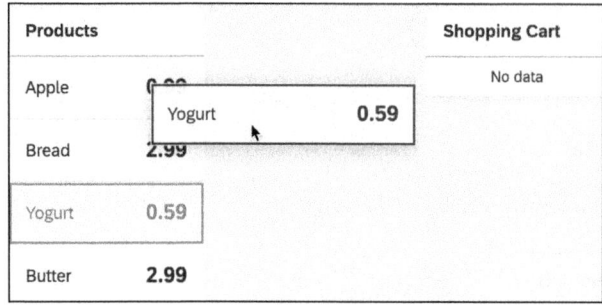

Figure 12.25 Dragging One List Item

We can't drop our item though. So, we need to define that our second list serves as a drop target, as shown in Listing 12.33.

```
        <List id="shoppingCartList" items="{/shoppingCart}" headerText=
"Shopping Cart" width="250px">
            <items>
                <ObjectListItem title="{name}" number="{price}" numberUnit=
"€" />
            </items>
            <dragDropConfig>
                <dnd:DropInfo
                    groupName="productsToShoppingCart"
                    targetAggregation="items"
                    drop="onProductDropped"
                    dropPosition="On"
                    />
            </dragDropConfig>
        </List>
```

Listing 12.33 Adding the DropInfo to the Second List

We add `DropInfo` to the list's `dragDropConfig` aggregation. This `DropInfo` again contains a `groupName`, the `targetAggregation` defining where items can be dropped, an event handler for the `drop` event, and the `dropPosition` defining how we arrange dropped items. This allows us now to register the second list as a drop target, as shown in Figure 12.26.

If you drop the product, however, nothing happens. Drag and drop is configured, but not the logic on what should happen with the dropped control. Therefore, we add the event handler from the drop event from Listing 12.33 into our controller.

Here, we need to remove the dragged item from the first list and add it to the second list in our model, as shown in Listing 12.34.

```
    public onProductDropped(event: Event){
        const draggedProduct: ObjectListItem =
event.getParameter("draggedControl"),
            model = (this.getView()?.getModel() as JSONModel);

        const product = model.getProperty((draggedProduct as
any).getBindingContext().getPath());
        let newProducts: any[] = [],
            shoppingList: any[] = model.getProperty("/shoppingCart");

        newProducts = model.getProperty("/products").filter((prod: any)=>
prod.key != product.key);
        model.setProperty("/products", newProducts);

        shoppingList.push(product);
        model.setProperty("/shoppingCart", shoppingList);
    }
```


Listing 12.34 Adding the Event Handler for the "drop" Event

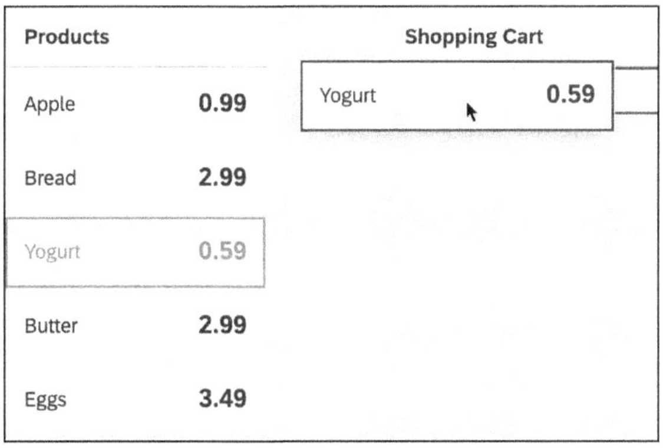

Products		Shopping Cart	
Apple	0.99	Yogurt	0.59
Bread	2.99		
Yogurt	0.59		
Butter	2.99		
Eggs	3.49		

Figure 12.26 Second List: Now a Drop Target

In this handler, we first get the dragged ObjectListItem from the event parameters. We then remove the item that was bound to this ObjectListItem from the JSON model's products list. After that, we add it to the JSON model's shoppingCart list. As this is registered as a change in the model, both lists' items aggregations are automatically re-rendered, and we see our dragged product in the shopping list, as shown in Figure 12.27.

This was a very simple sample, as in OData scenarios this goes hand in hand with sending OData calls to remove and add items. In addition, if you want to drop an item between items in a certain order, the new order must be determined manually.

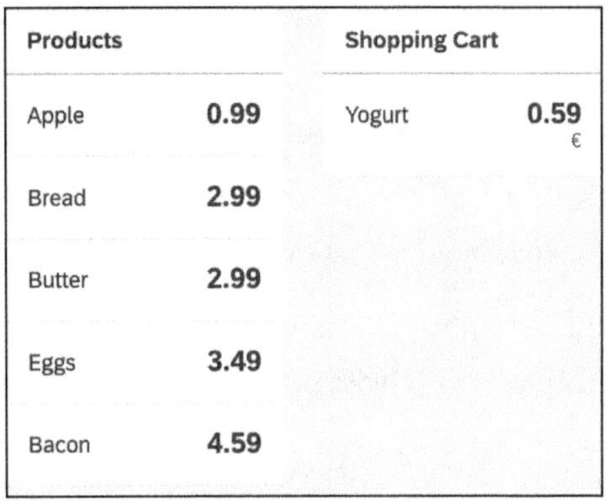

Products		Shopping Cart	
Apple	0.99	Yogurt	0.59 €
Bread	2.99		
Butter	2.99		
Eggs	3.49		
Bacon	4.59		

Figure 12.27 The Product Was Added to the Shopping Cart

Now you know how easy it is in general to enable dragging and dropping controls. The real challenge is implementing the behavior afterward.

12.5 Input and Form Validation

Input and form validation in SAPUI5 means validating data against certain requirements with specific methods. These validations can be one of the following items:

- Basic requirements
- Built-in datatypes
- Custom datatypes
- Custom validations
- Service-related validations

The validation goes hand in hand with the SAP Fiori design guidelines, as validation should be coherent through your applications and should be visualized using certain UI elements. We're also going to show you a few examples of how you can validate user input.

12.5.1 Basic Requirements

As most of the SAPUI5 input controls inherit their data from control `sap.m.InputBase`, they inherit property `required`. Setting this property to `true` might seem like the validation if the control holds any value, but this property is only needed for accessibility purposes, so checking if data has been inserted must be done manually in JavaScript or by using built-in datatypes.

12.5.2 Built-in Datatypes

SAPUI5 uses *simple types* to provide the definition of datatypes during data binding. These simple types are built-in to the framework itself and can serve as a base for specific custom datatypes. In addition, they define built-in constraints, which are automatically applied when a value has been provided to an input field that is bound with a specific simple type. The following simple types are available in SAPUI5 and are in namespace `sap.ui.model`:

- `Boolean`
  Represents Boolean datatypes and doesn't define any constraints.
- `Currency`
  Represents a composite type that consists of an amount and a currency code. It provides constraints for `minimum` and `maximum` amounts.
- `Date`
  Represents simple dates and provides constraints for `minimum` and `maximum` dates.
- `DateInterval`
  Represents two dates in an interval and provides constraints for `minimum` and `maximum` dates.
- `DateTime`
  Represents simple dates plus timestamps and provides constraints for `minimum` and `maximum` date plus time.
- `DateTimeInterval`
  Represents simple dates plus timestamps in an interval and provides constraints for `minimum` and `maximum` date plus time.
- `FileSize`
  Represents simple file sizes and provides constraints for `minimum` and `maximum` files sizes.
- `Float`
  Represents simple floats and provides constraints for `minimum` and `maximum` numbers.
- `String`
  Represents strings and provides constraints for `maxLength`, `minLength`, `startsWith`, `startsWithIgnoreCase`, `endsWith`, `endsWithIgnoreCase`, `contains`, `equals`, and `search` validations.
- `Integer`
  Represents simple integers and provides constraints for `minimum` and `maximum` numbers.
- `Time`
  Represents simple timestamps and provides constraints for `minimum` and `maximum` time values.

- `TimeInterval`
  Represents simple timestamps in an interval and provides constraints for `minimum` and `maximum` time values.
- `Unit`
  Can be used as a base type for your own unit types.

When used in some sort of data binding, these simple types provide three functions that are called during the lifecycle of a bound datatype:

- `formatValue`
  This method is called when the value of the property of the bound model is changed and formats data for output. This throws a `FormatException`.
- `parseValue`
  This method is called when a user inputs data, and the data then gets converted to be stored with the actual datatype provided in the binding. This throws a `ParseException`.
- `validateValue`
  This method is the main method this chapter covers. Here, the parsed data is validated against the basic constraints or against custom schemes. This throws a `ValidateException`.

12.5.3 Custom Datatypes

Custom datatypes are SAPUI5 modules that inherit one of the simple types or base class `sap.ui.model.SimpleType` and can be used to define custom datatypes such as a ZIP code, a sales order ID, or an ISO country code type.

12.5.4 Custom Validations

Custom validations can either be implemented with the help of custom datatypes or with the help of change events of the different input controls. There, you can do your own validations and define how constraints will be applied.

SAPUI5 controls inherit class `sap.ui.base.ManagedObject`, which provides event listeners for `validationSuccess` and `validationError`. There, you can define a method that will be called when the validation of the base or custom datatype fails.

12.5.5 Service-Related Validations

OData services should, depending on their implementation, return errors if nonvalid data was transmitted. These errors can be caught by using the OData model and its error callbacks.

12.5.6 Recommendations in SAP Fiori Design Guidelines

The SAP Fiori design guidelines provide detailed information on how to handle form validation and validation errors, which we'll discuss now. Validation errors of input controls should be made visible with the help of the value states from base class `sap.m.InputBase`. When one of the following value states is set to an input control, it's rendered according to the state:

- **None**
- **Error**
- **Warning**
- **Success**
- **Information**

Next, we'll take a closer look into these five value states, when they should be displayed, and how. Additionally, SAP Fiori provides a validation flow, indicating what you need to validate and how to prevent errors. These validation results can then be displayed in message popovers.

Value State: None

By default, input controls have the state **None** (see Figure 12.28), which means the user input hasn't yet been validated or the validation has been successful without issues. When messages are displayed, they should be of a noncritical nature.

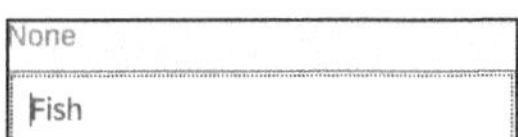

Figure 12.28 Value State: None

Value State: Error

The state **Error** (see Figure 12.29) marks an input control as being provided with an incorrect value, meaning that constraints have been violated. A user shouldn't be able to continue work without correcting the mistake first. In addition, a message should be shown hinting at how the mistake can be fixed.

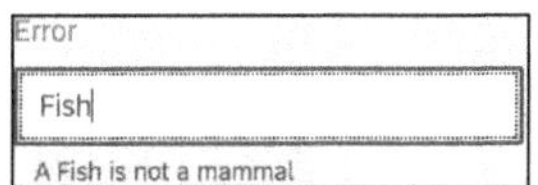

Figure 12.29 Value State: Error

Value State: Warning

Warning (see Figure 12.30) indicates that a minor problem occurred. Users can continue working, but an error might occur later. Again, a warning message needs to be presented.

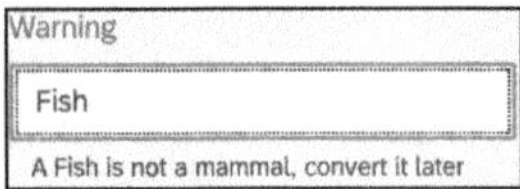

Figure 12.30 Value State: Warning

Value State: Success

When using the state **Success** (see Figure 12.31), the provided data has been validated and draws the attention of the user to it. This might come in handy when more data is required and should be copied. However, this state shouldn't be used for simple validation success messages but for process finalization purposes.

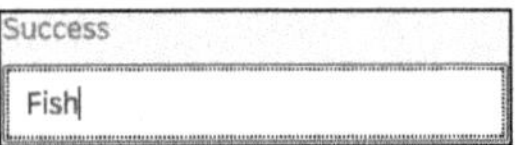

Figure 12.31 Value State: Success

Value State: Information

The state **Information** (see Figure 12.32) draws the attention of the user to the input field and shouldn't be used for validation purposes.

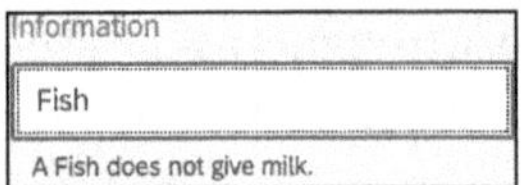

Figure 12.32 Value State: Information

Validation errors should be made visible with the value state **Error** and successful validations with the value state None. In addition, certain value states should be accompanied with a *value state text*, which is displayed when the user focuses the input field.

Validation Flows

Per SAP Fiori design guidelines, there are two validation guidelines that need to be applied when validating validation flows:

- **Global edit flow**
  When validating data with the global edit flow, the whole page will be validated. The validation is triggered by leaving the focus of an input control (field validation), pressing `Enter` in an input control (form validation), or clicking on an action button (page validation). Validation messages should be displayed on the field level and in a message popover on the page level.
- **Partial editing**
  Partial editing means validation on the form level. When all validators pass successfully, the form should go into display mode. When at least one field validation fails,

there should be a corresponding message displayed with the help of value state texts in the input field.

Message Popover Control

The message popover (control `sap.m.MessagePopover`) should be used in SAP Fiori object page floorplan scenarios for showing global edit flow messages (see Figure 12.33). With its help, you show validation and general error messages. The message popover must be available only when messages occur and need to be positioned to the left in a footer toolbar on the object page.

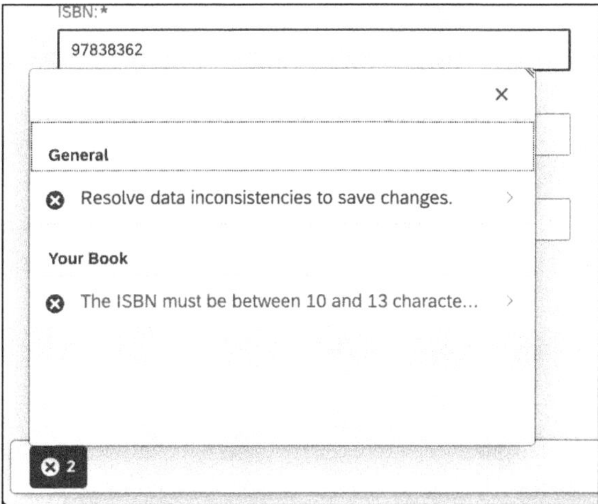

Figure 12.33 A Message Popover Containing Errors

Error Prevention

The best way is to prevent user errors before they occur. Input controls should always be labeled according to the data that needs to be entered and should provide helpers, either by different input types or, for example, mask inputs (control `sap.m.MaskInput`). In addition, always start with simple data input and get more complex during the input flow if necessary.

12.5.7 Implementation

In this section, we'll investigate the implementation approaches that can be used to provide validations, as follows:

- Use different datatypes.
- Use value states in input controls.
- Validate the whole form.
- Use the message popover.

Using Different Datatypes

We can use the simple types to get a basic validation flow into our application on a field level. For example, an input field for an ISBN expects a string with certain constraints, as follows:

- A maximum length of 13
- A minimum length of 10

We can implement this control by using a `sap.m.Input` control with simple type `sap.ui.model.type.String` and a binding to `sap.ui.model.json.JSONModel` in an XML view:

```
<Label text="Provide an ISBN"/>
<Input id="input" value="{path: 'bookModel>/isbn', type:
'sap.ui.model.type.String', constraints: {maxLength: 13, minLength: 10}}"
width="15rem"/>
```

This coding will add a new input control to our application that gets its value from a named model called `bookModel`. We also define the simple type for the binding with the additional constraints for `maxLength` and `minLength`.

Additionally, you need to set the model in the view controller:

```
onInit: function () {
 let oModel = new JSONModel({isbn: undefined});
 this.getView().setModel(oModel, "bookModel");
}
```

By adding a new model to our view, we can now store an ISBN. However, the constraint violations that might occur aren't automatically handled. To achieve this, SAPUI5 provides a message manager. We can provide the ID of our input control to the message manager to automatically set the right value state and value state text when the constraints of the binding are violated:

```
sap.ui.getCore().getMessageManager().registerObject(this.getView().byId("input"
), true);
```

Now, when the constraints are violated, a `ValidatorExeption` is thrown, and the exception message is shown below our input field (see Figure 12.34).

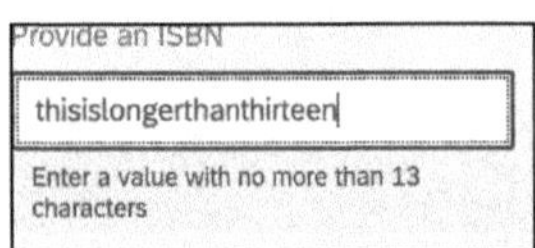

Figure 12.34 The Message Manager Showing an Error

SAPUI5 Message Manager

The singleton MessageManager, retrieved with sap.ui.getCore().getMessageManager(), offers a central management for messages. These can either be UI messages, generated by input validations, or server messages, generated by OData request responses.

To register controls or views for input validation message handling, method registerObject of the message manager can be used to automatically listen to value state changes of this control. Alternatively, automatic message handling can be added by defining property handleValidation: true of node sap.ui5 in the *manifest.json* file of the application.

All messages are stored in a message model, which can be retrieved via method getMessageModel.

12

Value States in Input Controls

Value states of input controls can also be set either directly in the view or by using JavaScript logic. This can be done when either dynamic validations will be used or when the message manager isn't suited for the task.

We'll enhance the input control from the preceding step to check whether the ISBN matches a regular expression we define. To achieve this, the input control needs to implement a listener method for the change event, where we're going to test the value against a regular expression:

```
<Input id="input" value="{path: 'bookModel>/isbn', type:
'sap.ui.model.type.String'}" width="15rem" change="onISBNChanged"/>
```

The event handler onISBNChanged now needs to be added in our JavaScript controller (see Listing 12.35).

```
onISBNChanged: function (oEvent) {
 let sValue = oEvent.getParameter("value"),
 oRegEx = new RegExp("^[0-9]{10,13}$"),
 oInput = oEvent.getSource();

 if (oRegEx.test(sValue)) {
  oInput.setValueState("None");
  oInput.setValueStateText("");
 } else {
   oInput.setValueState("Error");
   oInput.setValueStateText("Only numbers with a length between 10 and 13 are
```

```
allowed");
  }
}
```

Listing 12.35 Event Handler for the Change Event

The event handler is called whenever the user changes the value of the input field, which is then stored in variable sValue. In addition, a new regular expression in variable oRegEx is defined. When our value is tested against the regular expression, and the test is successful, we set the value state of the input field to None and clear any value state texts. When the test fails, we set the value state to Error and provide an accompanying value state text (see Figure 12.35).

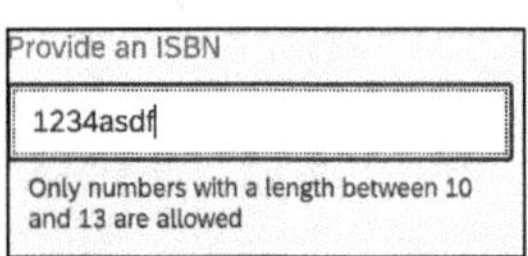

Figure 12.35 Custom Setting of the Value State

Validating a Whole Form

When you want to validate multiple input fields of a form such as a simple form (control sap.ui.layout.form.SimpleForm), we recommend adding all fields that should be validated to the message manager and writing specific validation methods for the different input fields.

Here, we're adding a simple form to our application that contains the ISBN input field and an additional input field for a book title, which will be required (see Listing 12.36).

```
<f:SimpleForm editable="true" layout="ResponsiveGridLayout" title="Book"
labelSpanXL="3" labelSpanL="3" labelSpanM="3" labelSpanS="12" adjustLabelSpan=
"false" emptySpanXL="4" emptySpanL="4" emptySpanM="4" emptySpanS="0" columnsXL=
"1" columnsL="1" columnsM="1" singleContainerFullSize="false">
 <f:content>
 <Label required="true" text="Title"/>
 <Input id="title" value="{path: 'bookModel>/title', type:
'sap.ui.model.type.String'}" change="onTitleChanged"/>
 <Label text="Provide an ISBN"/>
 <Input id="input" value="{path: 'bookModel>/isbn', type:
'sap.ui.model.type.String'}" change="onISBNChanged"/>
 </f:content>
</f:SimpleForm>
```

Listing 12.36 Adding a New SimpleForm to Validate

We also add a button somewhere in our view to start the form validation, as follows:

```
<Button text="Validate Form" press="onValidatePress"/>
```

Then, we need to add a new property to our JSONModel in our JavaScript controller and register the input field for the title in the message manager (see Listing 12.37).

```
onInit: function () {
 let oModel = new JSONModel({
  isbn: undefined,
  title: undefined
 });

 this.getView().setModel(oModel, "bookModel");
sap.ui.getCore().getMessageManager().registerObject(this.getView().byId("input"
), true);

sap.ui.getCore().getMessageManager().registerObject(this.getView().byId("title"
), true);
},
```

Listing 12.37 The onInit Method

Now we can define a method in our controller that will validate whether the book title has been provided (see Listing 12.38).

```
validateTitle: function (sValue) {
 let oInput = this.getView().byId("title");

 if (sValue && sValue.length > 0) {
  oInput.setValueState("None");
  oInput.setValueStateText("");
  return true;
 } else {
  oInput.setValueState("Error");
  oInput.setValueStateText("A Title needs to be provided");
  return false;
}
},
```

Listing 12.38 Method for Validating the Title

The method will handle the setting of the value state and the value state text if the book title is empty. After that, we add a method for validating the ISBN to be a certain length (see Listing 12.39).

```
validateISBN: function (sValue) {
 let oRegEx = new RegExp("^[0-9]{10,13}$"),
```

```
oInput = this.getView().byId("input")

if (oRegEx.test(sValue)) {
  oInput.setValueState("None");
  oInput.setValueStateText("");
  return true;
} else {
   oInput.setValueState("Error");
   oInput.setValueStateText("Only numbers with a length between 10 and 13 are
allowed");
  return false;
 }
},
```

Listing 12.39 Method for Validating the ISBN

We already know how this method works from the sample shown in Listing 12.38 earlier; we just define the validation logic in its own method.

Now we add the event handler for our button to trigger the start of the validation. Here, we'll call the two validation methods we just defined (see Listing 12.40).

```
onValidatePress: function (oEvent) {
 let oTitleInput = this.getView().byId("title"),
 oISBNInput = this.getView().byId("input");

 this.validateTitle(oTitleInput.getValue());
 this.validateISBN(oISBNInput.getValue());
},
```

Listing 12.40 Press Event Handler to Validate the SimpleForm

Now, the event handler methods for the change event of the input fields are missing. We're going to add them now to provide validation after the focus has left the field (see Listing 12.41).

```
onTitleChanged: function (oEvent) {
 let sValue = oEvent.getParameter("value");
 this.validateTitle(sValue);
},
onISBNChanged: function (oEvent) {
 let sValue = oEvent.getParameter("value");
 this.validateISBN(sValue);
}
```

Listing 12.41 Change Event Handler

Now, when we click the **Validate Form** button, we can automatically validate all fields of our simple form (see Figure 12.36).

Figure 12.36 Complete Form Validation

This is just a basic example. You can also retrieve the message model, bind its length to a button to disable it when messages occur, or implement further functionalities such as a message popover, which we'll show next.

Using the Message Popover

The control `sap.m.MessagePopover` provides an easy way to display messages in SAPUI5. Using it in combination with `MessageManager` and `MessageModel`, messages can be added and removed to show multiple errors in your application.

We'll further enhance our logic from the example in the previous section to have an additional `MessagePopover` available, showing all errors in one place.

First, we need to add a button (see Listing 12.42) that should open the message popover.

```
<Toolbar>
 <Button visible="{=!!${messageModel>/}.length}" icon="sap-icon://error" type=
"Reject" text="{= ${messageModel>/}.length}" press="onMessagePopoverPress"/>
 <ToolbarSpacer/>
</Toolbar>
```

Listing 12.42 Adding a Toolbar with a Button

This button should be made available in a toolbar on the bottom of the screen. This is bound to a named model called `messsageModel`, which will be set later. In addition, this button will only be visible when messages are in the `messageModel` and should display the number of stored messages. When clicked, the button calls method `onMessagePopoverPress`, which will handle the opening of the message popover.

After that, you need to import all message-relevant SAPUI5 modules (see Listing 12.43) into your controller.

```
sap.ui.define([
 ...
 'sap/m/MessagePopover',
```

```
 'sap/m/MessageItem',
 'sap/ui/core/MessageType',
 'sap/ui/core/message/Message'
], function (Controller, JSONModel, MessagePopover, MessageItem, MessageType,
Message)
...
```

Listing 12.43 Adding Imports to the Message Controls

The required modules are as follows:

- **MessagePopover**
  This SAPUI5 popover displays messages such as error, success, or warning messages, summarized in a list that also provides a detailed message view.
- **MessageItem**
  This is the item type of a message, which is required for displaying individual messages in the `MessagePopover`.
- **MessageType**
  SAPUI5 offers multiple `MessageTypes`, each representing a different level of message severity.
- **Message**
  The message itself is needed to add new messages to the message model, which will then be displayed as a `MessageItem` in the `MessagePopover`.

Then, the `onInit` method refactoring and two helper variables can be added (see Listing 12.44).

```
oMessagePopover: undefined,
oMessageManager: undefined,

onInit: function () {
 ...
 this.oMessageManager = sap.ui.getCore().getMessageManager();
 this.getView().setModel(this.oMessageManager.getMessageModel(),
"messageModel");
},
```

Listing 12.44 Setting Up the Message Model

Here, controller property `oMessageManager` is instantiated with the message manager, and its message model is set to the view as a named model called `messageModel`.

Now we implement the press event handler of the button (see Listing 12.45).

```
onMessagePopoverPress: function (oEvent) {
 if (!this.oMessagePopover) {
  this.createMessagePopover();
```

```
 }
 this.oMessagePopover.toggle(oEvent.getSource());
},
```

Listing 12.45 Adding a Press Handler

This event handler checks if we already have instantiated the `MessagePopover` or not. When we haven't, we'll call a method responsible for creating one; otherwise, we'll just open it.

This instantiating method also needs to be implemented (see Listing 12.46).

```
createMessagePopover: function () {
 this.oMessagePopover = new MessagePopover({
  items: {
   path: "messageModel>/",
   template: new MessageItem({
   title: "{messageModel>message}",
   subtitle: "{messageModel>additionalText}",
   type: "{messageModel>type}",
   description: "{messageModel>message}"
   })
 },
 groupItems: false
 });
 this.getView().addDependent(this.oMessagePopover);
},
```

Listing 12.46 Creating the Message Popover

The `createMessagePopover` method instantiates a new `MessagePopover`, which will be filled with the data of the message model. As a template for the items, we add a new `MessageItem` and fill its properties with the data from the message model.

Now we're ready to fill the message model with our own messages, so we'll enhance our validation methods (see Listing 12.47).

```
validateTitle: function (sValue) {
 ...
 this.removeMessageFromTarget("Title");

 if (sValue && sValue.length > 0) {
  ...
 } else {
  oInput.setValueState("Error");
  oInput.setValueStateText("A Title needs to be provided");
```

```
  this.oMessageManager.addMessages(new Message({
   message: "Field is required",
   type: MessageType.Error,
   additionalText: 'A Title needs to be provided',
   target: "Title",
   processor: this.oMessageManager.getMessageModel()
   }));
  return false;
  }
},
```

Listing 12.47 Adding a New Message

Let's manage our title validation first. We remove any old messages using method removeMessageFromTarget, which we're going to add later. When the validation fails, we add a new message to our message manager, providing all the necessary information. Then, we enhance the second validator (see Listing 12.48).

```
validateISBN: function (sValue) {
 ...
 this.removeMessageFromTarget("ISBN");

 if (oRegEx.test(sValue)) {
...
 } else {
  ...
  this.oMessageManager.addMessages(new Message({
   message: "ISBN length invalid",
   type: MessageType.Error,
   additionalText: 'Only numbers with a length between 10 and 13 are allowed',
   target: "ISBN",
   processor: this.oMessageManager.getMessageModel()
  }));
  return false;
  }
},
```

Listing 12.48 Adding a Second Message

In our ISBN validator, we reuse the same message insertion logic as before, only changing the error message and the target. The last thing still missing is the removal method of existing messages (see Listing 12.49).

```
removeMessageFromTarget: function (sTarget) {
 this.oMessageManager.getMessageModel().getData().forEach(function (oMessage) {
  if (oMessage.target === sTarget) {
```

```
    this.oMessageManager.removeMessages(oMessage);}
  }.bind(this));
},
```

Listing 12.49 Adding a Method to Remove Messages

The `removeMessageFromTarget` method gets a target as an import parameter, which will be provided individually by the validation methods. When a message with the target exists, we remove the message.

Now let's try to invalidate our form and check the message popover (see Figure 12.37).

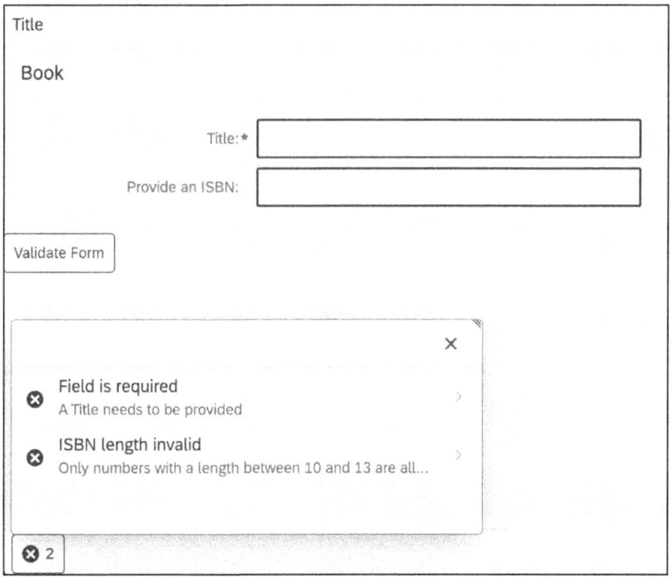

Figure 12.37 An Opened Message Popover

12.6 Error Handling

When implementing business logic, errors can happen. Because the business logic is implemented on the backend side, errors that occur there and that should have an impact on the user need to be forwarded to the UI as stated by the SAP Fiori design guidelines.

Error handling in SAP Fiori is closely coupled to general message handling. Carefully selected and written messages should help the user resolve the problem in an easy way. Error messages should therefore prevent the user from proceeding forward without solving the issue first. The best way to prevent errors in the first place is to add validators to your UI, minimizing the risk of needing error handling.

OData-related errors can be automatically caught in SAPUI5 by using the message manager. The SAPUI5 OData models have a `MessageParser` (class `sap.ui.model.odata.ODataMessageParser`) that is responsible for parsing OData messages and inserting them into the message model.

No additional coding is required for this to work because only the message handling must be made available through the manifest. Afterward, all OData messages are registered and can be shown in a message popover.

Alternatively, you can also attach an event handler for the following methods:

- `sap.ui.model.odata.v2.ODataModel`
  - `attachBatchRequestCompleted`
  - `attachBatchRequestFailed`
  - `attachRequestCompleted`
  - `attachRequestFailed`
- `sap.ui.model.odata.v4.ODataModel`
  - `attachRequestCompleted`
  - `attachRequestFailed`

Using these methods, event handlers for either batch requests (OData V2 model) or simple HTTP requests (OData V2 and OData V4 models) can be included to add custom messages to the message model when a request is either completed or simply fails. Here, additional custom messages can be set. These methods also might come in handy when providing event handlers when a request has run through and custom logic needs to be implemented, such as a custom refresh of certain data.

When multiple messages are returned upon request completion, we recommend displaying them either in a message popover or in a dialog (control `sap.m.Dialog`). When single messages are returned, either show them in a message toast (control `sap.m.MessageToast`) or in a message box (control `sap.m.MessageBox`).

DSAG Error Handler

The DSAG (Deutschsprachige SAP Anwendergruppe, the German equivalent to ASUG) developed an error handler specifically for OData-relevant messages that you can use and further extend. It's available on their GitHub at *https://1dsag.github.io/UI5-Best-Practice/error-handling/DSAG_Sample_ErrorHandler_EN.js.*

Whenever an OData service returns an error, this error catches and displays them accordingly. This means OData services can be registered using this error handler, and their messages are displayed in message views, message toasts, or message dialogs.

It's really handy to use this in your applications because you can then focus on the logic of your app and don't need to worry about missing displaying errors. So, a big shoutout goes to the developers behind this. We highly suggest that you also read their best practices on SAPUI5 development.

12.7 Smart Controls

Smart controls are SAPUI5 controls that are closely knit to OData V2 services. They are metadata driven, which means they behave like the metadata and annotations of an OData V2 rules. In addition, they offer an easy and fast way of developing UIs that are dependent of an OData service. They offer the SAP Fiori elements behavior broken down into dedicated controls, each in compliance with the SAP Fiori design guidelines.

A smart control always needs to have a binding to a certain OData entity or an `EntitySet` to load its configuration and behave as stated in the metadata document of such an OData service. So, if you're already familiar with OData annotations for SAP Fiori elements, these must be used to configure your smart controls.

Smart Controls and the OData Protocol Version

Smart controls are only available with a binding to an OData V2 service. They can't work with an OData V4 service. So, if you want to use this behavior in V4, you need the *flexible programming model.*

The flexible programming model uses the flexibility of freestyle SAPUI5 in combination with SAP Fiori elements for OData V4. This means, that SAP Fiori elements applications can be enhanced to contain freestyle SAPUI5 views, logic, and extensions. As a direct equivalent to the smart controls, the flexible programming model provides macros/building blocks for the most-used SAPUI5 controls that are metadata driven.

As OData V4 gains prominence and is becoming the default SAPUI5 OData model type, the SAP Fiori elements offer more functionality in this protocol version. We highly recommend checking this out.

SAPUI5 offers the following smart controls, bundled into the `sap.ui.comp` namespace:

- `SmartChart`
  The smart chart offers a chart based on the OData metadata.
- `SmartMicroChart`
  The smart micro chart offers a micro chart based on the OData metadata.
- `SmartField`
  The smart field offers a form field based on the OData metadata. It automatically renders either a display control or an input control based on its edit mode. It also provides different input controls based on the entity data model (EDM) datatype.

- `SmartForm`
  The smart form, combined with smart fields and smart labels, adds a form bound to an OData entity. This form is displayed either in display mode or edit mode.
- `SmartLabel`
  The smart label offers a label based on the OData metadata. It labels a smart field and gets its displayed text from the OData properties label that was bound to the smart field.
- `SmartFilterBar`
  The smart filter bar offers a filter bar for smart tables. It offers smart variant management and selection fields for each property of its bound OData entity that is filterable and annotated to be a selection field.
- `SmartTable`
  The smart table offers a table that gets its column configuration from the OData services annotations. It offers data export, search and grouping functionalities, and lazy loading.
- `SmartList`
  The smart list offers a list that is bound to and set up by an OData service `EntitySet`.

Let's look at these controls a little closer to understand how they work.

12.7.1 SmartField and SmartLabel

The `SmartField` has two modes. Its first mode is display mode. There, it either displays its bound OData entities property as a `Text`, a `Link`, or a `SmartLink`. When switched to edit mode, it displays different controls based on the EDM type of the bound property. These controls are `Input`, `Select`, `ComboBox`, `TextArea`, `Checkbox`, `DatePicker`, `DateTimePicker`, or `TimePicker`.

Depending on the bound property, it handles different visual representations, such as mandatory flags read-only, hide, or display.

The `SmartLabel` needs to have an association to a `SmartField`. It reads its binding and loads the properties label to display it. For example, if you have an `EntitySet` with a certain property that you want to use in the UI, you then need to look at your OData services metadata document and search for your entity set's property. We use this one as a sample:

```
<Property Name="PurchasingDate" Type="Edm.DateTime" Precision="0" sap:display-format="Date" sap:label="Date" sap:quickinfo="Field of type DATS"/>
```

You then add a `SmartField` and a `SmartLabel` to your XML view that reference this property in a relative property binding, as follows:

```
            <smartfield:SmartLabel labelFor="purchasingDate"/>
            <smartfield:SmartField value="{PurchasingDate}" id=
"purchasingDate"/>
```

Of course, for this to work, an entity must be bound to the view where your smart controls exist:

```
this.getView()?.bindElement("/YourEntitySet(keyProperties)");
```

This renders a label and a text control based on the bound property, as shown in Figure 12.38.

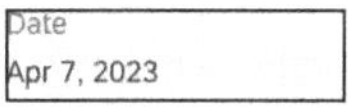

Date
Apr 7, 2023

Figure 12.38 The SmartField, Labeled by the SmartLabel

Let's say, for example, that you use a Boolean property for a smart field, as shown in the following:

```
<Property Name="ReadingFinished" Type="Edm.Boolean" sap:display-format=
"UpperCase" sap:label="Boolean Variable (X = True, - = False, Space = Unknown)"
sap:heading=""/>
```

You then add a `SmartField` and a `SmartLabel` to your XML view that reference this property in a relative property binding, as follows:

```
            <smartfield:SmartLabel labelFor="readingFinished"/>
            <smartfield:SmartField value="{ReadingFinished}" id=
"readingFinished"/>
```

You'll get a textual, translated representation for this Boolean value, as shown in Figure 12.39.

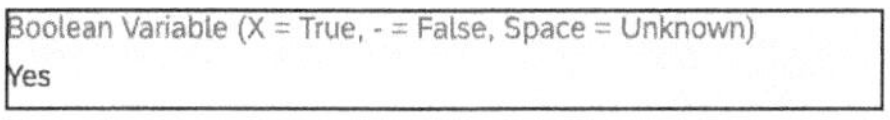

Boolean Variable (X = True, - = False, Space = Unknown)
Yes

Figure 12.39 A Boolean Property Translated

12.7.2 SmartForm

To add a form that behaves as a metadata-driven form, you can use `SmartForm`. It works similar to a normal form but uses smart fields as its form fields and offers the switch between display and edit mode. You can add a smart form as shown in Listing 12.50.

```
<smartform:SmartForm title="Book Data" entityType="<Your EntitySet> "
editTogglable="true" expandable="true" >
                <smartform:Group title="Reading Data">
                    <smartform:GroupElement >
```

```
                    <smartfield:SmartField value="{PurchasingDate}"/>
                </smartform:GroupElement>
                <smartform:GroupElement >
                    <smartfield:SmartField value="{ReadingFinished}"/>
                </smartform:GroupElement>
            </smartform:Group>
        </smartform:SmartForm>
```

Listing 12.50 Defining a SmartForm

Here, we add a SmartForm and define its entityType with our entity set's name. We can add a title and set the editTogglable property. This adds a button that allows us to switch between display and edit mode, meaning that editable smart fields will render input controls.

The SmartForm contains Groups. Each group works like a FormGroup, providing a dedicated container for group elements that contain the SmartFields. This produces an interactive result, as shown in Figure 12.40.

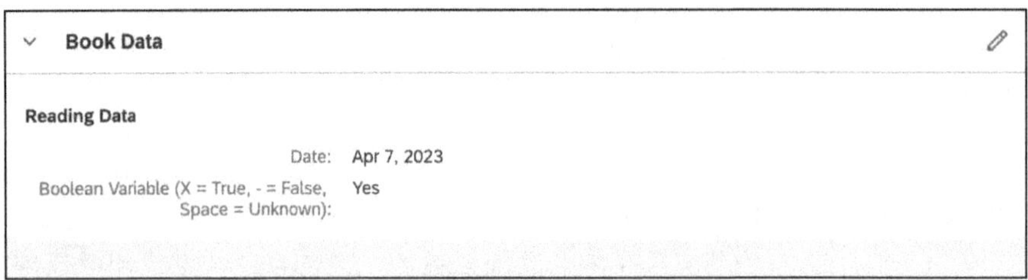

Figure 12.40 The SmartForm

12.7.3 SmartTable and SmartFilterBar

The SmartTable renders a table configured by annotations made to the entity set that is bound to it. It mostly comes in hand with the SmartFilterBar, offering variant management and filter options for a SmartTable.

The SmartTable offers personalization and export features. The personalization contains visible columns, sorting, and filtering capabilities. The SmartFilterBar provides optional variant management, and personalized search and filter capabilities

In this section, we'll do a step-by-step implementation of a SmartTable and a SmartFilterBar to show how they work. Let's start by defining both in a view and binding an entity set, as shown in Listing 12.51.

```
        <smartfilterbar:SmartFilterBar id="smartFilterBar" entitySet="<Your
Entity Set"/>
        <smarttable:SmartTable entitySet="<Your Entity Set> "
```

```
smartFilterId="smartFilterBar" enableAutoBinding="true">

            </smarttable:SmartTable>
```

Listing 12.51 Adding a SmartTable and a SmartFilterBar

Here, we add a SmartFilterBar and assign an entitySet to it. The SmartTable references our SmartFilterBar, is assigned an entitySet, and enables automatic binding, which is responsible for automatically loading the table data. Running this will render these two controls, which are empty for now, as shown in Figure 12.41.

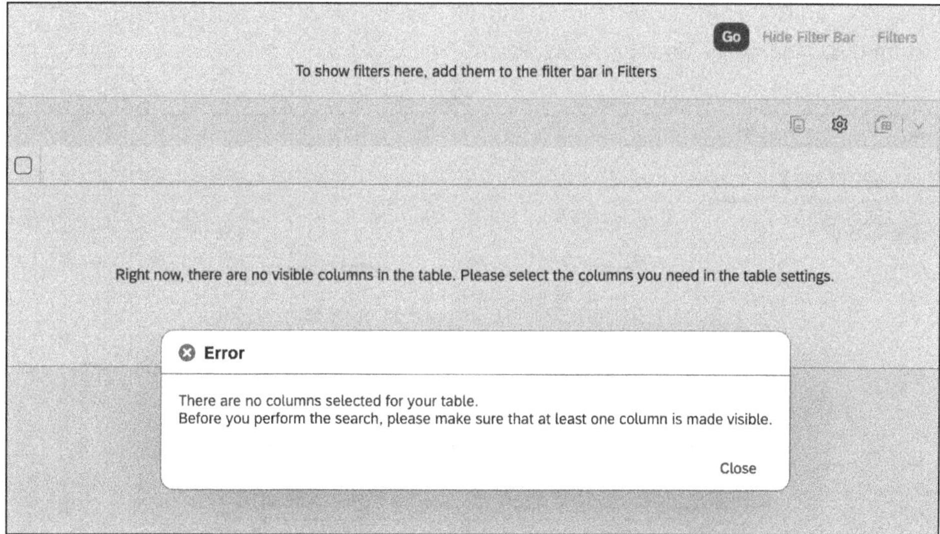

Figure 12.41 The Empty SmartTable in Combination with a SmartFilterBar

As we haven't annotated our OData service yet, the SmartTable shows an error, indicating that the column configuration is missing. You can either use remote annotations or use local annotations. The UI.lineItem annotation is the responsible annotation for displaying columns in tables. For adding local annotations, you can use the **Annotation File Manager** page, opened from the application information view of the SAP Fiori tools, as shown in Figure 12.42.

Here, you can add a file in the **Add local annotation file** area. You need to select your service and then add a new annotation file. This generates an XML annotation file, where you can add your UI.lineItem annotation, as shown in Listing 12.52.

```
<Annotations Target="SAP.<Your Entity Type> ">
            <Annotation Term="UI.LineItem">
                <Collection>
                    <Record Type="UI.DataField">
                        <PropertyValue Property="Value" Path="Title"/>
```

```
                        </Record>
                        <Record Type="UI.DataField">
                            <PropertyValue Property="Value" Path="Author"/>
                        </Record>
                        <Record Type="UI.DataField">
                            <PropertyValue Property="Value" Path=
"PurchasingDate"/>
                        </Record>
                    </Collection>
                </Annotation>
            </Annotations>
```

Listing 12.52 Adding the LineItem Annotation

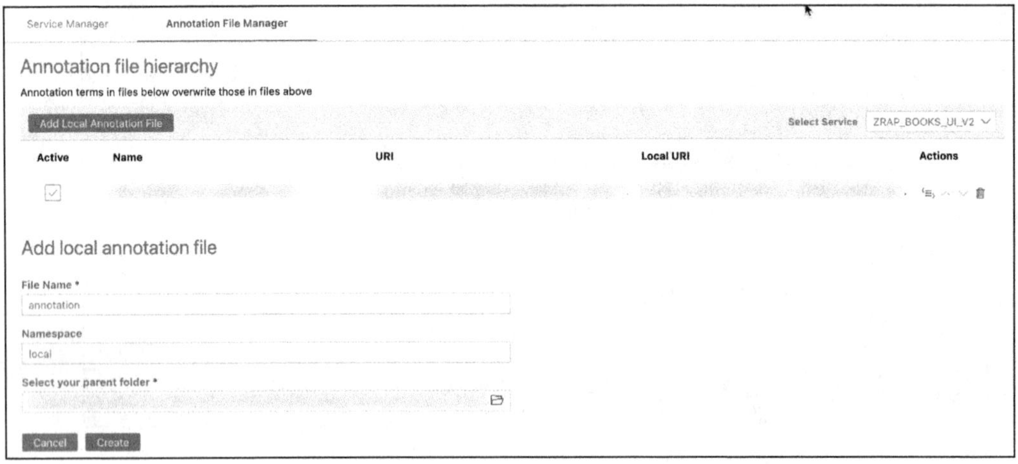

Figure 12.42 Annotation File Manager

We define that we want to display three properties of our entity type that is referenced in the entity set of our smart table. Therefore, we add `UI.DataField` annotations that are bundled into the `UI.LineItem` annotation. If you now reload your application, you see these three fields as dedicated columns, as shown in Figure 12.43.

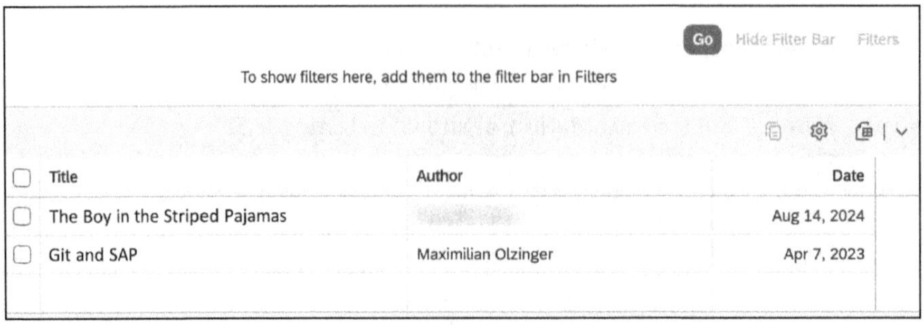

Figure 12.43 Your Columns

To add filters to our `SmartFilterBar`, we need to add the `UI.SelectionField` annotation to our entity reference in our annotation file, as shown in Listing 12.53.

```
<Annotations Target="<Your Entity Type>">
                <Annotation Term="UI.SelectionFields">
                    <Collection>
                        <PropertyPath>Title</PropertyPath>
                        <PropertyPath>PurchasingDate</PropertyPath>
                    </Collection>
                </Annotation>
                <Annotation Term="UI.LineItem">
                    ...
                </Annotation>
            </Annotations>
```

Listing 12.53 Adding SelectionField Annotations

Now we get the added filter fields as selection fields in our smart filter bar, as shown in Figure 12.44.

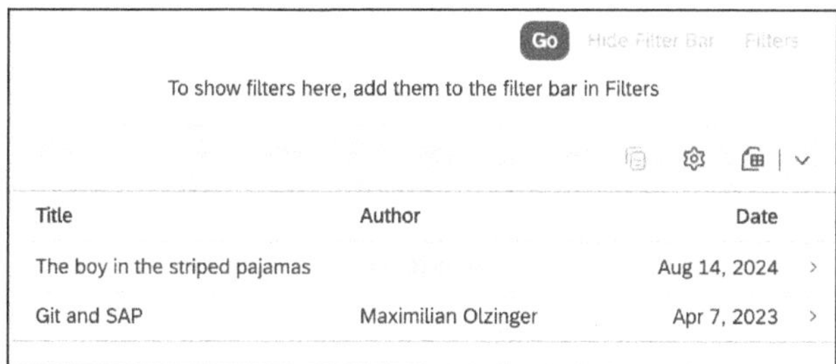

Figure 12.44 The Selection Fields

If you want to see what the `SmartTable` and `SmartFilterBar` are capable of besides displaying columns and selection fields, open the **View Settings** dialog of the smart table, as shown in Figure 12.45, accessible by clicking on the gear icon in the toolbar of the smart table.

This dialog is divided into a **Columns**, **Sort**, and **Filter** tabs. Here, you can select all properties of the entity set to add them as visible columns, add sorting parameters, and add simple filters.

The `SmartFilterBar` offers a similar personalization option, where you can select filterable OData entity set properties to be displayed as selection fields.

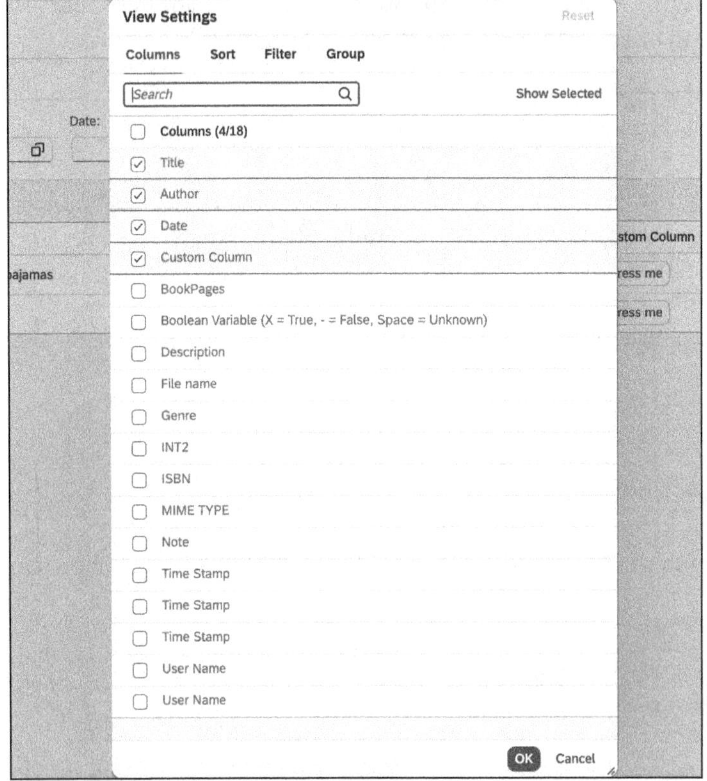

Figure 12.45 The View Settings Dialog

The `SmartTable` in addition offers the following properties to further define its behavior:

- `demandPopin`
  Enables the popin feature of `sap.m.Table`.
- `enableCopy`
  Enables copy-to-clipboard to copy columns and their values.
- `enableExport`
  Enables export to Microsoft Excel.
- `enablePaste`
  Enables pasting values as new rows.
- `requestAtLeastFields`
  Defines which properties of the entity set should be loaded.
- `showVariantManagement`
  Shows the variant management to store personalization.
- `tableType`
  Defines the type of the table, such as responsive table or grid table.

The `SmartFilterBar` offers the following properties to further enhance its behavior:

- **`enableBasicSearch`**
  Enables a search input to search over multiple columns.
- **`liveMode`**
  Enables automatic table reloading after changing a filter value.

We also want to show you how you can enhance the features of the `SmartTable`. It's possible to add a dedicated table control to the smart table to define custom columns and redefine the behavior of the row. To do so, add a table control to the smart table, as shown in Listing 12.54.

```
<smarttable:SmartTable entitySet="<Your Entity Set>" smartFilterId=
"smartFilterBar" enableAutoBinding="true"
                tableType="ResponsiveTable"
                >
                <Table >
                    <columns></columns>
                    <items></items>
                </Table>
            </smarttable:SmartTable>
```

Listing 12.54 Adding a Table

By adding the `tableType` `ResponsiveTable`, we can add a `sap.m.Table` instance inside the smart table.

During instantiation, the smart table creates a table instance itself. By providing one ourselves, the smart table uses ours to define its behavior. If we, for example, want to define the navigation behavior, we can add a `ColumnListItem` template to the `items` aggregation, as shown in Listing 12.55.

```
<smarttable:SmartTable entitySet="<Your Entity Set>" smartFilterId=
"smartFilterBar" enableAutoBinding="true"
                tableType="ResponsiveTable"
                >
                <Table >
                    <columns></columns>
                    <items>
                        <ColumnListItem type="Navigation" press=
"onItemPress"></ColumnListItem>
                    </items>
                </Table>
            </smarttable:SmartTable>
```

Listing 12.55 Adding a ColumnListItem to the Table

This adds a custom template to our smart table that should be used for displaying the items as navigable ones, as shown in Figure 12.46. You can also add a press event handler to manually implement which routing target you want to navigate to.

Figure 12.46 The ColumnListItem Type Set to Navigation

To add custom columns, you add those to the columns aggregation and your items template, as shown in Listing 12.56.

```
<smarttable:SmartTable entitySet="<Your Entity Set>" smartFilterId=
"smartFilterBar" enableAutoBinding="true"
                tableType="ResponsiveTable"
                >
                <Table >
                    <columns>
                        <Column>
                            <Text text="Custom Column"/>
                            <customData>
                                <core:CustomData key="p13nData" value='\
{"columnKey": "BookGuid", "leadingProperty": "BookGuid", "columnIndex": 4}'/>
                            </customData>
                        </Column>
                    </columns>
                    <items>
                        <ColumnListItem type="Navigation" press="onItemPress">
                            <Button text="Press me"/>
                        </ColumnListItem>
                    </items>
                </Table>
            </smarttable:SmartTable>
```

Listing 12.56 Adding Custom Columns to a SmartTable

We add a new column to our `columns` aggregation. There we define the `text` of the column header and add a `CustomData` instance. This is necessary because the `CustomData` needs a specific `key` with the value `p13nData` for the custom column to work. As the `value` of the custom data, we need to define a `columnKey` pointing to a property of our entity set, define its `leadingProperty`, and define the `index` of the new column.

Then, we're able to add the cell's template for the new custom column to the `ColumnListItem`. In our case this is a button. This produces a new column that is displayed as the fourth one in our smart table (see Figure 12.47).

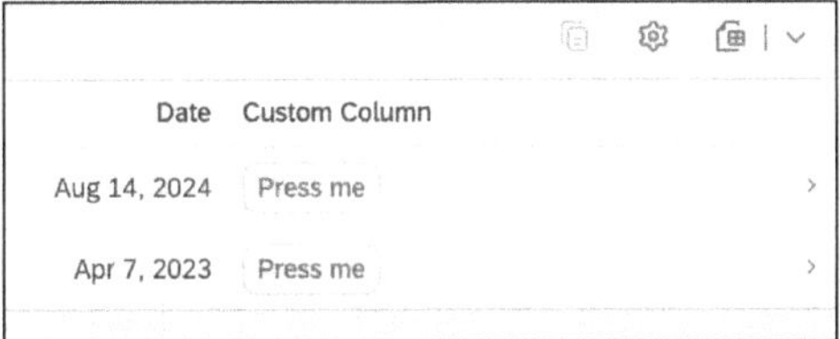

Figure 12.47 Custom Column Displayed

This was a short overview of the basic capabilities of smart controls. They are far more capable than that but this goes side by side with knowledge about the SAP Fiori elements annotations. But keep in mind that they help you build UIs very fast without the need of programming everything on your own. Building a table with personalization and variant management on your own, for example, isn't an easy task, especially keeping everything in compliance with the SAP Fiori design guidelines. This is where smart controls come in to play.

12.8 Summary

In this chapter, we looked at more advanced programming topics in SAPUI5. Dialogs and fragments allow you to add view parts that you can load on demand and reuse as modules. Base controllers offer modules in your controller, bringing often used and convenient functions into a single extendable class.

Localization, or internationalization, brings translatable texts into your application. Texts can be translated into multiple languages, where the current language is determined by SAPUI5, which provides these in a resource model. If working with file upload and download, SAPUI5 offers dedicated controls to work with media entities such as the `UploadSet`.

Working with modern, touch-enabled UIs enables SAPUI5 to offer drag-and-drop features to reorder lists or move items from one to another place.

Validating input is the first step to preventing errors. SAPUI5 provides value states, indicating that input wasn't provided in a valid state. Messages that occur either during validation or elsewhere can then be displayed in dedicated message boxes or popovers.

Smart controls are metadata-driven controls that conform to SAP Fiori design guidelines and behave as stated by an OData V2 service for easier implementation.

Chapter 13
Custom Control Development

Extending the base functionality and UI-relevant parts of SAPUI5 can enhance your applications to fit your specific needs. Whether it's the adaptation and extension of preexisting controls or the integration and development of your own controls, SAPUI5 offers the flexibility to seamlessly integrate and build your own ideas.

In this chapter, we'll take a deep dive into the development and extension of SAPUI5 controls. As SAPUI5 offers a huge variety of already existing controls, bundled into libraries, sometimes those control functionalities don't match our specific needs. Therefore, we can enhance already built and shipped controls as *extension controls*, as discussed in Section 13.1.

We'll also investigate the control metadata in Section 13.1.2. to understand how a control is defined. Once we understand this, we can enhance preexisting controls.

In the case of SAPUI5 not providing controls for specific use cases, we can develop and integrate our own controls, as discussed in Section 13.2. These *custom controls* have their own UI that you can define along with its behavior.

Extended or custom controls can then be bundled into libraries, as shown in Section 13.3, which provide the functionality to collect your controls based on their functionality and usage. You can then use the libraries in your applications.

13.1 Extending Controls

To understand what we're actually extending and developing, we must understand that a control is the SAPUI5 version of a UI element. In SAPUI5, we have two different kinds of UI elements, as follows:

- **Elements: `sap.ui.core.Element`**
  The element is the abstract base class for all SAPUI5 controls. It extends the `sap.ui.base.ManagedObject` class, which provides properties, events, and data binding features. Its ID is registered in SAPUI5 and can be written into the *Document Object Model* (DOM). Although this is a UI element, it has no renderer, which means that it lacks visual representation. An element can be used to add functionality to a control, for example, the `sap.ui.core.Item` element in the items aggregation of an

sap.m.Select control. The sap.ui.core.Item element provides entries for the sap.m.Select dropdown, which is responsible for the correct rendering. It's also responsible for listening to HTML5 events and registering event handlers.

- **Controls: `sap.ui.core.Control`**
  The control is the same as an element plus a dedicated renderer. The renderer method renders the control by using string-based rendering to generate HTML5 code, which will then be written into the DOM to be displayed. The main thing that separates a control from an HTML5 element is that the control not only has a visualization but also dedicated JavaScript or TypeScript code to define its behavior. During runtime, a control consists of the following parts:
  - A Control API to have an interface to the controls metadata
  - A control renderer to parse the control into HTML5 code
  - A control behavior, which is its JavaScript/TypeScript coding
  - The control style, which defines the styling by using cascading style sheets (CSS)

So, if you want to write your own UI element that should have visualization, you choose a control. Otherwise, you need to use an element. For the sake of simplicity, we'll also refer to an element as a control. First, you need to create an extension of an already-existing control and then add your own metadata to it.

13.1.1 Creating a Control Extension

We'll now create our own extension of a control: the sap.m.Button control. Let's say we want to have a button with which we can cancel something. Depending on the cancellation state, we want to display a corresponding text. Upon clicking the button, we want to trigger an event that opens a confirm dialog to confirm cancellation.

First, we need an application where we can add such a button. You can either generate a new application or choose one we've already created in this book. This sample will be implemented in TypeScript, but can also be done in JavaScript, by using JavaScript syntax.

We start with adding a new **TS** file called *CancellationButton.ts* in a new folder called *controls* under the *controller* folder, as shown in Figure 13.1.

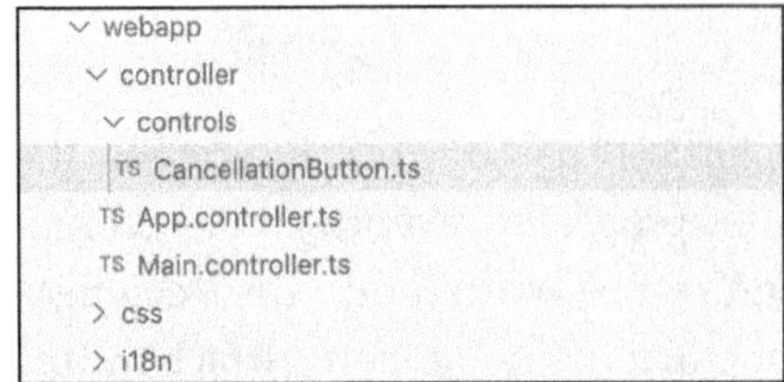

Figure 13.1 Add a New TS File

In this file, we'll add a new class called CancellationButton, which extends the sap.m.Button control, as shown in Listing 13.1.

```
import Button from "sap/m/Button";
/**
 * @namespace at.clouddna.customcontrol
 */
export default class CancellationButton extends Button {
renderer = {}
};
```

Listing 13.1 Adding a New Extension Control

This will be the base of our new custom button. In the following sections, we'll add functionality to our button. Later, we'll also talk about the renderer, but for now on, we want to use the button's renderer. First, we need to add properties and events in the form of metadata.

13.1.2 Control Metadata

To define the technical description of what a control consists of, we need to add metadata to it. The metadata of a control is generally available because it inherits its base functionality from the sap.ui.base.ManagedObject class. The sap.ui.base.ManagedObject.MetadataOptions serves as an interface, defining all possible configurations.

We'll start with defining our custom button's properties to represent its state. Then, we'll add events that are triggered at specific times.

Properties

A property, type sap.ui.base.ManagedObject.MetadataOptions.Property, stores simple values (string, Boolean, int, float, CSS, date, or custom types). They represent the state of the control and can be bound to via data binding (property binding, to be specific). Each property will automatically generate getter and setter methods, but can also be retrieved and set via the getProperty and setProperty methods. The base syntax of a property is shown in Listing 13.2.

```
properties: {
        <name>: {
            type: <type>,
            defaultValue: <optional default value>
        }
    }
```

Listing 13.2 The Property Definition

We'll now add a property to our button for storing the cancellationState, as shown in Listing 13.3.

```
import Button from "sap/m/Button";
//@ts-ignore
import type { MetadataOptions } from "sap/ui/core/Element";
/**
 * @namespace at.clouddna.customcontrol
 */
export default class CancellationButton extends Button {
    static readonly metadata: MetadataOptions = {
        properties: {
            cancellationState: {
                type: "boolean",
                defaultValue: false
            }
        }
    }
    renderer = {}
};
```

Listing 13.3 Adding a New Property to the Button

Here, we add new metadata by defining a static class variable called metadata. There, we can add new properties, namely our cancellationState. This property has the boolean datatype, and its defaultValue is false.

By adding this new property, the Control API now has two public methods: getCancellationState and setCancellationState. We're now going to overwrite the setCancellationState method to automatically set the text and the button's enablement based on the value of the property, as shown in Listing 13.4.

```
private setCancellationState(cancellationState: boolean){
        this.setProperty("cancellationState", cancellationState, true);
        if(cancellationState == true ){
            this.setText("Already cancelled");
            this.setEnabled(false);
        } else {
            this.setText("Cancel");
            this.setEnabled(true);
        }
    }
```

Listing 13.4 Overwriting the Setter Method

This method first sets the property value via setProperty. Based on the value, it also sets the text property of the button to a fixed text and enables or disables the button. We can now add our custom button in one of our views, as shown in Listing 13.5.

```
<mvc:View controllerName="at.clouddna.customcontrol.controller.Main"
    xmlns:mvc="sap.ui.core.mvc" displayBlock="true"
    xmlns="sap.m"
    xmlns:custom="at.clouddna.customcontrol.controller.controls">
    <Page id="page" title="{i18n>title}">
        <content>
            <custom:CancellationButton id="cancelButton" cancellationState=
"false"/>
        </content>
    </Page>
</mvc:View>
```

Listing 13.5 Including the Extended Button in a View

13

Here, we'll add a new namespace called custom that points to our *apps-resource* root and the folder where our extended button is defined. Then, using this namespace, we can add our CancellationButton, provide an id, and set a value to the property cancellationState. When we start our application, we should now see a button like the one in Figure 13.2 with a predefined text corresponding to the cancellation state.

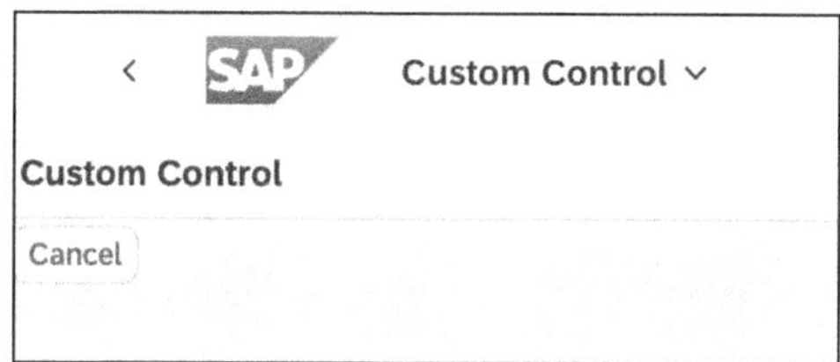

Figure 13.2 Our Extended Control

Next, we want to add functionality to our button by working with events.

Events

Events, type sap.ui.base.ManagedObject.MetadataOptions.Event, provide a way of communication between components based on a state change, which will notify event listeners. When declaring events, methods for adding, removing, and firing events are automatically added to the Control API. The base syntax of an event is as follows:

```
events: {
            <event name>: {}
        }
```

We're going to add an event called cancel, which should be triggered when the user clicks on the button and confirms a confirmation dialog. So, let's add a new event to the metadata object, as shown in Listing 13.6.

```
...
export default class CancellationButton extends Button {
    static readonly metadata: MetadataOptions = {
        properties: {
            ...
        },
        events: {
            cancel: {}
        },
    }

    private setCancellationState(cancellationState: boolean){
        ...
    }

    renderer = {}
};
```

Listing 13.6 Adding a New Event to the Control's Metadata

Here, we add the event cancel to the control's metadata object. Then, we'll overwrite the init method and open a confirm message, which will trigger our event upon confirmation, as shown in Listing 13.7.

```
import MessageBox, { Action } from "sap/m/MessageBox";
import Event from "sap/ui/base/Event";
...
export default class CancellationButton extends Button {
    static readonly metadata: MetadataOptions = {
        ...
    }

    init(): void {
        this.attachPress(this.onCancel.bind(this));
     }

    private setCancellationState(cancellationState: boolean){
        ...
    }

    private onCancel(event: Event){
```

```
        MessageBox.confirm("Do you want to cancel?",{
            onClose: (action: string)=>{
                if(action == Action.OK){
                    this.fireEvent("cancel", {
                        cancellationState:
this.getProperty("cancellationState")
                    })
                }
            }
        })
    }

    renderer = {}
};
```

Listing 13.7 Adding a Confirmation Dialog

Here, we add the `init` method, which is called upon the initialization of the control. We add a new method called `onCancel` to the `press` event registration of the button. The `onCancel` method opens a new `MessageBox`, showing a confirmation dialog. When the user selects `OK`, we trigger our `cancel` event, notifying all the listeners that are bound on this event.

Let's click on our button to see a confirmation message, as shown in Figure 13.3.

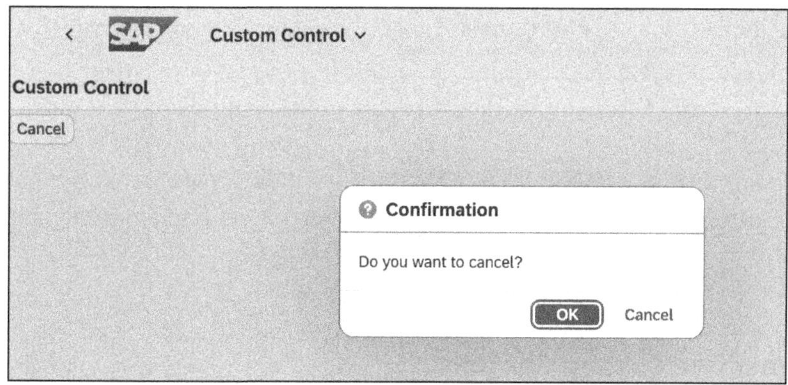

Figure 13.3 A Confirmation Dialog Appears

We can now add a listener to this event in the view, which uses our control, as follows:

```
<custom:CancellationButton id="cancelButton" cancellationState="false" cancel=
"onCancelPress"/>
```

The `onCancelPress` event listener must then be implemented in the corresponding controller, as shown in Listing 13.8.

```
import MessageToast from "sap/m/MessageToast";
...
export default class Main extends Controller {

    public onInit(): void {
      ...
    }

    public onCancelPress(event: Event){
        MessageToast.show("I was cancelled");
    }
}
```

Listing 13.8 Adding an Event Listener

Here, we want to just display a `MessageToast` when the event is triggered, as shown in Figure 13.4.

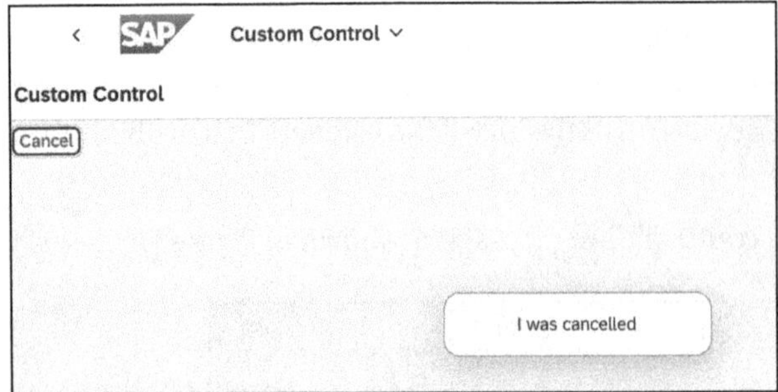

Figure 13.4 A MessageToast Appears

Our extended control is now finished. The only thing we didn't cover were aggregations and associations, which will be a part of our next section on implementing our custom controls.

13.2 Implementing Custom Controls

When implementing custom controls, the main difference is the need for a dedicated renderer method that generates HTML5 code that can be written into the DOM.

> **Control Renderer**
>
> The `sap.ui.core.Renderer` is responsible for creating the HTML structure of a control. It can be either defined as a method in the control definition or written separately in a dedicated SAPUI5 module.

The renderer provides methods for a string-based rendering based on the `sap.ui.core.RenderManager` API, which defines the structure of the control for the DOM. The `sap.ui.core.RenderManager` provides methods such as the following:

- `openStart` and `openEnd` to create a new opening HTML tag
- `close` to create a closing HTML tag
- `attr` to add HTML attributes in an opening tag
- `class` to add CSS classes in an opening tag
- `icon` to add rendered SAPUI5 icons
- `renderControl` to call a control's renderer
- `style` to add the style attribute in an opening tag
- `text` to add a text node in a HTML tag

The `sap.ui.core.RenderManager` also provides methods for HTML sanitizing and unsafe HTML that should be written into the DOM.

We're going to create a custom control for a property list, displaying a title and an HTML list where we can display properties. Then, we'll add aggregations for the items of our list and a renderer, where we can define the controls HTML5 representation.

13.2.1 Creating a Custom Control

Let's start by creating a new file `called` *PropertyList.ts* in our *control* folder, as shown in Listing 13.9.

```
import Control from "sap/ui/core/Control";
//@ts-ignore
import type { MetadataOptions } from "sap/ui/core/Element";
/**
 * @namespace at.clouddna.customcontrol
 */
export default class PropertyList extends Control {
    static readonly metadata: MetadataOptions = {
    }

    init(): void {
     }

};
```

Listing 13.9 Adding a New Control

Our `PropertyList` control extends the base `sap.ui.core.Control` class, making it a new custom control, where we need to set a new property for our title (see Listing 13.10).

```
...
export default class PropertyList extends Control {
    static readonly metadata: MetadataOptions = {
        properties: {
            mainLabel: {
                type: "string",
                defaultValue: ""
            }
        }
    }
...
};
```

Listing 13.10 Adding a New Property

The property `mainLabel`, type `string`, should be our label that will be displayed above our list. To now add a way of providing items to our list, we'll use aggregations.

13.2.2 Aggregations

An aggregation, type `sap.ui.base.ManagedObject.MetadataOptions.Aggregation`, can store one or multiple references to other controls or elements that are dependent on the lifecycle of the control. That means, when their parent is destroyed, they also will be deleted. An aggregation is defined via a name, a type, and a cardinality. Adding an aggregation to a control will also enable aggregation binding functionality and dedicated Control API methods for inserting and removing elements in the aggregation. The base syntax of an aggregation is as follows:

```
aggregations: {
            <aggregation name>: {
                type: <type>,
                multiple: <true or false>,
                singularName: <singular name>
            }
        }
```

Here, we add an aggregation to our custom control that stores items that should be displayed in the to-be rendered list, as shown in Listing 13.11.

```
...
export default class PropertyList extends Control {
    static readonly metadata: MetadataOptions = {
        properties: {
            ...
        },
```

```
        defaultAggregation: "propertyItems",
        aggregations: {
            propertyItems: {
                type: "sap.ui.core.Item",
                multiple: true,
                singularName: "propertyItem"
            }
        }
    }
    ...
};
```

Listing 13.11 Adding an Aggregation

Our aggregation `propertyItems` receives items from the `sap.ui.core.Item` element and has a cardinality of 0..N. We also set this aggregation as the `defaultAggregation` of our control.

Next, we need to add a dedicated renderer to our control.

13.2.3 Control Renderer

To implement a renderer for a custom control, either a dedicated SAPUI5 module or a method in the control implementation must be added. We create a dedicated module by adding a new file called *PropertyListRenderer.ts* to the *control* folder and then implement the renderer, as shown in Listing 13.12.

```
import RenderManager from "sap/ui/core/RenderManager";
import PropertyList from "./PropertyList";

/**
 * @namespace at.clouddna.customcontrol
 */
export default {
    apiVersion: 2,

    render: function (rm: RenderManager, control: PropertyList) {

    }
};
```

Listing 13.12 Creating a Dedicated Renderer

The renderer module consists of an object, defining the `apiVersion` of the renderer and the renderer method. This method will be called upon SAPUI5 rendering the control and receives a render manager and control instance. We can then implement the render method, as shown in Listing 13.13.

```
...
import Title from "sap/m/Title";
import Item from "sap/ui/core/Item";

...
export default {
    apiVersion: 2,

    render: function (rm: RenderManager, control: PropertyList) {
        rm.openStart("div", control);
        rm.openEnd();

        //@ts-ignore
        const mainLabel = control.getMainLabel(),
            label = new Title({
                text: mainLabel
            });
        rm.renderControl(label);

        rm.openStart("ul");
        rm.openEnd();
        //@ts-ignore
        const propertyItems: Item[] = control.getPropertyItems();

        propertyItems.forEach(propertyItem => {
            const text = `${propertyItem.getKey()}: ${propertyItem.getText()}`
            rm.openStart("li");
            rm.openEnd();
            rm.text(text)
            rm.close("li");
        });

        rm.close("ul");
        rm.close("div");
    }
};
```

Listing 13.13 The Renderer Implementation

The RenderManager first creates a new HTML div tag by using openStart and then writes basic control information such as the ID into the attributes of the div tag. Then, a new sap.m.Label is created, which gets its text from the getter method of the mainLabel property of the PropertyList control. The renderControl method then renders this label. Afterward, a new unordered HTML list is added. Then, the items of the PropertyList's propertyItems aggregation are read, and their key and text are added as new list items to the unordered list. At last, the closing div tag is created. The RenderManager will now insert this rendered control into the DOM.

This renderer module must now be referenced in the PropertyList control, as shown in Listing 13.14.

```
...
export default class PropertyList extends Control {
    ...
     static renderer: typeof PropertyListRenderer = PropertyListRenderer;

};
```

Listing 13.14 Adding the Renderer Reference

Now we can use this control in one of our views, like we did with the CancellationButton (see Listing 13.15).

```
<mvc:View controllerName="at.clouddna.customcontrol.controller.Main"
    xmlns:mvc="sap.ui.core.mvc" displayBlock="true"
    xmlns="sap.m"
    xmlns:core="sap.ui.core"
    xmlns:custom="at.clouddna.customcontrol.controller.controls">
    <Page id="page" title="{i18n>title}">
        <content>
            <custom:PropertyList id="myApple" mainLabel="An Apple">
                <custom:propertyItems>
                    <core:Item key="color" text="Red"/>
                    <core:Item key="type" text="Red Delicious"/>
                    <core:Item key="taste" text="sweet"/>
                </custom:propertyItems>
            </custom:PropertyList>
        </content>
    </Page>
</mvc:View>
```

Listing 13.15 Adding the Custom Control in an SAPUI5 View

Here, we add our PropertyList and provide a value for the mainLabel property. Then, we add new sap.ui.core.Item elements to the propertyItems aggregation. In this case, we want to display information about an apple. If the application runs, the PropertyList control will now be rendered and is displayed, as shown in Figure 13.5.

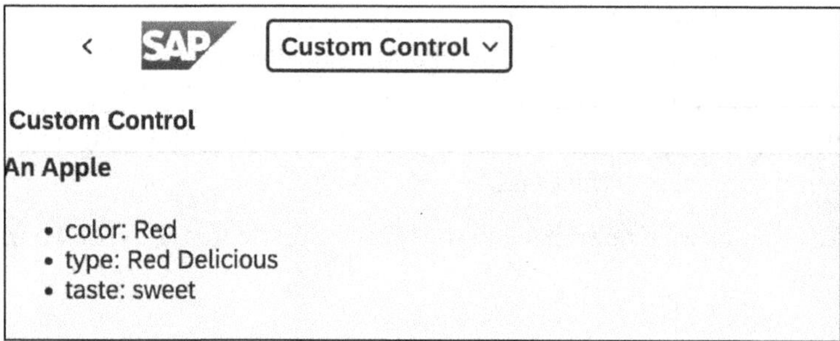

Figure 13.5 The Rendered Control

Now that we've created an extended control and a custom control, we'll look into how we can bundle them into a library to make them generally available in our other applications.

13.3 Libraries

Libraries help us to bundle custom code into one place. This enables us to reuse it systemwide in both our on-premise and cloud applications, get automatic theming support, and stay consistent with the SAP Fiori design.

A library itself can either be created by generating a new SAPUI5 application or by using a custom generator, which we'll talk about later. To create a library, you'll need a folder called *src*, which includes a *library.ts* file. This will serve as the descriptor of the library and as the equivalent to the *manifest.json* file of a SAPUI5 application (see Listing 13.16).

It's also possible to create a *manifest.json* file to be used as a library descriptor. This can then be used to include OData services, load additional references, and so on. But this isn't required, and dependencies to OData services and additional things can be done directly in your controls.

```
sap.ui.getCore().initLibrary({
    name: "<library id>",
    version: "${version}",
    dependencies: [
        "sap.ui.core",
        "sap.f",
        "sap.ushell",
        "sap.suite.ui.commons",
        "sap.ui.richtexteditor"
    ],
    types: [],
    interfaces: [],
    controls: []
```

```
    elements: [],
    noLibraryCSS: true
});

const thisLib: { [key: string]: unknown } = ObjectPath.get("<library id>") as {
[key: string]: unknown };

export default thisLib;
```

Listing 13.16 The library.ts File

This file initializes a new library with a certain ID (see the placeholder `<library id>`). A library can have cross-dependencies to other libraries such as `sap.ui.core`. These can be specified in the `dependencies` property to load them when the library is used somewhere.

TypeScript libraries can also define `types` that you can use as enums when you use custom types for your custom control properties, for example. This also goes for the `interfaces` property. The main property is the `controls` property where you define the custom or extended controls that should be included in your library. This is also done for `elements`, meaning non-UI elements. Then, you can specify if you also want to bundle custom CSS files. After that, the library must be created and exported.

CSS in Libraries

Libraries also support the definition of custom CSS logic via Leaner Style Sheets (Less), a stylesheet language for easier CSS definitions. By using the *themes* folder of your library, stylesheets for each SAP Fiori theme can be created and used to define custom style rules that can then used by your controls or renderers.

We'll now create a custom library that includes the controls we built in the previous sections to make them available for consumption in our other SAPUI5 apps.

13.3.1 Create a New Library

Unfortunately, there's no generator for SAPUI5 libraries included in the SAP Fiori tools. This was the case for SAP Web IDE, but hasn't been added to Visual Studio Code (VS Code) or SAP Business Application Studio. So, a big shout out goes to SAP Community and a few SAP employees for providing a sample library that you can use as a boilerplate to easily create your own library. We'll use this library, which can be found via the GitHub repository already tailored to our needs: *https://github.com/clouddnagmbh/samplelibrary*.

Start by opening a new workspace or folder in your development environment—this sample uses VS Code—and clone this repository. If you don't have Git installed, you can also download it. Use a new terminal to run the `git clone` command:

```
git clone https://github.com/clouddnagmbh/samplelibrary
```

This will download the repository. After that, open a terminal in the *samplelibrary* folder, and run the command `npm install`. Now this library is ready to be tested and adapted to our needs (see Figure 13.6).

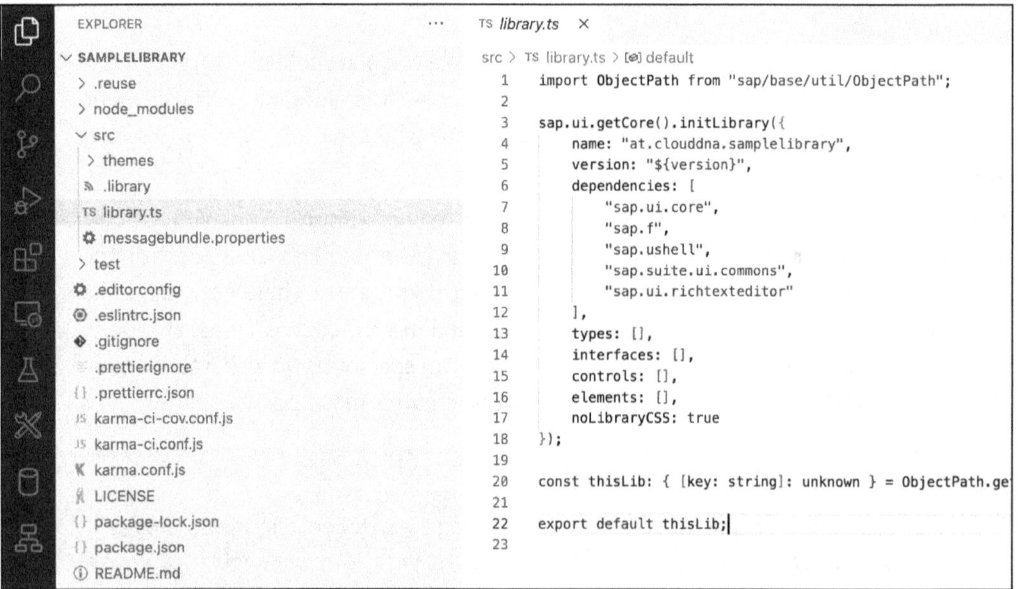

Figure 13.6 The Cloned Repository

This sample library comes with an option to directly test its contents by including *Example.html* and *Example.ts* files in the *test* folder. These files can be used to instantiate our controls to see if they work. Try it out by running the command `npm run start`. This will run the library that includes a custom control called `Example`, which renders a little colored box. You can make yourself familiar with the contents of the library if you want, but we'll now continue with making it our own.

Changing the ID of the Library

Each library, same as with SAPUI5 applications, needs an ID that is unique to the system it will be deployed on. To make this library your own, you can use the search and replace function of your IDE to replace the ID `at.clouddna.samplelibrary` with one you choose. Don't forget to replace the path *at/clouddna/samplelibrary* with a path to your ID.

13.3.2 Development

We now add the `CancellationButton` and `PropertyList` controls from the previous sections to our library. Therefore, copy the files *CancellationButton.ts*, *PropertyList.ts*, and *PropertyListRenderer.ts* from the previous exercise, and paste them in the *src* folder.

Your *src* folder should now look like Figure 13.7. The *.gen* files will be automatically generated after running the `npm run start` command in the terminal if not already done.

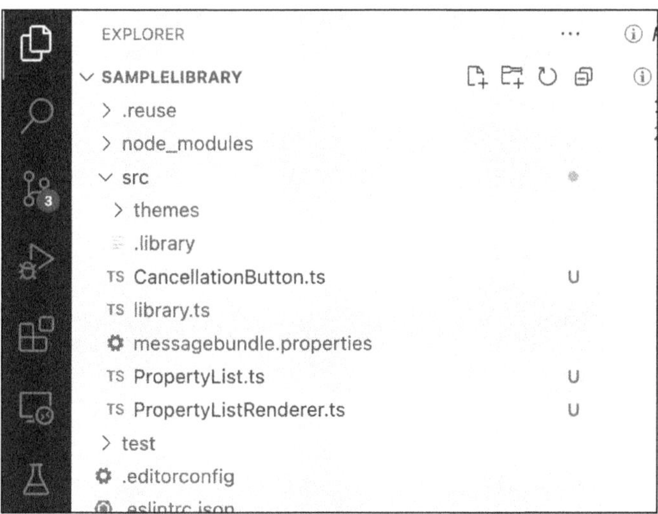

Figure 13.7 Added Custom Controls

For the library to work and to understand our controls, we need to make a few adjustments to the controls. We start by opening the *CancellationButton.ts* file. Here, we add a few lines of code, as shown in Listing 13.17.

```
...
/**
 * @namespace at.clouddna.samplelibrary
 * @name at.clouddna.samplelibrary.CancellationButton
 */
export default class CancellationButton extends Button {
    static readonly metadata: MetadataOptions = {
        ...
    }

    constructor(id?: string | $CancellationButtonSettings);
    constructor(id?: string, settings?: $CancellationButtonSettings);
    constructor(id?: string, settings?: $CancellationButtonSettings) {
        super(id, settings);
    }
```

```
    ...

    renderer = {}
};
```

Listing 13.17 Adjusting the Control in CancellationButton.ts

We start by adding the *.ts* doc annotations for the name and the namespace of our control. This needs to include our library's ID. Then, we add a constructor for TypeScript to be able to recognize out control's signature. Save this and make the same changes in the *PropertyList.ts* file, as shown in Listing 13.18.

```
...
import PropertyListRenderer from "at/clouddna/samplelibrary/
PropertyListRenderer";
/**
 * @namespace at.clouddna.samplelibrary
 * @name at.clouddna.samplelibrary.PropertyList
 */
export default class PropertyList extends Control {
    ...

    constructor(id?: string | $PropertyListSettings);
    constructor(id?: string, settings?: $PropertyListSettings);
    constructor(id?: string, settings?: $PropertyListSettings) {
        super(id, settings);
    }

    ...

};
```

Listing 13.18 Adjusting the Control in PropertyList.ts

Here, we also add a constructor and change the name and namespace. In addition, however, we need to change the import of the `PropertyListRenderer` to include our library path.

Save this and open the *PropertyListRenderer.ts* file, add the name and namespace annotation, and adjust the import of the `PropertyList`, as shown in Listing 13.19.

```
...
import PropertyList from "at/clouddna/samplelibrary/PropertyList";
...

/**
```

```
 * @namespace at.clouddna.samplelibrary
 * @name at.clouddna.samplelibrary.PropertyListRenderer
 */
export default {
    ...
    }
};
```

Listing 13.19 Adjusting the Renderer

Save this and open the *library.ts* file. Here, we need to include our custom controls, as shown in Listing 13.20.

```
import ObjectPath from "sap/base/util/ObjectPath";

sap.ui.getCore().initLibrary({
    name: "at.clouddna.samplelibrary",
    version: "${version}",
    dependencies: [
        ...
    ],
    types: [],
    interfaces: [],
    controls: [
        "at.clouddna.samplelibrary.CancellationButton",
        "at.clouddna.samplelibrary.PropertyList",
    ],
    elements: [
        "at.clouddna.samplelibrary.PropertyListRenderer",
    ],
    noLibraryCSS: true
});

const thisLib: { [key: string]: unknown } =
ObjectPath.get("at.clouddna.samplelibrary") as { [key: string]: unknown };

export default thisLib;
```

Listing 13.20 Adding the Controls to Their Respective Definitions

We add the `ProperyList` and the `CancellationButton` controls to the `controls` list of our library. The `ProperyListRenderer` gets added to the `elements` list.

Now our library is complete and can be deployed to a system. Before deployment though, it's highly recommended to test the contents of your library.

Internationalization and Translation in Libraries

SAPUI5 libraries support the creation of language-dependent property files. They can be found in the *messagebundle.property* file in the *src* folder. Here, similar to the i18n files of an SAPUI5 application, key-value pairs of language-dependent texts can be added.

To consume them inside your library, you can use the resource bundle as follows:

```
const resourceBundle = Core.getLibraryResourceBundle("at.clouddna.sampleli-
brary") as ResourceBundle;
const text = resourceBundle.getText('key of a text in the properties-file',[
'placeholders']);
```

As already mentioned, to test your library, an *Example.ts* file was added to the project. Open this file, and add a new `CancellationButton` instance, as shown in Listing 13.21.

```
import { CancellationButton } from "at/clouddna/samplelibrary/library";
// or import CancellationButton from "../src/CancellationButton";

// create a new instance of the Example control and

const cancellationButton = new CancellationButton({
    cancellationState: false
});

cancellationButton.placeAt("content");
```

Listing 13.21 Adding a New Instance

Now if you run `npm run start` in a terminal, you'll see that our `CancellationButton` is working, as shown in Figure 13.8.

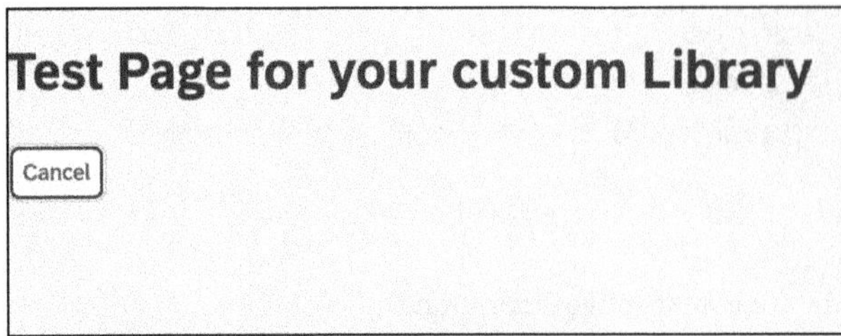

Figure 13.8 The Working CancellationButton

Wonderful! Now we're ready to deploy our library to make it usable in other applications.

13.3.3 Deployment

For simplicity purposes, a *ui5-deploy.yaml* file has already been added to the project. Here, you can adjust the properties to deploy a new application, as shown in Listing 13.22.

```
# yaml-language-server: $schema=https://sap.github.io/ui5-tooling/schema/
ui5.yaml.json

specVersion: "2.6"
metadata:
  name: at.clouddna.samplelibrary
type: library
builder:
  resources:
    configuration:
      paths:
        src: dist/resources/at/clouddna/samplelibrary/
    excludes:
      - /test-resources/**
      - /localService/**
  customTasks:
    - name: deploy-to-abap
      afterTask: generateCachebusterInfo
      configuration:
        target:
          url: <paste system url>
          client: "100"
        app:
          name: ZTSSAMPLELIB
          description: 'TypeScript Sample Library'
          package: $TMP
          transport:
        exclude:
          - /test/
```

Listing 13.22 Adjusting the Deployment File

Here, you need to provide the `url` and `client` of you SAP S/4HANA system. Then, provide a `name`, a `description`, a `package`, and a `transport` request number if not deployed as a local object.

Open a new terminal, and execute the command `npm run deploy`. Wait until the build process is finished, and then press [Y] to start the deployment to your system. Don't worry if the deployment states that all files are uploaded, but it finishes with the status code **400**. It still worked.

13

Deployment on Different Operating Systems

For the deployment to be executed on different operating systems, the `flatten` and `clean-after-flatten` scripts in *package.json* need to be adjusted:

- macOS

```
"flatten": "copy dist\\resources\\at\\clouddna\\samplelibrary\\* dist\\"
"clean-after-flatten": "rmdir /s /q dist\\resources dist\\test-resources"
```

- Windows

```
"flatten": "cp -R './dist/resources/at/clouddna/samplelibrary/' './dist'"
"clean-after-flatten": "rm -rf './dist/resources' './dist/test-
resources'"
```

If you log on to your SAP system and open Transaction SE80, you can view your deployed library, as shown in Figure 13.9.

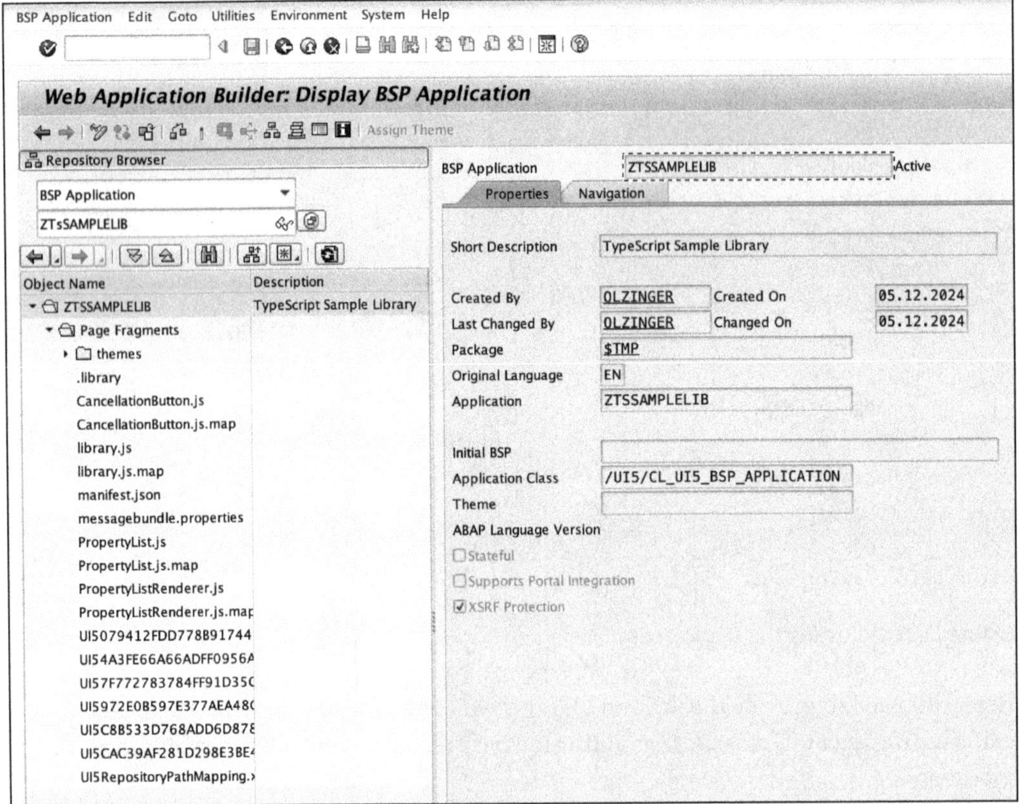

Figure 13.9 Your Library in Your System

Now we're ready to use the library in one of our applications.

13.3.4 Usage in Applications

To add a library to an existing application, it must be added to the *manifest.json* of your SAPUI5 application. Here, you add your library ID in the node sap.ui5 -> dependencies -> libs. For your application to be able to load your library, both must be deployed onto the same system. To integrate one of our controls of our library we just deployed, we can add its ID to one of our applications, as shown in Listing 13.23.

```
"sap.ui5": {
    ...
    "dependencies": {
      "minUI5Version": "1.126.0",
      "libs": {
        ...
        "at.clouddna.samplelibrary": {}
      }
    }
    ...
}
```

Listing 13.23 Adding a Library Reference

Then, all you need to do is deploy the application. However, in most scenarios, it's necessary to test the contents of the library that you've included during your development process. Therefore, the *ui5.yaml* file must be enhanced to know where it should load the library from when running the application locally.

You need to inform SAP Fiori tools where your library is by providing the path that your library should be routed through. You can do this by adding a new backend configuration in the *ui5.yaml* file pointing to your system, as shown in Listing 13.24.

```
server:
  customMiddleware:
    - name: fiori-tools-proxy
      ...
        ui5:
          ..
        backend:
          ...
          - path: /at
            url: https://<your system>
            client: <your client>
```

Listing 13.24 Adding a Backend Configuration

This will route all requests going to at to your system, and because your library's ID is at.clouddna.samplelibrary, this will load your library from your system.

Then, you can already use your library's content in views or controllers, as shown in Listing 13.25.

```
<mvc:View controllerName="at.clouddna.customcontrol.controller.Main"
    xmlns:mvc="sap.ui.core.mvc" displayBlock="true"
    xmlns="sap.m"
    xmlns:core="sap.ui.core"
    xmlns:custom="at.clouddna.samplelibrary">
    <Page id="page" title="{i18n>title}">
        <content>
            <custom:PropertyList id="myApple" mainLabel="An Apple">
                <custom:propertyItems>
                    <core:Item key="color" text="Red"/>
                    <core:Item key="type" text="Red Delicious"/>
                    <core:Item key="taste" text="sweet"/>
                </custom:propertyItems>
            </custom:PropertyList>
        </content>
    </Page>
</mvc:View>
```

Listing 13.25 Adding Content from the Library into a View

You can now test your included controls before you deploy your application.

13.4 Summary

In this chapter, we extended existing SAPUI5 controls and added new controls that previously didn't exists to adjust the SAPUI5 base functionality to our specific needs. Extension controls can be used to add new functionality and enhance existing controls. If we need a custom user experience, we can create custom controls with custom HTML5 rendering. These controls and other functionality can be bundled into libraries to reuse custom logic and controls in our applications.

All of these additional programming topics enhance your programming experience, tailor SAPUI5 to your needs, and enable reusability throughout your development lifecycle.

Chapter 14
SAP Fiori Elements

The robust SAP Fiori elements framework is designed to simplify and accelerate the development of SAPUI5 applications. This chapter begins by introducing the various floorplans—standardized layouts and navigation patterns tailored for common application scenarios. It then delves into the flexible programming model, enabling developers to create responsive and adaptive applications. Key topics include the edit flow for seamless data entry and validation, the use of macros for automating repetitive tasks, and the extensive options for customizing and extending standard applications to meet specific business needs.

From the most diverse requirements for business applications, comparable use cases crystallize again and again in which data must be presented in a certain way. To implement these use cases as simply as possible and with relatively little development effort, you can fall back on floorplans. Admittedly, you're more flexible if you develop your own freestyle SAPUI5 applications. However, freestyle development is more complex and requires in-depth knowledge of SAPUI5 development. If you rely on *SAP Fiori elements* instead, you can create applications with little knowledge of SAPUI5. At the same time, you automatically achieve a consistent UI design, and your applications comply with the SAP Fiori design guidelines.

The SAP Fiori elements framework offers you a range of floorplans. Whether you want to create an overview page with key figures or simply display a large amount of data in list form, there's a floorplan for every common use case. In addition to a list display, you can also create a detailed view for the individual objects within this list, which you can jump to when you select an object. You can use this detailed view to create new objects or edit existing ones. In addition, statistical data can also be displayed in various ways with a floorplan.

We briefly introduce the available floorplans and their functions in Section 14.1. Then, we'll look at the flexible programming model in Section 14.2, which allows seamless integration of SAPUI5 coding, controls, and extension points without forcing a strict choice between SAP Fiori elements and freestyle SAPUI5 development. You can leverage building blocks, controller extensions, or even develop a fully custom app within the SAP Fiori elements framework, treating every page as a custom page. Finally, we'll explain how applications can be extended in Section 14.3.

OData Versions

The individual floorplans sometimes differ depending on which OData version you're using. OData Versions 2 (V2) and 4 (V4) are available, so certain functions and annotations are only available in one of the two versions or work differently in the two versions. In the following examples we use OData V2. It's also important to note that both the flexible programming model and the edit flow are exclusively supported by OData V4. This ensures a more modern and robust framework for application development, leveraging the advanced capabilities and enhanced features of the latest OData standard to provide superior performance, scalability, and flexibility in SAPUI5 applications.

Further Reading

For more in-depth information on the topic of SAP Fiori elements, you can also refer to our other book, *SAP Fiori Elements: Development and Extensibility* (*www.sap-press.com/sap-fiori-elements_5641/*).

14.1 Floorplans

In SAP Fiori elements, you can choose from several predefined floorplans to create responsive and consistent user interfaces (UIs) for your SAP Fiori apps. These floorplans are designed to simplify app development by providing a standardized set of UI elements and behaviors. Probably the most commonly used floorplans are as follows:

- Overview page
- List report
- Object page

In the following, we'll look at them in detail. After having a look at the single floorplans, we'll discuss the generic annotations present for every floorplan.

14.1.1 List Report

The *list report* is mainly used when large amounts of data need to be displayed and the individual entries need to be handled. The presentation doesn't necessarily have to be in tabular or list form, as a diagram is also possible. In addition, this floorplan enables users to prepare the result set accordingly by means of filters, sorting, and grouping. In this way, large amounts of data can be reduced to the relevant records.

However, you shouldn't use the list report if you want to see the details of a record. In this case, the object page is suitable. The list reports is structured as follows:

- Dynamic page header
- Content area
- Footer toolbar

Figure 14.1 shows what a list report might look like. In this example, however, no footer toolbar is used—only the dynamic page header with the filter functions and the content area are used. At the top of the page, you'll find the *dynamic page header* ❶. This header contains the title of the report and tells you what information is displayed with the application. In addition, you'll find in the dynamic page header, for example, the filter line and buttons for general actions. These actions allow you to expand or compress the header. You can also pin or unpin it so that it's always displayed or only at the top when scrolling through the list.

The *content area* ❷ displays the actual content in table, list, or chart form. You can display data in different ways in this area. In addition, a title and an icon tab bar are displayed above the list. The latter can contain text buttons and/or icons.

In the *footer toolbar*, you'll find overarching actions that relate to the entire data overview, for example, saving or starting a simulation, as well as a notification icon, if applicable. This icon allows you to view notifications of errors that have occurred in a floating menu (popover). The footer toolbar is optional, so you don't have to use it.

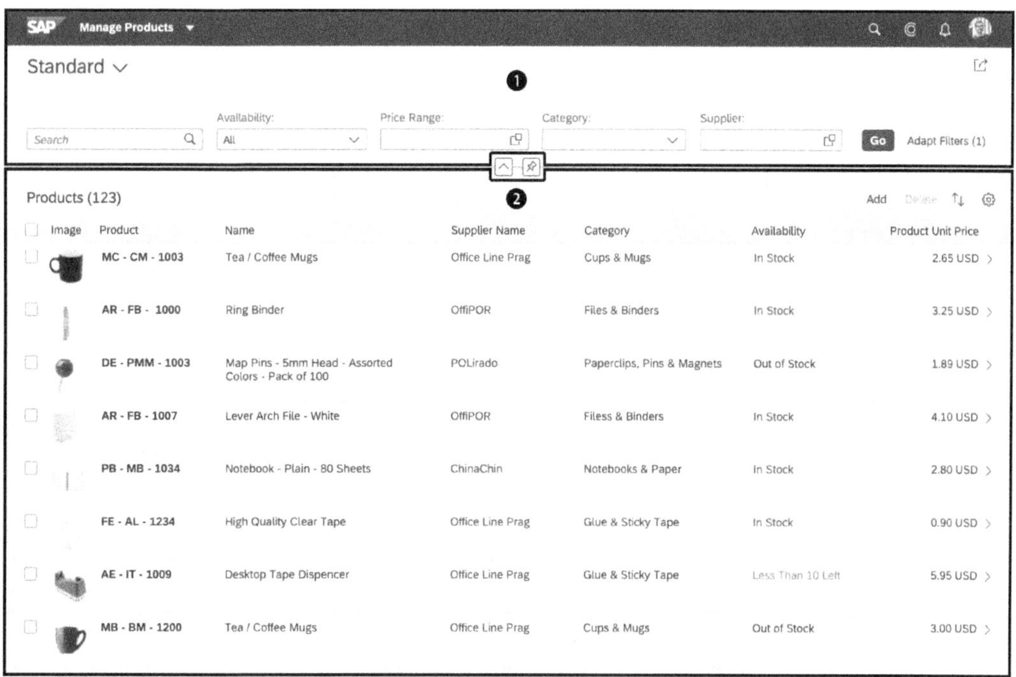

Figure 14.1 SAP Fiori Elements: List Report

The following are some use cases in which the list report is a good choice:

- Users need to efficiently locate and manage pertinent items from a vast collection of data. This is achieved by providing them with search, filter, sort, and group features.
- Your aim is to allow users to view the complete dataset using a range of visual representations such as tables or charts, without requiring any interactions between them. An example use case for this is generating reports.
- Users require the ability to interact with several views of the same content, such as open, in process, or completed items. To achieve this, you want to provide users with a way to switch between different views using features such as tabs, segmented buttons, or a select control.
- In general, drilldown is infrequently used or not available as a free or flexible option within the same page. Instead, it's typically accessible via navigation to another page.
- Users handle various types of items.

In the following, we'll see the most used annotations for the implementation of a list report:

- `UI.HeaderInfo`
  - Data type: `STRING(60)`
  - Example: `UI.HeaderInfo.typeName: 'Default'`

  This annotation basically refers to the content of a header. Various properties such as the title (e.g., name of the displayed object in singular and plural), a description, or an image URL can be defined with this annotation. The title can be a maximum of 60 characters long.
- `UI.LineItem`
  - Example: `UI.LineItem: {label: 'Name', position: 10}`

  With this annotation, you can define how the individual table rows should look. You can define properties such as the column titles, the column positions, the values, the data type of the respective list element, and so on.
- `UI.SelectionFields`
  - Example: `UI.SelectionFields: [{qualifier: 'Name', position: 10}]`

  With this annotation, you can define which filter properties you want to provide.

14.1.2 Object Page

In the previous section, we looked at floorplans that allow data to be displayed in tables or charts. What we're still missing is the detailed view of a business object. This is made possible by the *object page*. However, the functions of the object page go beyond those of a detailed view. You can also use it to create new objects, edit existing ones, or delete them. Depending on how complex your objects are, you can display the data in simple forms or in tables. The object page is made up of the following components:

- Dynamic page header
- Navigation bar (optional)
- Content area
- Footer toolbar (optional)

Figure 14.2 shows what an object page might look like. In the header, some important key performance indicators (KPIs) are shown.

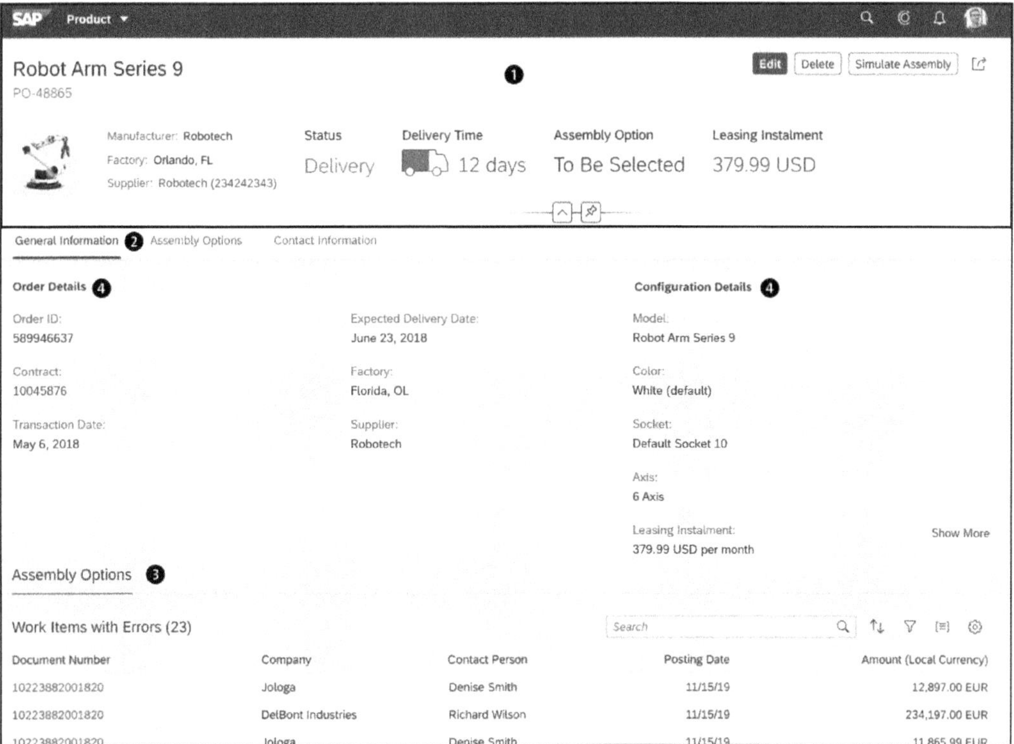

Figure 14.2 SAP Fiori Elements: Object Page

You'll also find actions such as editing or deleting an object, but also more specific functions. In the content area, the details of the selected object are displayed in forms and tables. The dynamic page header ❶ contains key information about the displayed object. You'll also find an edit button and a delete button here. In addition to the title, the header also contains a subtitle or breadcrumbs. You can use the navigation bar ❷ to control which content is displayed in the content area. This can be used in two different ways:

- **As an anchor bar**
  An anchor bar consists of a series of links that refer to different subsections. Clicking

on such a link takes you to the corresponding subsection. The anchor bar is mainly used when the contents of the individual sections belong together.

- **As a tab bar**
  The tab bar also contains links. However, these refer to other pages within the application. This form of navigation is used when different contents are displayed on the individual pages. The available tabs are displayed in the form of horizontally arranged links. Each of these links can be used to jump to a section or subsection.

The content area displays the detailed data of an object. This representation is done by means of sections ❸ and subsections ❹. A section can be seen as a kind of container that contains several subsections. Each section has a title. In the subsections, you'll find the actual contents. Each subsection also has a title and a toolbar. Whether you display tables, forms, or diagrams in the subsections is up to you. Below the content area is the footer toolbar. This contains, if available, final actions such as saving, canceling, or accepting changes.

The following are some use cases in which the object page is a good choice:

- Users need to view, generate, or modify an object.
- Users need to obtain a comprehensive understanding of an object and interact with its various components.

In the following, we see which annotations are relevant for the object page floorplan:

- **`UI.headerInfo.typeName`**
  - Type: `STRING(60)`
  - Example: `UI.headerInfo.typeName: 'Order'`

  Specifies the name of the object type and is displayed in the header of the SAP Fiori launchpad as the title of the application. This annotation must be assigned unless the `EndUserText.label` annotation has been assigned.
- **`UI.headerInfo.typeNamePlural`**
  - Type: `STRING(60)`
  - Example: `UI.headerInfo.typeNamePlural: 'Orders'`

  Represents the title of an object type above a list and must be mandatory. The value corresponds to the plural of the value from the `UI.headerInfo.typeName` annotation.
- **`UI.headerInfo.typeImageUrl`**
  - Type: `STRING`
  - Example: `UI.headerInfo.typeImageUrl: 'https://myurl/myimage.png'`

  References an image by URL that serves as a representation of an object type. Such an image is optional.
- **`UI.headerInfo.title.label`**
  - Type: `STRING(60)`

- Example: `UI.headerInfo.title.label: 'Order Nr.'`

Serves as a description of the element by which the object is identified. (The described value is taken from the `UI.headerInfo.title.value` annotation.)

- **`UI.headerInfo.title.value`**
  - Type: `elementRef`
  - Example: `UI.headerInfo.title.value: 'OrderNr'`

  Displays the value of an entity attribute as the title of the page. This value is usually the key attribute but should be understandable and readable by users. This annotation must be mandatory.

- **`UI.headerInfo.title.type`**
  - Type: `Enum`
  - Example: `UI.headerInfo.title.type: #STANDARD`

  Optionally, it allows you to define the type of header. The following values are allowed:
  - `STANDARD`: The page title is displayed as a normal data field, and the annotations for the label, value, and criticality attributes can be used.
  - `WITH_NAVIGATION_PATH`: The page title serves as a navigation element that points to an OData navigation attribute via the annotations for the `label`, `value`, and `targetElement` attributes.
  - `WITH_URL`: The page title serves as a navigation element that points to a URL via the annotations for the `label`, `value` and `url` attributes.
  - `WITH_INTENT_BASED_NAVIGATION`: If this type is used, an intent-based navigation can be implemented via the annotations for the value and `semanticObjectAction` attributes.

- **`UI.headerInfo.title.criticality`**
  - Type: `elementRef`
  - Example: `UI.headerInfo.title.criticality: 'CriticalValue'`

  Provided that the type annotation has the value `STANDARD`, the traffic light color of the title (standard, red, yellow, green) can optionally be controlled via the criticality annotations. The referenced element must contain a number (0, 1, 2, 3).

- **`UI.headerInfo.title.iconURL`**
  - Type: `STRING`
  - Example: `UI.headerInfo.title.iconUrl: '/img/cust.png'`

  Optionally points via URL to an image that displays a header icon.

- **`UI.headerInfo.title.targetElement`**
  - Type: `elementRef`
  - Example: `UI.headerInfo.title.targetElement: 'toSupplier'`

14

Optionally points to the navigation attribute that is triggered when the header is clicked (the type must be STANDARD).

- **UI.headerInfo.title.url**
  - Type: elementRef
  - Example: UI.headerInfo.title.url: 'CustomUrl'

  Optionally points to the URL that is triggered when the header is clicked (the type must be WITH_URL).

- **UI.headerInfo.description.label**
  - Type: STRING(60)
  - Example: UI.headerInfo.description.label: 'OrderDescLabel'

  Optionally serves as a label for the displayed object description. (The described value comes from the UI.headerInfo.description.value annotation.).

- **UI.headerInfo.description.value**
  - Type: elementRef
  - Example: UI.headerInfo.description.value: 'OrderDesc'

  Optionally displays the value of an entity attribute as a subtitle of the page. This value is usually an additional description and should be understandable and readable for users.

- **UI.headerInfo.description.type**
  - Type: Enum
  - Example: UI.headerInfo.description.type: #STANDARD

  Defines the type of the subtitle field. The following values are allowed:
  - STANDARD: The page description is displayed as a normal data field, and the annotations for the label, value, and criticality attributes can be used.
  - WITH_NAVIGATION_PATH: The page description serves as a navigation element that points to an OData navigation attribute via the annotations for the label, value, and targetElement attributes.
  - WITH_URL: The page description serves as a navigation element that points to a URL via the annotations for the label, value, and url attributes.
  - WITH_INTENT_BUSINESS APPLICATION STUDIOED_NAVIGATION: When this type is used, intent-based navigation can be implemented via the annotations for the value and semanticObjectAction attributes.

- **UI.headerInfo.description.criticality**
  - Type: elementRef
  - Example: UI.headerInfo.description.criticality: 'CriticalValue'

If the type annotation has the value STANDARD, the criticality annotations can be used to control the color of the description (standard, red, yellow, green). The referenced element must contain a number (0, 1, 2, 3).

- **UI.headerInfo.description.iconURL**
  - Type: STRING
  - Example: UI.headerInfo.description.iconUrl: '/img/cust.png'

  Optionally points via URL to an image that displays an icon next to the description.
- **UI.headerInfo.description.targetElement**
  - Type: elementRef
  - Example: UI.headerInfo.description.targetElement: 'toSupplier'

14.1.3 Overview Page

The SAP Fiori design guidelines define exactly what an SAP Fiori elements overview page should look like. The *overview page* floorplan basically serves as the central entry point. From here, you can jump to other floorplans and views. For example, you could use the overview page to give an overview of the information on a department from different applications. Subsequently, you could use a list report floorplan or an object page to present more detailed information.

If you want to display an overview or dashboard for a specific business area or topic, it's best to use the floorplan overview page for this purpose. The overview page is particularly suitable for displaying the following information:

- Key tasks to be completed
- Various notes on a task area
- Various KPIs

In addition, you can also display diagrams and graphics on an overview page. The data displayed on the overview page is usually compiled from a variety of sources and is intended to provide the user with an overall view. For example, you could show data on all your products, customers, suppliers, or similar on an overview page.

As cards are a fundamental design element when implementing an overview page, we'll first look at those in the following section. After that, we'll learn about use cases and best practices for the implementation of this floorplan. Finally, we'll see which annotations are relevant for this floorplan and how this floorplan can be implemented.

Cards

Figure 14.3 shows what an overview page can look like. Cards are used to display the data. *Cards* are design elements on which the most important information about an object can be clearly displayed. Similar to tiles, cards allow a UI that brings together information on several objects or from several applications to be structured dynamically and clearly. This floorplan is also characterized by this design element. What kind of content you place on the individual cards is up to you. You can display tables or lists, diagrams, or tiles of any kind within the individual cards.

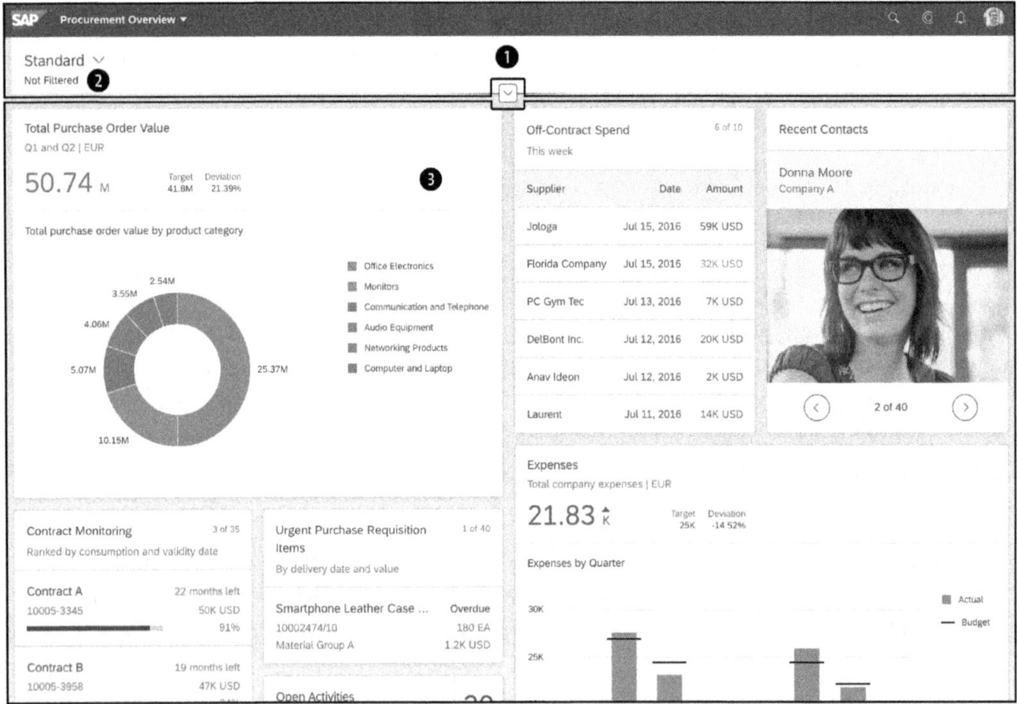

Figure 14.3 SAP Fiori Elements Overview Page

Basically, an overview page consists of two areas:

- Dynamic page header
- Dynamic page content

In the upper area of the application is the dynamic page header, as shown in Figure 14.3 ❶. This contains general information about the application. For example, the title describes the area for which the application provides data, such as purchasing or maintenance. You'll also find some user-specific settings in the header area. For example, you can call up and manage your user profile here. It's also possible to manage the displayed cards here.

If you want to further restrict the data displayed, you can do so using a smart filter ❷, which is located below the title in the dynamic page header. For example, you can filter the data displayed so that only entries that match your filter criteria are shown (e.g., only certain date values, totals, etc.).

The actual content is represented by the area with the cards, also called dynamic page content ❸. Cards belong to the components of the UI that can be configured via annotations. This makes them very versatile, as data can be displayed in many different ways. All kinds of charts, tables, or lists can be displayed within a card. The cards are displayed in five dynamic columns, whereby a card can also extend over several columns. They can be arranged as desired via drag and drop.

Note that a card is always bound to exactly one entity set. For example, if you want to display a product list on a card, this can be defined in a JavaScript Object Notation (JSON) object, as shown in Listing 14.1. This object can be found in the *manifest.json* file of your SAP Fiori elements application. When adding a card via one of the development environments, the corresponding definition is added at the code level.

```
"sap.ovp": {
   "cards": {
      "card01": {
         "model": "ZCP_PRODUCTS_SRV",
         "template": "sap.ovp.cards.list",
         "settings": {
            "sortBy": "Price",
            "sortOrder": "descending",
            "listFlavor": "bar",
            "annotationPath": "com.sap.vocabularies.UI.v1.LineItem#bar",
            "category": "{{card01_category}}",
            "entitySet": "ProductSet"
         }
      }
   }
}
```

Listing 14.1 Definition of a Card in manifest.json

Use Cases and Best Practices

You can use an overview page in the following situations:

- Users need a central entry point whose content is related to a specific role.
- Users want to be able to filter and act on information relevant to their role.
- Users need to present information that comes from two or more different applications.
- Users want to present data in different formats, such as lists, tables or charts.

Before you start implementing an overview page, you should think about what you want to achieve with the page and what look you're aiming for. The following best practices have proven to be effective in achieving the best possible user experience (UX):

- **Determine in advance how many cards you'll display.**
  Make sure only cards that add value to the users are displayed. For example, the cards available should always be relevant to the user's role or task.
- **Highlight the essential information.**
  Choose meaningful colors to highlight important information. This applies especially to texts and diagrams but also to individual pieces of information on large cards.
- **Try to mix types of cards that are as appropriate as possible.**
  To present information more clearly and to achieve a high recognition value, you should rely on a suitable mix of the different card types.
- **Go for a conscious order.**
  Users usually consider large cards, which are located at the top of the page, to be more important than other cards. This is what you should consider when ordering your cards.
- **Group related topics.**
  In addition to the order of the individual cards, it's also important to ensure that related information is displayed close together. This also contributes to a better overview.
- **Choose powerful texts.**
  Especially if you use cards with actions, you should rely on activating texts.
- **Pay attention to continuity.**
  End users are used to certain statuses being color-coded, for example: success messages should be displayed in green, warnings in orange, or errors in red. It's important to follow the same labeling consistently throughout the application.

Annotations

In this section, we'll introduce you to annotations that are essential for creating an overview page. The following annotations can be added directly in the core data services (CDS) view or locally in the SAPUI5 application. With regard to the reusability of the annotations, there's a lot to be said for adding them directly in the CDS view. The following annotations are relevant:

- `UI.SelectionVariant`
  - Example: `UI.SelectionVariant: {qualifier: 'Default'}`

  This annotation allows the data to be preselected using filters. In this way, filters defined for specific cards can be applied to both reduce the loading time and aggregate the data for the respective card. A unique identifier for the variant to be used can be specified via a string of up to 120 characters.
- `UI.HeaderInfo`
  - Example: `UI.headerInfo: {typeName: 'Demo', typeNamePlural: 'Demos', typeImageUrl: '/img/demo.png'}`

The `UI.HeaderInfo` annotation is used to define the properties of a title. For the overview page, this annotation defines the title area of a card. This allows you to define whether you want to display a title (for singular and plural), a description, and/or optionally an image. The titles can be up to 60 characters long.

- **`UI.FieldGroup`**
  - Example: `UI.FieldGroup.label: 'Demo'`

  Using this annotation, a group of data fields is defined where each data field includes a label (i.e., a display text) and a value. In this way, a form can be built. This display text has a maximum length of 60 characters.
- **`UI.Facets`**
  - Example: `UI.Facet: [{Label: 'Demo', targetElement: 'demoElement'}]`

  The `UI.Facets` annotation is used to establish a context between a card and another annotation. For example, it wouldn't be enough to just add the `UI.FieldGroup` annotation, but it must also be assigned to a card via a `UI.Facets` annotation. This mapping is done via the `Qualifier` property.
- **`UI.Chart`**
  - Example: `UI.Chart: {title: 'Chart Title', type: #COLUMN, dimensionAttributes: [{dimension: 'demoElement'}], measureAttributes: [{measure: 'demoElement'}]}`

  This annotation can be used to define which dimensions and measures will be used for the display of a chart. For this purpose, `UI.Chart.DimensionAttributes.Dimension` is used to define the dimension, and `UI.Chart.MeasureAttributes.Measure` is used to define the measure.

14

14.1.4 Generic Annotations

When you create apps using SAP Fiori elements, you'll notice that a lot is controlled by annotations within these apps. You can determine what content is displayed and how it's displayed, but functions such as filters can also be controlled via annotations.

The annotations can come from the OData service in the backend system, but they can also be stored locally in the SAPUI5 application in a separate file. To simplify the maintenance of this file for SAP Fiori elements applications, the annotation modeler is provided. This is available in SAP Web IDE.

In the following sections, we'll present specific UI annotations for each of the floorplans presented, which you can use to influence the display on the interface. The annotations listed in this section are generic and can thus be applied to any SAP Fiori elements floorplan.

UI.HeaderInfo

The following annotation is used to control the display of the respective headers:

- **UI.HeaderInfo.typeName**
  - Data type: `STRING(60)`
  - Example: `UI.HeaderInfo.typeName: 'Order'`

  Can contain up to 60 characters, is displayed in the header of the called application, and indicates the name of an object type. This annotation must be obligatory if the annotation `EndUserText.label` isn't used instead. This annotation is typically defined in a CDS view and contains a title for the CDS view that is as meaningful as possible.
- **UI.HeaderInfo.typeNamePlural**
  - Data type: `STRING(60)`
  - Example: `UI.HeaderInfo.typeNamePlural: 'Orders'`

  Represents the title of a list and is mandatory.
- **UI.HeaderInfo.typeImageUrl**
  - Data type: `STRING`
  - Example: `UI.HeaderInfo.typeName: '/img/order'`

  Contains the URL of an image used to represent the object type and can be specified optionally.

UI.selectionField

With these annotations, you can determine which filter criteria you want to provide. The following variants are available:

- **UI.selectionField.qualifier**
  - Data type: `STRING(120)`
  - Example: `UI.selectionField.qualifier: 'Ordernumber'`

  Specifies a unique name for a selection field with up to 120 characters.
- **UI.selectionField.position**
  - Data type: `Decimal`
  - Example: `UI.selectionField.position: 16`

  Controls the order of the individual filter criteria. The individual criteria are sorted in ascending order.

UI.lineItem

The annotation `UI.lineItem` allows you to control the presentation of data fields in tables and lists. For example, you can specify which columns should be present, how they should be arranged, and what the column heading should look like. The following specifications are possible here:

- **UI.lineItem.position**
  - Data type: `Decimal`
  - Example: `UI.lineItem.position: 16`

  Determines the position of a column by the value used. This annotation must be obligatory.
- **UI.lineItem.importance**
  - Data type: `STRING`
  - Possible values: `HIGH`, `MEDIUM`, `LOW`
  - Example: `UI.lineItem.importance: HIGH`

  Indicates the priority of a column. This specification plays a role if not all columns can be displayed due to the display size.
- **UI.lineItem.label**
  - Data type: `STRING(60)`
  - Example: `UI.lineItem.label: 'Ordernumber'`

  Indicates the title of the respective column in the table.

UI.hidden

With this annotation, you can control the visibility of UI elements. Possible values for this are `true` or `false`. A corresponding specification might look like `UI.hidden: false`.

14.2 Flexible Programming Model

The *flexible programming model* combines the flexibility of freestyle SAPUI5 apps with the low-code principle of SAP Fiori elements. SAP provides an online resource called the *flexible programming model explorer*, which provides an easy overview of the different capabilities of the flexible programming model. Here, all the different building blocks of the flexible programming model are shown in a sample implementation.

The implementation of such a flexible programming model application will generate an SAP Fiori elements app, which either defines views as SAPUI5 freestyle views or SAP Fiori elements pages. These freestyle views can combine standard SAPUI5 controls plus the provided building blocks, whereas the SAP Fiori elements pages can be included and extended.

The flexible programming model is the newest iteration of the SAP Fiori elements framework, offering a combination of both SAPUI5 freestyle and SAP Fiori elements application development approaches for OData V4. It can either be used as its own SAP Fiori elements application template or be integrated into already-existing SAPUI5 freestyle applications.

The flexible programming model offers the custom development flow of SAPUI5. All SAPUI5 controls can be used in the accustomed way. Views can be created, custom routes can be configured, and custom modules (e.g., formatters) can be defined. All technologies that freestyle apps offer can be used.

Additionally, flexible programming model projects are SAP Fiori elements application components on a core level. In other words, the page map and flow logic of SAP Fiori elements provides a low-code development approach that normally aren't customizable, except for adaptation projects and extensions based on the guided development. Here, the flexible programming model integrates the SAPUI5 freestyle approach to get the generic functionalities of SAP Fiori elements inside an application.

This provides the advantage of building apps, where generic blocks such as tables, forms, filter bars, and so on can be used in the SAP Fiori elements way, and everything else can be developed freestyle. A flexible programming model app incorporates three SAP Fiori elements base features:

- **Extension points**
  Extension points can be used to extend SAP Fiori elements floorplans and their different templates. They are defined in the *manifest.json* (application descriptor) of your app to add custom development controls to your SAP Fiori elements pages and sections such as custom columns in a table, custom sections in an object page, or custom pages. Extension points are the default way of extending SAP Fiori elements applications and are also available in other floorplans.
- **Building blocks**
  Building blocks offer SAP Fiori elements blocks that can be included in an SAPUI5 freestyle view to get the low-code functions of SAP Fiori elements such as tables, forms, fields, and so on. They have now been introduced with the flexible programming model and can also be integrated into preexisting SAPUI5 freestyle applications.
- **Controller extensions**
  Controller extensions work like normal controller extensions for SAP Fiori elements, extending the generic page controller to add custom JavaScript functionalities.

In the following, we'll take a closer look at the edit flow. Furthermore, we'll see how we can use macros, which are the meta-driven UI controls.

14.2.1 Edit Flow

The *edit flow* in SAPUI5 is a structured approach provided by SAP Fiori elements for managing data entry, updates, and validation in applications built with SAPUI5. It's designed to streamline user interactions for editing data while ensuring consistency and compliance with business logic. Here are its key components and characteristics:

- **Draft handling**
  - Uses the draft concept to allow users to make changes without immediately committing them to the database.
  - Supports autosave and explicit save actions, reducing the risk of data loss during editing.
- **Validation**
  - Ensures data integrity by performing client-side and server-side validations during data entry.
  - Provides immediate feedback to the user for correcting errors before the data is saved.
- **State management**
  - Tracks the state of data changes (e.g., unsaved, validated, or finalized).
  - Supports parallel user sessions, allowing multiple users to work on different drafts simultaneously.
- **OData V4 integration**
  - Fully leverages the capabilities of OData V4, enabling efficient handling of complex data structures and operations.
  - Relies on OData V4 to ensure advanced functionality such as reduced payloads and better metadata handling.
- **UX optimization**
  - Provides seamless navigation between edit and display modes.
  - Enables intuitive undo and redo operations to enhance user control over data changes.
- **Extensibility**
  - Supports custom logic during the save or validation steps through annotations or event hooks.
  - Allows developers to enhance or adapt the default behavior to specific business requirements.

At a basic level, the workflow of the edit flow runs as follows:

1. **Entering edit mode**
   The user initiates editing by selecting a record, which creates a draft or locks the data for editing.
2. **Editing data**
   Changes are made in the UI, and validations are triggered in real time.
3. **Saving or canceling**
   If the user saves, the draft data is committed to the backend system. If the user cancels, the draft is discarded, and the original data remains unchanged.

4. **Finalization**
   After a successful save, the system ensures data consistency and releases any locks or drafts.

The main benefits of the edit flow are as follows:

- **Improved user productivity**
  Provides a guided and intuitive process for editing data.
- **Data integrity**
  Ensures data is always validated and accurate before saving.
- **Reduced errors**
  Minimizes the risk of invalid or incomplete data submissions via real-time feedback.
- **Flexibility**
  Provides users with flexibility and control through the draft handling and undo capabilities.

The edit flow is an essential feature for modern SAP Fiori apps, particularly in scenarios where data quality, user productivity, and seamless interaction are critical. In addition, the edit flow is included in flexible programming model applications out of the box with a full range of functions.

The `EditFlow` is provided in the form of a class (`sap.fe.core.controllerextensions.EditFlow`). This contains various functions and handlers, the most important of which are explained in the following:

- `applyDocument`
  The call signature is `this.editFlow.applyDocument(oContext): Promise`:
  - `oContext` is the context, for which changes will be saved.
  - The returned promise resolves once the changes have been saved.

  This function submits the current set of changes and then navigates back.
- `cancelDocument`
  The call signature is `this.editFlow.cancelDocument(oContext, mParameters): Promise`:
  - `oContext` is the context of the editable document.
  - `mParameters` can contain attributes such as `control` (defines the control on which the discard popover will be opened) or `skipDiscardPopover` (suppresses the discard popover).
  - The returned promise resolves once the editable document has been discarded.

  This function discards the editable document.
- `createDocument`
  The call signature is `this.editFlow.createDocument(source, mInParameters): Promise`:

- source is an ObjectListBinding or a table building block for a temporary list binding which defines for which entity a document should be created.
- mInParameters defines several parameters:
  - createAtEnd: Specifies if the entry should be created at the top or bottom of the table.
  - creationDialogFields: Defines the list of properties displayed in the creation dialog when the creation mode is set to CreationnDialog.
  - creationMode: Defines the creation mode with supported values NewPage, Inline, External, CreationDialog.
  - data: Initial data for the created document.
  - outbound: Navigation target, where the document is created in case of creationMode "External".
  - selectedContexts: Contexts selected in the table initiating the creation and used in case of a TreeTable to determine the parent context of the created document.
  - singleDraftForCreate: When enabled, an extra request checks if any drafts exist that were never saved. If unsaved drafts are found, the newest one is opened. If no unsaved drafts are found, a new entity is created.
  - tableId: ID of the table.
- The returned promise resolves once the object has been created.

This function creates a new document.

- **customMassEditSave**

  The call signature is this.editFlow.customMassEditSave(_mParameters): Promise:

  - _mParameters is an object containing the parameters passed to the function customMassEditSave. It consists of two properties:
    - aContext: Array containing the selected contexts for mass edit.
    - oUpdateData: Dictionary containing the propertyPath and its value.
  - Returns a promise, which returns a Boolean. If it returns true, the default execution is prevented. If false is returned, the standard behavior is kept.

  This function is designed to execute code during the mass edit save process. It allows you to implement custom logic as needed. If you choose to handle the save operation yourself, you can return true to bypass the default save behavior. This function is intended to be overridden by consuming controllers for specific use cases and shouldn't be called directly.

- **deleteDocument**

  This call signature is this.editFlow.deleteDocument(oContext, mInParameters): Promise:

  - oContext defines the context of the document that should be deleted.

  - `mInParameters` is an optional parameter and can contain properties such as the following:
    - `description`: Description of the object being deleted.
    - `title`: Title of the document being deleted.
  - The returned promise is resolved once the document has been deleted.

  This function allows you to delete a document.

- **`editDocument`**
  The call signature is `this.editFlow.editDocument(context): Promise`:
  - `context` defines the context of the active document.
  - The returned promise resolves once the editable document is available with the editable context.

  This is used to create a draft for an active document.

- **`invokeAction`**
  This call signature is `this.editFlow.invokeAction(sActionName, mInParameters): Promise`:
  - `sActionName` defines the action that will be called.
  - `mInParameters` is optional and is an object containing attributes such as these:
    - `contexts`: Array of contexts for which the action will be executed.
    - `innvocationGrouping`: Defines the mode for how actions are called. Possible values are `ChangeSet` or `Isolated`. `ChangeSet` means that all action calls are bundled in one changeset, whereas `Isolated` means that they are put into separate changesets.
    - `label`: A human-readable label for the action. This property is needed if the action contains parameters and a parameter dialog is shown to the user. The label will be displayed as the title of the dialog.
    - `model`: For an unbound action, the instance of the OData V4 model must be provided.
    - `parameterValues`: Map of action parameter names and values.
    - `requiresNavigation`: Indicates whether navigation is required upon executing the action.
    - `skipParameterDialog`: Skips the parameter dialog if values are provided in `parameterValues`.
  - The returned promise resolves once the action has been executed.

  This can be used to call bound and unbound actions.

- **`onAfterCreate`**
  This call signature is `this.editFlow.onAfterCreate(_mParameters): Promise`:

 - `_mParameters` is an optional parameter represented as an object containing the parameters passed to `onAfterCreate`. The object consists of one parameter, `context`, which defines the context to be created.

 With this function, custom code can be executed after the `Create` action.

- **`onAfterDelete`**
  This call signature is `this.editFlow.onAfterDelete(_mParameters): Promise:`
  - `_mParameters` is an optional parameter represented as an object containing the parameters passed to `onAfterDelete`. The object consists of one parameter, `contexts`, which defines the contexts to be deleted.

  With this function, custom code can be executed after the `Delete` action.

- **`oAfterDiscard`**
  This call signature is `this.editFlow.onAfterDiscard(_mParameters): Promise:`
  - `_mParameters` is an optional parameter represented as an object containing the parameters passed to `onAfterDiscard`. The object consists of one parameter, `context`, which defines the context to be discarded.

  With this function, custom code can be executed after the `Discard` action.

- **`onAfterEdit`**
  This call signature is `this.editFlow.onAfterEdit(_mParameters): Promise:`
  - `_mParameters` is an optional parameter represented as an object containing the parameters passed to `onAfterEdit`. The object consists of one parameter, `context`, which defines the context to be edited.

  With this function, custom code can be executed after the `Edit` action.

- **`onAfterSave`**
  This call signature is `this.editFlow.onAfterSave(_mParameters): Promise:`
  - `_mParameters` is an optional parameter represented as an object containing the parameters passed to `onAfterSave`. The object consists of one parameter, `context`, which defines the context to be saved.

  With this function, custom code can be executed after the `Save` action.

- **`onBeforeCreate`**
  This call signature is `this.editFlow.onBeforeCreate(_mParameters): Promise:`
  - `_mParameters` is an optional parameter represented as an object containing the parameters passed to `onBeforeCreate`. The object consists of the following parameters:
    - `contextPath`: Path pointing to the context on which the create action is executed.
    - `createParameters`: Array of values that are filled via the action parameter dialog.

  With this function, the `Create` action can be intercepted and used to execute custom coding before calling the `create` action.

- `onBeforeDelete`
  This call signature is `this.editFlow.onBeforeDelete(_mParameters): Promise:`
  - `_mParameters` is an optional parameter represented as an object containing the parameters passed to `onBeforeDelete`. The object consists of the parameter, `contexts`, which defines the contexts to be deleted.

  With this function, the `Delete` action can be intercepted and used to execute custom coding before calling the `create` action.
- `onBeforeDiscard`
  This call signature is `this.editFlow.onBeforeDiscard(_mParameters): Promise:`
  - `_mParameters` is an optional parameter represented as an object containing the parameters passed to `onBeforeDiscard`. The object consists of the parameter `context`, which defines the context to be discarded.

  With this function, the `Discard` action can be intercepted and used to execute custom coding before calling the `create` action.
- `onBeforeEdit`
  This call signature: is `this.editFlow.onBeforeEdit(_mParameters): Promise:`
  - `_mParameters` is an optional parameter represented as an object containing the parameters passed to `onBeforeEdit`. The object consists of the parameter `context`, which defines the context to be edited.

  With this function, the `Edit` action can be intercepted and used to execute custom coding before calling the `create` action.
- `onBeforeSave`
  This call signature is `this.editFlow.onBeforeSave(_mParameters): Promise:`
  - `_mParameters` is an optional parameter represented as an object containing the parameters passed to `onBeforeSaved`. The object consists of the parameter `context`, which defines the context to be saved.

  With this function, the `Save` action can be intercepted and used to execute custom coding before calling the `create` action.
- `saveDocument`
  This call signature is `this.editFlow.saveDocument(oContext): Promise:`
  - `oContext` defines the context of the editable document.

  With this function, a new document can be saved.
- `securedExecution`
  This call signature is `this.editFlow.securedExecution(fnFunction, mParameters): Promise:`
  - `fnFunction` defines the function to be executed. This function should return a promise.

  - `mParameters` is an optional parameter that defines the preconditions to be checked before execution. It consists of the object `busy`, which contains the following properties:
    - `check`: Executes the function only if the application isn't in a busy state.
    - `set`: Triggers a busy indicator when the function is executed.
    - `updatesDocument`: Updates the current document without using the bound model and context.
  - The returned promise is resolved if returned by the `fnFunction`. If checks fail, the promise is rejected.

  This function can be used to trigger a secured execution of a given function. This ensures that the function will only be executed if certain conditions are met.

- **updateDocument**
  This call signature is `this.editFlow.updateDocument(updatedContext, updatePromise): Promise`:
  - `updatedContext` is the context of the update field.
  - `updatePromise` is a promise to determine when the update operation is completed. It should be resolved when the update operation is completed; subsequently, the draft status is updated.
  - The returned promise resolves once the draft status has been updated.

  This updates the draft status and displays error messages if errors occurred.

14.2.2 Macros

Macros or building blocks define template instructions on how controls should be created depending on the bound entity and its annotations. They work like smart controls, but don't represent a single SAPUI5 control and offer more metadata-driven approaches. The building blocks are bundled in namespace `sap.fe.macros`. The flexible programming model offers the following building blocks as discussed in their individual sections to follow:

- `Field`
- `Form`
- `FormElement`
- `Table`
- `FilterBar`
- `FilterField`
- `MicroChart`
- `Chart`
- `FlexibleCloumnLayoutActions`
- `Share`
- `Paginator`

Field

The `Field` building block can be used to display single properties of an OData entity and is controlled by the annotations of the following property:

```
<macros:Field xmlns:macro="sap.fe.macros" metaPath="AuthorName"/>
```

Building blocks should have a stable ID and their appearance can be determined in addition to the annotations via the Field Building Block API, which offers formatting options, display modes, and so on. This enables properties to not only be displayed as text fields but also as status objects, rating indicators, and more.

The `Field` building block can be used when the UI of a field and its behavior should be determined by OData. The most prominent properties are as follows:

- `metaPath`
  Serves as the property binding path of the OData entity property in use.
- `busy`
  Defines the busy-state of the field.
- `readonly`
  Redefines if a field is read-only or not.
- `formatOptions`
  Defines how the value of the field should be formatted.

The most prominent event is as follows:

- `change`
  Fires when the value gets changed by user input.

Form

The `Form` building block defines a form bound to the `FieldGroup` annotation that can be extended with additional form elements, as follows:

```
<macros:Form id="myCustomForm" metaPath=
"@com.sap.vocabularies.UI.v1.FieldGroup#BookInformation" title="Book
Information"></macros:Form>
```

When an OData service already defines `FieldGroup` facets, you can use the `Form` building block to display this facet at your given position in the UI and not only in an object page. The most prominent properties are as follows:

- `metaPath`
  Context binding path of the OData entity in use.
- `title`
  Title displayed above the form.

- `fieldGroupIds`
  Stable ID for validation purposes.

The most prominent event is as follows:

- `validateFieldGroup`
  Fired when the validation of a form should be triggered.

FormElement

The `FormElement` adds a new combination of label and input/display control of an OData property to an `sap.ui.layout.form.Form` control inside the `formElements` aggregation, as follows:

```
<f:formElements>
   <macros:FormElement xmlns:macro="sap.fe.macros" metaPath="Name" id="name" />
</f:formElements>
```

When you've included a custom form as a fragment in an SAP Fiori elements object page or in an SAPUI5 freestyle view, and you want to get automatic field control for an OData property, you should use the `FormElement` building block. The most prominent properties are as follows:

- `metaPath`
  Property binding path of the OData entity in use.
- `label`
  Defines a custom label for the `FormElement`. When not used, the label from the OData metadata is used.

The most prominent event is as follows:

- `validateFieldGroup`
  Fired when the validation of a form element should be triggered.

Table

The `Table` building block (see Figure 14.4) renders a table bound to an OData entity set. It's configured via the `LineItem` annotations and can be combined with a `FilterBar` building block. In addition, custom actions and menus can be added to the table itself, as follows:

```
<macros:Table xmlns:macro="sap.fe.macros" metaPath="MyNavProperty/
@com.sap.vocabularies.UI.v1.LineItem"/>
```

14

Search

ID	Integer Property	Number Pro...	Property With Contact annotation	String Property	Boolean Pro...	TextArrangementTextFirstProperty
1	89	78.777	Richard Wilson	Hello	No	Test First Description (Value)
2	39	75.41	Laurent Dubois	Good morning	No	Test First Description (Value)
3	9	8.12	John Miller	Bye	No	Test First Description (Value)
4	11	17.70	Ravi Mueller	Sup	Yes	Test First Description (Value)
5	15	13.145	Ravi Mueller	Sup	Yes	Test First Description (Value)

Figure 14.4 Example of the Table Building Block

When you want to display data inside a metadata-driven table, you should use the `Table` building block to minimize UI coding. The most prominent properties are as follows:

- `metaPath`
  List the binding path of the OData entity in use or a reference to the `LineItem` annotation.
- `enableExport`
  Enables the export functionalities of the table.
- `filterBar`
  Expects the ID of a `FilterBar` building block.
- `personalization`
  Configures the available table personalization options.
- `selectionMode`
  Defines the selection mode.
- `type`
  Defines if the table should be a grid, a responsive, or an analytical table.

The most prominent events are as follows:

- `rowPress`
  Fires when a row is clicked.
- `selectionChange`
  Fires when rows have been selected.

FilterBar

The `FilterBar` building block (see Figure 14.5) adds a filter bar for an entity set to the UI that is configured with the `SelectionFields` annotation, as follows:

```
<macros:FilterBar metaPath="@com.sap.vocabularies.UI.v1.SelectionFields#
MySelectionFields" id="myFilterBar" search=".onSearchPress" filtersChanged=
".onFiltersChanged"/>
```

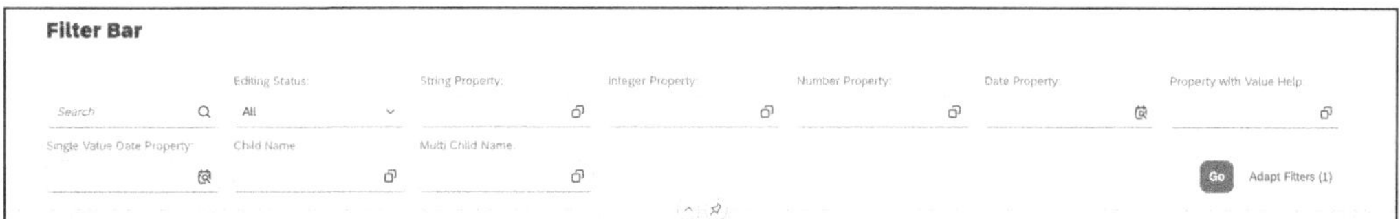

Figure 14.5 Example of the FilterBar Building Block

When you want to add filter functionalities to a page containing a table, you should use the `FilterBar` building block. The most prominent properties are as follows:

- `metaPath`
  List the binding path of the OData entity in use or a reference to the `SelectionFields` annotation.
- `liveMode`
  Enables the filtering after each change in the filter bar.
- `showClearButton`
  Shows a button to clear filters.
- `showMessages`
  Displays errors in a message box that occur during filtering.

The most prominent events are as follows:

- `afterClear`
  Fires when the **Clear** button is clicked.
- `filterChanged`
  Fires when the filters change.
- `search`
  Fired when the **Go** button is clicked or the live filter changes apply.

FilterField

The `FilterField` building block can be added to a `FilterBar` building block to extend it with fragment-based custom filter fields, as follows:

```
<macros:FilterBar … >
   <macros:filterFields>
      <macros:FilterField key="BookDescription" label="Description">
         <core:Fragment fragmentName="custom.fragments.MyFilter " type="XML" />
      </macros:FilterField>
   </macros:filterFields>
</macros:FilterBar>
```

The control that should be used as a filter is defined in a fragment. When the filtering is triggered, the value of the filter needs to be appended to the filter request.

When the normal SelectionFields offer less functionalities as required, or additional filters are required, you should use the FilterField building block. The most prominent properties are as follows:

- **key**
  Key of the FilterField.
- **label**
  Displayed label of the FilterField.
- **placement**
  Placement of the custom filter.

MicroChart

MicroChart can be used to add small charts (see Figure 14.6) to your UI. In SAP Fiori elements, they were limited to be displayed either in a table row or as a header facet in an object page. With this building block, you can use an existing chart annotation to display a micro chart wherever you want, as follows:

```
<macros:MicroChart id="soldBooks" metaPath="@com.sap.vocabularies.UI.v1.Chart#
SoldBooks" contextPath="/Books/_SoldBooks" batchGroupId="$auto.abc"/>
```

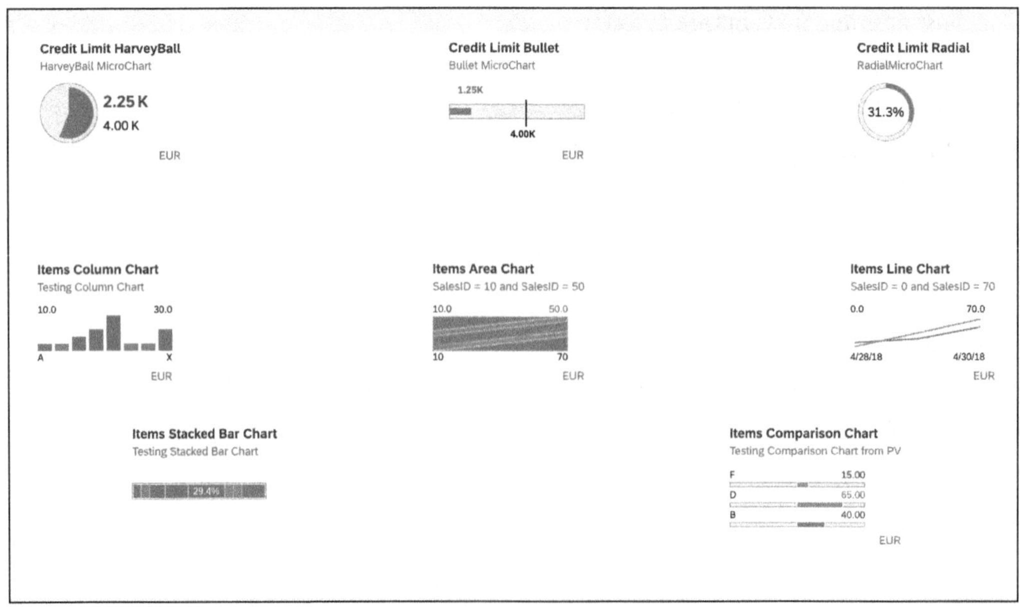

Figure 14.6 Examples of the MicroChart Building Block

As stated, they are completely configured with the UI.charts annotation, which provides several chart options. They are also used in combination with analytical CDS views to aggregate data.

When you already have an annotation-based chart, you can use the MicroChart building block to display it. The most prominent properties are as follows:

- metaPath
  Annotation path of the UI.chart annotation used.
- contextPath
  List binding path of the OData entity that provides the chart data.

Chart

The Chart building block (see Figure 14.7) works almost the same as MicroChart, but it provides more analytical functionalities and renders a bigger chart, as follows:

```
<macros:Chart id="totalBookSales" contextPath="/Books" metaPath=
"@com.sap.vocabularies.UI.v1.Chart" dataPointsSelected=
".onChartSelectionChanged" personalization="Sort,Type,Item" selectionMode=
"Multiple"></macros:Chart>
```


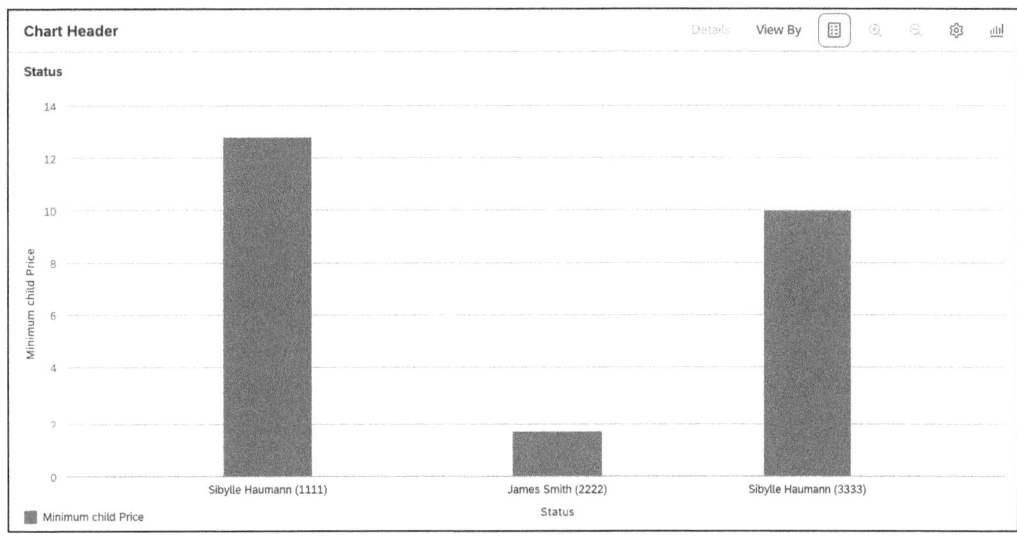

Figure 14.7 Example of the Chart Building Block

When you want to display data in a chart that offers analytical functionalities, you should use the Chart building block. The most prominent properties are as follows:

- metaPath
  Annotation path of the UI.chart annotation used.
- contextPath
  List binding path of the OData entity that provides the chart data.
- filterBar
  ID of a FilterBar building block to add filters to the chart.

- `header`
  Defines the header label.
- `personalization`
  Adds personalization to the chart.
- `selectionMode`
  Defines the chart selection mode.

The most prominent event is as follows:

- `selectionChange`
  Fires when chart data has been selected.

FlexibleColumnLayoutActions

When you use the `FlexibleColumnLayoutActions` as a root container of your application, you can add the actions for close, enter full screen, and exit full screen to your pages:

```
<macros:FlexibleColumnLayoutActions/>
```

Share

You can add the **Share** button of an object page to your custom pages in combination. This allows you to either save the page as a tile or send information as mail (in combination with the semantics annotations):

```
<macros:Share id="share" visible="true" />
```

Paginator

When you need to display a lot of data inside a list or table, you can use the `Paginator` building block to handle pagination for you. This renders paginator buttons (up and down) in a detail page to load either the next or the previous data:

```
<macros:Paginator id="paginator" />
```

This must be used in combination with the Paginator API in a controller to define how the paginator should work. When you want to traverse inside a detail page, you can use the `Paginator` building block.

14.3 Extensibility

A large number of applications are already delivered as standard. However, the use cases in individual companies are often such that the standard apps can't be adopted 100%. To avoid having to develop a completely new application from scratch in such cases, there are various ways to extend SAPUI5 applications. There are three options for extending applications:

- Adaptation projects
- Key user adaptations
- Adaptations via guided development

While adaptation projects can be used to extend entire applications, which of course also requires development effort, key user adaptations can be used to add or remove further UI elements, to create different views for applications, or to change the arrangement of UI elements. Key user adaptations can also be used to embed external content or change control identifiers. The guided development can be used, for example, to add custom columns to a list report.

14.3.1 Adaptation Projects

The different floorplans, such as list report, object page, or overview page, as discussed earlier, can be created using guided development. This guide also offers the option to create adaptation projects, especially SAPUI5 freestyle developers in particular at home.

An adaptation project is based on an application already deployed in your system. In the adaptation project, you can then use a GUI to select which parts of the existing application you want to change, hide, or even replace. Many modification options are offered for this purpose. We list the most common customizations here:

- Replace selection with your own XML fragment.
- Extend a controller.
- Change attributes of a control.
- Hide UI controls completely.
- Show available UI controls.
- Move UI controls to another position.
- Add additional content to an extension point.

An enhancement object doesn't copy the existing application, the source code, or the metadata definitions. The project always saves only the deviations from the original application, which are called changes. For this purpose, a project directory named *changes* is created, where the actual changes are stored in files with the extension *.change*. All other parts of the program code are loaded from the original application. Figure 14.8 outlines this operation of an adaptation project.

In the following sections, we'll see how adaptation projects can be created using SAP Business Application Studio, how extensions can be built using the graphical editor, and lastly how an adaptation project can be deployed.

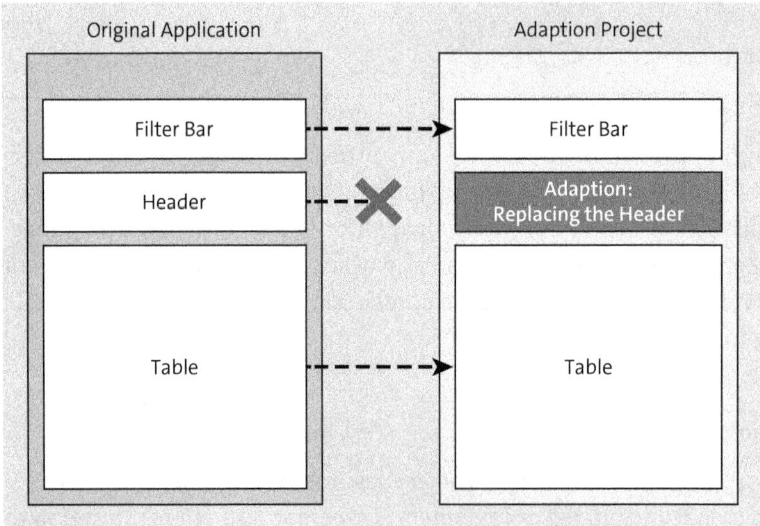

Figure 14.8 How Adaptation Projects Work

Creating an Adaptation Project

To create an adaptation project in SAP Business Application Studio, follow these steps:

1. Select **New Project From Template**, and then click on the **Adaptation Project** template (see Figure 14.9). Start the project creation by clicking **Start**.

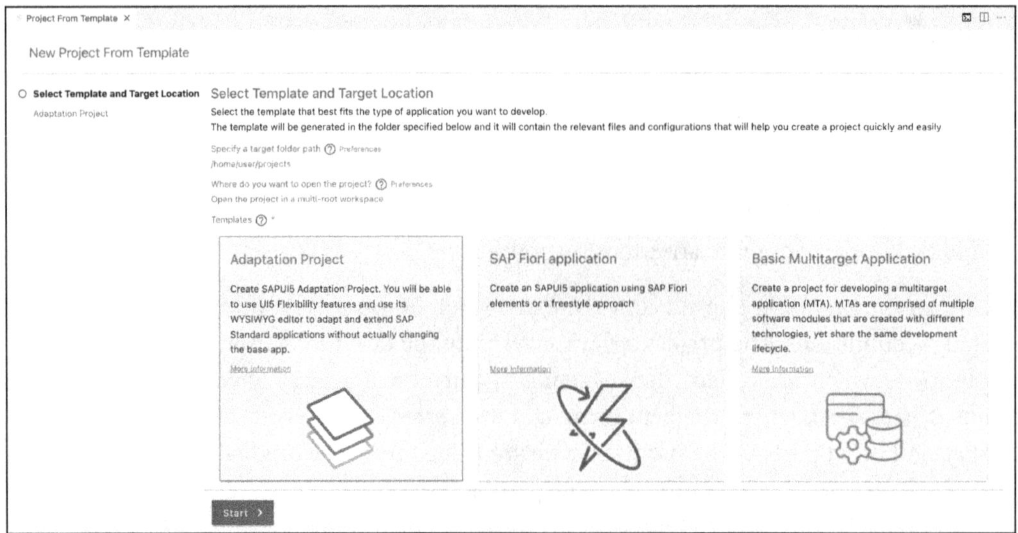

Figure 14.9 Creating an Adaptation Project

2. In the **Target environment** step, select the desired target platform on which to deploy the project (see Figure 14.10). Continue with **Next**.

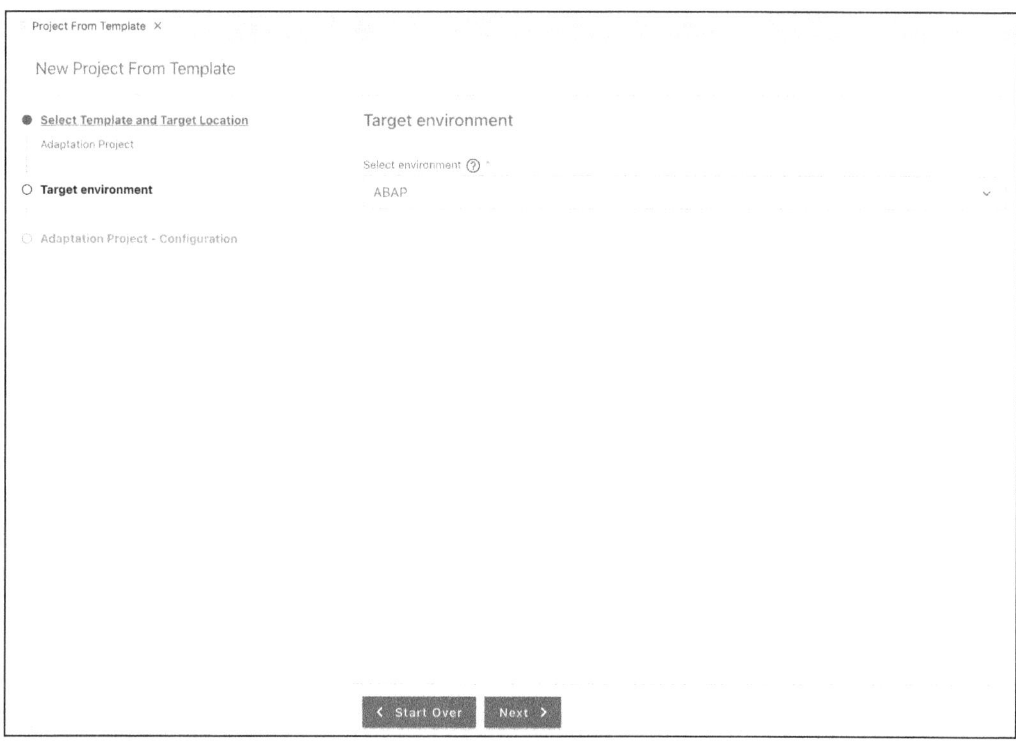

Figure 14.10 Selecting the Target Environment

3. Specify a **Project Name** for the extension application, an **Application Title**, and a **Namespace** (see Figure 14.11). Click **Next** again.

Figure 14.11 Adding Basic Information

4. Select the system, and enter your credentials. After that, select the original application that you want to extend (see Figure 14.12), and specify the SAPUI5 version. Close the project creation with **Finish**.

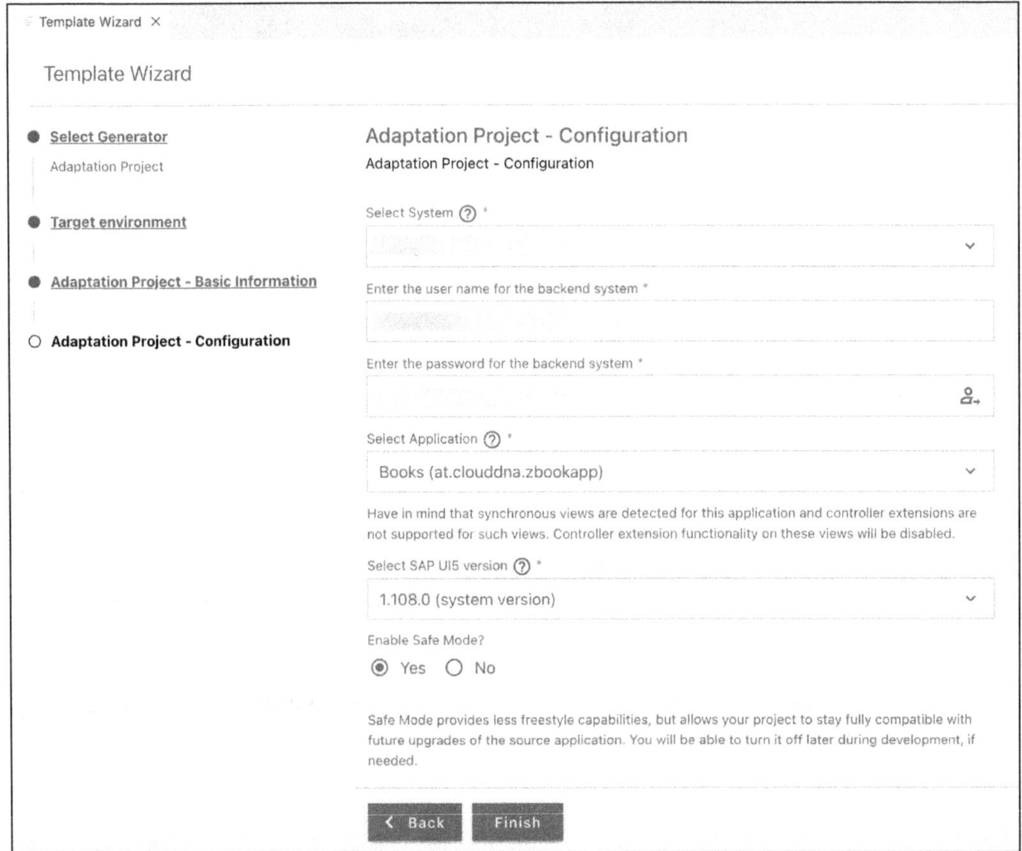

Figure 14.12 Selecting the Original Application

Building Extensions in the Graphical Editor

In Figure 14.13, you can see the project structure generated with the creation of the project. Right-click the **manifest.appdescr_variant** file to open the context menu, and select the **Open SAPUI5 Visual Editor** option to open the graphical editor with which you edit the adaptation project.

The editor now tries to open the app and display it in a preview and test view. You'll see not only the original app but also your customizations, if you've already made any, as shown in Figure 14.14.

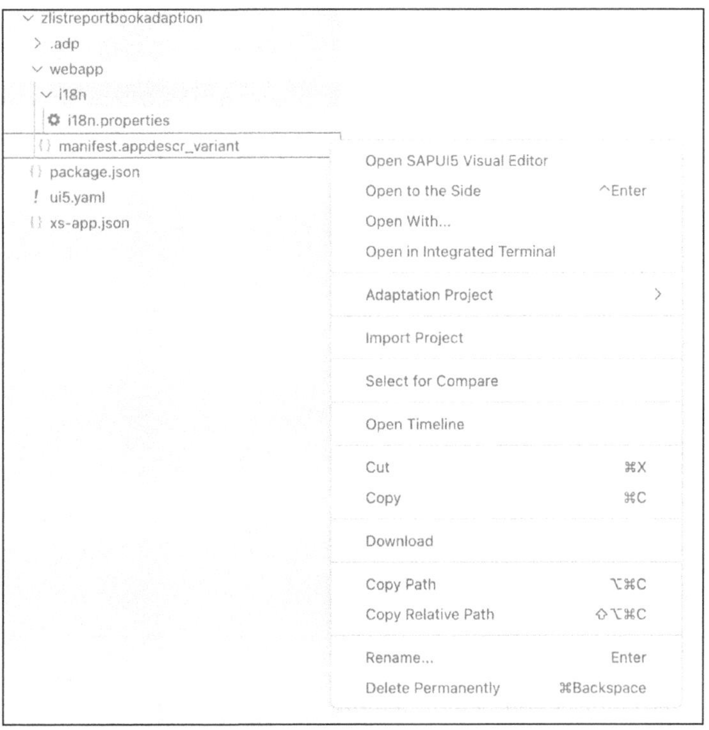

Figure 14.13 Opening the SAPUI5 Visual Editor

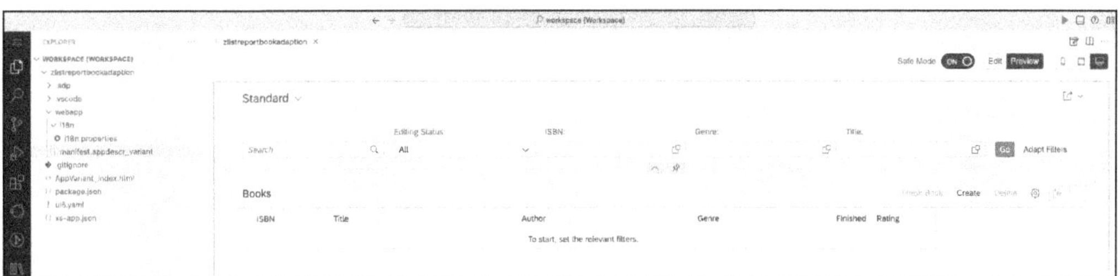

Figure 14.14 Preview of the Application

Now switch to edit mode by clicking on the **Edit** button in the upper-right corner. An UI divided into three parts opens (see Figure 14.15). This is the graphical editor with the following areas:

- On the left, you can see the hierarchical structure of the application's UI in the **Outline**. Here, you can select the elements you want to customize.
- In the middle, you can still see the preview of the application and the interaction options with the UI.
- On the right, you'll see the attributes of the UI element selected in the outline.

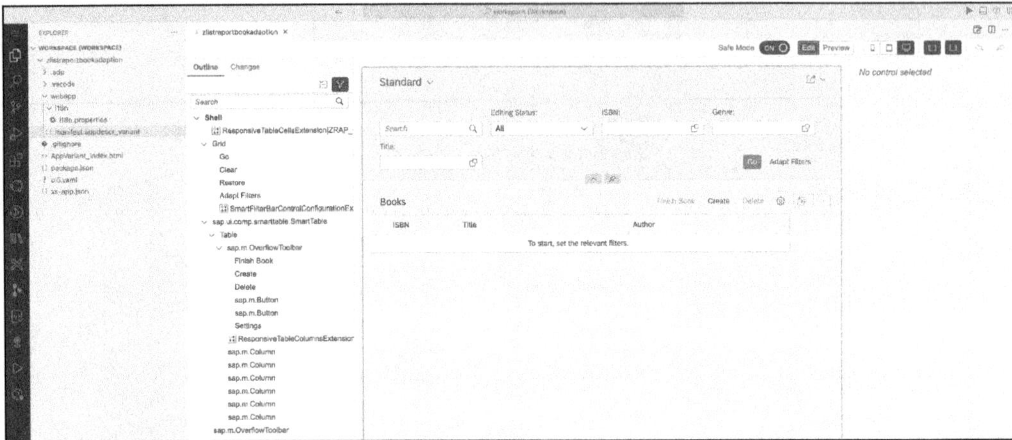

Figure 14.15 Editing the Application

In Figure 14.16, you can see that due to safe mode, the editing is limited. This only allows changes at designated points. Semantic changes are allowed.

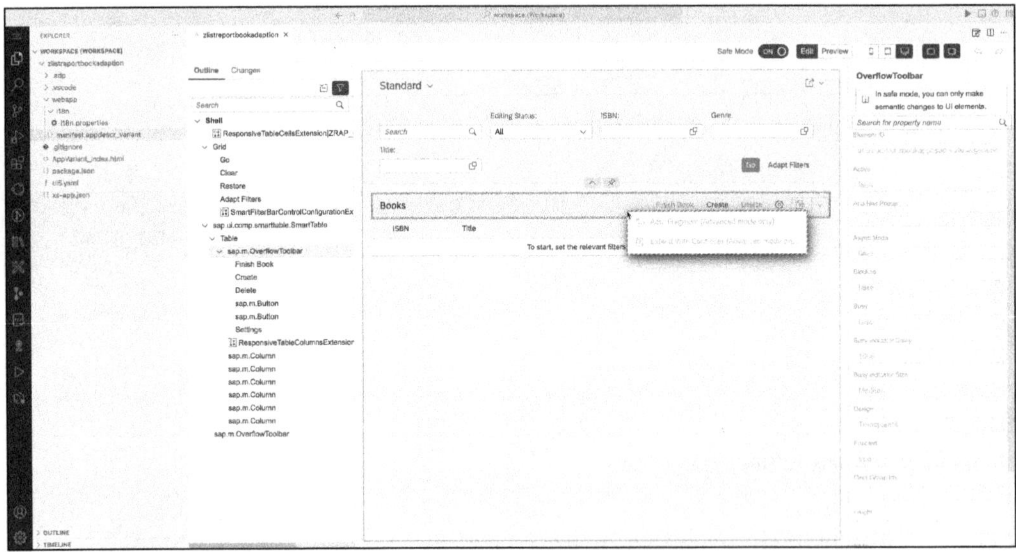

Figure 14.16 Trying to Edit with Safe Mode Enabled

To disable safe mode, change the **Safe Mode** switch in the top right from **ON** to **OFF**. After that, a dialog will pop up, where the deactivation needs to be confirmed with **Apply** (see Figure 14.17).

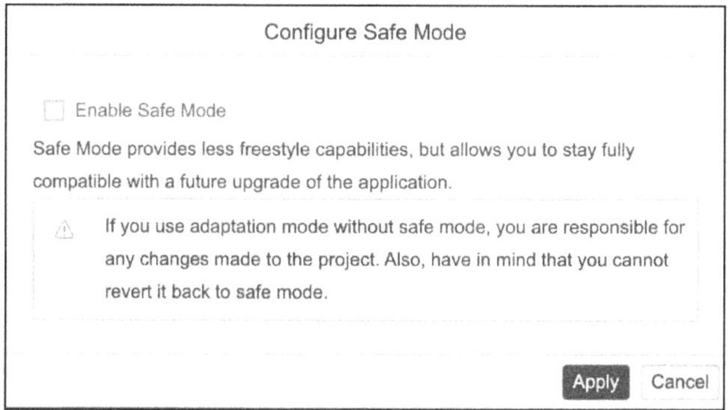

Figure 14.17 Confirming Deactivation of Safe Mode

We've now deactivated safe mode. If you want to add something in an area of the UI, select this area by clicking or right-clicking in edit mode. As you can see in Figure 14.18, you have two options for an extension in this case:

- **Add: Fragment**
- **Extend With Controller**

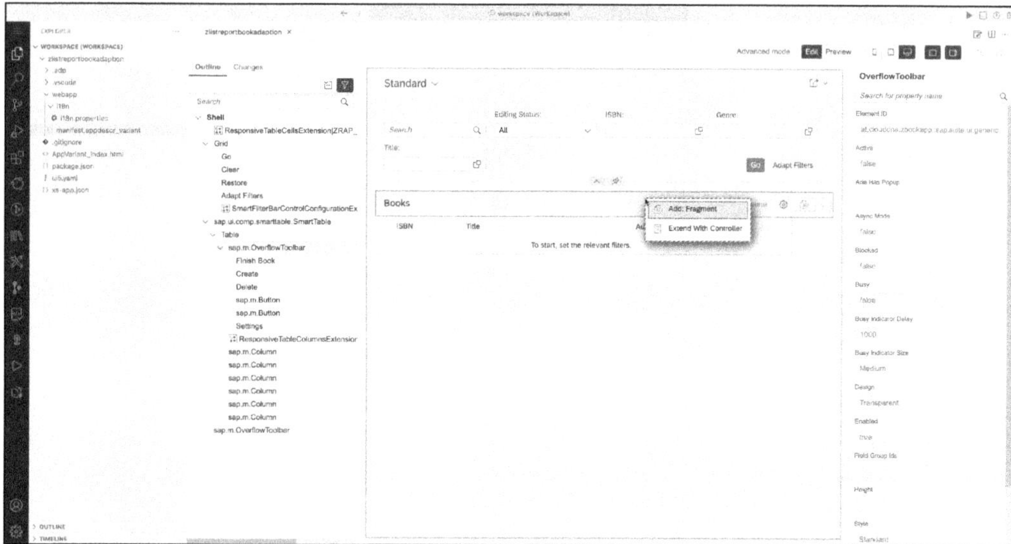

Figure 14.18 Editing an Area

If you select a specific element in the preview instead of a section, you'll have more options (see Figure 14.19):

- **Add: Fragment**
- **Extend With Controller**
- **Remove**

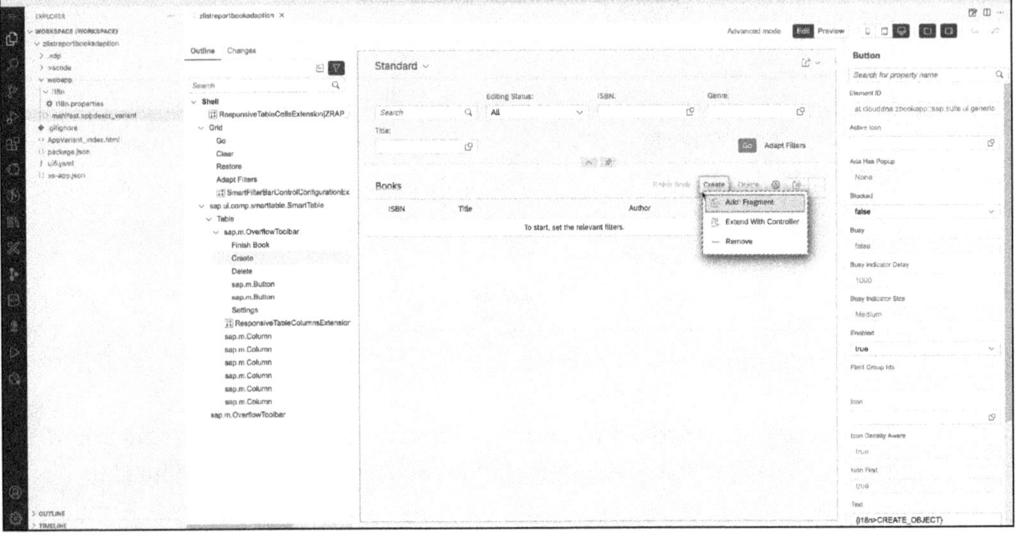

Figure 14.19 Editing One Element

Because there are various possibilities for extensions, we'll show you an example of how to add a fragment, (i.e., reusable UI elements), as follows:

1. Select the area you want to extend, for example, the table's toolbar, and choose the **Add Fragment** option.
2. For the extension of the table's toolbar, you'll be offered options for where exactly you want the fragment to be inserted in the **Target Aggregation** field (see Figure 14.20). You select the exact position within this area in the **Index** field.

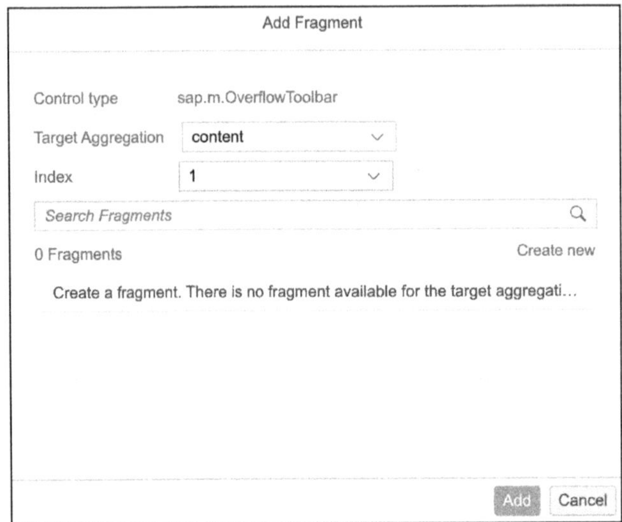

Figure 14.20 Adding a Fragment

3. You can select an existing fragment, or use **Create new** to create a new fragment. Here, select this option to create a new fragment.
4. Assign a name for the fragment, and click **Create** (see Figure 14.21).

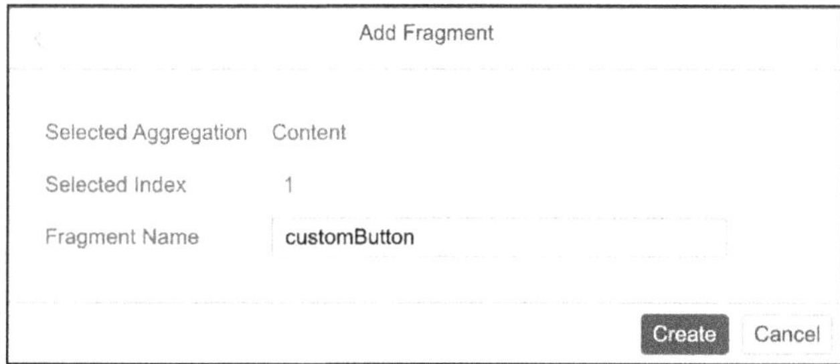

Figure 14.21 Creating a New Fragment

A directory fragment is now created with our new XML fragment *customButton.fragment.xml*. In this file, you can use all possible SAPUI5 controls with XML.

Assign IDs for Controls

It's important that you use IDs for all UI elements. If you haven't assigned an ID for an element, the element won't be displayed.

Our XML definition in this XML fragment looks like the following snippet:

```
<core:FragmentDefinition xmlns="sap.m" xmlns:core="sap.ui.core">
   <Button id="customBtn" text="Custom Button"/>
</core:FragmentDefinition>
```

If you now reload the preview of the application, a new button (**Custom Button**) will be visible in the table's toolbar (see Figure 14.22).

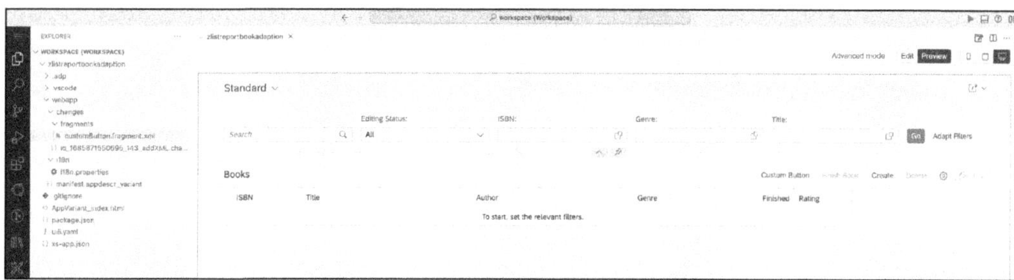

Figure 14.22 Custom Fragment Added: Custom Button

Deploying an Adaptation Project

Once you've made your extensions, you can deploy the extension application via the SAP Business Application Studio terminal using the deployment wizard (see Figure 14.23). You can find this wizard by right-clicking on *manifest.appdescr_variant* and choosing **Adaptation Project • Open Deployment Wizard**. You'll be guided through the deployment. Among other things, you must specify the project name and a package.

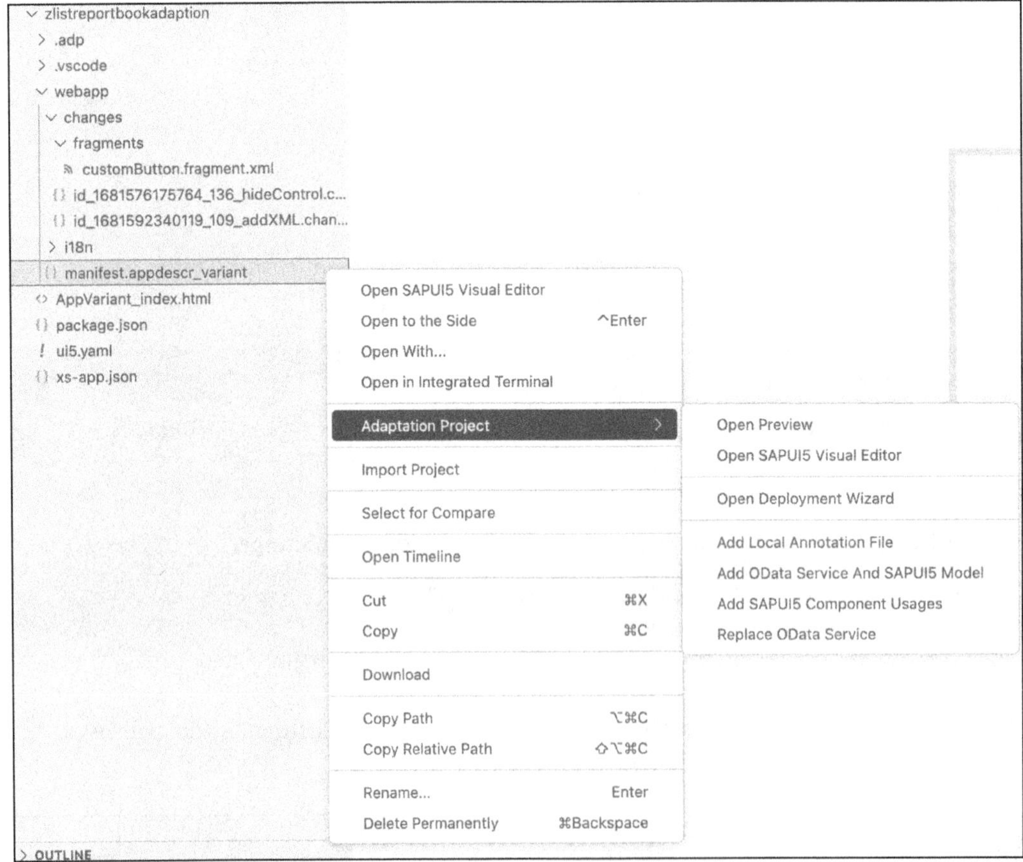

Figure 14.23 Open Deployment Wizard

In the next step, the target system needs to be selected. Furthermore, you may need to enter credentials (see Figure 14.24).

In the next step, you have to provide the **Package** information (see Figure 14.25). For a local object, select the **$tmp** package. If you don't choose a local package, you must also specify a transport request. Click **Next** to start the deployment.

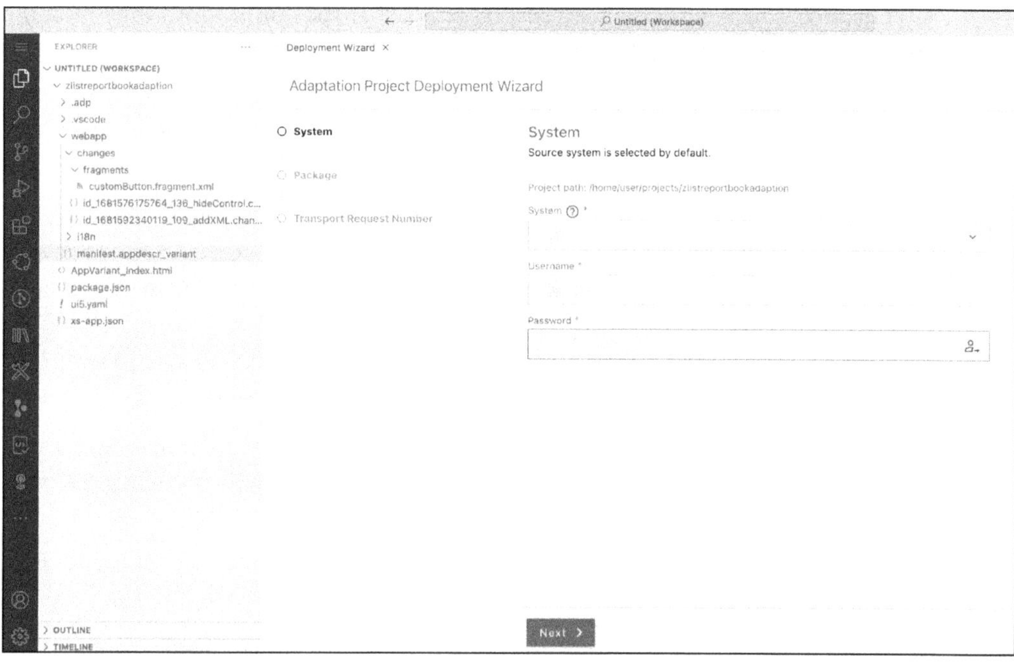

Figure 14.24 Selecting the Target System

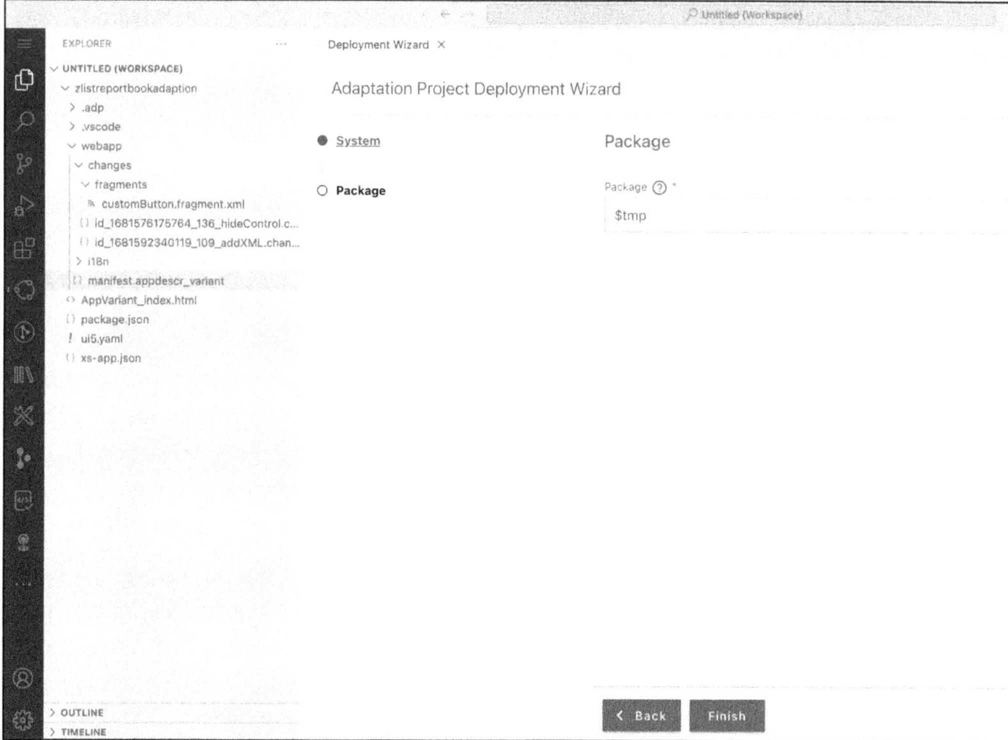

Figure 14.25 Specifying a Package

After deployment, you should see a success message in the notifications window at the bottom right (see Figure 14.26).

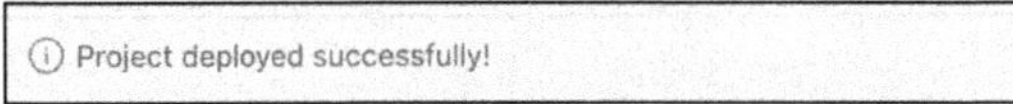

Figure 14.26 Deployment Successful

The deployed extension application can be searched for in the layered repository (LRep) via Transaction SUI_SUPPORT. In the LRep, applications can be managed layer by layer. This means that the original source code doesn't have to be manipulated if a new layer is added to this application via an adaptation project. There are several such layers. LRep stores the information about changes in a JSON-based file.

Using the **List repository files (support only)** option, you can start a search for extension implementations, for example (see Figure 14.27). You can search for either the ID or the namespace.

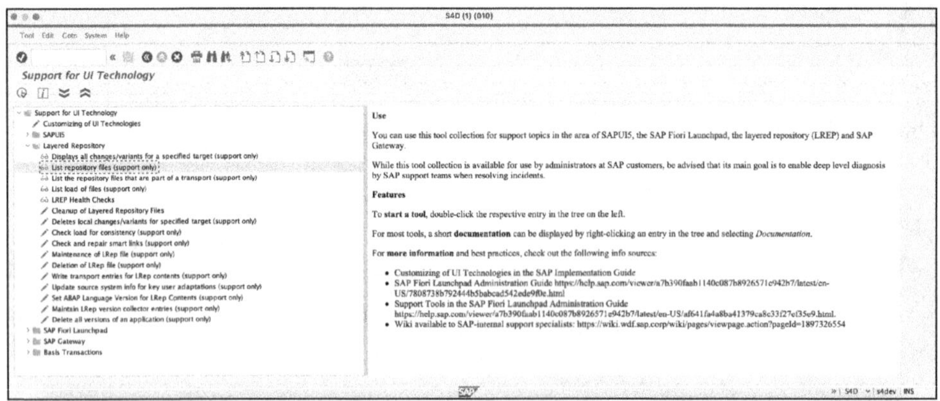

Figure 14.27 Layered Repository

14.3.2 Key User Adaptations

The SAPUI5 development framework can be used to create specific apps and offer them to your users or customers. But the UX of an app isn't always perfect for every user group from the start. Some want a more streamlined UI, others need additional UI elements, and still others want to save time by having certain filters applied automatically. Therefore, SAP enables you to make changes to the UI on runtime using *key user adaptations*. Key user adaptations are part of SAPUI5 flexibility.

With SAPUI5 flexibility, different user groups can design the UI of applications without having to touch the source code. It doesn't matter whether the application is a standard SAP application or a custom development. The ability to change the UI of an app with SAPUI5 flexibility is built-in to the SAPUI5 development framework. Therefore, if an app is developed with SAPUI5, it will support SAPUI5 flexibility in most cases.

The changes made to the UI with it don't change the source code of the original application. The changes are kept separate from the source code. So, if the original app changes due to an upgrade, it doesn't conflict; and the changes are still there afterward.

The customizations that are possible with SAPUI5 flexibility span the entire spectrum from simple adjustments, such as moving a field, to complex enhancements, such as adding sections with specific UI elements to an app. It's even possible to add custom fields end to end, from the UI to the database. The changes can be managed, reviewed, and enabled for specific user groups.

Availability of SAPUI5 Flexibility

SAPUI5 flexibility features are available on the following platforms:

- ABAP platform on premise (SAP NetWeaver, SAP S/4HANA)
- SAP S/4HANA Cloud
- SAP BTP, Cloud Foundry environment
- SAP BTP, ABAP environment

An application doesn't necessarily have to allow key user adaptations. To control this, there's a parameter in *manifest.json* called `flexEnabled`. If it's set to `true`, a key user adaptation can be carried out; if it's `false`, the application isn't available for key user adaptation (see Figure 14.28).

```
EXPLORER                              {} manifest.json ×
UNTITLED (WORKSPACE)                  zbookapp > webapp > {} manifest.json > ...
  zbookapp                                "sap.ui5": {
    > .vscode                                 "flexEnabled": true,
    > dist                                    "dependencies": {
    > node_modules                                "minUI5Version": "1.108.0",
    v webapp                                      "libs": {
      > annotations                                   "sap.m": {},
      > i18n                                          "sap.ui.core": {},
      > localService                                  "sap.ushell": {},
      > test                                          "sap.f": {},
      > utils                                         "sap.ui.comp": {},
      Component.js                                    "sap.ui.generic.app": {},
      index.html                                      "sap.suite.ui.generic.template": {}
      manifest.json                               }
    .gitignore                                },
    package-lock.json                         "contentDensities": {
    package.json                                  "compact": true,
    README.md                                     "cozy": true
    ui5-deploy.yaml                           },
    ui5-local.yaml                            "models": {
    ui5-mock.yaml                                 "i18n": {
    ui5.yaml                                          "type": "sap.ui.model.resource.ResourceModel",
                                                      "settings": {
                                                          "bundleName": "at.clouddna.zbookapp.i18n.i18n"
                                                      }
                                                  },
```

Figure 14.28 Enabling Key User Adaptation

Key users are typically business experts who coordinate one or more teams of end users. Their objective should be that all end users can work efficiently with the SAP Fiori apps. Key users therefore can perform key user customizations.

However, to be able to make such changes requires special authorizations. For this, the role `SAP_UI_FLEX_KEY_USER` must be assigned to the respective user. If this role is assigned, we first open an application in which the previously mentioned flag is set to `true` and thus key user adaptations can be carried out.

Next, we select **Adapt UI** (see Figure 14.29) in the user menu.

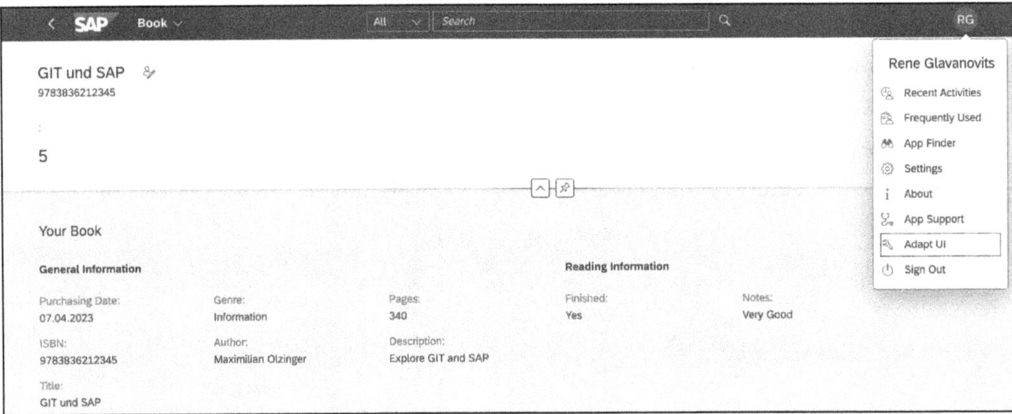

Figure 14.29 Adapt UI Action

Afterward, you can switch to the graphical editor where you can make various adjustments. For example, you can remove or add fields or entire blocks from the interface (see Figure 14.30).

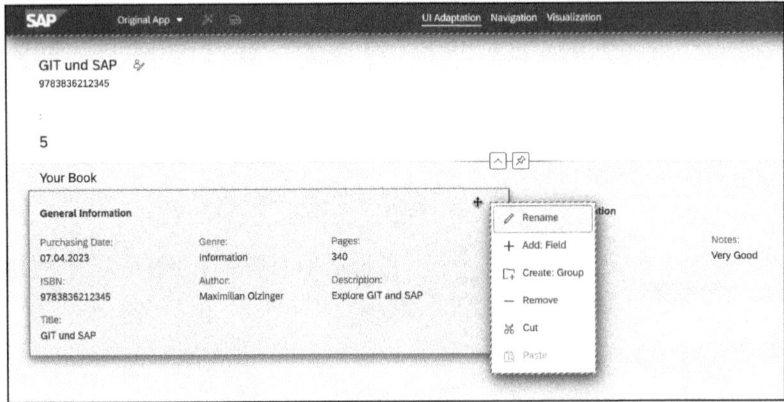

Figure 14.30 Customizing UI by Key Users

The customized application is then made available to users. Key users can make the following changes:

- Move controls.
- Add or remove controls.
- Rename controls.

- Change settings of controls.
- Combine individual fields into groups and ungroup them again.
- Integrate external content.

In addition to these changes, key users can also make changes concerning variant management. But keep in mind that this is only possible when using OData V2. Therefore, key users can create public variants, which will be delivered to all users or to users with specific roles. Additionally, they can perform the following:

- Modify existing public variants.
- Rename variants.
- Manage views by adding or removing favorites and modifying the visibility.

Depending on the environment in which an application runs, you can also save a customized application as a variant. In this way, you can provide different versions of an application to different user groups or teams. A variant looks like a custom application, but it's still linked to the original app. This means that if the original app changes due to an update, those changes will be reflected in your variants. However, the changes you make to the UI aren't affected.

To create a variant as a key user—that is, to be available to specific end users—follow these steps:

1. Switch to the graphical editor by selecting **Adapt UI** from the menu.
2. Specify any filter. In our case, we filter by **Title** (see Figure 14.31).

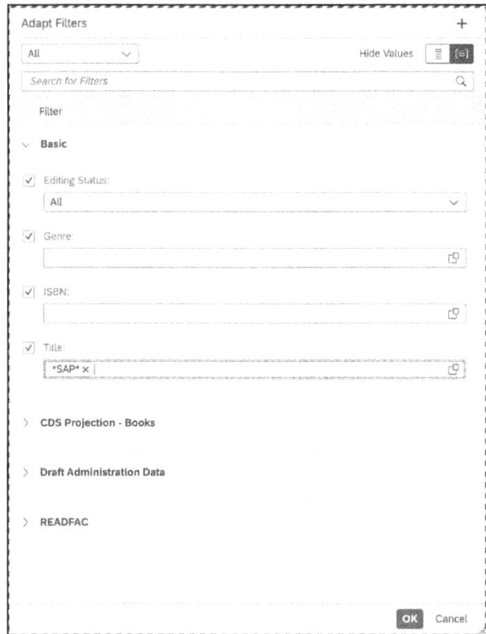

Figure 14.31 Adding the Filter for Title

3. After the filter has been applied, a dialog box appears, which you must confirm with **Save as New View** (see Figure 14.32).

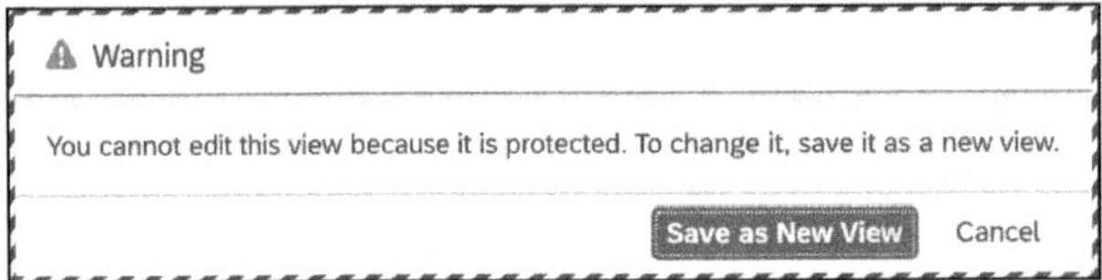

Figure 14.32 Confirmation Dialog: Save as New View

4. A dialog then appears in which a description and properties of the variant can be set. Among other things, a role assignment can also be made here. This means that only users with the assigned roles will have the variant available. Here, assign a name, and save the variant by clicking **Save** (see Figure 14.33).

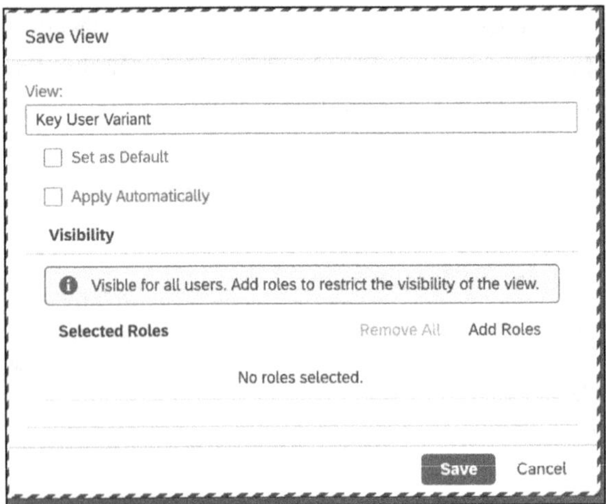

Figure 14.33 Saving the Variant

After the variant has been saved, you can validate the result as follows:

1. Select **Manage Views** in the variant management menu (see Figure 14.34).

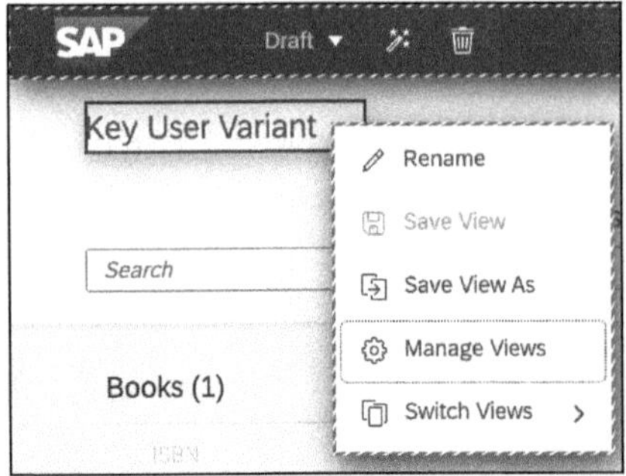

Figure 14.34 Variant Management Menu

2. Then, a dialog opens with the available variants. As shown in Figure 14.35, the previously created variant is also listed here. In addition, this is visible to all users without restriction.

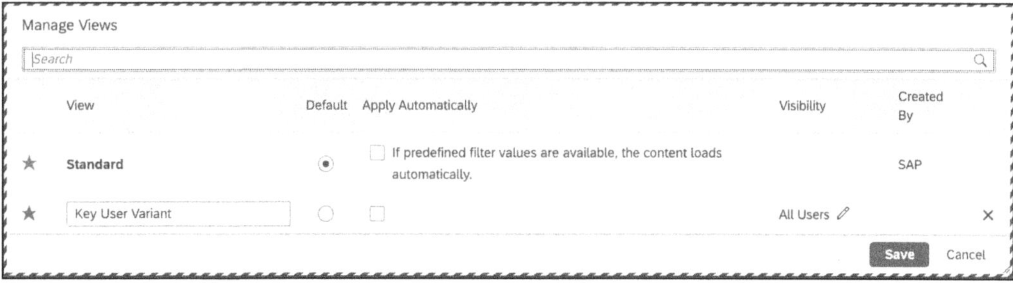

Figure 14.35 Available Variants

Embedding iFrames

The integration of iFrames is possible with UI adaptations, but this isn't recommended.

Best Practices

Following are a few things to consider in key user adaptations:

- **Use stable IDs.**
  Stable IDs are used to identify the controls that can be changed by UI customization or personalization. Therefore, each control and each view must have a stable ID assigned.
- **Use SAPUI5 controls supported by UI adaptation.**
  When integrating new SAPUI5 controls, care should be taken to ensure that the controls used are actually supported. To find this out, you can look in the samples. The UI adaptation feature can also be tested here.
- **Load your views asynchronously.**
  If stashing is used for loading views, care should be taken that views are loaded asynchronously. Otherwise, changes won't be recognized by key users during an app reload. Furthermore, asynchronous loading of views is an advantage for the performance of the application anyway. In any case, make sure that stashing is activated, especially when stashing is used.

14.3.3 Guided Development

Guided development is an SAP Fiori tools extension designed to simplify and broaden access to SAP Fiori development. Available in both SAP Business Application Studio and VS Code, it offers step-by-step instructions for adding functionality to your SAP Fiori projects. Whether adding new columns to a table, inserting charts, or creating cards for overview pages, guided development provides tailored guides for every SAP Fiori elements floorplan.

The guides offer an interactive approach to SAP Fiori tools documentation with the added capability to directly update your project. Simply open the desired guide, configure the relevant parameters, and click **Insert Snippet** to integrate the tailored code snippets into your project files. Because the parameters are project-specific, the snippets are ready to use without the need for placeholder modifications.

For newer developers, the wizard-like functionality simplifies implementing features they may be unfamiliar with. Experienced developers benefit from faster development, supported by auto-generated and updated code snippets.

To use guided development, the *SAP Fiori Tools - Guided Development* extension, which is part of the *SAP Fiori Tools - Extension Pack*, must be installed in VS Code. This installation step isn't required in SAP Business Application Studio, as the extension is already installed there by default.

To open guided development, you must first open an SAP Fiori elements app. Guided development can then be opened as follows:

1. Select **View • Command Palette** from the menu bar (see Figure 14.36).

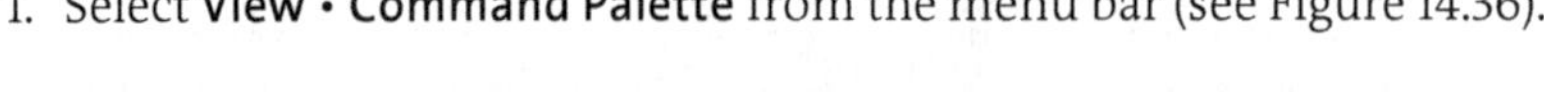

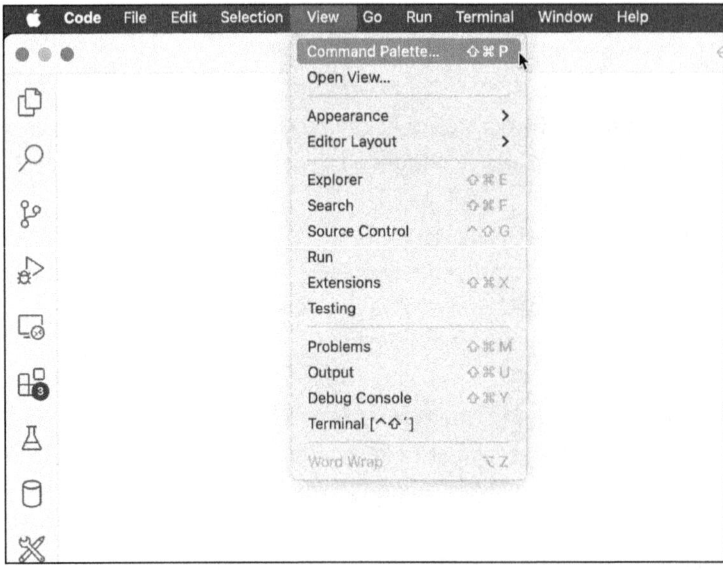

Figure 14.36 Opening the Command Palette Option in VS Code

2. Search for and open "Guided Development" (see Figure 14.37).
3. Once the guided development has been opened, the following view is displayed (see Figure 14.38).

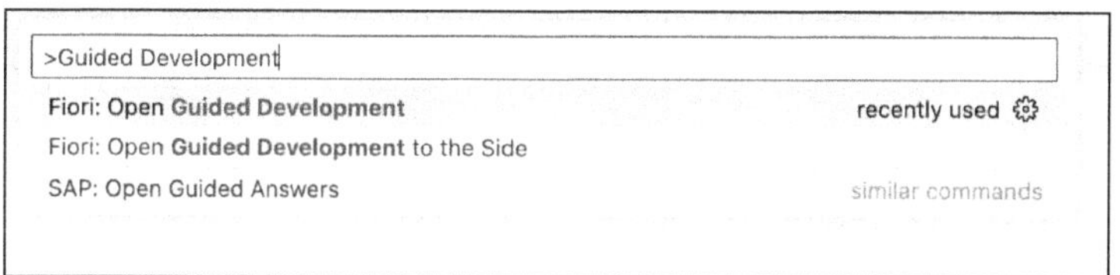

Figure 14.37 Searching for "Guided Development" in the Command Palette

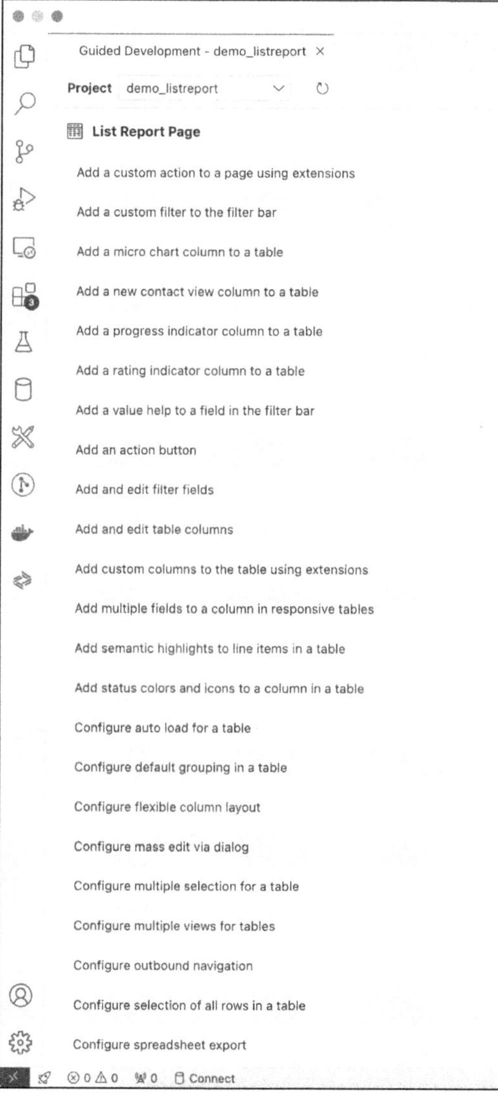

Figure 14.38 Guided Development

4. Various selection options can be applied in the upper area. You can define for which project the guided development is used, the list of guides can be grouped according

to various parameters (page type, OData version, application artifacts), and the guides can also be restricted so that only those relevant for the selected project are displayed.

Our example application is a list report. To give you an idea of how guided development works, we'll add a custom column to this list report with the following steps (for all other guides, this basically works in the same way):

1. Click on **Add and edit table columns**. A new window opens where the guide is described (see Figure 14.39). Click **Start Guide** to start the guide.

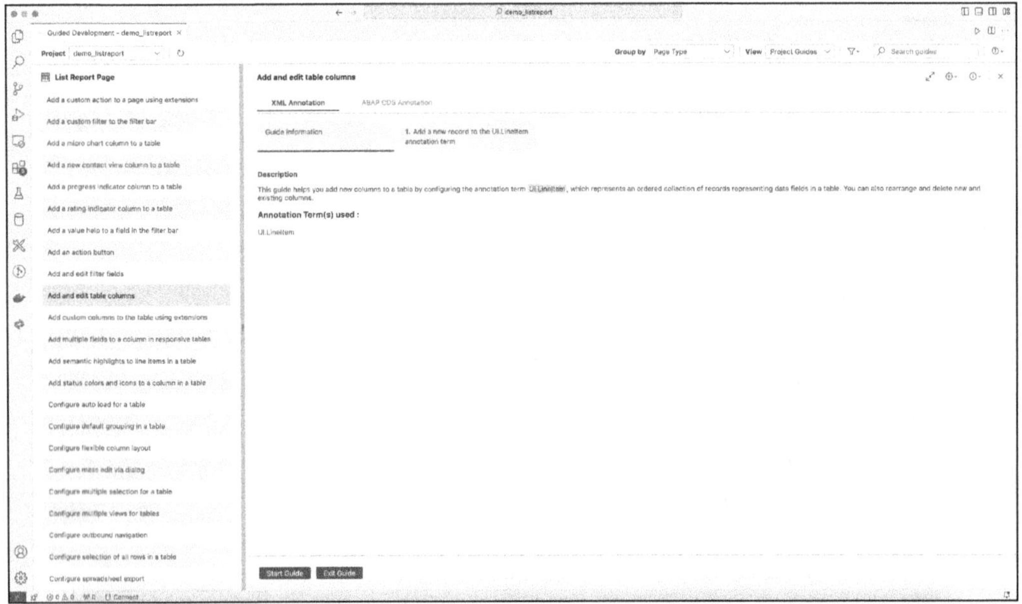

Figure 14.39 Opening the Add and Edit Table Columns Guide

2. Now the fields such as **Entity Type** and properties of the column (**label**, **property**, and **importance**) must be defined (see Figure 14.40). For this example, we chose the following values:
   - **Entity Type: CustomerType**
   - **Label: Demo Column**
   - **Property: CustomerId**
   - **UI Importance: High**
3. After all the values are entered, you can see the annotation generated based on the previous input. Click **Insert Snippet** to add the code snippet to the local annotation file.
4. Once that's done, close the guide with **Exit Guide**. To see whether the annotations were inserted correctly, we can check the local annotation file (see Figure 14.41).

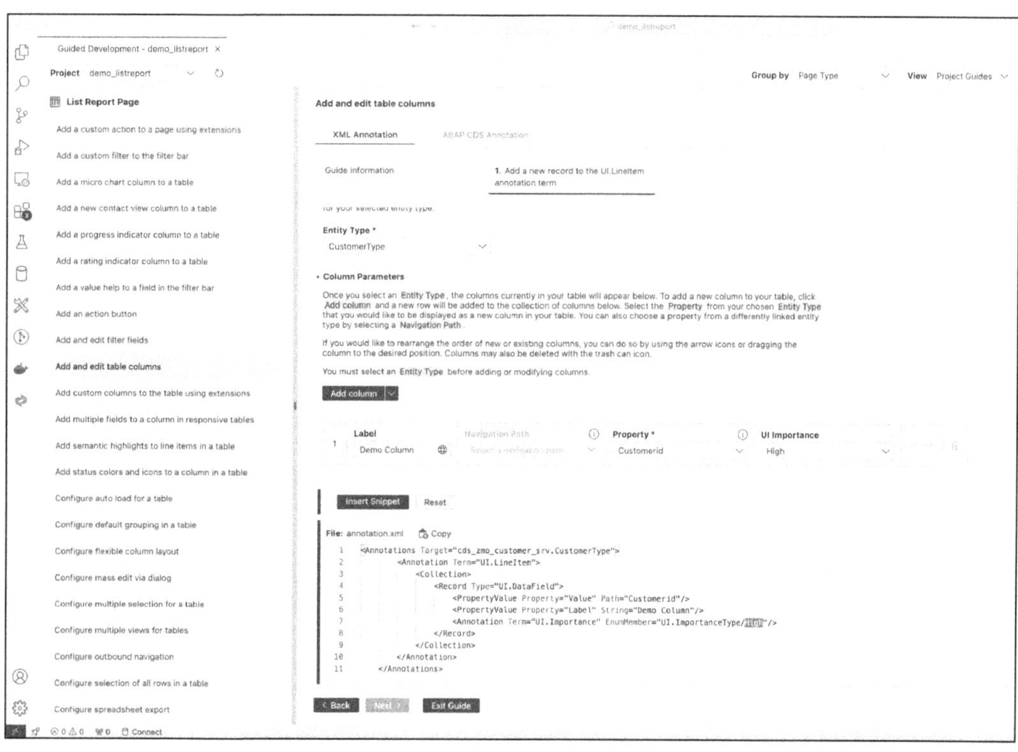

Figure 14.40 Entering Properties for the New Custom Column

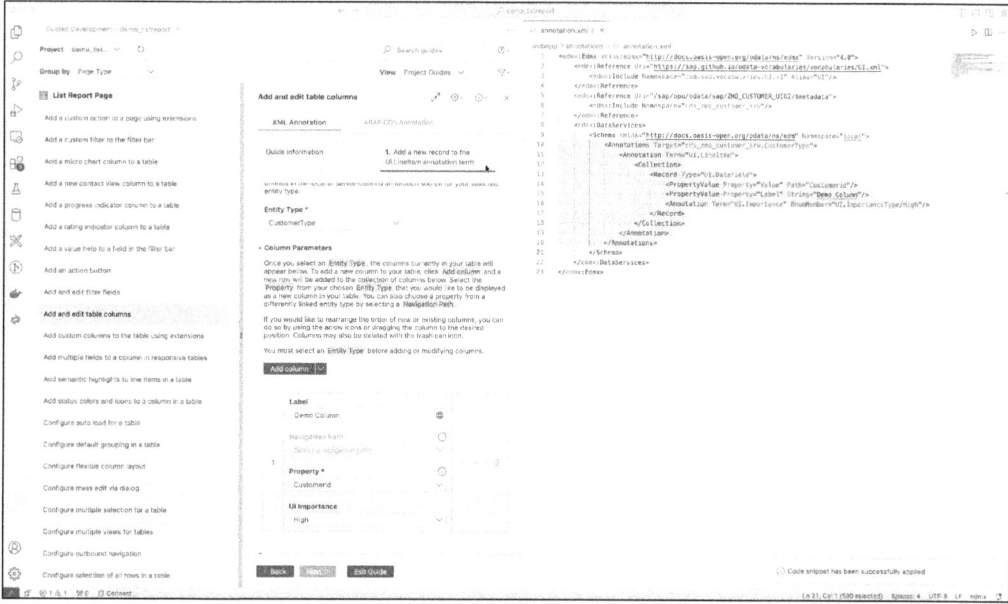

Figure 14.41 Local Annotation File after Inserting Code Snippet

14.4 Summary

In this chapter, you've seen which floorplans are available in SAP Fiori elements, when they should be used, and how they are structured, as well as what possibilities the flexible programming model offers and how applications can be extended. In the upcoming chapters, we'll now deal with topics that are more in the direction of administration and developer productivity. For example, we'll look at Git and its deployment process and deployment options, as well as the SAP Fiori launchpad configuration.

PART V

Administration and Developer Productivity

Chapter 15
Git

Git has become indispensable in modern software development and is the de facto standard when it comes to version management and source code distribution. This tool makes it easy to develop in distributed teams of any size.

"Why should I worry about version control?" This is a common question among software developers that can be answered very simply: as soon as you've introduced a major bug into one of your source codes, tested it insufficiently, and put it into productive use, you'll wish for nothing more than version control. *Version control* is a system that records changes to files chronologically, thus making it possible to retrieve historical versions at any time.

In the past, many developers started their careers by setting up local directory structures and storing backups or versions of their developments there. As soon as development was done in teams, however, this approach was no longer valid, and structured procedures and tools were needed to support the team. Today, version control systems are the de facto standard and an indispensable part of the curriculum for young software developers. This chapter is about *Git*, the most widely used tool in this environment.

But before we can dive into this powerful tool, we need a bit of history first. Git has its origins in Linux development and was developed out of necessity by Linux creator and mastermind Linus Torvalds. Until that point, kernel developers relied on the *BitKeeper* version control system to manage their source code. After BitKeeper's license was changed, free use was no longer possible. Another source code management system popular at the time, called *Subversion*, couldn't meet the high demands of the Linux release process. That was the birth of Git. Linus Torvalds, then head Linux developer, put his work aside and programmed the basic framework for Git in just two weeks. In doing so, he developed a completely new source code management system from scratch in 2005.

Note

Contrary to many assumptions, Git isn't an abbreviation. However, the word does refer to a "foolish person" in English.

Initially, Git's focus was on covering the functions of Linux kernel development. But other developers were also confronted with similar problems and found a suitable solution in Git. One of Git's objectives is to enable distributed development that can also be carried out offline.

Outside of Linux kernel development, the hosting service *GitHub*, which was launched in 2008, has contributed significantly to its success. Torvalds probably didn't plan for the centralized approach of hosting services to be the foundation of its success. His goal was merely to design a decentralized version management system. GitHub has quickly become the most widely used development platform for open-source projects worldwide. The reason for this is very simple: GitHub offers free hosting for open-source projects. The relevance of GitHub can also be seen from the fact that the service was bought by Microsoft in 2018 for around $7.5 billion. This is an impressive story for a tool that was born out of necessity and developed by a man who actually never wanted to build a source code management tool. With Git, Torvalds has started another revolution in addition to Linux, without which the world of software would look completely different. In an interview, Torvalds once said that, in addition to databases, he considers source control management to be "about the least interesting thing in the computing world" (*www.linux.com/news/10-years-git-interview-git-creator-linus-torvalds/*).

Git wasn't present in the SAP universe for a long time, and this is also in the nature of things, because the classic ABAP developments in an SAP system didn't require it either. In the classic ABAP world based on SAP NetWeaver, developers don't have to worry about managing the source code. Until SAP NetWeaver 7.4, developments were usually carried out in the ABAP Workbench (Transaction SE80), which is a complete programming environment delivered with every SAP NetWeaver AS ABAP system. In contrast to almost all other technologies, no separate development tool needs to be installed on the developers' computers. However, the ABAP Workbench also has a disadvantage: it requires a permanent connection between your local computer and the application server. The changes you make in the ABAP Workbench are stored in the repository tables of the SAP database. The content of these tables is referred to as repository data or repository objects. With a few exceptions, the repository data stored in these tables is cross-client.

With SAP NetWeaver 7.4, SAP has given the code-to-data approach significant importance. This means that the processing and preparation of data is carried out in the database, if possible, and not in the application server. With SAP NetWeaver 7.4, development objects such as *core data service* (CDS) or *ABAP Managed Database Procedures* (AMDP) were introduced, which can only be developed in the Eclipse development environment. To do this, it's necessary to install the *ABAP Development Tools* (ADT) for Eclipse. But even the partial change of the development environment in the direction of Eclipse doesn't require new version management.

Now, however, Git is becoming more important every day, and it's no longer possible to imagine many areas, such as SAPUI5 or SAP cloud development (referring to the two frameworks *SAP Cloud Application Programming Model* and *ABAP RESTful application programming model*), without it. For many SAP developers, SAPUI5 and SAP Fiori development is their first point of contact with version management outside of the SAP NetWeaver ABAP stack. SAP recommends managing SAPUI5 applications in an external version management system.

15.1 Local and Remote Repositories

Git is a *third-generation version control system*, while most of its predecessors belong to the second generation. The main difference is that third-generation version control systems are distributed, while second-generation systems are centralized. In the following sections, we'll look at the architecture of repositories and then briefly introduce you to forks and branches.

15.1.1 Architecture

15

In a centralized system, all developers have a copy of the data on their computers (see Figure 15.1). The historical data is managed on the server. Developers commit their changes directly to the remote repository.

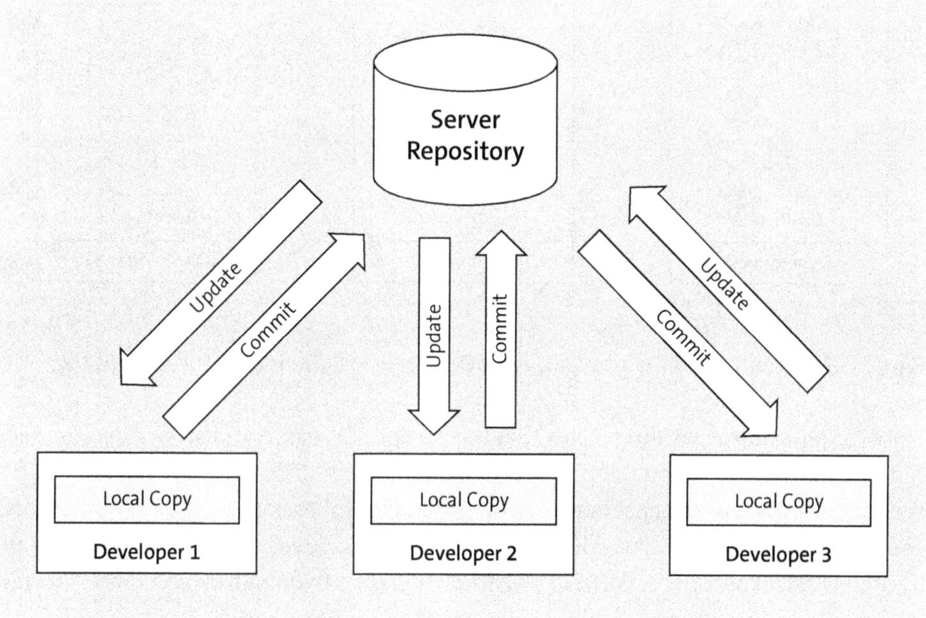

Figure 15.1 How Central Distribution Works

This means that a faulty change can destroy all builds, which is when the application is built or compiled. Furthermore, remote commits are very slow in the centralized system. We'll clarify what exactly a commit is, but for now it's sufficient to define a commit as a permanent change to the source code in the Git repository.

In a decentralized system such as Git, all developers have their local copies of the files (see Figure 15.2). These are stored in a local repository, and when they package changes in a commit, these are first stored only in the local repository. After that, they can be updated on the associated remote repository. The term *push* in this context means transferring the local changes to the remote repository on the server.

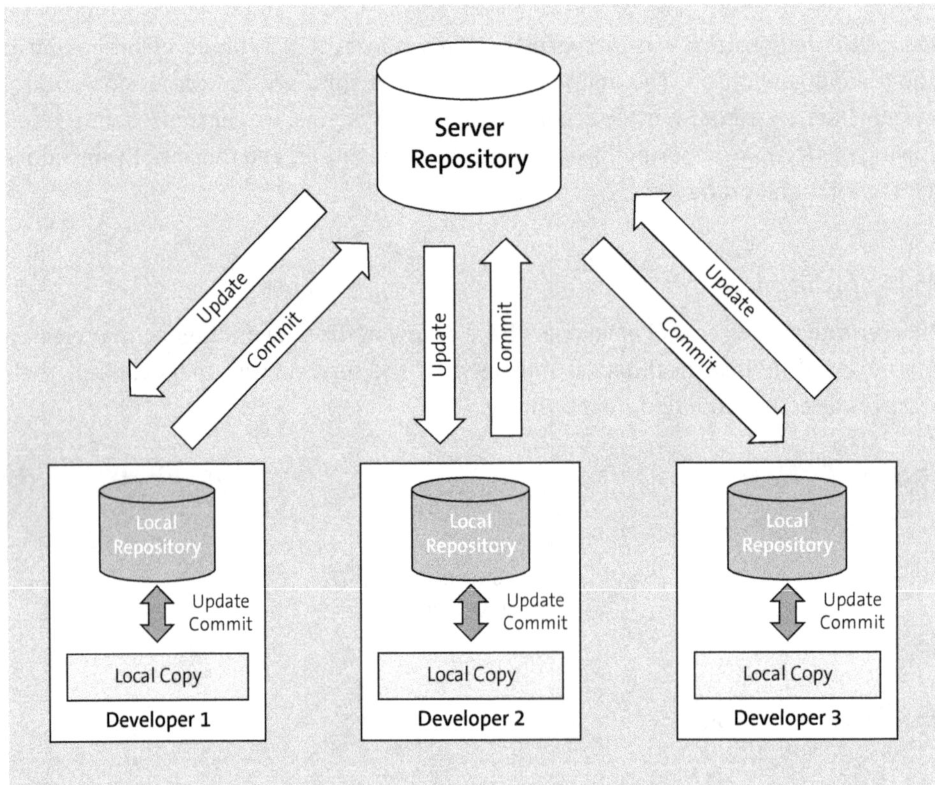

Figure 15.2 Decentralized Distribution: Every Developer Has Their Own Local Repository

Think of repositories as directories that have a specific name and store files. The files can be of various types and don't have to contain executable program code or a functional project. A repository remembers all the changes that have been made during the course of the project, so that you can also retrieve the developments of the files step-by-step in retrospect in the form of a history. There are two main types of repositories:

- **Local repositories**
  Repositories that are actually stored on the developer's computer, whether on a personal computer or in an integrated development environment (IDE) in the cloud.

This repository is only visible to that one person and isn't public or visible to other individuals or groups of people.

- **Remote repositories**
  Repositories that are either hosted within the corporate network or by an external provider in the cloud. These repositories can also be accessed by multiple people. Ideally, these repositories are either secured themselves by connecting an *Active Directory*, for example, or certain users in certain roles are entered in the cloud as contributors.

The design of the access is always dependent on the type of hosting. Either the remote repository is located in an internal network, and you can access the Git server directly—even if it requires a detour via a virtual private network (VPN) connection—or you need a URL because the remote repository is located at some provider. Nevertheless, you, as a tester, developer, or project manager, want to have access to it to work on this project.

If a remote repository already exists for a project, everyone who has access to and wants to use this repository must first clone it before they can work with it. With the command `git clone <path/url>`, you clone the remote repository. You may also be asked to authenticate to make sure that you're authorized to work with the repository. Where the copy of the remote repository is stored locally on your computer depends on the directory in which you executed the command.

Directory Structure

As mentioned earlier, cloning a repository creates a local copy and stores all files and directories that exist in the remote repository on your own computer. Because you're using the command line, you need to pay attention to which path in your directory structure you're currently on. You can usually see the current path at the beginning of the line, where you also enter commands.

If you're not in the right place, you can use the command `cd <path>` (stands for "change directory") to switch to a different directory. Whether you enter an absolute or a relative path as a parameter is important here. Absolute paths can always be processed, regardless of which hierarchy level you're on in a directory structure. Relative paths, on the other hand, can only be evaluated at the current level.

You already know one way of managing remote repositories. We've summarized the most important distributed workflows in the Git environment for you. For administrators and architects, there are three well-known ways of managing their Git repositories to organize team collaboration:

- **Subversion-style workflow**
  This is the classic case and is called a *centralized workflow*. There's one remote repository, which in this case is also called a *shared repository*. The developers clone this repository and thus have a copy on their computers (see Figure 15.3).

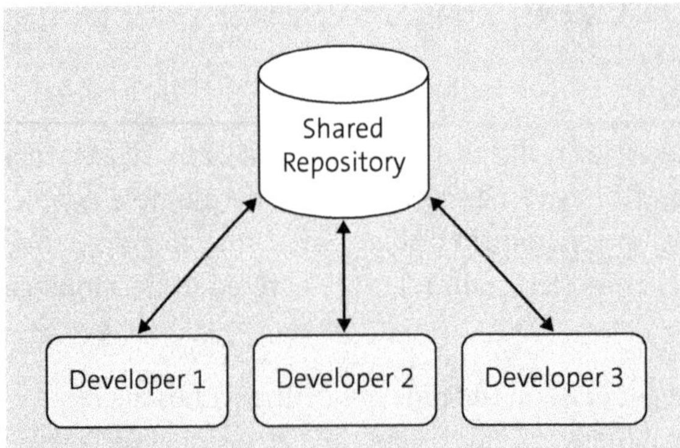

Figure 15.3 Subversion-Style Workflow

- **Integration manager workflow**
  In an integration manager workflow, which is a popular architecture for open-source projects, there are several separate repositories that developers work with. One or more *integration managers* ensure that only those contributions from the developers that add value to the project are taken from these public repositories. The changes that are taken over end up in the *blessed repository*, a repository that contains the public sources of the project (see Figure 15.4).

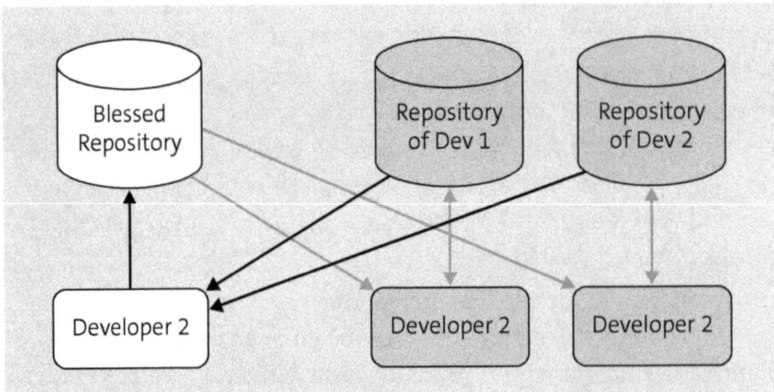

Figure 15.4 Integration Manager Workflow

- **Dictator and lieutenants workflow**
  This variant is very similar to the previous one, but here there are three different roles. This is translated as "workflow for dictator and lieutenants." The lieutenants are responsible for smaller parts of the project, which they monitor and for which they check and compile the developers' changes. The dictator then ensures that the compiled project parts are put together and made available in a blessed repository (see Figure 15.5).

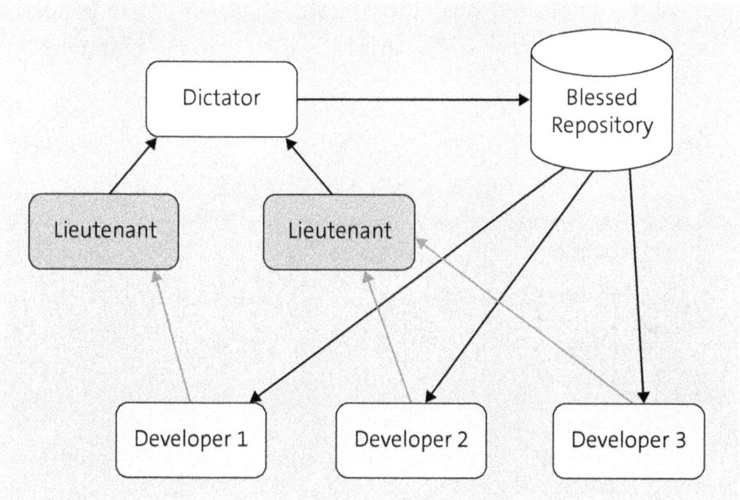

Figure 15.5 Dictator and Lieutenants Workflow

If no remote repository exists yet, someone has to take the first step and create one. You create a local repository locally on the computer using the command `git init <repo_name>`, where you should be in the directory that will represent the repository. After this command, a hidden folder named *.git* is created.

Now, you can link this local repository to a remote repository. This is necessary if you've created a local repository yourself and want to assign it to a remote repository, or because you want to change the associated remote repository for some other reason. To do this, use the command `git remote add <name> <your_repo_url>`. As the first parameter, enter a name for this remote repository, and, as the second parameter, enter the unique path that points to the remote repository.

15.1.2 Forks

To enable developers and teams to work together efficiently, including on open-source projects, thought has been given to how best to implement this collaboration from a technical point of view. In this context, *fork* (because, figuratively speaking, the project develops several forks) was invented.

With the help of a fork, you copy a remote repository created by another person as a separate remote repository for your own work (of course, you must have access to the corresponding remote repository to do this). Then, you clone this repository and create a local copy on your computer. This way, you initially only work against your own remote repository of the project. As soon as you think your changes are ready for the public, you submit a *pull request* to your contact person (integration manager, lieutenant, etc.) to have your changes incorporated into the blessed repository. This approach can help to keep the blessed repository clean, especially for large teams, because only the integration manager has write access, while all other team members work in their own forks (see Figure 15.6).

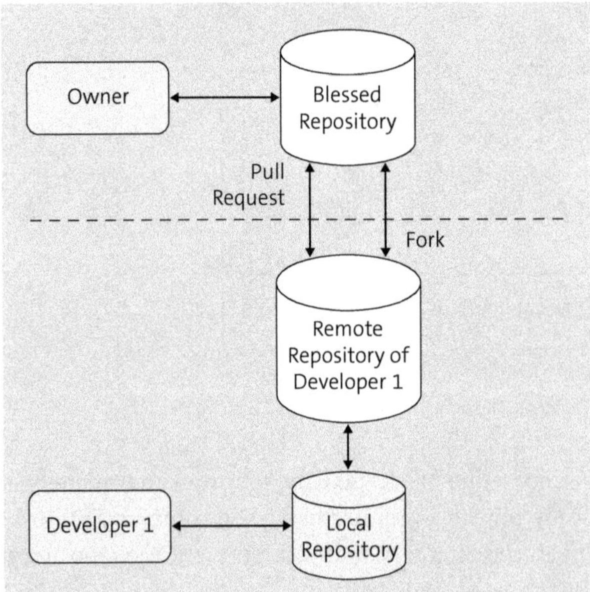

Figure 15.6 Forking: Best Practice for Open-Source Projects Especially

15.1.3 Branches

Git offers a special source code management and versioning option. You can split your development into several strands or *branches* and develop them separately, merge them back together again or even discard individual branches.

A repository can consist of several such branches, but there's always a `main` branch, which in most cases is called (who would have thought) the `main` branch. This name can be overridden or defined using the `init.defaultBranch` configuration option. For repositories that manage executable programs, there's an unwritten rule that the `main` branch must always contain executable code without errors.

Nondiscriminatory Language in IT

You'll also come across the term *master branch* in some places. However, you should no longer use this term. As you may have noticed, a movement began in 2020 that aims to make the language of IT, that is, technical terms, design patterns, architecture and reference models, and much more, nondiscriminatory and free of racist comments and expressions. For example, you should use terms such as *primary* and *replica* instead of *master* and *slave*, or *denylist* and *allowlist* instead of *blacklist* and *whitelist*. In our case, that results in *main branch* instead of *master branch*.

Branches fulfill several purposes, as follows:

- Multiple versions of the source code can be distinguished in the remote repository. This allows you, for example, to simulate a multisystem landscape and separate all development statuses from the test version to the productive status of the application.
- In a local repository, branches are often created for new developments (feature branches). They are branched off from the main branch to make changes and, in the best case, to merge them back into the main branch later.
- Another advantage of feature branches is that if the changes fail or are rejected, you don't have to laboriously search for the inserted lines; instead, you can simply delete the branch to return to the previous main branch, which can normally be executed.

There's a good reason why we speak of one or more branches. You can imagine the structure as branches that branch off from the trunk at certain points and move away, but which can also come together again (see Figure 15.7). The name of the remote branches is preceded by origin/ (or the name you've chosen to add your remote repository) to show that we aren't working locally, but that the resource is located elsewhere. The names of the branches themselves can be chosen freely.

15

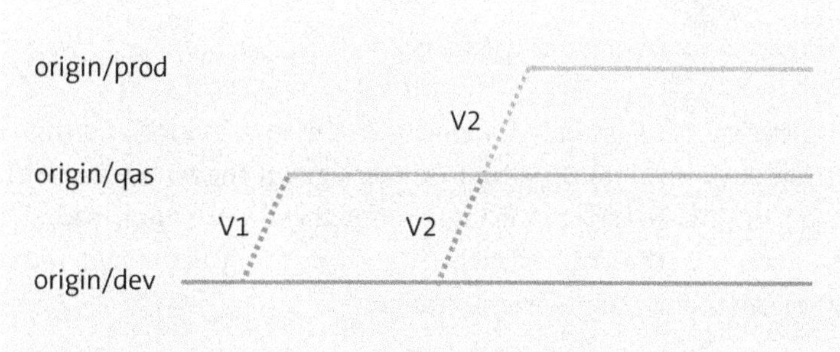

Figure 15.7 Three-System Landscape as a Simple Example for Branches

In the following example, we'll show you a simple scenario to illustrate how these branches might be imagined. We're mapping a three-system landscape, which is why we have a branch for development (dev), one for quality assurance (qas), and one for production (prod).

Furthermore, you can see that this repository starts with the dev branch, and then a qas branch for quality assurance with a first version (V1) was branched off at some point. After a feedback loop, change requests were implemented in the dev branch, and, after the second version was successfully accepted on the already-existing qas branch, a third branch called prod was created for production. This branch contains the tested and productive source code.

This example shows just one of many ways to manage your remote repository. Changing everything in one branch and having no further branches would be like always changing and implementing something in the production system. However, Git professionals among you know that it doesn't necessarily take a structure like this to map a three-system landscape. You can also tag the corresponding project states to mark the versions that were used for deployment. Again, it looks different if you're using *continuous integration*, *continuous delivery*, or even *continuous deployment*. Especially with the latter, you'll probably only be working on a single branch, where each change leads to the execution of the automated tests and subsequent deployment. As you can see, there are many factors that can influence the structure.

We've mentioned the distinction between branches in local repositories and remote repositories for specific purposes. Figure 15.8 shows a graphical representation of how such `feature` branches can be imagined in local repositories.

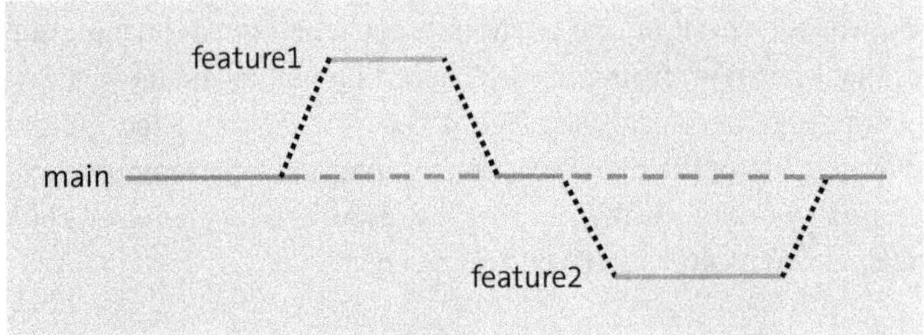

Figure 15.8 Using Local Branches for Feature Development

Here, you can see that development initially took place in the `main` branch and that a fork occurred at a certain point in time. This first fork was given the name `feature1`. Adjustments were made in this `feature` branch and, after these were approved, the changes were merged back into the `main` branch. The same thing happened when another feature was implemented in the `feature2` branch.

In this case, you're also in control of the distribution and use of local branches. What is very clear in this case, however, is that if the changes are revised, the `main` branch isn't affected, and only the `feature` branch needs to be deleted.

15.2 Essential Git Commands

You've already learned about repositories and branches, but these only provide a framework for the various functions and important commands that Git offers. In this section, we'll take a closer look at the most important commands that you'll typically execute on the command line. With these commands, you can edit the branches and repositories and breathe life into your project. In Section 15.5, you'll learn about graphical tools in SAP Business Application Studio and Visual Studio Code (VS Code) that almost eliminate the need for commands. These tools are very helpful, but even the best tool is of little use if you don't understand the logic behind it.

All of these commands have become very powerful over time, so if we wanted to cover all the command options and parameters, we could write several pages per command. As we don't have space for that, note that the commands you'll learn about will only be presented in as much detail as is necessary for the subsequent chapters and examples. See for yourself the wealth of possibilities and, after reading this introductory section, take a look at the official listing of all Git commands at *https://git-scm.com/docs/git#_git_commands*. But first, we want to introduce you to the most important commands that you'll definitely need in your daily work.

15.2.1 Commit

The smallest measurable units in the Git environment are called *commits* and are essentially a snapshot of the current state of development. A commit is triggered by a developer using a command (discussed later) and usually contains all changes made since the last commit. In addition, a commit stores who (`user.name` and `user.email` from the Git configuration file) made the changes at what time (date and time). This information is stored together with a unique commit ID, a checksum, and a comment (commit message), which the person making the commit can add when the commit command is executed.

The question naturally arises as to which changes made since the last commit will be saved with the next commit. These are all the changes that you have in the staging area. The *staging area* is an area that holds a selected number of changes that will be included in the next commit. You can add files to the staging area using the following command:

```
git add <filename>
```

If you want to include all changes in the staging area, use this command:

```
git add --all
```

Figure 15.9 shows the sequence of commands for ensuring that the changes first end up in the staging area and are then transferred to the local repository via commit.

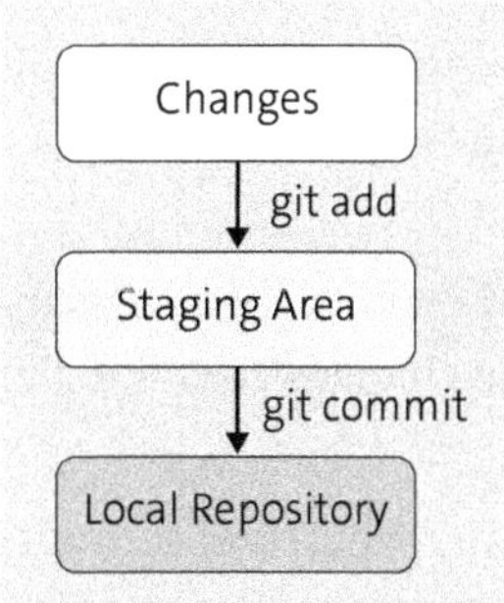

Figure 15.9 Adding Changes to the Staging Area before the Commit

To shorten the detour via the staging area, there's the command `git commit -a`. This command writes everything that is currently in the change state directly to the commit. Figure 15.10 shows this abbreviated method.

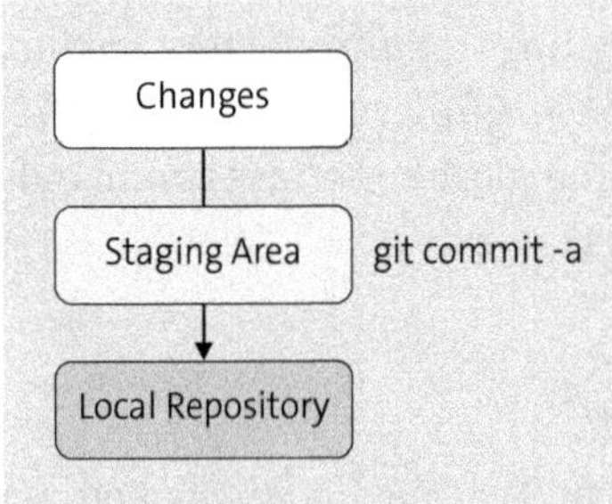

Figure 15.10 Abbreviation Method to Add All Changes to the Next Commit

Git also has its own file lifecycle to keep track of which files have been newly added, edited, or deleted. There are basically two types of files: files that are tracked and files that aren't tracked. The name says it all: a distinction is made here between whether changes to a file are recognized and documented or not.

If you decide that Git shouldn't track certain files or file types, you need to create a file named *.gitignore* in the project's root directory, if it doesn't already exist. In this file, you write line by line the paths to certain files, complete directories, or—using wildcards—files with certain extensions. Here are three of these examples:

- `node_modules`
- `data/mock.json`
- `*.zip`

Otherwise, trackable files can be distinguished between unmodified, modified, or staged (the state when they are in the staging area). Because, in our opinion, these lifecycles are very well described in the official documentation, we refer you to the Git documentation at this point (*https://git-scm.com/*), but we don't want to withhold Figure 15.11 from you, which describes these cycles and shows which functions the cycle changes effect.

After the changed files have been brought into the staging area, execute a commit using `git commit -m <comment>`, and you can add a comment using the `-m` option. We've identified the comment parameter with a placeholder `<comment>`.

Commit comments should be written in the present tense and contain the necessary information—but no unnecessary ramblings. Simply write the changes contained in the respective commit in a human-readable form.

Figure 15.12 shows how the relationships between repositories, branches, and commits are mapped in a healthy system. We speak of a healthy system because, of course, you could also create an empty repository, but that makes no sense.

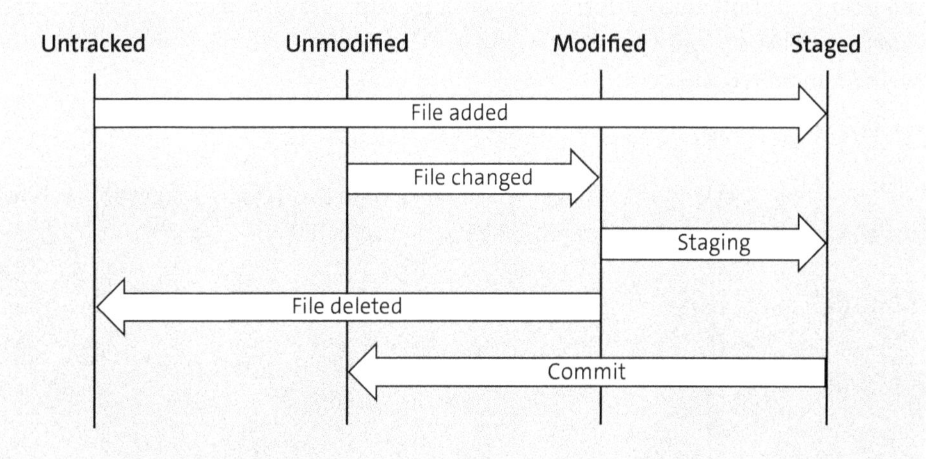

Figure 15.11 Lifecycle of Trackable Files within Git (Source: Git SCM)

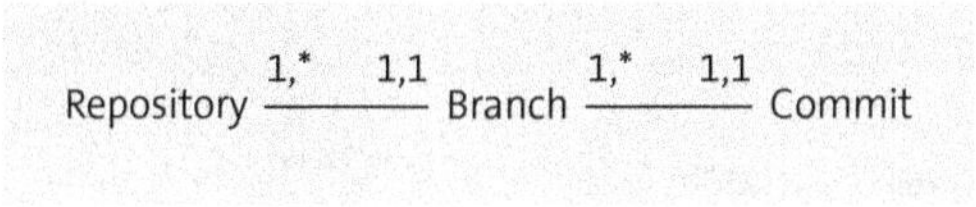

Repository 1,* —— 1,1 Branch 1,* —— 1,1 Commit

Figure 15.12 Relationship between Repositories, Branches, and Commits

Experience has shown that visual explanations of Git functions and commands are always helpful for understanding what commands do and how they affect the current state of the source code and the Git repository, both locally and remotely. We'll cover the next few functions in a continuous example.

15.2.2 Clone

Figure 15.13 shows the initial situation. The circle indicates that there's a commit with the unique identifier Ⓐ on `origin/main` in the remote repository.

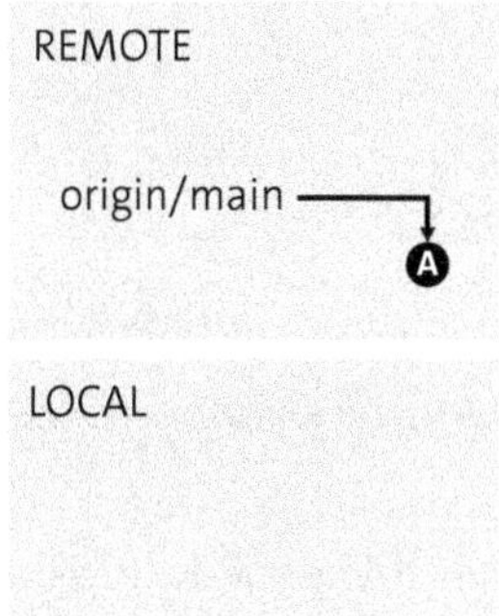

Figure 15.13 Starting Point: Initial Commit Is Located in the Remote Repository

The following command performs a clone, indicating that you want to have an exact local copy of the remote repository you're working against. In other words, you're cloning the remote repository:

```
git clone <your_repo_url>
```

You'll see that a local repository and a `main` branch have been created. Figure 15.14 shows that a local copy of the remote repository has been created.

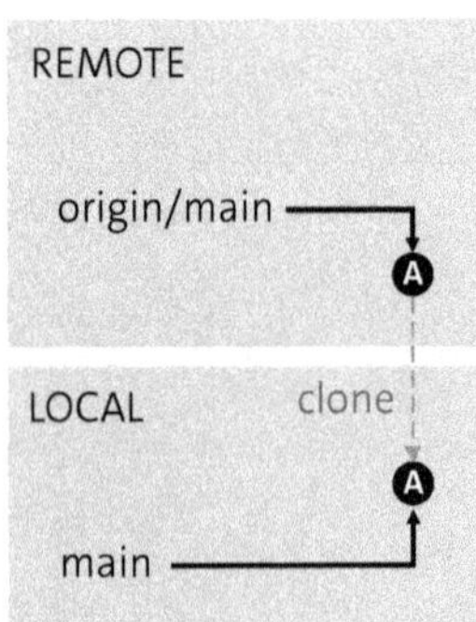

Figure 15.14 Cloning the Remote Repository Leads to a Local Copy

The result of a clone on your local repository is shown in Figure 15.15. After this process is complete, the `main` branch points to the current commit, which in this case has the ID Ⓐ.

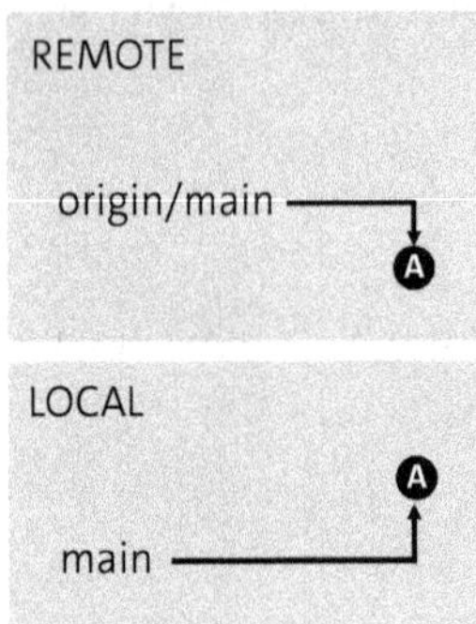

Figure 15.15 Identical State Both of Local and Remote Repository after Clone

Assuming you now make changes and there are two further commits with the IDs Ⓑ and Ⓒ, which you make in the `main` branch of your local repository, the example develops, as shown in Figure 15.16.

You can think of the name associated with the `main` branch, in this case, `main`, as a pointer that always points to a commit, but not necessarily the current one. In addition to these pointers to branches, there are also other pointers, which we'll discuss in more detail later.

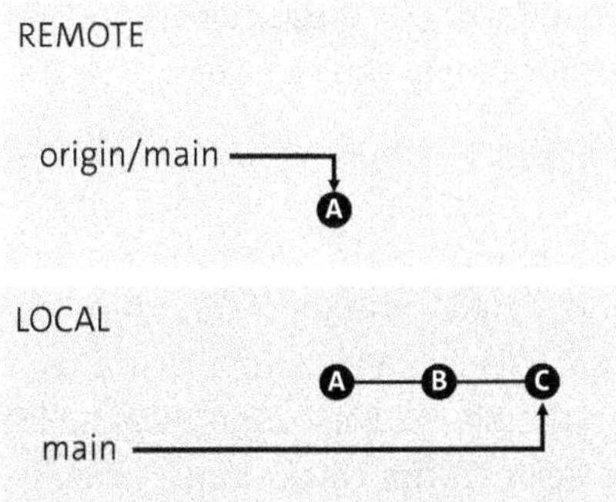

Figure 15.16 Some New Commits at the Local Main Branch

15.2.3 Fetch

In most cases, you aren't working alone, of course. Your colleagues are also working on the project and sharing their changes with you and everyone else in the remote repository. As shown in Figure 15.17, your local repository doesn't yet know about these changes. Until you check in, you don't know if there have been any changes, what they are, or how many new commits are in the remote repository.

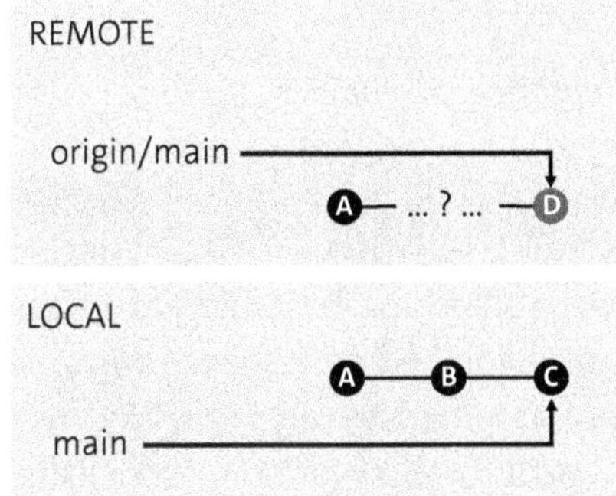

Figure 15.17 Several Changes Made on the Remote Origin/Main Branch

You have to instruct Git yourself when your local repository should check to see if something has changed in the remote repository. If you execute the following command, your local repository checks to see if something has changed in all connected remote repositories:

```
git fetch --all
```

If you only want to check a specific remote repository, you can do so with this command:

```
git fetch <repository>
```

If you only have one remote repository, the command without the remote repository name is sufficient. If you only want to fetch a specific branch, you can execute the following command:

```
git fetch <repository> <branch>
```

In Figure 15.18, you can see that the fetch has created a circle with the ID **D**, which represents the fetch and also shows that something has been found. Note that a fetch doesn't actually create a new commit, as shown in the figure. The commit was added merely for the sake of clarity, and you'll see how we deal with this commit when we look at merges.

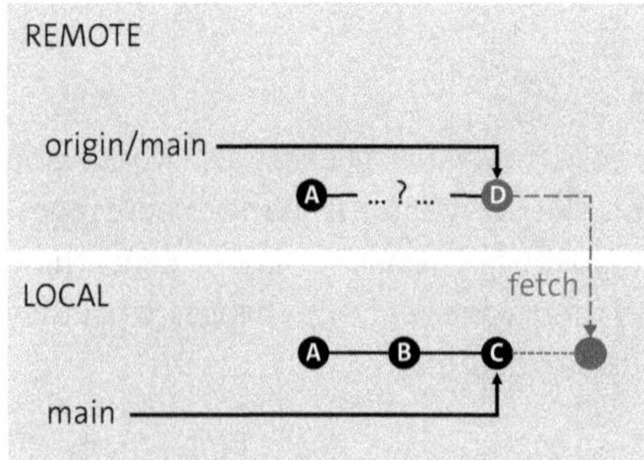

Figure 15.18 Fetching the Commits from the Remote Repository and Branch

15.2.4 Merge

Thanks to your fetch request, you now know that something has changed in the remote repository. The next step is to merge the files you just downloaded with the files you already have. This process is called a *merge*. Merging files can be a tricky business, especially when conflicts arise. We won't deal with conflicts for now, as we'll discuss these issues in more detail in Section 15.4.

The main thing to understand is what characterizes a merge. As you can see in Figure 15.19, the main pointer points to an older commit **C**. So you have to merge commit **C** with the new commit you downloaded from the remote repository. To merge, you must first select which branch is to be compared with which one, that is, transferred. In our case, you're currently in the local `main` branch and want to merge the changes from the `origin/main` branch. There are basically two types of merge operations:

- **Three-way merge**
  A three-way merge occurs when new commits have been made since the last merge both in the base branch and in the branch selected for merging. Here, the files are merged according to specific merge strategies and procedures, and a new merge commit symbolizes the merge.

- **Fast-forward merge**
  If no changes have been made to the base branch since the last merge, that is, no new commits have been added, no merge commit needs to be created. It's sufficient to simply set your pointer to the new commit because, in this case, there's nothing to merge.

The following command represents a merge request:

```
git merge <branch_name>
```

The result after a merge with the `origin/main` branch is shown in Figure 15.19.

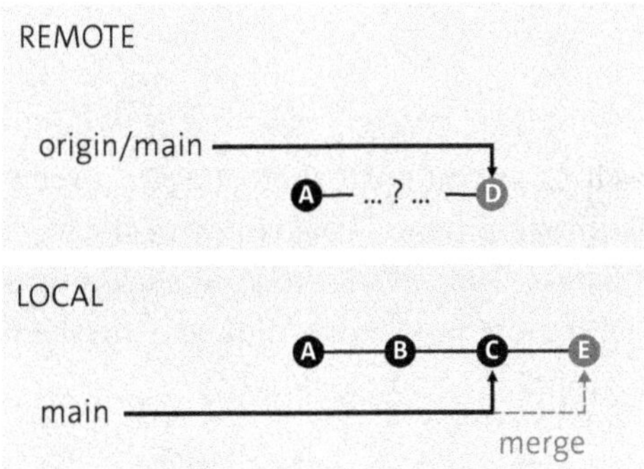

Figure 15.19 Performing a Merge

You can clearly see here that a merge commit is created that represents the merge, which is why we had already set the fetch for a simpler understanding. The pointer from the local `main` branch must, of course, also be set, but this happens automatically during or after the merge.

15.2.5 Checkout

In the background, and almost unnoticed by us, Git always sets a `HEAD` pointer that marks the place where we are right now. The changes we make are applied to the place where this pointer points. We'll see the `HEAD` pointer in the examples later on.

To move this pointer, you need to know another important command called `checkout`:

```
git checkout <id>
```

The placeholder `<id>` can take several values. These values can be either filenames, commits, or branches. We'll need to use this command frequently to switch between individual branches.

Detached Changes

We recommend that you don't use this command to check out commits unless you're sure of the consequences. Such actions increase the risk of detached HEADs occurring. We won't discuss detached HEADs further here.

To prevent such detached changes from occurring, you can also use the command `git switch <branch>` to switch between branches. The advantage is that you can only specify a branch here and thus can't accidentally specify a file or a commit. Here's an example of this command:

```
git switch main
```

15.2.6 Push

A merged version has no real added value as long as you don't share it again, except that you're locally up-to-date with the latest developments. To share your development status with others, you have to execute a `push`, which tries to send all commits that have occurred since the last push to the remote repository, thus making them available to others. So, you push your changes into the remote repository:

```
git push
```

You don't need to worry if you forget to do a fetch and merge beforehand because Git will give you an error message in this case. Everything you want to push should be mergeable at the remote repository with a fast-forward merge. You can see the result after the push in Figure 15.20.

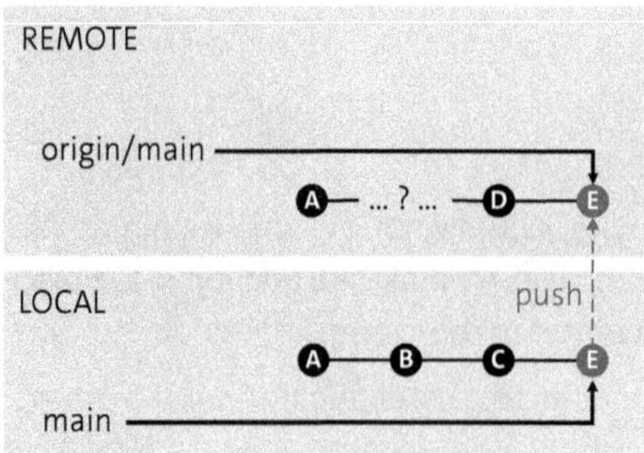

Figure 15.20 Pushing the Commits to the Remote Branch

15.2.7 Pull

You probably already realize that if you have to execute fetch and merge every time before a push, then these two commands almost always occur in conjunction. Yes, that's right, and you can use the `pull` command to remedy the situation. Figure 15.21

shows the initial situation. Both you and your team members have been busy working. There are new changes in the local and remote repositories that you don't yet have in your local environment.

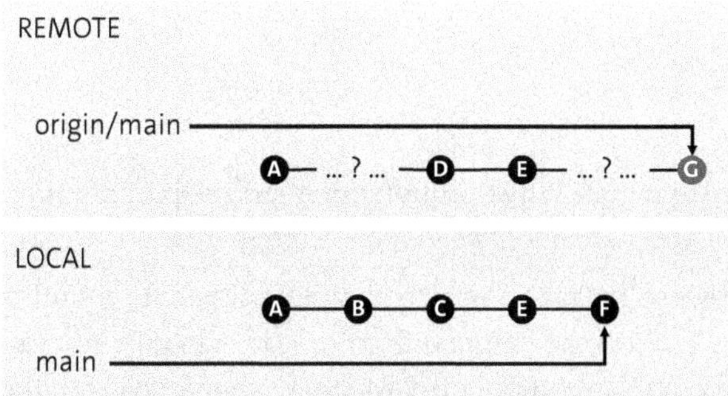

Figure 15.21 Again, Changes Both Remotely and Locally

If you now execute the `git pull` command, a fetch and merge are carried out together: it checks for changes in one step and, if there are any, it immediately tries to merge the changes using a merge query.

15

Pull = Fetch + Merge?

We'll discuss later that a pull doesn't always result in a merge commit. For the sake of simplicity, however, we'll initially only address this option.

As usual, we've visualized the result after the merge request (see Figure 15.22).

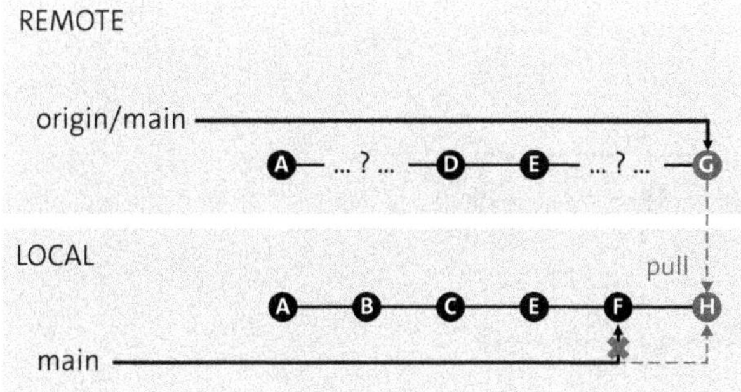

Figure 15.22 Pull Leads to a Fetch and Merge

Stay Up-to-Date with Pull

You should get into the habit of always pulling as soon as you want to share local developments. This not only saves you time but also avoids annoying error messages.

Even if you don't always want to share your current state, it pays to check from time to time with a pull whether the current state of the local repository—after you've reached a certain local state, of course—can also be merged with the current developments at the remote repository.

15.2.8 Rebase

Rebase is basically an "alternative" to merge. Why the quotation marks? Opinions differ when it comes to discussing the better variant between rebase and merge.

The short version is that with a rebase, we take a selected branch with all its commits and attach them to another branch. With the command `git rebase <branch>`, you can attach the branch you're currently in to another branch, which we've marked with the placeholder `<branch>`. This way, it looks as if these commits were made in the respective branch and a separate branch was never created and used.

The rest of this section shows the differences between merge and rebase in a concrete example with a step-by-step description. Let's summarize the properties of the merge request again in a quick run-through, so that we can then go into more detail about rebase and the differences. Figure 15.23 shows the initial situation again, that is, a `feature` branch has been created from the commit with ID **H**. You create a new branch with the following command:

```
git branch <branch_name>
```

The first state of the `feature` branch thus corresponds to the state of commit **H**, and two further commits **X** and **Y** have been added from this commit.

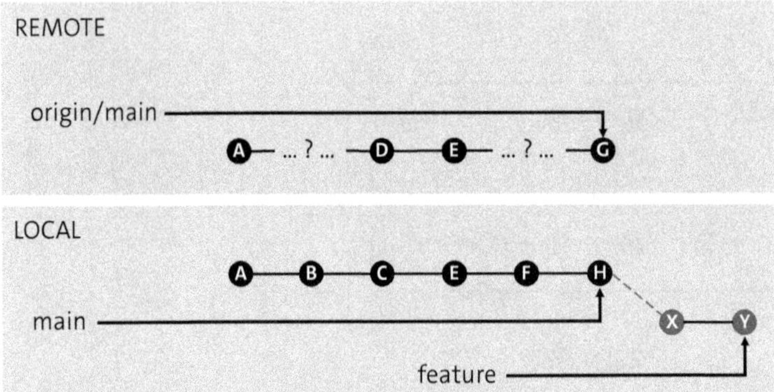

Figure 15.23 Two New Commits on a Feature Branch

In the next step, you determine that the developments on the `feature` branch are finished and worthy of being merged back into the `main` branch. As soon as you execute a merge request, a fast-forward merge is performed because no commits have been added to the base branch (in our case, the `main` branch), since then. This is also visible in Figure 15.24: no new merge commit is created here, as you would expect with a normal merge, but the pointer from the `main` branch can simply be moved forward.

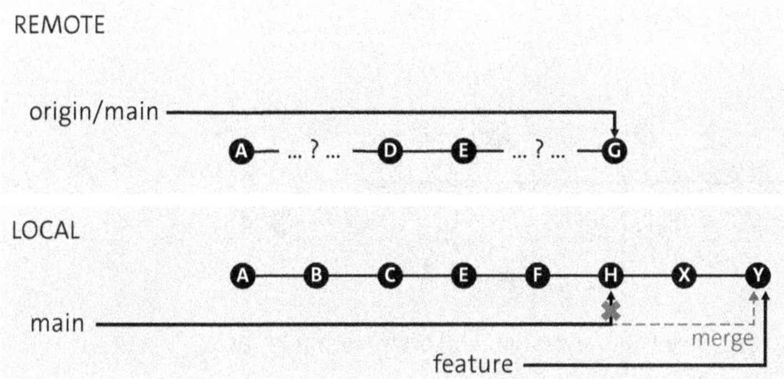

Figure 15.24 Merge Leads to a Fast-Forward Merge

As shown in Figure 15.25, after the deletion of the `feature` branch, the pointer no longer exists, but the `main` branch is aware of these new commits, and they have become part of the history.

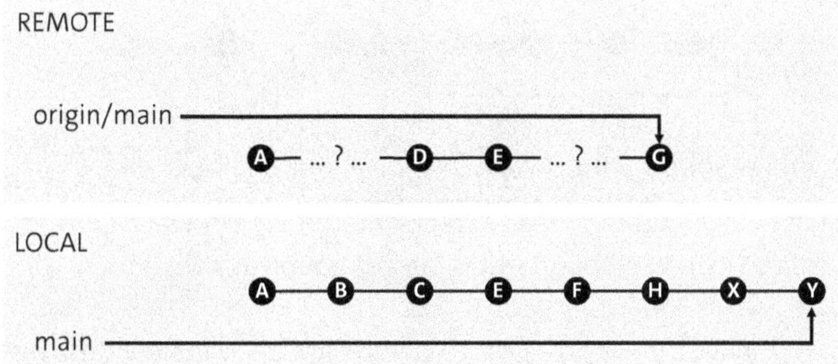

Figure 15.25 After Deleting the Feature Branch, No Trace of This One

Rewind to the point where two new changes were made to the `feature` branch. Now assume that a new commit has also been added to the `main` branch (here with ID ❶). This means that you have two different project statuses that have moved away from each other and need to be merged again (see Figure 15.26).

In this case, no fast-forward merge is possible, so a three-way merge must take place. Git does this for you, searching for the smallest common denominator of the two branches, so to speak. Recursive methods are used to merge both branches. The result is a common state in the form of a new commit ❷. This commit is also called a merge commit (see Figure 15.27).

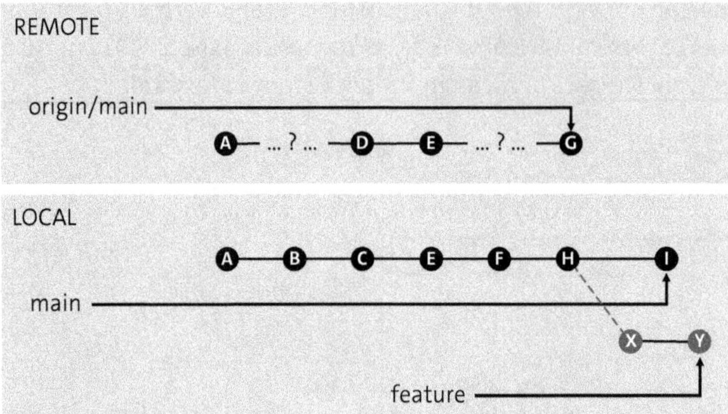

Figure 15.26 Again, Changes on a Feature Branch and on the Main Branch

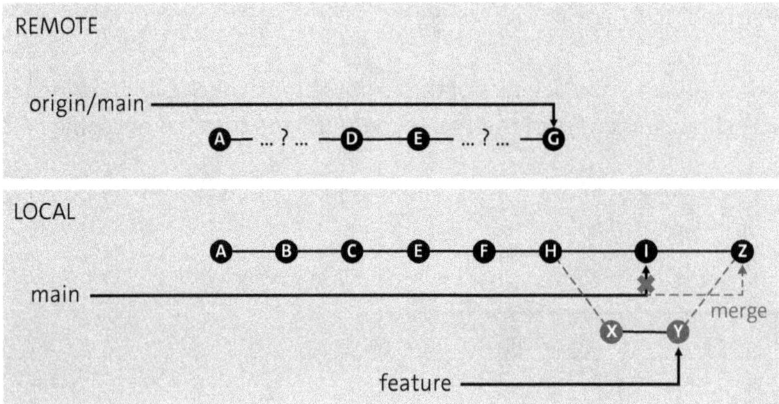

Figure 15.27 Merge Leads to a Three-Way Merge with a Merge Commit as a Result

After deleting the `feature` branch, you realize that a detour has been made in the history of the project and the two commits X and Y are out of line. Even after deleting the `feature` branch, this detour remains and becomes visible and detectable through graphical tools when you read out the predecessors of the individual commits (see Figure 15.28).

At this point, opinions differ, because for some, this state of development doesn't count as a clean line. This is only a cosmetic problem for our human eye, but by no means a problem or an obstacle for Git.

Let's back up a few steps and imagine that you created a `feature` branch and added the two commits X and Y to that `feature` branch (see Figure 15.29). This time, however, you want to integrate the complete development history from the `feature` branch into the `main` branch.

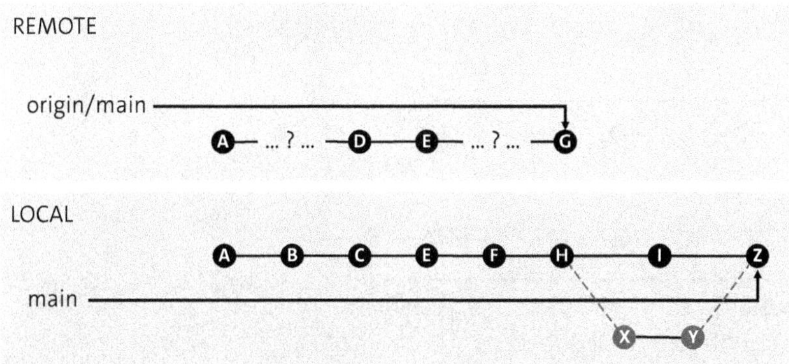

Figure 15.28 After Deleting the Feature Branch, the History Remains

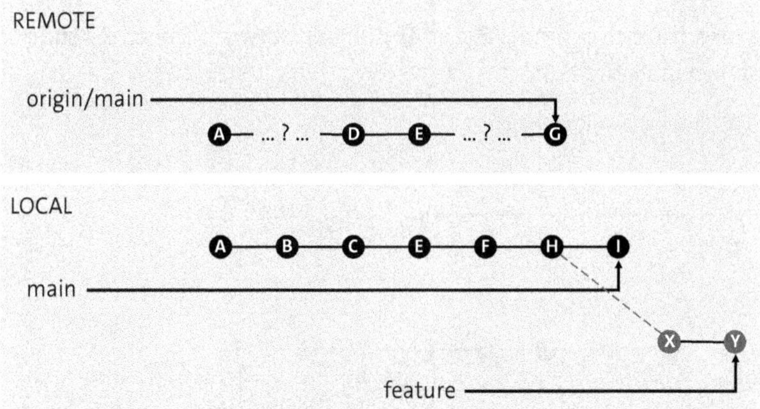

Figure 15.29 Again, Some Changes on the Feature Branch and on the Main Branch

15

When doing a merge, you need to be in the branch you want to merge with another one. When doing a rebase, on the other hand, you need to be in the branch you want to integrate into another one. The following command shows how to check out the branch using `checkout`:

```
git switch feature
```

Now, you're in the right branch and can do a `rebase`. Figuratively, you can think of it as integrating your `feature` branch, with all the commits that have been made, into the `main` branch. With the following command, you can start a rebase in the direction of the `main` branch, starting from the current branch:

```
git rebase main
```

In this process, all the commits on the `feature` branch are integrated into the history of the `main` branch (see Figure 15.30).

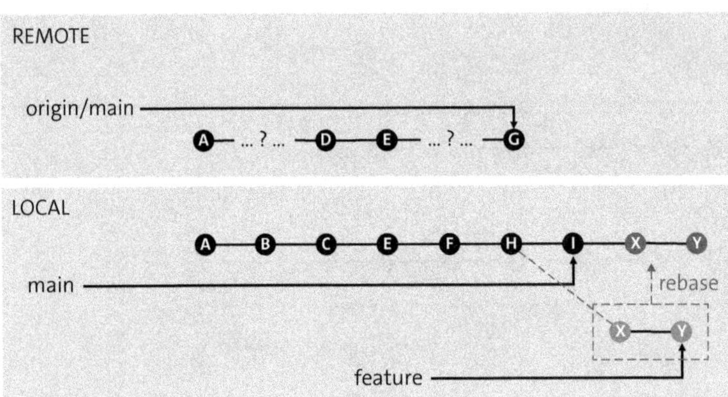

Figure 15.30 Rebase Feature on to Main

Of course, in this case, the two commits X and Y don't exist twice. The actual state of development is shown in Figure 15.31.

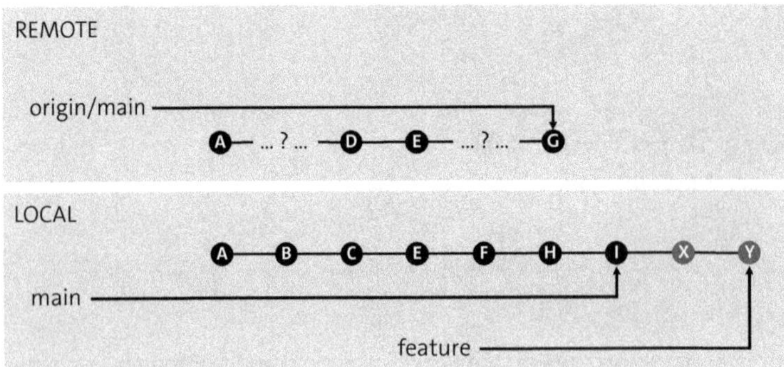

Figure 15.31 Feature Branch Docking at the Main Branch

Now, in the last step, a fast-forward merge can be done so that the pointer from the `main` branch can point to the current commit (see Figure 15.32). In this case, there's no trace of a merge commit.

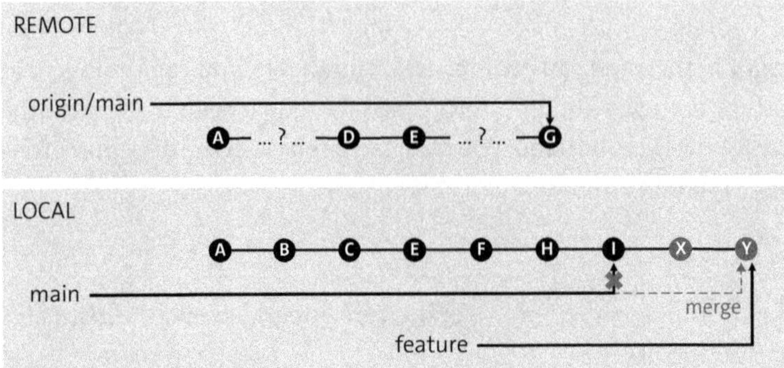

Figure 15.32 The Merge Leads to a Fast-Forward Merge

Figure 15.33 shows the current state of the project after the `feature` branch has been deleted.

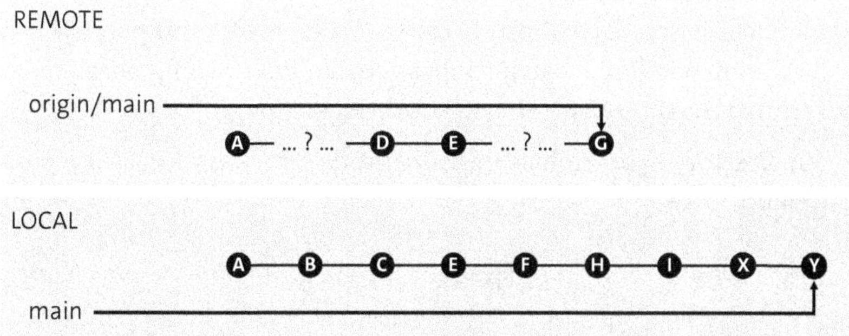

Figure 15.33 Again after Deleting the Feature Branch, No Trace

If we do a direct comparison, we can show that the following differences arise between merge and rebase:

- In contrast to a merge, a rebase usually doesn't result in another commit (there are exceptions here again when it comes to conflicts).
- When you rebase, there are no detours in the history and no commits that step out of line in the development branch.

In a final step, you push the current state into the remote repository so that the newly created commits H, I, X, and Y are also available to other developers (see Figure 15.34).

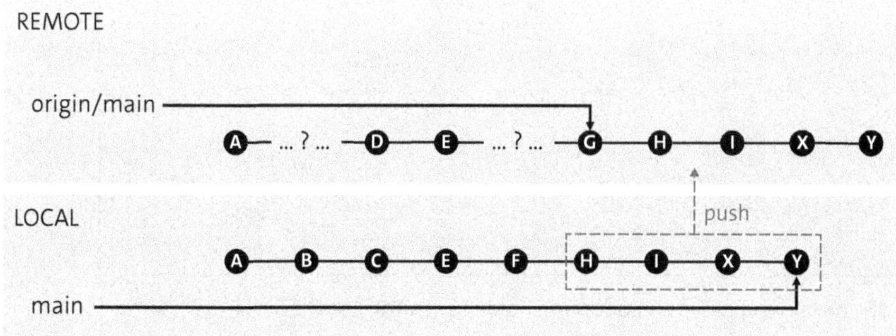

Figure 15.34 Pushing the Last Changes to the Remote Main Branch

15.3 Working with the Git Command-Line Interface

Before you can explore a number of graphical tools in the following chapters, we want to show you how to work with the in-house command-line interface (CLI) and Git commands. We've already mentioned how important it is that you also work with the CLI and see the functionalities here. This knowledge serves as a basis for you to better understand later what the graphical tools do for you in the background.

In this section, you'll see the end-to-end example that you learned about in the preceding section in action. We've created a directory called *GitExercise* for this purpose and will show you the commands step-by-step so that you can follow along once you've installed Git and created the *GitExercise* directory on your computer.

After opening your directory in a terminal, initialize the directory as a local Git repository, as shown in Figure 15.35:

```
git init
```

If you realize that the initial branch with your Git version has been named master, you can override this setting for all future branches with the following command:

```
git config --global init.defaultBranch main
```

Nevertheless, you can rename a branch at any time:

```
git branch -m main
```

At first glance, nothing has changed in your directory, but you now have a hidden folder named *.git*. To see this folder, you have press `Cmd`+`Shift`+`.` on a Mac or select **View • Options • Folder and Search Options • Show Hidden Files and Folders** in the top menu bar on Windows.

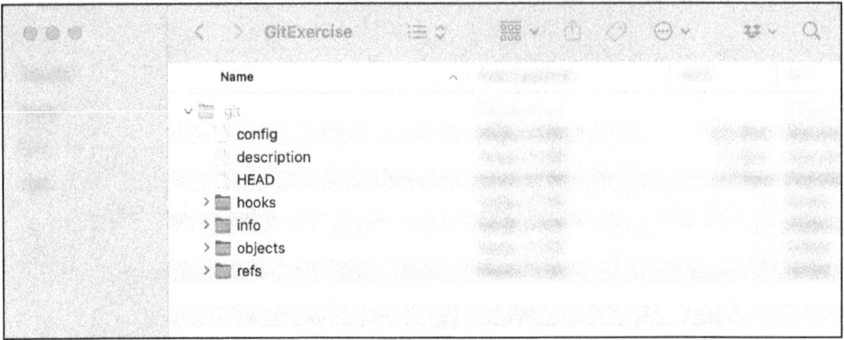

Figure 15.35 Initialize Creates a Hidden .git Folder with the Local Config File

Now create a text file directly in the CLI, and fill it with a simple word/sentence:

```
echo "Welcome" > file.txt
```

To include the changes to this file, that is, the creation and the addition of this sentence, in a commit, we must first put the changes into the staging area. There are two commands that you can use to put either all current changes or just one or more selected files into the staging area:

- `git add -A`
- `git add file.txt`

If you want to remove the file *file.txt* from the staging area, you can use the following command:

```
git rm --cached file.txt
```

With the following command, you summarize the current staging area in a commit and assign a comment:

```
git commit -m "Initial commit"
```

After each commit, you'll receive a summary that might look like Figure 15.36 and contains information such as the unique ID of the commit, the comment, and a list of all changes with information om whether these changes have been added, changed, or removed.

```
GitExercise — -zsh — 87×7
(base) danielkrancz@MacBook-Pro-von-Daniel GitExercise % git add file.txt
(base) danielkrancz@MacBook-Pro-von-Daniel GitExercise % git commit -m "Initial commit"
[main (root-commit) f5bf15c] Initial commit
 2 files changed, 1 insertion(+)
 create mode 100644 .DS_Store
 create mode 100644 file.txt
(base) danielkrancz@MacBook-Pro-von-Daniel GitExercise %
```

Figure 15.36 Summary of a Commit

Now edit your file by customizing the first line and then adding a second line. The following command displays all currently open changes in a summary (by default in detail):

```
git status
```

If you think that these changes or the change in a particular file should be discarded, execute the following command:

```
git restore file.txt
```

Now add your changes again, and view the short summary of all changes recorded by Git with the following command:

```
git status --short
```

Because, in this case, we want to include all changes in the next commit and get to the result faster, use the command that skips the staging area:

```
git commit -a -m "Second line"
```

To display the work history and view the history that Git is continuously capturing, use the following command:

```
git log
```

With the `--oneline` option, you can also request a more compact view of the history.

Figure 15.37 shows the timeline of when each commit was made, along with all the important information. It also clearly shows that the current pointer of your branch `HEAD ->` with the name main points to the current commit.

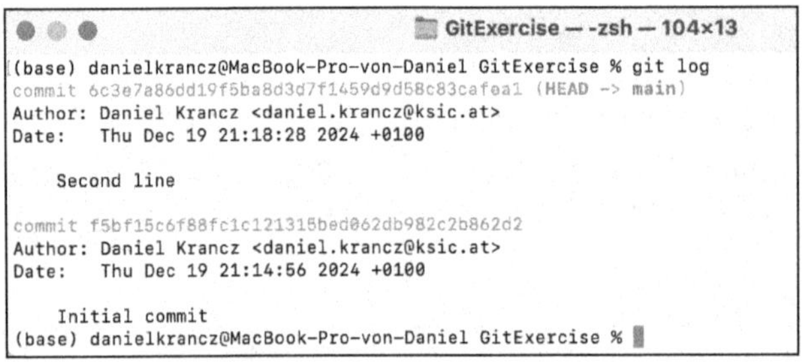

```
GitExercise — -zsh — 104×13
(base) danielkrancz@MacBook-Pro-von-Daniel GitExercise % git log
commit 6c3e7a86dd19f5ba8d3d7f1459d9d58c83cafea1 (HEAD -> main)
Author: Daniel Krancz <daniel.krancz@ksic.at>
Date:   Thu Dec 19 21:18:28 2024 +0100

    Second line

commit f5bf15c6f88fc1c121315bed062db982c2b862d2
Author: Daniel Krancz <daniel.krancz@ksic.at>
Date:   Thu Dec 19 21:14:56 2024 +0100

    Initial commit
(base) danielkrancz@MacBook-Pro-von-Daniel GitExercise %
```

Figure 15.37 Git Log Showing the History and Which Commits the Branches Are Pointing To

If you want to take a closer look at a commit and also view a summary of all changes that were recorded in it, use the following command:

```
git show 6c3e7a86dd19f5ba8d3d7f1459d9d58c83cafea1
```

Note the parameter (in our case, a long SHA-1 checksum) because you have to enter the checksum of the commit you want to view here. The checksum we've provided can be found in Figure 15.37, in the second line starting with `commit`. This checksum is made up of several factors, including the ID of the commit. It's also possible to just provide the ID. The output of this call is shown in Figure 15.38.

```
GitExercise — -zsh — 111×19
(base) danielkrancz@MacBook-Pro-von-Daniel GitExercise % git show 6c3e7a86dd19f5ba8d3d7f1459d9d58c83cafea1
commit 6c3e7a86dd19f5ba8d3d7f1459d9d58c83cafea1 (HEAD -> main)
Author: Daniel Krancz <daniel.krancz@ksic.at>
Date:   Thu Dec 19 21:18:28 2024 +0100

    Second line

diff --git a/.DS_Store b/.DS_Store
index c7d51a9..4c889e0 100644
Binary files a/.DS_Store and b/.DS_Store differ
diff --git a/file.txt b/file.txt
index ba4c514..3c05686 100644
--- a/file.txt
+++ b/file.txt
@@ -1 +1,2 @@
 Welcome
+This is the second line!
\ No newline at end of file
(base) danielkrancz@MacBook-Pro-von-Daniel GitExercise %
```

Figure 15.38 Git Showing More Details for One Commit

Especially in the development of applications, you'll often need auxiliary information, such as development and interface keys (e.g., API keys), which are essential and must also be stored in a file. You already know that Git tracks and records all changes made in this *GitExercise* directory.

In the next step, you'll create a directory with another file that stores information that only concerns your local development and isn't of interest to others. This directory, named *private*, will store fictitious key `apiKey=1234` in a file named *data.txt*. In the future, we want this folder and all the files in it to be excluded from Git tracking. Create the directory with the file, and fill it with data via the following:

```
mkdir private
cd private
echo "apiKey=1234" > data
```

Now switch to the main folder *GitExercise*, and create the file *.gitignore*. This file contains line by line the paths of the directories and underlying files that should be explicitly excluded from change tracking:

```
cd ..
echo "private" >.gitignore
```

Now, you need files of the *.properties* file type for your development, for example. However, you don't want all files with *.properties* extension to be tracked. To do this, add the following command to the *.gitignore* file, and create a *development.properties* file at the same time to test the ignoring:

```
echo "key=value" > development.properties
echo "*.properties" >>.gitignore
```

In Figure 15.39, you can see that the changes are no longer listed under the nonversioned files.

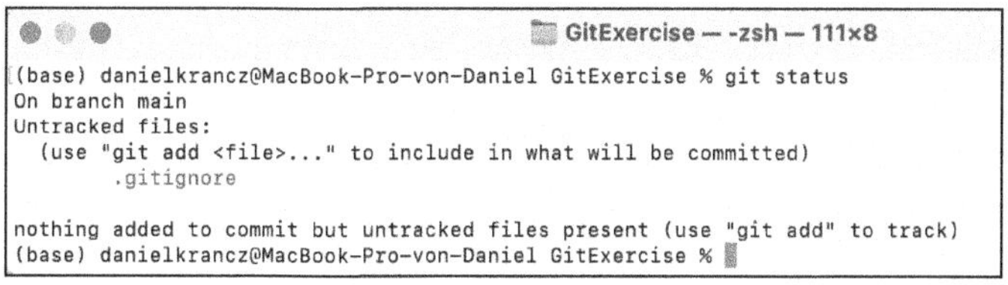

```
GitExercise — -zsh — 111×8
(base) danielkrancz@MacBook-Pro-von-Daniel GitExercise % git status
On branch main
Untracked files:
  (use "git add <file>..." to include in what will be committed)
        .gitignore

nothing added to commit but untracked files present (use "git add" to track)
(base) danielkrancz@MacBook-Pro-von-Daniel GitExercise %
```

Figure 15.39 Files and Directories in the .gitignore File Aren't Tracked

.DS_Store File on macOS

Don't be confused if you see a nonversioned file called *.DS_Store* on a macOS. This file has nothing to do with Git, but rather with macOS. It's automatically created by the operating system when you navigate to a folder or open a file using the Finder. This file doesn't exist on Windows and Linux.

Use the following command to disable change tracking for *.DS_Store* globally and for all future folders tracked by Git:

```
echo .DS_Store >> ~/.gitignore_global
git config --global core.excludesfile ~/.gitignore_global
```

Alternatively, you can add the file to the *.gitignore* file just for this project:

```
echo .DS_Store >>.gitignore
```

Now compare the entries in the Git status report with the actual directories and files in your *GitExercise* directory (see Figure 15.40). You'll see that neither the *development.properties* file nor the *private* directory with its files appears in the Git status report, but they are still physically stored on your machine and can be used for development purposes.

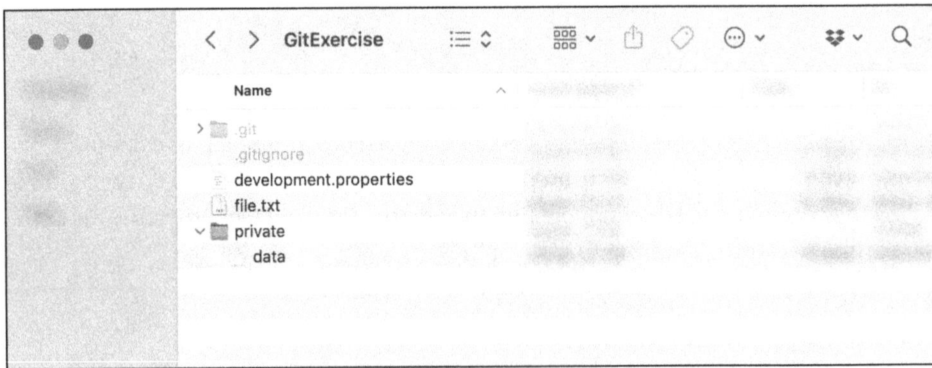

Figure 15.40 Untracked Files Are Still on Your Machine

Finally, add the creation of the *.gitignore* file to a commit with the message "Gitignore extended". Use the knowledge you've gained so far to do this.

Now, let's assume that you want to add a second file. However, because you don't know whether this new requirement is the right way to go, you should put the related changes into a separate branch. You won't suffer any disadvantages from this procedure, but if the changes have to be discarded, it will be easier for you to return to the original state. Create a new branch called `feature`, as follows:

```
git branch feature
```

If you were to make changes at this point, they would affect the `main` branch. That's why you first must switch to the `feature` branch using a `checkout`:

```
git checkout feature / git switch feature
```

Note that you can use the `checkout` command not only to switch to different branches but also to switch to individual commits. Switching to individual commits can be useful for restoring or creating new branches from a specific and older commit.

Now check the state of the project, and you'll see that the pointer `HEAD ->` points to the same commit for both `main` and `feature`.

Pointer HEAD

We've already briefly discussed this pointer, but because this is the first time we've encountered it, here's a brief explanation. In the first line of Figure 15.37, you see a pointer named `HEAD ->`. This pointer currently points to the branch that you've just checked out. In other words, it points to the branch in which you're currently located and in which changes and commits are applied.

Now create a new file called *file2.txt*, and fill it with data. You can see these steps in Listing 15.1.

```
echo "Second file:" > file2.txt
echo "Another line" >> file2.txt
git add .
git commit -m "Second file"
```

Listing 15.1 Make Some Changes within file2.txt and Commit These

Over time, you may conclude, after careful consideration, that you're working in the wrong direction with your current `feature` branch and that development doesn't make sense this way. Therefore, you want to discard everything you've done in this branch. This is done with the following command:

```
git switch main
git branch -d feature
```

Git will now say that you have at least one commit in the `feature` branch, but that the changes haven't yet been merged into your `main` branch. Because you're sure you want to delete this branch, run the following command:

```
git branch -D feature
```

Create a new branch named `feature` with the Git branch feature and switch to this branch with `git switch feature`. If you now look at the status report, you'll see that no changes are currently registered and there's nothing to commit. Now execute the following command:

```
echo "One line" > file2.txt
git add .
git commit -m "Second file"
```

This will create a new file and insert a line. You'll immediately put these changes into the staging area and create a commit.

If you would have added the *file2.txt* manually via the Explorer/Finder and filled it with data, Git would have recognized these changes. This means that Git is working in the background whenever a repository gets initialized in a folder.

Run `git log` again to see the history. You can see that the pointer for your `main` branch is still pointing to the older commit with the message "`Gitignore extended`". However, the `feature` branch is already further along and has a new commit that isn't yet visible to the `main` branch.

Figure 15.41 shows a graphical representation of how the pointers currently stand in your project. The pointer of the `feature` branch is further ahead, and the pointer of the `main` branch is one commit behind, so it doesn't yet know the changes in the "`Second file`" commit.

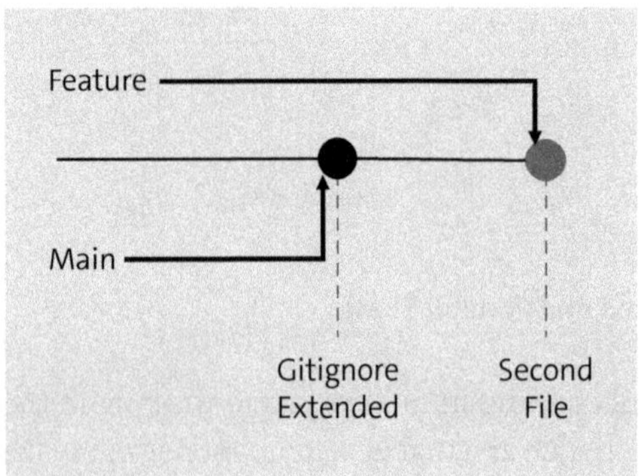

Figure 15.41 Feature Is One Commit ahead of Main

To merge the `feature` branch into the `main` branch, use a merge. Before each merge, you have to switch to the branch you want to merge with. We want to load all the changes that have been made on the `feature` branch since the last common commit and make them known to the `main` branch. To do this, execute the following commands:

```
git switch main
git merge feature
```

With each merge, you also get a summary of what happened and how the request went (see Figure 15.42). In this figure, you can see that a file named *file2.txt* was transferred. It also shows which method was used to merge these two branches.

We've already described the two methods in detail, but here's a brief description of what "fast-forward" technically means in this case. Because the `feature` branch originated from the `main` branch at some point, and no further changes have been added to the `main` branch since that point, it's sufficient to simply move the pointer from the `main` branch to the commit from the `feature` branch. This method of merging, which Git chose, requires no further commit, which we would normally expect in a merge. The fast-forward method is also very clear when you look at your history and view the pointers. You can see that the pointer of `main` was simply set to the last commit, or rather to the commit that the `feature` branch pointed to. Compare the unique keys of the commits between before and after the merge. This way, you can be sure that no new commit was created.

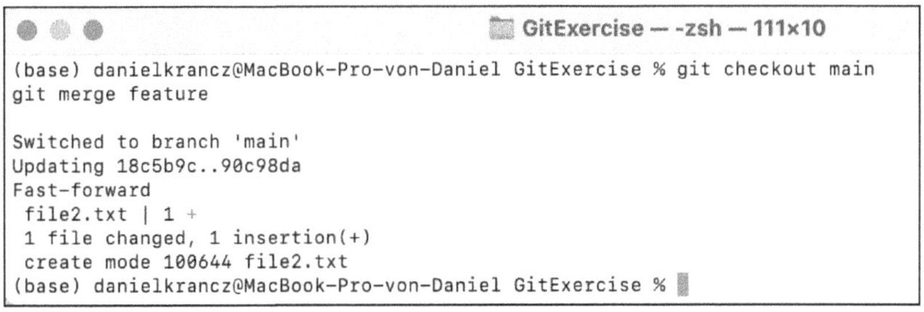

```
GitExercise — -zsh — 111×10
(base) danielkrancz@MacBook-Pro-von-Daniel GitExercise % git checkout main
git merge feature

Switched to branch 'main'
Updating 18c5b9c..90c98da
Fast-forward
 file2.txt | 1 +
 1 file changed, 1 insertion(+)
 create mode 100644 file2.txt
(base) danielkrancz@MacBook-Pro-von-Daniel GitExercise %
```

Figure 15.42 Merge Successfully Done via a Fast-Forward Approach

In the next step, we want to find out what happens when the `main` branch and the `feature` branch are merged, and both branches contain new commits. Make sure you start with the `main` branch with the changes. First, add another line of data to the file *file.txt*, and package this change in a commit. Then, switch to the `feature` branch, add a line to the file *file2.txt*, and make a commit as well. Then, display the history of the current branch, that is, the `feature` branch (see Listing 15.2).

```
git switch main
echo "Another line in the main branch" >> file.txt
git add .
git commit -m "Change to the main branch"
git switch feature
echo 'A new line is also added here' >> file2.txt
git add .
git commit -m "Change to the feature branch"
git log
```

Listing 15.2 Make Some Changes Both on Feature and on Main

Because you're currently on the `feature` branch, the history on this branch is of course displayed. All changes made before the fork on the `main` branch are displayed and that the commit with the comment `"Change on the feature branch"` is also displayed. You don't see one commit here: namely the one you created when you were on the `main` branch. You also don't see the `main` branch pointer here because that pointer points to a commit that isn't visible for the `feature` branch.

You can also check the history of the `main` branch. Except for the last commit, this history looks exactly like the history of the `feature` branch. Here, you can see all the past commits that have taken place over time, but in contrast to the feature branch, you can also see the main branch pointer. The main branch pointer points to the last commit with the comment "Main branch changes". You can't see the feature branch pointer.

Figure 15.43 shows these branches again graphically. You can see the commits on the main branch in the bottom line. A branch-off occurred at the commit named "Second File". This results in the feature branch. The main branch continues without consideration of the feature branch.

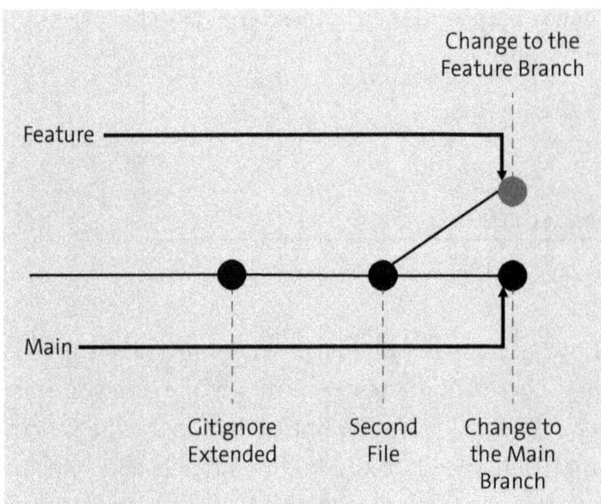

Figure 15.43 New Commits Both on the Feature and the Main Branch

To merge these two branches again, you must switch to the main branch and merge with the feature branch. In this case, a three-way merge is done. This merge commit requires you to provide a description of what happens during the merge and which decisions you've made (see Figure 15.44).

GitExercise — vi ◂ git merge feature — 132×13

```
Merge branch 'feature'
# Please enter a commit message to explain why this merge is necessary,
# especially if it merges an updated upstream into a topic branch.
#
# Lines starting with '#' will be ignored, and an empty message aborts
# the commit.
~
~
~
~
~
~
```

Figure 15.44 Three-Way Merge Results in a Merge Commit Requiring a Corresponding Message

Once you've entered your message and closed the input window by pressing [Esc] + [ZZ], the merge process begins. As usual, you receive a summary after the operation is completed. If you look at the history again, you can see that a new commit has actually been created. This commit looks a bit different from all the other commits: on the second line, you see the keyword merge and two links after it. These links point to the two predecessors based on the short keys. Because one commit has emerged from two forks, Git also stores these two links for documentation purposes and so that you can see who the predecessors were.

Figure 15.45 shows the current state of the branches. When the "Second File" commit is made, the two branches move further apart and then meet again at a merge commit.

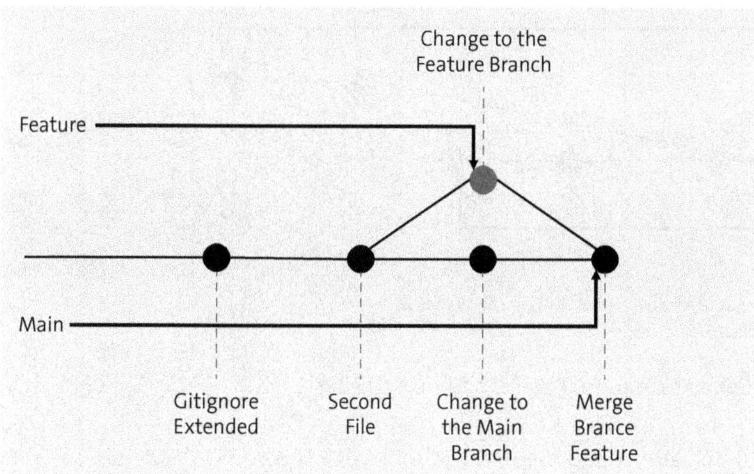

Figure 15.45 Merge Commit Created via the Three-Way Merge

There's a second way to merge two branches. Create a new branch again (we named this branch feature2), and make two commits on both branches (see Listing 15.3). We've illustrated these individual changes in Figure 15.46.

```
git checkout -b feature2
echo "New line1" >> file2.txt
git commit -a -m "New Line1 Feature"
echo "New line2" >> file2.txt
git commit -a -m "New Line2 Feature"
git checkout main
echo "New line1" >> file.txt
git commit -a -m "New Line1"
echo "New line2" >> file.txt
git commit -a -m "New Line2"
```

Listing 15.3 Two New Commits Both on the feature2 Branch and the main Branch

To merge the two branches, you can use rebase. Rebase transfers commits made on one branch to another branch and integrates them into its history. Unlike a merge, you need to be on the branch that you want to integrate into another branch. In our case, we want to get the two commits from the feature2 branch on the main branch, so we switch to the feature2 branch. Then, we execute a rebase on main, as follows:

```
git checkout feature2
git rebase main
```

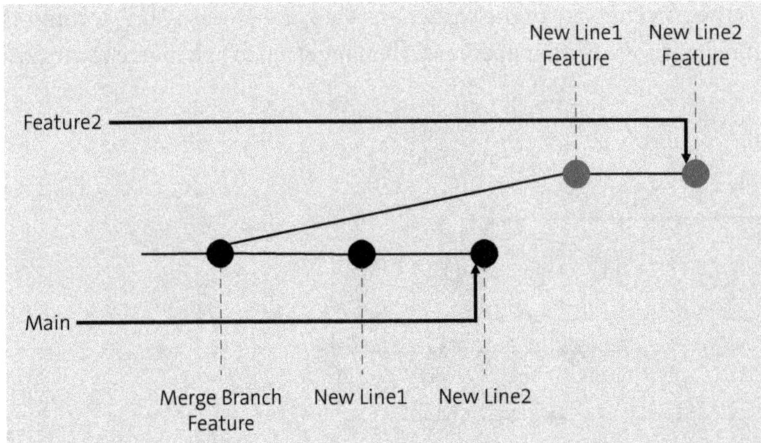

Figure 15.46 Two Commits Made in Parallel on Two Branches

After a rebase, our visually edited history now looks like Figure 15.47. You can see that the two commits have now become part of the main branch. Even though these commits happened later in time, it looks as if they were made directly in the main branch.

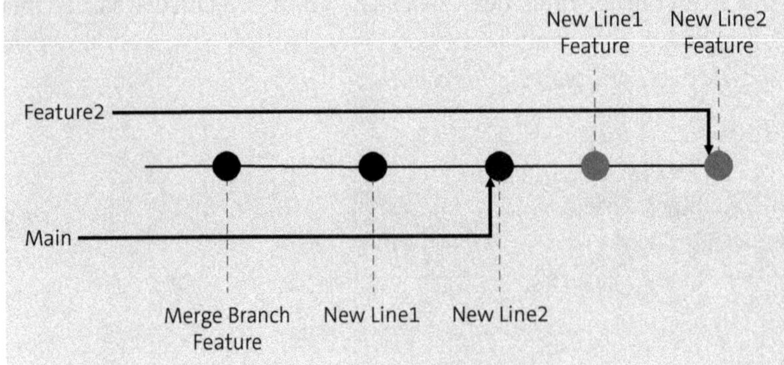

Figure 15.47 Rebase Leads to a Concatenation of the Commits of feature2 with main

For the main branch pointer to point to the current commit, a merge must be performed. With this visual history, you can already guess that we won't need a merge commit because a fast-forward merge can be done. Figure 15.48 shows the current state and that there's no detour to be seen in this history. Everything is neatly in one branch, and the visual representation is easy to read.

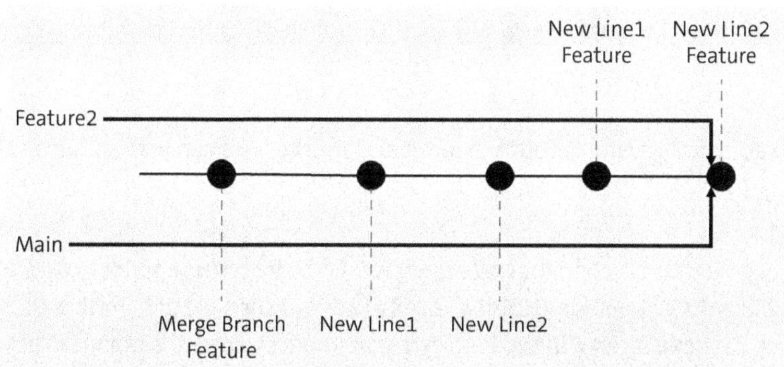

Figure 15.48 Merge Leads to a Fast-Forward Merge as Changes Are Made in Different Files

Of course, you haven't gained much here just because you used the `rebase` command instead of the `merge` command. You can achieve the same result with either command, namely, to have two branches come together again, but with the one difference that the process is clearer here.

Because you've created this project and now want to share it with others, you need to connect it to a remote repository. Cloning wouldn't be possible at this point because the remote repository doesn't yet exist.

Use the following command to connect to a remote repository named origin that you've created or been assigned:

```
git remote add origin <your_repo_url>
```

To get your changes onto the remote repository, execute a `push` command and specify the name of the remote branch as the first parameter and the source as the second optional parameter. After the execution of this command, you'll see a summary:

```
git push origin main
```

With the following command, you can check whether changes have been made to the remote repository:

```
git fetch
```

You can also merge directly with the remote branch:

```
git merge origin/main
```

Merge When Using Several Remote and Local Branches

As you know, a single fetch isn't enough to merge two branches. It's up to you whether you first load the latest state of the remote branch with a `git fetch` and then do a `git merge` with the loaded local branch, or whether you do a `git merge` with the remote branch right away. If you choose the locally loaded branch, it's important that it's always up-to-date.

You don't have to go through this detour via fetch and merge; you can also do a pull:

```
git pull origin/main
```

It will rarely be the case that no changes have been made to the remote repository and remote branch while you've been developing locally on your own branch. In this case, a pull request when first executed will show an error message because it's not clear how the changes should be merged. You can make a setting in the configuration and determine whether a merge, a rebase, or just a pull should be done when a fast-forward approach can be taken. In our case, we want a merge to be done always and accept that a merge commit may arise under certain circumstances. We set this setting globally for all repositories on this system:

```
git config --global pull.rebase false
```

15.4 Conflict Resolution

Resolving conflicts can be a tricky business, as you must make important decisions about the further course of the project that is being managed and versioned in Git.

First, let's clarify what's behind the term *conflict*. In the context of Git, we speak of a conflict whenever complications arise when merging two states. When we speak of merging, we specifically mean the merge functionality, which we've already covered in detail. Complications arise from the fact that changes have been made in different branches in the same file at the same place. If you think of source code, you could say that a change has been made on the same line. In this case, Git alone can't decide and will need your help.

In the previous sections, you learned that a merge can be used both to merge two local branches (e.g., `main` and `feature`) and to merge a remote branch with a local branch. It's precisely in these two cases (local-local and remote-local) that conflicts can arise.

In Figure 15.49, you can see how a conflict can arise if one developer pulls the changes from the remote repository at time `t1` and another developer pulls the same changes at time `t2`. Both change the value of the same variable `A`, but to a different one. Developer 1 changes the value from `1` to `2` and developer 2 changes the value from `1` to `3`. You've learned in the earlier sections that before you push, you should check whether anything has changed in the remote repository. Developer 1 does this and pulls before pushing. Developer 2 does the same, but a conflict occurs during his pull or in the merge part of the pull. Git doesn't know at this point whether variable A should have the value `2` or `3` because two people made a change at this point. After the developer has stuck with his variant, the fixed version with `A = 3` is pushed to the remote repository.

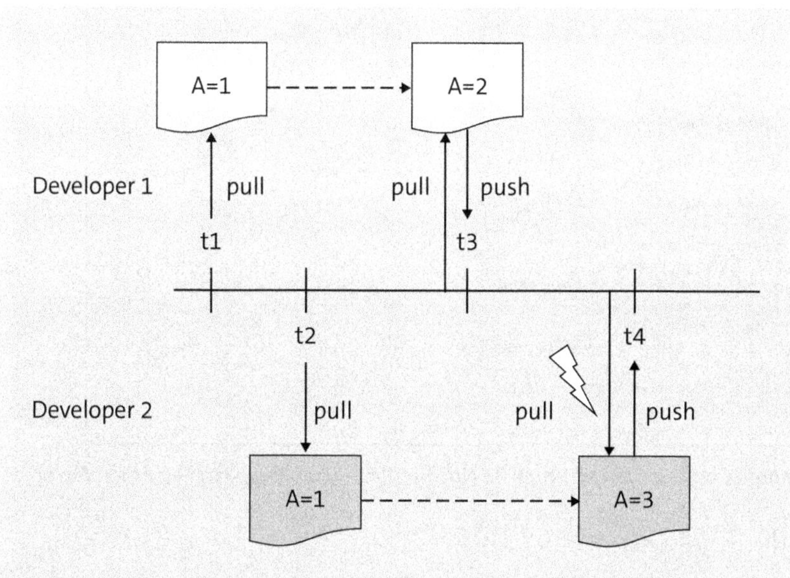

Figure 15.49 Conflict When Two Commits Have Changes for the Same Line of a File

15

The creators of Git considered which method would be best for resolving conflicts that can't be automatically resolved. In this case, the familiar method is used: manual resolution by humans.

For testing purposes, create a local branch called `feature`. For now, stay on the `main` branch, and add a line to the file *file.txt*. Then, switch to the `feature` branch, and add a change to the first line of the file there too (see Listing 15.4).

```
git checkout -b feature3
echo "Change Feature" > file.txt
git add .
git commit -m "Change Feature"
git checkout main
echo "Change Main" > file.txt
git add .
git commit -m "Change Main"
```

Listing 15.4 Make Some Changes on Different Branches of the Same File

As a final step, try to run a merge with `git merge feature`. During this merge, Git notices that both branches have a change in the same file at the same line. Git doesn't want to decide in this case, so it marks the respective locations for you and gives you a hint that something went wrong during the merge (see Figure 15.50). The status display shows you the file in which the conflict occurred.

```
GitExercise — -zsh — 132×19
(base) danielkrancz@MacBook-Pro-von-Daniel GitExercise % git checkout -b feature3
echo "Change Feature" > file.txt
git add .
git commit -m "Change Feature"
git checkout main
echo "Change Main" > file.txt
git add .
git commit -m "Change Main"
git merge feature3
Switched to a new branch 'feature3'
[feature3 6102260] Change Feature
 1 file changed, 1 insertion(+), 4 deletions(-)
Switched to branch 'main'
[main 08c6cda] Change Main
 1 file changed, 1 insertion(+), 4 deletions(-)
Auto-merging file.txt
CONFLICT (content): Merge conflict in file.txt
Automatic merge failed; fix conflicts and then commit the result.
(base) danielkrancz@MacBook-Pro-von-Daniel GitExercise %
```

Figure 15.50 Merge Conflict Occurs: You'll Be Notified That the Merge Isn't Handled Automatically

If you want to abort this merge at any point, you can use the following command:

```
git merge --abort
```

To make it easier for you to decide how to resolve the conflict, the affected location is not only marked but both versions are also offered. Figure 15.51 shows the current state of the file. The first part of this conflict shows the state of the current HEAD. Remember, HEAD marks the branch you've just checked out. The equals signs separate the state of HEAD from the state of the other source. The source in question is identified by the name of the respective branch at the very end of the marked passage. If there's any data left in the file outside the conflict area, it will also be displayed but will remain unaffected by the conflict. In our case, we deleted everything and only had one line in the file in both branches.

Figure 15.51 Conflict Area Highlighted and Both States Presented

Now, it's up to you to modify this *file.txt* file so that it contains the new, merged state. You can edit the file directly in the CLI or in a file editor of your choice, and because you now have everything under control, you can even adopt both versions. From now on, everything follows the regulated process; the change can be brought into the staging area so that it can then be summarized in a commit.

Resolving a conflict when merging with a remote branch works the same way. Looking at a file, you'll see that the markings are like those of a local conflict. There, you can again modify the file as usual and then package the merged version in a commit. Don't forget that you're also sharing the change with others when you push the commits.

15.5 IDE and Git Integration

It's good to know the most important Git commands and what effects they have because modern IDE integrations do this for us entirely and encapsulate the functionalities behind buttons and popups. You'll see that this is no different in SAP Business Application Studio and VS Code as well. In this section, we'll now rediscover the most important functionalities, which we've already learned in theory and tried out on the Git CLI, in visual form and try them out directly in SAP Business Application Studio.

We've already created a simple SAPUI5 application. This application hasn't yet been checked in anywhere and hasn't been shared with any colleagues. It's only stored in our dev space. We want to change that, which is why we're selecting the **Source Control** entry on the left-hand side. This gives us the option to choose **Initialize Repository** in this project (see Figure 15.52).

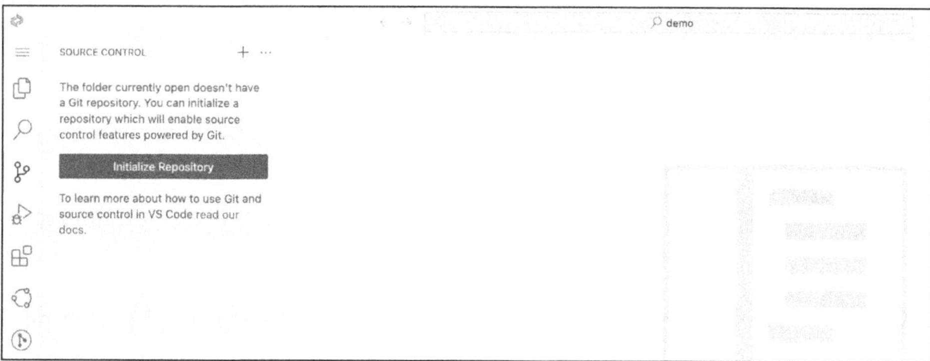

Figure 15.52 Initialize Git in a Project in SAP Business Application Studio

After Git has been initialized, the directory *.git* is created, and a file called *.gitignore* is added to your project. All files that aren't excluded from tracking are immediately registered as a change (see Figure 15.53).

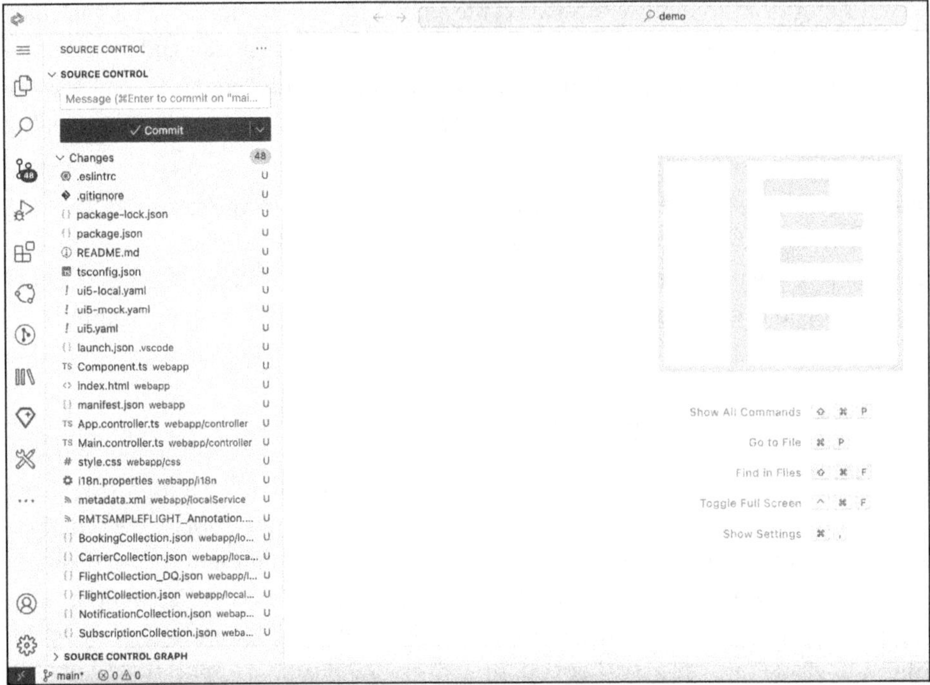

Figure 15.53 Changes Are Tracked in the Source Control Pane

We could now push all the changes to the staging area with **Stage All Changes** (see Figure 15.54).

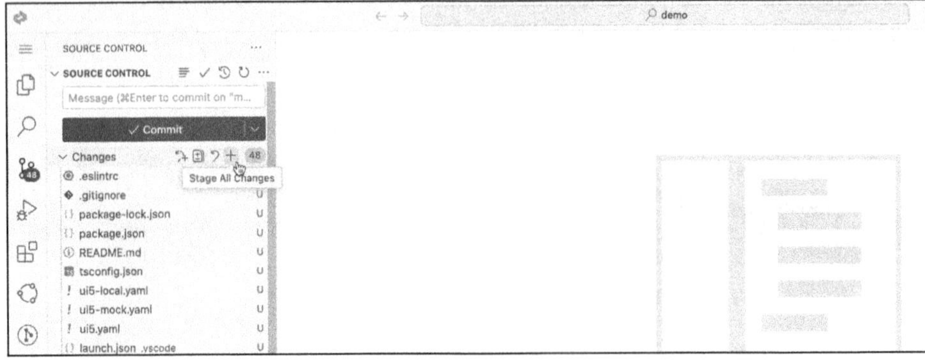

Figure 15.54 Stage All Changes within a Project

However, it's also possible to move the files individually and only selectively into the staging area (see Figure 15.55).

If we switch back to the **Explorer** on the left side, we'll see that the files now have markers (see Figure 15.56). There, we can see the current state of the file (tracked, untracked, changed, staged, etc.). In this figure, you can also see the file *.gitignore* open. The files and folders that are irrelevant for distributed work are listed there.

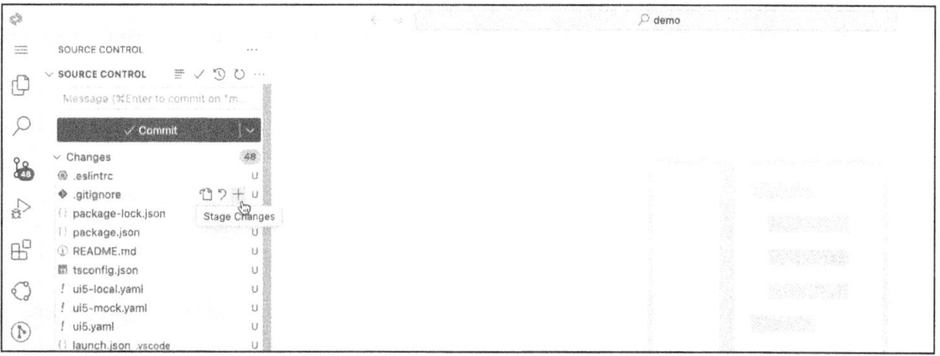

Figure 15.55 Move Changes Selectively into the Staging Area

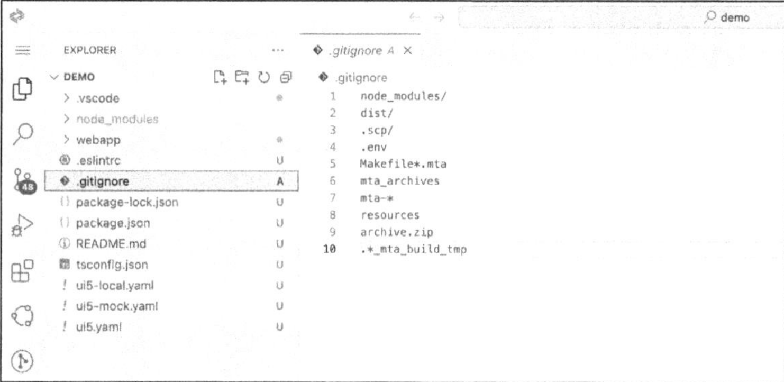

Figure 15.56 Markers Are Added to the Files in the Explorer

If we've now made one or more changes and brought these changes into the staging area, we can combine the changes into a commit in the source control by clicking the **Commit** button (see Figure 15.57). The commit message needs to be entered in the input field above the button.

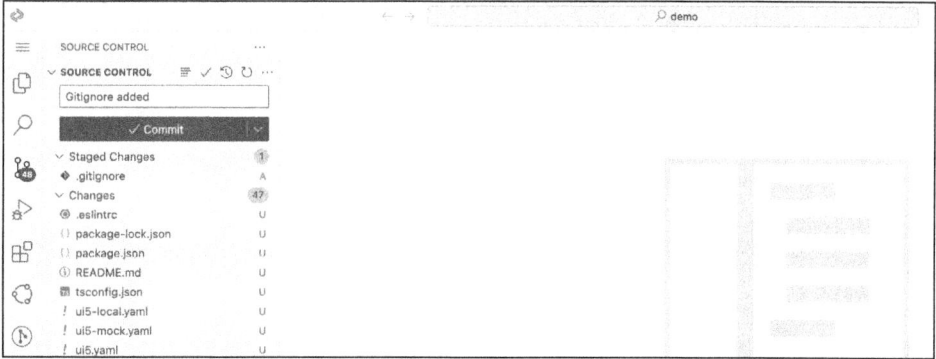

Figure 15.57 Creating a New Commit in the Source Control Pane

If you made this commit by mistake and haven't yet pushed to a remote repository, you can still undo it. To do so, click on the settings in the **SOURCE CONTROL** area, and select **Commit • Undo Last Commit** (see Figure 15.58).

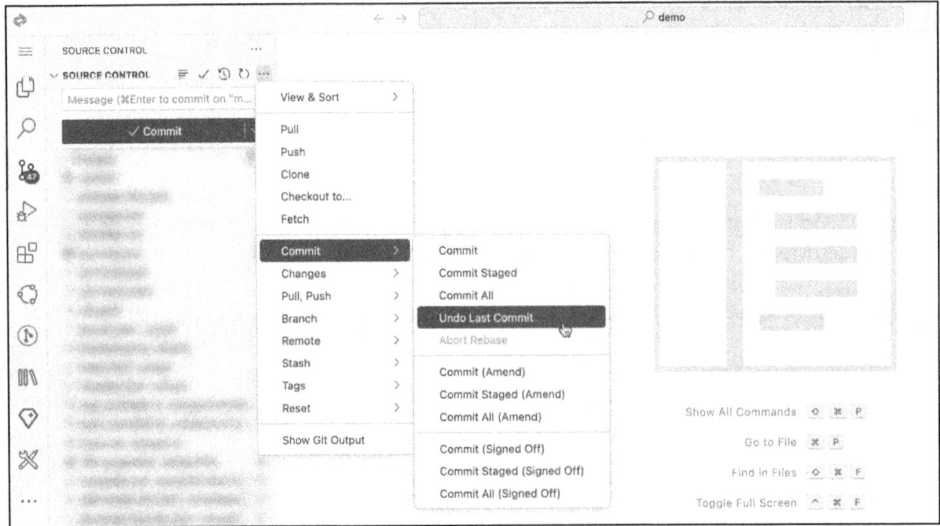

Figure 15.58 Undo the Last Commit with Ease

Even if you've now initialized a Git locally and made a number of commits, it's still of no use because it's all still in your dev space. To share these changes with others, you first have to connect a remote repository, if you haven't already done so. Go to the settings again, and select **Remote • Add Remote** (see Figure 15.59).

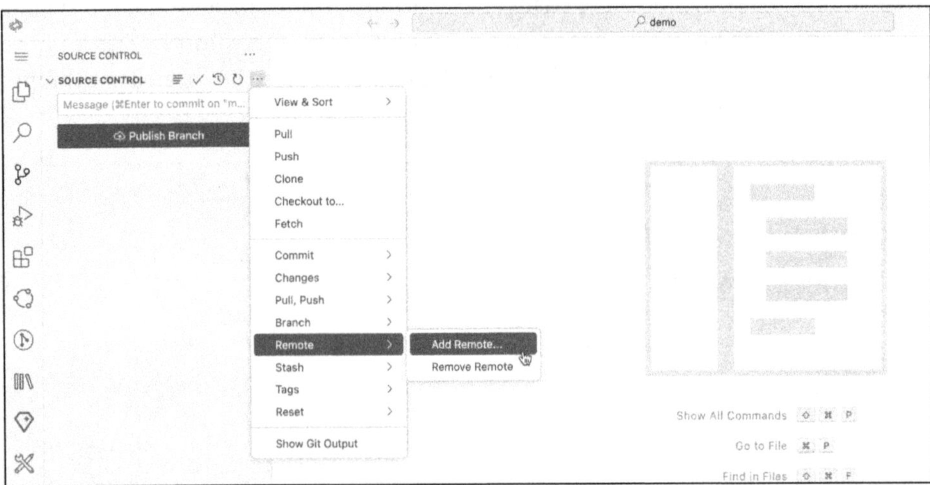

Figure 15.59 Adding a Connection to a Remote Repository

In the upper middle of the screen, a dialog box opens where you can connect a remote repository using a URL (see Figure 15.60).

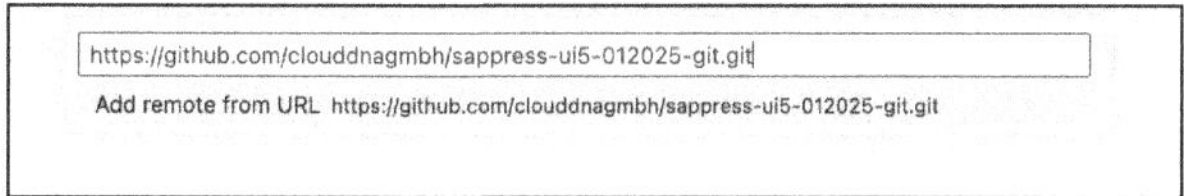

Figure 15.60 Connecting to a Remote Repository from a URL

Each remote repository gets a name so that we can recognize it (in case we have more than one). Usually, you name it "origin", but you could choose any name (see Figure 15.61).

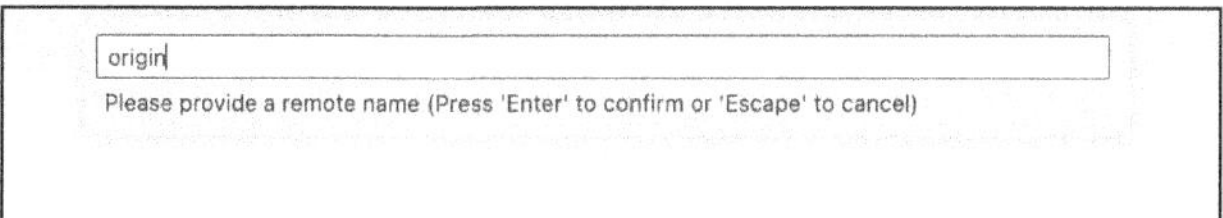

Figure 15.61 Assign a Name to the Remote Repository Connection

Now that the remote repository is connected and still empty, we can publish our local `main` branch with the commits we've collected so far using **Publish Branch** and then push them (see Figure 15.62).

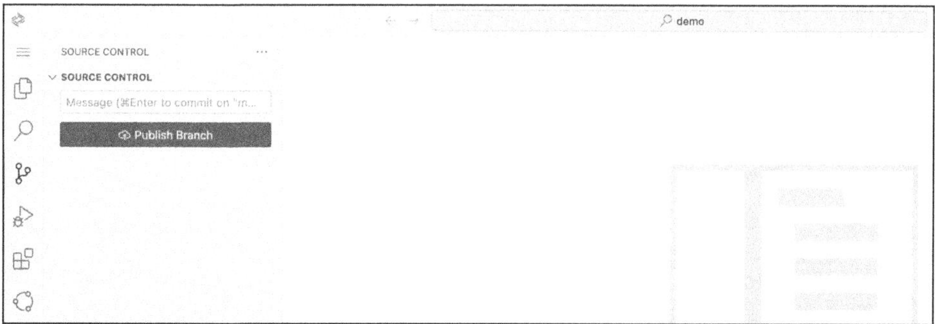

Figure 15.62 Publish Your Local Branches

Depending on the settings and the connection to your remote repository, you may need to authenticate. Once you've done this, you'll be asked in SAP Business Application Studio whether these credentials should only be stored for this session or in the dev space (see Figure 15.63).

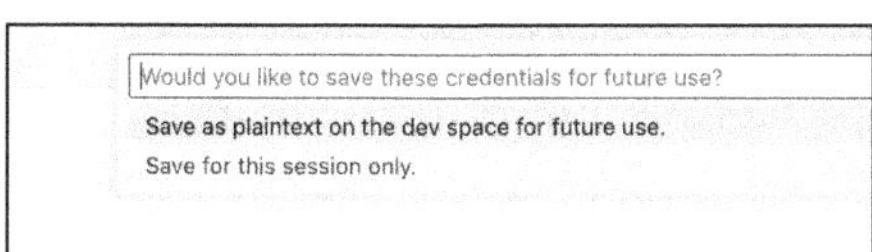

Figure 15.63 Store the Git Credentials in SAP Business Application Studio

In the left-hand **SOURCE CONTROL** tab, there are several views that you can expand and collapse. For example, you can expand the **SOURCE CONTROL GRAPH** and see the last commits there in a small summary (see Figure 15.64). If you click on a commit, the differences of each file from that commit are briefly listed in a new tab on the right.

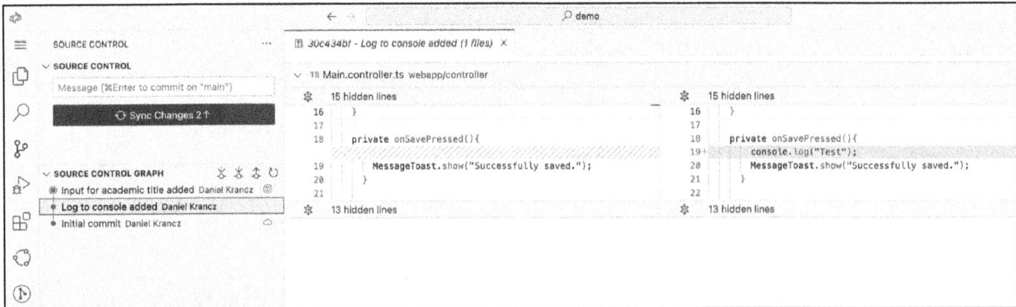

Figure 15.64 Source Control Graph to Show the Last Commits

However, if you want to access the familiar `git log`, you can use the **Git: View History** button (see Figure 15.65) to open it separately. The entire history is visually displayed in a new tab, which you can even filter at the top. For each commit, you have the option to view the differences, to check them out, or to create a new branch based on a specific commit. You can also easily see which pointer of every branch is currently pointing to which commit.

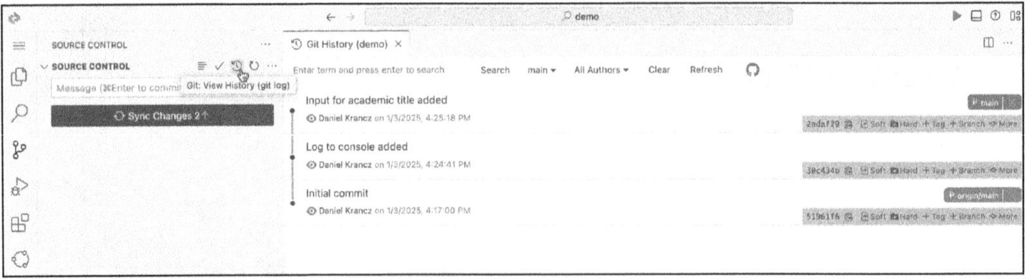

Figure 15.65 Git History Showing the Well-Known git log

We've recognized that we've made two new commits since the initial commit that haven't yet been shared with our colleagues; that is, they haven't yet been pushed to the remote repository. We can do this by opening the settings again and selecting **Push** or, because we have no open changes at the moment, by selecting the **Sync Changes** button (see Figure 15.66).

It's important to note that clicking **Sync Changes** does nothing other than first doing a pull to make sure that no changes since our last pull are missing locally. After that, a push is triggered, and because our local `main`, on which we're currently working, is linked to the `origin/main`, our commits are made available there (see Figure 15.67).

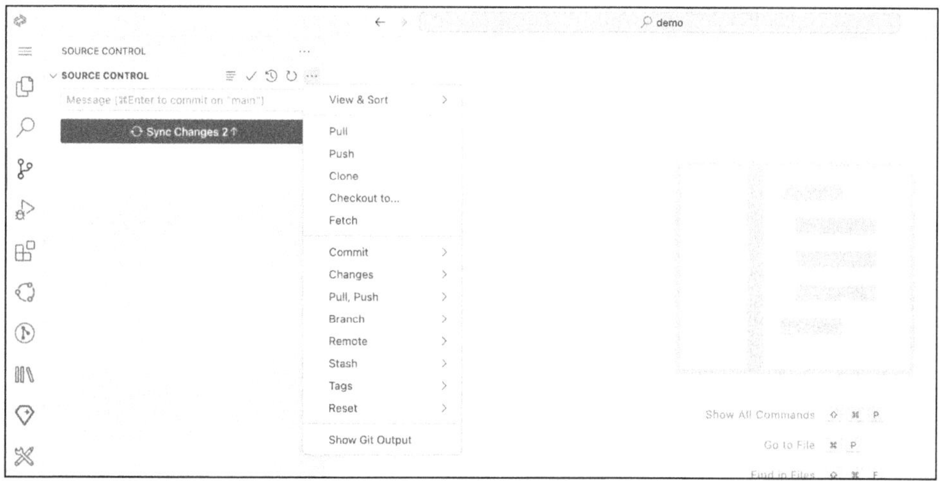

Figure 15.66 Sync/Push Your Changes

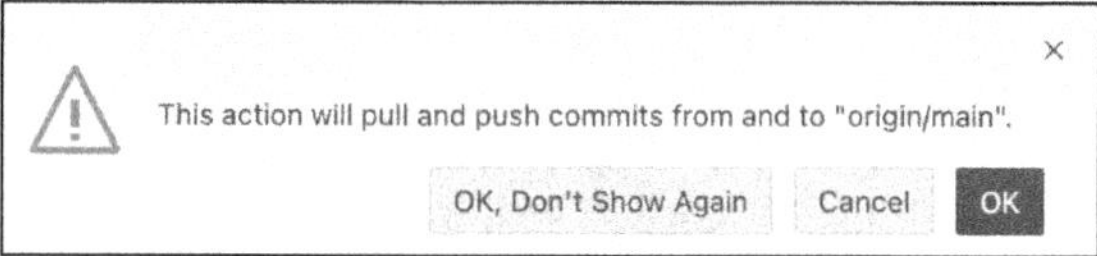

Figure 15.67 A Sync Results in the Actions Pull and Push

Let's create a new branch because we're not sure yet whether the next changes we make to the app should really be merged into the `main` branch. To do that, we'll open the settings again and select **Branch • Create Branch** (see Figure 15.68).

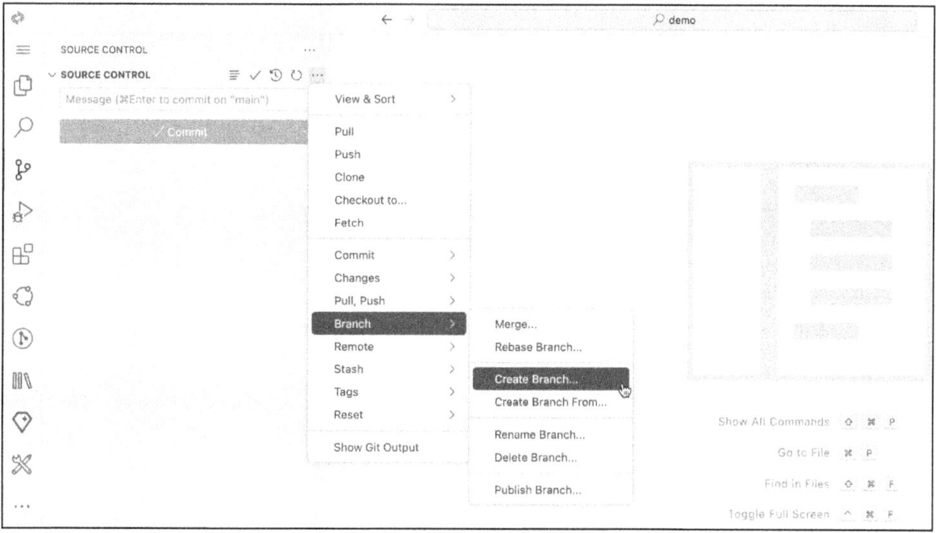

Figure 15.68 Creating a New Local Branch

After you've entered the name for the new branch in a small dialog that will appear again at the top center of the screen, you'll immediately be switched to that branch. We've entered the name "feature/saveForm" and can check at the bottom left which branch we currently checked out (see Figure 15.69).

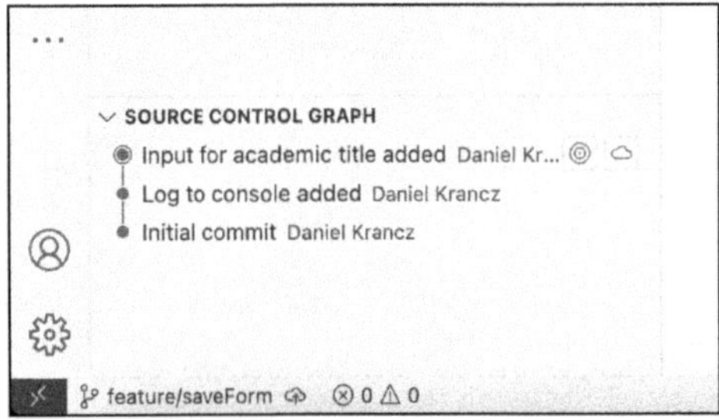

Figure 15.69 Viewing Which Branch You Currently Checked Out

If you click on this name, you'll be given the option in the dialog box at the top to check out another branch or to create a new one (see Figure 15.70).

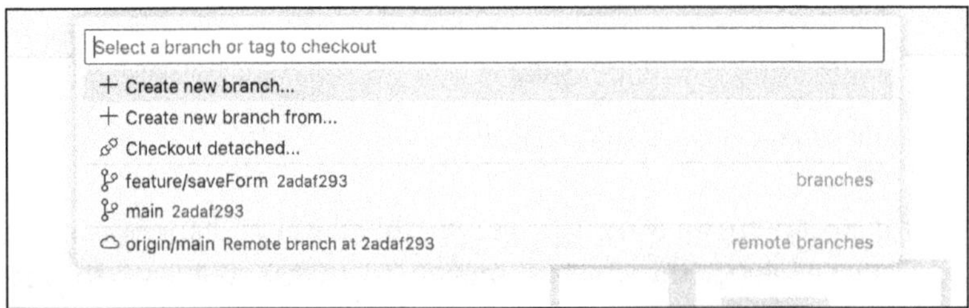

Figure 15.70 Check Out Another Branch or Create a New One

We linger on our branch, so we close this dialog by pressing Esc. We make a change in our main controller, which we also immediately pack in a commit. In the source code, we now always see the information about commits inline on the line where our cursor is located (see Figure 15.71).

```
private onSavePressed(){
    console.log("Test2");
    MessageToast.show("Successfully saved.");
}
```

Figure 15.71 Inline Information about the Git Status of the Current Line

Our **Git History** tab also shows that our two branches **main** and **feature/saveForm** have drifted apart. Commits were created on both branches (see Figure 15.72).

Now that our change is relevant to the feature, we want to merge it with the `main` branch. To do this, we first switch to the `main` branch and go to **Settings** to select **Branch • Merge** (see Figure 15.73). A dialog will open again, where we have to select the branch with which we want to merge. We've selected our **feature/saveForm** branch there.

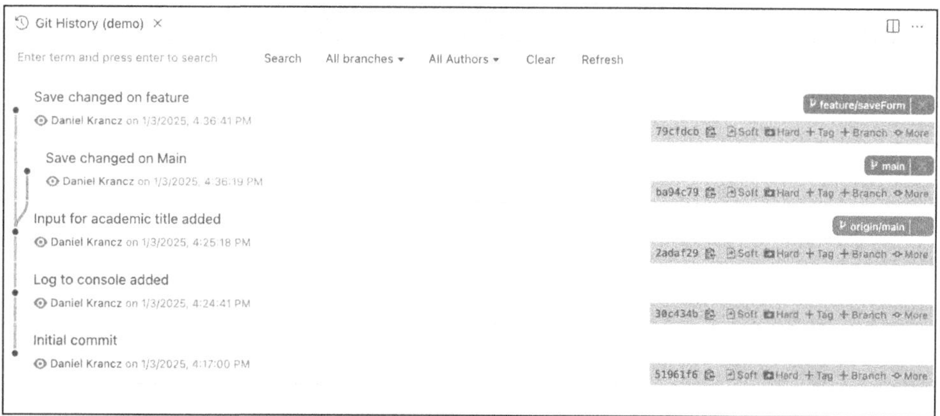

Figure 15.72 Commits at Both Branches Were Created

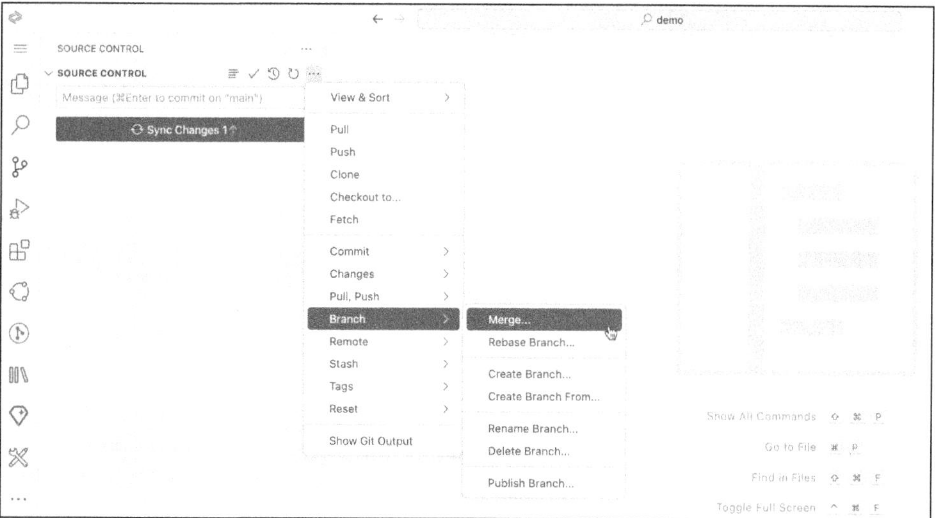

Figure 15.73 Merging the Current Branch with Another Local or Remote Branch

Now, we realize that there was a conflict during the merge process. No fast-forward merge can be done because something was changed in the same file and line on both branches. Now, we can select the files in **SOURCE CONTROL** on the left side that are marked with an exclamation point (see Figure 15.74). These files, as well as our *Main.controller.ts*, have a conflict. Now, we can see visually in which lines a conflict arises and decide whether to edit the file manually or to use the links above the color-coded blocks. These blocks represent the changes from the `main` branch and the `feature` branch, respectively. In our case, we decided on the **Accept Both Changes** link because we felt that both changes were necessary for the application. After we've fixed the files one after the other, we then have to manually complete the merge commit by packaging the merge changes into a merge commit using the **Commit** button.

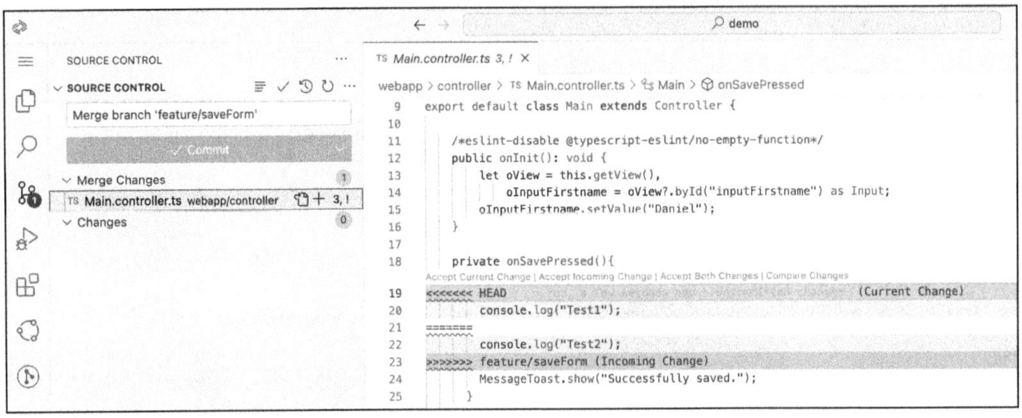

Figure 15.74 Resolving a Merge Conflict Directly in the File Editor

Of course, such merge procedures are also visualized in the **Git History** tab (see Figure 15.75).

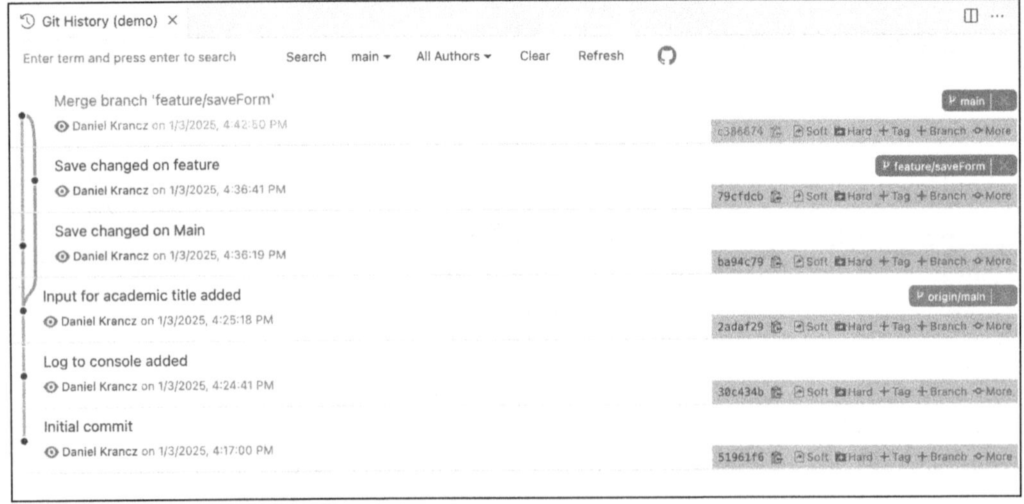

Figure 15.75 Merge Process Shown in the Git History Tab

If you're not the person who created the project and initially wants to share it, you can also clone an existing repository. There are several ways to do this. If you don't have a folder open, you'll see a **Clone Repository** button on the left (see Figure 15.76). There's also a **Clone from Git** button on the **Get Started** welcome screen. In any case, you can switch to the **SOURCE CONTROL** menu on the left and go to **Clone** in the settings. Then, enter the URL of the repository as usual, complete the authentication, and select the folder in your local filesystem where you want to clone the repository. By doing so, not only is the source code cloned into your folder but also the remote repository is added automatically as a remote source.

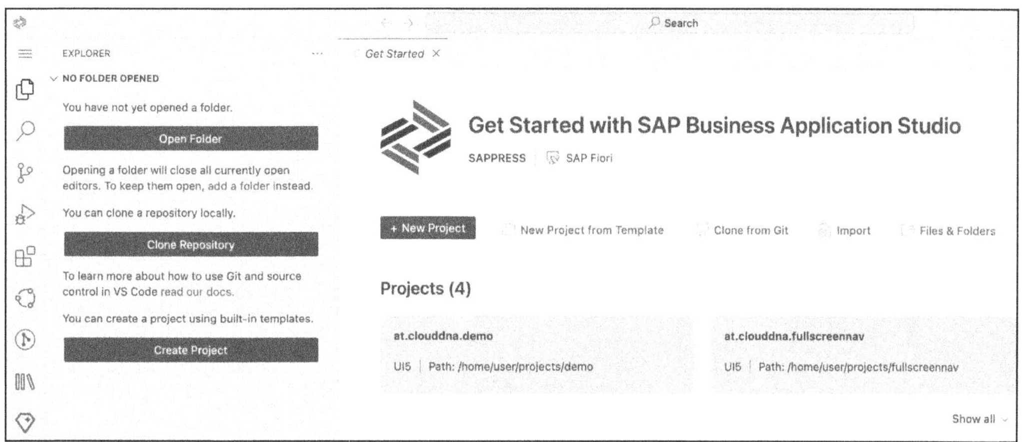

Figure 15.76 Cloning a Project

Don't Forget about the Other Tasks after Cloning a Project

After cloning the project, you have to manually run `npm install` because this isn't picked up by Git. As you already know, only then will the necessary `node_modules` be downloaded, which are necessary for execution, previewing, deployment, and other tasks within the development environment.

15

15.6 Summary

In this chapter, you've learned a great deal about Git. Although we weren't able to cover every detail, the aim was to show you the most important commands and options so that you can also manage SAPUI5 applications in a Git repository or participate in their development in the future. We started with a few definitions of terms related to Git: repositories, branches, commits, and changes. After that, we looked at a number of functions which ensure that you can store your changes in a bundled and versioned form but also share them with others (clone, commit, fetch, merge, rebase, pull, push, etc.).

Finally, we clarified that conflicts can arise when merging multiple snapshots if two changes affect the same location in a file. In this case, the person who encountered the conflict must intervene. Fortunately, we don't have to memorize all the commands that we can also execute in the terminal or in the Git CLI because most development environments are shipped with graphical tools. This is also the case in SAP Business Application Studio, which has source control built in by default. This tool takes care of the correct application of these commands for us, but we still have to be aware of what we're doing.

Chapter 16
Deployment

The development of an app is usually completed by deployment. This makes the app available to other users. Deployment can be done in different ways. Essentially, it depends on the target environment to which the app must be deployed.

After covering the development of SAPUI5 apps in detail in the previous chapters, this chapter is dedicated to making these applications available to end users—known as deployment. *Deployment* of software refers to the process of installing, configuring, and deploying a software application to a specific target environment. It doesn't matter whether it's a local infrastructure or a cloud infrastructure. Deployment is a critical step in moving software from a development or test environment to a production environment so that it can be used by users.

For cloud deployment, you use SAP Business Technology Platform (SAP BTP), which provides a scalable and flexible environment for modern applications. In the on-premise environment, on the other hand, there are several options, such as SAP S/4HANA systems or SAP NetWeaver Application Server for ABAP (SAP NetWeaver AS ABAP). Because the deployment steps on SAP S/4HANA and SAP NetWeaver AS ABAP are largely identical, we'll focus on deployment in an SAP S/4HANA environment in this chapter.

Deployment can be either manual or automated. In the case of manual deployment, the developer initiates the process by clicking a button or executing a corresponding command. In contrast, automated deployment is realized using the SAP Continuous Integration and Delivery service, which is available in SAP BTP and enables continuous integration and deployment.

In the rest of this chapter, we'll explain manual deployment step-by-step in Section 16.1, while Section 16.2 will cover automated deployment in detail. In addition, we'll discuss the advantages and specific uses of both deployment methods to give you a solid basis for deciding on your specific project requirements.

After an app has been deployed, it can be integrated into the SAP Fiori launchpad to create a central and user-friendly interface for end users. This integration will be covered in detail in the next chapter of this book. Alternatively, the app can also be started as a standalone application, depending on the specific requirements and use cases.

16.1 Manual Deployment

Manual deployment of SAPUI5 apps is a simpler and more immediately accessible option compared to automated procedures, as it only requires an integrated development environment (IDE) and no additional services or extensive infrastructure. This approach is particularly suitable for development and test environments where fast results and direct control over the deployment process are required.

In Section 16.1.1, we'll introduce the deployment of SAPUI5 apps to SAP BTP. SAP BTP provides a robust cloud environment that enables SAPUI5 apps to be deployed directly in the cloud and made available to users. We'll go through the steps and configurations required to successfully deploy your app on SAP BTP. This includes creating a corresponding HTML5 repository and configuring the target environment to ensure that the app is correctly available to end users.

Section 16.1.2 then deals with deployment to an SAP S/4HANA on-premise system. This option is particularly relevant for companies that want to operate their SAPUI5 apps within a local system landscape and without an external cloud connection. Deployment in an on-premise environment requires that the SAP S/4HANA system is correctly configured to accept and run HTML5 applications. In this section, we'll explain the necessary steps and prerequisites in detail to ensure a smooth deployment and optimal performance of the application in the SAP S/4HANA context.

In principle, it's also possible to deploy SAPUI5 apps as static HTML pages on a conventional web server such as nginx or Apache Tomcat. This can be a cost-effective and flexible alternative if the app is to be operated independently of SAP systems or if fast, standalone deployment is required. This method is particularly suitable for scenarios in which no special SAP backend functionalities are required. We'll also cover the basic steps on how to configure an SAPUI5 app as a static website and host it on a web server, including the necessary customizations in the configuration file to ensure that the app loads and runs correctly.

16.1.1 SAP Business Technology Platform

To deploy an SAPUI5 app to SAP BTP, a few preparatory steps are necessary. First, an *mta.yaml* file is added to the project, which describes the concept of the multi-target application (MTA). In this file, you define which modules and resources belong to your application, how they are connected to each other, and which dependencies exist.

The *mta.yaml* file is to a certain extent the heart of the deployment configuration and ensures that all components of the application can be provided together and coherently. In addition to *mta.yaml*, you need an *xs-app.json* and an *xs-security.json* file. The *xs-app.json* file defines important routes and access control for your application. It ensures that the app can be reached via the correct URLs and that access to the various parts of the application is regulated. The *xs-security.json* in turn is responsible for the security configuration. This is where the authentication and authorization rules are defined to ensure that only authorized users have access to the application and its data.

A deployment configuration is created to ensure that these files are correctly and completely available in the project. This configuration helps to automatically create and adapt all necessary files such as *mta.yaml*, *xs-app.json*, and *xs-security.json*. This structures and simplifies the entire deployment process, and you can be sure that your SAPUI5 app is optimally prepared to be deployed as an MTA in SAP BTP.

In the following, we'll show you how to create these steps using the application information. In the first step, right-click to open the context menu in the *webapp* folder in your project, as shown in Figure 16.1, and click on the menu item **Open Application Info**.

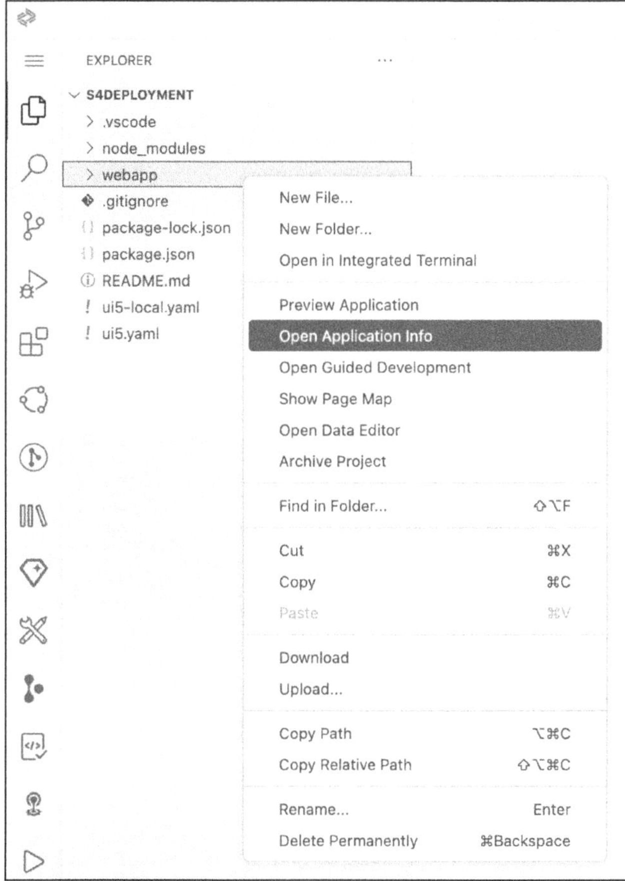

Figure 16.1 Open Application Info

The application info provides you with an overview of your app. In addition, you can perform common operations such as adding an OData service at this point. It's also possible to create a *deploy config*. To do this, click **Add Deploy Config**, as shown in Figure 16.2.

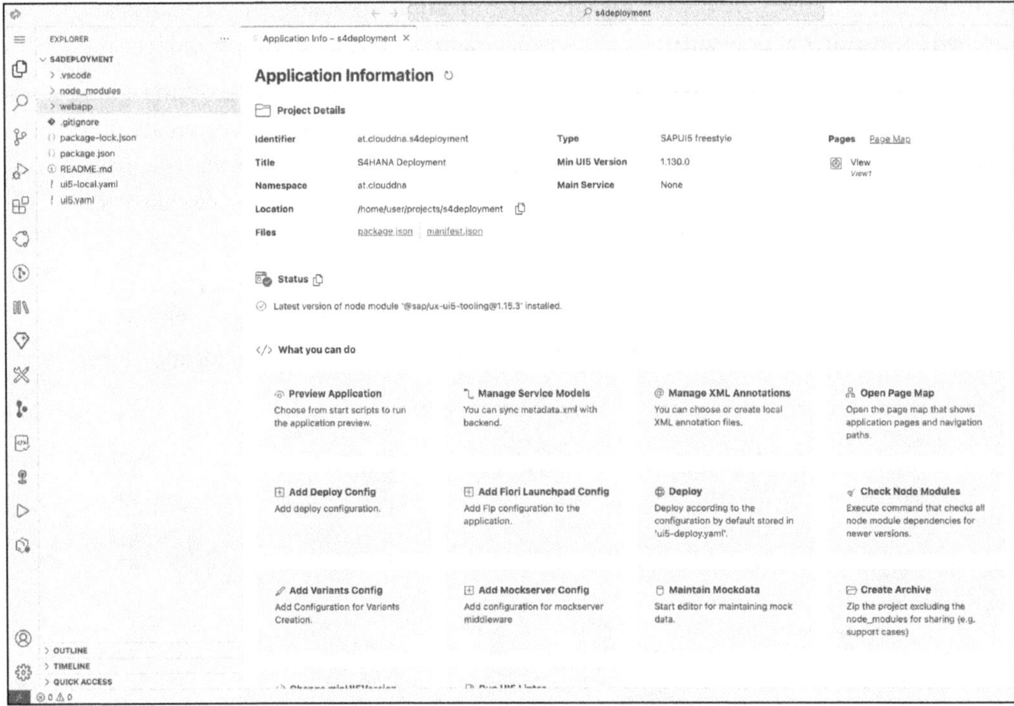

Figure 16.2 Add Deploy Config

In the next step, select **Cloud Foundry** as the target environment. In the **Destination Name** field, you can select the **None** option. At this point, you can also specify a particular destination to be used for the deployment. Then, select the **Yes** option to add the **Managed Application Router** and confirm by also selecting **Yes** to overwrite the existing configuration. Finally, click **Finish** (see Figure 16.3).

Deployment Configuration Generator

Deploy Configuration
Target Type: Cloud Foundry
Destination Name: None
Add To Router: Yes

Deploy Configuration
Create/Update deployment configuration - cfdeployment

Please choose the target *
Cloud Foundry

Destination name *
None

Add application to managed application router?
Yes No

Editing the deployment configuration will overwrite existing configuration, are you sure you want to continue?
Yes No

Finish

Figure 16.3 Cloud Foundry Deploy Configuration

Managed Application Router

The managed application router enables access to and execution of HTML5 applications in a cloud environment without the need to operate a proprietary runtime infrastructure. This runtime environment for HTML5 applications is provided by the following products:

- SAP Build Work Zone, standard edition
- SAP Build Work Zone, advanced edition
- SAP Cloud Portal

In the next step, the files **mta.yaml**, **xs-app.json**, **xs-security.json**, and **ui5-deploy.yaml** are automatically added to the project, as shown in Figure 16.4.

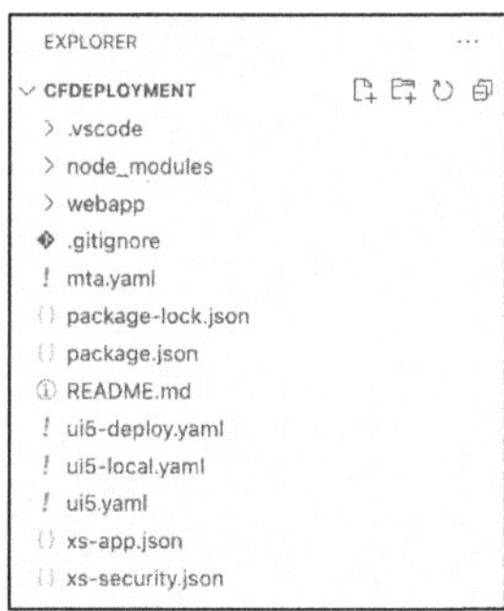

Figure 16.4 Files Added for Cloud Foundry Deployment

The deployment is controlled by the *ui5-deploy.yaml* file. Open this to familiarize yourself with the structure of the file. The `name` attribute in the `metadata` area and the `type` that has the `application` value are important (see Figure 16.5).

Application Info – cfdeployment   ! ui5-deploy.yaml ×

! ui5-deploy.yaml > ...

```
# yaml-language-server: $schema=https://sap.github.io/ui5-tooling/schema/ui5.yaml.json
specVersion: '2.4'
metadata:
  name: at.clouddna.cfdeployment
type: application
resources:
  configuration:
    propertiesFileSourceEncoding: UTF-8
builder:
  resources:
    excludes:
      - "/test/**"
      - "/localService/**"
  customTasks:
  - name: webide-extension-task-updateManifestJson
    afterTask: replaceVersion
    configuration:
      appFolder: webapp
      destDir: dist
  - name: ui5-task-zipper
    afterTask: generateCachebusterInfo
    configuration:
      archiveName: atclouddnacfdeployment
      additionalFiles:
      - xs-app.json

```

Figure 16.5 Generated ui5-deploy.yaml File

16

Before an app can be deployed, a build must be carried out. To do this, open the context menu of the **mta.yaml** file, and select the entry **Build MTA Project** (see Figure 16.6).

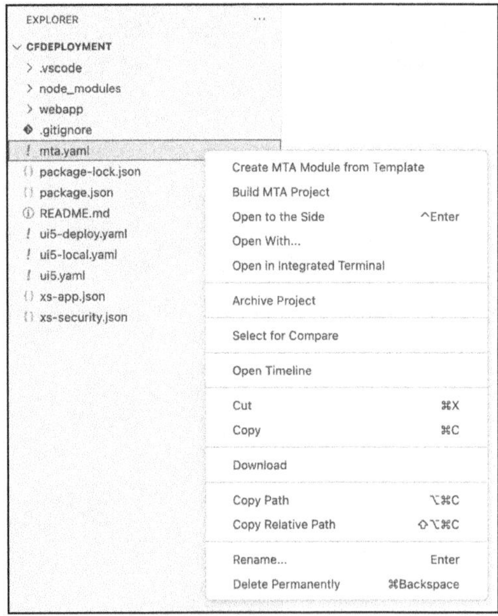

Figure 16.6 Build MTA

This adds a directory called **mta_archives** to the project, as shown in Figure 16.7. The result of the build is stored in this directory in the form of a MTA archive (*.mtar*). Open the context menu on this file, and select **Deploy MTA Archive** to start the deployment.

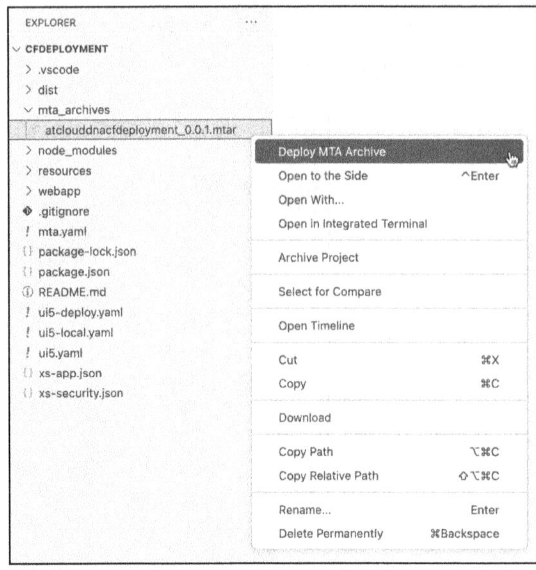

Figure 16.7 Trigger Deployment

Before the deployment is carried out, however, you still must register with the **Cloud Foundry Endpoint** via the API. The subaccount in which SAP Business Application Studio is located is proposed as the Cloud Foundry endpoint. You can either register with **Credentials** or with an **SSO Passcode**. We've chosen the **SSO Passcode** option. To do this, click on the link shown in Figure 16.8.

Cloud Foundry Sign In ×

Cloud Foundry Sign In and Targets

Provide your Cloud Foundry parameters to sign in to the Cloud Foundry enviroment

Cloud Foundry Sign In

Enter Cloud Foundry Endpoint *

https://api.cf.eu10-004.hana.ondemand.com

Select authentication method

Credentials SSO Passcode

Open a new browser page to generate your SSO passcode

Enter your SSO Passcode *

Enter your passcode

Sign in

Figure 16.8 Selecting the Authentication Method

This opens a new window in which you must select the desired **identity provider** (see Figure 16.9). You'll then be authenticated against this identity provider.

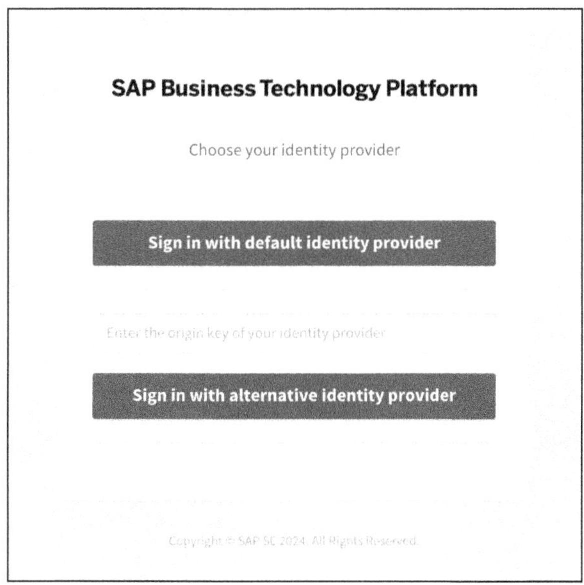

Figure 16.9 Selecting an Identity Provider

After successful authentication, the **Temporary Authentication Code** is displayed, as shown in Figure 16.10. Copy this to the clipboard.

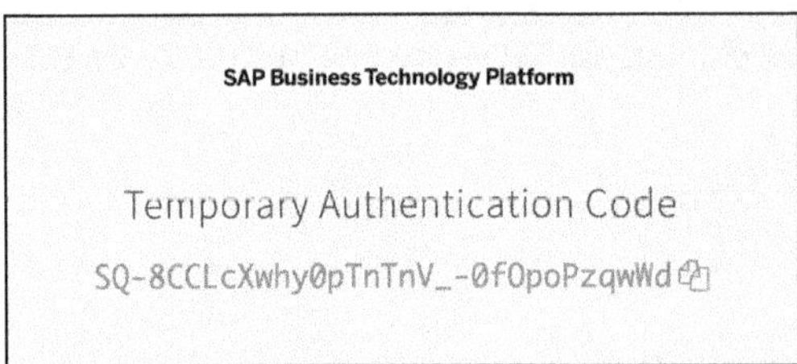

Figure 16.10 Temporary Authentication Code

Now insert the generated authentication code into the **Enter your SSO Passcode** field, and click on **Sign In** (see Figure 16.11).

Cloud Foundry Sign In ×

Cloud Foundry Sign In and Targets

Provide your Cloud Foundry parameters to sign in to the Cloud Foundry enviroment

Cloud Foundry Sign In

Enter Cloud Foundry Endpoint *

https://api.cf.eu10-004.hana.ondemand.com

Select authentication method

Credentials SSO Passcode

Open a new browser page to generate your SSO passcode

Enter your SSO Passcode *

SQ-8CCLcXwhy0pTnTnV_-0fOpoPzqwWd

Sign in

Figure 16.11 Entering the SSO Passcode

As shown in Figure 16.12, you can now see that you're logged in to the Cloud Foundry environment. You now need to select a **Cloud Foundry Organization** and then a **Cloud Foundry Space** within that organization. Click **Apply**.

The deployment is now started. This may take some time. The terminal will display the deployment status and also its successful execution, as shown in Figure 16.13.

Now open the SAP BTP cockpit for the subaccount, and navigate to the **HTML5 Applications** area in the side menu, as shown in Figure 16.14. To your surprise, you won't find any applications there. You'll only see a note telling you that you need to subscribe to an edition of SAP Build Work Zone or the SAP Cloud Portal service. This is because, as mentioned earlier, we've decided to use the managed application router. You'll learn how to set up SAP Build Work Zone in Chapter 17.

Cloud Foundry Sign In and Targets

Provide your Cloud Foundry parameters to sign in to the Cloud Foundry enviroment

Cloud Foundry Sign In You are signed in to Cloud Foundry.

Cloud Foundry Endpoint

https://api.cf.eu10-004.hana.ondemand.com

Sign Out

Cloud Foundry Target

Select Cloud Foundry Organization *

CloudDNA_sapui5-a29wqi23

Select Cloud Foundry Space *

dev

Apply

Figure 16.12 Selecting the Cloud Foundry Organization and Space

PROBLEMS OUTPUT DEBUG CONSOLE TERMINAL

```
Creating service key "atclouddnacfdeployment-destination-content-atclouddnacfdeployment-destination-service-credentials" for service instance "atclouddnacfdepl
oyment-destination-service"...
Creating service key "atclouddnacfdeployment-uaa-key" for service instance "atclouddnacfdeployment-xsuaa-srv"...
Creating service key "atclouddnacfdeployment-repo-host-key" for service instance "atclouddnacfdeployment-html5-srv"...
Deploying content module "atclouddnacfdeployment-destination-content" in target service "atclouddnacfdeployment-destination-service"...
Creating service key "atclouddnacfdeployment-app-content-atclouddnacfdeployment-repo-host-credentials" for service instance "atclouddnacfdeployment-html5-srv".
..
Uploading content module "atclouddnacfdeployment-app-content" in target service "atclouddnacfdeployment-repo-host"...
Deploying content module "atclouddnacfdeployment-app-content" in target service "atclouddnacfdeployment-repo-host"...
Skipping deletion of services, because the command line option "--delete-services" is not specified.
Process finished.
Use "cf dmol -i 5f8b1b12-9c3b-11ef-b9ee-eeee0a889a02" to download the logs of the process.

 * Terminal will be reused by tasks, press any key to close it.
```

Build MTA Project Task

Deploy atclouddnacfdeploy...

The Build/Deploy task has completed!

Figure 16.13 Deployment Competed Information

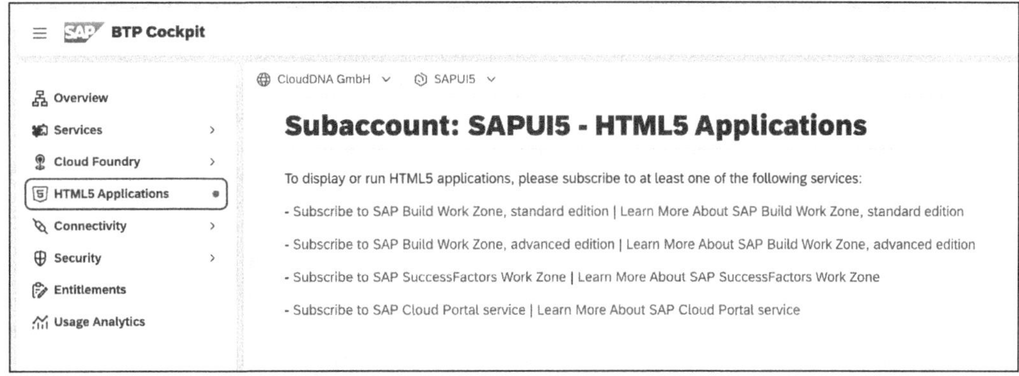

Figure 16.14 Missing Subscriptions

16.1.2 SAP S/4HANA

After learning how to deploy to SAP BTP in the previous section, this section shows how to deploy an SAPUI5 app to an on-premise SAP S/4HANA system. The deployment to an SAP NetWeaver AS ABAP system is identical. The prerequisite for this is that a destination is created in the subaccount in which SAP Business Application Studio is also running. The virtual hostname with the appropriate port must be stored in the URL field. You can view this in the cloud connector. How the cloud connector is connected to SAP BTP and how the mappings are created was covered in Chapter 2.

Select **OnPremise** as the **Proxy Type**, and maintain the desired type of authentication. We've decided to use **NoAuthentication**, as shown in Figure 16.15. This means that the developer is prompted to enter a username and password during deployment. In addition, create the following **Additional Properties**:

- **HTML5.DynamicDestination** = "true"
- **HTML5.Timout** = "30000"
- **WebIDEEnabled** = "true"
- **WebIDEUsage** = "odata_abap,dev_abap"

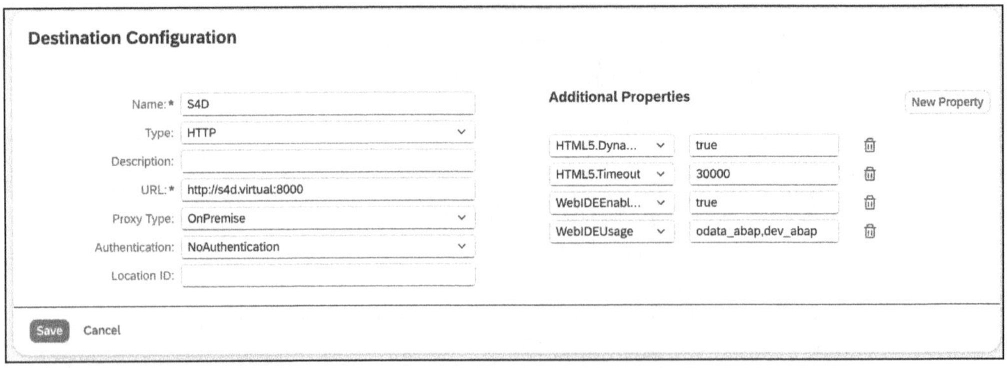

Figure 16.15 Adding the Destination

You now must create a deploy configuration for your project. To do this, open the **Application Information** page (refer to Figure 16.1) as we've already done. On this page, click on **Add Deploy Configuration** (see Figure 16.2).

In the **Please choose the target** field, select **ABAP** (see Figure 16.16). In the **Destination** field, select the destination you created earlier. Now enter a username and the corresponding password for the SAP S/4HANA system. Then, click the button next to the password to log in.

After that, you must enter the **SAPUI5 ABAP Repository**. This is the name of the Business Server Pages (BSP) application that is created for your app. Note that this must be in the customer namespace or a namespace that you've registered. Then, optionally assign a **Deployment Description**, and enter a **Package** and a **Transport Request**. In the example shown, we've decided on the "$TMP" package. Therefore, no transport request needs to be specified. Finally, click **Finish** to create the configuration artifacts.

As you can see in Figure 16.17, a file named *ui5-deploy.yaml* has been added to your app. Open it and take a look at the content. You'll find all the entries you made previously there.

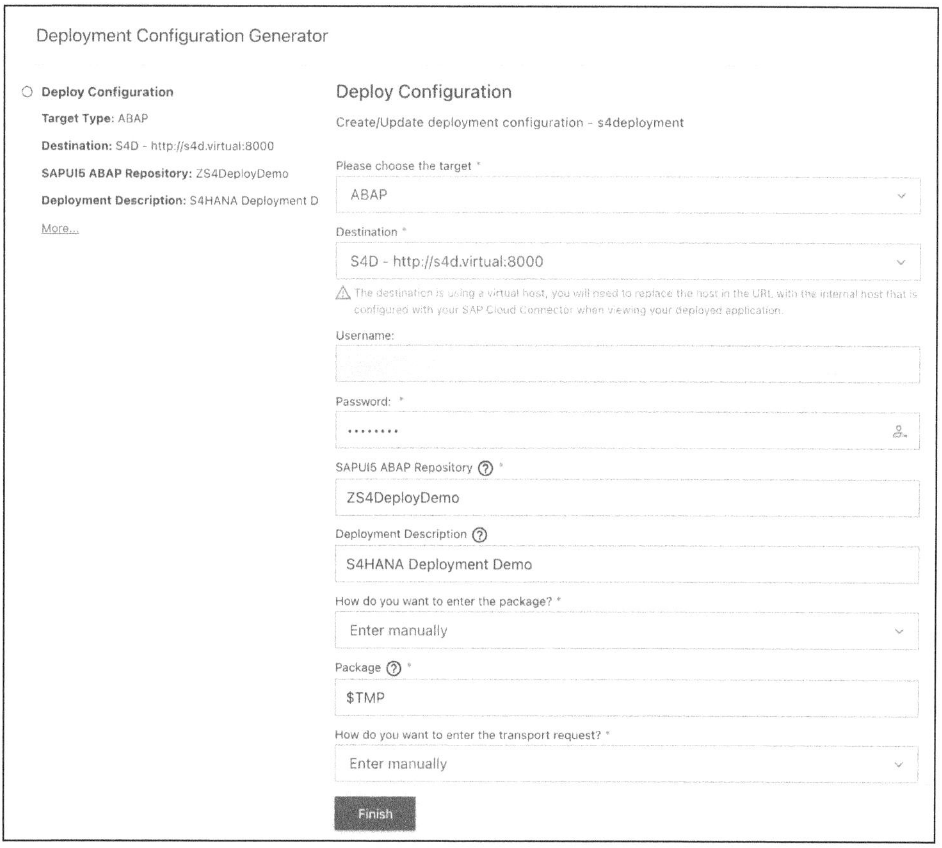

Figure 16.16 Adding a Deploy Configuration

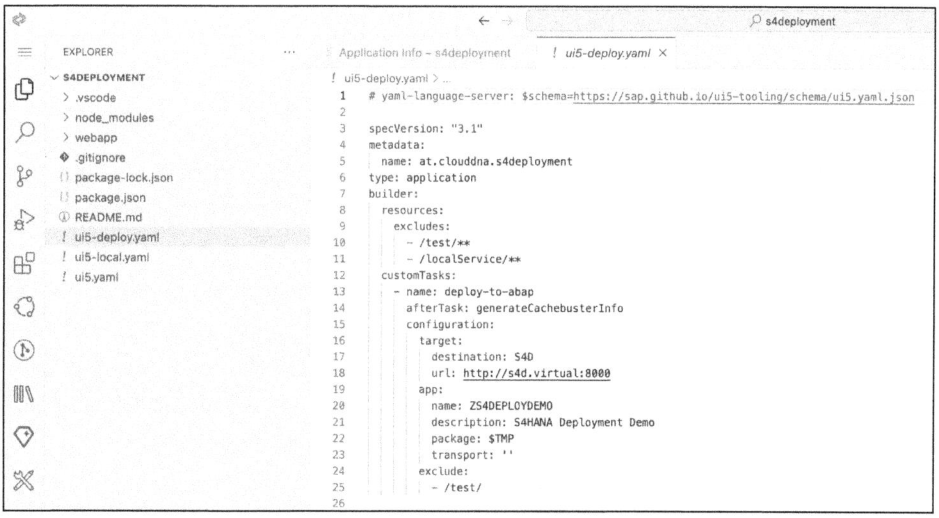

Figure 16.17 Generated ui5-deploy.yaml

The deployment is started via a terminal. Therefore, as shown in Figure 16.18, open a new terminal window using the context menu, and select **Terminal • New Terminal**.

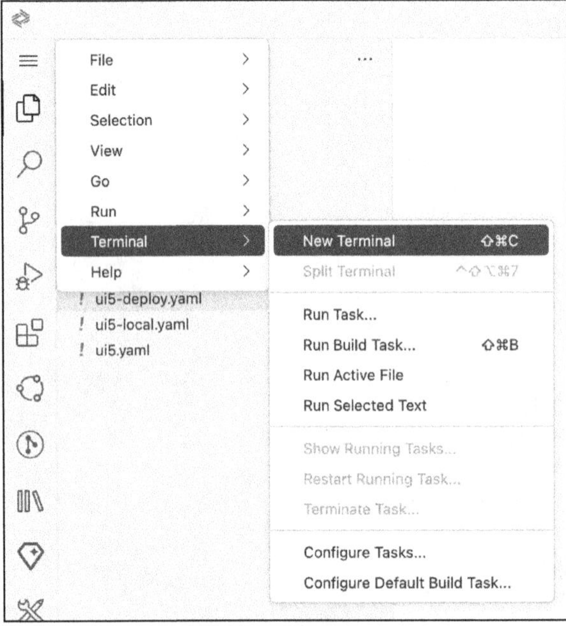

Figure 16.18 Opening a New Terminal

Run the command `npm run deploy` in the terminal (see Figure 16.19).

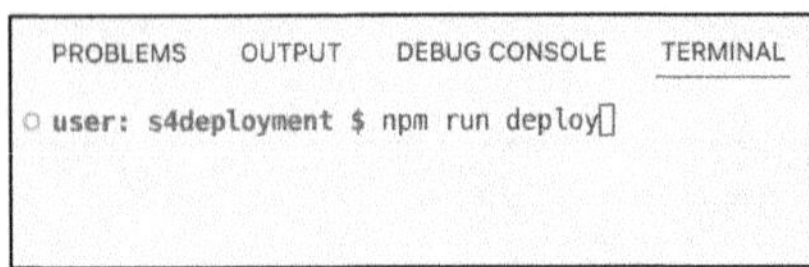

Figure 16.19 Deploying the Application

As shown in Figure 16.20, you'll now see a series of tasks that are carried out. You must then confirm that you want to start a deployment with the configuration shown by entering the letter "Y".

The deployment can take some time. The terminal displays the deployment status (see Figure 16.21). If the deployment was successful, you should see the output **Deployment Successful**. However, errors may also occur. For example, the name of the application may already be in use, or you may have referenced a transport request that has already been released. In all cases, you'll see an informative error message.

```
PROBLEMS    OUTPUT    DEBUG CONSOLE    TERMINAL

user: s4deployment $ npm run deploy

> s4deployment@0.0.1 deploy
> npm run build && fiori deploy --config ui5-deploy.yaml && rimraf archive.zip


> s4deployment@0.0.1 build
> ui5 build --config=ui5.yaml --clean-dest --dest dist

info ProjectBuilder Preparing build for project at.clouddna.s4deployment
info ProjectBuilder   Target directory: dist
info ProjectBuilder Cleaning target directory...
info Project 1 of 1: ❯ Building application project at.clouddna.s4deployment..
info at.clouddna.s4deployment › Running task escapeNonAsciiCharacters...
info at.clouddna.s4deployment › Running task replaceCopyright...
info at.clouddna.s4deployment › Running task replaceVersion...
info at.clouddna.s4deployment › Running task minify...
info at.clouddna.s4deployment › Running task generateFlexChangesBundle...
info at.clouddna.s4deployment › Running task generateComponentPreload...
info ProjectBuilder Build succeeded in 344 ms
info ProjectBuilder Executing cleanup tasks...

Confirmation is required to deploy the app:

    Application Name: ZS4DEPLOYDEMO
    Package: $TMP
    Transport Request:
    Destination: S4D
    SCP: false
    Target System SAPUI5 version: 1.108.0

Target system's SAPUI5 version is lower than the local minUI5Version. Testing
9/Latest/en-US/09171c8bc3a64ec7848f0ef31770a793.html

? Start deployment (Y/n)?

  › (Y/n)
```

Figure 16.20 Confirming the Deployment Start


```
✓ Start deployment (Y/n)?

 … yes
info abap-deploy-task ZS4DEPLOYDEMO Creating archive with UI5 build result.
info abap-deploy-task ZS4DEPLOYDEMO Archive created.
info abap-deploy-task ZS4DEPLOYDEMO Starting to deploy.
warn abap-deploy-task ZS4DEPLOYDEMO Deployment failed with authentication error.
info abap-deploy-task ZS4DEPLOYDEMO Please maintain correct credentials to avoid seeing this error
info abap-deploy-task ZS4DEPLOYDEMO     (see help: https://www.npmjs.com/package/@sap/ux-ui5-tooling#setting-environment-variables-in-a-env-file)
info abap-deploy-task ZS4DEPLOYDEMO Please enter your credentials.
✓ Username: …
✓ Password: …
info abap-deploy-task ZS4DEPLOYDEMO ZS4DEPLOYDEMO found on target system: false
info abap-deploy-task ZS4DEPLOYDEMO SAPUI5-Anwendung ZS4DEPLOYDEMO erfolgreich hochgeladen und registriert
info abap-deploy-task ZS4DEPLOYDEMO ***** Upload von SAPUI5-App oder -Bibliothek aus ZIP-Archiv in SAPUI5-ABAP-Repository *****
info abap-deploy-task ZS4DEPLOYDEMO Standardmodus aktiv: Kurzprotokoll
info abap-deploy-task ZS4DEPLOYDEMO 20 Dateien in Archiv gefunden.
info abap-deploy-task ZS4DEPLOYDEMO * Parameter *
info abap-deploy-task ZS4DEPLOYDEMO Die Textdateien werden mit Hilfe der Standardeinstellungen identifiziert.
info abap-deploy-task ZS4DEPLOYDEMO Die Binärdateien werden mit Hilfe der Standardeinstellungen identifiziert.
info abap-deploy-task ZS4DEPLOYDEMO Nicht zu berücksichtigende Dateien und Ordner werden mit Hilfe der eingebauten Standardeinstellungen identifiziert.
info abap-deploy-task ZS4DEPLOYDEMO Name des SAPUI5-Repository "ZS4DEPLOYDEMO" wurde aus dem entsprechenden Importparameter identifiziert.
info abap-deploy-task ZS4DEPLOYDEMO Beschreibung des SAPUI5-Repository wurde aus dem entsprechenden Importparameter identifiziert.
info abap-deploy-task ZS4DEPLOYDEMO Paket der SAPUI5-Anwendung "$TMP" wurde aus dem entsprechenden Importparameter identifiziert.
info abap-deploy-task ZS4DEPLOYDEMO Name der externen Codepage "UTF8" wurde aus dem entsprechenden Importparameter identifiziert.
info abap-deploy-task ZS4DEPLOYDEMO Akzeptanz Unix-artiger Zeilenendemarker in Textdateien wurde aus dem entsprechenden Importparameter identifiziert.
info abap-deploy-task ZS4DEPLOYDEMO Unix-artige Zeilenende-Marker in Textdateien werden akzeptiert.
info abap-deploy-task ZS4DEPLOYDEMO Deltamodus wurde aktiviert.
info abap-deploy-task ZS4DEPLOYDEMO Wird im sicheren Modus ausgeführt
info abap-deploy-task ZS4DEPLOYDEMO * Neues SAPUI5-ABAP-Repository ZS4DEPLOYDEMO wird angelegt *
info abap-deploy-task ZS4DEPLOYDEMO UPLOAD FILE   : Component-dbg.js (Text)
info abap-deploy-task ZS4DEPLOYDEMO UPLOAD FILE   : Component-preload.js (Text)
info abap-deploy-task ZS4DEPLOYDEMO UPLOAD FILE   : Component-preload.js.map (Text)
info abap-deploy-task ZS4DEPLOYDEMO UPLOAD FILE   : Component.js (Text)
info abap-deploy-task ZS4DEPLOYDEMO UPLOAD FILE   : Component.js.map (Text)
info abap-deploy-task ZS4DEPLOYDEMO UPLOAD FILE   : controller/App-dbg.controller.js (Text)
info abap-deploy-task ZS4DEPLOYDEMO ... 14  Nachrichten ausgelassen ...
info abap-deploy-task ZS4DEPLOYDEMO * Anwendungsindex wird aktualisiert *
info abap-deploy-task ZS4DEPLOYDEMO SAPUI5-Anwendung  erfolgreich hochgeladen und registriert
info abap-deploy-task ZS4DEPLOYDEMO * Erledigt *
info abap-deploy-task ZS4DEPLOYDEMO App available at http://s4d.virtual:8000/sap/bc/ui5_ui5/sap/zs4deploydemo
info abap-deploy-task ZS4DEPLOYDEMO Deployment Successful.
info abap-deploy-task ZS4DEPLOYDEMO (Note: As the destination is configured using an On-Premise SAP Cloud Connector, you will need to replace the host
```

Figure 16.21 Deployment Success Message

16.2 Automated Deployment with the SAP Continuous Integration and Delivery Service

In addition to manual deployment, there's also the option of performing an automated deployment. Therefore, SAP provides organizations with the SAP Continuous Integration and Delivery service in SAP BTP. The service is provided with different continuous integration/continuous delivery (CI/CD) pipelines for different use cases. The service is delivered exclusively in the Cloud Foundry environment.

To use the service, it must first be activated. In Chapter 2, we showed how to install SAP Business Application Studio using boosters. In the process, we also activated and configured the SAP Continuous Integration and Delivery service by using the booster.

Before we jump into deploying apps using the SAP Continuous Integration and Delivery service to a Cloud Foundry environment or an ABAP system, we first need to finish setting up the SAP Continuous Integration and Delivery service by assigning role collections.

16.2.1 Assign Role Collections

Open the SAP BTP cockpit of the subaccount in which the services have been activated, and navigate in the side menu, as shown in Figure 16.22, to the **Services • Instances and Subscriptions** area. You'll find the **Continuous Integration and Delivery Service** in the **Subscriptions** area.

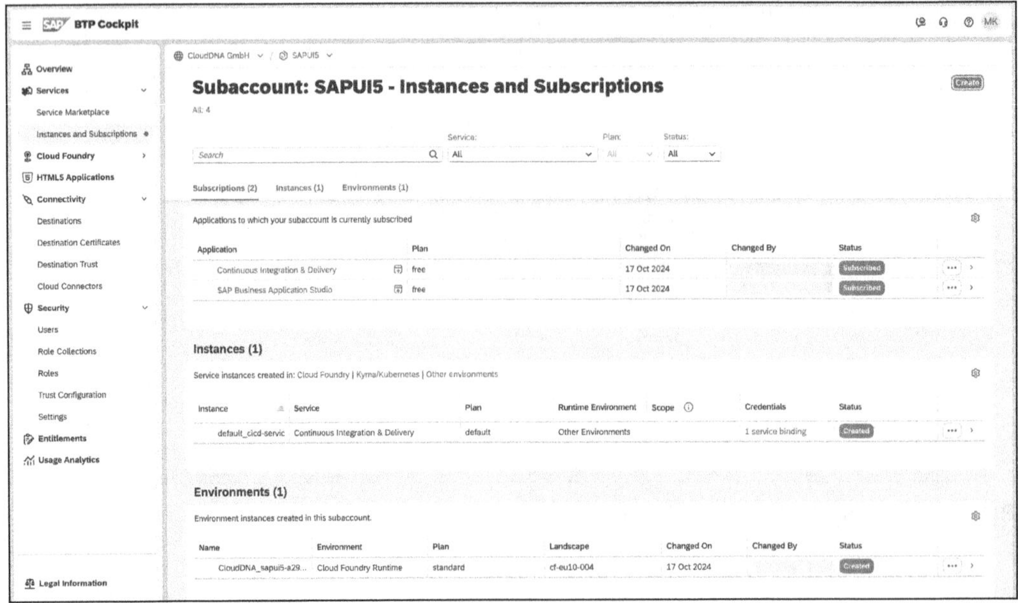

Figure 16.22 Subaccount Services and Instances Overview

The service comes with two role collections: one for a **CICD Service Administrator** and one for a **CICD Service Developer**. You must assign the desired role to the user based on their function. Navigate to the **Security • Role Collections** area in the side menu, as shown in Figure 16.23.

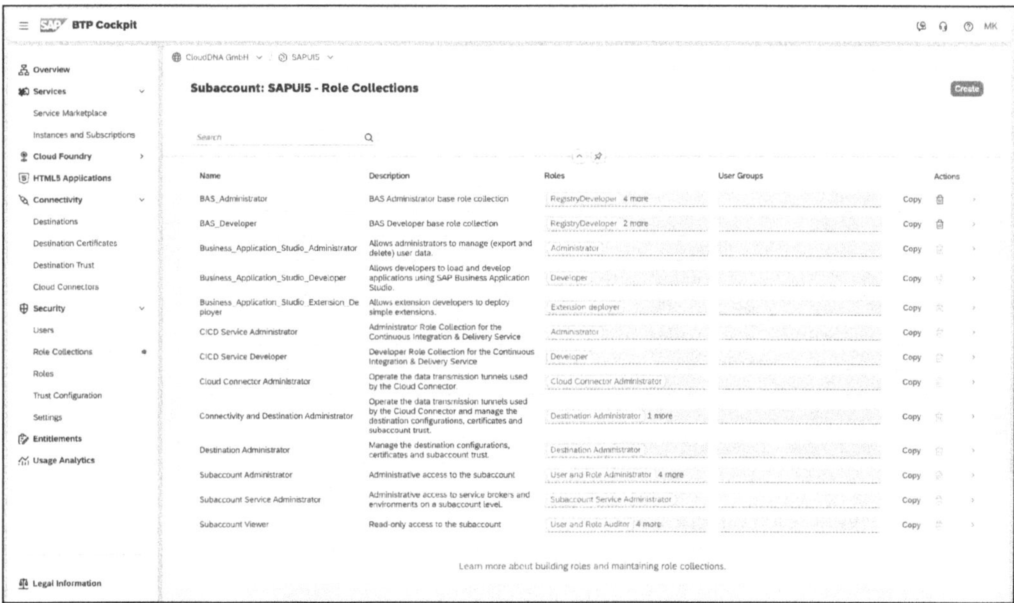

Figure 16.23 Continuous Integration and Delivery Role Collections

The differences between the administrator and developer role are shown in Table 16.1.

Category	Permission	Administrator	Developer
Credentials	Create	Yes	No
	Modify	Yes	No
	Delete	Yes	No
	View	Yes	Yes
Jobs	Create	Yes	No
	Modify	Yes	No
	Delete	Yes	No
	View	Yes	Yes

Table 16.1 Role Collection Permissions

Category	Permission	Administrator	Developer
Repositories	Create	Yes	No
	Modify	Yes	No
	Delete	Yes	No
	View	Yes	Yes
Builds	Trigger a build	Yes	Yes
	View a build	Yes	Yes
	Delete a build	Yes	Yes
	View a build log	Yes	Yes
Allowed spaces	View	Yes	No
	Modify	Yes	No

Table 16.1 Role Collection Permissions (Cont.)

To assign a role collection to a user, navigate to **Security • Role Collections** in the side menu as usual, and select the desired role in the details area. Figure 16.24 shows this using the **CICD Service Administrator** role collection as an example. Then, click the **Edit** button and enter the email address of the desired user in the **Users** area.

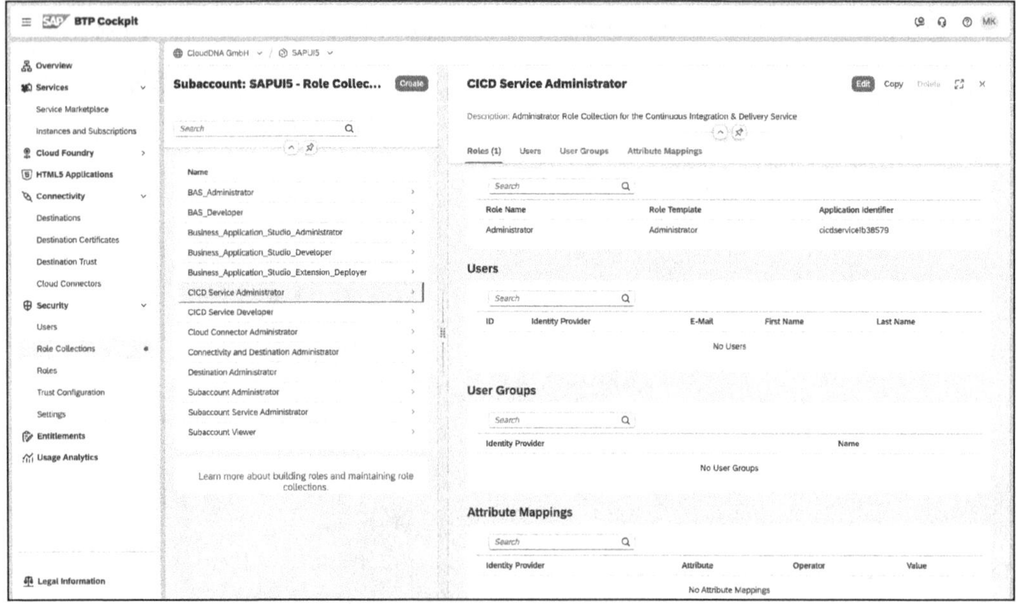

Figure 16.24 Assigning the CICD Service Administrator Role Collection

As shown in Figure 16.25, don't forget that you have to click on the **Save** button after you've made changes.

We generally recommend that you never assign users directly to role collections and instead derive the role collections from user groups in the underlying identity provider, such as SAP Cloud Identity Services or Microsoft Entra ID.

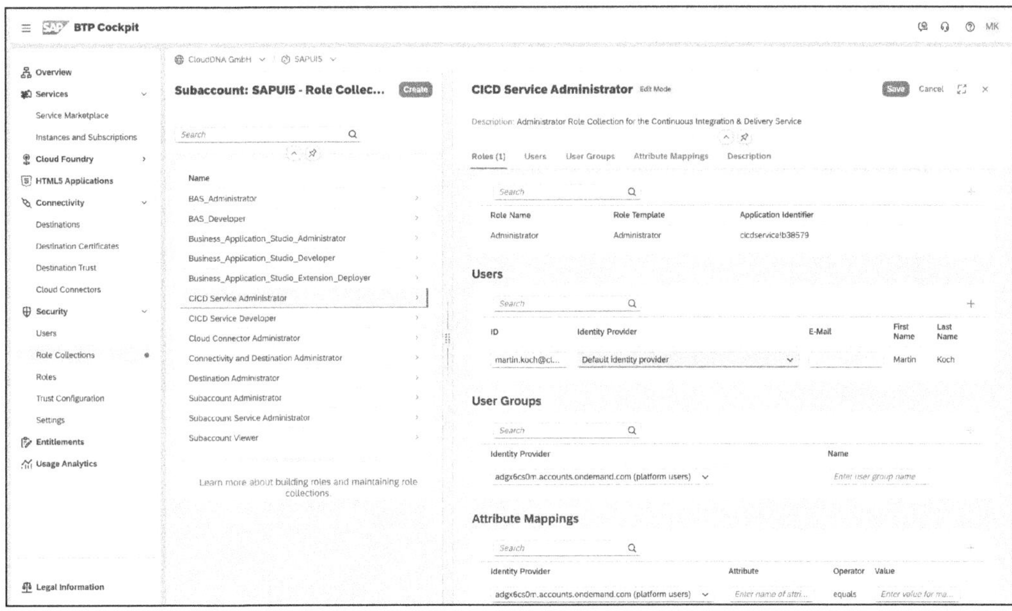

Figure 16.25 CICD Service Administrator Role Collection Edit Mode

Then, open the SAP Continuous Integration and Delivery service by clicking on the service link in the subscriptions area, as shown earlier in Figure 16.22. You should then see the screen shown in Figure 16.26.

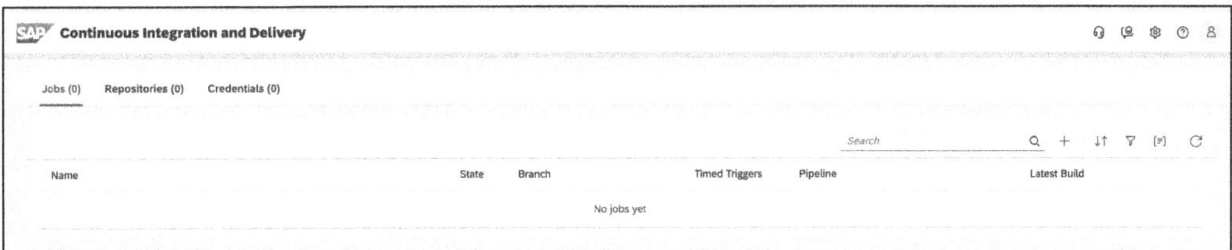

Figure 16.26 Opening the SAP Continuous Integration and Delivery Service

16.2.2 Cloud Foundry Deployment

As already mentioned, you have the option of deploying to the Cloud Foundry environment. The necessary procedure is described in this section.

You can trigger a build manually through the service user interface or automatically in response to changes in your source code repository. To automate your pipeline and initiate a build with every commit, you must set up a webhook between SAP Continuous Integration and Delivery and your source repository. This webhook sends a push event whenever changes are pushed, triggering a build for the associated job. We'll focus on triggering the build via a webhook, as this is the more challenging way from a configuration perspective. In our scenario, we work with a Git repository hosted on GitHub. Therefore, it's necessary to create credentials for both the webhook and the Git repository.

Let's get started with the credential for the webhook. Open the **Credentials** section, as shown in Figure 16.27, and click the **+** button to create a new credential. Use a meaningful **Name** and select **Webhook Secret** as **Type**. Click on the **Generate** button beside the **Secret** field to create the secret. Note that the secret is only visible during this step. It's not possible to receive the secret at any other point in time. Finally, click the **Create** button in the footer of the **Create Credentials** section.

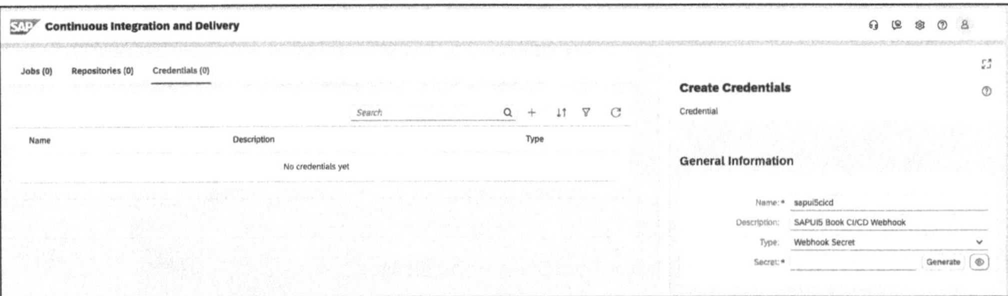

Figure 16.27 Creating Webhook Credentials

In the next step, you'll create the credentials for accessing the Git repository. To do this, click the **+** button again in the **Credentials** section. Assign a meaningful name, and select **Basic Authentication** as the **Type** (see Figure 16.28). Enter the desired username in the **Username** field. It's important to note that GitHub no longer supports basic authentication and instead uses access tokens, which provide more options for managing permissions. We'll subsequently explain how to create an access token.

To create an access token, you must open the developer settings for your user in GitHub. Then navigate to **Personal access tokens • Tokens (classic)** in the side menu and click on **Generate new token** (see Figure 16.29) on the details screen. Select the **Generate new token (classic)** option in the popover.

Enter a meaningful description for the use of the access token in the **Note** field (see Figure 16.30). Select how long the token should be valid (in days) in the **Expiration** field, and select the **repo** checkbox in the **Select scopes** table. Save the access token.

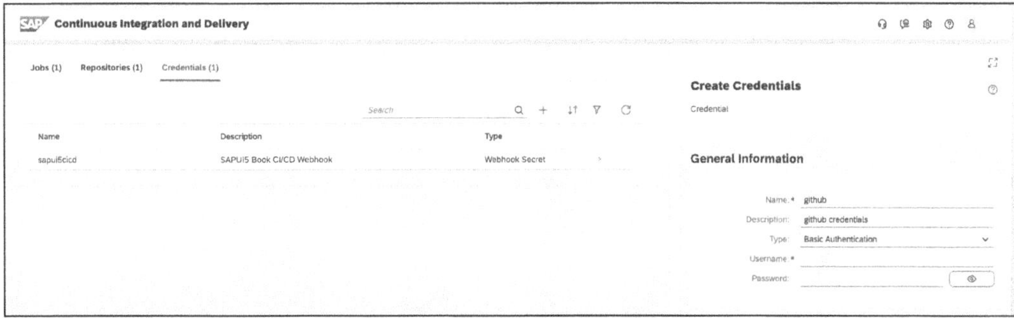

Figure 16.28 Adding GitHub Credentials

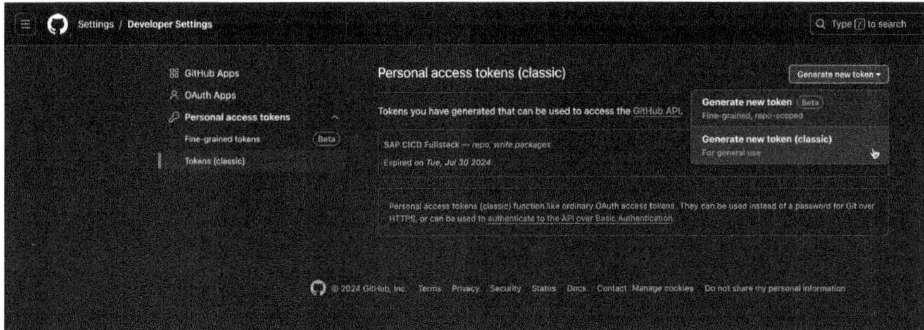

Figure 16.29 Creating an Access Token

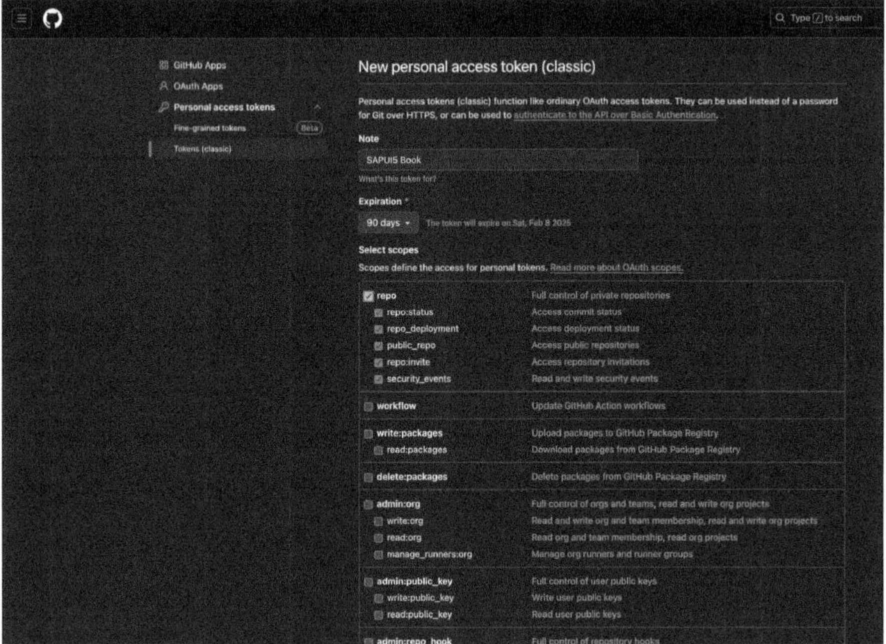

Figure 16.30 Configuring the Scope

After saving the new personal access token, you'll be presented with the access token, as shown in Figure 16.31. Copy the access token, and enter it in the **Password** field in the GitHub credentials (Figure 16.28).

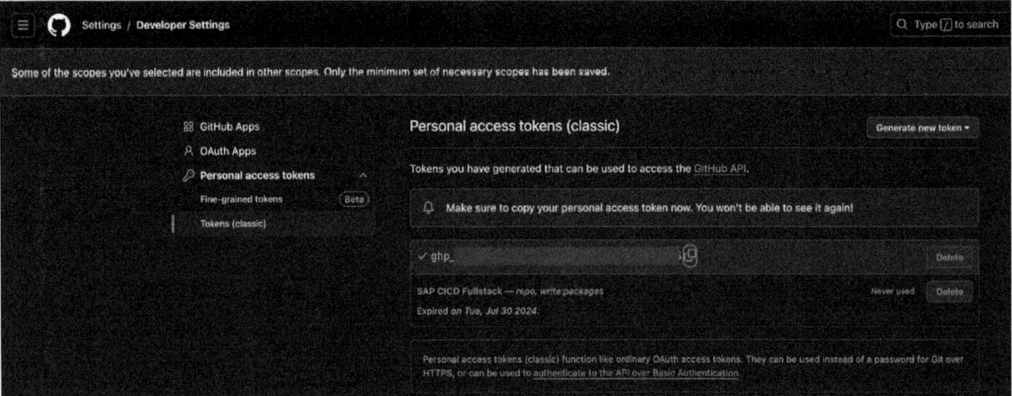

Figure 16.31 Copying the Access Token

> **Warning**
> You can't retrieve the access token at any other time, so be sure to note or copy it carefully.

After the build has been successfully executed, the generated build artifact should usually be deployed, in our case, to a subaccount with an activated SAP BTP, Cloud Foundry runtime. To do this, a credential must be created. Proceed identically to the other two credentials. Enter a descriptive **Name**, and select **Basic Authentication**. Enter the **Username** and the corresponding **Password** (see Figure 16.32).

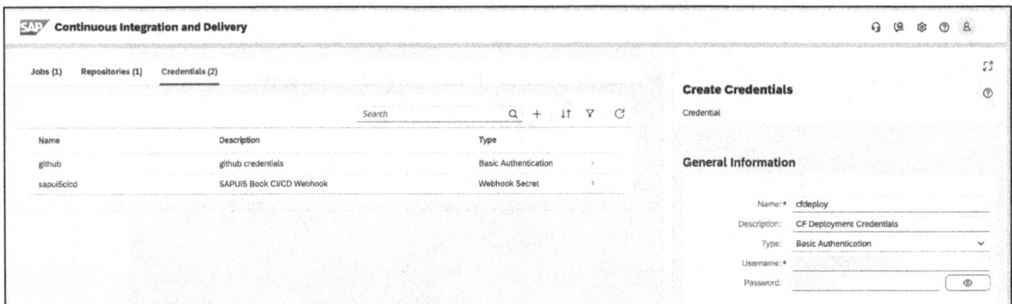

Figure 16.32 Adding Cloud Foundry Deployment Credentials

The next step is to configure the repository from which the source code is to be pulled. To do this, open the **Repositories** tab, as shown in Figure 16.33, and then click the **+** button. Enter a **name** for the repository, and enter the URL of the Git repository in the

Clone URL field. Select the previously created Git credentials as the value for the **Credential** field.

In the **WEBHOOK EVENT RECEIVER** area, select the **GitHub** type, and select the previously created webhook credential in the **Webhook Credential** field. Then, click the **Save** button.

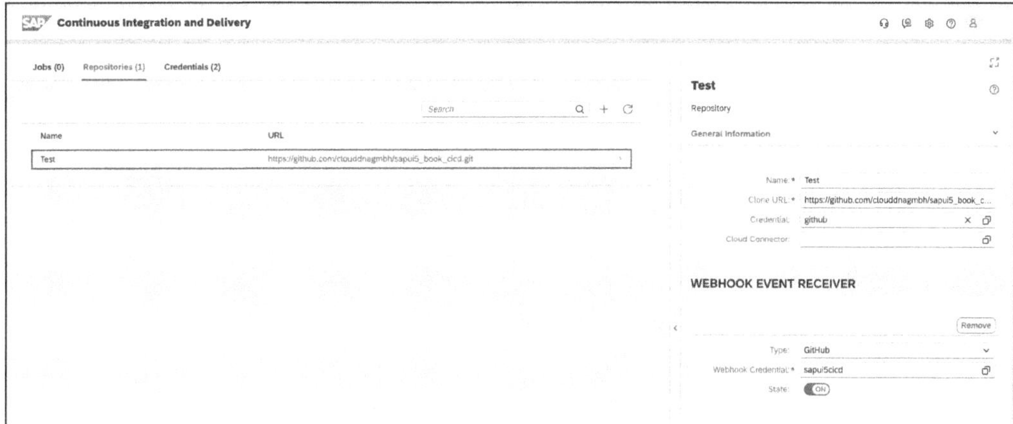

Figure 16.33 Configuring the Repository

You can now create a job. To do this, open the **Jobs** tab, as shown in Figure 16.34, and then click on the **+** button.

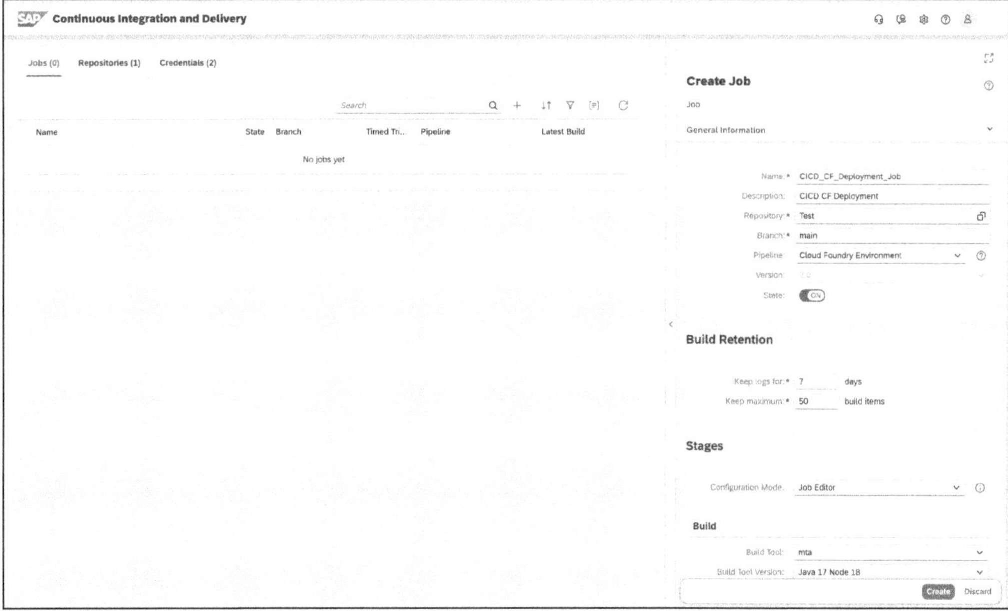

Figure 16.34 Configuroing the Job: General Information

Configure the job in the **General Information** area. To do this, select the GitHub repository you created earlier as the **Repository**, and enter the name of the branch you want to clone in the **Branch** field. For the **Pipeline** field, select the **Cloud Foundry Environment** entry.

You now only need to configure the deployment target. To do this, scroll to the **Acceptance** area, as shown in Figure 16.35, and activate the corresponding switch. You now need to specify the **API Endpoint**, the **Org Name**, and the **Space**. The easiest way to get this information is from the desired subaccount. To do this, open the SAP BTP cockpit of the subaccount, as shown in Figure 16.36, and navigate to the **Overview** area in the side menu. You can find the information you need there. Then, enter it in the corresponding fields in the **Job** configuration (see Figure 16.35). Select **Standard** as **Deploy Type**, and select the **cfdeploy** you created previously for the **Credentials** attribute.

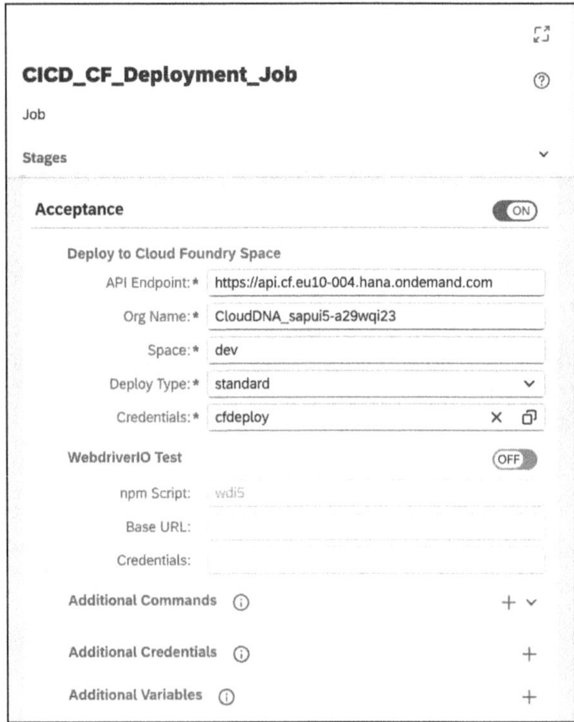

Figure 16.35 Configuring the Job: Add Acceptance Stage

You've now successfully completed the configuration of the job, so you can test it very easily. Make a change in the desired project, and push the changes into the Git repository. This triggers the job via the webhook. Then, open the associated job in the SAP Continuous Integration and Delivery service, and open the **Builds** tab (see Figure 16.37). There, you can see the individual stages of the build. In the example shown, you can see that the build, including deployment, was successful.

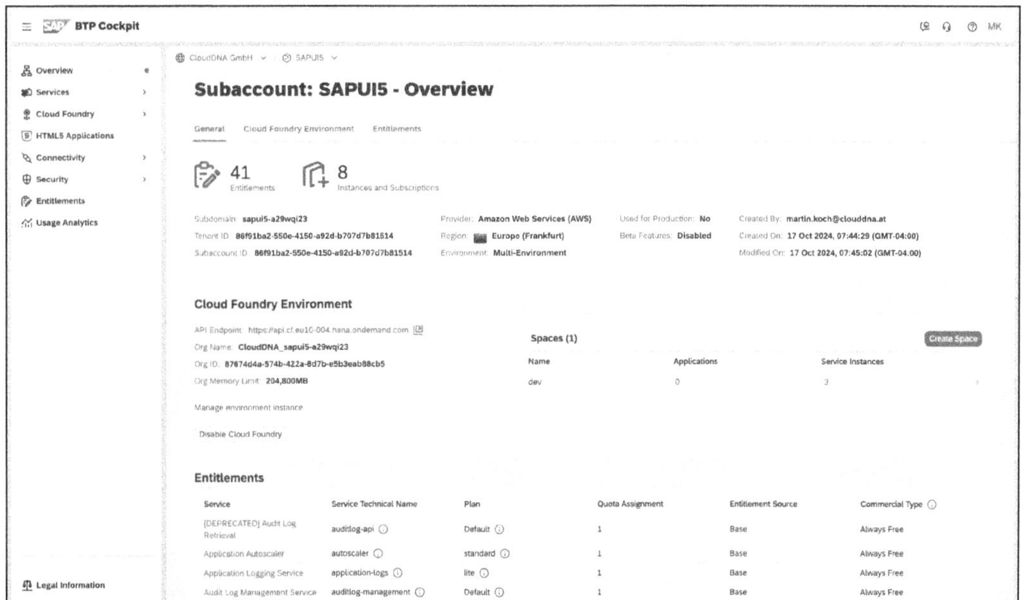

Figure 16.36 Retrieve Cloud Foundry Deployment Information

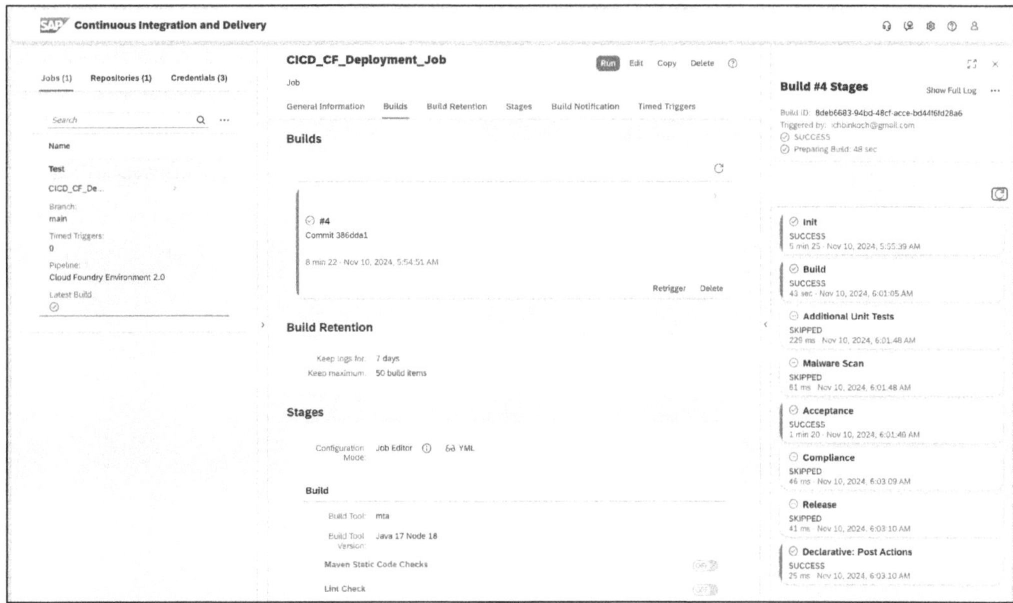

Figure 16.37 Checking the Build Status

You can then check in the subaccount to see if the app has been deployed. To do this, open the SAP BTP cockpit for the subaccount, and navigate to the **HTML5 Applications** area, as shown in Figure 16.38. You'll see the deployed SAPUI5 app there.

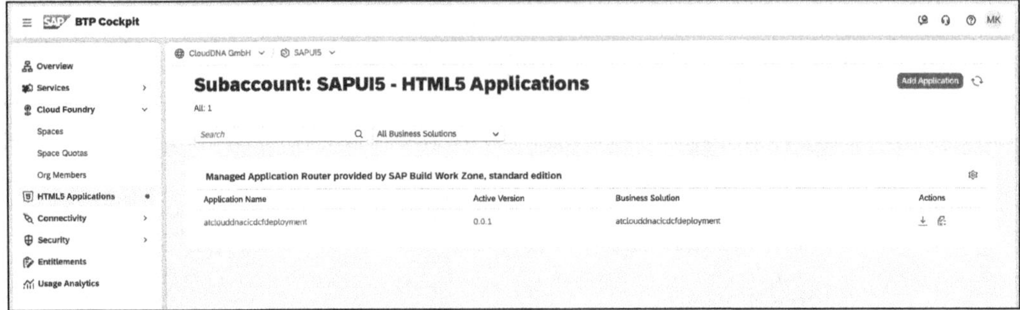

Figure 16.38 Deployed HTML5 Application

16.2.3 ABAP Deployment

After learning how to deploy to the SAP BTP, Cloud Foundry runtime in the previous section, this section will teach you how to deploy an SAPUI5 app to an ABAP system. It doesn't matter whether you're deploying to a modern SAP S/4HANA system or to an older SAP Business Suite system. The basic configurations, such as setting up the Git credentials and the webhook, won't be covered again here, as they are identical to the Cloud Foundry deployment.

To deploy the app to an ABAP system, you need to create the necessary ABAP system credentials. This requires a user that has the necessary authorizations for the deployment. In the SAP Continuous Integration and Delivery service, go to the **Credentials** area, and click the **+** button to create a new credential.

Enter a descriptive name, as shown in Figure 16.39, and select the value **Basic Authentication** for the **Type** attribute. Then, enter the username of the desired user in the **Username** field and the corresponding password in the **Password** field. Once all fields have been filled in, save your entries by clicking the **Create** button.

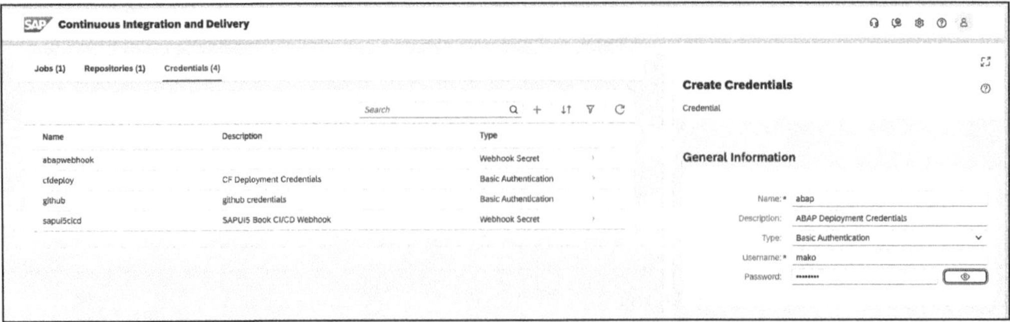

Figure 16.39 Adding the ABAP Credential

You now need to create a credential of **Type Cloud Connector** (see Figure 16.40). This is mandatory for an on-premise deployment. If necessary, enter the **Location ID** of the cloud connector connection.

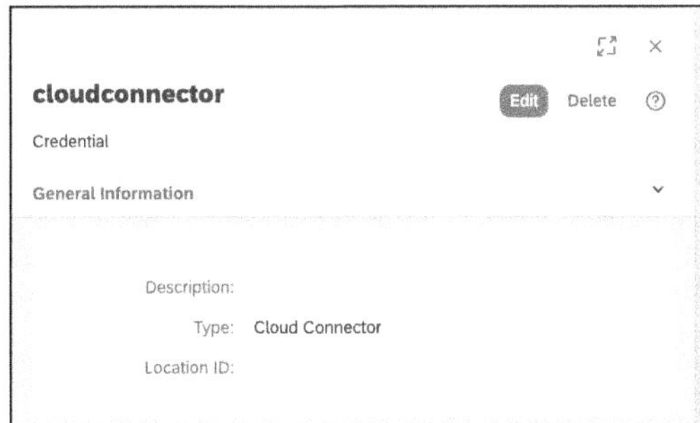

Figure 16.40 Adding the Cloud Connector Credential

In the next step, you have to create a connection to the Git repository, identical to the Cloud Foundry deployment. To do this, navigate to the **Repositories** area (see Figure 16.41) and create the connection, including the **Webhook Event Receiver**, as before.

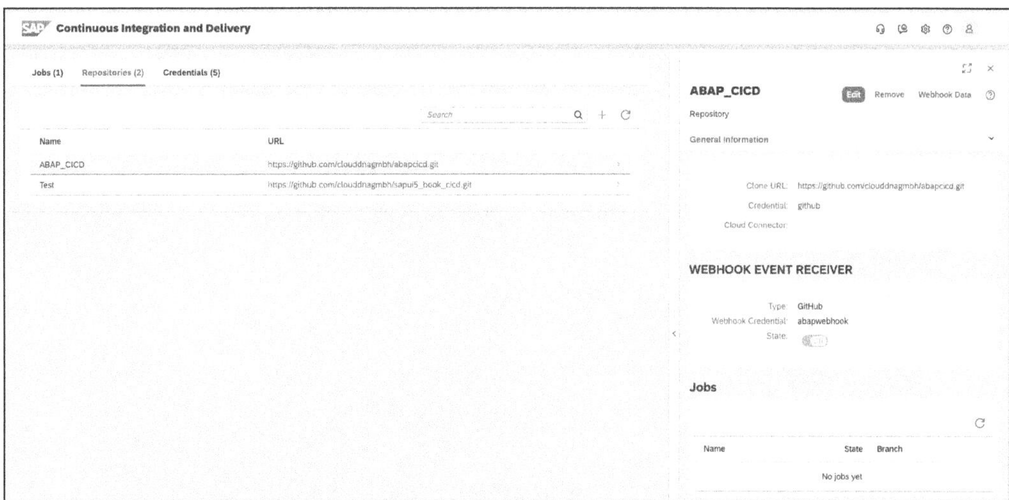

Figure 16.41 Adding the ABAP Git Repository

In the last step, you must create a job, as shown in Figure 16.42. To do this, navigate to the **Jobs** area, and click the **+** button. Select the **Repository** you created earlier, and select the value **SAP Fiori for ABAP platform** for the **Pipeline** attribute. Make sure that the **State** is set to **ON**.

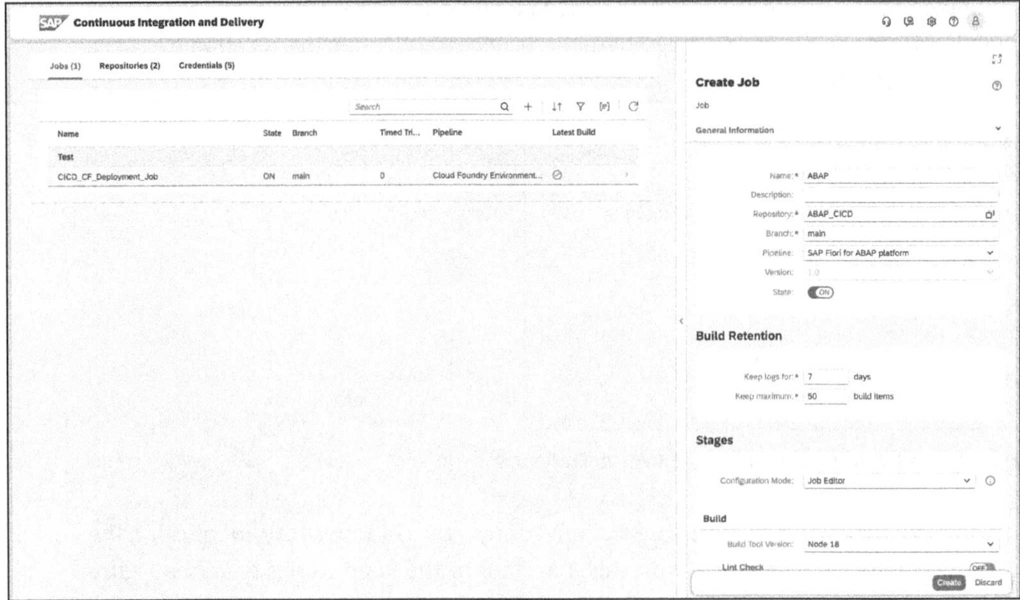

Figure 16.42 Job Configuration: General Information

Now scroll to the **Stages** area, as shown in Figure 16.43, and then to the **Release Stage** in particular. Set the corresponding switch to **ON** to activate the deployment in this stage. Then, maintain the following attributes:

- **Upload Credentials**
  The previously created credentials of the ABAP system.
- **Cloud Connector**
  The previously created cloud connector.
- **ABAP Platform Endpoint**
  Virtual host and port of the ABAP system, as maintained in the cloud connector in the mapping.
- **ABAP Package**
  Name of the ABAP package in which the app is to be added.
- **Application Name**
  Name of the BSP application that is to be created and is later available as a node in the Internet Communication Framework (ICF).
- **Configuration Mode of Transport Request ID**
  Here you can select **Job Editor**.
- **Transport Request ID**
  Number of the transport request.

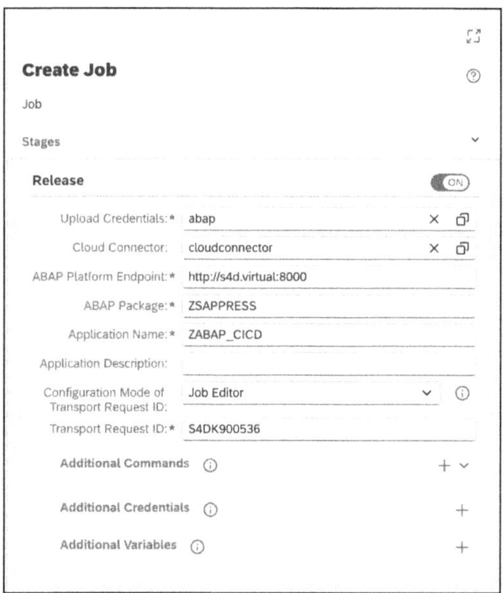

Figure 16.43 Job Configuration: Release

You can now test the job by pushing to the corresponding Git repository on the branch configured in the job. After that, the webhook will trigger the job. Then, open the job, and navigate to the **Builds** area, as shown in Figure 16.44. There, you can see the status of the build and deployment. In case of an error, you'll usually find a self-explanatory error messages.

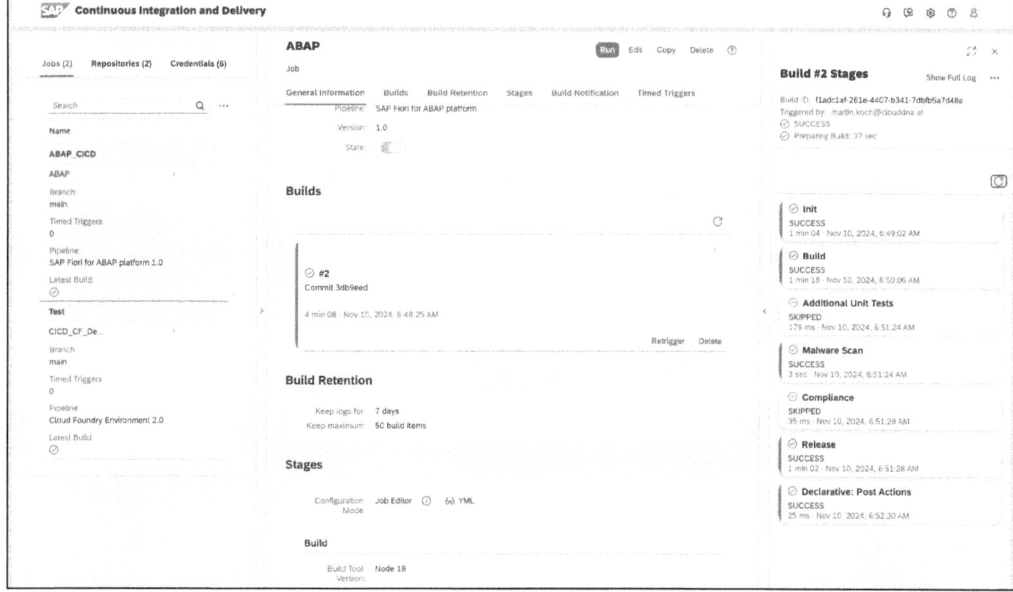

Figure 16.44 Build Status

16.3 Summary

In Section 16.1, the process of manually deploying SAPUI5 applications was explained step-by-step. This approach enables developers to transfer applications to a target environment in a targeted and controlled manner, whether in the cloud via SAP BTP or on-premise in an SAP S/4HANA system. It was also pointed out that the deployment steps for SAP S/4HANA and SAP NetWeaver AS ABAP are largely identical.

Section 16.2 was dedicated to automated deployment using the SAP Continuous Integration and Delivery service. This approach was highlighted as the key to efficient and continuous integration and deployment, optimizing deployment processes and supporting quality assurance.

Throughout the chapter, the importance of choosing between manual and automated deployment in relation to project-specific requirements was emphasized, with the specific advantages and use cases of both methods described in detail.

An application can either be used directly as a standalone application or as described in the next chapter, integrated into the SAP Fiori launchpad. The latter creates a central and user-friendly interface for end users. With the content covered in this chapter, you're well prepared to successfully deploy and integrate SAPUI5 applications into your IT landscape.

Chapter 17
SAP Fiori Launchpad Configuration

The SAP Fiori launchpad is the central entry point for SAP applications, providing a personalized, role-based, and unified user experience that significantly increases productivity and adoption in organizations.

The *SAP Fiori launchpad* is a central component of the SAP ecosystem that provides a unified and consistent user experience (UX) across different enterprise solutions. It uses the capabilities of user roles to combine specialized apps in one interface, greatly simplifying access to the applications and data needed. Each app is designed for a specific step and role, and combining several apps makes it possible to map complete business processes that often span multiple enterprise solutions and systems. This development marks the transition from monolithic software solutions to flexible, activity-based apps and represents a fundamental simplification and optimization of business processes.

The SAP Fiori launchpad was originally developed for the on-premise landscape. It was designed to serve as a central entry point for users within classic SAP environments. There are some basic concepts such as groups, catalogs, spaces, and pages. We'll discuss these in Section 17.1.

The configuration of the SAP Fiori launchpad in an on-premise SAP S/4HANA system will be covered in Section 17.2. With the further development of IT landscapes and the increasing shift to the cloud, the SAP Fiori launchpad was later made available in SAP Business Technology Platform (SAP BTP) to meet the requirements of modern, hybrid IT architectures.

In SAP BTP, a launchpad was initially provided as a standalone SAP Launchpad service. This service offered companies the opportunity to create a centralized and personalized UX for their cloud-based applications and services, while retaining the familiar features of the SAP Fiori launchpad. The SAP Launchpad service made it possible to seamlessly integrate cloud and on-premise apps into a single interface, bridging the gap between classic and modern architectures.

With the further development of SAP BTP, the SAP Launchpad service was eventually replaced by *SAP Build Work Zone, standard edition*. This further development expands the concept of the launchpad to include additional collaboration and personalization features. In addition to the usual app integration, SAP Build Work Zone, standard edition, also offers options for integrating widgets, dashboards, and other useful tools that further optimize the UX. This makes SAP Build Work Zone, standard edition, a central hub that not only focuses on apps but also promotes collaboration and access to relevant information. The launchpad configuration for SAP Build Work Zone, standard edition, will be covered in Section 17.3.

17.1 General Concepts

The most important artifact in the SAP Fiori launchpad is the *tile*. A tile serves as the visual representation of an app within the SAP Fiori launchpad, making it an essential component for intuitive navigation. The tile definition specifies how the tile appears on the launchpad's homepage or in the app finder. It encompasses various design elements such as texts, icons, and, optionally, dynamic information. This dynamic information can include real-time data such as counts, percentages, or aggregated metrics—for example, "85% budget used" or "5 open tasks." These features help users quickly assess relevant information without needing to open the app.

Tiles can also be configured as links instead of full visual elements. Links provide a streamlined alternative, offering direct access to the app while omitting additional details that a traditional tile might display. This can be particularly useful for less critical apps or scenarios where minimal screen space is desired.

There are various types of tiles available, each serving different purposes, as follows:

- *Static tiles* provide a fixed visual representation, displaying consistent information such as app names or icons, without any real-time updates.
- *Dynamic tiles* offer a more interactive experience by displaying real-time or periodically updated data, such as key performance indicators (KPIs), counts, or other aggregated metrics.
- *News tiles* are designed to highlight messages or announcements, making them ideal for showcasing updates, company news, or critical alerts.
- *Custom tiles*, on the other hand, allow customers to tailor the UX by developing tiles to meet their specific requirements. These tiles can be programmed to include unique functionality or integrate with custom data sources, providing flexibility and adaptability to business needs.

Intent-based navigation in the SAP Fiori launchpad provides a powerful mechanism to ensure that users can access the appropriate application views or modes based on their specific roles and requirements. This approach enables dynamic and context-sensitive navigation, abstracting the underlying technical details while ensuring flexibility and adaptability in application interaction.

The SAP Fiori launchpad uses *intents* to resolve navigation targets rather than directly encoding the technical details of an application into the URL hash. Each application in the launchpad has a resource locator (URL) to load it. However, instead of hardcoding these technical identifiers, intents introduce a layer of abstraction, allowing seamless navigation between applications.

An intent is essentially an abstraction layer that defines what action a user wants to perform on a business entity (semantic object). This decouples the navigation process from the UI technology or the specific technical implementation, offering greater flexibility and user-centric design.

Key use cases for intent-based navigation are as follows:

- **Role-based navigation**
  Depending on the end user's role, the same application can be launched in different views or modes. For example, a sales manager might access an analytics dashboard, while a salesperson views transactional details within the same application.
- **Scenario extensions and customization**
  Intents allow modifications to navigation targets without altering the underlying application code, enabling easy extension and customization of SAP Fiori scenarios.
- **App lifecycle independence**
  Applications have different deployment and lifecycle phases. With intent-based navigation, one app can operate independently of others, ensuring stability even if certain apps are unavailable in the production environment.
- **Smooth transition between apps**
  Organizations can seamlessly integrate legacy applications and later migrate to SAPUI5 apps without impacting the UX. The use of intents ensures users remain unaware of underlying technical changes.

An intent is constructed using *semantic objects*, *actions*, and *semantic object parameters*. A semantic object represents a business entity such as a customer, sales order, or product. These objects group applications into scenarios by referring to entities in a standardized, implementation-independent way. SAP provides predefined semantic objects, but custom objects can also be created to meet specific requirements. For example, a semantic object could be a sales order or product. The action defines the operation to be performed on the semantic object. Examples include display, approve, or edit. For instance, `sales order-display` could refer to displaying a sales order's details. Semantic object parameters provide additional details to specify the instance of the semantic object. For example:

- Semantic object: `Employee`
- Action: `display`
- Parameter: `EmployeeID=12345`

The syntax for intents is `#<semantic object>-<action>?<semantic object parameter>=<value>`. For the example shown, it would be `#Employee-display?EmployeeID=12345`.

Target mapping serves as the bridge between a defined intent and the actual navigation target. It links the intent, comprising a semantic object and an action, to the specific application or resource that should be launched. This connection is essential for enabling navigation within the SAP Fiori launchpad. Target mapping acts as a prerequisite for navigation to a target application. Without it, the launchpad can't resolve an intent to its corresponding application, preventing successful navigation. It defines not only what the user intends to do but also where that action should lead within the system.

To grant launchpad users access to applications, administrators must configure several key entities. Figure 17.1 illustrates an overview of these entities and their relationships. The arrows in the diagram represent the direction in which the following entities are assigned and connected:

- *Business roles* determine the authorizations required to access the SAP Fiori launchpad and its content. These roles are assigned to users, granting them the necessary permissions to run the launchpad and interact with specific applications and features.
- A *space* is a structural unit within the SAP Fiori launchpad that contains one or more pages. Users can access multiple spaces, which are displayed in the launchpad's navigation bar, enabling seamless switching between different spaces. Each space is designed to organize related content and applications, providing a clear and user-friendly way to navigate specific business areas or workflows.
- A *page* is a component within a space that organizes apps into various sections, grouping related applications for easier navigation. Pages are displayed in the main area of the SAP Fiori launchpad, providing users with a structured and intuitive interface to access the applications and information relevant to their tasks.
- A *group* defines how applications are organized and displayed on the launchpad homepage. It determines the grouping, sort order, and appearance (as tiles or links) of a subset of apps visible to users. Groups consist of tiles and links, offering a curated selection of applications for specific roles or tasks. Users with the appropriate roles can view the group and its contained apps on their launchpad homepage. Additionally, users can personalize their experience by adding or removing apps from predefined groups or creating their own custom groups, ensuring the launchpad meets their individual needs and preferences.
- A *catalog* is a curated collection of applications associated with specific roles or multiple roles. Users assigned a particular role gain access to all the apps within the catalogs linked to that role. Catalogs serve as the smallest unit for defining and managing the apps available to users. They streamline app selection and authorization processes, ensuring that users can access only the applications relevant to their responsibilities while maintaining system security and efficiency.

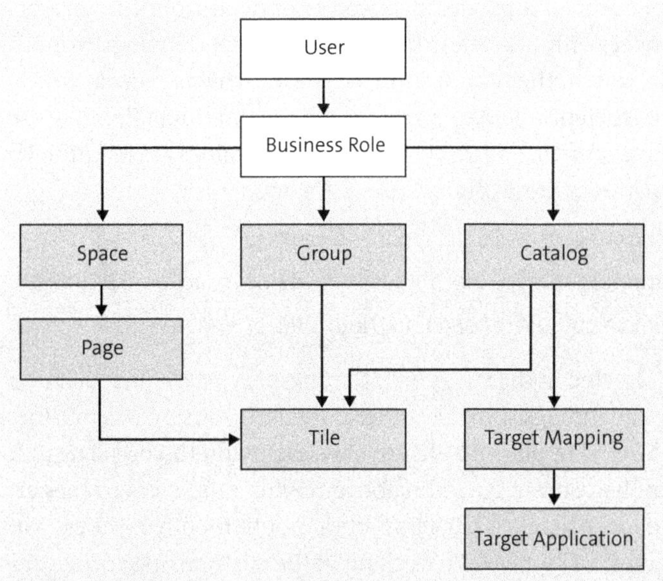

Figure 17.1 SAP Fiori Launchpad Content

SAP distinguishes between *technical catalogs* and *business catalogs*, each serving a specific purpose in managing app assignments and navigation within the SAP Fiori launchpad.

17

Technical catalogs are the source repositories that contain the original definitions of tiles and target mappings. They are designed to store all available app configurations in a centralized and structured manner. Key characteristics of technical catalogs are as follows:

- Serve as a reference for creating business catalogs
- Not intended for direct assignment to end users
- Provide flexibility and reusability by allowing tiles and target mappings to be referenced in multiple business catalogs

You can leverage predefined catalogs that SAP delivers to simplify implementation. SAP-provided technical catalogs are identified by the naming convention SAP_TC_<...>.

Custom technical catalogs, developed in the configuration scope, are primarily intended for customer-specific apps. It's recommended to rely on SAP-delivered technical catalogs for SAP apps, adapting them only when necessary. Adjustments to SAP-delivered content can be made using the adaptation mode for minor changes, such as modifying titles or descriptions. For more significant alterations, items can be copied to a custom technical catalog and modified as required.

Business catalogs contain a selection of tiles and target mappings tailored for a specific business role. They are designed to meet the functional requirements of individual business users. The content of a business catalog is derived from the technical catalog. Administrators select only the relevant apps and target mappings from the technical catalog to create a streamlined collection for specific roles or scenarios. This layered approach ensures efficient role-based access while maintaining flexibility for app configuration and management. SAP-provided business catalogs are identified by SAP_BC_<...>. Key characteristics of these catalogs are as follows:

- Represent a subset of the content from the technical catalogs
- Include only the tiles and target mappings relevant to the user's responsibilities
- Intended for assignment to end users based on their roles

Custom business catalogs, created in the customizing scope, provide client-specific collections of tiles and target mappings. Typically, these are derived from SAP business catalogs, which serve as a starting point to ensure consistent app-to-app navigation. Administrators can refine the content by adding or removing references to tiles and target mappings. To avoid inconsistencies, it's important not to modify business catalogs created in the configuration scope while working in the customizing scope. Additionally, maintaining tiles and their corresponding target mappings within the same business catalog ensures navigation consistency.

In some cases, apps such as SAP GUI for HTML applications aren't included in any business catalog and are only available in technical catalogs. These apps need to be manually added to custom business catalogs. To optimize performance, it's advisable to limit the number of tiles and target mappings within a catalog. By following these principles, organizations can efficiently structure their SAP Fiori launchpad catalogs to meet business needs while maintaining flexibility and system performance.

17.2 On-Premise Configuration for SAP S/4HANA

In an on-premise landscape, the SAP Fiori launchpad has become the central entry point with the introduction of SAP S/4HANA. Over time, it's expected to replace the traditional SAP GUI as the primary interface. Therefore, understanding how to configure the SAP Fiori launchpad on an SAP S/4HANA system in an on-premise environment is critical.

In the previous section, you became familiar with the core elements of the launchpad, such as spaces, pages, business roles, groups, catalogs, target mappings, and tiles. In this section, we'll guide you through the configuration process required to set up these components effectively.

To facilitate this, SAP provides various tools to streamline and simplify the configuration process. These tools allow administrators to create and manage the structural components of the launchpad, ensuring a user-friendly and role-specific interface. By leveraging these tools, you can tailor the SAP Fiori launchpad to align with your organization's requirements and pave the way for a smooth transition from SAP GUI to a modern and efficient UX.

We'll follow this sequence for this task:

1. Create a technical catalog.
2. Add tiles and create target mappings.
3. Create a business catalog.
4. Assign tiles to the business catalog.
5. Create a PFCG role, and assign business catalog.
6. Create spaces and pages.
7. Add spaces to the PFCG role.
8. Define page content and structure.
9. Test.

Use the Launchpad App Manager app to create your own technical catalogs within the cross-client scope CONF. This allows you to define custom tiles and target mappings to represent specific content that extends beyond the functionalities of standard SAP Fiori apps.

In addition, you can customize *launchpad app descriptor items* from SAP's technical catalogs directly. This approach is particularly useful for making minor adjustments, saving both time and effort. It's generally recommended to explore this customization option first before developing your own content or technical catalogs.

To get started, open the Launchpad App Manager app. You can access it via SAP GUI using Transaction /UI2/FLPAM, or alternatively, through the tile depicted in Figure 17.2 within the SAP Fiori launchpad.

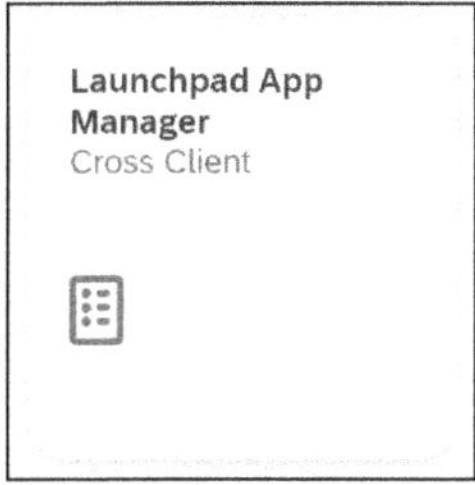

Figure 17.2 SAP Fiori: Launchpad App Manager Tile

In the **SAP Fiori Launchpad Application Manager** page, as shown in Figure 17.3, you'll find an overview of all technical catalogs. This includes both the catalogs delivered by SAP as part of the standard and those you've created yourself. Each catalog is displayed with its **Technical Catalog ID**, the **Technical Catalog Title**, and other metadata to help you distinguish between the different catalogs available in your system.

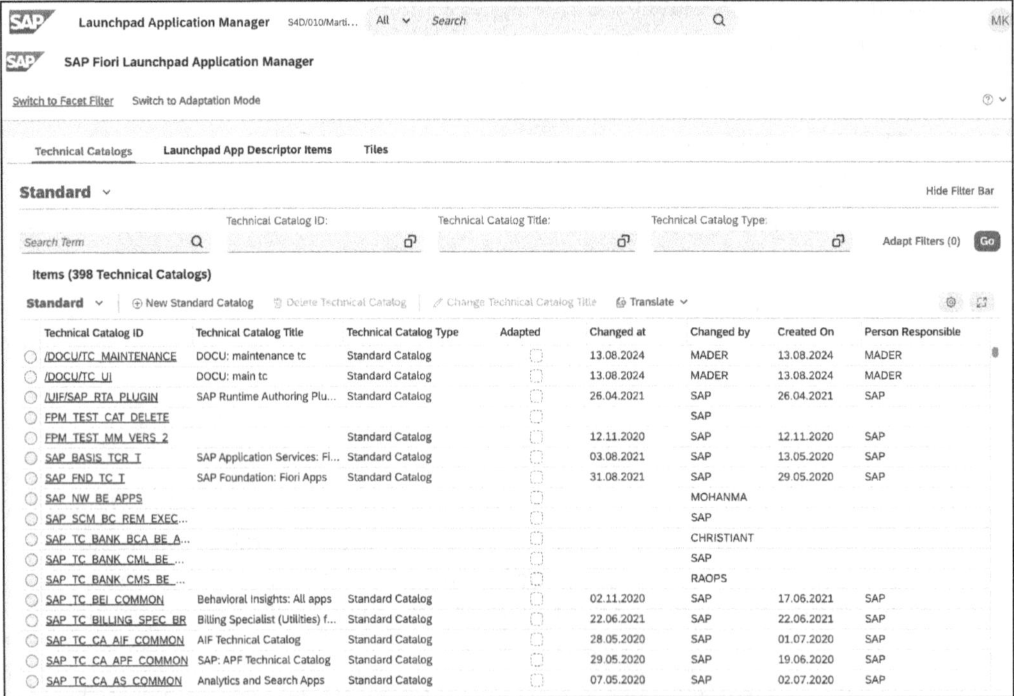

Figure 17.3 Launchpad Application Manager Overview

To create a new catalog, click on **+ New Standard Catalog**. This action will open a dialog shown in Figure 17.4 where you can define the basic properties of the catalog, such as the **Technical Catalog ID** and the **Technical Catalog Title**. You can also assign a **Package** and a **Transport Request**, which allows you to transport the catalog to the next stage in your system landscape. Finally, click the **Save** button.

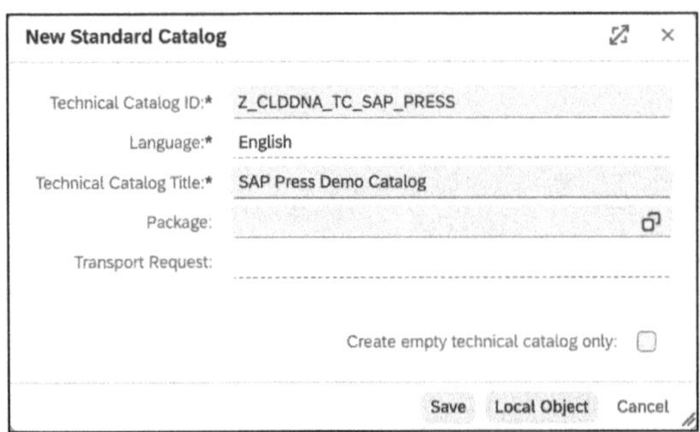

Figure 17.4 New Catalog Dialog

You can now add apps to the catalog. To do so, click on **Add App**, as shown in Figure 17.5, and then select **SAPUI5 Fiori App**. Before adding the app, ensure that it has been successfully deployed to the system. Without a properly deployed app, it won't be available for selection in the catalog.

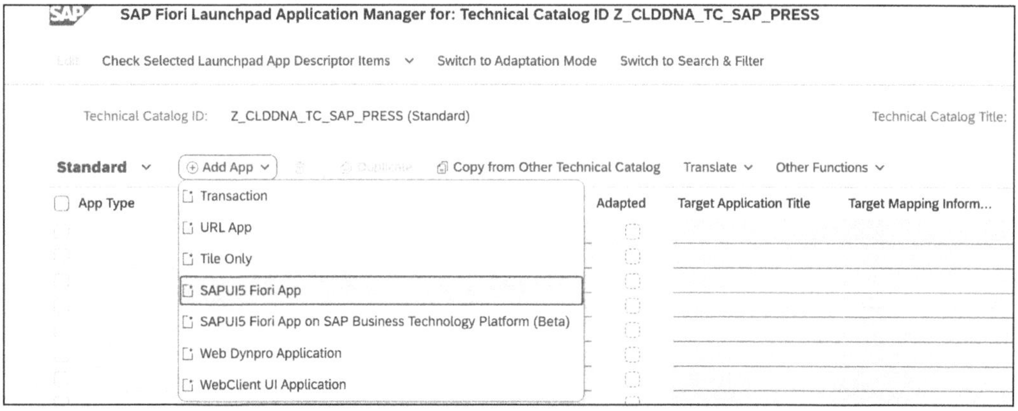

Figure 17.5 Add SAPUI5 Fiori App to the Catalog

In the first step, you must maintain the basic data of the app in the **Target Application Fields** tab page. The **App Type** is already preselected, as shown in Figure 17.6. Assign a **Semantic Object** and an **Action** to clearly define the target assignment of the app. If no suitable semantic objects are available, you can create your own using Transaction /UI2/SEMOBJ. Alternatively, you can use the semantic objects delivered by SAP. This saves customizing effort and makes it easier to standardize your content.

You also need to specify the **SAPUI5 Component ID** of the app. Use the input help (F4) to select the correct component from the list of available components. As soon as you've selected a component, the **ICF Path** field is automatically prepopulated. This path defines the app's access point and ensures that the app is accessed correctly. After entering this basic data, you can make further settings to optimally integrate the app into the catalog and the SAP Fiori launchpad.

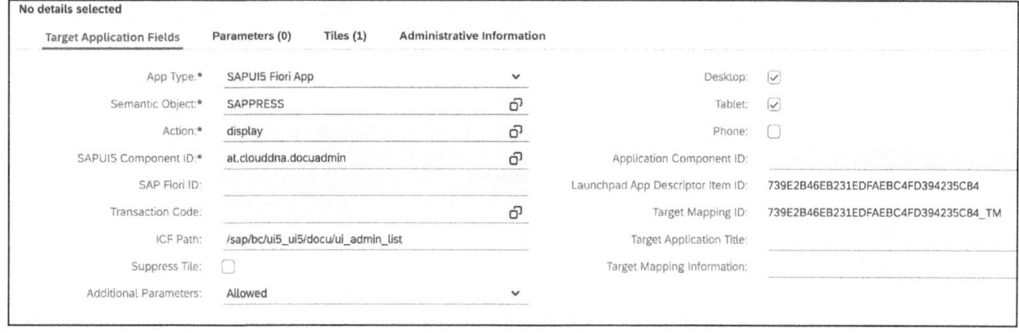

Figure 17.6 Defining the Target Application Fields

You can now maintain the **Title** and **Subtitle** in the **Tiles** tab and assign a **Tile Icon** (see Figure 17.7). This allows you to control the visual appearance of the tile within the SAP Fiori launchpad. The title represents the main text displayed on the tile, while the subtitle provides additional context or information to help users understand the purpose of the app. The tile icon serves as a visual indicator, making the tile easily recognizable and improving user navigation. You can select the icon from the available *SAP icon library*.

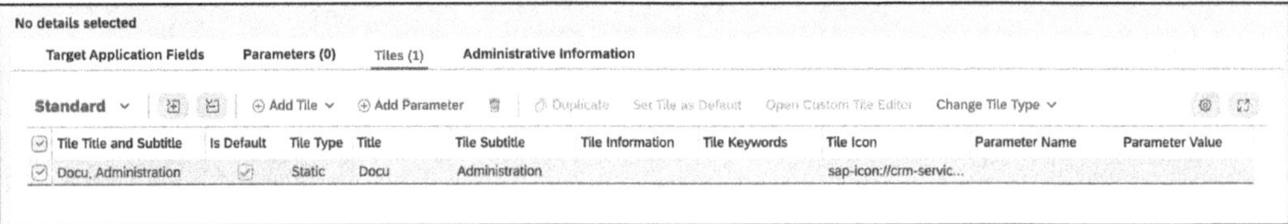

Figure 17.7 Configuring the Tile

After creating the technical catalog, you can proceed to create the business catalog. This is done using the Launchpad Content Manager app. You can access the content manager either via Transaction /UI2/FLPCM_CONF in SAP GUI or through the SAP Fiori launchpad (see Figure 17.8).

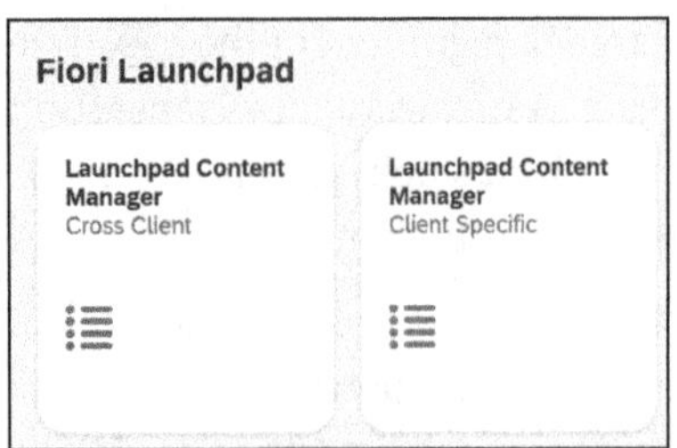

Figure 17.8 Content Manager Tiles

In the Launchpad Content Manager app, as illustrated in Figure 17.9, you'll find a comprehensive overview of all available catalogs. When you select a specific catalog, the content assigned to it is displayed in a tabular format. In the content manager, you can now create business catalogs. To do this, navigate to the **Catalogs** table, and click on the **Create** button.

This will open a dialog (see Figure 17.10) where you can define the basic properties of the new business catalog, such as the following:

- **New ID**
  A unique identifier for the catalog.
- **New Title**
  A descriptive name for easy identification.

After providing these details, click the **Continue** button.

Next, select a **Package** to which the catalog will be assigned (see Figure 17.11). If you don't intend to transport the catalog, for example, if it's created for testing purposes, you can choose **Local Object** instead.

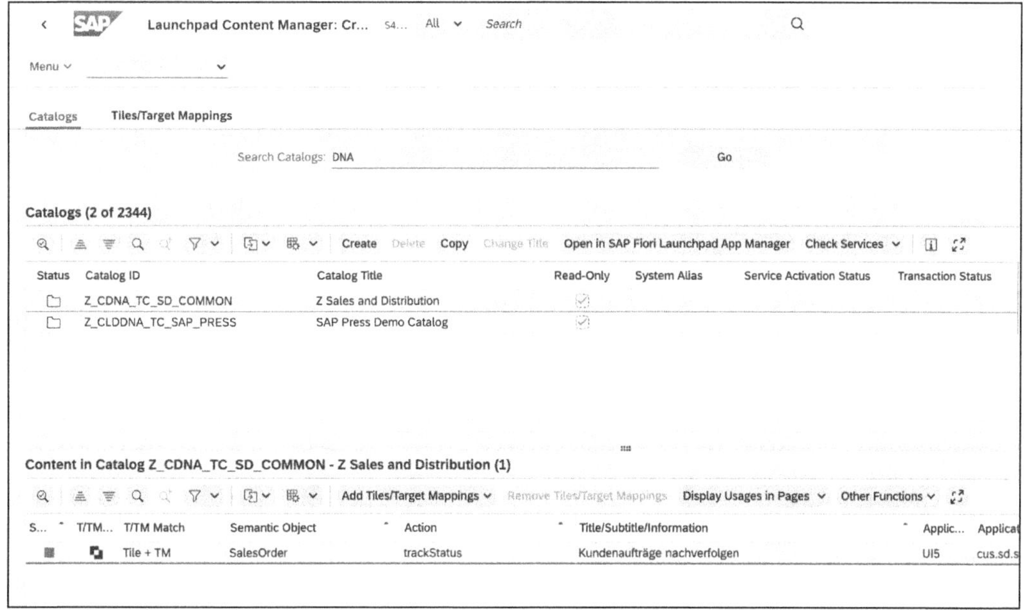

Figure 17.9 Content Manager Overview

Figure 17.10 Create Business Catalog

Figure 17.11 Assigning a Package

The newly created business catalog will appear in the **Catalogs** table and can be further customized by assigning content from technical catalogs.

The easiest way to add content to a business catalog is to open the technical catalog containing the desired tiles and target mappings, and then transfer them directly. In the **Content in Catalog** section, select the desired tile, and click on **Add Tiles/Target Mappings** in the toolbar. Then, as shown in Figure 17.12, choose the **Add Selected Tiles/TMs to Other Catalog** option.

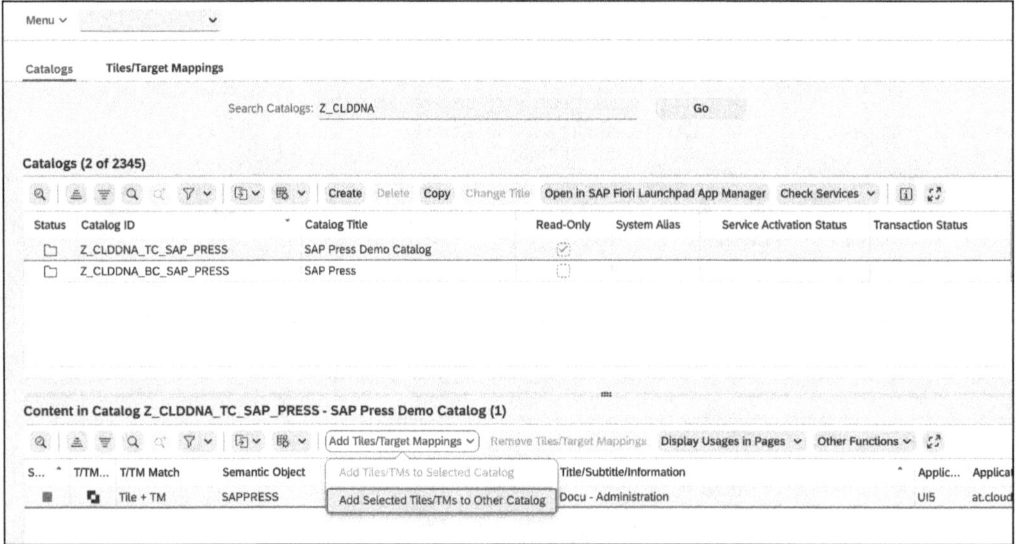

Figure 17.12 Adding a Tile to a Business Catalog

Next, as shown in Figure 17.13, select the business catalog into which you want to transfer the tile and target mapping.

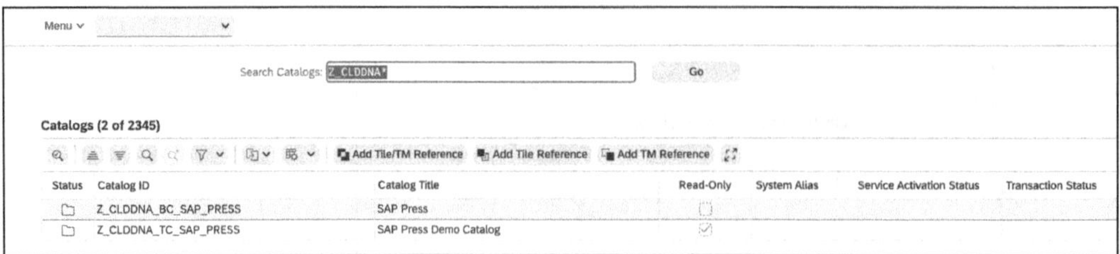

Figure 17.13 Select Target Catalog

Now, it's necessary to create a role in the SAP system to include the business catalog. To do this, start the role maintenance using Transaction PFCG. Enter a name for the **Role**, and click on the **Single Role** button (see Figure 17.14).

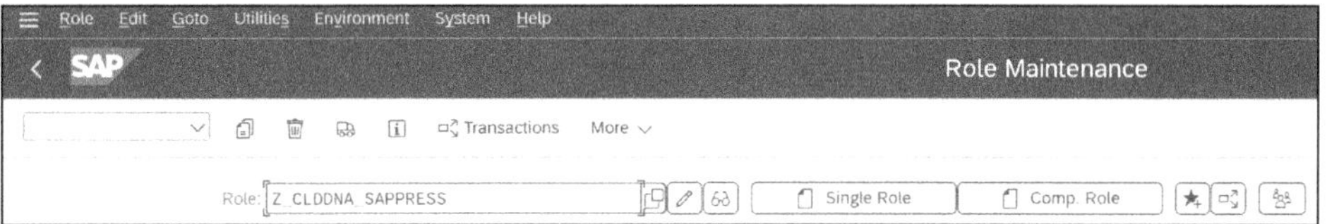

Figure 17.14 Creating the PFCG Role

The business role can be configured in the **Menu** tab by selecting the required catalog through the **Transaction** button. As shown in Figure 17.15, click on **SAP Fiori Launchpad** in the menu, and choose the **Launchpad Catalog** option.

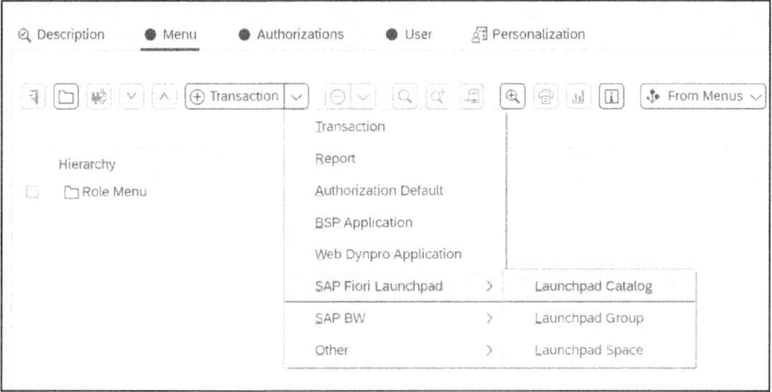

Figure 17.15 Adding the Launchpad Catalog to the Role

In the dialog, select **Fiori Launchpad Catalogs** as the **Catalog Provider**. Then, search for the desired business catalog or directly enter the **Catalog ID** of the previously created business catalog in the respective field (see Figure 17.16). It's important to note that assigning the technical catalog to the role isn't necessary.

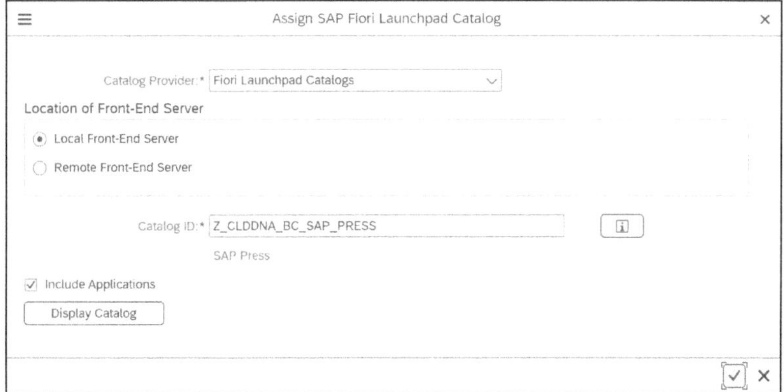

Figure 17.16 Assigning the Catalog

All prerequisites are now in place to proceed with creating spaces and pages. The process always begins with a space, which can contain one or more pages. To get started, open the Manage Launchpad Spaces app from the SAP Fiori launchpad (see Figure 17.17).

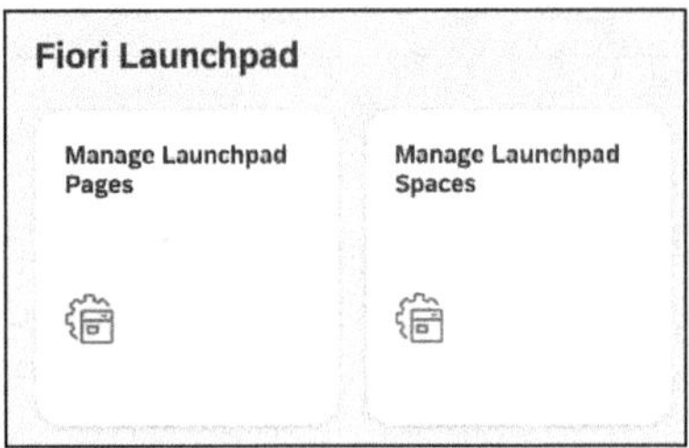

Figure 17.17 Spaces and Pages Tiles

This will take you to the spaces overview, as shown in Figure 17.18. To create a new space, click on the **Create** button in the toolbar.

Manage Launchpad Spaces

Customer-Created Predefined

Spaces (20)

Search Create Copy Delete

	ID / Description	Title	Roles	Created By / Created On	Changed By / Changed On
○	Z_SP_BP		1	Thorsten REISINGER 11.12.2024	Thorsten REISINGER 11.12.2024
○	Z_SP_THURNER_INNO		1	Tobias Thurner 22.11.2024	Tobias Thurner 22.11.2024
○	Z_SP_ZANKL_INNO		1	Julian Zankl 22.11.2024	Julian Zankl 22.11.2024
○	Z_SP_OLZINGER_INNO		1	Maximilian Olzinger 22.11.2024	Maximilian Olzinger 22.11.2024
○	Z_SP_CURIAC_INNO Innovation Day Curiac	Innovation Day Curiac	1	Marian Curiac 22.11.2024	Marian Curiac 22.11.2024

Figure 17.18 Spaces Overview

This opens the dialog for creating a space, as shown in Figure 17.19. Enter a system-wide unique **Space ID**, a **Space Description**, and a **Space Title**. To create a page linked to the space in the same step, activate the **Also create a page** checkbox. Then, provide a unique **Page ID**, a **Page Description**, and a **Page Title** for the new page. In addition, you must select a transport request in the mandatory **Transport** field to ensure that the space and page can be transported to other systems if needed.

Afterward, you'll be directed to the details of the newly created space. Here, you can add additional pages to the space if needed (see Figure 17.20). As shown in the figure, the **Role Assignment** tab indicates that the space isn't yet assigned to any role. Therefore, in the next step, you'll assign the space to a Transaction PFCG role to make it accessible to the appropriate users.

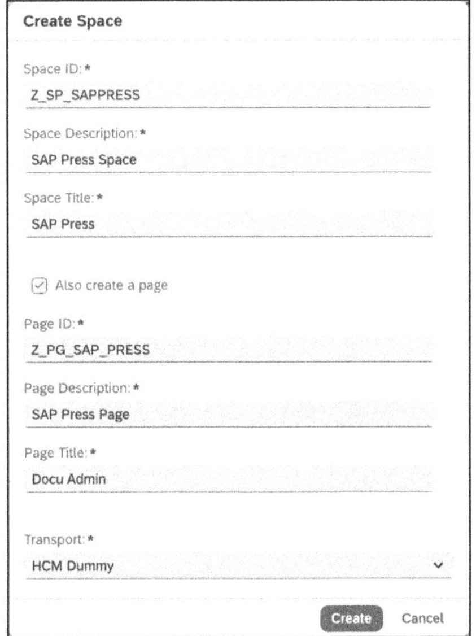

Figure 17.19 Creating the Space and Page

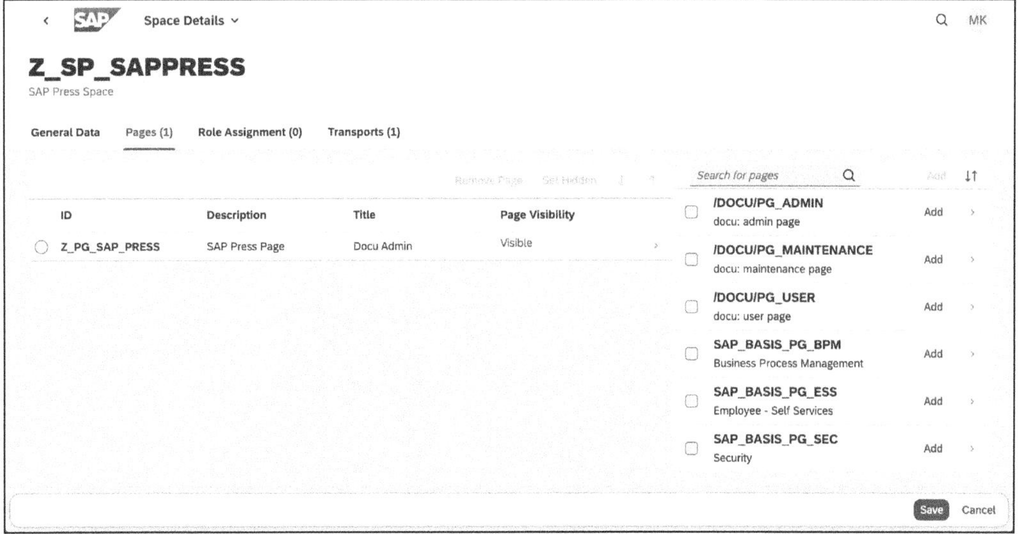

Figure 17.20 Page Details

To do this, open the previously created role again in role maintenance (Transaction PFCG). Navigate to the **Menu** tab, and click on **Transaction** (see Figure 17.21). Then, select **SAP Fiori Launchpad**, and choose **Launchpad Space** to assign the space to the role.

In the dialog that appears, enter the **Space ID** of the previously created space in the corresponding field (see Figure 17.22). Make sure to save your entries to finalize the assignment.

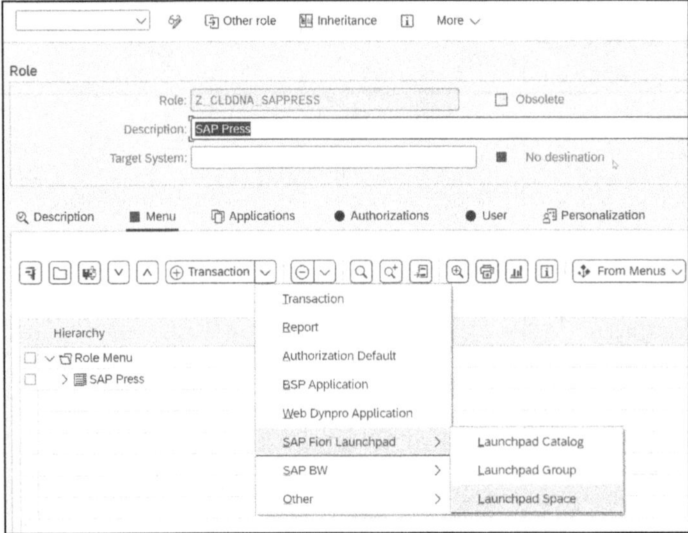

Figure 17.21 Assigning a Space to the Role

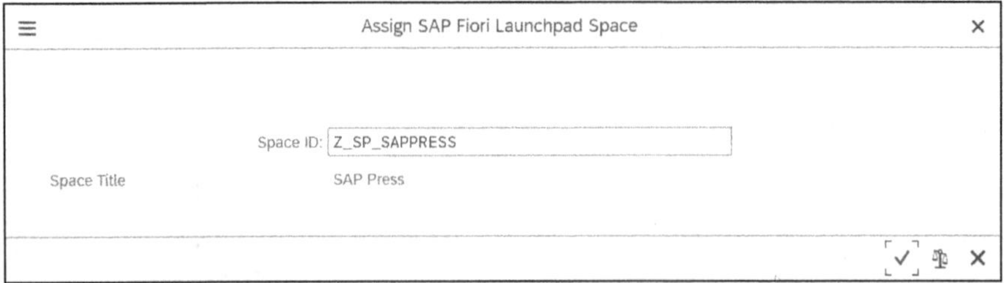

Figure 17.22 Selecting the Launchpad Space

In the next step, the content of the page can be configured. To do this, open the Manage Launchpad Pages app, as shown in Figure 17.17. Then, navigate to the **Page Content** tab, and click on the **Edit** button in the toolbar (see Figure 17.23) to begin modifying the page's content.

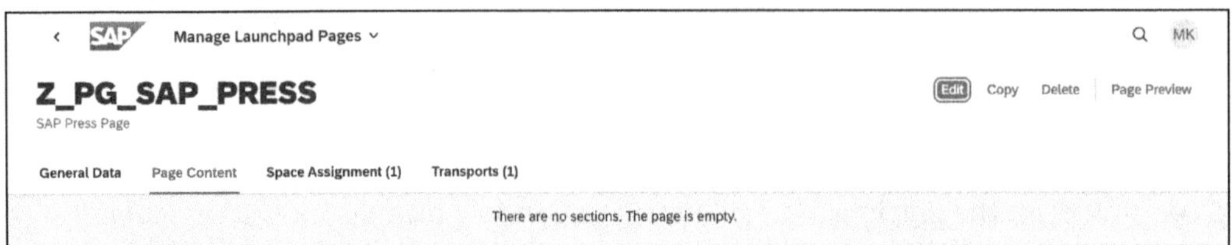

Figure 17.23 Page Overview

On the right side of the screen, you'll see a table titled **Derived from Roles**, which lists all the apps included in the business catalogs assigned to the role (see Figure 17.24). Click on the **Add** button for the desired app, and then select the menu entry **Add as Tile (preferred)** to include the app as a tile on the page.

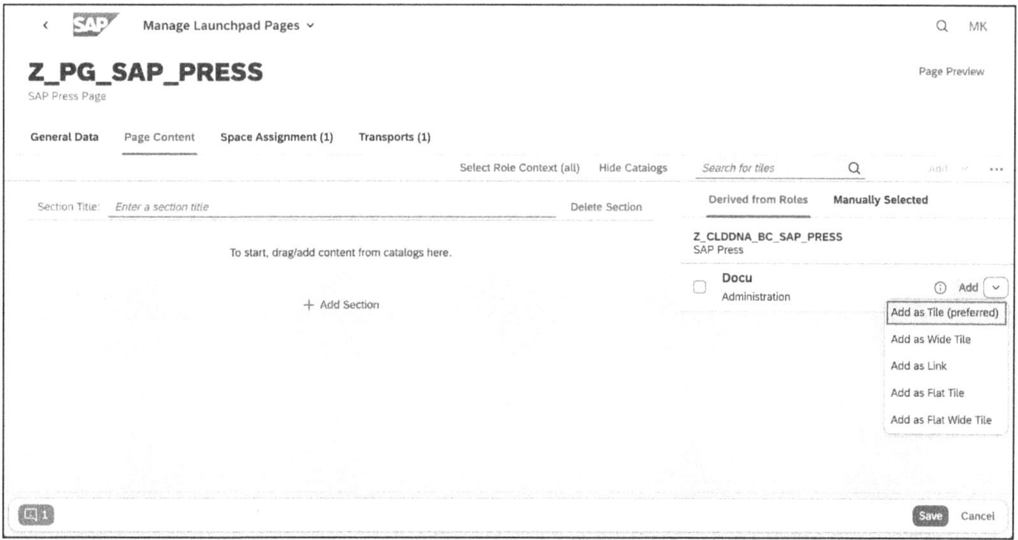

Figure 17.24 Adding a Tile to the Page

Pages themselves don't directly contain tiles; instead, they are organized into one or more sections, where tiles are displayed. As shown in Figure 17.25, assign a **Section Title** to define the purpose or context of the section. Once done, click on the **Save** button to finalize the changes.

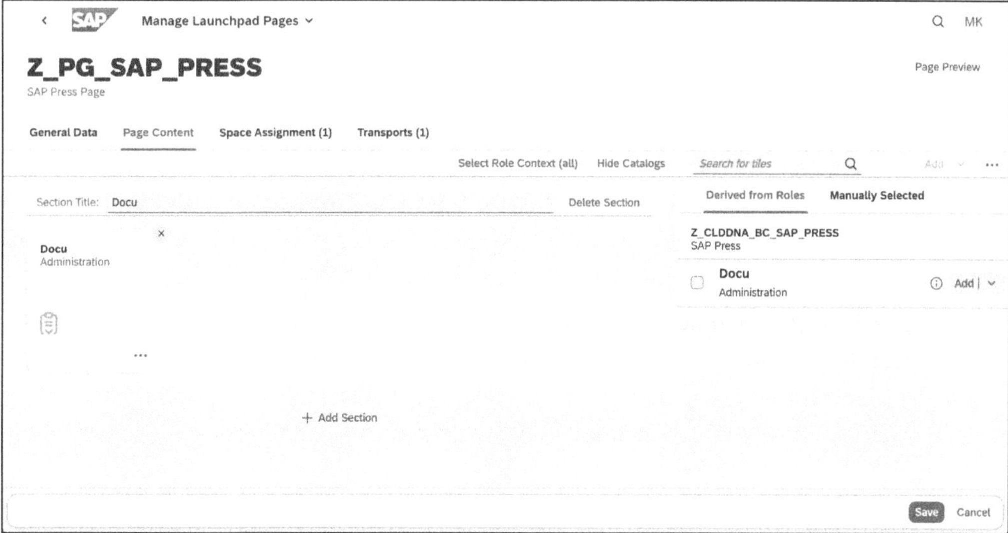

Figure 17.25 Assigning a Section Title

You've now completed all the required steps. To test the configuration, assign the previously created Transaction PFCG role to the desired user. Afterward, log in with this user, and open the SAP Fiori launchpad. You'll now see the newly created space containing the page, the section, and the added tile (see Figure 17.26).

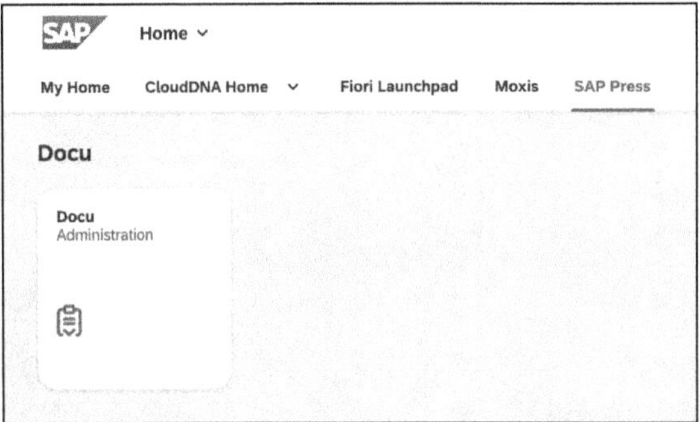

Figure 17.26 Testing

17.3 SAP Build Work Zone, Standard Edition

In the previous section, we focused on configuring the SAP Fiori launchpad in an on-premise landscape. In this section, we'll demonstrate how to set up a launchpad in SAP BTP. For this purpose, SAP provides SAP Build Work Zone, standard edition. In many cases, the service provisioning can be facilitated through a booster at the global account level. However, there's no specific booster for SAP Build Work Zone, standard edition, unless you also intend to enable SAP Task Center alongside it.

17.3.1 Configuration

As a first step, you need to create a subaccount. Because we assume you're already familiar with this process, we won't cover it in detail here. Once the subaccount is created, you must assign the SAP Build Work Zone, standard edition entitlement to it. To do this, navigate to the **Entitlements • Entity Assignments** section in your global account, as shown in Figure 17.27, and click the **Edit** button.

In the **Subaccount Build Workzone: Add Service Plans** dialog, as shown in Figure 17.28, search for **SAP Build Work Zone, Standard Edition**. Select the two available plans: **standard** and **standard (Application)**. Next, click on the **Add 2 Service Plans** button to close the dialog. Finally, save the entity assignment to complete the process.

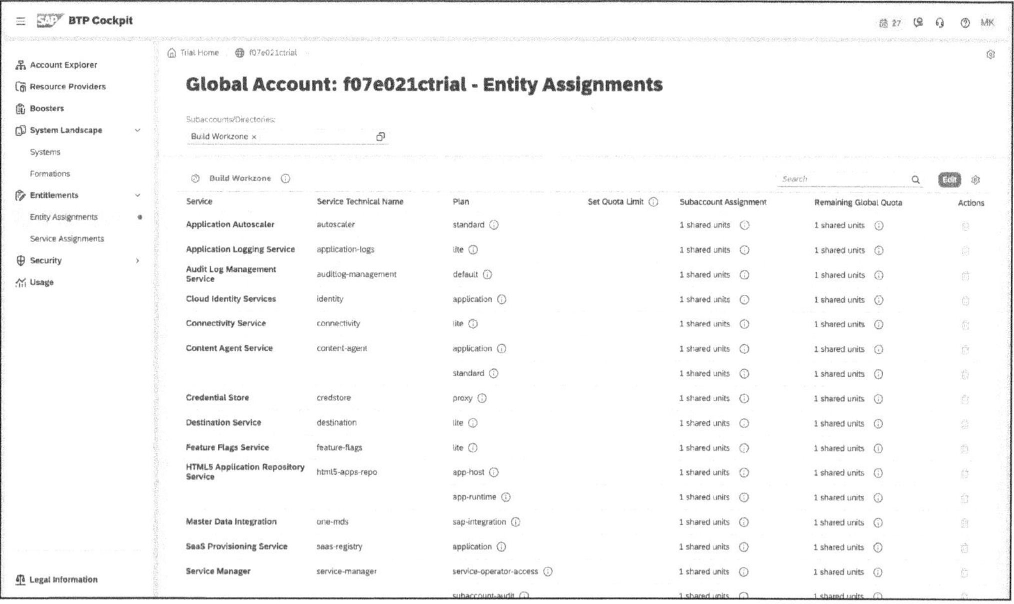

Figure 17.27 Entity Assignments

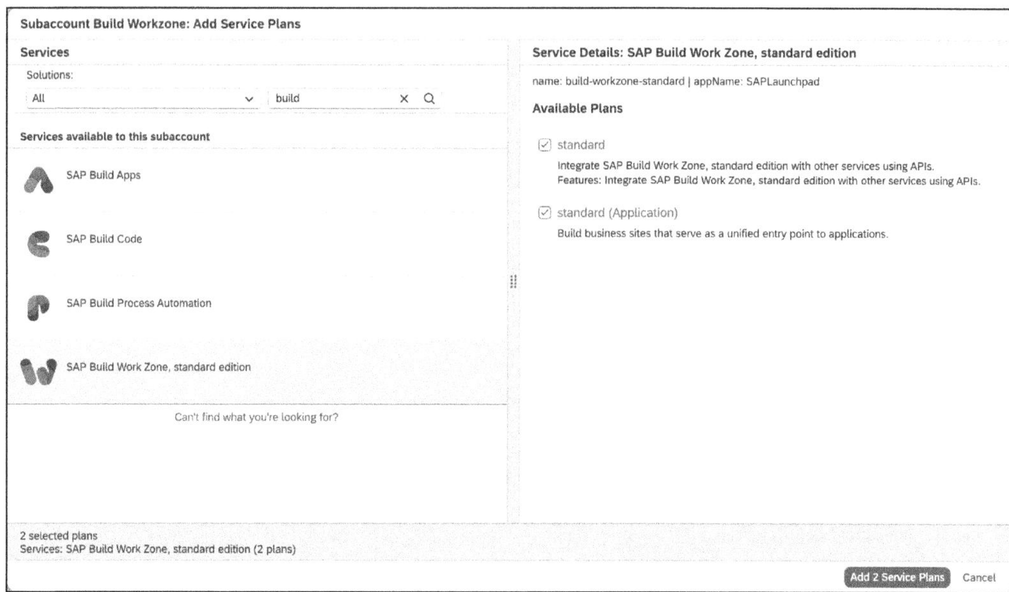

Figure 17.28 Adding a Service Plan

Now, navigate to the subaccount. Here, you need to activate the Cloud Foundry environment. To do this, click on **Enable Cloud Foundry**, as shown in Figure 17.29.

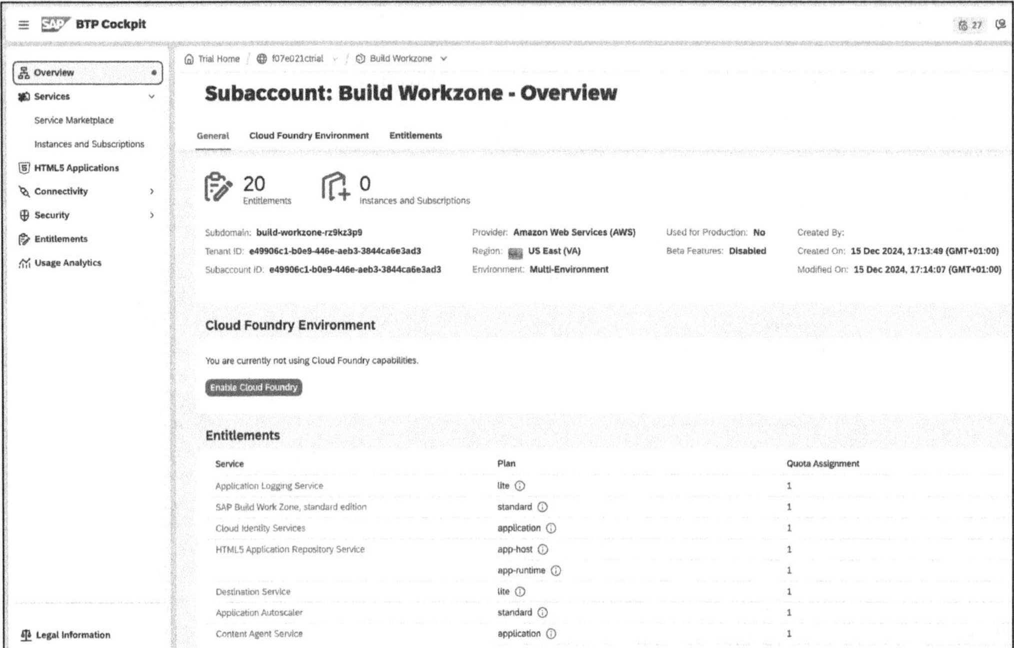

Figure 17.29 Enable Cloud Foundry

Select an appropriate plan. Because we're working with a trial account in this example, we've selected the **trial** plan, as shown in Figure 17.30. The system suggests the **Instance Name** and the **Org Name**, which can be adjusted if needed. Once the configuration is set, click on **Create** to activate the Cloud Foundry environment with the specified settings.

Enable Cloud Foundry

1
Basic Info

Enter basic info for your environment instance.

Environment: * Can't find what you're looking for?

Cloud Foundry Runtime

Plan: *

trial

Instance Name:

build-workzone-rz9kz3p9_Build-Workzone

Org Name: *

f07e021ctrial_build-workzone-rz9kz3p9

Cloud Foundry API endpoints may differ between subaccounts in the same region.
The API endpoint for this subaccount is shown in the Cloud Foundry environment instance details.

Create Cancel

Figure 17.30 Cloud Foundry Runtime Configuration

In the next step, you have the optional ability to create a space. However, if you're configuring access to an SAP Fiori app deployed within the same subaccount, creating a space is mandatory. To do this, navigate to the **Services** section in the side menu, and click on the **Create Space** button, as shown in Figure 17.31.

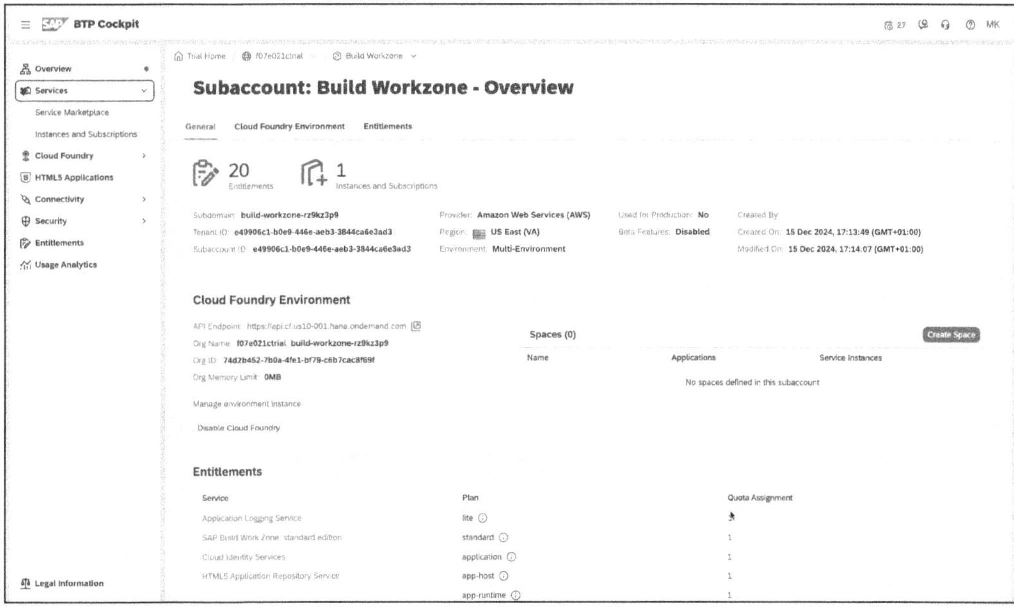

Figure 17.31 Create Space

Enter an appropriate name in the **Space Name** field. Additionally, select the space roles to be assigned to the user creating the space. By default, the roles **Space Developer** and **Space Manager** are selected. Once the configuration is complete, click on **Create** to finalize the space creation (see Figure 17.32).

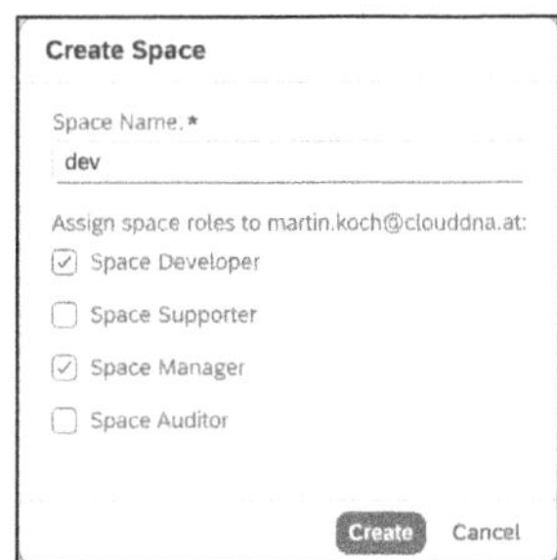

Figure 17.32 Space Configuration

You can now create the SAP Build Work Zone, standard edition subscription. To do this, navigate to **Services • Service Marketplace** in the side menu (see Figure 17.33). Search for **SAP Build Work Zone, standard edition** service, and click on the **...** button. In the popover menu, select **Create** to initiate the subscription.

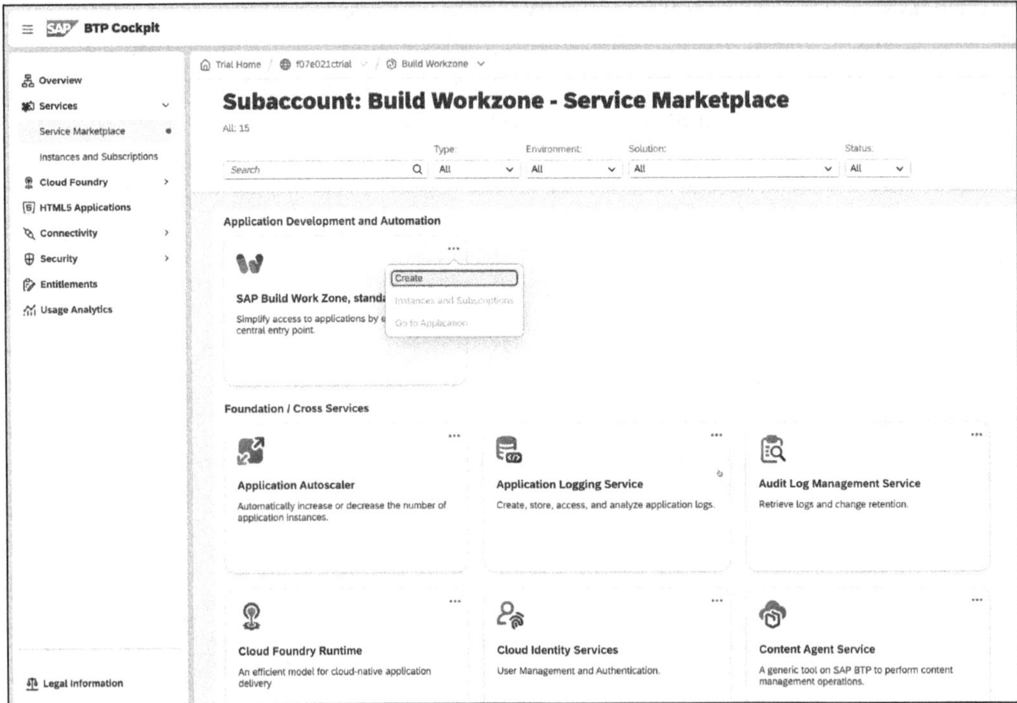

Figure 17.33 Creating an SAP Build Work Zone Instance

In the **Subscriptions** section, as shown in Figure 17.34, select the **standard Plan**. Then, click on the **Create** button to finalize the subscription setup.

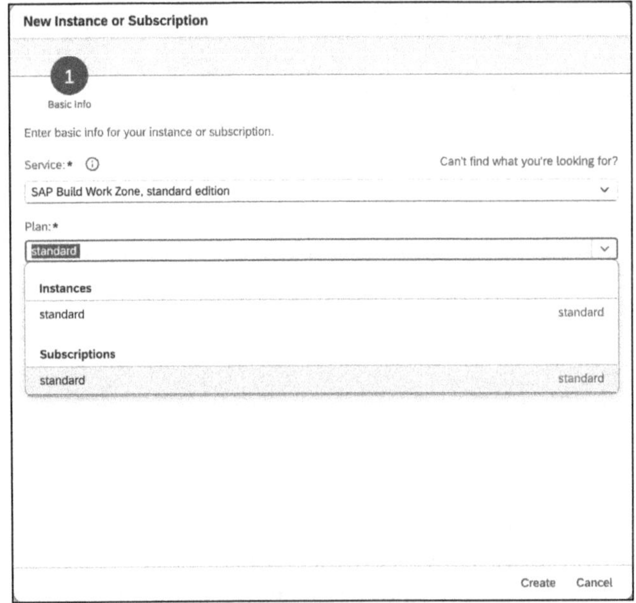

Figure 17.34 Instance Details

A popup will appear, providing updates on the progress. Once the process is complete, click on **Close**, as shown in Figure 17.35, to dismiss the dialog.

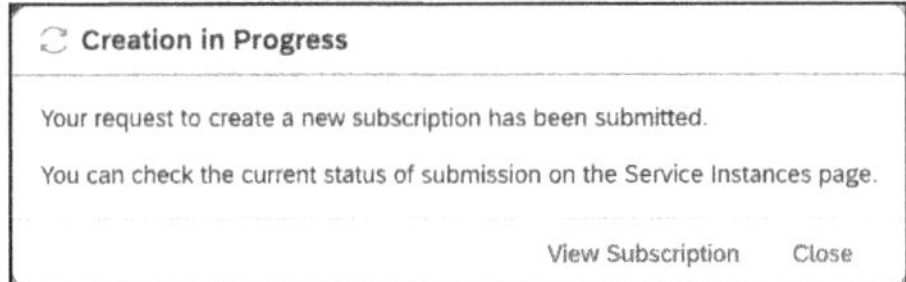

Figure 17.35 Instance Creation Progress

When the subscription is created, the corresponding roles and role collections are automatically added to the Subaccount. To view all available role collections, navigate to **Security • Role Collections** in the side menu. The subscription includes role collections such as **Launchpad_Admin, Launchpad_Advanced_Theming,** and **Launchpad_External_User** (see Figure 17.36). To work with SAP Build Work Zone, standard edition, these role collections must be assigned to the appropriate users.

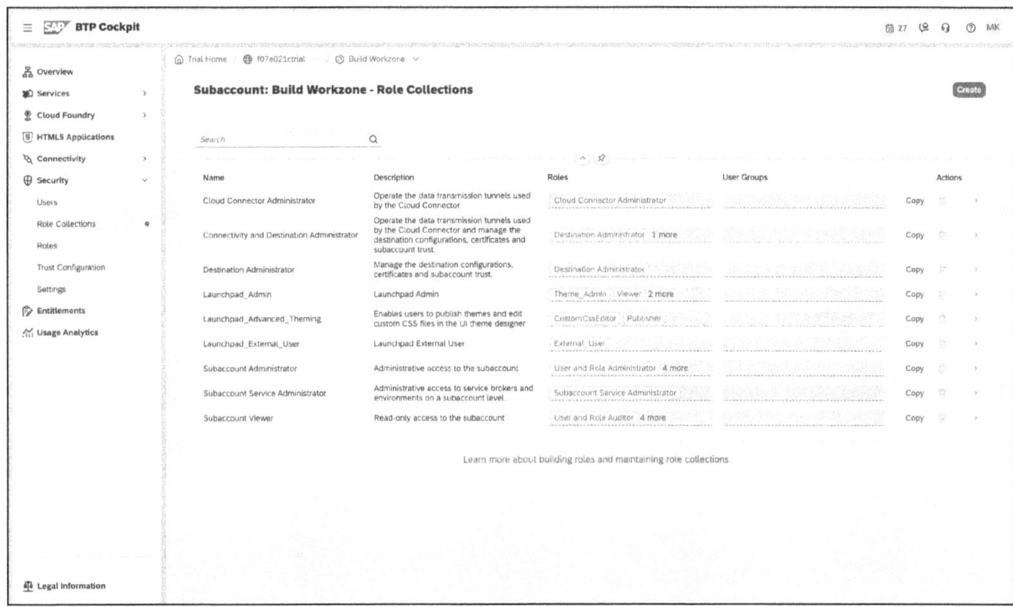

Figure 17.36 Role Collection Overview

Open the **Launchpad_Admin** role collection. Switch to **Edit** mode, and in the **User** section, add the email addresses of the desired users. Once the entries are complete, click on **Save** (see Figure 17.37) to finalize the changes.

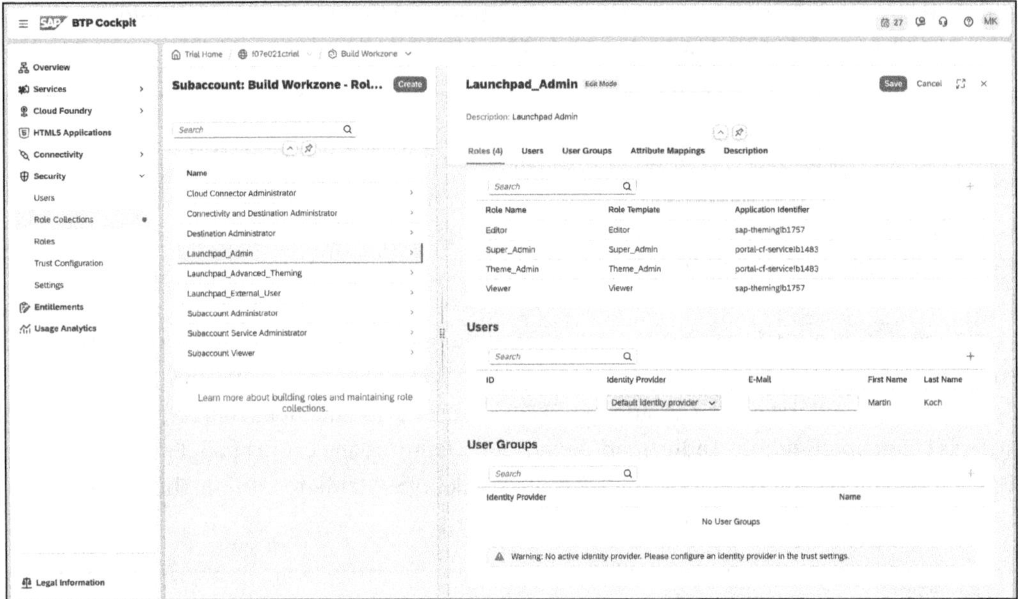

Figure 17.37 Assigning the Role Collection

17.3.2 Launchpad Creation

You can now open the SAP Build Work Zone, standard edition service. Applications are integrated into SAP Build Work Zone using *content managers*. For all applications deployed in the same subaccount as SAP Build Work Zone, a default *channel* is already available in the content manager. This channel is essential for adding apps to catalogs. As shown in Figure 17.38, the default channel, named **HTML5 Apps**, is displayed as a tile. Click on this tile to access its details.

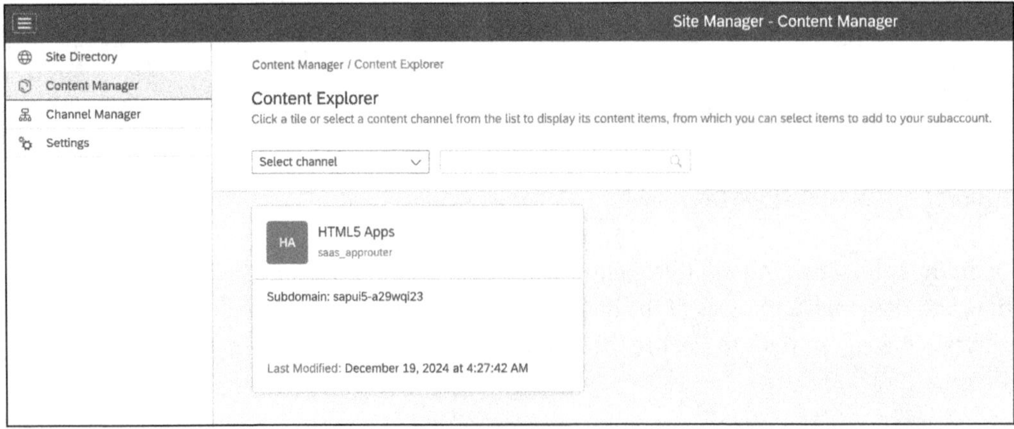

Figure 17.38 Content Manager Overview

In the details, you'll find a list of all available apps (see Figure 17.39). Unlike deployment in an on-premise landscape, you must first add the app before it can be used in a catalog. To do this, select the desired apps from the list, and then click on the **Add** button. In the example shown, we'll use an app called **SAP Press**.

Figure 17.39 Adding the App

The next step involves creating a *catalog*. To do this, return to the content manager, and, as shown in Figure 17.40, click on the **Create** button. From the dropdown menu, select the option **Catalog** to proceed.

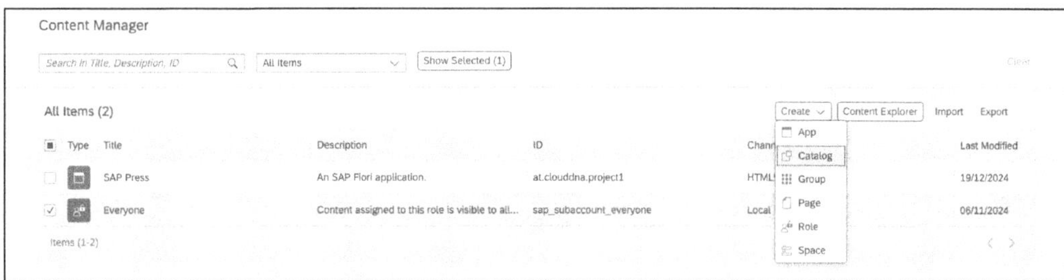

Figure 17.40 Creating the Catalog

This action takes you to the catalog details screen. Enter a meaningful name in the **Catalog Title** field, and optionally provide a description in the **Description** field. In the **Apps** tab, you'll see a list of all available apps. To add an app to the catalog, toggle the **Assignment Status** switch in the respective column. Figure 17.41 illustrates this process using the example of the SAP Press app.

You can now create a new *page*. Navigate to the content manager, and, as previously shown for creating a catalog, click on the **Create** button. This time, select the **Page** option from the dropdown menu (see Figure 17.40). Enter a meaningful name in the **Page Title** field, and, optionally, provide a description in the **Description** field. Apps can't be directly assigned to a page; instead, they must be added to a *section*. A page can contain multiple sections to organize content effectively. To create a new section within the page, click on **Add Section**, as shown in Figure 17.42. This allows you to structure the page by grouping related apps into distinct sections.

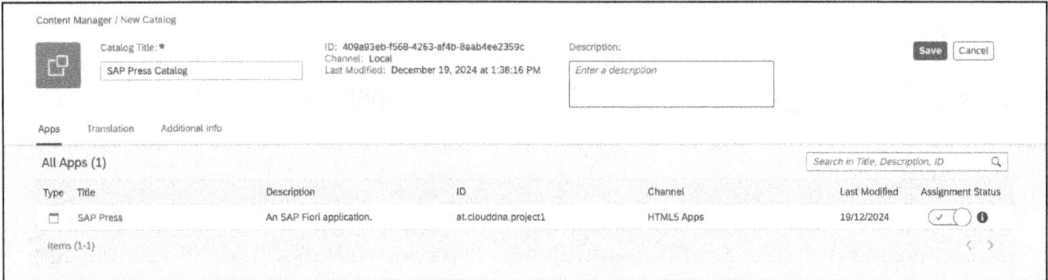

Figure 17.41 Catalog Details

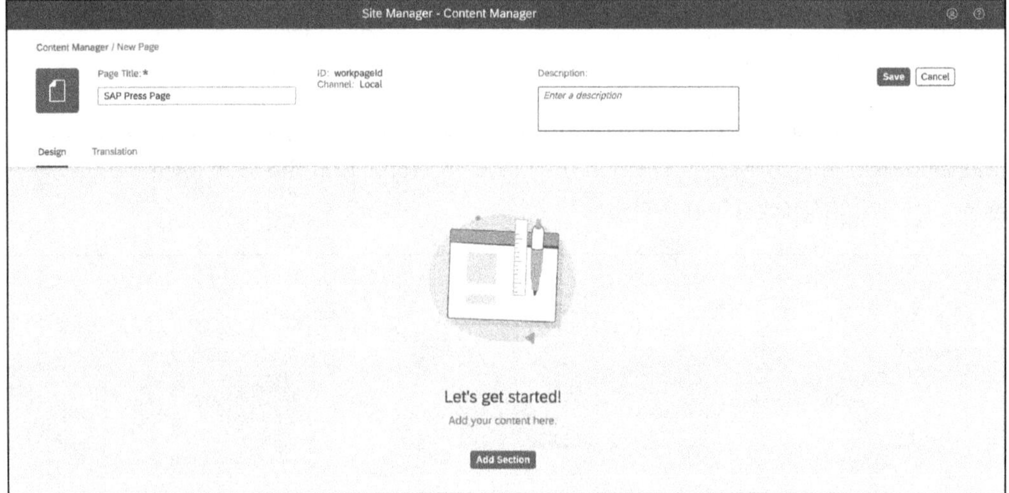

Figure 17.42 Creating a Page

You can now assign a title to the section by entering it in the **Header** field (see Figure 17.43). Apps are added to the section as *widgets*. To proceed, click on **Add Widget** in the next step.

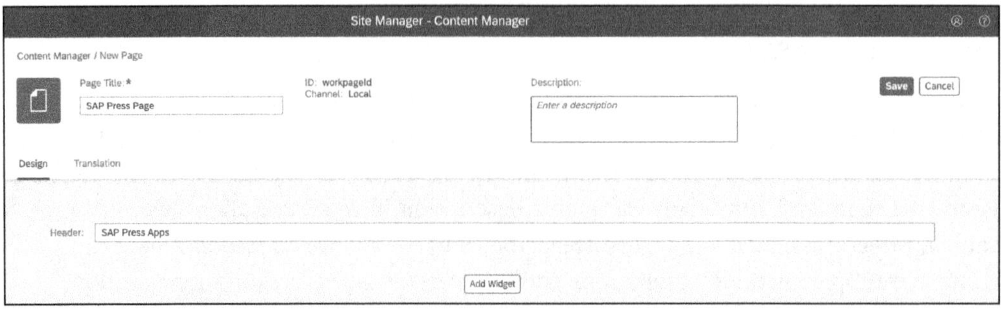

Figure 17.43 Adding a Section

Widgets are offered as **Tiles** or **Cards**, as shown in Figure 17.44. To add an SAPUI5 app, select the **Tiles** option.

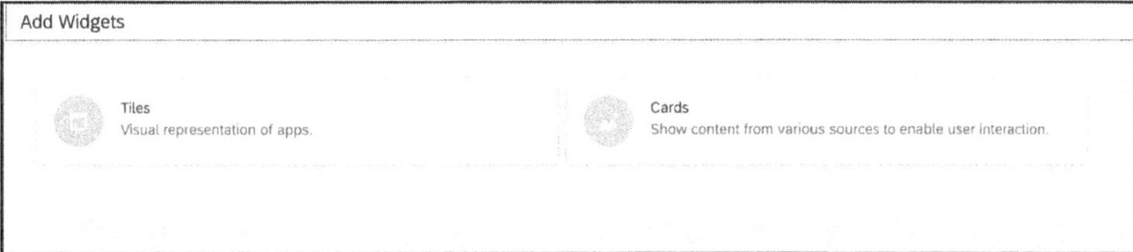

Figure 17.44 Adding a Widget

You can now select the desired apps. To do this, click the checkbox next to each app (see Figure 17.45), and then click on the **Add** button to include them in the section.

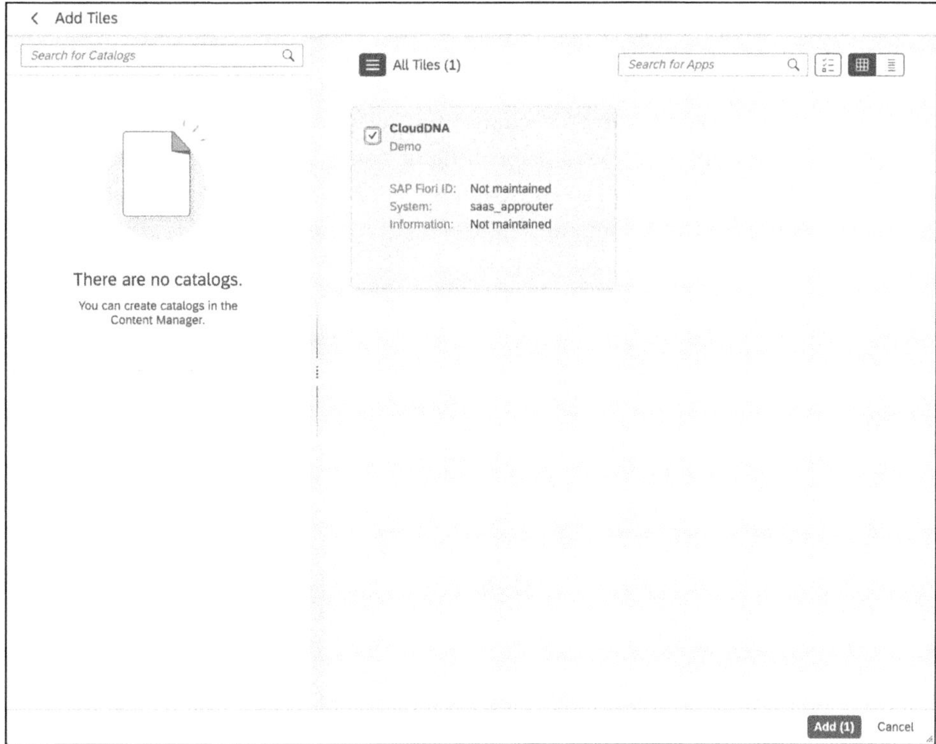

Figure 17.45 Assigning an App to a Page/Section

You can now view the page, including its sections and widgets. At this point, you can rearrange the order of the apps if needed. Finally, click on **Save** to confirm your changes (see Figure 17.46).

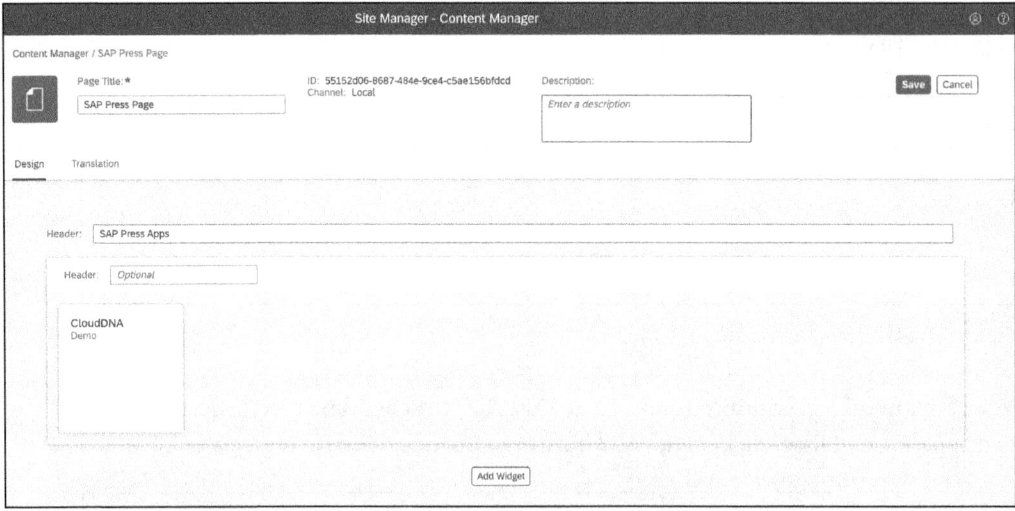

Figure 17.46 Page Design Mode

After creating the page, you can proceed with setting up the *space*. In the content manager, create a new space. Provide a meaningful name in the **Space Title** field and, optionally, add a description in the **Description** field. Next, assign the desired pages to the space by toggling the **Assignment Status** switch in the corresponding column (see Figure 17.47). Finally, click the **Save** button.

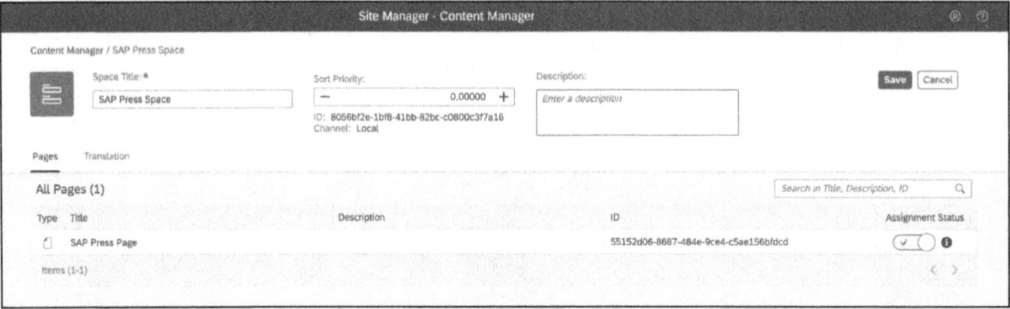

Figure 17.47 Create a Space

You now need to create a *role* to combine the apps and spaces. Navigate to the content manager, and create a new role by selecting the **Role** option from the dropdown menu, as shown in Figure 17.40. This will take you to the role details screen (see Figure 17.48). Provide a **Role Title** and, optionally, a **Description**. In the **Apps** tab, assign the desired apps to the role by toggling the **Assignment Status** switch in the corresponding column. This step ensures that the selected apps are associated with the role for streamlined access.

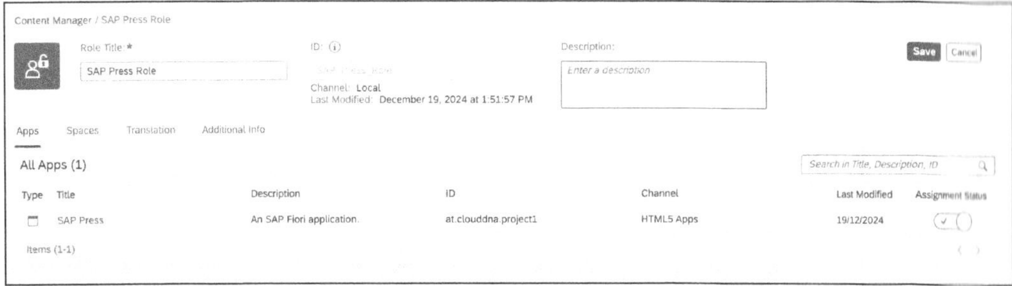

Figure 17.48 Creating a Role

Next, you need to assign the spaces. Open the **Spaces** tab. and, as shown in Figure 17.49, activate the **Assignment Status** switch for the desired spaces.

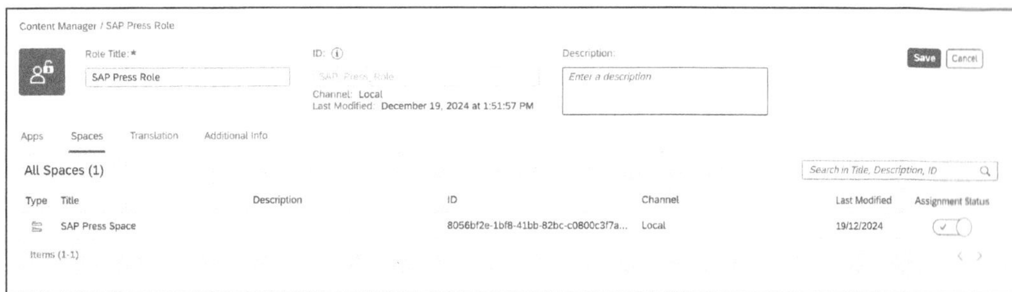

Figure 17.49 Assigning Spaces to a Role

The tasks in the content manager are now complete. You can proceed to create a *site*. To do this, navigate to the **Site Directory** in the side menu, and click on **+ Create Site** (see Figure 17.50).

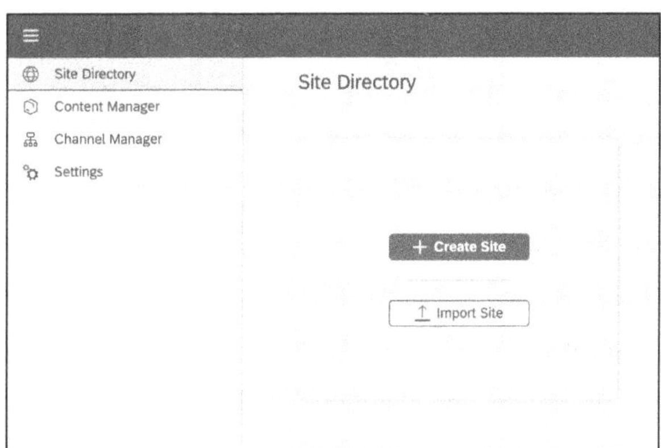

Figure 17.50 Create Site

In the dialog, enter a **Site Name**, and then click on **Create** (see Figure 17.51).

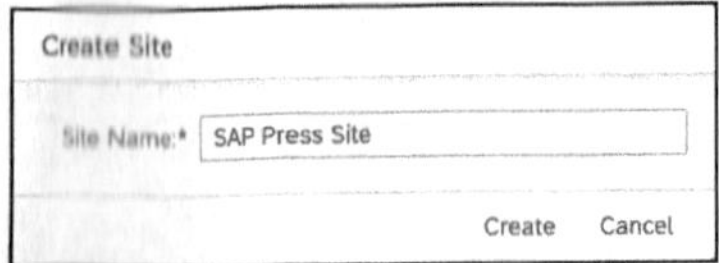

Figure 17.51 Providing a Site Name

You can now view the site properties in display mode, as shown in Figure 17.52. As the figure illustrates, there are various settings available. To make changes, click on **Edit** to switch to edit mode.

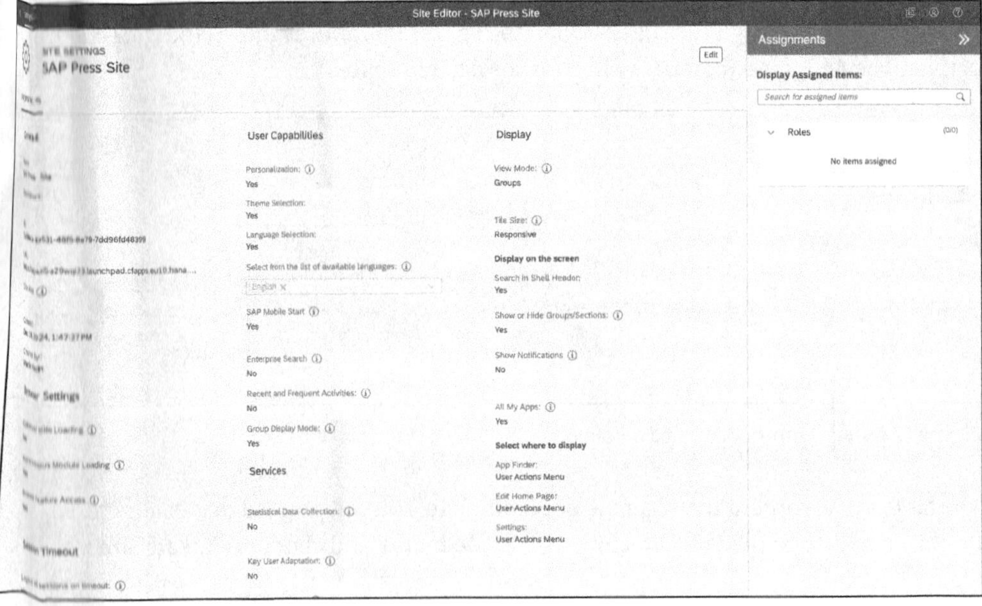

Figure 17.52 Site Settings

First, in the **Display** section, select **Spaces and Pages - New Experience** for the **View Mode** property. Next, in the right-hand section, click into the search field under **Assign Items**. This will display all available roles. Select the **SAP Press Role**, and then click **Save** to finalize the configuration (see Figure 17.53).

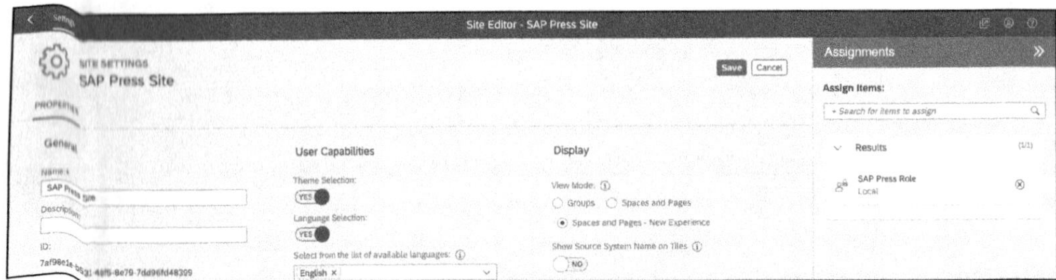

Figure 17.53 Adding a Role to the Site

For each role created in the content manager, a corresponding role collection is automatically generated in the subaccount. To test the site, you need to assign this role collection to the desired users. This is done in the SAP BTP cockpit of the subaccount. Navigate to **Security • Role Collection** in the side menu, as shown in Figure 17.54. Open the **SAP_Press_Role** role collection, and assign the users to it as usual.

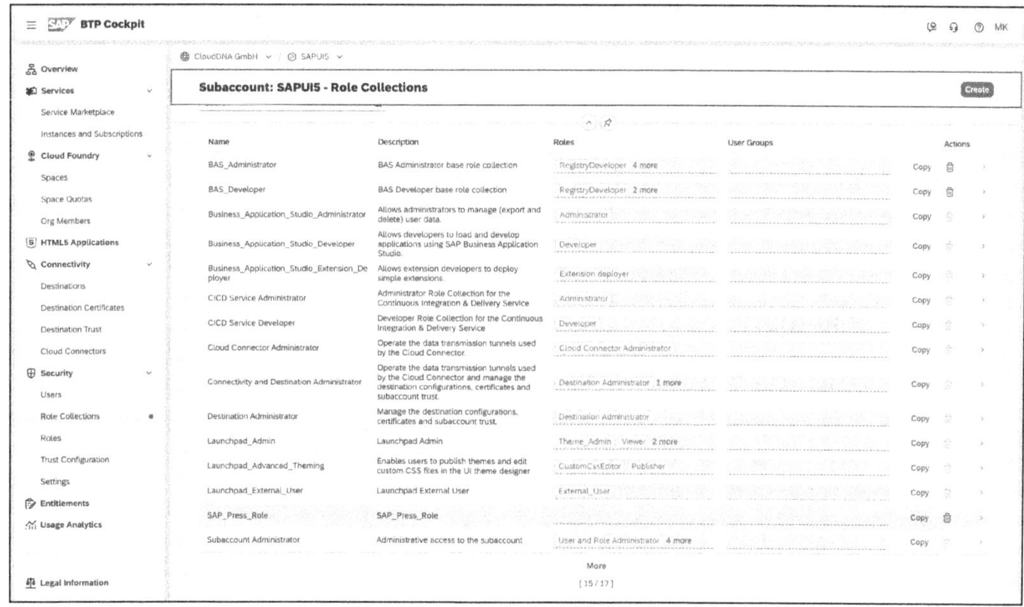

Figure 17.54 Opening the Role Collection in the SAP BTP Subaccount

Reopen the site directory. Click on [icon] to open the site (see Figure 17.55).

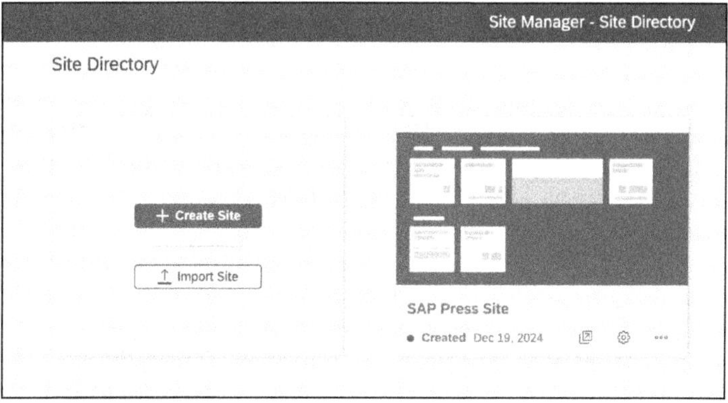

Figure 17.55 Site Directory

You can now see the launchpad, as shown in Figure 17.56, displaying the configured space, the page, the sections, and the tiles.

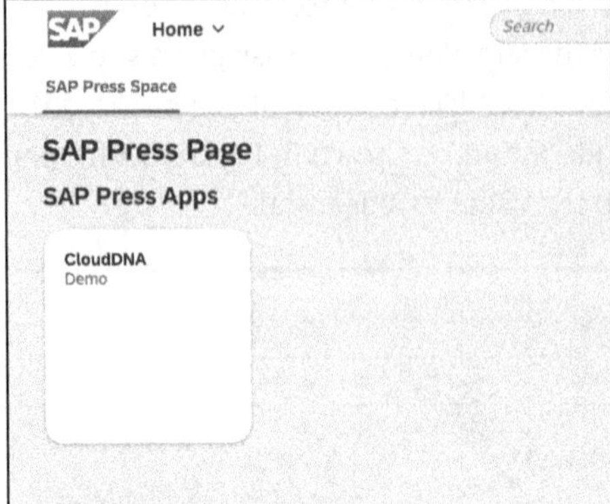

Figure 17.56 Site Preview

17.4 Summary

This chapter taught you to configure the SAP Fiori launchpad for access to SAPUI5 applications. It includes setups for both on-premise (SAP S/4HANA) environments and cloud deployments (by leveraging SAP Build Work Zone, standard edition). In the on-premise section, you learned to personalize the launchpad experience by organizing tiles, groups, and catalogs within SAP S/4HANA landscapes. The chapter also explains the concepts of pages and spaces, which replace groups from previous releases. The SAP Build Work Zone, standard edition, section demonstrates how to use SAP Build Work Zone, standard edition, to design and deploy a customized launchpad experience in the cloud.

Chapter 18
Debugging and Code Quality

Debugging and software testing are essential for JavaScript-based business applications, as errors can have fatal consequences for business processes. Without robust testing and efficient debugging, companies risk leaving their applications unstable and prone to error, which in the worst case can lead to data loss and expensive downtime. Developers must not only have the technical skills to use test frameworks and debugging tools but also the courage to consistently question unused code and potential sources of error.

Debugging and software testing are crucial for the quality and reliability of JavaScript-based web applications, especially when these are used as business applications. Business applications must offer the highest reliability as they often support mission-critical processes. Errors in these applications can lead to production downtime, data loss, or security breaches, which can have a significant impact on the business. Regular debugging and testing help to catch and fix bugs early, increasing application stability and user satisfaction. Testing ensures that all features work as expected and guards against regressions that might occur after changes or updates. 18

Effective debugging and testing require in-depth knowledge of various techniques and tools. Developers must be familiar with JavaScript debugging tools such as browser developer tools to quickly identify and fix errors in the code. In addition, it's important to create automated tests to ensure that the application remains stable with every change. This requires skills in test frameworks such as QUnit. Testers should also be able to simulate the behavior of the application and consider different user interactions and edge cases. Another important aspect is the ability to correctly test asynchronous processes because many modern web applications rely heavily on asynchronous communication. To prevent overlooking defects, it's also important to follow a systematic testing approach that includes both unit testing and integration testing, and to perform code reviews regularly.

In this chapter, we'll get to know the most important tools and techniques used in the SAPUI5 environment. We'll start in Section 18.1 with the browser developer tools, an out-of-the-box debugging and analysis tool available in all common browsers. We'll limit ourselves to the DevTools in Google Chrome. One of the reasons for this is that in Section 18.3, we'll install an extension for Google Chrome that enables even more specific debugging and error analysis of SAPUI5 applications. We're talking about the UI5

Inspector, which is only available in this form in the Google Chrome Web Store. However, Section 18.2 will show what the Support Assistant can do first because here we'll get to know the standard rulesets that indicate design and accessibility requirements. In Section 18.4, we'll briefly discuss SAPUI5 Diagnostics, a built-in window that also allows you to read information or record an end-to-end process. In Section 18.5, we discuss *static code analysis*, a process in which source code is examined without being executed to identify potential errors, security vulnerabilities, or violations of coding guidelines. Finally, we'll talk about testing and test-driven development in Section 18.5. Two possible test types are delivered with the SAPUI5 framework by default. We can use QUnit to write unit tests and OPA5 to perform integration tests.

18.1 Browser Developer Tools

The *browser developer tools* (DevTools, for short) are integrated tools in modern web browsers such as Google Chrome, Firefox, Microsoft Edge. and Safari that help developers and users analyze, debug, and optimize websites and web applications. These tools offer a variety of functions that are indispensable for frontend and web developers.

The DevTools are an essential tool for web developers to efficiently analyze, debug, and optimize web pages and web applications. One of the key features is the ability to inspect the *Document Object Model* (DOM). Developers can examine and modify the HTML and cascading style sheets (CSS) code of a website directly in the browser and adjust in real time. This makes it easier to identify layout and design issues and test solutions immediately without changing the actual files. Particularly useful here is the visual representation of the box model, which shows how elements are positioned by padding, margin, and border.

Another important feature is JavaScript debugging, which makes it possible to find and fix errors in the code. Developers can set breakpoints to stop the code execution at specific points and check variable values at runtime. Together with the console, which displays error messages and debug output, this provides an effective way to identify problems in the application logic. Analyzing network activity is also a central component of DevTools. With the network overview, developers can see which resources are loading, how long this takes, and whether errors such as **404** (file not found) occur. This feature helps to optimize loading times and ensure that all necessary resources are correctly integrated.

In addition to debugging and performance analysis, the DevTools also support the optimization of responsive design. They offer a simulation of different screen sizes and device types, allowing developers to check whether a website is displayed correctly on smartphones, tablets, and desktops. Together with performance optimization tools, such as the loading time measurement or the Lighthouse integration in Chrome, bottlenecks can be quickly identified and eliminated. In addition, the DevTools allow a detailed analysis of memory and security aspects. Developers can inspect local storage and cookies, as well as view information about HTTPS certificates or potential security risks.

In our case, we'll use the DevTools in Google Chrome. You can open it with a keyboard shortcut `Ctrl` + `Shift` + `I` or `F12`/`Fn` + `F12` (Windows) or `Cmd` + `Option` + `I` or `Fn` + `F12` (macOS), or you can simply go to **More Tools • Developer Tools** at the browser settings. By default, this opens docked to your current browser window with the last tab used from the DevTools (see Figure 18.1).

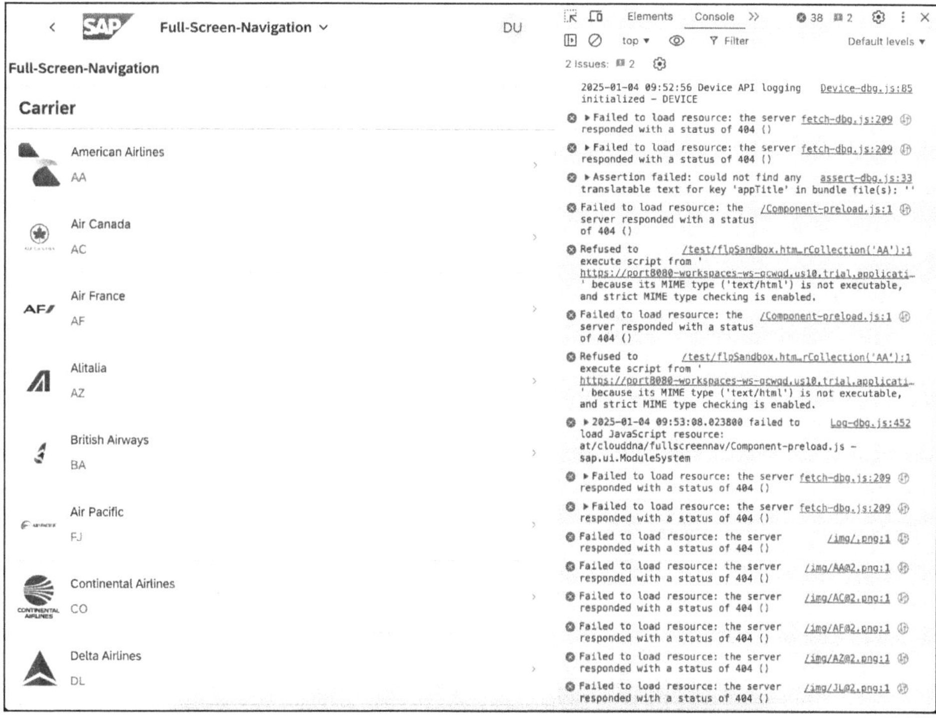

Figure 18.1 Developer Tools as a Built-in Functionality of Common Browsers

You can dock the DevTools on the left, bottom, or right as you like, or even open it as a separate window on a second screen, for example (see Figure 18.2).

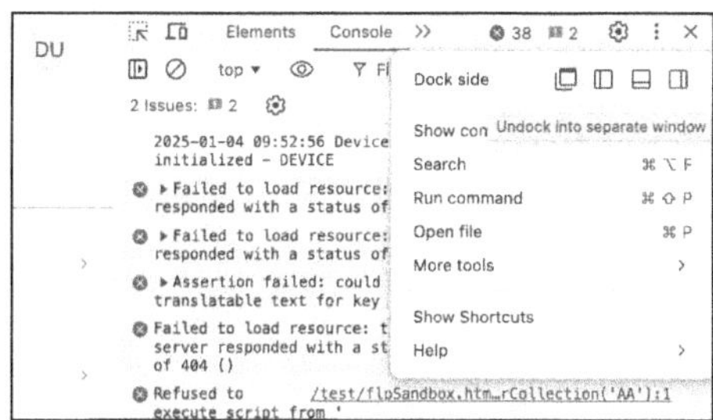

Figure 18.2 Undocking the DevTools into a Separate Window

18

We'll now introduce you to the most important tabs for SAPUI5 development and that you'll sometimes need to search for information or errors in more or less detail. Speaking of errors, we first open the **Console** tab (see Figure 18.3). This is where logger statements are displayed, but also, very importantly, the error messages that were thrown at runtime. Don't be misled because there can be a lot of error messages even by default (due to missing translation files, but also from the SAP Fiori launchpad itself), which can make the search more difficult. Unfortunately, only experience in troubleshooting can help here to localize the right ones.

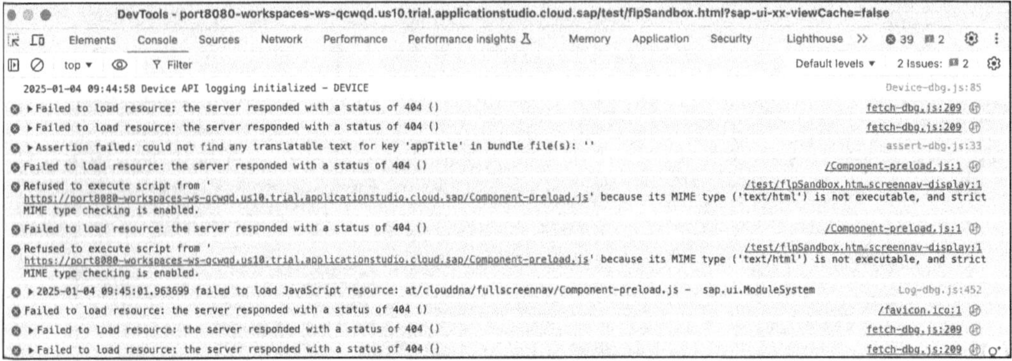

Figure 18.3 Runtime Errors and Logs Shown in the Console

Let's try out how we can output information to the console ourselves both to log information but also, for example, to support us in developing the application. Do you remember our full-screen navigation example from Chapter 7, where we have an event handler when a table entry is selected? We place a `console.log()` statement there and use string concatenation to output the binding context path of the currently selected entry before navigating to the **Detail** view (see Listing 18.1).

```
private onNavToDetail(oEvent: Event) :  void {
  let oRouter = (this.getOwnerComponent() as UIComponent).getRouter(),
    oBindingContext = (oEvent as any).getSource().getBindingContext(),
    oObject = oBindingContext?.getObject(),
    sPath = oBindingContext?.getPath();
  console.log("Path: " + sPath);
  oRouter.navTo("Detail", { path: encodeURIComponent(sPath) });
}
```

Listing 18.1 Using console.log to Display Information on the Console

As soon as we click on a table entry, we can see our output in the console (see Figure 18.4). On the right, we can also see which line in *Main.controller.ts* triggered this message. Click on this link to go to the method and thus to the source.

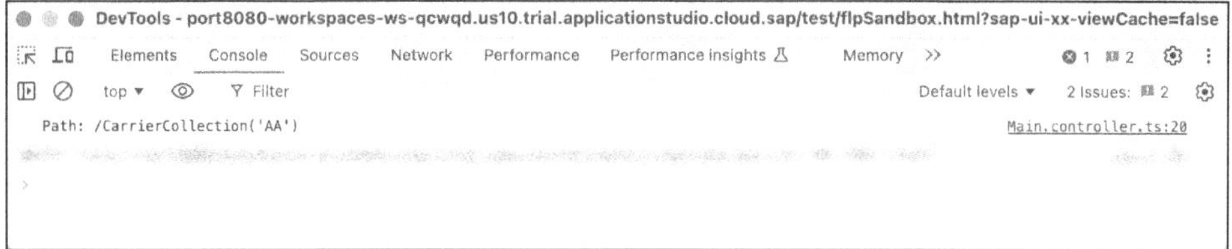

Figure 18.4 Log Displayed the Value of a Local Variable

Now, without having to search for the file for long, we've landed in the **Sources** tab (see Figure 18.5). In this tab, we can see all the app-specific and general files with content, all of which have already been loaded at runtime. If we want to debug a specific file, we can click on the line counter on the left to set a breakpoint. This breakpoint is then also visible on the right in the list of all breakpoints and can be activated and deactivated again.

Figure 18.5 Using Breakpoints in the Sources Tab

If we now click on a table entry again, we'll stop at our breakpoint, and we can debug the application. We can switch back to the **Console** tab so that we can also make entries, but you can also use Esc to show the console again (if it's hidden). As you can also see in Figure 18.6, we've already executed a number of commands to elicit the details from our event object. We can access all variables and methods here that are also accessible and visible to the application at this point.

On the right, you also have the **Locals** (local variables and functions) as well as the **Globals** listed (member variables, member functions, global variables, and global functions). So, if you aren't sure how to get a value, you can play with getters and setters in the console. However, debugging doesn't eliminate the asynchrony. Remember that asynchrony still exists even during debugging, and you can't always test all functions and methods directly with a single statement but may have to deal with promises or callbacks.

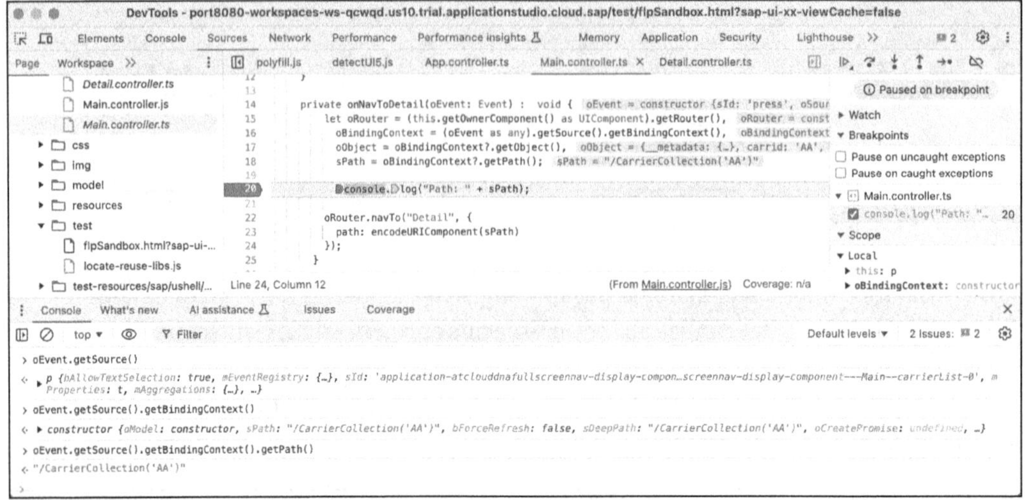

Figure 18.6 Debugging Allows Access to All the Local and Global Variables in Scope

An alternative for debugging an application is the debugger statement. The advantage here is that you simply write this keyword into the source code, and with DevTools open, the application automatically stops (see Figure 18.7). The disadvantage is that if you forget to remove this statement before going live, the debugger in the SAP Fiori launchpad will also stop with DevTools open for the user.

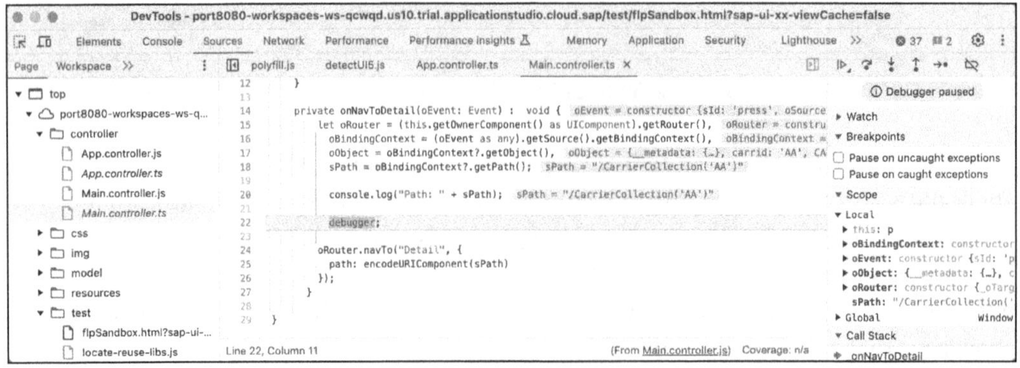

Figure 18.7 Application Stops at the Debugger Statement If the DevTools Window Is Opened

Another important tab for us is the **Network** tab. Here, you can monitor all outgoing requests and read the replies and their status. In Figure 18.8, you can see that all JavaScript Object Notation (JSON), XML, and JavaScript files for our application have to be loaded, as well as the read and manipulate calls to the connected OData service.

Name	Status	Type	Initiator	Size	Time
Component.js	200	script	ui5loader-dbg.js:1355	1.7 kB	514 ms
models.js	200	script	ui5loader-dbg.js:1355	910 B	144 ms
manifest.json	200	xhr	fetch-dbg.js:209	1.2 kB	254 ms
i18n_en.properties	404	xhr	fetch-dbg.js:209	205 B	157 ms
i18n.properties	200	xhr	fetch-dbg.js:209	337 B	183 ms
style.css	200	stylesheet	includeStylesheet-dbg.js:77	158 B	147 ms
[illegible]	200	xhr	ui5loader-dbg.js:1283	26.8 kB	163 ms
i18n_en.properties	404	xhr	fetch-dbg.js:209	202 B	148 ms
i18n.properties	200	xhr	fetch-dbg.js:209	327 B	148 ms
$metadata?sap-language=EN	200	xhr	datajs-dbg.js:2592	4.0 kB	523 ms
$value/?sap-language=EN	200	xhr	jquery-dbg.js:10224	1.9 kB	388 ms
App.view.xml	200	xhr	fetch-dbg.js:209	251 B	136 ms
Main.view.xml	200	xhr	fetch-dbg.js:209	387 B	136 ms

133 requests 3.8 MB transferred 13.5 MB resources Finish: 10.08 s DOMContentLoaded: 437 ms Load: 3.37 s

Figure 18.8 Outgoing HTTP Requests and Their Replies in the Network Tab

If we click on a request, we can see the URL, that is, where this request was sent, the status code (if the reply is already available), and other request and reply headers (see Figure 18.9).

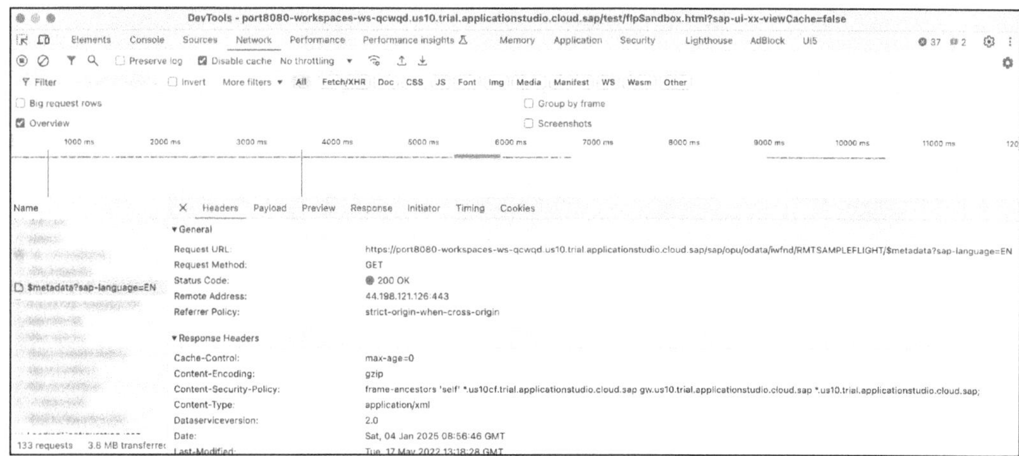

Figure 18.9 URL and Headers of a Request

In our case, we've opened the exact request that was sent to the OData service to request the metadata file. In the preview, we can see that the metadata was returned as XML (see Figure 18.10).

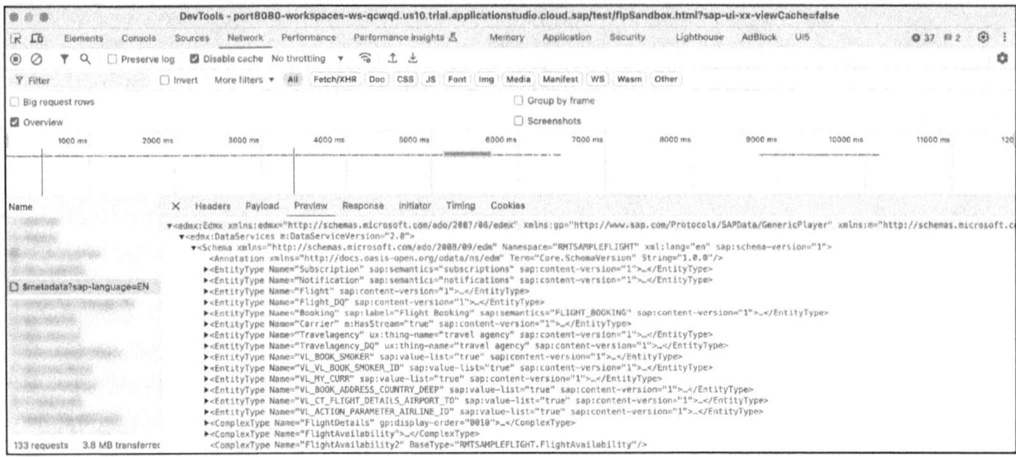

Figure 18.10 Preview of the Response

However, we can often see little in this small preview window, and it's difficult for us to access the data. In this case, however, we can double-click on the entry in the list of requests on the left, and the same call will open for us in a new tab (see Figure 18.11). This makes it much easier for us to search for entity sets, entity types, properties, or navigation properties.

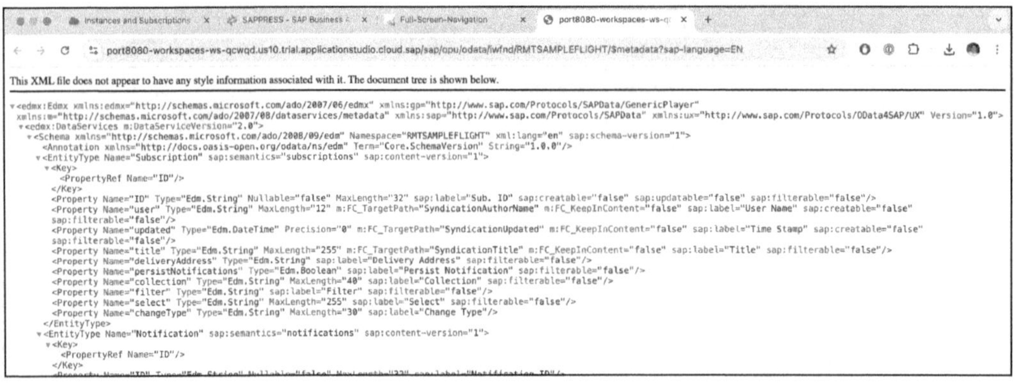

Figure 18.11 Opening the Request in a New Tab

The same applies to all other requests sent to the OData service. Here, too, we often want to open the individual requests in a new tab to test the primary key, and try out the navigation properties, filters, sorters, or other things. However, the individual requests sent to an OData service are no longer sent individually but bundled as batch requests. A batch request can contain one-to-many (1:M) HTTP requests and receives the responses of all 1:M requests in a single reply. You can see such a `$btach` request, as illustrated in Figure 18.12.

Figure 18.12 Batch Request May Be Used Instead of Single HTTP Requests Sent to the SAP System

If you were to double-click on this `$batch` entry, a new tab would open, but with an error message and not, as expected, with a request to the */CarrierCollection* API of the OData service (see Figure 18.13).

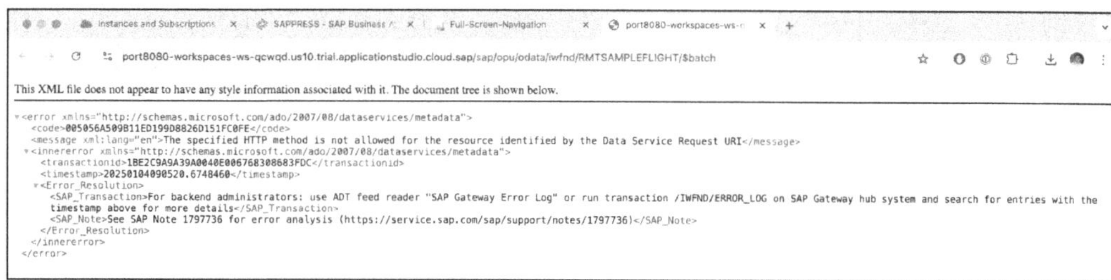

Figure 18.13 Batch Request Can't Be Opened in a New Tab as Easily

This is because, on one hand, the individual requests, like the `GET` request for */CarrierCollection*, are bundled together in the payload (body) of the batch request (see Figure 18.14 ❶ and ❷), and, on the other hand, batch requests are sent as `POST` requests instead of `GET`. If we're opening a URL/request in the browser, this request automatically results in a `GET` request.

We also have a tip for you here. Batch requests were invented to send individual requests to the same SAP system in a bundled form, thus (if you do everything right) saving time but also resources. You can switch off this batch handling in the application descriptor under `settings` for an OData model (see Listing 18.2). If you set the property `useBatch` to `false`, the application will no longer send batch requests via the OData model, but individual requests instead. You can remove this parameter again before going live, or even leave it in, at your discretion, as it wouldn't be applied in the preview.

```
"": {
  "dataSource": "mainService",
  "preload": true,
  "settings": { "useBatch": false }
}
```

Listing 18.2 Deactivate Batch Handling in the Configuration of the OData Model

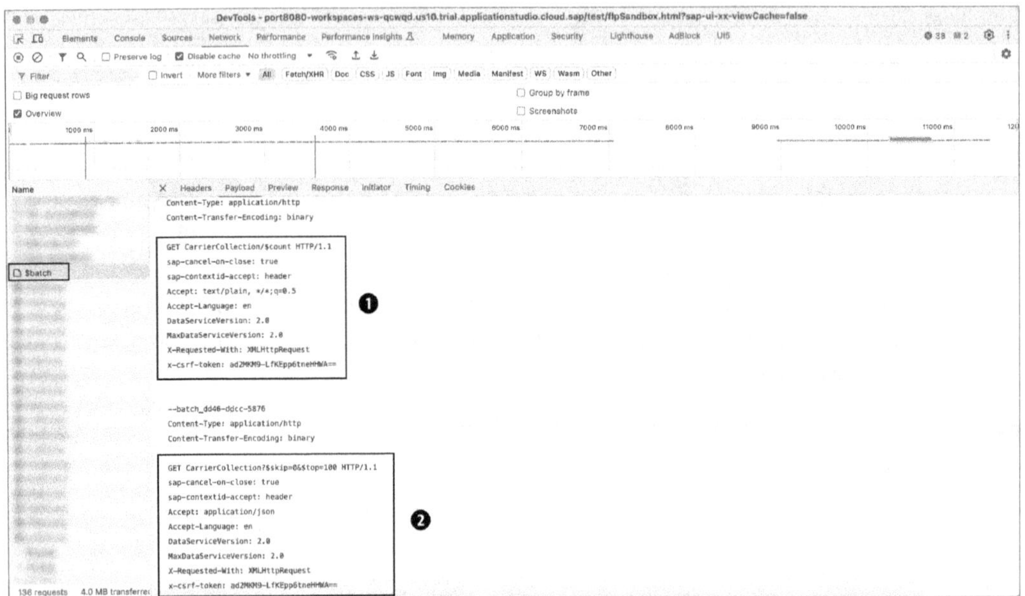

Figure 18.14 Batch Requests Bundle Single HTTP Requests in the Body and Are Sent as POST Requests

We won't discuss any of the other tabs from the DevTools in this section because they offer little support in the SAPUI5 framework. To access individual elements in the DOM and change certain properties and layout options or to find out information about data binding, we'll get to know other tools in this chapter.

18.2 Support Assistant

The *Support Assistant* is a tool that developers can use to check whether their applications have been developed in accordance with SAPUI5 best practices and guidelines. The aim of the tool is to reduce maintenance and consulting times and to optimize the development of SAPUI5 applications. With the help of a predefined *ruleset*, the Support Assistant checks various aspects of an application, such as accessibility, performance, data binding, and usability. With a single click, the current state of the app can be analyzed, and corrective action can be taken based on the results.

The easiest way to open the Support Assistant is to add the URL parameter `sap-ui-support=true`. This parameter causes the tool to appear as a toolbar in the footer of the app:

```
https://port8080-workspaces-ws-qcwqd.us10.trial.applicationstudio.cloud.sap/
test/flpSandbox.html?sap-ui-support=true&...
```

The tool allows you to select specific rulesets and start the analysis of the app with them. It also offers the option of saving rules and settings in the browser's local storage so that work can continue even after the browser window is closed. The **Available Rulesets** tab shows the currently loaded rulesets that are used in the application (see Figure 18.15). On the left, there's a list of the available rules by library, while on the right, detailed information about the selected rule is displayed. You can select the rules to be executed during the analysis by checking the checkboxes in front of the rules. The list of rulesets dynamically adjusts to the libraries used in the current state of the application. By clicking on the Analyze button, you can execute the analysis run.

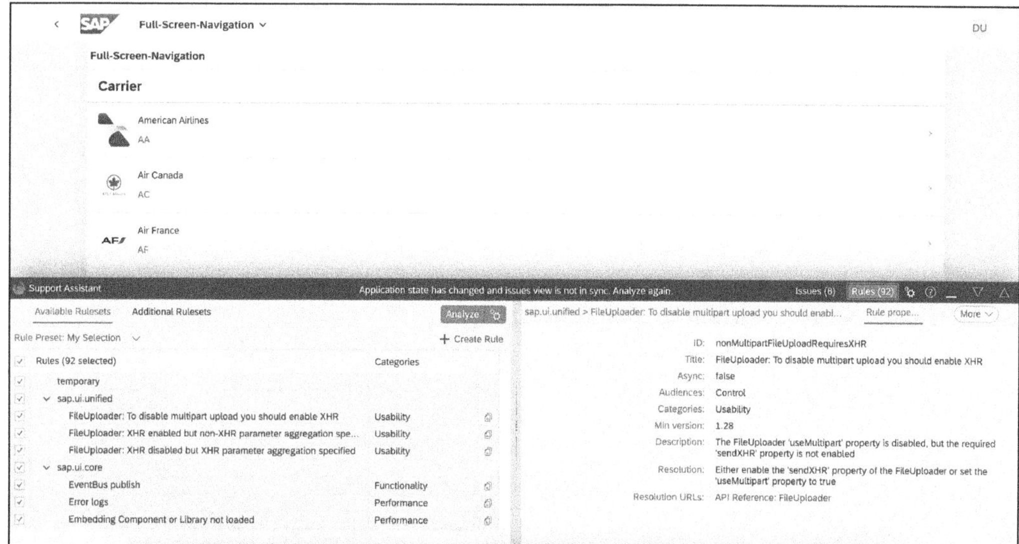

Figure 18.15 Support Assistant Showing the Available Rules for the UI Libraries Used in the App

The **Additional Rulesets** tab shows rules for libraries that aren't currently used by the application (see Figure 18.16). Selected rules can be loaded via this tab to move them to the **Available** Rulesets and use them in the analysis.

After an analysis, the Support Assistant displays an overview of all triggered rules, including their description, solution steps, and a control tree with highlighted problematic elements (see Figure 18.17).

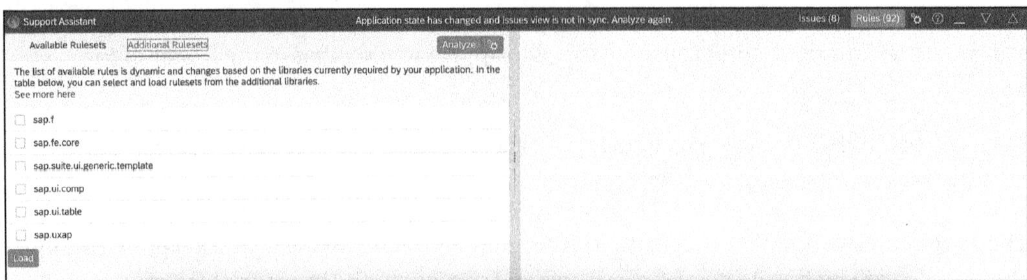

Figure 18.16 Additional Rulesets Can Be Included

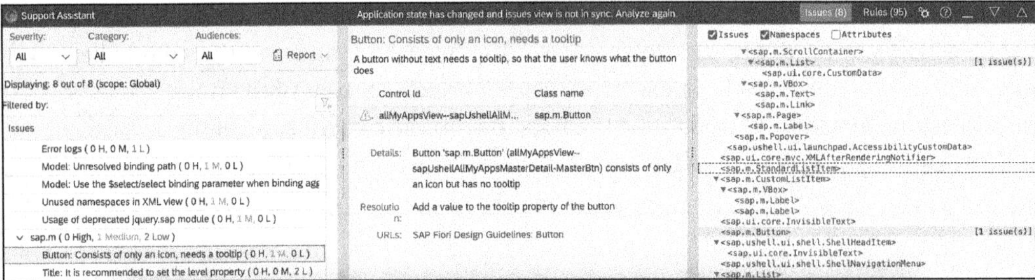

Figure 18.17 Analysis Showing the Issues, Severities, Resolutions, and the Effected UI Elements

On the left, a list of triggered rules is sorted by severity (**High**, **Medium**, and **Low**). Drop-down menus allow filtering by severity, category, or target. The **Report** option generates an HTML report with the current rules and the analysis area, which can be displayed in a new tab or downloaded as a ZIP file. Detailed information about the selected rule is displayed in the center, including the following:

- General description of the rule
- Steps to resolve the problem
- Useful links such as API references or documentation
- Severity of the problem
- Namespace and ID of the affected element

On the right, the element tree of the application is displayed, whose root is designated as `<WEBPAGE>`. Rules that don't target specific controls are assigned to this element. When such rules are triggered, the corresponding problems are also assigned to the `<WEBPAGE>` element.

Hovering over an element in the tree highlights it in the application (if visible). Selecting an element with problems loads them into the issues list. In addition, the table in the issues list shows all other elements that have triggered the same rule, along with the respective severity level.

18.3 UI5 Inspector

The UI5 Inspector is an open-source extension for the Chrome DevTools that allows developers of SAPUI5 applications to inspect, analyze, and support apps. The extension is available for SAPUI5 versions 1.28 and later. This extension can be searched for and installed via the Chrome Web Store (see Figure 18.18).

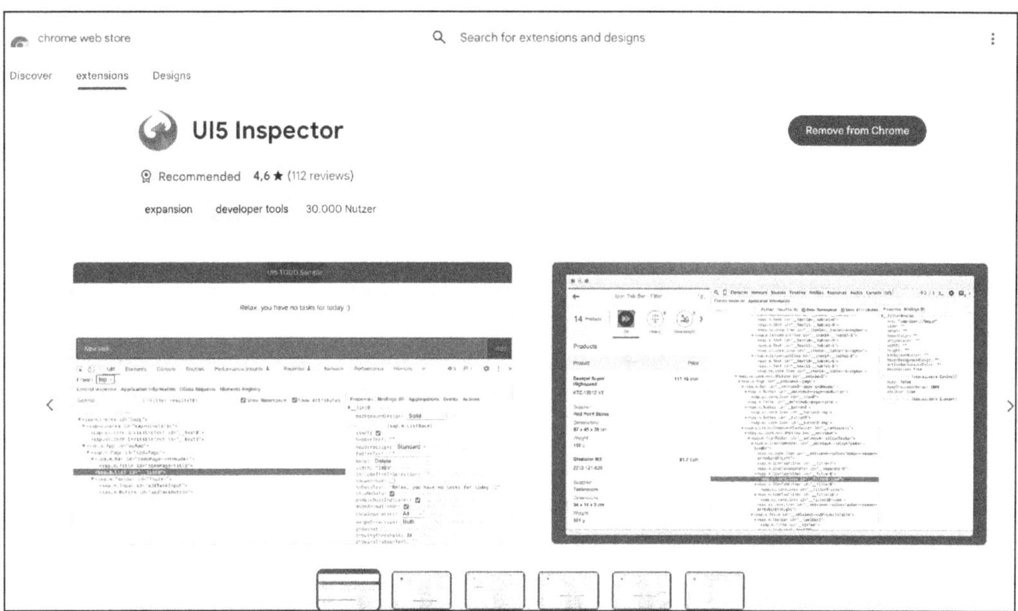

Figure 18.18 UI5 Inspector as a Chrome Extension in the Chrome Web Store

The UI5 Inspector is part of the DevTools and can be found in the DevTools on the far right, next to the tabs by default. The structure and nesting of the SAPUI5 controls are displayed there first. Developers can search and filter for specific controls and show or hide their namespace and attributes in the tree. Hovering over a tree branch highlights the corresponding control in the app (see Figure 18.19). Alternatively, right-click on a control in the app, and select **Inspect UI5** control to select and inspect it directly in the tree (to do so, the DevTools must be opened).

On the right, the **Properties** tab lists the set and inherited properties for the selected control. Values can be changed directly, with the changes being validated by the framework and rendered immediately. Errors for incorrect values are displayed in the console in the DevTools, as already presented.

Furthermore, there is the **Bindings** tab (see Figure 18.20). The data bindings of a control are displayed here, including the number of data bindings specified in the title. The tab provides details on models, paths, and values. A link to the corresponding binding file leads to further information in the **Model Information** section.

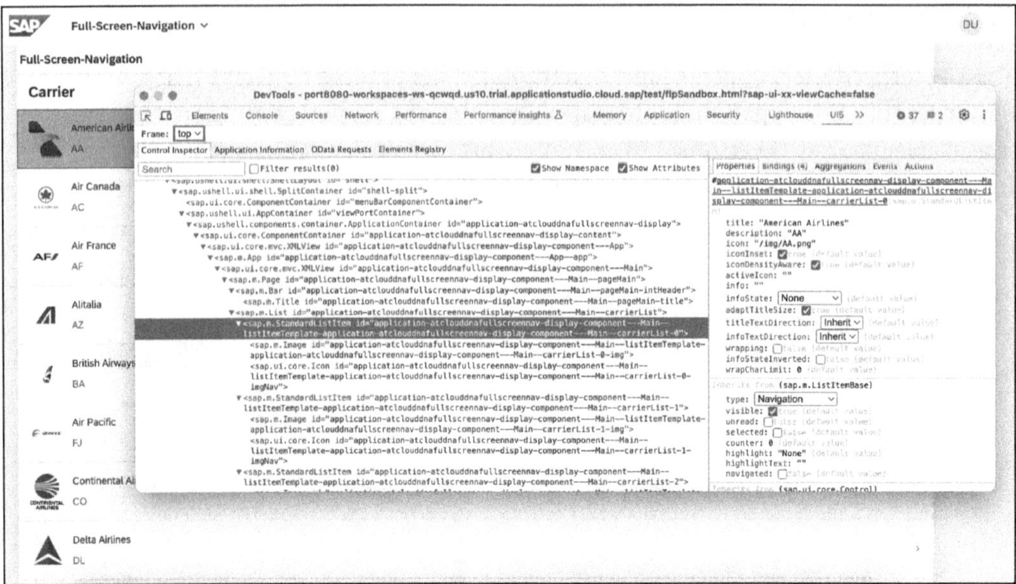

Figure 18.19 Searching for UI Elements with the Component Inspector and Change Properties at Runtime

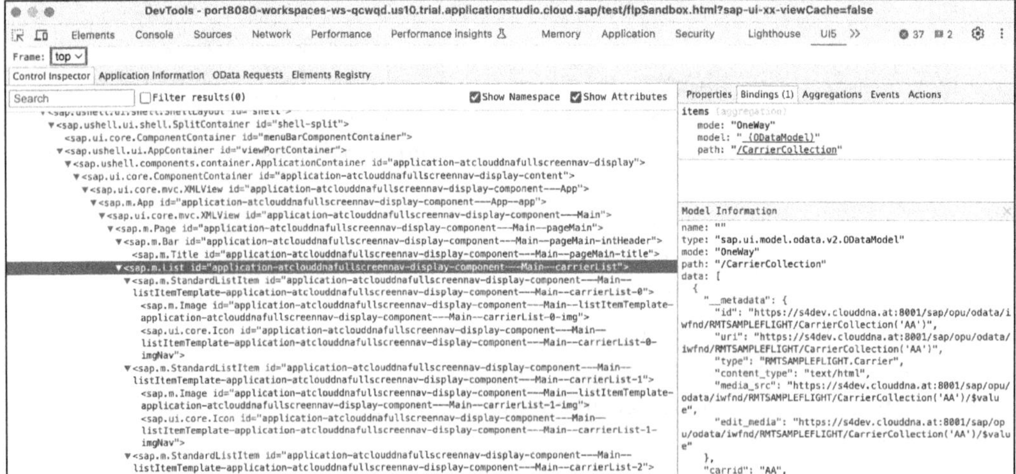

Figure 18.20 Bindings Can Be Examined If Element, Property, and Aggregation Binding Was Used

The **Aggregations** tab displays all aggregations of the selected UI element, including which data and other UI elements they contain (see Figure 18.21).

The Events tab lists the events and, if an event handler/listener is assigned, the corresponding function (see Figure 18.22).

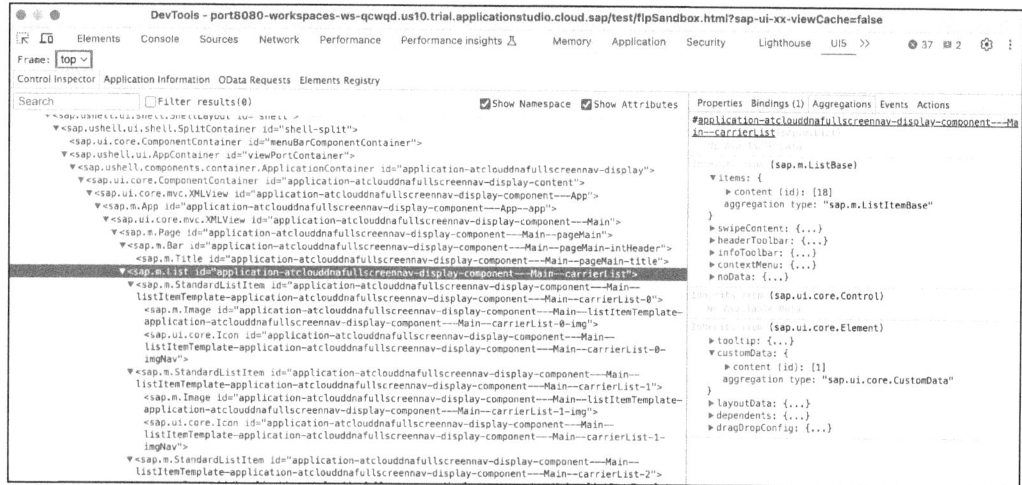

Figure 18.21 Aggregations Tab Showing the Used Aggregations with Their Data

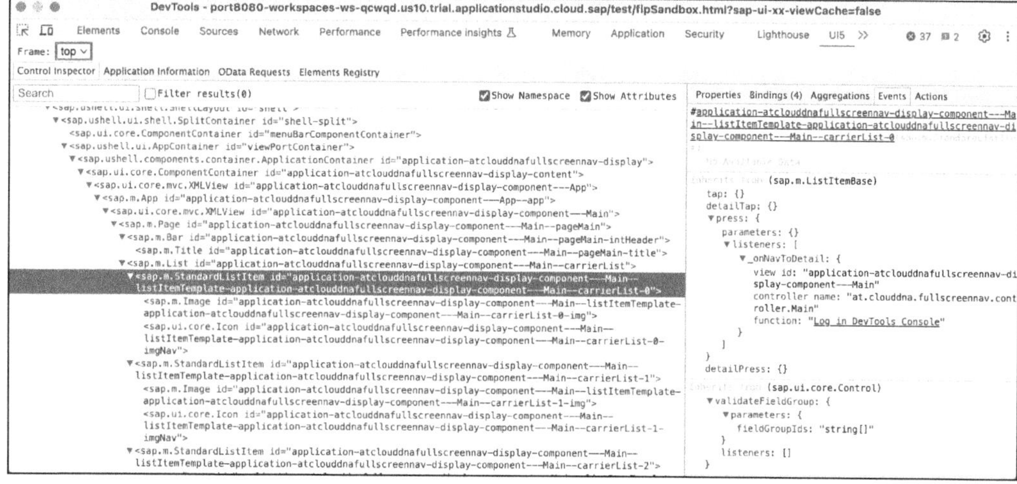

Figure 18.22 All the Used Events and Their Registered Listeners in the Events Tab

The upper tab, **Application Information**, contains general information about the app, such as the exact SAPUI5 version, the browser version, and the app URL (Figure 18.23). Information about loaded libraries and modules is hidden by default but can be viewed if required.

The **OData Requests** tab is located next to **Application Information**. Here, you can search for OData requests, as you can in the **Network** tab of the DevTools. The difference here is that you get a list that is really filtered only for OData requests and that the batch requests can be resolved and broken down into the individual requests (see Figure 18.24).

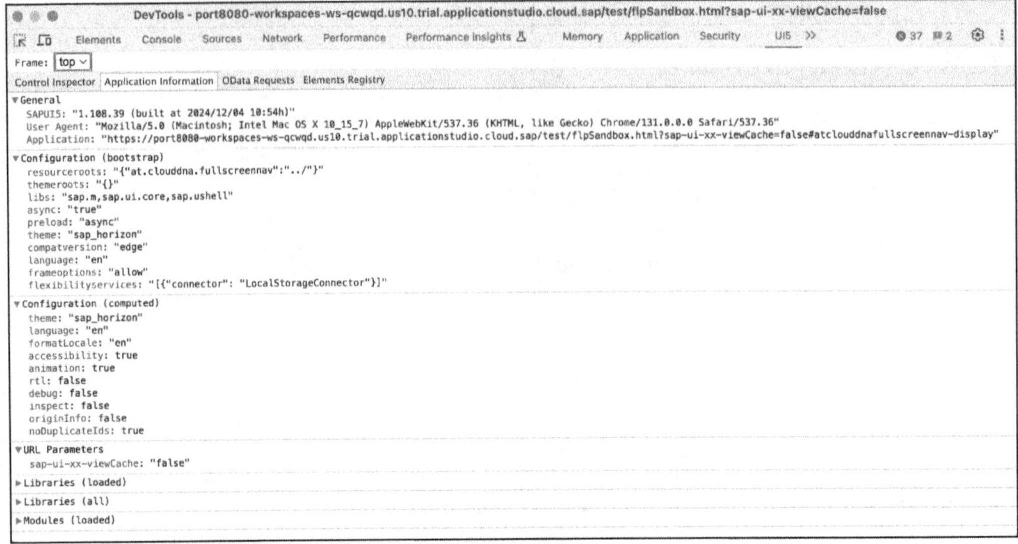

Figure 18.23 See All the Relevant Application Information and Libraries Loaded by the Framework

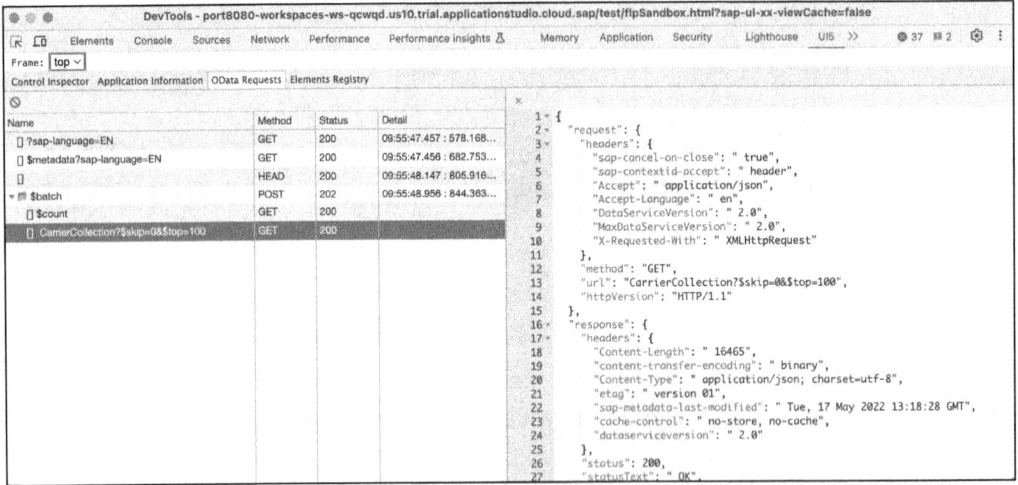

Figure 18.24 More Detailed Overview of All OData Requests Sent by the App

18.4 Diagnostics

The *Diagnostics Window* in SAPUI5 is a supportive tool that runs within an existing SAPUI5 app. The window is opened with the keyboard shortcut [Shift]+[Ctrl]+[Alt]+[S] (Windows) or [Shift]+[Ctrl]+[Option]+[S] (macOS) (see Figure 18.25). It provides **Technical Information** about the app, such as the bootstrap configuration and loaded libraries. In addition, debugging can be activated, and end-to-end traces can be created to analyze communication within the app.

The **Control Tree** shows all controls used in the app. Developers can change properties, update bindings, or set breakpoints to check the behavior of individual methods or controls. The export function can be used to export content to a ZIP archive and import it again, for example, for external analysis.

The window enables breakpoints at object and class level. It also allows you to temporarily change the SAPUI5 version of the app to perform compatibility tests or error checks. Developers are given a detailed insight into class methods and their execution using the browser debugger console (**Debugging** and **JavaScript Trace**).

A special feature is the collection and display of performance data for interaction steps. Developers can analyze load times, data transfer sizes, and OData statistics. This is activated via URL parameters or directly in the interface (**Interaction** and **Performance**). The results are detailed and show, for example, end-to-end duration and server processing times.

Captured performance data can be exported as a ZIP archive and reimported. This makes it possible to pass it on to other teams or experts for detailed analysis.

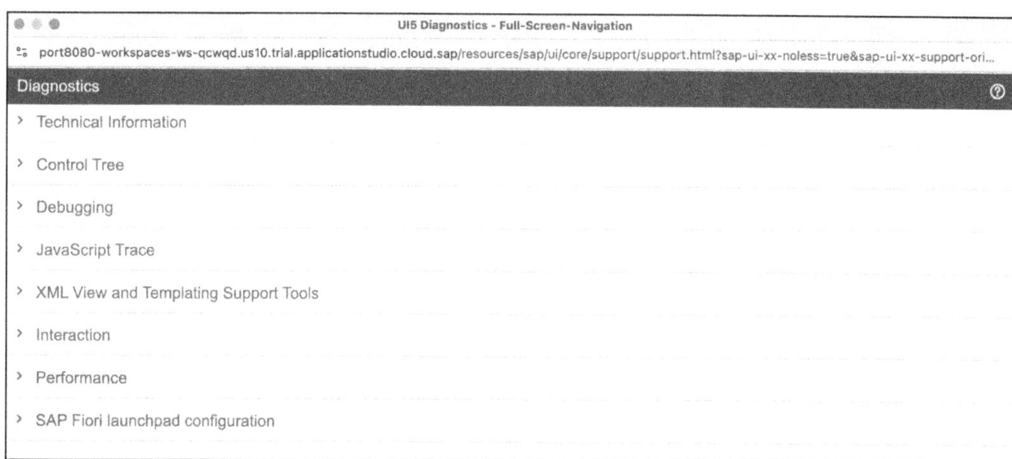

Figure 18.25 Diagnostics Window with a Few More Options Compared to the UI5 Inspector

18.5 Static Code Analysis

Static code analysis is a process in which source code is examined without being executed to identify potential errors, security vulnerabilities, or violations of coding guidelines. They help to improve code quality, ensure maintainability, and identify bugs early in the development process. In software development, static analysis is frequently used in continuous integration/continuous delivery (CI/CD) pipelines to perform automated quality control, ensuring more reliable and secure software.

ESLint is an open-source project that helps find and fix problems in JavaScript code, whether it's running in the browser or on the server, no matter if the project is using a framework or not (like SAPUI5, Angular, and others). ESLint analyzes code statically to quickly catch errors. It integrates with most text editors and can also be run in a continuous integration pipeline. Many of the problems ESLint detects can be fixed automatically, and it's syntax-aware, so it doesn't introduce errors through traditional search and replace algorithms. ESLint allows you to preprocess code, use custom parsers, and write your own rules to work alongside ESLint's built-in rules. This means that ESLint can be configured exactly according to the project's requirements.

Rules are the core component of ESLint. They check whether code meets certain expectations and define actions if it doesn't. Many rules offer additional configuration options, such as whether JavaScript statements should end with a semicolon (the semi rule). ESLint comes with hundreds of built-in rules, but you can also create your own or use rules from plugins. Some rules can automatically fix violations using the `--fix` command or via editor extensions, without changing the application's logic. In addition, rules can make suggestions that cause changes to the logic but aren't automatically applied. The ESLint configuration file contains the settings for rules, plugins, shared configurations, and target files. An ESLint plugin is an npm module that can contain rules, configurations, processors, and languages, often including custom rules. Plugins support style guides, JavaScript extensions (e.g., TypeScript), libraries (e.g., React), and frameworks (e.g., Angular). A parser converts code into an abstract syntax tree that ESLint can analyze. By default, ESLint uses the Espree parser for compatible JavaScript. Custom parsers allow the analysis of nonstandard syntax and are often integrated into plugins or configurations so that they don't have to be used directly. The CLI allows you to run linting commands in the terminal, while the Node.js API can be used to integrate ESLint programmatically into Node.js applications.

SAPUI5 uses ESLint to check JavaScript but also TypeScript sources, that is, controllers and libraries. The activated rules are defined at the root level in the project directory in the .eslintrc.json file and apply to all JavaScript and TypeScript code in the project. When creating a project, you can specify whether you want to configure advanced options 1 or to use ESLint in the last step of the wizard 2 (see Figure 18.26).

To run an ESLint check, we open the terminal, switch to the project's root level, and execute the following command:

```
npm run lint
```

The defined rules are run through, and both warnings and errors are output to the console (see Figure 18.27).

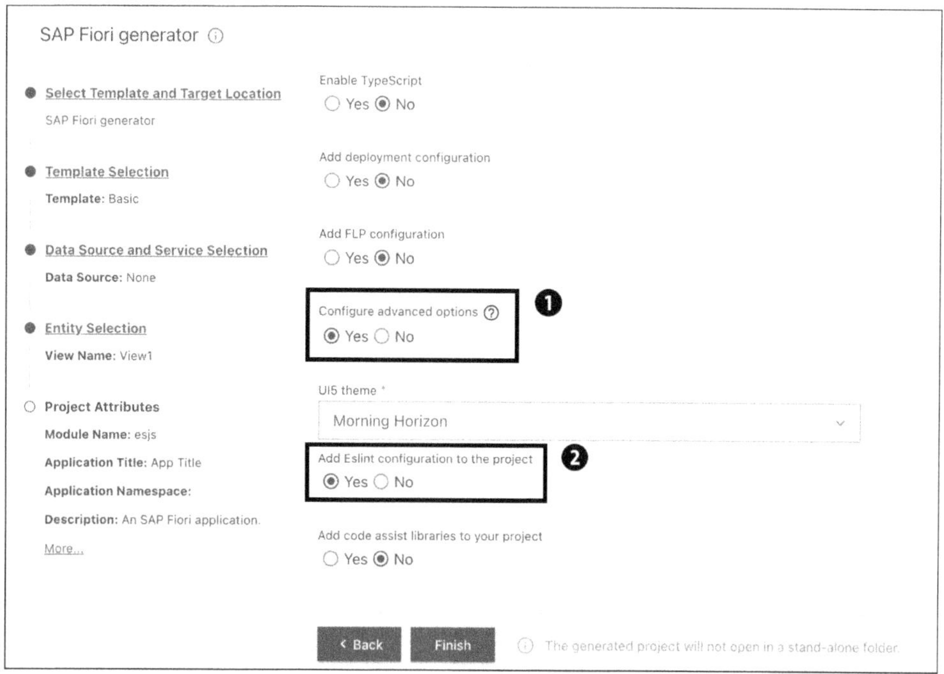

Figure 18.26 Activating ESLint via the Wizard in the Application Generator

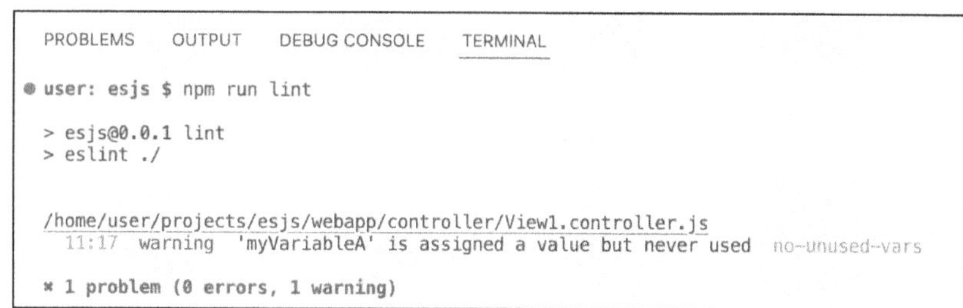

Figure 18.27 ESLint Checking If Your Source Code Breaks the Rules

You can find out which rules are applied and which additional rules you can add from the set of predefined rules in the official OpenUI5 documentation: https://github.com/SAP/openui5/blob/master/.eslintrc.json.

These ESLint rules can be configured with different levels of severity:

- **off or 0**
  Rule is deactivated and won't be checked.
- **warn or 1**
  Active rule, but only with a warning as a result—no exit code.
- **error or 2**
  If the rule is broken, the lint command will exit with an error.

These levels provide fine-grained control over how ESLint applies rules. Further configuration options are described in the documentation.

This static code analysis is not only executed explicitly but also implicitly when a build process (npm run build) is carried out in the project, and of course when the application is deployed (npm run deploy), because of the fact that the deployment contains the build at first or when the application is tested in SAP Business Application Studio or VS Code (e.g., with npm run start).

18.6 Testing

When developing applications, we place a high value on clean, testable code to ensure good test coverage and high app quality. It's important to examine test methods and implement new features using *test-driven development* (TDD) and best practices. This focus on testing pays off in the long run. Automated tests ensure app quality, avoid frequent manual testing, and minimize errors, for example after code changes or updates to the SAPUI5 framework. In addition, test coverage facilitates future refactorings, as we can ensure that existing functions continue to work as expected.

SAPUI5 offers various testing options such as unit tests, integration tests, and the mock server, which we've already covered in Chapter 4. Before implementing tests, it's important to consider the various aspects of the application and suitable test methods. The agile test pyramid illustrates the use of appropriate test tools for different test levels (see Figure 18.28). If we wanted to classify the previously presented ESLint, it would be at level 0 (lower than **Unit Tests**) because static code analysis isn't part of the classic agile testing.

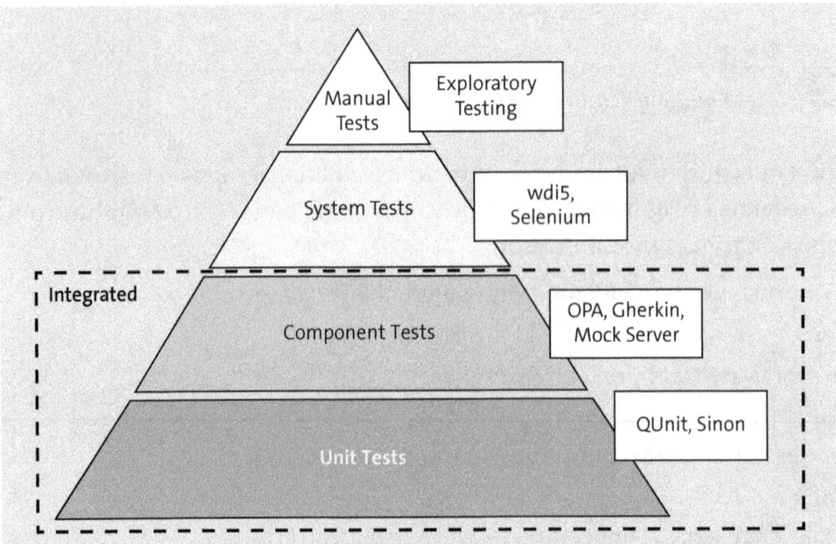

Figure 18.28 Testing Pyramid and the Supported Tools in SAPUI5 (Source: UI5 Demo Kit)

The following should be tested:

- Public functions that have been introduced by your development team.
- Overridden getters and setters.
- Events of the control and their call frequency.
- All possible user interactions (e.g., typing, keyboard, focus).
- Re-rendering the control on interactions (only if inefficient or missing rendering is suspected).
- Right-to-left (RTL) mode if JavaScript adjustments are done.
- Integration tests for composite controls (but no need to cover all the child controls).
- Default property values.
- Interactions of the control with models (OData, JSON).
- Destruction of the control: Check that dependencies/events have been correctly removed.

What shouldn't be tested:

- Nonoverridden getters and setters (tested by the framework).
- Complete CSS (only check if the correct classes are set)—focus on testing JavaScript and TypeScript instead.
- Generic framework functionality—focus on your own controls.

As just mentioned, instead of testing the framework, we should try to test our own content. So, in the next two subsections, we'll look at how we can test the controller and its JavaScript/TypeScript coding, but also how we can test the interactions with our view and the integration between view and controller. Section 18.6.1 covers QUnits that we can use to write unit tests for the methods programmed in the controller. With OPA5 tests, we can simulate interaction with the UI as if we were the end user. We'll discuss this in Section 18.6.2.

18.6.1 QUnit

QUnit tests provide good support for the asynchronous testing that is often required in UI functional testing, such as waiting for rendering, animations, or backend calls to complete. A QUnit test page can be executed independently in the browser without an additional tool, which makes it easier to create and execute individual tests. Furthermore, a test suite already exists by default and was created by the wizard when the project was created.

When writing QUnit tests, they should always be atomic, that is, independent of other tests and without global dependencies. Each test function should only check one aspect, and created controls must be destroyed after use. Tests should be clear and readable, with descriptive variable names, and shouldn't unnecessarily reuse or test too many internal aspects. It's important to design tests in a way that they can fail at first before code is fixed, and make sure all assertions are executed, especially for asynchronous behavior. For this, the use of `expect` in QUnit helps. Complexity should be split to make tests and production code easier. General SAPUI5 functionality shouldn't be tested; the focus is on the specific functionality of the respective control. Additionally, tests should be as small as possible, with no unnecessary code. The use of modules and `beforeEach`/`afterEach` can help to better organize tests, but setup shouldn't be too global or complex. Asynchronous tests can often be replaced by fake timers, and duplicate tests of the same behavior should be avoided.

Remember our Split app from Chapter 7? In the **Detail** view, we had a table of flights for which we used a formatter, as shown in Listing 18.3. Since then, the app has been further developed by several people, and the formatter may have been adapted. Nevertheless, we now want to write a test so that the original definition of the color setting is applied correctly.

```
public stateAvSeats(seats: int, seatsocc: int): string {
        let iDiff = seats - seatsocc;
        if (iDiff <= 15) {
            return "Error";
        } else if (iDiff < 25) {
            return "Warning";
        } else {
            return "Success";
        }
    }
```

Listing 18.3 Formatter Used in the Split App, Which Needs to Be Tested

As already mentioned, the test suite already exists, so we don't have to make any configurations at this point. It's in the *test* directory under *unit*. We'll see how to start the tests in a moment, but here's a hint that all the tests are handled thanks to the file *unitTests.qunit.ts* (see Figure 18.29). Here, all tests are run one after the other in a `Promise.all`. You can see that there's currently only one import on line `7`, namely the import of the tests for the *Main.controller.ts*. These tests are also located in this *unit* folder. If you create new views and controllers in your project and test the functions of the new controller, you have to make a new entry in this *unit/controller* directory and complete the reference in *qunitTests.qunit.ts*.

In Listing 18.4, let's look at the source code of the *MainPage.controller.ts* file, that is, the file that contains the tests for the *Main.controller.ts*. We see that a new test block begins with `QUnit.module`, marking the start of all tests for the formatter in question. After that comes `QUnit.test`, a method that will execute a test. This test has a descriptive sentence and, as a second parameter, the test itself. In this test, the *arrange-act-assert pattern* ensures a structured approach. It consists of the following steps:

- **Arrange**
  Prepare the dependencies and options of the system under test (e.g., constructor objects, spies/stubs/mocks, models). In this step, the system under test is created and, if necessary, rendered.
- **Act**
  Execute the function to be tested, ideally in a single line of code.
- **Assert**
  Execute QUnit assertions to check the expected results. Not too many assertions should be used per test, and unexpected scenarios should also be checked.
- **Cleanup (optional)**
  Clean up the created controls/models and reset spies/stubs/mocks.

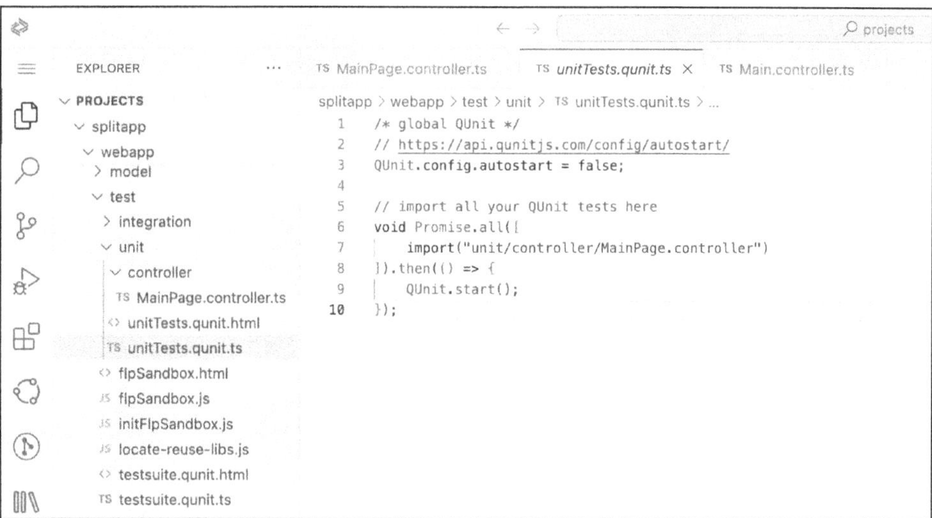

Figure 18.29 Tests That Will Be Executed When Starting the Test Suite

In our first test, we now check whether the formatter returns the value `Error` if there are only `15` (`16` maximum seats minus `1` occupied seat) seats available. We check this with a `strictEqual` (i.e., in JavaScript language with `===`).

```
/*global QUnit*/
import Controller from "at/clouddna/splitapp/controller/Main.controller";
import { MessageType } from "sap/ui/core/library";
QUnit.module("Available Seats State");
QUnit.test("Should format the seats with available seats lower than 16 to
Error", function (assert: Assert) {
    //Arrange
    const oAppController = new Controller("Main");
```

```
    //Act
    const sState = oAppController.stateAvSeats(16, 1);
    //Assert
    oTestData.assert.strictEqual(sState, MessageType.Error);
    //Cleanup (optional)
    oAppController.destroy();
});
```

Listing 18.4 Test Checks If the Desired State Is Returned by the Formatter

A script for running the tests has also already been prepared and stored in *package.json*. You can find the unit tests in the preview options under `unit-tests` (see Figure 18.30), or you can start them manually in the terminal using `npm run unit-tests`.

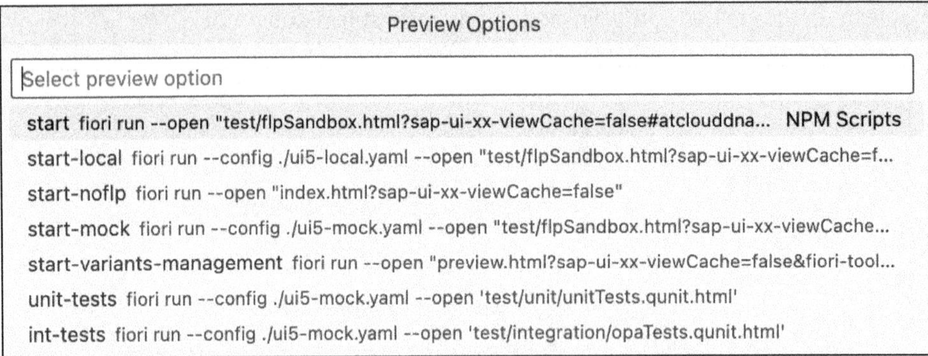

Figure 18.30 Preview Options Already Include the Scripts for Testing

In the new tab that opens, all tests bundled per controller in *qunitTests.qunit.ts* are run. As you can see in Figure 18.31, our first test was also run, and the result was a true positive.

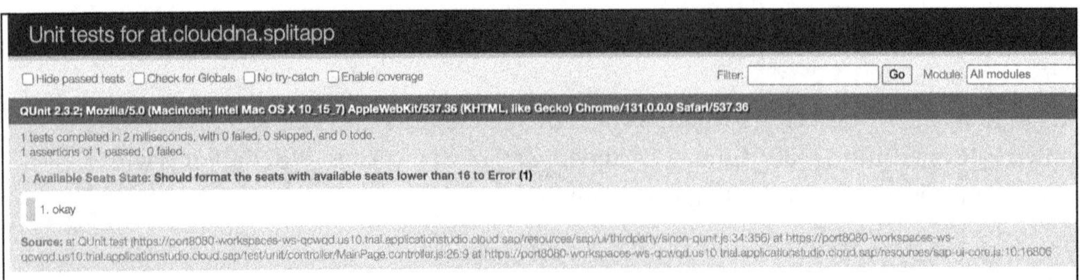

Figure 18.31 Testing Results Will Be Displayed in a Separate Tab as an HTML Page

In the next step, we have to consider how we can best test the formatter for all combinations. So, in this construct, we'll use the boundary values for each colored marking because we know from the original definition that this is desired, as shown in Table 18.1.

Range	State
<= 15	Error
15 < and <= 25	Warning
25 <	Success

Table 18.1 Ranges with the Desired State We're Expecting to be Returned by the Formatter

Using *boundary value analysis* (BVA), we find out that it would be sufficient for us to test the boundary values 15, 16, 25, and 26 because then we've covered all areas and the correct change of sate within these areas. In addition, we could also test inner values such as 14, 20, and 30, but we'll refrain from doing so for our example. Nevertheless, testing inner values alongside boundaries ensures comprehensive coverage, detects hidden logic errors, and validates typical use cases beyond edge conditions.

We must not forget the principle of *don't repeat yourself* (DRY) when writing the tests either. We won't copy all the tests and insert the same coding in each one. As shown in Listing 18.5, we've written a function `availableSeatsStateTestCase` that receives the `oTestdata` and executes the actual test. We've defined this test data as a separate TypeScript type. The tests can now be executed cleanly one after the other by calling this generic method.

```
type TestData = {seatsMax: int, seatsOcc: int, assert: Assert, expected:
MessageType};
function availableSeatsStateTestCase (oTestData: TestData){
  //Arrange
  const oAppController = new Controller("Main");
  //Act
  const sState = oAppController.stateAvSeats(oTestData.seatsMax,
oTestData.seatsOcc);
  //Assert
  oTestData.assert.strictEqual(sState, oTestData.expected);
  //Cleanup (optional)
  oAppController.destroy();
}
QUnit.test("Should format the seats with available seats lower than 16 to
Error", function (assert: Assert) {
  avSeatsTestCase({ seatsMax: 16, seatsOcc: 1, assert: assert, expected:
MessageType.Error });
});
QUnit.test("Should format the seats with available seats higher than 15 to
Warning", function (assert: Assert) {
```

18

```
  avSeatsTestCase({ seatsMax: 17, seatsOcc: 1, assert: assert, expected:
MessageType.Warning });
});
QUnit.test("Should format the seats with available seats lower than 26 to
Warning", function (assert: Assert) {
  avSeatsTestCase({ seatsMax: 26, seatsOcc: 1, assert: assert, expected:
MessageType.Warning });
});
QUnit.test("Should format the seats with available seats higher than 25 to
Success", function (assert: Assert) {
  avSeatsTestCase({ seatsMax: 27, seatsOcc: 1, assert: assert, expected:
MessageType.Success });
});
```

Listing 18.5 DRY Implementation of the Tests Testing the Boundary Values

When running the tests, we now see that something can't be right with the range `iDiff <= 25` (see Figure 18.32).

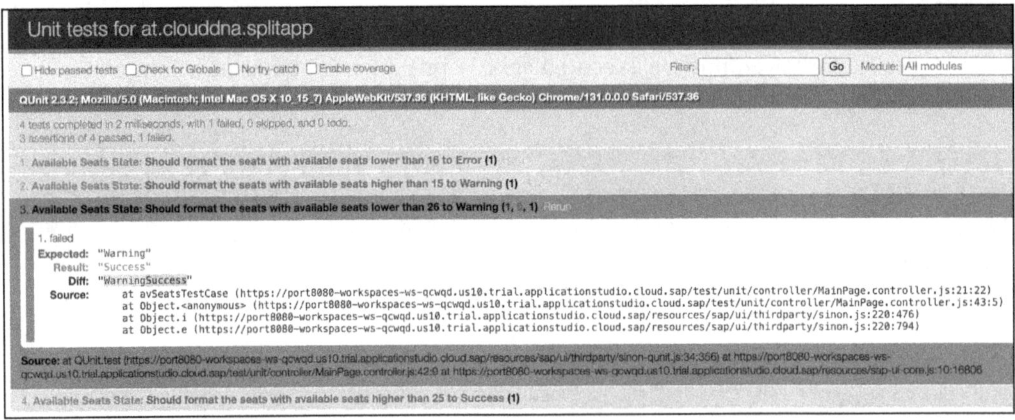

Figure 18.32 Test Failed Because the Result Didn't Match the Expected Value

This is because either we made a mistake when writing the formatter or someone unintentionally changed it. We mistakenly queried for `< 25` and not for a `<= 25`. We'll correct this immediately in our formatter in *Main.controller.ts*:

```
} else if (iDiff <= 25) { return "Warning"; }
```

Now all tests are run, and we can be sure that the states returned by the formatter match the definition (see Figure 18.33). Even if several people in *Main.controller.ts* have work to do in the future, we can ensure that our formatter continues to work by running *regression tests*.

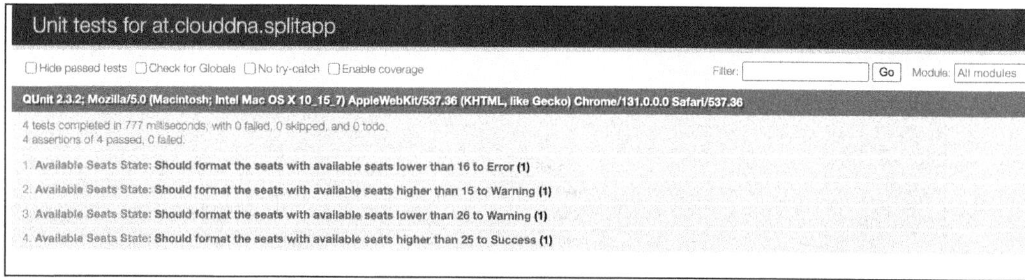

Figure 18.33 All Tests Finished True Positive

18.6.2 OPA5

OPA5 provides *one-page acceptance tests* via an API for SAPUI5 controls that abstracts asynchrony and facilitates access to SAPUI5 elements. OPA5 is particularly suitable for testing user interaction, navigation, data binding, and integration. Because OPA5 is JavaScript-based, tests can be written in the same language as the app, enabling quick access to functions, easy familiarization, and debugging.

The advantages of OPA5 include tight integration with SAPUI5, fast feedback, the use of polling instead of timeouts for asynchrony, and support for TDD. Tests follow the arrange-act-assert pattern, which increases readability and comprehensibility. In addition, tests can be executed directly in the browser without additional plugins or apps. This saves time because regressions decrease, and the tests become more stable. However, OPA5 has limitations: it doesn't support screen capturing, testing across multiple pages, or remote test execution. End-to-end testing isn't recommended due to authentication issues and unstable test data.

We started with a demo app in Chapter 4, which we built up step-by-step. In this app, we had a `SimpleForm` with two input fields (first name and last name), as well as two buttons in the footer for save and cancel. We extended the event handler of the `press` event of the save button and output a `MessageToast` there, which concatenates a static text with the input of the first name. You can see this event handler in Listing 18.6.

```
private onSavePressed(){
  let sValueFirstName = (this.getView()?.byId("inputFirstname") as
Input).getValue();
  MessageToast.show("Successfully saved, " + sValueFirstName);
}
```

Listing 18.6 Event Handler of the Press Event

How can we test it now? A test suite has already been prepared for OPA5 tests too. This is also located in the *test* directory and in the *integration* directory. Here, everything is controlled via *NavigationJourney.ts* (see Figure 18.34). As with a user journey, the journey of the tests runs through this one after the other. We see between lines 24 and 27 that four tests are carried out. Two of them apparently check whether a view has been opened; we've added the other two methods, which we'll explain in the next step.

```
/*global QUnit*/
import opaTest from "sap/ui/test/opaQunit";
import AppPage from "./pages/AppPage";
import ViewPage from "./pages/MainPage";
import Opa5 from "sap/ui/test/Opa5";

QUnit.module("Navigation Journey");

const onTheAppPage = new AppPage();
const onTheViewPage = new ViewPage();

Opa5.extendConfig({ viewNamespace: "at.clouddna.demo.view.", autoWait: true });

opaTest("Should see the message toast after save button press", function () {
    // Arrangements = Given
    // eslint-disable-next-line @typescript-eslint/no-floating-promises
    onTheAppPage.iStartMyUIComponent({
        componentConfig: {
            name: "at.clouddna.demo"
        }
    });

    // Assertions = When
    onTheAppPage.iShouldSeeTheApp();
    onTheViewPage.iShouldSeeThePageView();
    onTheViewPage.iPressOnTheButton();
    onTheViewPage.iShouldSeeTheToast();

    // Cleanup
    // eslint-disable-next-line @typescript-eslint/no-floating-promises
    onTheAppPage.iTeardownMyApp();
});
```

Figure 18.34 NavigationJourney.ts Includes All the Tests That Need to Be Executed

The source code for these methods can be found in the test files provided for each view in the *pages* directory. We're now in *MainPage.ts*, which will test *Main.view.xml*. As you can see in Figure 18.35, we have three methods there. The method `iShouldSeeThePageView` has already been created in the standard. The other two methods contain our test cases, which we'll explain in a moment.

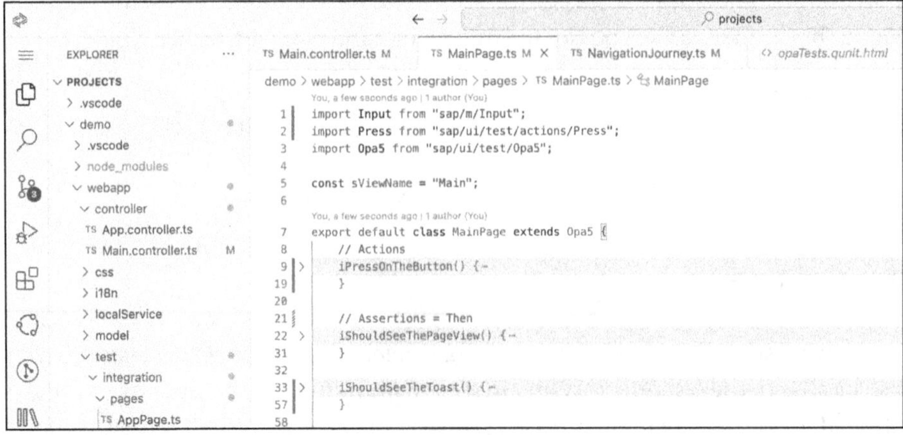

Figure 18.35 Each View Has Its Own Test Page Where Actions and Assertions Are Done

As shown in Listing 18.7, the iPressOnTheButton method triggers an action, namely clicking on the button with the ID saveButton. Thanks to the waitFor method, it waits for this to happen until a timeout is triggered in case of doubt. If the button has been clicked, thanks to the actions definition, we end up in the success callback and can confirm the click with assert.ok.

```
iPressOnTheButton() {
  this.waitFor({
    id: "saveButton",
    viewName: sViewName,
    actions: new Press(),
    success: () => {
      Opa5.assert.ok(true, "The button was pressed");
    },
    errorMessage: "Did not find the save button"
  });
}
```

Listing 18.7 User Inputs and Actions Executed/Simulated in the Actions Section

With the button, you've seen how actions, such as a press, can be triggered. With the MessageToast, we've chosen a more complex case because a message toast has the peculiarity that it's briefly displayed using static methods and then disappears again. We haven't initialized this with an ID, which is why we can't easily access it using the ID. What OPA5 offers here is a way to disable your own autoWait and write your own check method. In this check method, we search for a UI control that has the CSS class sapMMessageToast. As soon as this is found, we extract the text and compare it (for now) with a static string. If this matches, we make another assert.ok (see Listing 18.8).

```
iShouldSeeTheToast() {
  <placeholder>
  return this.waitFor({
    viewName: sViewName,
    autoWait: false,
    check: () => {
      let oMessageToast = Opa5.getJQuery().find(".sapMMessageToast") as any;
      return oMessageToast[0].innerText === "Successfully saved, Daniel";
    },
    success: () => {
      Opa5.assert.ok(true, "The toast message is displayed");
    },
    errorMessage: "Did not find the toast"
  });
}
```

Listing 18.8 Expected Result Compared with the Returned Result within the Assertion

Of course, it's not clean to assert with a static string. For this reason, we've prepared something for you in Listing 18.9 that you should replace instead of the `<placeholder>` in the preceding method. Here, we show you how you can access individual UI controls. Again, using the `waitFor` method, we can search for a UI control with the ID `inputFirstname`, and in the `success` handler, using the first parameter, read this control and save it in a local variable.

```
let oInput: Input;
this.waitFor({
  id: "inputFirstname",
  viewName: sViewName,
  success: (oControl: any) => {
    oInput = oControl;
  }
});
```

Listing 18.9 Accessing All the UI Elements via IDs If Stable IDs Are Used in Each View

Now, we adjust the check method and compare the text in the `MessageToast` with the static text "`Successfully saved,`" concatenated with the first name from the input field:

```
return oMessageToast[0].innerText === "Successfully saved, " + oInput.getValue();
```

If we run the test with the `init-tests` option from the **Preview Options** or start it manually in the terminal with `npm run init-tests`, the *NavigationJourney.ts* is run (see Figure 18.36).

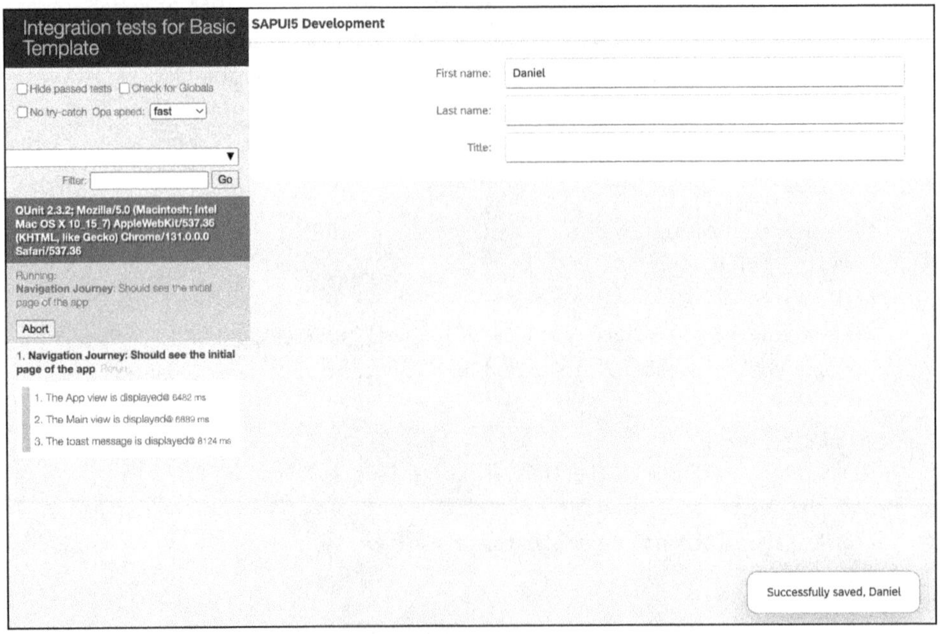

Figure 18.36 OPA5 Tests Can Be Observed during Execution in a New Tab

In contrast to the QUnit tests, you can follow the individual test steps directly in the UI here if you set the **Opa speed**. You don't have to click or type anything, just watch and follow the outcome of the tests on the left.

18.7 Summary

In this chapter, we've learned a few things about debugging and software testing in the SAPUI5 environment. We also looked at how we can debug our source code to access local and global variables and methods that help us develop the application during this debugging process. There, we can test the effect of individual calls or see how we can get the values of certain variables or instances. The built-in developer tools of every common browser are a great help here. In the console, we can see the errors that have occurred, debug the sources as already mentioned, and view the HTTP requests and their replies. The UI5 Inspector extends this circle so that SAPUI5-specific information can also be read, properties of UI elements can be adjusted, bindings can be viewed, or events and event handlers can be examined.

With the Support Assistant, we can run prepared rulesets and check our application for deprecated UI controls and method calls, and we can see if we're violating design guidelines or accessibility requirements. We can also write our own rules. This tool also gives us tips on how to fix the errors. A lot of information can be viewed in the diagnostics window, which is also made possible with the UI5 Inspector. In addition, end-to-end logs and recordings can be created to gain more insights into runtime and performance.

Regarding testing, we've looked at the two most important tools that are delivered with a fully functional test suite by default. In the standard, QUnit tests can be written to test individual functions in the controllers. Here, we are in the test pyramid at the lowest level, where we want to test atomic parts of the application. One level higher are the integration tests, of which OPA5 is an example. Here, we can go up a level, although the complexity can also increase. We're trying to test the functionalities that are offered via the view. OPA5 and the underlying connection to the UI elements via APIs enable us to simulate and test a user journey.

Chapter 19
Security

The security of client-side JavaScript frameworks such as SAPUI5 is crucial, as they act as an interface between users and systems but are vulnerable to attacks such as cross-site scripting or manipulation due to their execution in the browser. Although a framework has only limited influence on the execution environment, it provides important tools such as input sanitization and secure default configurations to minimize risks. However, the main responsibility for security lies with the application, which must make sensible use of these functions and supplement them with server-side measures.

SAPUI5 is a client-side JavaScript library that has been developed and tested securely but doesn't guarantee the security of the entire application. Client-side application generation means that the entire files and scripts that contribute to the functionality of a framework must be loaded and executed on the client. Unlike Web Dynpro, where the application is generated on the server side, and the framework handles the HTML display and communication with the browser, with SAPUI5, the application itself controls the HTML output process and provides its own JavaScript code, which is executed on the client side and regulates communication between the client and server.

This freedom brings many opportunities but also a great deal of responsibility in terms of security. Developers must be aware of security risks and actively take measures to prevent vulnerabilities. In addition, the correct configuration of the HTTP server used is of great importance. Although this isn't part of the typical responsibilities of a developer, it has a major influence on the security of the overall process. Important security functions such as user authentication, session management, authorization checks, and encryption aren't part of SAPUI5. These must be implemented by server-side frameworks or individual adaptations.

Here's one specific example: When it comes to authentication, SAPUI5 can't determine the identity of a particular user itself. To do this, it requires an Identity Provider (IdP) that is familiar with both the SAPUI5 application and the backend. This provider issues a token after a user has successfully authenticated. With this token in hand, the SAPUI5 framework must send this token with every call that communicates with a backend via a service. The server takes the token from the call and asks the IdP whether the token is valid or not. If it's valid, the token can be used to determine which user it is. After that, the process continues on the server side, and the server has to check whether the user is authorized to perform the action at all. Either an unauthorized/forbidden message or a success message can be sent back to the SAPUI5 application. As you can see from this example, SAPUI5 has a small role to play here, but a mechanism like this has a significant impact on the security of an SAPUI5 application. SAPUI5 isn't tied to a specific server implementation or programming language and can be used with SAP NetWeaver Application Server for ABAP, SAP NetWeaver Application Server for Java, the SAP HANA XS engine, or other standard non-SAP web servers. Therefore, the respective security guidelines of the corresponding servers also apply to SAPUI5.

For this reason, we can only offer you a little content in this chapter that specifically concerns the security of an SAPUI5 application. In this chapter, we'll provide an insight into the individual dependent software components and communication partners of SAPUI5. Section 19.1 covers the confidentiality and integrity of the provided and loaded files. In Section 19.2, we'll discuss the browser and its vulnerabilities because it has a great influence on security as the runtime environment on the client. Transport security concerns the communication of the SAPUI5 application with a web server or content delivery network (CDN), or how secure the communication itself is between the communication partners. This will be covered in Section 19.2.3. Section 19.4 briefly discusses server security because SAPUI5 is a client-side framework and only has a small number of server-side configurations.

19.1 Content Security

The *content security policy* offers an additional layer of security to detect and prevent certain attacks such as *cross-site scripting* (XSS) and data injections. The content security policy restricts the sources from which the browser may load resources such as scripts, fonts, and images. Enforcing the use of HTTPS helps to prevent packet sniffing. The content security policy can be activated via the web server configuration (HTTP header `Content-Security-Policy`) or an HTML meta tag in *index.html*. Further generic information can be found in the content security policy specification (*www.w3.org/TR/CSP2*). This defines the policy language used to declare a series of content restrictions for a web resource. It also describes a mechanism for transferring the policy from a server to a client, which is used to enforce the policy.

XSS is a web application vulnerability in which attackers inject malicious code, usually in the form of JavaScript, into a web page that is then viewed by other users. This vulnerability arises when an application fails to properly validate or sanitize user input before inserting it into an HTML page. This allows the attacker to steal sensitive data such as cookies or session IDs, execute malicious code, or manipulate user actions. XSS is divided into three main categories: *stored XSS* (the code is permanently stored on the server), *reflected XSS* (the code is mirrored back via a link or request), and *DOM-based XSS* (the vulnerability arises from manipulation of the Document Object Model [DOM] by client-side JavaScript).

Data injection in JavaScript-based web applications is a security vulnerability in which attackers feed malicious or manipulated data into the application, which is then processed unchecked. Such attacks can occur when user input is incorporated directly into database queries, API calls, or dynamic JavaScript functions without prior validation or sanitization. An attacker could use insecure input to insert SQL injections, command injections, or malicious JavaScript code. This can lead to the disclosure of sensitive data, manipulation of application logic, or the execution of malicious code in the user's browser. To prevent this, input should always be validated and encoded, and secure methods such as prepared SQL statements and secure JavaScript functions should be used. SAPUI5 has a parser for the standard user interface (UI) controls that are delivered, which not only parses the input data before it's written to the model, but also partially validates it. Nevertheless, such checks must always be carried out on the server side because skilled users can also bypass the built-in functions of the individual controls and falsify requests or submit them themselves.

Packet sniffing in HTTP refers to the interception and analysis of data packets that are transmitted between a client (e.g., a web browser) and a server over the network. Because HTTP is unencrypted, attackers who have access to the network (e.g., in public wireless local area networks [WLANs]) can read the data traffic and intercept sensitive information such as passwords, usernames, or cookies. This is made easier by tools such as Wireshark, which can monitor and display network packets in real time. To prevent packet sniffing, data traffic should be encrypted with HTTPS, as Secure Sockets Layer/Transport Layer Security (SSL/TLS) secures the transmitted data and makes it unreadable for third parties.

19.2 Browser Security

Browser security includes topics such as XSS, clickjacking, and the built-in functionalities of HTML5 applications. A browser is considered an unreliable client because users can manipulate client-side code or modify data requests. Therefore, the server must validate all incoming data.

19.2.1 Cross-Site Scripting

As mentioned earlier, XSS is one of the most dangerous vulnerabilities in web applications. Attackers can use it to steal session cookies, execute actions in the current session context, and exploit browser vulnerabilities to execute native code. SAPUI5 protects against XSS when processing and outputting content in controls through input validation and output encoding. However, applications are responsible for the secure encoding of HTML content and data (JavaScript Object Notation [JSON]/XML), the secure processing of JSON/XML data, and the security of custom controls.

Ajax frameworks are a popular target for XSS attacks because not only the HTML initially sent from the server to the browser but also the client-side code that visualizes content can contain vulnerabilities. An additional risk factor with Ajax applications is that injected scripts often remain active for a long period of time. Because Ajax frameworks usually don't reload the entire page but only perform delta updates, running JavaScript code isn't automatically cleaned up.

In the context of SAPUI5, it's important to understand that the framework itself isn't responsible for creating the HTML page that is sent to the client. Therefore, SAPUI5 can't prevent XSS vulnerabilities that are already present in this HTML page. The application is responsible for the correct escaping of user data on the server side. Developers should ensure that user information is prepared in such a way that no JavaScript injections are possible and should refer to the documentation of the server-side rendering framework used. SAPUI5, on the other hand, takes over the correct escaping of all content that is created using the framework's UI controls and displayed on the screen. The application itself must pass user data to the API of a UI control unescaped because SAPUI5 automatically applies the necessary escaping for the respective context in which the data is to be displayed.

An *HTML sanitizer* is a tool or function that is used to remove harmful or unwanted code from HTML documents before they are processed or displayed. It filters out unsafe content such as scripts, inline JavaScript, potentially dangerous tags such as `<script>` or `<iframe>`, or harmful attributes such as `onclick` to avoid security risks such as XSS. The sanitizer usually only allows a defined whitelist of safe HTML tags and attributes, so that the functionality and design of the content are preserved without compromising the security of the application or its users.

SAPUI5 uses a customized version of Google's HTML4 Sanitizer to support HTML5 code. The Google Sanitizer also supports CSS3 code. The HTML5 Sanitizer uses a URL whitelist to check embedded URLs for correct formatting or against a predefined whitelist. To adapt the sanitizer to HTML5, the HTML attributes and elements were reorganized according to the current HTML5 specification of the W3C. All types and flags were carefully reviewed, and HTML4 elements that are no longer used in HTML5 were removed. However, they are still visible as comments. New or changed rules for HTML5 are marked as `new` in the comments. These comments also explain which attributes and elements are assigned to the respective types and flags. Rules that weren't completely clear were analyzed by similarity; for example, audio and video content behave like images.

In addition, there are concepts such as `URIEFFECTS`, which describe whether a URL is loaded within a tag that controls the type of content (e.g., an image) or whether a new document is loaded (e.g., a `href`). `LOADERTYPES` indicate whether content is loaded in a sandbox (e.g., video content) or without restrictions. In SAPUI5, controls that accept arbitrary HTML content, such as `sap.ui.richtexteditor.RichTextEditor` or `sap.ui.core.HTML`, use the HTML5 Sanitizer to sanitize the HTML code of their content and values to avoid dangerous code. The option for cleaning can be activated or deactivated for the respective controls using the corresponding property: RichTextEditor.sanitizeValue or HTML.sanitizeContent. This option is disabled by default for the HTML control and enabled by default for the RichTextEditor.

Additionally, XSS can be prevented by ensuring that it's not possible to insert script code into an application page that is executed in the browser. To do this, controls must prevent scripts from being written to the page that originates from the application itself or from business data stored by other users. To ensure this, two central countermeasures must be combined: the validation of typed control properties and escaping.

The validation of typed control properties is carried out by the SAPUI5 core. This involves checking the value of properties set by the application against the type of the respective property. This ensures that, for example, an integer always remains an integer and an SAP/ui/core/library/CSSSize is a character string that represents a CSS size without containing a script tag. This validation also applies to enumerations or control IDs. Renderers can rely on this check when writing HTML. Typed property values can therefore be written without additional escaping.

Escaping is necessary, however, for strings and other values that originate from the application and aren't typed enough to completely exclude script tags. Developers of custom controls must ensure that such values have been escaped correctly before writing to HTML. The sap/ui/core/RenderManager and SAPUI5 core provide various auxiliary methods for this. To maximize security with a new renderer, certain best practices should be observed. First, the most specific types available should be used for control properties.

For example, sap.ui.core.CSSSize is more suitable than a generic string when dealing with CSS sizes. Developers should also use the auxiliary methods of the RenderManager to correctly escape string properties. Instead of writing values directly to the HTML using write, use writeEscaped. Similarly, for attributes, use writeAttributeEscaped instead of writeAttribute. If these methods aren't applicable, then sap/base/security/encodeXML can be used to escape the strings before they are processed further.

Furthermore, it's crucial to check the HTML coding carefully to ensure that application values can't enter the HTML unsecured. Developers should examine where variable values come from and whether the application can set values directly or whether it can only choose between predefined, hard-coded values. Parameters in method calls of controls must also be escaped, as they aren't automatically validated by the SAPUI5 core. Finally, always keep in mind that XSS can potentially occur anywhere, be it in cascading style sheets (CSS) classes, styles, or other areas.

19.2.2 Clickjacking

Clickjacking manipulates user clicks, for example, by using invisible *iFrames* over a UI. SAPUI5 protects against clickjacking by configuring frame options from version 1.28.0.

SAP has assessed the Common Vulnerability Scoring System (CVSS) v3 base score for this vulnerability at 4.3 out of 10 points. This value represents an estimate of the risk but doesn't consider the specific system configuration or the user's operational environment. SAP recommends that you perform your own risk analysis to evaluate the relevance for your scenario.

Clickjacking, also known as *UI redressing* or *UI redress attacks*, is a technique in which a user's mouse clicks in a web browser are manipulated to trick them into taking unwanted actions on a website. This is usually done by visual deception, where the user doesn't see the actual element they are clicking on and instead believe they are clicking on something else. An example of this is an attacker using HTML frames to embed and overlay a target page in their own page. The frame options configuration of SAPUI5 is a client-side feature designed to prevent security risks such as clickjacking. This configuration is different from the `X-Frame-Options` HTTP header.

This type of attack defeats the usual countermeasures against *cross-site request forgery* (CSRF) by feigning an apparent user interaction with the attacked application. What is particularly problematic is that such an attack can't be detected or traced on the backend side only. This is of interest to attackers if they can get users to execute business-critical tasks via an SAP system that gives the attacker an advantage.

SAP has thoroughly investigated the criticality of clickjacking, considering existing countermeasures. However, there is currently no general standard recommendation from committees such as the W3C that covers broad browser support. The most common protections are based on three approaches: *frame busting*, HTTP response headers (`X-FRAME-OPTIONS`), and *inter-frame communication*. However, due to the complex technical architecture of SAP solutions, none of these measures alone is sufficient to cover all scenarios, especially considering browser support and the costs for customers.

SAP is therefore pursuing a combined approach based on inter-frame communication. For scenarios in which only one domain is used, a frame-busting method based on domain relaxation is used. This solution requires only minimal configuration effort while providing protection for trusted domains. For multi-origin scenarios, the `postMessage` API is used to ensure security. This involves validation against a list of trusted hosts and domains on the backend side. SAP doesn't recommend combining this solution with alternatives such as `X-FRAME-OPTIONS` in an application. To simplify the protection of web applications based on SAP NetWeaver AS ABAP and SAP NetWeaver AS Java, SAP has integrated its own solution into the SAPUI5 framework (and other frameworks too). Applications that use the SAPUI5 framework are protected without code changes, unless otherwise specified in the associated SAP Note 2319727.

SAPUI5's frame-options are a JavaScript-based frontend feature that works in all browsers that are listed in the SAPUI5 compatibility list. When enabled with the values `deny` or `trusted`, an invisible blocking layer is placed over the page, preventing user input such as mouse and keyboard events, while the page content remains visible.

In contrast, the X-Frame-Options header is a server-side feature sent via HTTP response headers. It prevents a page from loading at the browser level when embedding isn't allowed. While this header supports the DENY and SAMEORIGIN options, it lacks broad support for ALLOW-FROM, which is now deprecated in most browsers. This feature must be configured by the backend and isn't fully supported by all browsers.

A modern approach is the content security policy header, which is also sent from the backend. With the frame-ancestors directive, it provides more precise control over trusted pages and is recommended as the preferred method over X-Frame-Options for implementing embedding restrictions.

SAPUI5 provides the following configuration options for the frame options to determine whether an application can be embedded in a separate frame:

- **allow (default)**
  Allows interaction with the application regardless of the origin of the parent frame.
- **deny**
  Denies all interaction with the application.
- **trusted**
  Only allows the interaction if the application is embedded in accordance with the same-origin policy and the trusted origins defined by an allowlist service.

An allowlist service can be configured for embedding an SAPUI5 application in cross-domain frames. This checks whether the application is allowed to be executed in the origin domain of the parent frame.

The *cache buster* mechanism in SAPUI5, on the other hand, makes it possible to instruct the browser to update resources only if the SAPUI5 resources have changed. As long as no changes have been made, these resources can be loaded from the browser cache, which significantly reduces the loading time. SAPUI5 supports this concept for Java and ABAP servers as well as SAP Business Technology Platform (SAP BTP). However, it's not available for SAP HANA XS. 19

To use the cache permanently, the URL in the SAPUI5 bootstrap tag can be changed from resources/sap-ui-core.js to resources/sap-ui-cachebuster/sap-ui-core.js. With this mechanism, SAPUI5 resources remain in the browser cache until a change is made to a UI library or web application. In contrast to conventional methods, in which resources are cached for a fixed period or checked in development mode with the 304 NOT MODIFIED mechanism, the cache buster is specifically optimized for productive scenarios. However, applications must explicitly decide to use the cache buster.

The cache buster is implemented as part of the ResourceServlet. For JavaScript resources such as the bootstrap script, the request is made using the URL resources/sap-ui-cachebuster/sap-ui-core.js. When loading, a JavaScript code is executed that ensures the global SAPUI5 configuration variable exists and the resource targets are defined according to the cache buster. In addition, a timestamp is integrated into the request, which ensures that the resources are updated correctly. All cache buster resources are cached for one year by default.

The process of a request with cache buster works as follows: the first request (resources/sap-ui-cachebuster/sap-ui-core.js) must not be cached because it determines the timestamp and forwards it to the actual resource. Subsequent requests, such as resources/~201207160201~/sap-ui-core.js, are then loaded from the cache. If necessary, the timestamp of the cache buster can be determined by means of the request to resources/sap-ui-cachebuster. The response is in text format and contains the value as ~20120716-0201~, for example.

The cache buster is particularly useful in scenarios where SAP NetWeaver 7.30 or higher is used to run web applications with SAPUI5. A common problem is that users must manually clear their browser cache for software updates to take effect. The cache buster solves this problem by ensuring that updated resources are loaded automatically. It also allows the configuration of private cache controls so that resources are only cached on the user's device.

The *application cache buster* (AppCacheBuster for short) is similar to the cache buster but is used specifically for application resources. SAPUI5 supports the AppCacheBuster exclusively on SAP NetWeaver AS ABAP. This concept isn't available for Java applications on SAP NetWeaver AS Java and SAP HANA XS.

Applications provide an index file called *sap-ui-cachebuster-info.json*, which is created dynamically and contains the last-modified timestamps of all contained files, such as scripts, properties, or other files programmatically loaded via XMLHttpRequest (XHR). Technically, this file maps the request path (below the application's context path) to the corresponding timestamps. The server instructs the client to cache all the listed resources without using the 304 Not Modified mechanism. However, for the index file, the 304 Not Modified mechanism is used to avoid unnecessary loading if the file remains unchanged.

On the client side, this index file is loaded on startup if the configuration option sap-ui-app-cache-buster is activated. The file is used for XHR requests by adding the timestamp as the leading segment to the request path. As long as the timestamp remains unchanged, the URL remains unique, and the resource is loaded from the cache. As soon as the file is changed, the URL parameter is updated, and the resource is retrieved again from the backend. On the server side, the timestamp must be removed from the URL to locate the file correctly. For SAP NetWeaver AS ABAP, this logic is implemented in the Internet Communications Framework (ICF) handler. For both SAP NetWeaver AS Java and SAP NetWeaver AS ABAP, the index file is generated dynamically as and when required.

However, the AppCacheBuster doesn't work across application boundaries. Resources from other applications aren't loaded using this mechanism. The index file includes all files that are to use the cache buster. Unlike the cache buster for runtime resources, each application file has a unique timestamp that is provided in *sap-ui-cache-buster-info.json*.

To activate the AppCacheBuster, the configuration `data-sap-ui-app-cache-buster="./"` must be added to the application's bootstrap script *index.html*. This parameter is an array of strings that can contain a comma-separated list of base URLs pointing to other applications that the AppCacheBuster should consider. By default, this list should include the base path of the local application. These base URLs are used to load the corresponding index files.

19.2.3 HTML5 Security Risks

HTML5 brings numerous new features, which also introduce potential new security risks. The following is an overview of some of the new features and the possible security issues that may arise from their use:

- All modern browsers now offer a *Local Storage API* that can be used to store a limited amount of data in the browser. Access to this data is restricted to JavaScript code that comes from the same domain in which the data was stored. SAPUI5 provides helper functions to make it easier to access local storage in different browsers. However, the browser's local storage isn't a secure store. While it can be used for static data such as enumerations, neither user data nor application data should be stored there. SAPUI5 uses the browser's local storage, for example, for the history capacity of dropdown and combo boxes.
- The growing support for *web graphics library* (WEBGL) in modern browsers enables direct access to the computer's graphics API. However, this can also pose low-level security risks, which is why some browsers don't support WEBGL at all. SAPUI5 doesn't currently use WEBGL functionality.
- *WebSockets* open new possibilities for client-server communication in web applications. However, numerous security problems arose during the first implementations by the browser manufacturers. With the standardization through RFC 6455, WebSockets have now reached a stable state and are supported by Chrome 16 and Firefox 11, for example. Even if the browser implementations are considered secure, the use of WebSockets requires additional security measures on the client side. SAPUI5 doesn't currently use WebSockets.
- Another feature in HTML5 is `postMessage`, which can cause massive security problems if not used correctly. This function enables inter-window communication between windows of different domains and thus opens a gap in the currently implemented same-origin policy of the browsers. As soon as an application listens for the `onMessage` event, it can receive messages from any other browser window. It's the application's responsibility to check the origin domain and only process messages from trusted domains. SAPUI5 uses `postMessage` for debugging and tracing purposes only.

19.3 Transport Security

Transport security includes topics such as encryption and session security. Client- and server-side security alone isn't enough if attackers can read, intercept, or manipulate the data transport between them. By default, HTTP is stateless and unencrypted, which is why it's necessary to configure encrypted connections (HTTPS) and add session management via cookies or URL rewriting. The use of HTTPS (HTTP over SSL/TLS) is not only standardized but also required for SAP applications. SAPUI5 fully supports HTTPS; however, there are restrictions when using the SAPUI5 library over a CDN with HTTPS. Activate and test HTTPS early in the application development process, as switching to HTTPS often causes problems. If the application uses HTTPS, the SAPUI5 library must also be loaded from an HTTPS server.

Even with secure data transport (SSL/TLS), attackers can hijack sessions and execute malicious requests. Typical attacks are CSRF or session fixation. SAPUI5 only offers CSRF protection for data sent via the OData model. This involves reading a CSRF token from the server and using it for write requests. The application itself is responsible for using the CSRF header and implementing further protective mechanisms for other types of server communication.

19.4 Server Security

Server-side security includes topics such as cross-origin resource sharing (CORS) and *resource handlers*. SAPUI5 has only a small server-side component that supports resource loading. Using the resource handler is optional, but SAPUI5 also provides static libraries that can be used with arbitrary HTTP servers. By default, for security reasons, the `XMLHttpRequest` API only allows access to resources in the same domain as the source document. However, CORS makes it possible to load data from other domains. This is important for using web services, OData services, or feeds (RSS/Atom), for example. The server sets special CORS headers in its responses, which tell the `XMLHttpRequest` API whether the requested data can be processed. For SAPUI5 applications that load data from a different server than the application, the CORS headers must be configured correctly on the data server to allow access from the application server. SAPUI5 also uses CORS headers on its CDN so that additional scripts, styles, and resources can be loaded from there. SAPUI5 is delivered with third-party libraries, although security-related aspects must be observed. The included libraries include jQuery and DataJS, which are essential for SAPUI5 to operate.

jQuery is a mandatory library because SAPUI5 is based on it. Although the jQuery website doesn't provide any specific security-related documentation, the jQuery team is known to be security-aware and generally responds quickly to vulnerabilities. SAPUI5 ships jQuery along with its own libraries, giving it the ability to make its own security updates or adjustments to jQuery as needed.

The *DataJS* library is required if OData services are to be used. Here, too, there is a lack of explicit security-related documentation on the official website. However, SAPUI5 integrates the DataJS library and can make its own security adjustments if necessary.

Technically, applications based on SAPUI5 can integrate any custom libraries into their application. However, SAPUI5 can't make any statements about the security of third-party libraries and can't guarantee their security. The responsibility for the security assessment, secure integration, and use of such third-party libraries lies entirely with the respective application.

The SAPUI5 framework provides a client-side API for managing allowed URLs using the `URLListValidator`. This API allows the validation of any URL and is used internally by various controls. One example of how the `URLListValidator` is used is in controls that accept any HTML content, such as `sap.ui.richtexteditor.RichTextEditor` and `sap.ui.core.HTML`. These controls use the `URLListValidator` to check content (sanitization). URLs within their content are automatically removed unless they have been explicitly allowed by the `URLListValidator`. When adding a path to the allowed URLs in the `URLListValidator`, you should ensure that the path is prefixed with a /, for example, */index.epx* instead of *index.epx*. The allowed URLs can be managed with the following methods of the `URLListValidator`:

- `add`
  Adds a new URL to the list of allowed URLs.
- `clear`
  Deletes all entries from the list of allowed URLs.
- `entries`
  Returns all current entries in the list of allowed URLs.

19

URLs are validated using the method `validate`. If no URLs have been explicitly added as allowed, the `URLListValidator` checks by default whether the URL is defined in a valid format.

19.5 Summary

In this chapter, you've seen that ensuring application security has a major impact on the SAPUI5 application, but that the framework itself can't contribute much. Because this is a client-side framework and we must always stick to the motto "never trust a client," additional server-side validation is always required. This may be for authentication, authorization, or user input, which could infiltrate unwanted malicious code/script or data into the system. This framework is already preconfigured to cover all important known and common countermeasures and represents the basic configuration for a secure framework. Especially when the SAPUI5 applications are running on an SAP S/4HANA system, you as a developer will have little to do with the configuration on the client or server side. However, if the application is to be made available on a standard web server, you may be involved in the implementation as well as in some of the configuration options that we've presented in this chapter.

Chapter 20
SAP Build Code

SAP Build Code uses generative AI to further simplify application development by providing developers with suggestions for code, automation workflows, and user interfaces.

Artificial intelligence (AI) is no longer a future trend, but a central component of modern corporate strategies. SAP, as one of the leading providers of enterprise software, is actively driving the integration of AI into its products and platforms. With solutions such as *Joule* and tools such as *SAP Build Code*, AI is not only becoming more accessible to companies but also more practical.

In 2023, SAP introduced Joule, an AI-powered capability that helps organizations make complex processes more efficient. Joule integrates seamlessly with existing SAP solutions such as SAP S/4HANA, SAP SuccessFactors, and SAP Business Technology Platform (SAP BTP). In doing so, Joule provides contextual insights, recommendations for action, and automations based on an organization's unique business needs.

What makes Joule special in the SAP context is its ability to analyze large amounts of data in real time and to provide intelligent recommendations for action based on that analysis. This enables companies to act not only faster but also on a more informed basis. One example of this is supply chain management, where Joule can identify potential supply bottlenecks at an early stage and suggest proactive measures. Human resource management is another area in which Joule can play to its strengths. Joule supports you in the early stages of the recruiting process in identifying talent, optimizing your HR processes, and providing data-based insights that can serve as a basis for strategic decisions.

While Joule provides support primarily at the operational level, SAP Build Code plays a central role in the development of new applications that use AI. SAP Build Code is part of the *SAP Build* suite and enables developers to quickly and efficiently create software solutions using a low-code or no-code environment.

SAP has a clear vision on AI. It should not only be integrated into existing applications but also be used to develop new solutions. The combination of Joule with SAP Build Code enables companies to create AI-powered applications that are customized and perfectly match their specific business needs.

This chapter offers you a comprehensive introduction to the world of *generative artificial intelligence* and its application in SAPUI5 development. At a time when technology is changing the way we develop software and design business processes, understanding these new tools and approaches is vital for being successful and competitive.

Section 20.1 offers you a basic introduction to generative AI. Here, you'll learn what's behind this technology, how it works, and what potential it offers for software development. Section 20.2 then guides you through the installation and configuration of SAP Build Code, a tool specifically designed to make the development of SAPUI5 applications more efficient and flexible. In Section 20.3, we'll create a basic project in SAP Build Code.

Note

At the time of writing, January 2025, SAP Build Code is still a work in progress and does contain a number of bugs. You may not be able to reproduce the results in the chapters due to its current instability.

Finally, in Section 20.4, we discuss specific use cases in SAPUI5 development. You'll learn how generative AI can be used in practice to overcome typical development challenges and create innovative solutions and modernize existing apps.

This combination of theoretical knowledge, practical instructions, and real-world examples will help you make the most of the possibilities offered by generative AI and take your skills as an SAPUI5 developer to a new level.

20.1 Introduction to Generative AI

Before we get into the topic of generative AI, we first need to explain the term artificial intelligence (AI). *Artificial intelligence* refers to the ability of machines to perform tasks that typically require human intelligence. This includes understanding natural language, making decisions, solving complex problems, and learning from experience.

A particularly exciting and dynamic field of AI is generative AI. This specialized form of AI goes beyond data analysis and can independently create content such as text, images, videos, music, or even software code. It works based on large neural networks, in particular transformer models such as *GPT* (generative pretrained transformer), which are trained on gigantic amounts of data. This enables them to learn patterns and structures that they can use to generate new content. In software development, generative AI is becoming increasingly relevant by generating program code or supporting developers with intelligent suggestions.

Generative AI is based on complex mathematical and algorithmic approaches that enable machines to create new content. Its success is based on a combination of large amounts of data, powerful computing resources, and advanced model architectures. Generative AI uses neural networks, deep learning in particular, to process data. One of the most advanced and successful architectures in this area is transformer models, which are designed specifically for sequential data such as text, images, and speech. These models consist of multiple layers of *encoders* and *decoders* that extract and process information from input data and translate it into understandable outputs. A core principle of transformer models is the *attention mechanism*. It enables the underlying model to prioritize relevant parts of the input. It doesn't matter where they are in the sequence. This allows complex dependencies between data to be captured efficiently.

For generative AI to create high-quality content, it undergoes an intensive training process that consists of several phases. In the first step, the model is trained with large amounts of data from a wide variety of sources, such as books, websites, images, music, and—particularly relevant in our case—source code. This data serves as the basis for understanding language or other patterns. For software development, large amounts of program code must be collected to serve as training data. This can come from open-source projects, documentation, or code snippets, for example. In the case of SAP, there are millions of lines of source code available from the company's own software solutions such as SAP S/4HANA, SAP S/4HANA Cloud, SAP SuccessFactors, and SAP Ariba. If you don't want to use the model for just one specific programming language such as ABAP, then it's important to collect data for different programming languages (e.g., Python, Java, JavaScript, TypeScript, ABAP). Both simple and difficult code examples are needed to cover different levels of complexity. A model that works well in one language might have difficulty transferring to other languages.

In the next step, the raw data must be prepared before training. This involves cleaning the data. This includes removing faulty or redundant code to ensure the quality of the data. After that, comments and documentation in the code are recognized and preserved to maintain semantic and explanatory relationships. Next, tokenization takes place. This involves breaking down the source code into individual components, called tokens, for example, keywords, variables, operators, and brackets. Finally, the source code is classified by language, paradigm (e.g., object-oriented, functional), and specific frameworks (ABAP RESTful application programming model, SAP Cloud Application Programming Model, Spring, etc.).

The next step is pretraining. During this process, the model learns to recognize general patterns and structures in the data. The model is familiarized with general programming patterns by being trained on large, mixed datasets. The model learns the syntax rules and keywords of the various programming languages. It recognizes how logical constructs such as loops, conditions, and functions are structured. Common programming patterns such as singleton, factory, and observer are learned through repeated examples. Poor or insufficient training data can lead to incorrect or inefficient results. During the training process, the quality of the generated results is continuously monitored.

After pretraining, fine-tuning takes place. This involves adapting the model to specific tasks or programming languages. The model is optimized separately for each programming language. Additional training data can cover specific frameworks or libraries such as TensorFlow, Spring, ABAP RESTful application programming model, and SAP Cloud Application Programming Model. After that, the model is trained to solve specific tasks, such as creating OData APIs, core data services (CDS) views, and user interface (UI) components based on SAPUI5, SAP Fiori, or SAP Fiori elements.

The model is continuously improved by learning from mistakes. Incorrect generations are analyzed, and the underlying weaknesses of the model are remedied by targeted data enrichment. After that, a bias removal is carried out. This is to minimize distortions caused by dominant language or framework preferences. Optimizing the training process also reduces resource consumption.

After the training is complete, the model can generate code. In doing so, it's important to master the art of prompting. The model works exclusively based on the input instructions (prompts). A clear, specific, and well-formulated prompt gives the model the context, structure, and direction it needs for the generation. Imprecise or vague prompts often lead to irrelevant, erroneous, or superficial results, as the model can't provide its own interpretation or correction. However, precise prompting can be used to generate targeted and high-quality content that is precisely tailored to the user's requirements, significantly increasing the efficiency and usefulness of generative AI.

However, these potentials of AI also entail risks. The automatic generation of content could raise copyright issues, especially if the content created resembles existing works. Another problem is the quality of the generated content: AI systems aren't infallible and occasionally produce erroneous results. Even if the generated code appears plausible at first glance and the accompanying explanations are convincingly formulated, AI-generated suggestions may be flawed. The range of possible problems is diverse, from logic errors to the use of nonexistent functions to references to variables that are named differently. Sometimes the code may run without problems, but it may prove inefficient or potentially unsafe. Additionally, there are challenges of technological dependency and the high energy consumption associated with training large models.

20.2 Installation and Configuration of SAP Build Code

SAP Build Code provides an AI-powered cloud development environment tailored to meet the needs of developers working with SAP Cloud Application Programming Model, SAP Fiori, mobile applications, and SAPUI5. By unifying coding, testing, and application management within a single platform, SAP Build Code simplifies and accelerates the application development process on SAP BTP. It integrates SAP Business Application Studio with the most essential services and software development kits (SDKs) available on SAP BTP. With Joule, SAP's AI copilot, SAP Build Code empowers developers to automatically generate application code, data models, services, and sample data. This ensures that applications are built following the best practices and recommendations outlined in the SAP BTP developer's guide.

SAP Build Code is composed of several SAP services, as shown in Figure 20.1. These services facilitate application development for all key extension use cases in SAP environments.

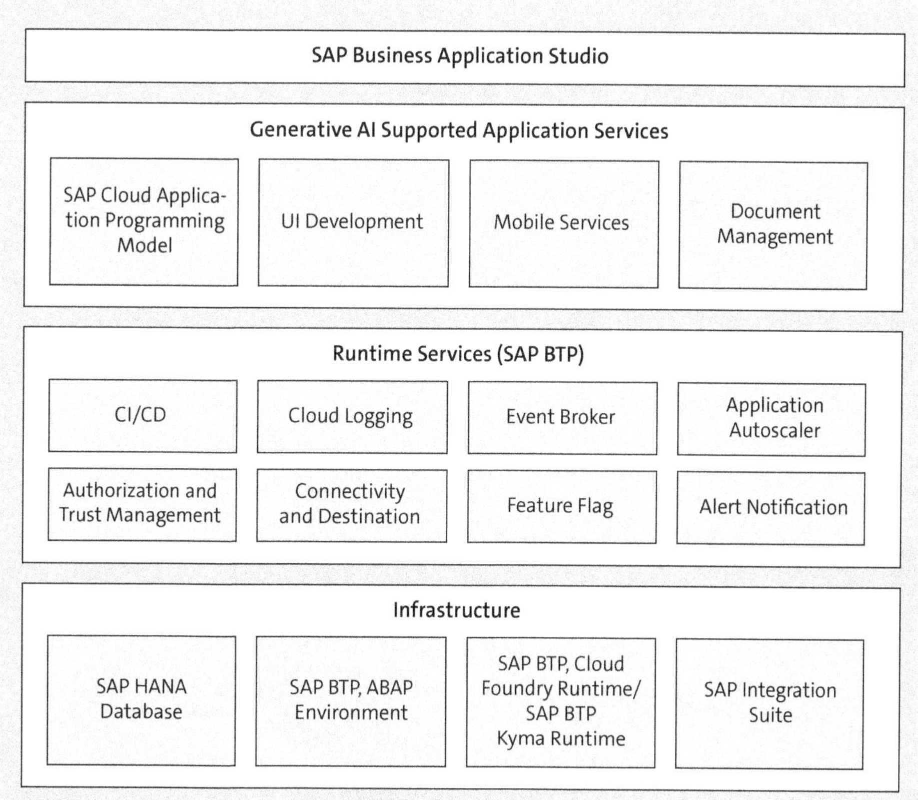

Figure 20.1 SAP Build Code Components

SAP Build Code is provided by SAP in two commercial models, the standard plan and the free plan. You can find details in SAP Discovery Center (see Figure 20.2). The free plan allows you to evaluate SAP Build Code. The standard plan is intended for productive use.

Capacity units are used as the billing metric. The SAP Build Code capacity unit calculator (see Figure 20.3) is provided by SAP so that you can estimate in advance how much the solution you've generated will cost. You can jump directly from SAP Discovery Center to the calculator.

20

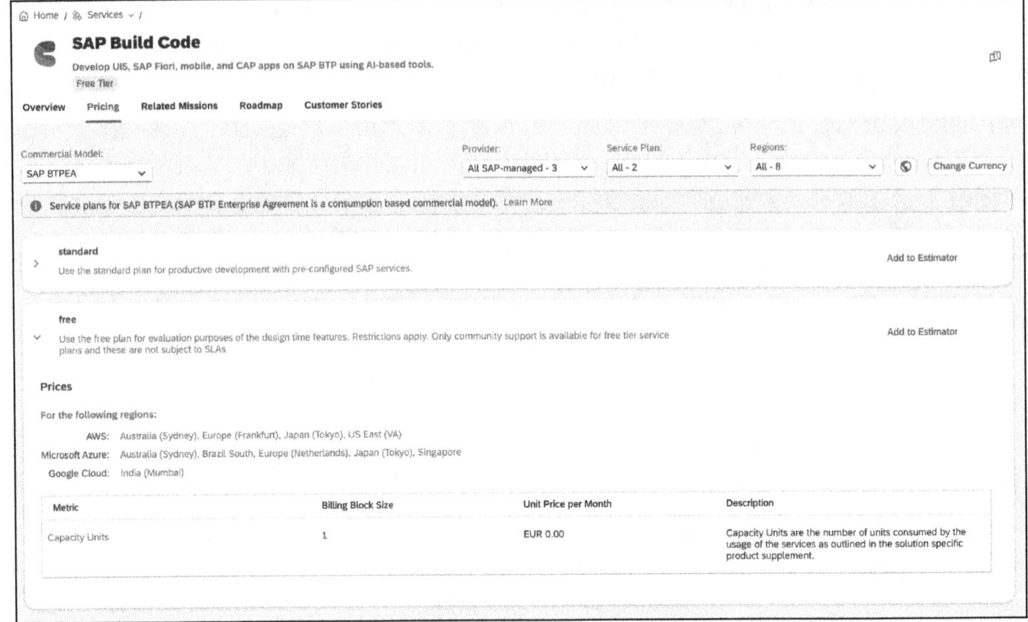

Figure 20.2 SAP Build Code Service Plans

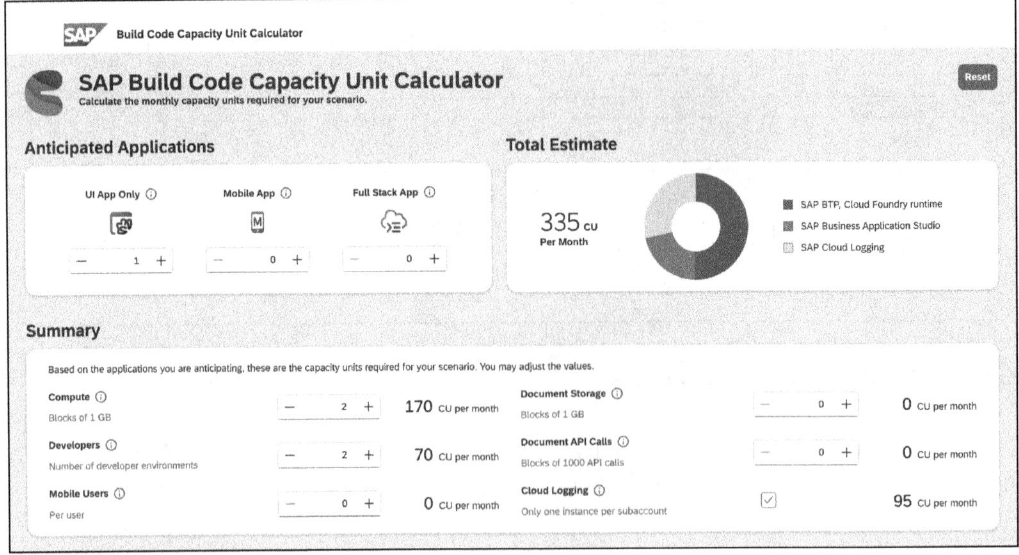

Figure 20.3 SAP Build Code Capacity Unit Calculator

We'll now show you how to set up SAP Build Code in SAP BTP. We'll use the free plan. Alternatively, you can also set this up in a trial account. For most services, including

SAP Build Code, SAP provides a booster that makes setup and initial configuration much easier for you. To do this, navigate to the global account level in the SAP BTP cockpit, and select **Boosters** from the side menu. As shown in Figure 20.4, look for the booster named **Get Started with SAP Build Code (Free Plan)**.

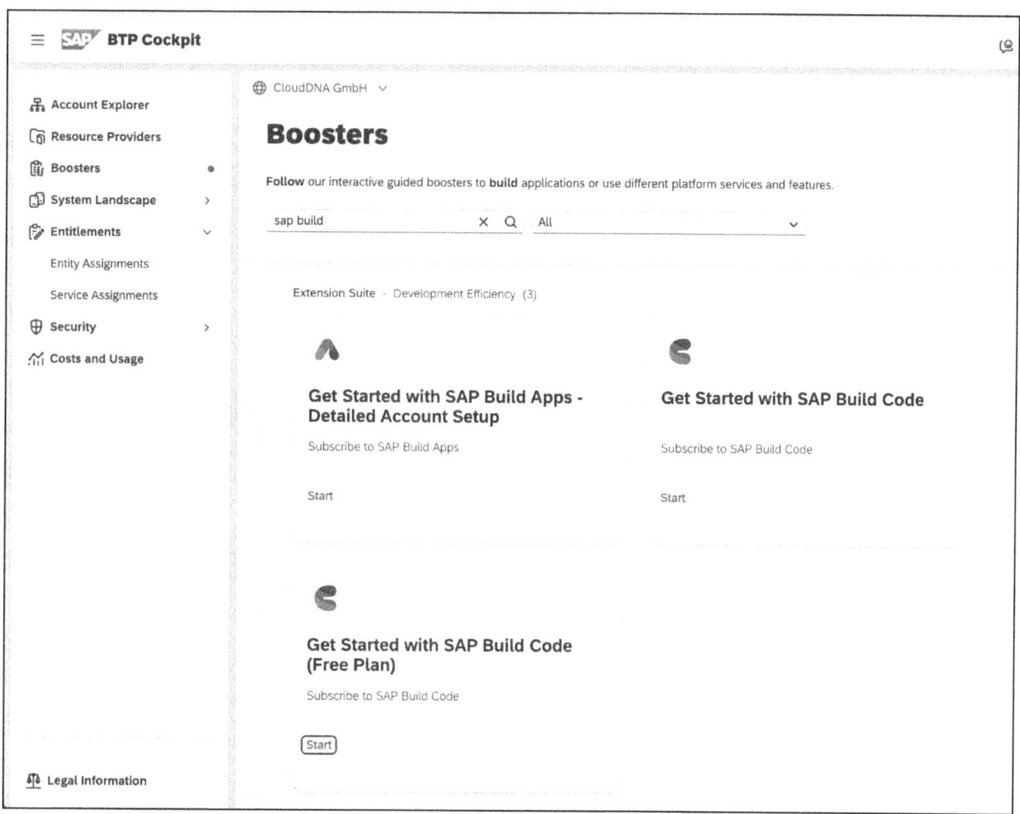

Figure 20.4 Finding the Booster

In the **Overview** of the booster, you can see an architecture diagram, as shown in Figure 20.5, which graphically displays the components used and how they are related. It's recommended that you familiarize yourself with these components, as this will give you a better understanding of the computing units metric in the pricing model. Once you're familiar with the details, click **Start** to launch the booster.

In the first step, the booster checks whether you meet the requirements for its execution. Among other things, this includes checking your authorizations and the available entitlements. This means that it ensures that you have the necessary licenses, which should be the case if you're using the free quota (see Figure 20.6).

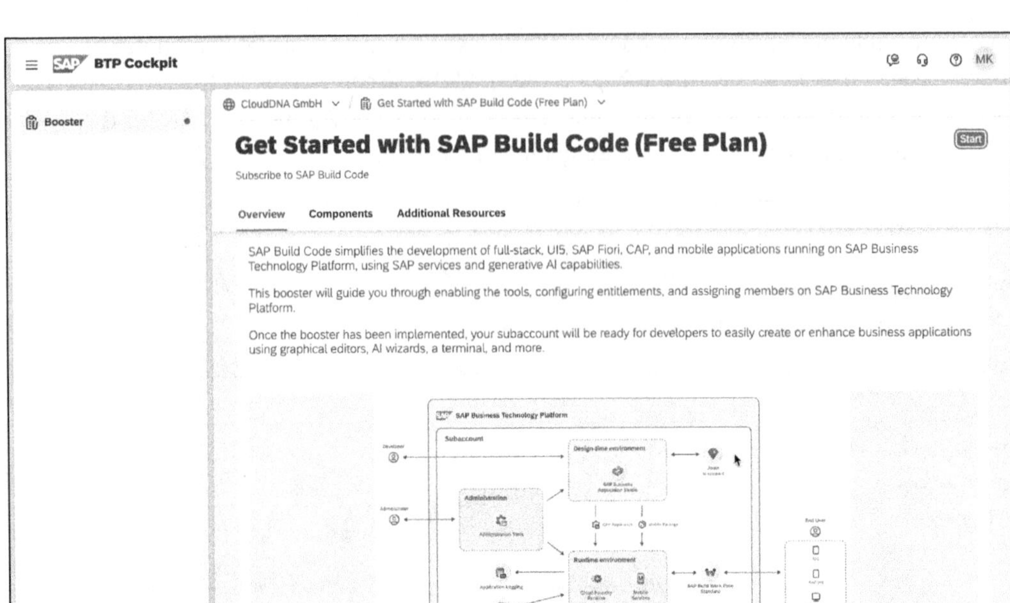

Figure 20.5 Booster Overview

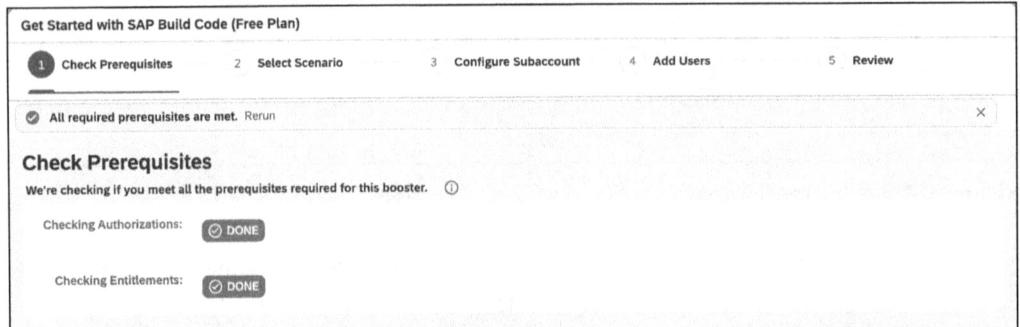

Figure 20.6 Booster Prerequisite Check

In the next step, select the subaccount mode, as shown in Figure 20.7. Here, you determine whether SAP Build Code is to be instantiated in a new subaccount or installed in an existing subaccount. In the example shown, we've decided on the first option of creating a new subaccount. However, if you've already installed and configured SAP Business Application Studio, we recommend that you choose the second option and install SAP Build Code in the same subaccount.

The next step is to configure the subaccount by assigning the required entitlements. As shown in Figure 20.8, you can customize the **Subaccount Name**, **Provider**, and associated

Region, and edit the **Subdomain**. The **Provider** and **Region** are derived from the default settings of your global account.

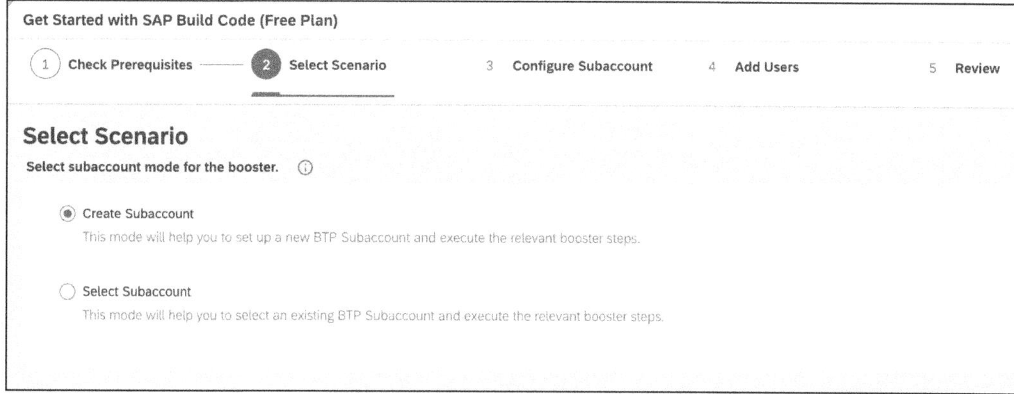

Figure 20.7 Selecting the Subaccount

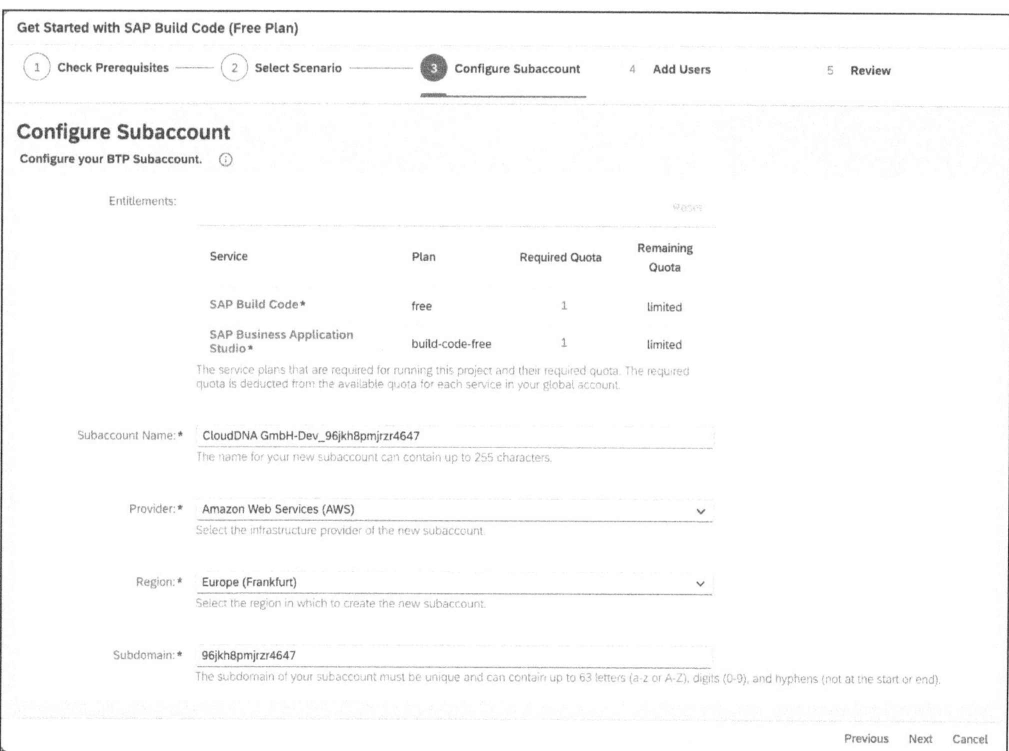

Figure 20.8 Configuring the Subaccount

In the next step, you have the option of adding other users as administrators or developers in addition to the user who is performing the configuration. As shown in Figure 20.9, you can also define which identity provider is to be used for the platform users and which for the application users. By default, the SAP ID service (**accounts.sap.com**) is used as the identity provider.

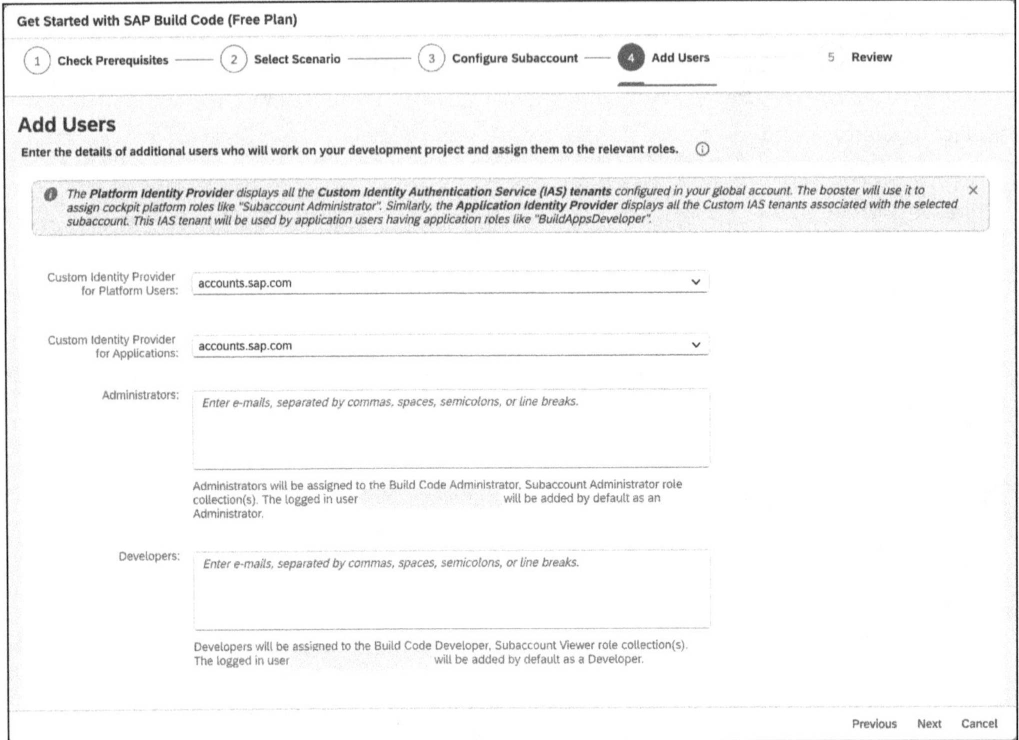

Figure 20.9 Adding Users to the Subaccount

Finally, you'll receive a summary of the settings you've made so far. Check them carefully. If the settings aren't correct, you can navigate back and make the necessary adjustments. After you've completed the configuration by clicking **Finish**, the progress of the steps performed is displayed (see Figure 20.10). This process may take a few minutes.

Progress

Creating Subaccount	PROCESSING
Assigning Service Quotas	PENDING
Subscribing SaaS Applications	PENDING
Creating Service Instances	PENDING
Creating Role Collection	PENDING
Assigning Role Collection	PENDING

Figure 20.10 Installation and Configuration Progress

As soon as the booster has successfully completed the configuration, you'll receive a success message, as shown in Figure 20.11. From there, you can jump directly to the corresponding subaccount to continue your work.

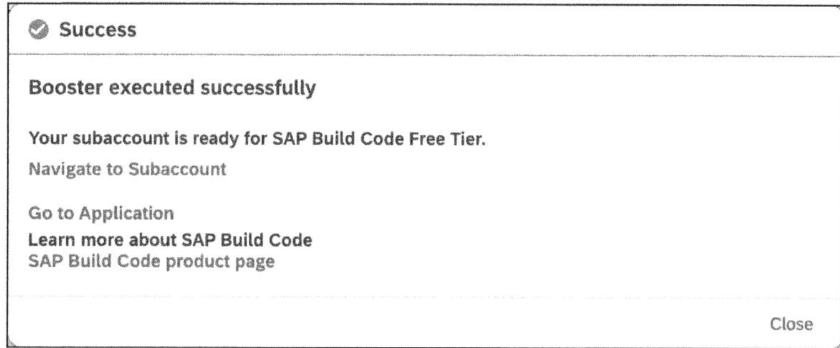

Figure 20.11 Booster Success Message

You can now check the configuration. To do this, navigate to the corresponding subaccount, and open the **Services • Instances and Subscriptions** area in the side menu. There, you'll see that both **SAP Build Code** and **SAP Business Application Studio** have been subscribed to and set up by the booster (see Figure 20.12).

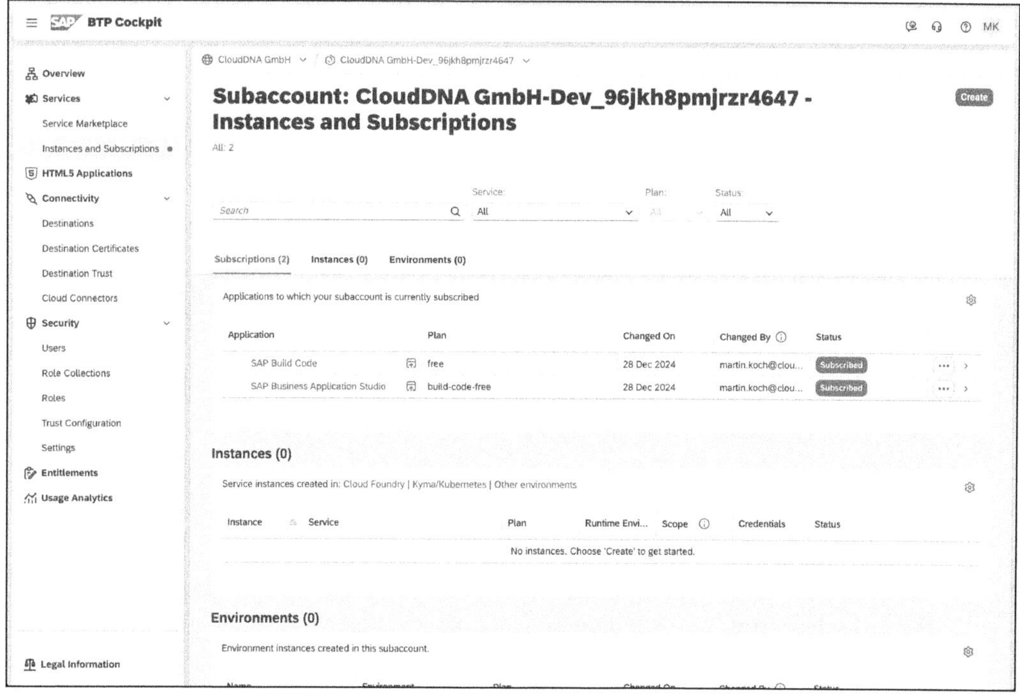

Figure 20.12 Subscriptions at the Subaccount Level

Now navigate to the **Security • Role Collections** area. As shown in Figure 20.13, you can see that the following specific role collections have been created:

- **Build Code - Lobby Admin**
- **Build Code - Lobby Developer**
- **Build Code Administrator**
- **Build Code Developer**
- **Business_Application_Studio_Administrator**
- **Business_Application_Studio_Developer**
- **Business_Application_Studio_Extension_Deployer**

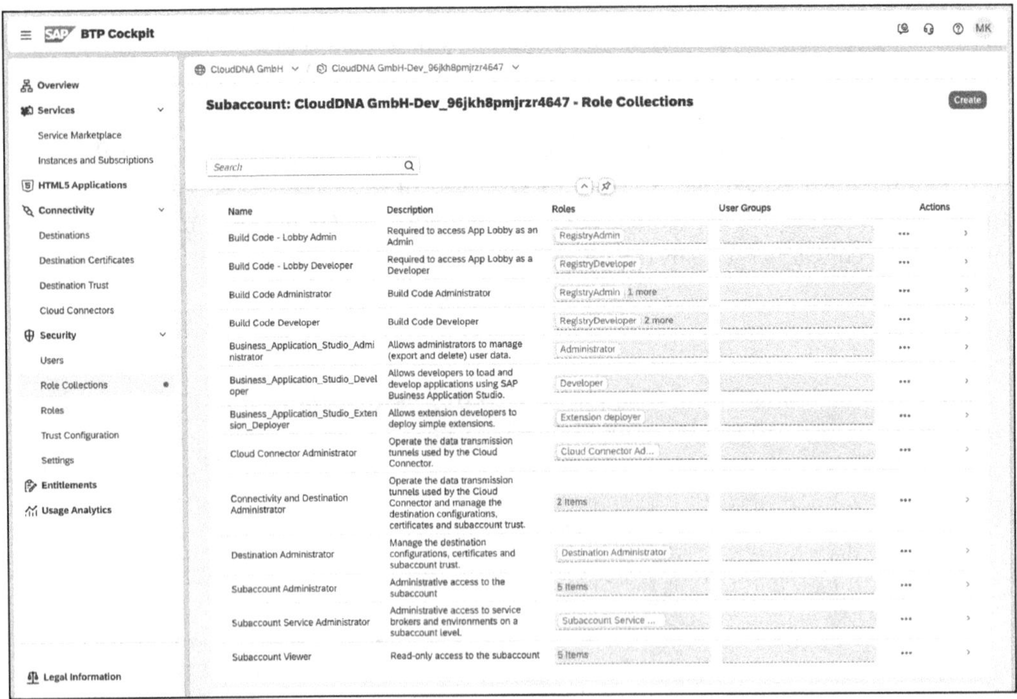

Figure 20.13 Checking the Role Collections

20.3 Create a Basic Project

Now that you've successfully set up SAP Build Code, this section shows you how to create a new project in SAP Build Code. To do this, open SAP Build Code, which will take you to the lobby (see Figure 20.14). In the lobby, you can create a new project by clicking on **Create**.

In this step, you can decide whether you want to build an application, an automated process for SAP Build Process Automation, or a business site for SAP Build Work Zone (see Figure 20.15). We'll choose the **Build an Application** option.

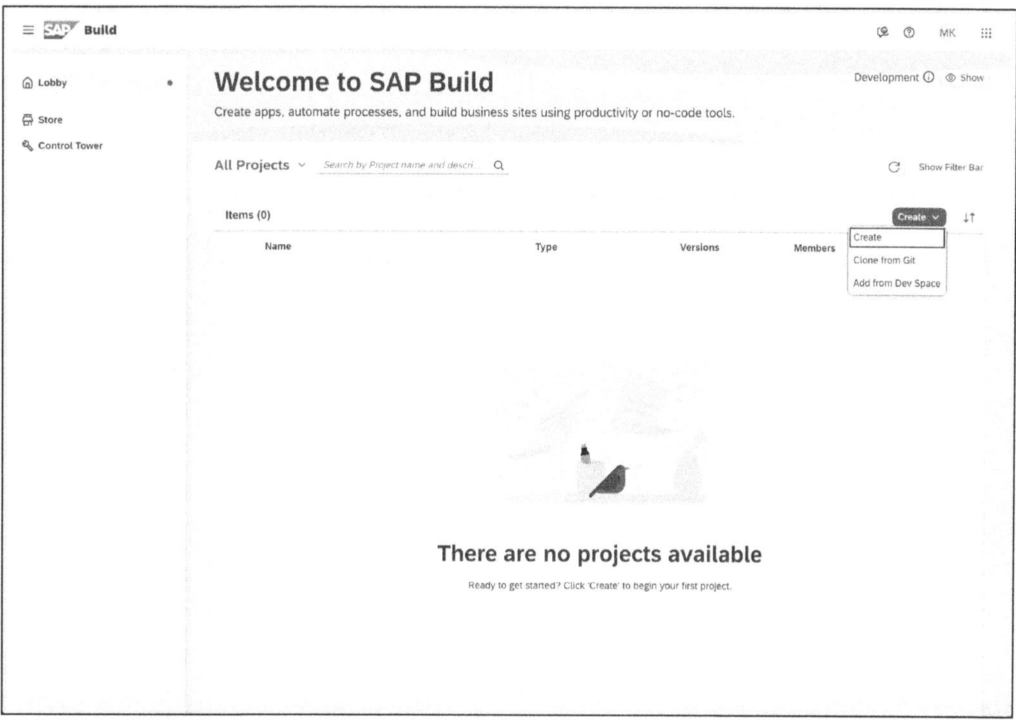

Figure 20.14 Creating a Project from the Lobby

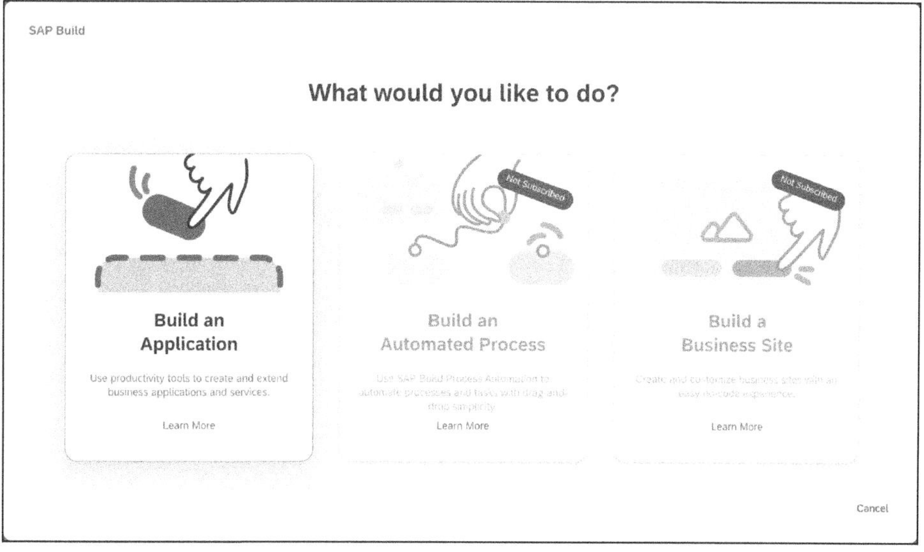

Figure 20.15 Selecting the Application Type

In the next step, you select whether you want to create a project with **SAP Build Apps**, **SAP Build Code**, or **ABAP Cloud** (see Figure 20.16). In this example, we've opted for the **SAP Build Code** option.

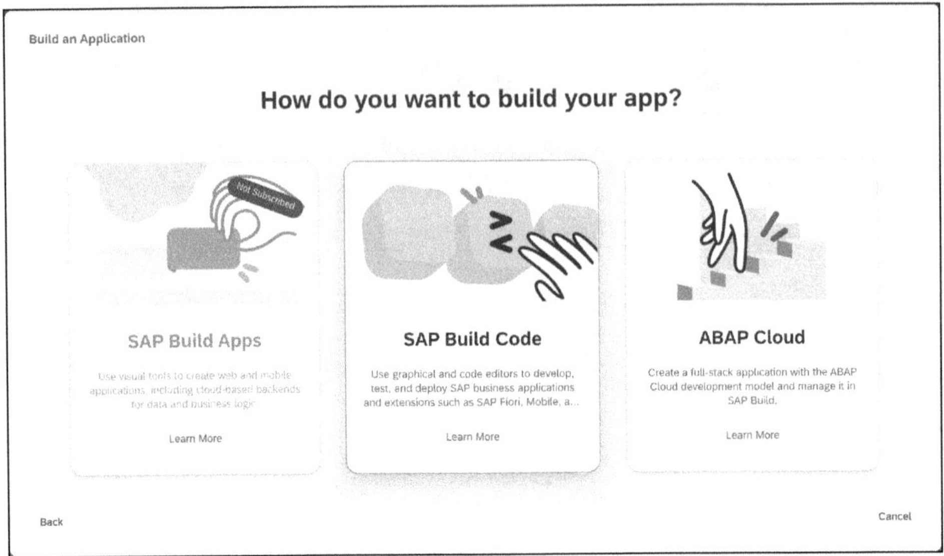

Figure 20.16 Selecting the Technology

Then, select whether you want to create an **SAP Fiori Application**, a **Full-Stack Application**, or a **Mobile Application** (see Figure 20.17).

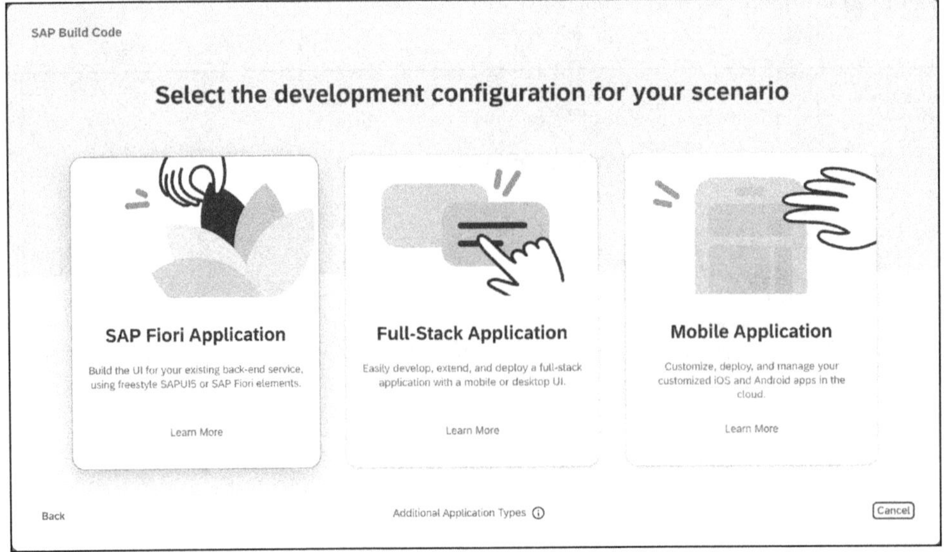

Figure 20.17 Choosing the Development Configuration

In the next step, you must specify a **Project Name**, a **Description** and a **Dev Space**, as shown in Figure 20.18.

Create an SAP Fiori project

Give your project a name

Project Name: *
CloudDNA

Description:
CloudDNA SAP Build Code Demo

Dev Space:*
SAP_Fiori

You can only run one dev space at a time. For more options, go to the Dev Space Manager

Back
Create
Cancel

Figure 20.18 Providing the Project Details

After completing the entry, you'll return to the SAP Build lobby (see Figure 20.19). Creating the project may take a few minutes. As soon as the project has been created, its name appears as a clickable link. Clicking on the link will take you directly to SAP Build Code and open the corresponding project there.

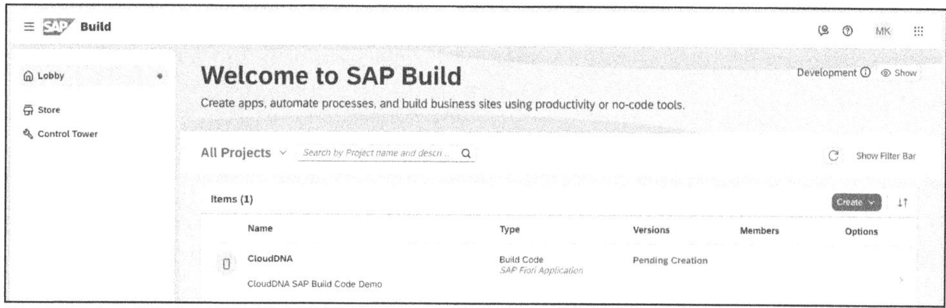

Figure 20.19 Project Overview

20.4 Use Cases in SAPUI5 Development

With SAP Build Code, you can address a wide range of use cases in SAPUI5 and SAP Fiori development, making it a versatile tool for modern application creation and maintenance. The key capabilities include the following:

- **Application development**
  - Create new SAPUI5 and SAP Fiori apps from scratch.
  - Convert existing applications or modules to TypeScript for improved type safety and maintainability.
  - Add advanced logic to enhance application functionality.
  - Seamlessly integrate new UI controls to enrich UIs.
- **Code documentation**
  Automatically generate comprehensive documentation to improve code readability and maintainability.
- **Code explanation**
  Leverage AI to understand complex code sections through clear and concise explanations, aiding both new developers and experienced teams.
- **Code refactoring**
  Optimize and restructure code for better performance, clarity, and adherence to best practices without altering functionality.
- **Unit test creation**
  Generate robust and reliable unit tests to ensure application quality and reduce the risk of errors during future updates.

In the following, we'll look at a few examples of how you can use SAP Build Code to efficiently advance your development projects. We'll start with the conversion of JavaScript to TypeScript, a challenge that many developers are currently facing. This is mainly because SAPUI5 applications were originally developed based on JavaScript, while TypeScript has since established itself as the de facto standard. We won't go into the numerous advantages of TypeScript in detail here. However, it's advisable to convert existing applications that are undergoing refactoring to TypeScript at the same time to benefit from its type safety and better development options.

As shown in Figure 20.20, you can conveniently trigger the conversion with the command `/ui5 convert the details.controller.js to TypeScript`. This process is automatically supported by SAP Build Code and helps you to quickly and efficiently convert existing JavaScript components into the more modern TypeScript format.

SAP Build Code also allows you to add new UI controls to a view to enhance the user experience. An example of this is the inclusion of a toolbar in the footer area, as shown in Figure 20.21. This process is simplified by SAP Build Code, enabling you to quickly and efficiently insert and customize new elements in your application. In the example, we've used the following command to insert a toolbar in the footer area of the page:

```
/ui5 add a footer toolbar containing an edit button to the details.view.xml
```

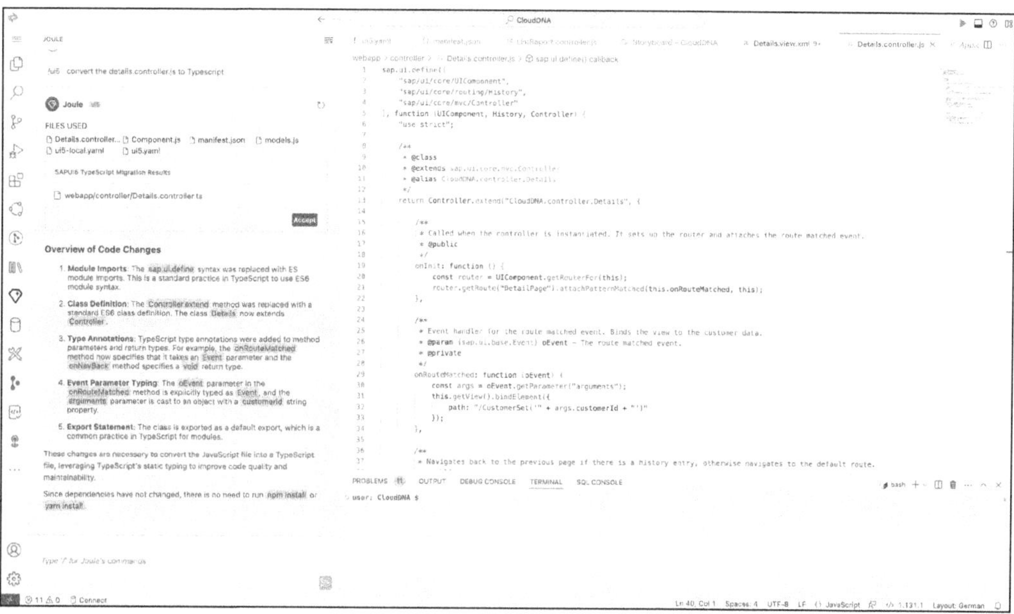

Figure 20.20 Converting to TypeScript

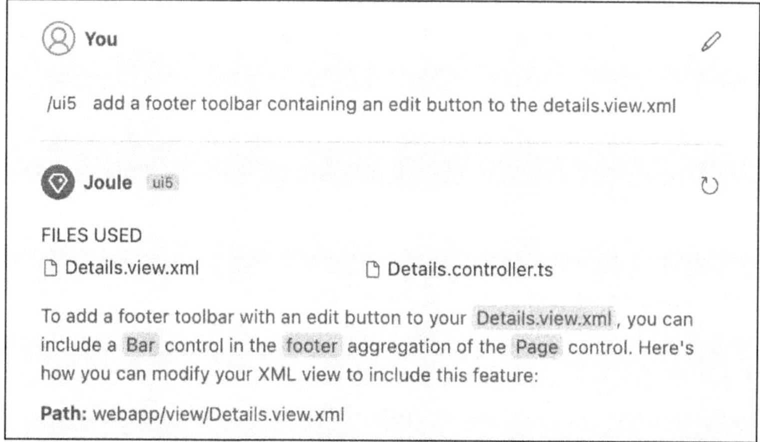

Figure 20.21 Adding UI Controls

You can see the result in Figure 20.22. By clicking on the **Accept** button, the suggested changes can be directly incorporated into the source code. The **Explanation** function is particularly helpful at this point, as it describes all the changes made in detail. This not only provides you with transparency regarding the adjustments but also a better understanding of the changes in the context of your application.

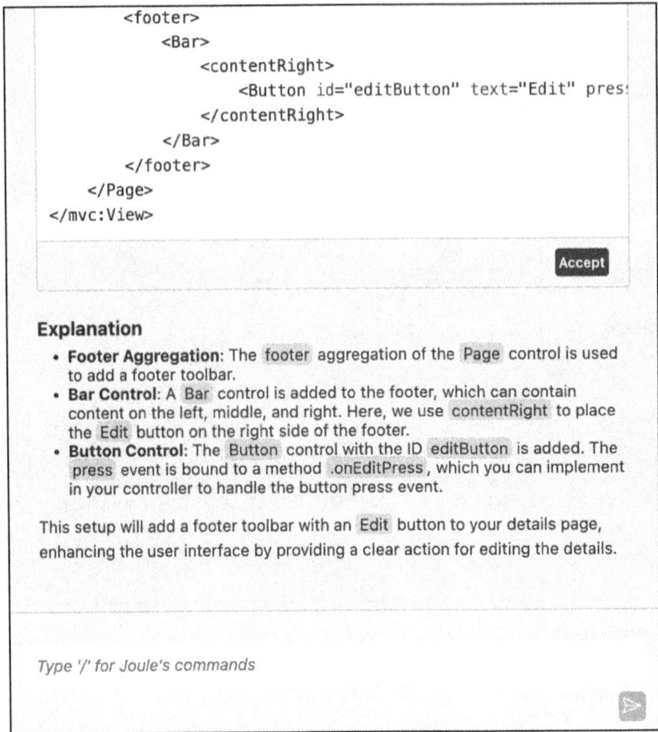

Figure 20.22 Adding UI Controls: Result

SAP Build Code also allows you to generate business logic directly in the controller. Figure 20.23 shows an example of a formatter function being created. Following is the command used:

`/ui5` add a `formatter` `function` to `Details.controller.js` `concatenating` `Firstname` and `Lastname` `separated` by a `space` and `remove` `leading` and `trailing` `spaces`

In Figure 20.24, you can see the generated function. It has been implemented in modern and efficient JavaScript and includes a JSDoc comment structure. This documentation describes the function in detail and makes it easy for other developers to quickly understand its purpose and use. The use of contemporary JavaScript practices ensures that the generated code is not only functional but also easy to maintain and future-proof.

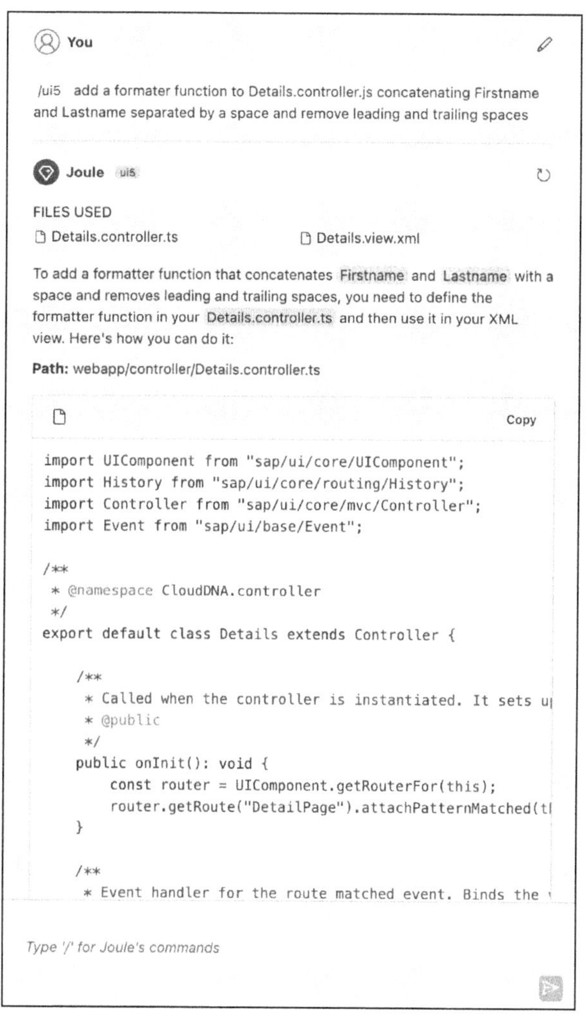

Figure 20.23 Adding Controller Logic

```
    /**
     * Formatter function to concatenate Firstname and Lastname with a space and trim s
     * @param {string} firstName - The first name of the customer.
     * @param {string} lastName - The last name of the customer.
     * @returns {string} The concatenated and trimmed full name.
     * @public
     */
    public formatFullName(firstName: string, lastName: string): string {
        return `${firstName} ${lastName}`.trim();
    }
}
```

Accept

Figure 20.24 Adding Controller Logic

Another scenario in which SAP Build Code can be used is to explain source code. As shown in Figure 20.25, SAP Build Code can analyze complex code sections and explain their function in simple and understandable language. This is particularly useful for understanding existing applications more quickly, training new developers on a project, or improving documentation.

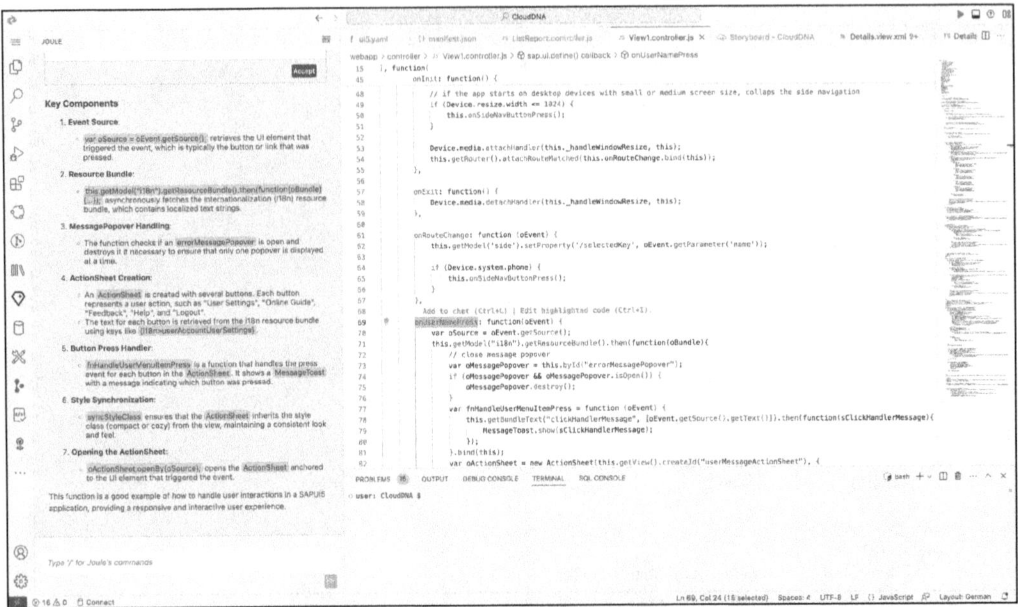

Figure 20.25 Explaining Existing Code

SAP Build Code can also be used for refactoring, as shown in Figure 20.26. In this example, the prompt `/ui5 refactor the View1.controller.js using modern JavaScript` was used to optimize the code in the *View1.controller.js* file. This refactoring converts the existing code to modern JavaScript practices, improving both readability and maintainability. This automated refactoring saves time and ensures that the code meets current standards.

Figure 20.27 shows the result of the refactoring. The key changes are clearly summarized and concisely described. This summary provides a clear overview of the adjustments made, such as the transition to modern JavaScript practices, the simplification of structures, and the removal of redundant code. This ensures that the changes are comprehensible and support the development process.

SAP Build Code also allows you to automatically generate code documentation based on JSDoc. As shown in Figure 20.28, the prompt is `/ui5 write a JSDoc for Details.controller.js`. The tool analyzes the source code of the *Details.controller.js* file and generates concise JSDoc comments that document functions, parameters, and return values. This automation saves a significant amount of time, improves code readability, and helps ensure development best practices.

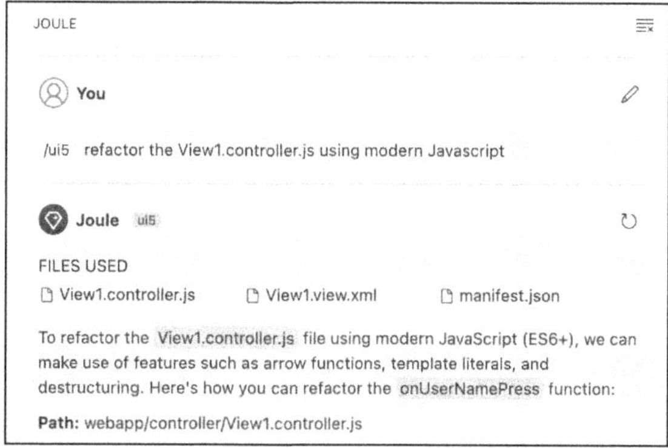

Figure 20.26 Refactoring

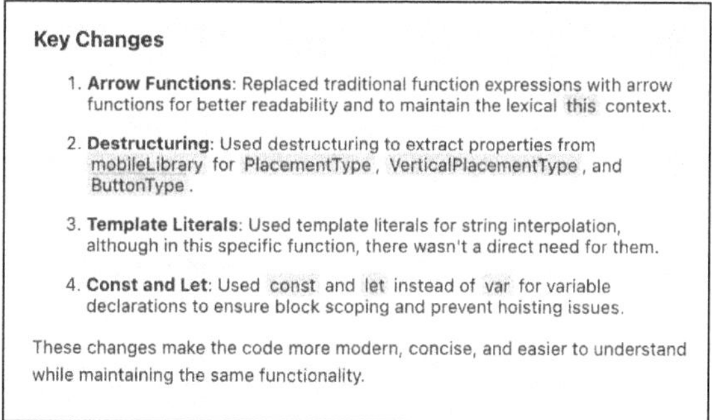

Figure 20.27 Refactoring Result

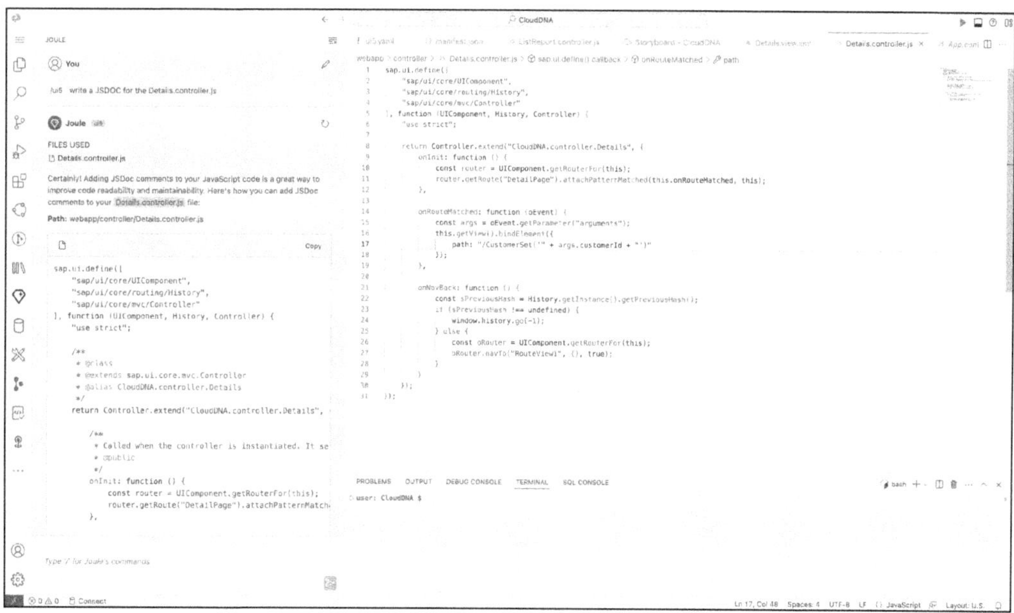

Figure 20.28 Adding JSDoc Documentation

20

SAP Build Code also allows you to create unit tests quickly and easily. Figure 20.29 shows an example. The tool automatically generates unit tests based on the functions defined in the source code. This makes it much easier to ensure code quality by catching potential errors early. Automated testing also promotes maintainability and reliability of your applications.

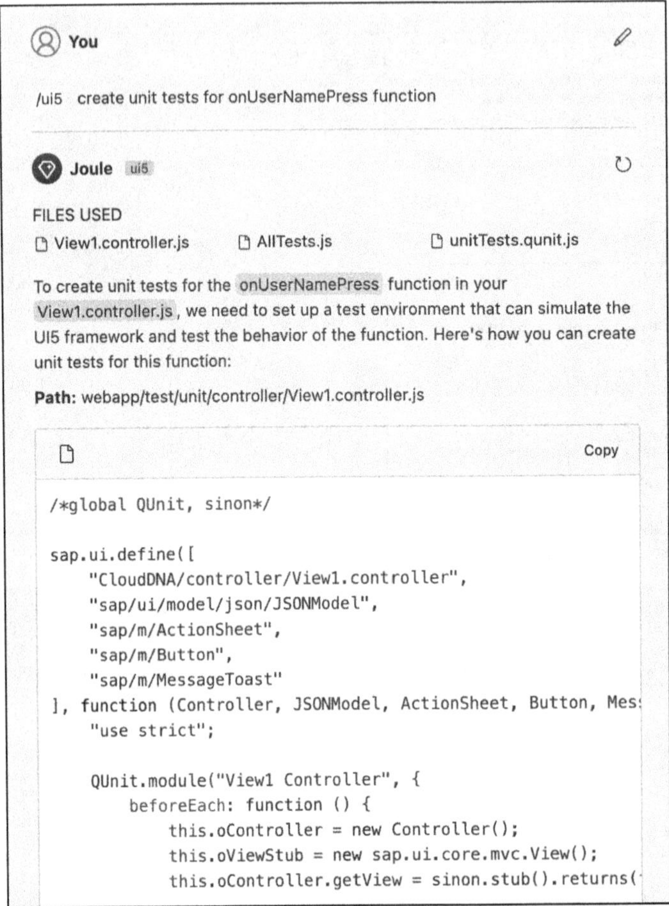

Figure 20.29 Creating Unit Tests

In Figure 20.30, you can see, among other things, the explanation of the created unit tests as a result of the prompt. This explanation describes in detail the structure and functioning of the generated tests, including the tested scenarios, the test data used, and the expected results. This transparency makes it easier for developers to understand and customize the tests and to ensure that they meet the application's requirements.

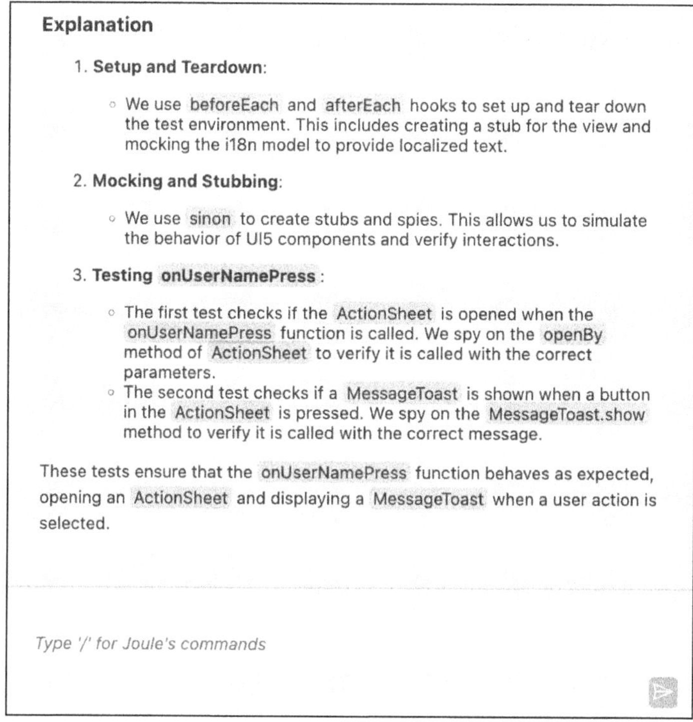

Figure 20.30 Unit Test Explanation

20.5 Summary

This chapter provides a comprehensive overview of SAP Build Code and shows how it simplifies and modernizes the development and maintenance of applications in the SAP environment. We started by explaining the basic functioning of generative AI and presented its application in the context of SAP. SAP Build Code uses the capabilities of generative AI to automate development tasks, increase efficiency, and support developers in recurring activities.

After that, we explained how to set up SAP Build Code in SAP BTP. This included the use of a booster, the configuration of subaccounts, and the assignment of the necessary authorizations. The process was described in detail for both the free plan and for specific requirements, including customization options such as selecting the region and subdomain.

We covered the process for creating a new project in SAP Build Code next. From choosing the project type—for example, SAP Fiori apps, full-stack applications, or mobile apps—to defining project names, descriptions, and dev spaces, all the important steps were described. We also discussed how developers can navigate directly to SAP Business Application Studio to seamlessly continue their work.

Finally, specific use cases were presented that illustrate the versatility of SAP Build Code in SAPUI5 development. These included automated conversion from JavaScript to TypeScript, integration of new UI controls, generation of specific business logic such as formatter functions, refactoring of existing applications using modern JavaScript practices, automatic creation of code documentation with JSDoc, and generation and explanation of unit tests.

The Authors

Rene Glavanovits is an SAP consultant and developer at CloudDNA GmbH, an SAP partner in Austria. He specializes in the latest SAP technologies, specifically in the development of full-stack applications with SAP Fiori, SAPUI5, OData, CDS, and SAP Cloud Application Programming Model.

Martin Koch is the managing director of CloudDNA GmbH. He and his team conduct training for SAP and have developed four of their own training courses on the topics of SAPUI5, SAP Fiori, cloud integration, and cloud security, which are listed in the SAP training catalogue. In addition to his work as a trainer, Martin works as an architect, consultant, and developer for international SAP customers of all company sizes.

Daniel Krancz is a software developer and consultant at CloudDNA GmbH. His focus is on full-stack development with SAPUI5, SAP Fiori, OData, SAP Cloud Application Programming Model, as well as mobile development.

Maximilian Olzinger is a software developer and consultant at CloudDNA GmbH. He is an SAP Certified Development Associate and has handled projects for companies in all industries with great success. He conducts trainings in the areas of SAP Fiori, ABAP and web development. He has developed several SAP standard trainings for SAP Fiori.

Index

D

E

F

G

H

I

J

K

L

T